Praise from Reviewers for This Edition

"Williams-Sawyer . . . is the most readable textbook that deals with computer terminology in a meaningful way without getting into tech jargon. The concepts are clearly presented and the [photos], illustrations, and graphics become part of the reading and enhance the ability of the reader to comprehend the material. . . . I think the level of difficulty is perfect. I find very few students, even international students, who have difficulty comprehending the book."

—*Beverly Bohn, Park University, Parkville, Missouri*

"[UIT is] geared toward a generation that grew up with computers but never thought about how they work. Should appeal to a younger audience."

—*Leleh Kalantari, Western Illinois University, Macomb*

"The treatment of MP3 players, satellite radio, digital photography, SDTV, HDTV, and cellphones [in Chapter 7, the new Personal Technology chapter] will enhance my classes."

—*Charles Brown, Plymouth State University, New Hampshire*

"I really liked the fact that you updated the text with items that would be important to students when they are looking to purchase a PC."

—*Stephanie Anderson, Southwestern Community College, Creston, Iowa*

"I like the authors' writing style very much. I found it to be almost conversational, which is good, in my opinion. . . . I truly looked for unclear areas and did not find any at all."

—*Laurie Eakins, East Carolina University, Greenville, North Carolina*

"I like how [the writing] is personalized. It seems as if the writer is speaking directly to the student—not the normal textbook emphasis."

—*Tammy Potter, Western Kentucky Community & Technical College, Paducah*

"[The authors'] writing style is clear and concise. [They have] taken some very technical topics and explained them in everyday language while not 'dumbing down' the material. The text flows smoothly. The inclusion of quotes from real people lends a conversational tone to the chapter [Chapter 6], making it easier to read and comprehend."

—*Robert Caruso, Santa Rosa Junior College, California*

"The level of difficulty is perfect for an intro level computer applications course taught at a 2- or 4-year college."

—*Jami Cotler, Siena College, Loudonville, New York*

"Chapter 2 is written in a readable, motivating style. I found it to be concise, and introducing topics in a proper sequence, defining terms accurately and effectively. I found myself thinking of topics to be added, and then THERE THEY WERE!"

—*Mike Michaelson, Palomar College, San Marcos, California*

"Strong writing style. This chapter [Chapter 8] was extremely thorough. And covered many subjects in depth. . . . Writing style has always been quite clear and concise with these two authors."

—*Rebecca Mundy, UCLA and University of Southern California*

"As a user of the sixth edition already, I find that the authors have the right level of difficulty presented for an introductory computer course."

—*Jerry Matejka, Adelphi University, Garden City, New York*

"I think the writing style is good and will work well with the students."

—*Michelle Parker, Indiana Purdue University, Fort Wayne*

"This text is written at a level that is fine for most of my students. I have many students for whom English is a second language. These students may have difficulty with certain phrasing.... As I read this chapter [Chapter 3], however, I found very little that I thought might cause confusion, even for those ESL students.... I have selected previous editions of this text in large part because it is very 'readable.'"

—*Valerie Anderson, Marymount College, Palos Verdes, California*

Praise for Previous Editions

"Williams and Sawyer do a consistently good job of explaining material. The graphics and examples are well done."

—*David Burris, Sam Houston State University, Huntsville, Texas*

"Practicality is in the title of the book and is definitely practiced in each chapter. Readability means clear writing, and that is also evident in the text."

—*Nancy Webb, San Francisco City College*

"The practical approach to information technology, along with the book's superior readability, make this a strong text. The book's emphasis on being current and a three-level learning system are great."

—*DeLyse Totten, Portland Community College, Oregon*

"I would rate the writing style as superior to the book I am currently using and most of the books I have reviewed.... I found this book much easier to read than most books on the market."

—*Susan Fry, Boise State University, Idaho*

"The easy-to-understand way of speaking to the readers is excellent. You put computer terminology into an easily understandable way to read. It's excellent."

—*Ralph Caputo, Manhattan College, New York*

"The major difference that I notice between your text and other texts is the informal tone of the writing. This is one of the main reasons we adopted your book—the colloquial feel."

—*Todd McLeod, Fresno City College, California*

"[The text] is written in a clear and non-threatening manner, keeping the student's interest through the use of real, colorful anecdotes and interesting observations. The authors' emphasis on the practical in the early chapters gets the students' interest by centering on real-life questions that would face everyone purchasing a new personal computer."

—*Donald Robertson, Florida Community College–Jacksonville*

"I enjoyed the writing style. It was clear and casual, without trivializing. I think the examples and explanations of Williams and Sawyer are excellent."

—*Martha Tillman, College of San Mateo, California*

"Ethics topics are far superior to many other textbooks."

—*Maryann Dorn, Southern Illinois University*

"[The critical thinking emphasis is important because] the facts will change, the underlying concepts will not. Students need to know what the technology is capable of and what is not possible . . ."

—*Joseph DeLibero, Arizona State University*

Seventh Edition

Using
Information
Technology

A Practical Introduction to Computers & Communications

Brian K. Williams

Stacey C. Sawyer

McGraw-Hill
Irwin

Boston   Burr Ridge, IL   Dubuque, IA   Madison, WI   New York   San Francisco   St. Louis
Bangkok   Bogotá   Caracas   Kuala Lumpur   Lisbon   London   Madrid   Mexico City
Milan   Montreal   New Delhi   Santiago   Seoul   Singapore   Sydney   Taipei   Toronto

McGraw-Hill
Irwin

USING INFORMATION TECHNOLOGY:
A PRACTICAL INTRODUCTION TO COMPUTERS & COMMUNICATIONS
Published by McGraw-Hill/Irwin, a business unit of The McGraw-Hill Companies, Inc., 1221
Avenue of the Americas, New York, NY, 10020. Copyright © 2007 by The McGraw-Hill
Companies, Inc. All rights reserved. No part of this publication may be reproduced or distributed
in any form or by any means, or stored in a database or retrieval system, without the prior written
consent of The McGraw-Hill Companies, Inc., including, but not limited to, in any network or
other electronic storage or transmission, or broadcast for distance learning.

Some ancillaries, including electronic and print components, may not be available to customers
outside the United States.

This book is printed on acid-free paper.

1 2 3 4 5 6 7 8 9 0 QPD/QPD 0 9 8 7 6

ISBN-13: 978-0-07-110768-6
ISBN-10: 0-07-110768-1

www.mhhe.com

Brief Contents

1 INTRODUCTION TO INFORMATION TECHNOLOGY: Your Digital World   1

2 THE INTERNET & THE WORLD WIDE WEB: Exploring Cyberspace   49

3 SOFTWARE: Tools for Productivity & Creativity   117

4 HARDWARE: THE CPU & STORAGE: How to Choose a Multimedia Computer System   189

5 HARDWARE: INPUT & OUTPUT: Taking Charge of Computing & Communications   251

6 COMMUNICATIONS, NETWORKS, & SAFEGUARDS: The Wired & Wireless World   309

7 PERSONAL TECHNOLOGY: The Future Is You   367

8 DATABASES & INFORMATION SYSTEMS: Digital Engines for Today's Economy   408

9 THE CHALLENGES OF THE DIGITAL AGE: Society & Information Technology Today   463

10 SYSTEMS ANALYSIS & PROGRAMMING: Software Development, Programming, & Languages   497

To the Instructor

Introduction: Teaching the "Always On" Generation

If there is anything we have learned from our 25 years of writing computer concepts books, it is this: *the landscape of computer education can change quickly, and it's our responsibility to try to anticipate it.*

USING INFORMATION TECHNOLOGY's First Edition was the first text to foresee and define the impact of digital convergence—the fusion of computers and communications—as the new and broader foundation for the computer concepts course. *UIT's* Fourth Edition was the first text to acknowledge the new priorities imposed by the internet and World Wide Web and bring discussion of them from late in the course to near the beginning (to Chapter 2).

Now, with this Seventh Edition we address another paradigm change: because of the mobility and hybridization of digital devices, **an "Always On" generation of students has come of age that's at ease with digital technology but—and it's an important "but"—not always savvy about computer processes, possibilities, and liabilities.** This development imposes additional challenges on professors: **instructors are expected to make the course interesting and challenging to students already at least somewhat familiar with information technology but with widely varying levels of computer sophistication.**

What's New in the Seventh Edition

To address these challenges, this edition includes the following new features:

1. *Now a 10-chapter book.* By making judicious cuts and combining topics, we have reduced chapters to make the book **a better fit for many instructors and students,** without sacrificing much in the way of coverage. (Changes are shown on page viii.)

2. *New chapter on personal technology.* This brand-new chapter (Chapter 7) describes **fast-changing personal technologies**—from podcasting to smartphones—and their effects.

3. *More "What's in it for me?" student questions:* More "I" and "me" questions, of the type students ask, have been added—and answered in the book.

4. *Expanded coverage of communications:* Because the subject of communications and networks has exploded, we have greatly increased topic coverage—on everything from Bluetooth to ZigBee, from EV-DO to Z-Wave (Chapters 2, 6).

5. *More on computer self-defense:* Computer threats have soared. We help readers stay safe by explaining spoofing, phishing, pharming, and other evils, all the way to online bullies, botnets, and zombies (Chapters 2, 6, 9).

Addressing Instructors' Two Most Important Challenges

Quotes

What instructors say is the most significant challenge in teaching this course

"Keeping the students interested."
—Evelyn Lulis, DePaul University

"Keeping a wide variety of students on the same page."
—Donald Robertson, Florida Community College–Jacksonville

As we embark on our twelfth year of publication, we are extremely pleased at the continued reception to *USING INFORMATION TECHNOLOGY*, which has been used by well more than a half million students and adopted by instructors in over 700 schools. One reason for this enthusiastic response may be that we've tried hard to address professors' needs. We've often asked instructors—in reviews, surveys, and focus groups—**"What is your most significant challenge in teaching this course?"**

The First Most Frequent Answer: "Trying to Make the Course Interesting and Challenging"

One professor at a state university seems to speak for most when she says: "Making the course interesting and challenging." Others echo her with remarks such as "Keeping students interested in the material enough to study" and "Keeping the students engaged who know some, but not all, of the material." Said one professor, "Many students take the course because they must, instead of because the material interests them." Another speaks about the need to address a "variety of skill/knowledge levels while keeping the course challenging and interesting"—which brings us to the second response.

"This will always be a difficult course to teach, since the students in any given class come from very different backgrounds and have vastly different levels of computer expertise."
—Laurie Eakins, East Carolina University

The Second Most Frequent Answer: "Trying to Teach to Students with a Variety of Computer Backgrounds"

The most significant challenge in teaching this course "is trying to provide material to the varied levels of students in the class," says an instructor at a large Midwestern university. Another says the course gets students from all backgrounds, ranging from "Which button do you push on the mouse?" to "Already built and maintain a web page with html." Says a third, "mixed-ability classes [make] it difficult to appeal to all students at the same time." And a fourth: "How do you keep the 'techies' interested without losing the beginners?"

As authors, we find information technology tremendously exciting, but we recognize that many students take the course reluctantly. And we also recognize that many students come to the subject with attitudes ranging from complete apathy and unfamiliarity to a high degree of experience and technical understanding.

To address the problem of **motivating the unmotivated and teaching to a disparity of backgrounds,** *UIT* offers unequaled treatment of the following:

1. **Practicality**
2. **Readability**
3. **Currentness**
4. **Three-level critical thinking system.**

We explain these features on the following pages.

Sixth & Seventh Editions of *UIT* Compared

This edition constitutes a major revision of UIT. The Sixth and Seventh editions are compared below.

UIT Sixth Edition (2005)	*UIT* Seventh Edition (2007)
1. Introduction to Information Technology	1. **Introduction to Information Technology** • New section: "How Becoming Computer Savvy Benefits You" • New section (moved from old Chap. 10): "Information Technology & Your Life" (effects on education, careers, etc.)
2. The Internet & the World Wide Web	2. **The Internet & the World Wide Web** • More on web portals, search engines, audio & video searching, desktop search, tagging, VoIP, blogging, RSS, podcasting. • New section: "The Intrusive Internet: Snooping, Spamming, Spoofing, Phishing, Pharming, Cookies, & Spyware"
3. Application Software	3. **Software** • Two former software chapters combined into one. • System software now discussed before application software. • History of DOS & Windows reduced. Updates on Apple and Windows OSs.
4. System Software	
5. Hardware: The CPU & Storage	4. **Hardware: The CPU & Storage** • Updates on chips, cache, hard-disk controllers, Blu-ray, smart cards, multicore processors.
6. Hardware: Input & Output	5. **Hardware: Input & Output** • More on specialty keyboards, flat-panel display, digital cameras.
7. Telecommunications	6. **Communications, Networks, & Safeguards** • New section (moved from old Chap. 9 and expanded): "Cyberthreats, Hackers, & Safeguards" • Moved compression/decompression to Chap. 8. • Moved smart TV to Chap. 7. • Resequenced discussion of networks, added material (HANs, PANs), more on firewalls. • Added material on Ethernet, HomePNA, HomePlug. • Discussion of GPRS, EDGE, EV-DO, UMTS, WiMax, Bluetooth 2.0, Ultra Wideband, Wireless USB, Insteon, ZigBee, Z-Wave, more on Wi-Fi. • Reduced/removed discussion of line configurations, transmission mode, multiplexing, OSI.
	7. **Personal Technology—BRAND NEW CHAPTER!** • Discussion of MP3 players; satellite, hi-def, internet radios; point-and-shoot & SLR digital cameras; PDAs and tablet PCs; DTV, HDTV, SDTV; smartphones (SMS, ringtones, TV & video, etc.); videogame systems. • Box on "Always On" generation.
8. Files, Databases, & E-Commerce	8. **Databases & Information Systems** • Combined databases and e-commerce from old Chap. 8, information systems from old Chap. 11, and expert systems and AI from old Chap. 10. • Moved "Concerns about Accuracy & Privacy" to Chap. 9. • Slightly reduced material on databases and MIS. • Resequenced section on computer-based information systems.
9. The Challenges of the Digital Age	9. **The Challenges of the Digital Age** • Section from old Chap. 8: "Truth Issues: Manipulating Digital Data" • New material on Evil Twin attack, zombies, botnets, blackmail, crashing internet. • Revision of section on computer criminals, adding terrorists, corporate spies, etc. • New section: "Protecting Children: Pornography, Sexual Predators, & Online Bullies"
10. The Promises of the Digital Age	Old Chap. 10 topics moved to Chaps. 1 and 8.
11. Information Systems	Old Chap. 11 topics moved to Chaps. 8 and 10.
Appendix A: Software Development	10. **Systems Analysis & Programming** • Combined systems from old Chap. 11 with programming & languages from old App. A.

Feature #1: Emphasis on Practicality

This popular feature received overwhelming acceptance by both students and instructors in past editions. **Practical advice**, of the sort found in computer magazines, newspaper technology sections, and general-interest computer books, is expressed not only in the text but also in the following:

The Experience Box

Appearing at the end of each chapter, the Experience Box is optional material that may be assigned at the instructor's discretion. However, students will find the subjects covered are of immediate value. *Examples:* "Web Research, Term Papers, & Plagiarism." "The Mysteries of Tech Support." "How to Buy a Laptop." "Preventing Your Identity from Getting Stolen." "Virtual Meetings: Linking Up Electronically."

New to this edition: "The 'Always On' Generation."

See the list of Experience Boxes and Practical Action Boxes on the inside front cover.

Practical Action Box

This box consists of optional material on practical matters. *Examples:* "Serious Web Search Techniques." "Preventing Problems from Too Much or Too Little Power to Your Computer." "When the Internet Isn't Productive: Online Addiction & Other Timewasters."

New to this edition: "Evaluating & Sourcing Information Found on the Web." "Tips for Fighting Spam." "Tips for Avoiding Spyware." "Utility Programs." "Get a PC or Get a Mac? Dealing with Security Issues." "Help in Building Your Web Page." "Starting Over with Your Hard Drive: Erasing, Reformatting, & Reloading." "Buying a Printer." "Telecommuting & Telework: The Nontraditional Workplace." "Ways to Minimize Virus Attacks." "How to Deal with Passwords." "Online Viewing & Sharing of Digital Photos." "Buying the Right HDTV."

See the list of Survival Tips on the inside front cover.

Survival Tips

In the margins throughout we present utilitarian **Survival Tips** to aid exploration of infotech.

New to this edition: Examples: "Test the Speed of Your Internet Connection." "Some Free ISPs." "Do Home Pages Endure?" "Look for the Padlock Icon." "Keeping Windows Security Features Updated." "New Software & Compatibility." "Where Do I Get a Boot Disk?" "Is Your Password Guessable?" "Update Your Drivers." "Service Packs 1 & 2." "Two Versions of Windows XP."

Survival Tip
Look for the Padlock Icon

To avoid having people spying on you when you are sending information over the web, use a secure connection. This is indicated at the bottom of your browser window by an icon that looks like a padlock or key.

"Compressing Web & Audio Files." "Try Before You Buy." "Setting Mouse Properties." "Digital Subscriptions." "Cellphone Minutes." "Reformat Your Memory Card to Avoid Losing Your Photos." "Keeping Track of Your Cellphone." "Fraud Baiters." "Alleviating Info-Mania."

How to Understand a Computer Ad

In the hardware chapters (Chapters 4 and 5), we explain important concepts by showing students **how to understand the hardware components in a hypothetical PC ad** (see p. 195).

Great PC Buy!

- 7-Bay Mid-Tower Case
- Intel Pentium 4 Processor 2.80 GHz
- 512 MB 533 MHz DDR2 SDRAM
- 1 MB L2 Cache
- 6 USB 2.0 Ports
- 56 Kbps Internal Modem
- 3D AGP Graphics Card (64 MB)
- Sound Blaster Digital Sound Card
- 160 GB SATA 7200 RPM Hard Drive
- 24X DVD/CD-RW Combo Drive
- 104-Keyboard
- Microsoft IntelliMouse
- 17" Flat Panel Display
- HP Business Inkjet 1000 Printer

Details of this ad are explained throughout this chapter and the next. See the little magnifying glass:

Feature #2: Emphasis on Readability & Reinforcement for Learning

We offer the following features for reinforcing student learning:

Interesting Writing

Where is it written that textbooks have to be boring? Can't a text have personality?

Actually, studies have found that textbooks **written in an imaginative style** significantly improve students' ability to retain information. Both instructors and students have commented on the distinctiveness of the writing in this book. We employ a number of journalistic devices—colorful anecdotes, short biographical sketches, interesting observations in direct quotes—to make the material as interesting as possible. We also use real anecdotes and examples rather than fictionalized ones.

Finally, unlike most computer concepts books, **we provide references for our sources—991 references in the back of the book, 40% of them from the year preceding publication.** We see no reason why introductory computer books shouldn't practice good scholarship by revealing their sources of information. And we see no reason why good scholarship can't go hand in hand with good writing. That is, scholarship need not mean stuffiness.

Key Terms AND Definitions Emphasized

To help readers avoid any confusion about which terms are important and what they actually mean, we print each key term in ***bold italic underscore*** and its definition in **boldface.** *Example* (from Chapter 1): "***Data* consists of raw facts and figures that are processed into information.**"

Material in Easily Manageable Portions

Major ideas are presented in **bite-size form,** with generous use of advance organizers, bulleted lists, and new paragraphing when a new idea is introduced. Most **sentences have been kept short,** the majority not exceeding 22–25 words in length.

"What's in It for Me?" Questions—to Help Students Read with Purpose

We have **crafted the learning objectives as Key Questions** to help readers focus on essentials. These are expressed as "I" and "me" questions, of the type students ask.

New to this edition! We have also added more "I" and "me" questions following both first-level and second-level headings throughout the book. (To save space, these replace the QuickChecks used in prior editions.)

Summary of Terms and Why They're Important

Each chapter ends with a **Summary** of important terms, with an explanation of **what they are and why they are important.** The terms are accompanied, when appropriate, by a picture. Each concept or term is also given a cross-reference page number that refers the reader to the main discussion within the text.

Eight Timelines to Provide Historical Perspective

Some instructors like to see coverage of the history of computing. Not wishing to add greatly to the length of the book, we decided on **a student-friendly approach: the presentation of eight pictorial timelines showing the most significant historical IT events.** These timelines, which occur in most chapters, appear along the bottom page margin. Each timeline repeats certain "benchmark" events to keep students oriented, but **each one is modified to feature the landmark discoveries and inventions appropriate to the different chapter material.** *Examples:* In Chapter 2, about the internet, the timeline features innovations in telecommunications, the internet, and the World Wide Web (see below).

See timelines beginning on pp. 14, 50, 162, 192, 258, 310, 416, 522

1995	1996	2000	2001	2002	2003	2004
NSFNET reverts to research project; internet now in commercial hands; the Vatican goes online	Microsoft releases Internet Explorer; 56 K modem invented; cable modem introduced; 12,881,000 hosts on internet (488,000 domains)	Web surpasses 1 billion indexable pages: 93,047,785 hosts on internet	AOL membership surpasses 28 million; Napster goes to court	Blogs become popular	First official Swiss online election; flash mobs start in New York City	More than 285,000,000 hosts on internet

See Ethics examples on pp. 37, 91, 99, 236, 276, 319, 348, 448, 449, 465, 480

Emphasis Throughout on Ethics

Many texts discuss ethics in isolation, usually in one of the final chapters. We believe this topic is too important to be treated last or lightly, and users have agreed. Thus, **we cover ethical matters throughout the book,** as indicated by the icon shown at right. *Example:* We discuss such all-important questions as copying of internet files, online plagiarism, privacy, computer crime, and netiquette.

Emphasis Throughout on Security

In the post 9-11 era, security concerns are of gravest importance. Although we devote several pages (in Chapters 2, 6, and 9) to security matters, we also reinforce student awareness by **highlighting with page-margin Security icons instances of security-related material throughout the book.** *Example:* On p. 99, we use the icon shown at right to highlight the advice that one should pretend that every email message one sends "is a postcard that can be read by anyone."

SECURITY

See Security icons on pp. 37, 99, 277, 291, 341, 344, 468

Feature #3: Currentness

About *UIT*'s currentness

"Very knowledge-able, very good research."
—Maryann Dorn, Southern Illinois University

Reviewers have applauded previous editions of *UIT* for being **more up to date than other texts.** For example, we have traditionally ended many chapters with a forward-looking section that offers a preview of technologies to come—some of which are realized even as students are using the book.

Among the new topics and terms covered in this edition are: *Abilene, adware, AMD Athlon 64 X2, Apple video iPod, black-hat hackers, blog-osphere, Bluetooth 2.0, Blu-ray, botnet, browser hijackers, business-to-consumer (B2C) systems, consumer-to-consumer (C2C) systems, contactless smart cards, cyberbullies, cyberterrorists, desktop search engines, Desk-topLX, downloaded ringtones, EDGE, employee internet management (EIM) software, entertainment PCs, EV-DO, EV-DV, Evil Twin attacks, exabytes, Fibre Channel, foldable PDA keyboards, GPRS, hacktivists, high-definition radio, home area networks, home automation networks, home networks, HomePlug, HomePNA, hotspots, ICANN 2.0, Insteon, Intel Itanium 2 Mon-tecito, Intel Pentium EE 840, Internet Protocol Television, internet radio, IrDA ports, key loggers, keyword index, L3 cache, Linspire, Macintosh OS X Tiger, malware, mash-ups, megapixels, mesh technologies, micropay-ments, Microsoft Xbox 360, MIMO, moblogs, MP3 players, multicore proces-sors, national ID card, National LamdaRail, Nintendo Revolution, one-hand PDA keyboards, online sexual predators, perpendicular recording technol-ogy, personal area networks, pharming, phishing, podcasting, point-and-shoot digital cameras, polymer memory, pornography, PowerPC chips, RSS aggregators, satellite radio, search hijackers, Short Message Service, SLR dig-ital cameras, smart TVs, smartphones, Sony PlayStation 3, spoofing, spy-ware, subject directory, Symbian OS, tagging, text messaging, thrill-seeker hackers, UltraCard, ultra wideband, UMTS, vblogs, videogame ratings, VoIP phoning, web-based email, white-hat hackers, WiMax, Windows Media Player 10 Mobile, Windows Vista, wireless USB, Xen, ZigBee, zombies, Z-Wave.*

See inside <u>back</u> cover for pages on which MoreInfo! icons appear.

"MoreInfo!" Icons Help Students Find Their Own Answers to Questions

In addition, **we have taken the notion of currentness to another level through the use of the "MoreInfo!" feature to encourage students to obtain their own updates** about material. *Examples:* "Finding Wi-Fi Hot Spots." "Finding ISPs." "Do Home Pages Endure?" "Do You Need to Know HTML to Build a Website?" "Urban Legends & Lies on the Internet." "Blog Search Engines." "Some Online Communities." "Links to Security Software." "Where to Learn More about Freeware & Shareware." "More about Watermarks." See the pages listed on the inside back cover.

Finding Wi-Fi Hot Spots

www.wififreespot.com/
www.wifihotspotlist.com/
www.wifinder.com/

Feature #4: Three-Level System to Help Students Think Critically about Information Technology

This feature, which has been in place for the preceding three editions, has been warmly received. More and more instructors seem to have become familiar with **Benjamin Bloom's *Taxonomy of Educational Objectives,*** describing a hierarchy of six critical-thinking skills: (a) two lower-order skills—*memorization* and *comprehension*; and (b) four higher-order skills—*application, analysis, synthesis,* and *evaluation.* Drawing on our experience in writing books to guide students to college success, we have implemented Bloom's ideas in a three-stage pedagogical approach, using the following hierarchical approach in the Chapter Review at the end of every chapter:

Stage 1 Learning—Memorization: "I Can Recognize and Recall Information"

Using self-test questions, multiple-choice questions, and true/false questions, we enable students to test how well they recall basic terms and concepts.

Stage 2 Learning—Comprehension: "I Can Recall Information in My Own Terms and Explain It to a Friend"

Using open-ended short-answer questions, we enable students to re-express terms and concepts in their own words.

Stage 3 Learning—Applying, Analyzing, Synthesizing, Evaluating: "I Can Apply What I've Learned, Relate These Ideas to Other Concepts, Build on Other Knowledge, and Use All These Thinking Skills to Form a Judgment"

In this part of the Chapter Review, we ask students to put the ideas into effect using the activities described. The purpose is to help students take possession of the concepts, make them their own, and apply them realistically to their own ideas. **Our web exercises are also intended to spur discussion in classroom and other contexts.** *Examples:* "Using Text Messaging in Emergencies." "What's Wrong with Using Supermarket Loyalty Cards?" "Are You in the Homeland Security Database?"

Because of the extensive text revisions, the supplements have been extensively revised as well. **Two key supplements—the Instructor's Manual and the Testbank—were done by the authors, Stacey Sawyer and Brian Williams, and were carefully checked for accuracy.**

Instructor's Manual

The electronic Instructor's Manual, available as part of the Instructor's Resource CD, helps instructors to create effective lectures. The Instructor's Manual is easy to navigate and simple to understand. Each chapter contains a chapter overview, a lecture outline, teaching tips, additional information, and answers to end-of-chapter questions and exercises.

Testbank

The Testbank format allows instructors to effectively pinpoint areas of content within each chapter on which to test students. The test questions include learning level, answers, and text page numbers.

EZ Test

McGraw-Hill's EZ Test is a flexible and easy-to-use electronic testing program. The program allows instructors to create tests from book-specific items. It accommodates a wide range of question types and instructors may add their own questions. Multiple versions of the test can be created, and any test can be exported for use with course management systems such as WebCT, Blackboard, or PageOut. EZ Test Online is a new service and gives you a place to easily administer your EZ Test–created exams and quizzes online. The program is available for Windows and Macintosh environments.

PowerPoint Presentation

The PowerPoint presentation includes additional material that expands on important topics from the text, allowing instructors to create interesting and engaging classroom presentations. Each chapter of the presentation includes illustrations to enable instructors to emphasize important concepts in memorable ways.

Figures from the Book

All photos, illustrations, screenshots, and tables are available electronically for use in presentations, transparencies, or handouts.

Online Learning Center

(www.mhhe.com/uit7e) Designed to provide students with additional learning opportunities, the website includes PowerPoint presentations for each chapter. For the convenience of instructors, all Instructor's Resource CD material is available for download.

PageOut

PageOut is our Course Web Site Development Center and offers a syllabus page, URL, McGraw-Hill Online Learning Center content, online exercises and quizzes, gradebook, discussion board, and an area for student web pages.

PageOut requires no prior knowledge of HTML, no long hours of coding, and a way for course coordinators and professors to provide a full-course website. PageOut offers a series of templates—simply fill them out with your course information and click on one of 16 designs. The process takes under an hour and leaves you with a professionally designed website. We'll even get you started with sample websites, or enter your syllabus for you! Page-Out is so straightforward and intuitive, it's little wonder why over 12,000 college professors are using it. For more information, visit the PageOut website at *www.pageout.net*

The Online Learning Center can be delivered through any of these platforms:

- **Blackboard.com**
- **WebCT (a product of Universal Learning Technology)**

Web CT and Blackboard Partnerships

McGraw-Hill has partnerships with WebCT and Blackboard to make it even easier to take your course online and have McGraw-Hill content delivered through the leading internet-based learning tools for higher education.

McGraw-Hill has the following service agreements with WebCT and Blackboard:

- **SimNet Concepts:** This is the TOTAL solution for training and assessment in computer concepts. SimNet Concepts offers a unique, graphic-intensive environment for assessing student understanding of computer concepts. It includes interactive labs for 77 different computer concepts and 160 corresponding assessment questions. The content menus parallel the contents of the McGraw-Hill text being used for the class, so that students can cover topics for each chapter of the text you are using.

  SimNet Concepts also offers the only truly integrated learning and assessment program available today. After a student has completed any SimNet Concepts Exam, he or she can simply click on one button to have SimNet assemble a custom menu that covers just those concepts that the student answered incorrectly or did not attempt. These custom lessons can be saved to disk and loaded at any time for review.

  Assessment Remediation records and reports what the student did incorrectly for each question on an exam that was answered incorrectly.

Acknowledgments

Two names are on the front of this book, but a great many others are important contributors to its development. First, we wish to thank our sponsoring editor, Paul Ducham, for his support during this fast-moving revision process. Thanks also go to our marketing champion, Sankha Basu, for his enthusiasm and ideas. Trina Hauger deserves our special thanks for her help with the supplements program. Everyone in production provided support and direction: Christine Vaughan, Shesha Bolisetty, and Artemio Ortiz. We also thank our media technology producers, Rose Range and Victor Chiu, for helping us develop an outstanding Instructor's Resource Kit.

Outside McGraw-Hill we were fortunate to have the services of the development house Content Connections, with David Brake and Jenn Erickson. Brian Kaspar helped on the Instructor's Manual. Scott Wallace provided creative ideas for early Chapter Review exercises. Photo researcher Judy Mason, copyeditor Sue Gottfried, proofreader Martha Ghent, and indexer James Minkin all gave us valuable assistance. Thanks also to all the extremely knowledgeable and hard-working professionals at TechBooks/GTS who provided so many of the prepress services.

Finally, we are grateful to the following reviewers for helping to make this the most market-driven book possible.

Stephanie Anderson
Southwestern Community College, Creston, Iowa

Valerie Anderson
Marymount College, Palos Verdes, California

Hossein Bidgoli
California State University, Bakersfield

Beverly Bohn
Park University, Parkville, Missouri

Charles Brown
Plymouth State College, Plymouth, New Hampshire

Robert Caruso
Santa Rosa Junior College, Santa Rosa, California

Jami Cotler
Siena College, Loudonville, New York

Laura A. Eakins
East Carolina University, Greenville, North Carolina

Janos Fustos
Metropolitan State College, Denver

Yaping Gao
College of Mount St. Joseph, Cincinnati

David German
Cerro Coso Community College, Bishop, California

Fillmore Guinn
Odessa College, Odessa, Texas

Norman P. Hahn
Thomas Nelson Community College, Hampton, Virginia

Laleh Kalantari
Western Illinois University, Macomb

Gina Long
Southwestern Community College, Creston, Iowa

Pamela Luckett
Barry University, Miami Shores, Florida

Thomas Martin
Shasta College, Redding, California

Jerry Matejka
Adelphi University, Garden City, New York

Jennifer Merritt
Park University, Parkville, Missouri

Mike Michaelson
Palomar College, San Marcos, California

Cindy Minor
John A. Logan College, Carterville, Illinois

Rebecca Mundy
University of California, Los Angeles, and University of Southern California

Kathleen Murray
Drexel University, Philadelphia

Michelle Parker
Indiana Purdue University, Fort Wayne

Tammy Potter
West Kentucky Community & Technical College, Paducah

Morgan Shepherd
University of Colorado at Colorado Springs

Anita Whitehill
Foothill College, Los Altos Hills, California

Alfred Zimermann
Hawai'i Pacific University, Honolulu

Contents

Chapter 1

INTRODUCTION TO INFORMATION TECHNOLOGY: YOUR DIGITAL WORLD 1

1.1 The Practical User: How Becoming Computer Savvy Benefits You   3

1.2 Information Technology & Your Life: The Future Now   4

The Two Parts of IT: Computers & Communications   4
Education: The Promise of More Interactive & Individualized Learning   5
Health: High Tech for Wellness   6
Money: Toward the Cashless Society   8
Leisure: Infotech in Entertainment & the Arts   9
Government & Electronic Democracy: Participating in the Civic Realm   10
Jobs & Careers   11

1.3 Infotech Is All Pervasive: Cellphones, Email, the Internet, & the E-World   13

The Telephone Grows Up   14
"You've Got Mail!" Email's Mass Impact   15
The Internet, the World Wide Web, & the "Plumbing of Cyberspace"   16
College Students & the E-World   18
Practical Action Box: *Managing Your Email   19*

1.4 The "All-Purpose Machine": The Varieties of Computers   20

All Computers, Great & Small: The Categories of Machines   20
Supercomputers   21
Mainframe Computers   22
Workstations   22
Microcomputers   22
Microcontrollers   23
Servers   24

1.5 Understanding Your Computer: How Can You Customize (or Build) Your Own PC?   24

How Computers Work: Three Key Concepts   24
Pretending to Order (or Build) a Custom-Built Desktop Computer: Basic Knowledge of How a Computer Works   27
Input Hardware: Keyboard & Mouse   27
Processing & Memory Hardware: Inside the System Cabinet   28
Storage Hardware: Floppy Drive, Hard Drive, & CD/DVD Drive   30
Output Hardware: Video & Sound Cards, Monitor, Speakers, & Printer   31
Communications Hardware: Modem   32
Software   32
Is Getting a Custom-Built PC Worth the Effort?   33

1.6 Where Is Information Technology Headed?   34

Three Directions of Computer Development: Miniaturization, Speed, & Affordability   34
Three Directions of Communications Development: Connectivity, Interactivity, & Multimedia   35
When Computers & Communications Combine: Convergence, Portability, & Personalization   35
"E" Also Stands for Ethics   37
Experience Box: *Better Organization & Time Management: Dealing with the Information Deluge in College—& in Life   38*

Chapter 2

THE INTERNET & THE WORLD WIDE WEB: EXPLORING CYBERSPACE   49

2.1 Connecting to the Internet: Narrowband, Broadband, & Access Providers   52

Narrowband (Dial-Up) Modem: Low Speed but Inexpensive & Widely Available   53

High-Speed Phone Lines: More Expensive but Available in Most Cities 55

Problem for Telephone Internet Connections: The Last Mile 56

Cable Modem: Close Competitor to DSL 56

Satellite Wireless Connections 57

Other Wireless Connections: Wi-Fi & 3G 58

Internet Access Providers: Three Kinds 58

Practical Action Box: *Looking for an Internet Access Provider: Questions to Ask at the Beginning* 60

2.2 How Does the Internet Work? 60

Internet Connections: POPs, NAPs, Backbone, & Internet2 61

Internet Communications: Protocols, Packets, & Addresses 62

Who Runs the Internet? 63

2.3 The World Wide Web 64

The Face of the Web: Browsers, Websites, & Web Pages 64

How the Browser Finds Things: URLs 66

The Nuts & Bolts of the Web: HTML & Hyperlinks 68

Using Your Browser to Get around the Web 69

Web Portals: Starting Points for Finding Information 73

Search Services & Search Engines, & How They Work 74

Four Web Search Tools: Keyword Indexes, Subject Directories, Metasearch Engines, & Specialized Search Engines 75

Smart Searching: Three General Strategies 76

Multimedia Search Tools: Image, Audio, & Video Searching 77

Practical Action Box: *Serious Web Search Techniques* 78

Desktop Search: Tools for Searching Your Computer's Hard Disk 79

Practical Action Box: *Evaluating & Sourcing Information Found on the Web* 80

Tagging: Saving Links for Easier Retrieval Later 81

2.4 Email & Other Ways of Communicating over the Net 81

Two Ways to Send & Receive Email 81

How to Use Email 83

Sorting Your Email 85

Attachments 85

Instant Messaging 86

FTP—for Copying All the Free Files You Want 88

Newsgroups—for Online Typed Discussions on Specific Topics 88

Listservs: Email-Based Discussion Groups 90

Real-Time Chat—Typed Discussions Among Online Participants 90

Netiquette: Appropriate Online Behavior 91

2.5 The Online Gold Mine: Telephony, Multimedia, Webcasting, Blogs, E-Commerce, & Other Resources 92

Telephony: The Internet Telephone & Videophone 92

Multimedia on the Web 93

The Web Automatically Comes to You: Webcasting, Blogging, & Podcasting 95

E-Commerce: B2B Commerce, Online Finance, Auctions, & Job Hunting 97

Relationships: Online Matchmaking 98

2.6 The Intrusive Internet: Snooping, Spamming, Spoofing, Phishing, Pharming, Cookies, & Spyware 98

Snooping on Your Email: Your Messages Are Open to Anyone 99

Spam: Electronic Junk Mail 99

Practical Action Box: *Tips for Fighting Spam* 100

Spoofing, Phishing, & Pharming: Phony Email Senders & Websites 101

Cookies: Convenience or Hindrance? 101

Spyware—Adware, Browser & Search Hijackers, & Key Loggers: Intruders to Track Your Habits & Steal Your Data 102

Practical Action Box: *Tips for Avoiding Spyware* 103

Experience Box: *Web Research, Term Papers, & Plagiarism* 104

Chapter 3
SOFTWARE: TOOLS FOR PRODUCTIVITY & CREATIVITY 117

3.1 System Software: The Power Behind the Power 118

3.2 The Operating System: What It Does 119

Booting 120

CPU Management 121

File Management 121

Task Management 122

Security Management 124

3.3 Other System Software: Device Drivers & Utility Programs 124

Device Drivers: Running Peripheral Hardware 124
Utilities: Service Programs 124
Practical Action Box: *Utility Programs* 125

3.4 Common Features of the User Interface 127

Using Keyboard & Mouse 127
The GUI: The Graphical User Interface 130
The Help Command 134

3.5 Common Operating Systems 135

Macintosh Operating System 136
Microsoft Windows 137
Network Operating Systems: NetWare, Windows NT/2000/2003, Microsoft.NET, Unix, & Linux 139
Practical Action Box: *Get a PC or Get a Mac? Dealing with Security Issues* 140
Operating Systems for Handhelds: Palm OS & Windows CE/Pocket PC/Windows CE. NET 143

3.6 Application Software: Getting Started 145

Application Software: For Sale, for Free, or for Rent? 145
Tutorials & Documentation 147
A Few Facts about Files & the Usefulness of Importing & Exporting 148
Types of Application Software 148

3.7 Word Processing 149

Creating Documents 150
Editing Documents 150
Formatting Documents with the Help of Templates & Wizards 153
Output Options: Printing, Faxing, or Emailing Documents 155
Saving Documents 155
Tracking Changes & Inserting Comments 155
Web Document Creation 155

3.8 Spreadsheets 156

The Basics: How Spreadsheets Work 156
Analytical Graphics: Creating Charts 158

3.9 Database Software 159

The Benefits of Database Software 159
The Basics: How Databases Work 160
Personal Information Managers 162

3.10 Specialty Software 163

Presentation Graphics Software 163
Financial Software 166
Desktop Publishing 168
Drawing & Painting Programs 170
Video/Audio Editing Software 171
Multimedia Authoring Software 173
Animation Software 172
Web Page Design Software 173
Practical Action Box: *Help in Building Your Web Page* 174
Project Management Software 174
Computer-Aided Design 175
Experience Box: *The Mysteries of Tech Support* 176

Chapter 4

HARDWARE: THE CPU & STORAGE: HOW TO CHOOSE A MULTIMEDIA COMPUTER SYSTEM 189

4.1 Microchips, Miniaturization, & Mobility 190

From Vacuum Tubes to Transistors to Microchips 190
Miniaturization Miracles: Microchips, Microprocessors, & Micromachines 192
Mobility 192
Choosing an Inexpensive Personal Computer: Understanding Computer Ads 194

4.2 The System Unit: The Basics 194

The Binary System: Using On/Off Electrical States to Represent Data & Instructions 195
The Parity Bit 198
Machine Language 199
The Computer Case: Bays, Buttons, & Boards 199
Power Supply 201
The Motherboard & the Microprocessor Chip 202
Practical Action Box: *Preventing Problems from Too Much or Too Little Electrical Power to Your Computer* 203
Processing Speeds: From Megahertz to Picoseconds 206

4.3 More on the System Unit 207

How the Processor or CPU Works: Control Unit, ALU, Registers, & Buses 207
How Memory Works: RAM, ROM, CMOS, & Flash 209

How Cache Works 211
Other Methods of Speeding Up Processing 212
Ports & Cables 213
Expandability: Buses & Cards 217

4.4 Secondary Storage 220

Floppy Disks & Zip Disks 221
Hard Disks 222
Optical Disks: CDs & DVDs 225
Magnetic Tape 229
Smart Cards 230
Flash Memory 231
Online Secondary Storage 232

4.5 Future Developments in Processing & Storage 232

Practical Action Box: *Starting Over with Your Hard Drive: Erasing, Reformatting, & Reloading 233*
Future Developments in Processing 235
Future Developments in Secondary Storage 236
Experience Box: *How to Buy a Notebook 238*

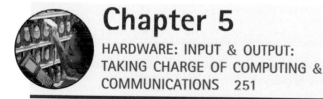

Chapter 5

HARDWARE: INPUT & OUTPUT: TAKING CHARGE OF COMPUTING & COMMUNICATIONS 251

5.1 Input & Output 253

5.2 Input Hardware 254

Keyboards 254
Pointing Devices 258
Scanning & Reading Devices 265
Audio-Input Devices 270
Webcams & Video-Input Cards 271
Digital Cameras 271
Speech-Recognition Systems 274
Sensors 275
Radio-Frequency Identification Tags 276
Human-Biology-Input Devices 277

5.3 Output Hardware 277

Traditional Softcopy Output: Display Screens 278
Traditional Hardcopy Output: Printers 282
Practical Action Box: *Buying a Printer 286*
Mixed Output: Sound, Voice, & Video 288

5.4 Input & Output Technology & Quality of Life: Health & Ergonomics 289

Health Matters 289
Ergonomics: Design with People in Mind 291

5.5 The Future of Input & Output 291

Toward More Input from Remote Locations 292
Toward More Source Data Automation 292
Toward More Output in Remote Locations 294
Toward More Realistic Output 294
Experience Box: *Good Habits: Protecting Your Computer System, Your Data, & Your Health 296*

Chapter 6

COMMUNICATIONS, NETWORKS, & SAFEGUARDS: THE WIRED & WIRELESS WORLD 309

6.1 From the Analog to the Digital Age 311

The Digital Basis of Computers: Electrical Signals as Discontinuous Bursts 311
The Analog Basis of Life: Electrical Signals as Continuous Waves 312
Purpose of the Dial-Up Modem: Converting Digital Signals to Analog Signals & Back 313
Converting Reality to Digital Form 314

6.2 Networks 315

The Benefits of Networks 316
Types of Networks: WANs, MANs, LANs, HANs, PANs, & Others 316
How Networks Are Structured: Client/Server & Peer-to-Peer 318
Intranets, Extranets, & VPNs 319
Components of a Network 320
Network Topologies: Bus, Ring, & Star 322
Two Ways to Prevent Messages from Colliding: Ethernet & Token Ring 324

6.3 Wired Communications Media 325

Wired Communications Media: Wires & Cables 325
Wired Communications Media for Homes: Ethernet, HomePNA, & HomePlug 326
Practical Action Box: *Telecommuting & Telework: The Nontraditional Workplace 327*

6.4 Wireless Communications Media 328

The Electromagnetic Spectrum, the Radio-Frequency (RF) Spectrum, & Bandwidth 328

Four Types of Wireless Communications Media 330
Long-Distance Wireless: One-Way Communication 333
Long-Distance Wireless: Two-Way Communication 337
Short-Range Wireless: Two-Way Communications 340

6.5 Cyber Threats, Hackers, & Safeguards 344

Cyber Threats: Denial-of-Service Attacks, Worms, Viruses, & Trojan Horses 344
Some Cyber Villains: Hackers & Crackers 348
Practical Action Box: *Ways to Minimize Virus Attacks 348*
Online Safety: Antivirus Software, Firewalls, Passwords, Biometric Authentication, & Encryption 350
Practical Action Box: *How to Deal with Passwords 352*

6.6 The Future of Communications 355

Satellite-Based Systems 355
Beyond 3G to 4G 355
Photonics: Optical Technologies at Warp Speed 355
Software-Defined Radio 356
A New Way to Compute: The Grid 356
Experience Box: *Virtual Meetings: Linking Up Electronically 357*

Chapter 7

PERSONAL TECHNOLOGY: THE FUTURE IS YOU 367

7.1 Convergence, Portability, & Personalization 368

Convergence 369
Portability 370
Personalization 371
Popular Personal Technologies 372

7.2 MP3 Players 372

How MP3 Players Work 372
The Societal Effects of MP3 Players 374
Using MP3 Players in College 374

7.3 High-Tech Radio: Satellite, High-Definition, & Internet 375

Satellite Radio 375
High-Definition Radio 377

Internet Radio 378
Podcasting 378

7.4 Digital Cameras: Changing Photography 379

How Digital Cameras Work 379
Practical Action Box: *Online Viewing & Sharing of Digital Photos 383*
The Societal Effects of Digital Cameras 385

7.5 Personal Digital Assistants & Tablet PCs 385

How a PDA Works 386
The Future of PDAs 387
Tablet PCs 387

7.6 The New Television 387

Interactive, Personalized, Internet, & Smart TVs & Entertainment PCs 388
Three Kinds of Television: DTV, HDTV, SDTV 388
Practical Action Box: *Buying the Right HDTV 390*
The Societal Effects of the New TV 390

7.7 Smartphones: More Than Talk 391

How a Mobile Phone Works 391
Smartphone Services 391
The Societal Effects of Cellphones 397

7.8 Videogame Systems: The Ultimate Convergence Machine? 397

Microsoft's Xbox 360 398
Sony's PlayStation 3 399
Nintendo's Revolution 399
The Results of Personal Technology: The "Always On" Generation 399
The Experience Box: *The "Always On" Generation 400*

Chapter 8

DATABASES & INFORMATION SYSTEMS: DIGITAL ENGINES FOR TODAY'S ECONOMY 407

8.1 Managing Files: Basic Concepts 408

How Data Is Organized: The Data Storage Hierarchy 409
The Key Field 410
Types of Files: Program Files & Data Files 411
Compression & Decompression: Putting More Data in Less Space 412

Contents

8.2 Database Management Systems 412

The Benefits of Database Management Systems 414

Three Database Components 414

Practical Action Box: Storing Your Stuff: How Long Will Digitized Data Last? 415

The Database Administrator 416

8.3 Database Models 416

Hierarchical Database 417

Network Database 418

Relational Database 419

Object-Oriented Database 422

Multidimensional Database 422

8.4 Data Mining 424

The Process of Data Mining 424

Some Applications of Data Mining 426

8.5 Databases & the Digital Economy: E-Business & E-Commerce 427

E-Commerce: Online Buying & Selling 427

Types of E-Commerce Systems: B2B, B2C, & C2C 429

8.6 Information Systems in Organizations: Using Databases to Help Make Decisions 431

The Qualities of Good Information 431

Information Flow within an Organization: Horizontally between Departments & Vertically between Management Levels 432

Computer-Based Information Systems 435

Office Information Systems 435

Transaction Processing Systems 435

Management Information Systems 436

Decision Support Systems 437

Executive Support Systems 438

Expert Systems 439

8.7 Artificial Intelligence 439

Expert Systems 440

Natural Language Processing 442

Intelligent Agents 442

Pattern Recognition 442

Fuzzy Logic 442

Virtual Reality & Simulation Devices 443

Robotics 443

Two Approaches to Artificial Intelligence: Weak versus Strong AI 446

Artificial Life, the Turing Test, & AI Ethics 448

8.8 The Ethics of Using Databases: Concerns about Privacy & Identity Theft 449

The Threat to Privacy 450

Identity Theft 452

Experience Box: Preventing Your Identity from Getting Stolen 453

Chapter 9
THE CHALLENGES OF THE DIGITAL AGE: SOCIETY & INFORMATION TECHNOLOGY TODAY 463

9.1 Truth Issues: Manipulating Digital Data 465

Manipulation of Sound 465

Manipulation of Photos 465

Manipulation of Video & Television 467

Accuracy & Completeness 467

9.2 Security Issues: Threats to Computers & Communications Systems 468

Errors & Accidents 468

Natural Hazards 470

Computer Crimes 471

Computer Criminals 475

Practical Action Box: Is the Boss Watching You? Trust in the Workplace 477

9.3 Security: Safeguarding Computers & Communications 478

Deterrents to Computer Crime 478

Identification & Access 479

Encryption 480

Protection of Software & Data 481

Disaster-Recovery Plans 481

9.4 Quality-of-Life Issues: The Environment, Mental Health, Child Protection, & the Workplace 482

Environmental Problems 482

Mental-Health Problems 484

Protecting Children: Pornography, Sexual Predators, & Online Bullies 485

Workplace Problems: Impediments to Productivity 487

Practical Action: When the Internet Isn't Productive: Online Addiction & Other Time Wasters 489

Contents

9.5 Economic & Political Issues: Employment & the Haves/Have-Nots 490

Technology, the Job Killer? 490
Gap between Rich & Poor 491
Whom Does the Internet Serve? 491
In a World of Breakneck Change, Can You Still Thrive? 492
Experience Box: *Student Use of Computers: Some Controversies* 493

Chapter 10

SYSTEMS ANALYSIS & PROGRAMMING: SOFTWARE DEVELOPMENT, PROGRAMMING, & LANGUAGES 497

10.1 Systems Development: The Six Phases of Systems Analysis & Design 498

The Purpose of a System 498
Getting the Project Going: How It Starts, Who's Involved 499
The Six Phases of Systems Analysis & Design 499
The First Phase: Conduct a Preliminary Investigation 500
The Second Phase: Do an Analysis of the System 500
The Third Phase: Design the System 502
The Fourth Phase: Develop the System 503
The Fifth Phase: Implement the System 504
The Sixth Phase: Maintain the System 505

10.2 Programming: A Five-Step Procedure 505

The First Step: Clarify the Programming Needs 506
The Second Step: Design the Program 507
The Third Step: Code the Program 513
The Fourth Step: Test the Program 514
The Fifth Step: Document & Maintain the Program 514

10.3 Five Generations of Programming Languages 515

First Generation: Machine Language 517
Second Generation: Assembly Language 517
Third Generation: High-Level or Procedural Languages 519
Fourth Generation: Very-High-Level or Problem-Oriented Languages 520
Fifth Generation: Natural Languages 522

10.4 Programming Languages Used Today 522

FORTRAN: The Language of Mathematics & the First High-Level Language 522
COBOL: The Language of Business 523
BASIC: The Easy Language 524
Pascal: The Simple Language 524
C: For Portability & Scientific Use 525
LISP: For Artificial Intelligence Programs 525

10.5 Object-Oriented & Visual Programming 525

Object-Oriented Programming: Block by Block 526
Three Important Concepts of OOP 526
Visual Programming: The Example of Visual BASIC 528

10.6 Markup & Scripting Languages 528

HTML: For Creating 2-D Web Documents & Links 529
VRML: For Creating 3-D Web Pages 529
XML: For Making the Web Work Better 530
JavaScript: For Dynamic Web Pages 530
ActiveX: For Creating Interactive Web Pages 531
Perl: For CGI Scripts 531
Experience Box: *Critical Thinking Tools* 532

Notes 541
Credits 552
Index I-1

chapter 1

Introduction to Information Technology
Your Digital World

Chapter Topics & Key Questions

1.1   **The Practical User: How Becoming Computer Savvy Benefits You** What does being *computer savvy* mean, and what are its practical payoffs?

1.2   **Information Technology & Your Life: The Future Now** What is information technology, and how does it affect education, health, money, leisure, government, and careers?

1.3   **Infotech Is All-Pervasive: Cellphones, Email, the Internet, & the E-World** How does information technology facilitate email, networks, and the use of the internet and the web; what is the meaning of the term *cyberspace*?

1.4   **The "All-Purpose Machine": The Varieties of Computers** What are the five sizes of computers, and what are clients and servers?

1.5   **Understanding Your Computer: How Can You Customize (or Build) Your PC?** What four basic operations do all computers use, and what are some of the devices associated with each operation? How does communications affect these operations?

1.6   **Where Is Information Technology Headed?** What are three directions of computer development and three directions of communications development?

Are there any surprises left?

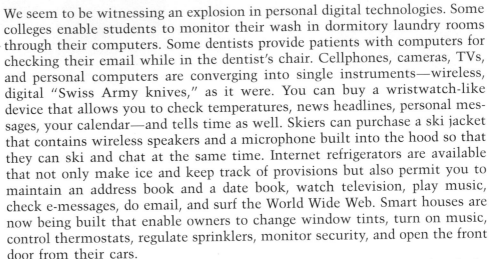

We seem to be witnessing an explosion in personal digital technologies. Some colleges enable students to monitor their wash in dormitory laundry rooms through their computers. Some dentists provide patients with computers for checking their email while in the dentist's chair. Cellphones, cameras, TVs, and personal computers are converging into single instruments—wireless, digital "Swiss Army knives," as it were. You can buy a wristwatch-like device that allows you to check temperatures, news headlines, personal messages, your calendar—and tells time as well. Skiers can purchase a ski jacket that contains wireless speakers and a microphone built into the hood so that they can ski and chat at the same time. Internet refrigerators are available that not only make ice and keep track of provisions but also permit you to maintain an address book and a date book, watch television, play music, check e-messages, do email, and surf the World Wide Web. Smart houses are now being built that enable owners to change window tints, turn on music, control thermostats, regulate sprinklers, monitor security, and open the front door from their cars.

Some of these devices might seem merely silly. Who really needs a high-tech fridge, for example? However, in some electricity-starved countries, such as Italy, networked appliances can monitor and adjust to power demands. More importantly, these devices serve to show that we live in the era of *pervasive computing* or *ubiquitous computing.* The world has moved on beyond boxy computers that sit on desks or even on laps. Today handheld wireless devices and smart cellphones (not to mention terminals everywhere—libraries, airports, cafés) let us access information anytime, anywhere. And not just general information but personal information—electronic correspondence, documents, appointments, photos, songs, money matters, and other data important to us.

Central to this concept is the internet—the "Net, or net," that sprawling collection of data residing on computers around the world and accessible by high-speed connections. Everything that presently exists on a personal computer, experts suggest, will move onto the internet, giving us greater mobility and wrapping the internet around our lives.[1] So central is the internet to our existence, in fact, that many writers are now spelling it without the capital "I"—*Internet* becomes *internet,* just as *Telephone* became *telephone*—because both systems belong not to just one owner but to the world. We will follow this convention in this book.

With so much information available everywhere all the time, what will this do to us as human beings? We can already see the outlines of the future. One result is *information overload:* a report from the University of California, Berkeley, estimated that 30% more information existed in 2003 than 2 years earlier.[2] Another is *less use of our brains for memorizing:* familiar phone numbers and other facts are being stored on speed-dial cellphones, pocket computers, and electronic databases, increasing our dependence on technology.[3] A third result is a *surge in "multitasking" activity:* people have become highly skilled in performing several tasks at once, such as doing homework while talking on the phone, watching TV, answering email, and surfing the web, although efficiency is diminished, since the brain has limits and can do only so much at one time. A fourth result is that many people, especially younger ones, *aren't careful about protecting privacy,* since they are used to providing information online and don't care about electronic surveillance.[4] The fifth is that *smart mobile devices could produce "smart mobs,"* groups of people who can cooperate in doing things together in ways

"Just keeping busy." Multiple electronic devices allow people to do multiple tasks simultaneously—multitasking.

never before possible, even if they don't know one another—*but they could also produce "dumb mobs" manipulated by the government or corporate marketers.*[5] These trends pose unique challenges to how you learn and manage information. An important purpose of this book is to give you the tools for doing so, as we explain at the end of this chapter.

In this chapter, we begin by discussing how becoming computer savvy can benefit you and how computing and the internet affect your life. We then discuss cellphones, the internet, the World Wide Web, and other aspects of the e-world. We next describe the varieties of computers that exist. We then explain the three key concepts behind how a computer works and what goes into a personal computer, both hardware and software. We conclude by describing three directions of computer development and three directions of communications development.

1.1

The Practical User: How Becoming Computer Savvy Benefits You

What does being computer savvy mean, and what are its practical payoffs?

There is no doubt now that for most of us information technology is becoming like a second skin—an extension of our intellects and even emotions, creating almost a parallel universe of "digital selves." Perhaps you have been using computers a long time and in a multitude of ways, or perhaps not. Either way, this book hopes to deliver important practical rewards by helping you become "computer streetwise"—that is, computer savvy. **Being computer savvy means knowing what computers can do and what they can't, knowing how they can benefit you and how they can harm you, knowing when you can solve computer problems and when you have to call for help.**

Among the practical payoffs are the following:

YOU WILL KNOW HOW TO MAKE BETTER BUYING DECISIONS No matter how much computer prices come down, you will always have to make judgments about quality and usefulness when buying equipment and software. In fact, we start you off right in this chapter by identifying the constituent parts of a computer system, what they do, and how much they cost.

YOU WILL KNOW HOW TO FIX ORDINARY COMPUTER PROBLEMS Whether it's replacing a printer cartridge, obtaining a software improvement ("patch"), or pulling photos from your digital camera or camera cellphone, we hope this book will give you the confidence to deal with the continual challenges that arise with computers—and know when and how to call for help.

YOU WILL KNOW HOW TO UPGRADE YOUR EQUIPMENT & INTEGRATE IT WITH NEW PRODUCTS New gadgetry and software are constantly being developed. A knowledgeable user learns under what conditions to upgrade, how to do so, and when to start over by buying a new machine.

Competence. To be able to choose a computer system or the components to build one, you need to be computer savvy.

3

YOU WILL KNOW HOW TO USE THE INTERNET MOST EFFECTIVELY
The sea of data that exists on the internet and other online sources is so great that finding what's best can be a hugely time-consuming activity. We hope to show you the most workable ways to approach this problem.

YOU WILL KNOW HOW TO PROTECT YOURSELF AGAINST ONLINE VILLAINS The online world poses real risks to your time, your privacy, your finances, and your peace of mind—spammers, hackers, virus senders, identity thieves, and companies and agencies constructing giant databases of personal profiles, as we will explain. This book aims to make you streetwise about these threats.

YOU WILL KNOW WHAT KINDS OF COMPUTER USES CAN ADVANCE YOUR CAREER Even top executives now use computers, as do people in careers ranging from police work to politics, from medicine to music, from retail to recreation. We hope you will come away from this book with ideas about how the technology can benefit you in whatever work you choose.

Along the way—in the Experience Boxes, Practical Action Boxes, Survival Tips, and More Infos—we offer many kinds of practical advice that we hope will help you become truly computer savvy in a variety of ways, large and small.

info! **More Info!**

From now on, whenever you see the More Info! icon in the margin, you'll find information about internet sites to visit and how to search for terms related to the topic just discussed.

1.2

Information Technology & Your Life: The Future Now

What is information technology, and how does it affect education, health, money, leisure, government, and careers?

This book is about computers, of course. But not just about computers. It is also about the way computers communicate with one another. When computer and communications technologies are combined, the result is *information technology*, or "infotech." **Information technology (IT) is a general term that describes any technology that helps to produce, manipulate, store, communicate, and/or disseminate information.** IT merges computing with high-speed communications links carrying data, sound, and video. Examples of information technology include personal computers but also new forms of telephones, televisions, appliances, and various handheld devices.

The Two Parts of IT: Computers & Communications

How do I distinguish computer technology and communications technology?

Note that there are two important parts to information technology—computers and communications.

COMPUTER TECHNOLOGY You have certainly seen and, we would guess, used a computer. Nevertheless, let's define what it is. **A _computer_ is a programmable, multiuse machine that accepts data—raw facts and figures— and processes, or manipulates, it into information we can use,** such as summaries, totals, or reports. Its purpose is to speed up problem solving and increase productivity.

COMMUNICATIONS TECHNOLOGY Unquestionably you've been using communications technology for years. **_Communications technology_, also called _telecommunications technology_, consists of electromagnetic devices and systems for communicating over long distances.** The principal examples are telephone, radio, broadcast television, and cable TV. In more recent times, there has been the addition of communication among computers—which is what happens when people "go online" on the internet. In this context, **_online_ means using a computer or some other information device, connected through a network, to access information and services from another computer or information device. A _network_ is a communications system connecting two or more computers; the internet is the largest such network.**

Information technology is already affecting your life in exciting ways and will do so even more in the future. Let's consider how.

Education: The Promise of More Interactive & Individualized Learning

How is information technology being used in education?

In her sociology classes at Indiana University, professor Melissa Wilde uses a small wireless keypad linked to a computer to enable students to answer questions not by raising their hands but by pressing buttons, with the results appearing on a screen in the front of the room. Wilde has her students answer multiple-choice questions to see whether they understand her lecture points and to make necessary adjustments. "I can instantly see that three-quarters of the class doesn't get it," she says.[6] She also uses the technology to get students to answer questions about themselves—race, income, political affiliation—to show that compared to the national average the classroom is skewed, for example, toward wealthier or poorer students, an event that fired up a half hour of excited class discussion.

Maybe the classrooms at your school haven't reached this level of interactivity yet, but there's no question that information technology is universal on college campuses, and at lower levels the internet has penetrated 99% of schools.[7] Most college students have been exposed to computers since the lower grades; indeed, one-fifth of college students report they were using computers between ages 5 and 8, and all had begun using computers by the time they were 16–18 years old. At the college level, the great majority (85%) of students have their own computer, and two-thirds use at least two email addresses. As you no doubt know, **_email_ is "electronic mail," messages transmitted over a computer network, most often the internet.** Three quarters of college students use the internet 4 or more hours a week, and about one-fifth use it 12 or more hours a week. About half are required to use email in their classes. For academic purposes, most students use email to set up appointments (62%) with professors, discuss grades (58%), or get clarification of an assignment (75%). (But nearly three-fourths of students say most of their online communication is with friends.)[8]

Besides using the internet to help in teaching, today's college instructors also use _presentation graphics software_ such as PowerPoint to show their lecture outlines and other materials on classroom screens (as we discuss in Chapter 3). In addition, they use Blackboard, WebCT, and other **_course-management software_ for administering online assignments, schedules, examinations, and grades.**[9] One of the most intriguing developments in education at all levels, however, is the rise of **_distance learning_, or _e-learning_, the name given to online education programs,** which has nearly 1 million online students enrolled nationwide and grows more than 20% a year.[10] E-learning has had some interesting effects: for example, the home-schooling movement, whereby children are taught at home, usually by parents, has come of age thanks to internet resources.[11] E-learning has also propelled the rise of for-profit institutions, such as DeVry and the University of Phoenix,

■more
info!

Online Colleges

The following websites provide detailed information about getting college degrees online:
www.classesusa.com/featured-schools/fos/index.cfm?gcid=S1
5196x002&keyword=distance
%20learning%20education

www.guidetoonlineschools.com/
?source=google_gtos_distlrn2

www.petersons.com/
distancelearning/

www.usdla.org

which 1 in 12 college students now attends.[12] More than a third of institutions of higher education—and 97% of public universities—offer online courses, and many have attracted on-campus students, who say they like the flexibility of not having to attend their classes at a set time.[13]

E-learning has been put to such varied uses as bringing career and technical courses to high school students in remote prairie towns, pairing gifted science students with master teachers in other parts of the country, and helping busy professionals obtain further credentials outside business hours. But the reach of information technology into education has only begun. In the future, we will see software called "intelligent tutoring systems" that give students individualized instruction when personal attention is scarce—such as the software Cognitive Tutor, which not only helps high school students to improve their performance in math but also sparks them to enjoy a subject they might have once hated. In colleges, more students may use interactive simulation games, such as McGraw-Hill's Business Strategy Game, to apply their knowledge to real-world kinds of problems. And employees in company training programs may find themselves engaged in mock conversations with **_avatars_—computer depictions of humans,** as are often found in online videogames—that represent imaginary customers and coworkers, combining the best parts of computer-based learning with face-to-face interaction.[14]

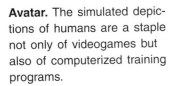

Avatar. The simulated depictions of humans are a staple not only of videogames but also of computerized training programs.

Health: High Tech for Wellness

How are computers being used in health and medicine?

Viktor Yazykov, competing in the perilous Around Alone solo sailing competition, found himself in the stormy South Atlantic with a seriously infected arm that needed emergency surgery. So, with the help of step-by-step instructions sent by email from Boston-based Dr. Daniel Carlin to his solar-powered laptop computer, Yazykov operated on his own arm.[15]

Yazykov's story is a dramatic example of **_telemedicine_—medical care delivered via telecommunications.** For some time, physicians in rural areas lacking local access to radiologists have used "teleradiology" to exchange computerized images such as X-rays via telephone-linked networks with expert physicians in metropolitan areas. Now telemedicine is moving to an exciting new level, as the use of digital cameras and sound, in effect, moves patients to doctors rather than the reverse. Already telemedicine is being

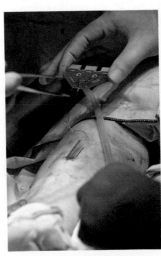
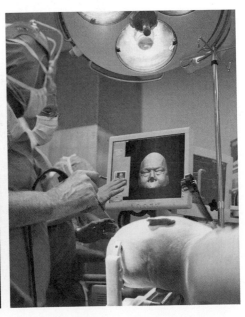

High-tech medicine. (Left) British university professor Ken Warwick having electrodes implanted in his head, enabling his nervous system to be connected to a computer, which will record his brain signals. He is the world's first cyborg—half human, half machine. (Right) Brain surgeons plot the surface of a patient's head into a 3D computer model, enabling them to accurately locate a tumor.

info!

more

Health Websites

Some reliable sources are these:

www.medlineplus.gov
www.healthweb.org
www.nimh.nih.gov
www.4woman.gov
www.mayoclinic.com
www.nationalhealthcouncil.org
www.yourdiseaserisk.
  harvard.edu

embraced by administrators in the American prison system, where by law inmates are guaranteed medical treatment—and where the increase in prisoners every year has led to the need to control health costs.

Computer technology is also radically changing the tools of medicine. All medical information, including that generated by X-ray, lab test, and pulse monitor, can now be transmitted to a doctor in digital format. And image transfer technology allows radiologic images such as CT scans and MRIs to be immediately transmitted to electronic charts and physicians' offices.[16] Patients in intensive care, who are usually monitored by nurses during off-times, can also be watched over by doctors in remote "control towers" miles away.[17] Electronic medical records and other computerized tools enable heart attack patients to get follow-up drug treatment and diabetics to have their blood sugar measured.[18] Software can compute a woman's breast cancer risk.[19] Patients can use email to query their doctors about their records (although there are still privacy and security issues).[20] Various **_robots_—automatic devices that perform functions ordinarily performed by human beings,** with names such as robo doc, RoboCart, TUG, and HelpMate—help free medical workers for more critical tasks; the four-armed da Vinci surgical robot, for instance, can do cuts and stitches deep inside the body, so that surgery is less traumatic and recovery time faster.[21] Hydraulics and computers are being used to help artificial limbs get "smarter."[22] And a patient paralyzed by a stroke has received an implant that allows communication between his brain and a computer; as a result, he can move a cursor across a screen by brainpower and convey simple messages—as in _Star Trek._[23]

Want to calculate how long you will live? Go to _www.livingto100.com,_ an online calculator developed by longevity researchers at Harvard Medical School and Boston Medical Center. Want to go about gathering your family health history to see if you're at risk for particular inherited diseases? Go to _www.hhs.gov/familyhistory_ to find out how. These are only two examples of health websites available to patients and health consumers. Although online health information can be misleading and even dangerous, many people now tap into health care databases, email health professionals, or communicate with people who have similar conditions. For instance, hours after 10-year-old Robert Lord of San Diego fractured his spine in a fall from a tree, his

Robots. (Left) A new humanoid robot, HRP-2 Promet, developed by the National Institute of Advanced Industrial Science and Technology and Kawada Industries, Inc. Five feet tall, it performs traditional Japanese dancing. Priced at $365,000, the robot can help workers at construction sites and also drive a car. (Right) This sea bream is about 5½ pounds and can swim up to 38 minutes—before recharging. The robot fish, created by Mitsubishi, looks and swims exactly like the real thing.

father found an experimental drug on the internet, saving the boy from life-time paralysis.[24] Often patients are already steeped in information about their conditions when they arrive in the offices of health care professionals. "It's a fundamental shift of knowledge, and therefore power, from physicians to patients," says one consultant.[25] In addition, health care consumers are able to share experiences and information with one another. Young parents, for example, can find an online gathering spot (chat room) at pediatrician Alan Greene's website at *www.drgreene.com.* Finally, if you want to begin storing your medical records in electronic form, visit *healthmanager .webmd.com* or, to put them on an electronic keychain, *med-infochip.com.* And a Florida firm has been cleared to market implantable microchips, worn under the skin, that would provide easy access to individual medical records.[26]

Money: Toward the Cashless Society

How will computers affect my financial matters?

"The future of money is increasingly digital, likely virtual, and possibly universal," says one writer.[27] ___Virtual___ **means that something is created, simulated, or carried on by means of a computer or a computer network,** and we certainly have come a long way toward becoming a cashless society. Indeed, the percentage of all financial transactions done electronically, both phone-initiated and computer-initiated, was projected to rise to 18.4% in 2013, up from 0.9% in 1993.[28] Besides currency, paper checks, and credit and debit cards, the things that serve as "money" include cash-value cards (such as subway fare cards), automatic transfers (such as direct-deposit paychecks), and digital money ("electronic wallet" accounts such as PayPal).

Many readers of this book will probably already have engaged in online buying and selling, purchasing CDs and DVDs, books, airline tickets, or computers. But what about groceries? After all, you can't exactly squeeze the cantaloupes through your keyboard. Even so, online groceries are expected to reach $6.5 billion in U.S. sales by 2008.[29] Despite spectacular failures (such as Webvan in 2001), e-grocers such as Safeway, Albertsons, PeaPod, and Freshdirect.com have steadily expanded their operations in cities with large populations and heavy internet use. To change decades of shopping habits, such grocers keep their delivery charges low and delivery times convenient, and they take great pains in filling orders, knowing that a single bad piece of fruit will produce a devastating word-of-mouth backlash.

Only about 46% of U.S. workers have their paychecks electronically deposited into their bank accounts (as opposed to 95% or more in Japan, Norway, and Germany, for example), but this is sure to change as Americans discover that direct deposit is actually safer and faster. Online bill paying is also picking up steam. For more than two decades, it has been possible to pay bills online, such as those from phone and utility companies, with special software and online connections to your bank. About 19 million American households do some bill paying online, and the number is expected to reach 61 million households by 2008.[30]

Some banks and other businesses are backing an electronic-payment system that allows internet users to buy goods and services with ___micropayments___, **electronic payments of as little as 25 cents in transactions for which it is uneconomical to use a credit card.** Micropayment, suggests futurist Paul Saffo, "allows you to buy things by the sip rather than the gulp."[31] The success of Apple Computer's iTunes online music service, which sells songs for 99 cents each, suggests that micro sales are now feasible. All kinds of businesses and organizations, from independent songwriters to comic book writers to the Legal Aid Society of Cleveland, now accept micropayments, using intermediaries such as BitPass and Peppercoin.[32] Thus, you could set up your own small business simply by constructing a website (we show you how later in the book) and accepting micropayments.

Leisure: Infotech in Entertainment & the Arts

How will my leisure activities be affected by information technology?

Information technology is being used for all kinds of entertainment, ranging from videogames to telegambling. It is also being used in the arts, from painting to photography. Let's consider just two examples, music and film.

Computers, the internet, and the World Wide Web are standing the system of music recording and distribution on its head—and in the process are changing the financial underpinnings of the music industry. Because of their high overhead, major record labels typically need a band to sell half a million CDs in order to be profitable, but independent bands, using online marketing, can be reasonably successful selling 20,000 or 30,000 albums. Team Love, a small music label established in 2003, found it could promote its first two bands, Tilly and the Wall and Willy Mason, by offering songs online free for <u>**downloading**</u>—**transferring data from a remote computer to one's own computer**—so that people could listen to them before paying $12 for a CD. It also puts videos online for sharing and uses quirky websites to reach fans. "There's something exponential going on," says one of Team Love's founders. "The more music that's downloaded, the more it sells."[33] Many independent musicians are also using the internet to get their music heard, hoping that giving away songs will help them build audiences.[34]

The web also offers sources for instantly downloadable sheet music (see *www.everynote.com*, *www.musicnotes.com*, *www.sheetmusicdirect.com*, and *www.sunhawk.com*). One research engineer has devised a computerized scoring system for judging musical competitions that overcomes the traditional human-jury approach, which can be swayed by personalities and politics.[35] And a Spanish company, PolyphonicHMI, has created Hit Song Science software, which they say can analyze the hit potential of new songs by, according to one description, "reference to a finely parsed universe of attributes derived from millions of past songs."[36]

As for movies, now that blockbuster movies routinely meld live action and animation, computer artists are in big demand. The 1999 film *Star Wars: Episode I*, for instance, had fully 1,965 digital shots out of about 2,200 shots. Even when film was used, it was scanned into computers to be tweaked with animated effects, lighting, and the like. Entire beings were created on computers by artists working on designs developed by producer George Lucas and his chief artist.[37]

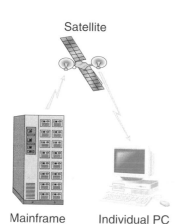

Satellite

Mainframe   Individual PC

Download
(reverse the direction of data transmission to **upload**)

more info!

Free Music Online

Places to look for free—and legal—music online:
www.webjay.org
www.epitonic.com
www.furthurnet.com
http://memory.loc.gov/ammem/audio.html
www.archive.org
www.garageband.com

Entertainment. (Left) Computer-based special effects in *Harry Potter and the Prisoner of Azkaban* turned a woman into a balloon. (Right) An indoor "winter" sports facility in Japan; the system uses microprocessors to keep lifts running, snow falling, and temperature at 26 degrees.

more info!

Online Movie Tickets

Three sites offer movie tickets, as well as reviews and other materials. In some cities you can print out tickets at home.
www.fandango.com
www.moviefone.com
www.movietickets.com

What is driving the demand for computer artists? One factor is that animation, though not cheap, looks more and more like a bargain, because hiring movie actors costs so much—some make $20 million a film. Moreover, special effects are readily understood by audiences in other countries, and major studios increasingly count on revenues from foreign markets to make a film profitable. Digital manipulation also allows a crowd of extras to be multiplied into an army of thousands. It can also be used to create settings: in the 2004 film *Sky Captain and the World of Tomorrow*, the actors—Gwyneth Paltrow, Angelina Jolie, and Jude Law—shot all their scenes in front of a blue screen, and computer-generated imagery was then used to transport them into an imaginary world of 1939.[38] Computer techniques have even been used to develop digitally created actors—called "synthespians." (Thespis was the founder of ancient Greek drama; thus, a thespian works in drama as an actor.) Actors ranging from the late James Dean to the late John Wayne, for instance, have been recruited for television commercials. And computerized animation is now so popular that Hollywood studios and movie directors are finding they can make as much money from creating videogames as from making movies.[39]

But animation is not the only area in which computers are revolutionizing movies. Digital editing has radically transformed the way films are assembled. Whereas traditional film editing involved reeling and unreeling spools of film and cutting and gluing pieces of highly scratchable celluloid together, nearly burying the editor in film, today an editor can access 150 miles of film stored on a computer and instantly find any visual or audio moment, allowing hundreds of variations of a scene to be called up for review. Even nonprofessionals can get into movie making as new computer-related products come to market. Now that digital video capture-and-edit systems are available for under $1,000, amateurs can turn home videos into digital data and edit them. Also, digital camcorders, which offer outstanding picture and sound quality, have steadily dropped in price.

Government & Electronic Democracy: Participating in the Civic Realm

What ways are computers changing government and politics?

A Rutgers University study suggests that the internet has great potential for civic betterment because it is fast and cheap for users (once they are connected) and facilitates communication among citizens better than do mass media such as radio and TV.[40] And a study by the Pew Internet & American Life Project found that internet users are much more likely to contact government than are nonusers because of the ease of finding information online and of contacting officials through email.[41]

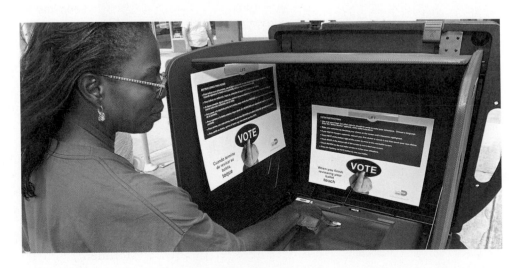

Electronic voting. Voting using computer technology and a touch screen to vote.

Some cities have adopted Neighborhood Link (*www.neighborhoodlink .com*), a free, easy-to-use system of neighborhood websites in which residents can communicate among themselves and with local governments. In Denver, for instance, the system serves 402 neighborhoods. In Austin, Texas, an entrepreneur formed E-The People (*www.e-thepeople.org*), which describes itself as "America's Interactive Town Hall" and is designed to connect citizens with their government officials, local and national, everywhere in the United States. In Nevada, citizens visit the state legislature's hearings and floor voting sessions by accessing the legislature's website, where they can either listen to internet broadcasts or read the text of legislation. In Seattle, citizens can go to their city's website to deal with everything from absentee ballots to youth and family services. Besides cutting expenses for stamps, paper, and employees, e-government helps reduce lines and offers people more convenience in paying taxes and parking tickets, renewing vehicle registration and driver's licenses, viewing birth and marriage certificates, and applying for public sector jobs.[42]

The internet and other information technology have also changed much of the political process, both for good and for ill. On the one hand, the net has enabled political candidates and political interest groups to connect with voters in new ways, to raise money from multiple small donors instead of just rich fat cats, and (using cellphones and text messaging) to organize street protests. On the other hand, computers have allowed incumbent legislators to design (gerrymander) voting districts that make it nearly impossible for them to be dislodged; electronic tools have also made it easier than ever for political parties to skirt or break campaign laws, and computerized voting machines still don't always count votes as they are supposed to. Still, websites and bloggers have become important watchdogs on government. The website eDemocracy (*www.e-democracy.org*), for instance, can help citizens dig up government conflicts of interest, and websites such as Project Vote Smart (*www.vote-smart.org*) outline candidates' positions.

Jobs & Careers

How could I use computers to advance my career?

Today almost every job and profession requires computer skills of some sort. Some are ordinary jobs in which computers are used as ordinary tools. Others are specialized jobs in which advanced computer training combined with professional training gives people dramatically new kinds of careers. Consider:

Careers. Front-desk workers at many hotels use computers to check guests in.

- In the hotel business, even front-desk clerks need to know how to deal with computerized reservation systems. Some hotels, however, also have a so-called computer concierge, someone with knowledge of computer systems who can help computer-carrying guests with online and other problems.

- In law enforcement, police officers need to know how to use computers while on patrol or at their desks to check out stolen cars, criminal records, outstanding arrest warrants, and the like. However, investigators with specialized computer backgrounds are also required to help solve fraud, computer break-ins, accounting illegalities, and other high-tech crimes.

- In entertainment, computers are used for such ordinary purposes as budgets, payroll, and ticketing. However, there are also new careers in virtual set design, combining training in architecture and 3-D computer modeling, and in creating cinematic special effects.

Clearly, information technology is changing old jobs and inventing new ones. To prosper in this environment, you need to combine a traditional

Police work. Many police officers now have computers in their cars; this officer is checking a license plate number.

education with training in computers and communications. You also need to be savvy about job searching, résumé writing, interviewing, and postings of employment opportunities. Advice about careers, job hunting, occupational trends, and employment laws is available at Yahoo!, Google, and other websites. Some starting annual salaries for recent college graduates are shown below; note that jobs involving degrees in computers and information systems occupy four of the seven top-paying starting salaries. (● *See Panel 1.1.*)

Computers can be used both for you to find employers and for employers to find you.

WAYS FOR YOU TO FIND EMPLOYERS   As you might expect, the first to use cyberspace as a job bazaar were companies seeking people with technical

● PANEL 1.1
Entering the job market
Starting annual salaries for 2004 graduates.

Discipline (bachelor's degree)	Current Average
Computer Engineering	$51,297
Electrical Engineering	$51,124
Computer Science	$49,036
Mechanical Engineering	$48,578
Aerospace/Aeronautical/Astronautical Engineering	$48,220
Information Sciences and Systems	$42,375
Management Information Systems/ Business Data Processing	$41,579
Accounting	$41,058
Economics/Finance	$40,630
Business Administration/Management	$38,254
Marketing/Marketing Management	$34,712
Political Science/Government	$32,296
English Language and Literature	$31,113
History	$30,344
Communications	$29,763
Liberal Arts and Sciences/General Studies	$29,713
Biological Sciences/Life Sciences	$29,629
Criminal Justice and Corrections	$29,428
Sociology	$29,168
Psychology	$28,230

Source: Adapted from the Fall 2004 *Salary Survey*, with the permission of the National Association of Colleges and Employers, copyright holder. © National Association of Colleges and Employers. All rights reserved. 62 Highland Ave. Bethlehem, PA 18017, *www.naceweb.org.*

America's Job Bank: **www.ajb.dni.us**

Career Builder: **www.careerbuilder.com**

College Grad Job Hunter: **www.collegegrad.com**

Erecruiting: **www.erecruiting.com**

FedWorld (U.S. Government jobs): **www.fedworld.gov**

NationJob Network: **www.nationjob.com**

Hot Jobs: **www.hotjobs.com**

Jobs.com: **www.jobs.com**

Jobs on Line: **www.jobsonline.com**

MonsterTrak.com: **www.jobtrak.com**

JobWeb: **www.jobweb.org**

Monster.com: **www.monster.com**

● **PANEL 1.2**
Some websites that post job listings

backgrounds and technical people seeking employment. However, as the public's interest in commercial services and the internet has exploded, the focus of online job exchanges has broadened. Now, interspersed among ads for programmers on the internet are openings for forest rangers in Idaho, physical therapists in Atlanta, models in Florida, and English teachers in China. Most websites are free to job seekers, although many require that you fill out an online registration form. (● *See Panel 1.2.*)

WAYS FOR EMPLOYERS TO FIND YOU Posting your résumé online for prospective employers to view is attractive because of its low (or zero) cost and wide reach. But does it have any disadvantages? Certainly it might if the employer who sees your posting happens to be the one you're already working for. In addition, you have to be aware that you lose control over anything broadcast into cyberspace. You're putting your credentials out there for the whole world to see, and you need to be somewhat concerned about who might gain access to them.

If you have a technical background, it's definitely worth posting your résumé with an electronic jobs registry, since technology companies in particular find this an efficient way of screening and hiring. However, posting may also benefit people with less technical backgrounds. Online recruitment is popular with companies because it prescreens applicants for at least basic computer skills. If you've mastered the internet, you're likely to know something about word processing, spreadsheets, and database searching as well, knowledge required in most good jobs these days.

One wrinkle in job seeking is to prepare a résumé with web links and/or clever graphics and multimedia effects and then put it on a website to entice employers to chase after you. If you don't know how to do this, there are many companies that—for a fee—can convert your résumé and publish it on their own websites. Some of these services can't dress it up with fancy graphics or multimedia, but since complex pages take longer for employers to download anyway, the extra pizzazz is probably not worth the effort. A number of websites allow you to post your résumé for free. Another wrinkle is to pay extra to move your résumé higher in the listings so that it will stand out compared with competing résumés. For example, for an extra $20–$150 apiece, Careerbuilder.com will move your listing toward the top of the search heap, and the company says that employers click on upgraded résumés 200% more often than regular ones.[43]

1.3

Infotech Is All-Pervasive: Cellphones, Email, the Internet, & the E-World

How does information technology facilitate email, networks, and the use of the internet and the web; what is the meaning of the term cyberspace?

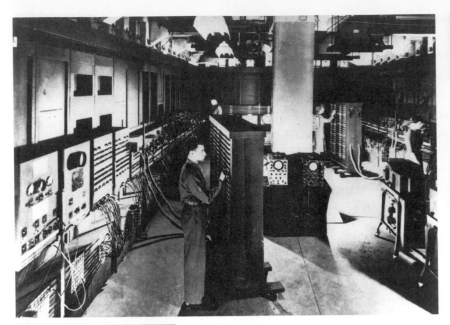

● **PANEL 1.3**
Grandparent and offspring
ENIAC (left) is the grandparent of today's smartphones (right).

One of the first computers, the outcome of military-related research, was delivered to the U.S. Army in 1946. ENIAC (short for "Electronic Numerical Integrator And Calculator") weighed 30 tons and was 80 feet long and two stories high, but it could multiply a pair of numbers in the then-remarkable time of three-thousandths of a second. (● *See Panel 1.3.*) This was the first general-purpose, programmable electronic computer, the grandparent of today's lightweight handheld machines—including the smart cellphone. Some of the principal historical developments are illustrated in the timeline below. (● *See Panel 1.4.*)

The Telephone Grows Up

How has the telephone changed?

● **PANEL 1.4**
Timeline
Overview of some of the historical developments in information technology. The timeline is modified in upcoming chapters to show you more about the people and advancements contributing to developments in information technology.

Cellphone mania has swept the world. All across the globe, people have acquired the portable gift of gab, with some users making 45 or more calls a day. Phone maker Nokia estimated that there will be 2 billion mobile phone subscriptions worldwide in 2006.[44] It has taken more than 100 years for the telephone to get to this point—getting smaller, acquiring push buttons, losing its cord connection. In 1964, the ★ and # keys were added to the keypad. In 1973, the first cellphone call was processed.

In its most basic form, the telephone is still so simply designed that even a young child can use it. However, it is now becoming more versatile and complex—a way of connecting to the internet and the World Wide Web. Indeed, internet smartphones—such as the Treo 650, Audiovox SMT 5600,

4000–1200 BCE	3500 BCE–2900 BCE	3000 BCE	1270 BCE	900 BCE	530 BCE	100 CE
Inhabitants of the first known civilization in Sumer keep records of commericial transactions on clay tablets	Phoenicians develop an alphabet; Sumerians develop cuneiform writing; Egyptians develop hierogylphic writing	Abacus is invented in Babylonia	First encyclopedia (Syria)	First postal service (China)	Greeks start the first library	First bound books

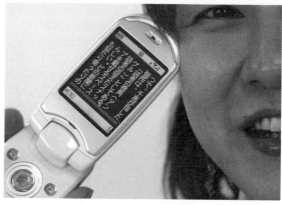

Smartphones. (Left) Motorola E398 smartphone. (Middle) Ojo videophone, which allows face-to-face chats long distance. (Right) Bandai Networks' mobile phone that displays a novel. Bandai offers 150 books on its site for phone owners to download and read on their phones.

Motorola E398, and LG VX8000—represent another giant step for information technology. Now you no longer need a personal computer to get on the internet. Smartphones in their various forms enable you not only to make voice calls but also to send and receive text messages, browse the World Wide Web, and obtain news, research, music, photos, movies, and TV programs. (And with camera and camcorder cellphones, you can send images, too.) According to one survey, the percentage of people who use nonvoice applications for text messages is 27%; email 11%; internet 9%; and photography 6%—and the numbers of users for these options are growing all the time.[45]

"You've Got Mail!" Email's Mass Impact

What makes email distinctive from earlier technologies?

It took the telephone 40 years to reach 10 million customers, and fax machines 20 years. Personal computers made it into that many American homes 5 years after they were introduced. Email, which appeared in 1981, became popular far more quickly, reaching 10 million users in little more than a year.[46] No technology has ever become so universal so fast. Thus, one of the first things new computer and internet users generally learn is how to send and receive email.

Until 1998, hand-delivered mail was still the main means of correspondence. But in that year, the volume of email in the United States surpassed the volume of hand-delivered mail. By 2006, the total number of email messages sent daily was expected to exceed 60 billion worldwide.[47] Already, in

700–800	1049	1450	1455	1621	1642	1666
Arabic numbers introduced to Europe	First moveable type (clay) invented in China	Newspapers appear in Europe	Printing press (J. Gutenberg, Germany)	Slide rule invented (Edmund Gunther)	First mechanical adding machine (Blaise Pascal)	First mechanical calculator that can add and subtract (Samuel Morland)

fact, email is the leading use of PCs. Because of this explosion in usage, suggests a *BusinessWeek* report, "email ranks with such pivotal advances as the printing press, the telephone, and television in mass impact."[48]

Using electronic mail clearly is different from calling on a telephone or writing a conventional letter. As one writer puts it, email "occupies a psychological space all its own. It's almost as immediate as a phone call, but if you need to, you can think about what you're going to say for days and reply when it's convenient."[49] Email has blossomed, points out another writer, not because it gives us more immediacy but because it gives us *less*. "The new appeal of email is the old appeal of print," he says. "It isn't instant; it isn't immediate; it isn't in your face." Email has succeeded for the same reason that the videophone—which allows callers to see each other while talking—has been so slow to catch on: because "what we actually want from our exchanges is the minimum human contact commensurate with the need to connect with other people."[50] It will be interesting to see, however, whether this observation holds up during the next few years if marketers roll out more videophones.

What is interesting, though, is that in these times when images often seem to overwhelm words, email is actually *reactionary*. "The internet is the first new medium to move decisively backward," points out one writer, because it essentially involves writing. "Ten years ago, even the most literate of us wrote maybe a half a dozen letters a year; the rest of our lives took place on the telephone."[51] Email has changed all that—and put pressure on businesspeople in particular to sharpen their writing skills.

The Internet, the World Wide Web, & the "Plumbing of Cyberspace"

What's the difference between the net, the web, and cyberspace?

As the success of the cellphone shows, communications has extended into every nook and cranny of civilization (with poorer nations actually the leaders in cellphone growth), a development called the "plumbing of cyberspace." The term *cyberspace* was coined by William Gibson in his novel *Neuromancer* (1984) to describe a futuristic computer network into which users plug their brains. (*Cyber* comes from "cybernetics," a term coined in 1948 to apply to the comparative study of automatic control systems, such as the brain/nervous system and mechanical-electrical communication systems.) In everyday use, this term has a rather different meaning.

Today many people equate cyberspace with the internet. But it is much more than that. Cyberspace includes not only the web, chat rooms, online

1714	1801	1820	1829	1833	1843
First patent for a typewriter (England)	A linked sequence of punched cards controls the weaving patterns in Jacquard's loom	The first mass-produced calculator, the Thomas Arithnometer	William Austin patents the first workable typewriter in America	Babbage's difference engine (automatic calculator)	World's first computer programmer, Ada Lovelace, publishes her notes

Always on. Most of today's students don't remember a time before the existence of cyberspace.

diaries (blogs), and member-based services such as America Online—all features we explain in this book—"but also such things as conference calls and automatic teller machines," says David Whittler.[52] We may say, then, that _cyberspace_ **encompasses not only the online world and the internet in particular but also the whole wired and wireless world of communications in general**—the nonphysical terrain created by computer and communications systems. Cyberspace is where you go when you go online with your computer.

THE NET & WEB DEFINED   The two most important aspects of cyberspace are the internet and that part of the internet known as the World Wide Web. To give them formal definition:

- **The internet—"the mother of all networks":** The internet is at the heart of the Information Age. Called "the mother of all networks," the _internet_ **(the "net") is a worldwide computer network that connects hundreds of thousands of smaller networks. These networks link educational, commercial, nonprofit, and military entities, as well as individuals.**

1844	1854	1876	1890	1895	1907	1920–1921
Samuel Morse sends a telegraph message from Washington to Baltimore	George Boole publishes "An Investigation on the Laws of Thought," a system for symbolic and logical reasoning that will become the basis for computer design	Alexander Graham Bell patents the telephone	Electricity used for first time in a data-processing project (punched cards) Hollerith's automatic census-tabulating machine (used punched cards)	First radio signal transmitted	First regular radio broadcasts, from New York	The word "robot," derived from the Czech word for compulsory labor, is first used to mean a humanlike machine

- **The World Wide Web—the multimedia part of the internet:** The internet has been around for more than 35 years. But what made it popular, apart from email, was the development in the early 1990s of the <u>*World Wide Web*</u>, **often called simply the "Web" or the "web"—an interconnected system of internet computers (called *servers*) that support specially formatted documents in multimedia form.** The word *multimedia*, from "multiple media," refers to technology that presents information in more than one medium, such as text, still images, moving images, and sound. In other words, the web provides information in more than one way.

THE INTERNET'S INFLUENCE   There is no doubt that the influence of the net and the web is tremendous. At present, more than 128 million Americans use the internet, according to the Pew Internet & American Life Project.[53] Between 51% and 69%, depending on the type of internet connections (dial-up versus broadband, as we'll describe), of all Americans go online in a typical day.[54] But just how revolutionary is the internet? Is it equivalent to the invention of television, as some technologists say? Or is it even more important—equivalent to the invention of the printing press? "Television turned out to be a powerful force that changed a lot about society," says *USA Today* technology reporter Kevin Maney. "But the printing press changed everything—religion, government, science, global distribution of wealth, and much more. If the internet equals the printing press, no amount of hype could possibly overdo it."[55] No massive study was ever done of the influence of the last great electronic revolution to touch us, namely, television. But the Center for Communication Policy at the University of California, Los Angeles, in conjunction with other international universities, has been looking at the effects of information technology—and at how people's behavior and attitudes toward it are changing. We will report on some of these findings throughout the book.[56]

College Students & the E-World

How does my use of information technology compare with that of other students?

One thing we know already is that cyberspace is saturating our lives. The worldwide internet population was projected to be 1.21 billion in 2006, with 185 million of that number American.[57] While the average age of users is rising, there's no doubt that people ages 18–27 (the "millennials," Gen Y, or the "Net Generation") love information technology, with 85% using computers and 78% using the net.[58] Among college students, 99% use email, four out of five carry cellphones (which more than a third use for text as well as voice messages), and more than 80% of on-campus students access the net through high-speed lines, which make it easier to obtain music and videos.[59] Most students multitask—switching between listening to music,

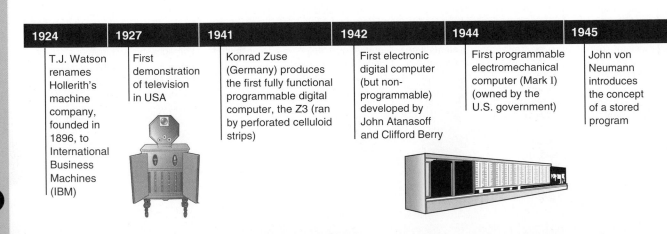

1924	1927	1941	1942	1944	1945
T.J. Watson renames Hollerith's machine company, founded in 1896, to International Business Machines (IBM)	First demonstration of television in USA	Konrad Zuse (Germany) produces the first fully functional programmable digital computer, the Z3 (ran by perforated celluloid strips)	First electronic digital computer (but non-programmable) developed by John Atanasoff and Clifford Berry	First programmable electromechanical computer (Mark I) (owned by the U.S. government)	John von Neumann introduces the concept of a stored program

>>

PRACTICAL ACTION
Managing Your Email

For many people, email *is* the online environment, more so than the World Wide Web. According to one study, 60% of people who do emailing at work average 10 or fewer messages a day, 23% receive more than 20, and 6% receive more than 50.[60] But some people receive as many as 300 emails a day—with perhaps 200 being junk email (spam), bad jokes, or irrelevant memos (the "cc," previously "carbon copy," now "courtesy copy").

It's clear, then, that email will increase productivity only if it is used properly. Overuse or misuse just causes more problems and wastes time. The following are some ideas to keep in mind when using email:

- *Do your part to curb the email deluge:* Put short messages in the subject line so that recipients don't have to open the email to read the note. Don't reply to every email message you get. Avoid "cc:ing" (copying to) people unless absolutely necessary. Don't send chain letters or lists of jokes, which just clog mail systems.

- *Be helpful in sending attachments:* Attachments—computer files of long documents or images attached to an email—are supposed to be a convenience, but often they can be an annoyance. Sending a 1-megabyte file to a 500-person mailing list creates 500 copies of that file—and that many megabytes can clog the mail system. (A 1-megabyte file is about the size of a 300-page double-spaced term paper.) Ask your recipients beforehand if they want the attachment.

- *Be careful about opening attachments you don't recognize:* Some dangerous computer viruses—renegade programs that can damage your computer—have been spread by email attachments, that automatically activate the virus when they are opened.

- *Use discretion about the emails you send:* Email should not be treated as informally as a phone call.

Don't send a message electronically that you don't want some third party to read. Email messages are not written with disappearing ink; they remain in a computer system long after they have been sent. Worse, recipients can easily copy and even alter your messages and forward them to others without your knowledge.

- *Make sure emails to bosses, coworkers, and customers are literate:* It's okay to be informal when emailing friends, but employers and customers expect a higher standard. Pay attention to spelling and grammar.

- *Don't use email to express criticism and sarcasm:* Because email carries no tone or inflection, it's hard to convey emotional nuances. Avoid criticism and sarcasm in electronic messaging. Nevertheless, you can use email to provide quick praise, even though doing it in person will take on greater significance.

- *Be aware that email you receive at work is the property of your employer:* Be careful of what you save, send, and back up.

- *Realize that deleting email messages doesn't totally get rid of them:* "Delete" moves the email from the visible list, but the messages remain on your hard disk and can be retrieved by experts. Special software, such as Spytech Eradicator and Window Washer, will completely erase email from the hard disk.

- *Don't neglect real personal contact:* More companies are asking employees to trade email for more in-person contact with the people they work with, through such practices as banning the use of email on Fridays. This has come about because so many employees complain that they have to leave multiple messages when trying to get answers since coworkers don't respond in timely fashion.

1946	1947	1951	1952	1958	1962	1969
First programmable electronic computer in United States (ENIAC)	Invention of the transistor (enabled miniaturization of electronic devices)	Computers are first sold commercially	UNIVAC computer correctly predicts election of Eisenhower as U.S. President	Integrated circuit; first modem	The first computer game is invented (Spacewar)	ARPANet established by U.S. Advanced Research Project Agency, led to internet

Introduction to Information Technology

19

watching TV, trolling the internet, talking on the phone, and e-messaging friends—and still somehow are able to do some studying. They are also big participants in e-commerce, e-shopping, and e-business. For the Net Generation, the digital media are "like air."[61] The electronic world is everywhere. The net and the web are everywhere. Cyberspace permeates everything.

1.4

The "All-Purpose Machine": The Varieties of Computers

What are the five sizes of computers, and what are clients and servers?

When the ★alarm clock blasts you awake, you leap out of bed and head for the kitchen, where you check the ★coffee maker. After using your ★electronic toothbrush and showering and dressing, you stick a bagel in the ★microwave, and then pick up the ★TV remote and click on the ★TV to catch the weather forecast. Later, after putting dishes in the ★dishwasher, you go out and start up the ★car and head toward campus or work. Pausing en route at a ★traffic light, you turn on your ★CD player to listen to some music.

You haven't yet touched a PC, a personal computer, but you've already dealt with at least 10 computers—as you probably guessed from the ★s. All these familiar appliances rely on tiny "computers on chips" called *microprocessors*. Maybe, then, the name "computer" is inadequate. As computer pioneer John von Neumann has said, the device should not be called the computer but rather the "all-purpose machine." It is not, after all, just a machine for doing calculations. The most striking thing about it is that it can be put to *any number of uses*.

What are the various types of computers? Let's take a look.

All Computers, Great & Small: The Categories of Machines

What are the five sizes of computers?

At one time, the idea of having your own computer was almost like having your own personal nuclear reactor. In those days, in the 1950s and 1960s, computers were enormous machines affordable only by large institutions. Now they come in a variety of shapes and sizes, which can be classified according to their processing power: *supercomputers, mainframe computers, workstations, microcomputers,* and *microcontrollers*. We also consider *servers*.

1970	1972	1975	1976	1978	1981	1982
Micro-processor chips come into use; floppy disk introduced for storing data; first dynamic RAM chip	First video-game (Pong)	First micro-computer (MIT's Altair 8800)	Apple I computer (first personal computer sold in assembled form)	5¼" floppy disk; Atari home videogame	IBM introduces personal computer; mouse becomes regular part of a computer	Portable computers

Computer/ Maker or Lab	Top Speed, Teraflops	Location
Blue Gene/L IBM	280.6	U.S.
Molecular Dynamics Machine Riken	78	Japan
Grape-6 U. Tokyo	64	Japan
Columbia SGI	61	U.S.
Earth Simulator NEC	41	Japan

Source: Data from Rikert, *www.top500.org*, reported in Otis Port, "Holy Screaming Teraflops," table, *BusinessWeek,* January 17, 2005, p. 62; *www.newsfactor.com/story.xhtml? story_id=31771*; and *www.supercomputingonline.com/print.php?sid=8879.*

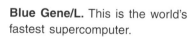

● more info!

FLOPS

In computing, FLOPS is an abbreviation of Floating-point Operations Per Second. FLOPS is used as a measure of a computer's performance, especially in fields of scientific calculations. Using floating-point encoding, extremely long numbers can be handled relatively easily. Computers operate in the trillions of FLOPS; for comparison, any response time below 0.1 second is experienced as instantaneous by a human operator, so a simple pocket calculator could be said to operate at about 10 FLOPS. Humans are even worse floating-point processors. If it takes a person a quarter of an hour to carry out a pencil-and-paper long division with 10 significant digits, that person would be calculating in the milliFLOPS range.

Supercomputers

Is there a chance I might use a supercomputer?

Typically priced from $1 million to more than $350 million, **_supercomputers_ are high-capacity machines with thousands of processors that can perform more than several trillion calculations per second.** These are the most expensive and fastest computers available. "Supers," as they are called, have been used for tasks requiring the processing of enormous volumes of data, such as doing the U.S. census count, forecasting weather, designing aircraft, modeling molecules, and breaking encryption codes. More recently they have been employed for business purposes—for instance, sifting demographic marketing information—and for creating film animation. The fastest computer in the world, costing $100 million and 10 times faster than its fastest predecessor (the Earth Simulator from Japan's NEC Corp.), is IBM's Blue Gene/L supercomputer, developed for nuclear weapons research for the U.S. Department of Energy. Blue Gene's present speed is 280.6 tcraflops (280.6 trillion floating-point calculations per second) but is expected to hit a mind-blowing 360 teraflops in short order. (● *See Panel 1.5.*)

Blue Gene/L. This is the world's fastest supercomputer.

1984	1993	1994	1998	2000	2001	2004
Apple Macintosh; first personal laser printer	Multimedia desktop computers	Apple and IBM introduce PCs with full-motion video built in; wireless data transmission for small portable computers; first web browser invented	PayPal is founded	The "Y2K" nonproblem; the first U.S. presidential webcast	Dell computers becomes the largest PC maker	IBM PC sold to Lenovo Group

Supercomputers are still the most powerful computers, but a new generation may be coming that relies on **_nanotechnology_, in which molecule-size nanostructures are used to create tiny machines for holding data or performing tasks.** A biological *nanocomputer*, which would be made of DNA and could fit into a single human cell, would use DNA as its software and enzymes as its hardware; its molecular-sized circuits would be viewable only through a microscope. (*Nano* means "one-billionth.") Some believe that within 10 years computers the size of a pencil eraser will be available that work 10 times faster than today's fastest supercomputer.[62] Eventually nanotech could show up in every device and appliance in your life.

Mainframe Computers

What kind of services am I apt to get from a mainframe?

The only type of computer available until the late 1960s, **_mainframes_ are water- or air-cooled computers that cost $5,000–$5 million and vary in size from small, to medium, to large, depending on their use.** Small mainframes ($5,000–$200,000) are often called *midsize computers;* they used to be called *minicomputers,* although today the term is seldom used. Mainframes are used by large organizations—such as banks, airlines, insurance companies, and colleges—for processing millions of transactions. Often users access a mainframe by means of a **_terminal_, which has a display screen and a keyboard and can input and output data but cannot by itself process data.** Mainframes process billions of instructions per second.

Workstations

What are some uses of workstations?

Introduced in the early 1980s, **_workstations_ are expensive, powerful personal computers usually used for complex scientific, mathematical, and engineering calculations and for computer-aided design and computer-aided manufacturing.** Providing many capabilities comparable to those of midsize mainframes, workstations are used for such tasks as designing airplane fuselages, developing prescription drugs, and creating movie special effects. Workstations have caught the eye of the public mainly for their graphics capabilities, which are used to breathe three-dimensional life into movies such as *The Lord of the Rings* and *Harry Potter.* The capabilities of low-end workstations overlap those of high-end desktop microcomputers.

Microcomputers

How does a microcomputer differ from a workstation?

Microcomputers, also called *personal computers* (*PCs*), which cost $500 to over $5,000, can fit next to a desk or on a desktop or can be carried around. They either are stand-alone machines or are connected to a computer network, such as a local area network. **A _local area network (LAN)_ connects, usually by special cable, a group of desktop PCs and other devices, such as printers, in an office or a building.**

Microcomputers are of several types: desktop PCs, tower PCs, notebooks (laptops), and personal digital assistants—handheld computers or palmtops.

DESKTOP PCs **_Desktop PCs_ are microcomputers whose case or main housing sits on a desk, with keyboard in front and monitor (screen) often on top.**

Mainframe computer

Workstation

Desktop microcomputer

Small. The Mac Mini has the smallest desktop microcomputer case, just 6.5 inches square and 2 inches tall.

TOWER PCs *Tower PCs* **are microcomputers whose case sits as a "tower," often on the floor beside a desk, thus freeing up desk surface space.** Some desktop computers, such as Apple's 2004 iMac, no longer have a boxy housing; most of the actual computer components are built into the back of the flat-panel display screen.

Tower PC

NOTEBOOKS *Notebook computers,* **also called** *laptop computers,* **are lightweight portable computers with built-in monitor, keyboard, hard-disk drive, battery, and AC adapter that can be plugged into an electrical outlet; they weigh anywhere from 1.8 to 9 pounds.**

Notebook computer

PERSONAL DIGITAL ASSISTANTS *Personal digital assistants (PDAs),* **also called** *handheld computers* **or** *palmtops,* **combine personal organization tools— schedule planners, address books, to-do lists—with the ability in some cases to send email and faxes.** Some PDAs have touch-sensitive screens. Some also connect to desktop computers for sending or receiving information. (For now, we are using the word *digital* to mean "computer based.") The range of handheld wireless devices, such as multipurpose cellphones, has surged in recent years, and we consider these later in the book (Chapter 7).

Personal digital assistant
(PDA)

Microcontrollers

What gadgets do I have that might contain microcontrollers?

Microcontrollers, **also called** *embedded computers,* **are the tiny, specialized microprocessors installed in "smart" appliances and automobiles.** These microcontrollers enable microwave ovens, for example, to store data about how long to cook your potatoes and at what power setting. Recently microcontrollers have been used to develop a new universe of experimental electronic appliances—e-pliances. For example, they are behind the new single-function products such as digital cameras, MP3 players, and organizers, which have been developed into hybrid forms such as gadgets that store photos and videos as well as music. They also help run tiny web servers embedded in clothing, jewelry, and household appliances such as refrigerators. In addition, microcontrollers are used in blood-pressure monitors, air bag sensors, gas and chemical sensors for water and air, and vibration sensors.

Microcontroller. The MPXY8020A pressure sensor from Motorola reduces tire blowouts and improves gas mileage. This embedded computer notifies drivers, via a dashboard display, when tire pressure is not optimal.

Cellphone microcontroller

Servers. A group of networked servers that are housed in one location is called a *server farm* or a *server cluster*.

Servers

How do servers work, and what do they do?

The word *server* describes not a size of computer but rather a particular way in which a computer is used. Nevertheless, because servers have become so important to telecommunications, especially with the rise of the internet and the web, they deserve mention here. (Servers are discussed in detail in Chapters 2, 6, and 7.)

A **_server_, or *network server*, is a central computer that holds collections of data (databases) and programs for connecting or supplying services to PCs, workstations, and other devices, which are called _clients_. These clients are linked by a wired or wireless network. The entire network is called a *client/server network.*** In small organizations, servers can store files, provide printing stations, and transmit email. In large organizations, servers may also house enormous libraries of financial, sales, and product information.

You may never lay eyes on a supercomputer or mainframe or server or even a tiny microcontroller. But most readers of this book will already have laid eyes and hands on a personal computer. We consider this machine next.

1.5

Understanding Your Computer: How Can You Customize (or Build) Your Own PC?

What four basic operations do all computers use, and what are some of the devices associated with each operation? How does communications affect these operations?

Perhaps you know how to drive a car. But do you know what to do when it runs badly? Similarly, you've probably been using a personal computer. But do you know what to do when it doesn't act right—when, for example, it suddenly "crashes" (shuts down)?

Cars are now so complicated that professional mechanics are often required for even the smallest problems. With personal computers, however, there are still many things you can do yourself—and should learn to do, so that, as we've suggested, you can be effective, efficient, and employable. To do so, you first need to know how computers work.

How Computers Work: Three Key Concepts

What are the three fundamental principles everyone should understand about how computers work?

Could you build your own personal computer? Some people do, putting together bare-bones systems for just a few hundred dollars. "If you have a logical mind, are fairly good with your hands, and possess the patience of Job, there's no reason you can't . . . build a PC," says science writer David Einstein. And, if you do it right, "it will probably take only a couple of hours," because industry-standard connections allow components to go together fairly easily.[63]

Actually, probably only techies would consider building their own PCs. But many ordinary users *order* their own custom-built PCs. Let's consider how you might do this.

We're not going to ask you to build or order a PC—just to pretend to do so. The purpose of this exercise is to give you a basic overview of how a computer works. That information will help you when you go shopping for a new system or, especially, if you order a custom-built system. It will also help you understand how your existing system works, if you have one.

Before you begin, you will need to understand three key concepts.

1. PURPOSE OF A COMPUTER: TURN DATA INTO INFORMATION
Very simply, the purpose of a computer is to process data into information.

- Data: **_Data_ consists of the raw facts and figures that are processed into information**—for example, the votes for different candidates being elected to student government office.
- Information: **_Information_ is data that has been summarized or otherwise manipulated for use in decision making**—for example, the total votes for each candidate, which are used to decide who won.

2. DIFFERENCE BETWEEN HARDWARE & SOFTWARE You should know the difference between hardware and software.

- Hardware: **_Hardware_ consists of all the machinery and equipment in a computer system.** The hardware includes, among other devices, the keyboard, the screen, the printer, and the "box"—the computer or processing device itself. Hardware is useless without software.
- Software: **_Software_, or _programs_, consists of all the electronic instructions that tell the computer how to perform a task.** These instructions come from a software developer in a form (such as a CD, or compact disk) that will be accepted by the computer. Examples are Microsoft Windows and Office XP.

3. THE BASIC OPERATIONS OF A COMPUTER Regardless of type and size, all computers use the same four basic operations: (1) input, (2) processing, (3) storage, and (4) output. To this we add (5) communications.

S u r v i v a l   T i p

Input is covered in detail in Chapter 5.

S u r v i v a l   T i p

Processing is covered in detail in Chapter 4.

- Input operation: **_Input_ is whatever is put in ("input") to a computer system.** Input can be nearly any kind of data—letters, numbers, symbols, shapes, colors, temperatures, sounds, pressure, light beams, or whatever raw material needs processing. When you type some words or numbers on a keyboard, those words are considered input data.
- Processing operation: **_Processing_ is the manipulation a computer does to transform data into information.** When the computer adds 2 + 2 to get 4, that is the act of processing. The processing is done by the *central processing unit*—frequently called just the *CPU*—a device consisting of electronic circuitry that executes instructions to process data.
- Storage operation: Storage is of two types—temporary storage and permanent storage, or primary storage and secondary storage. **_Primary storage_, or _memory_, is the internal computer circuitry that temporarily**

S u r v i v a l   T i p

Storage is covered in detail in
Chapter 4.

S u r v i v a l   T i p

Output is covered in detail in
Chapter 5.

S u r v i v a l   T i p

Communications is covered in
detail in Chapters 2, 6, and 7.

holds data waiting to be processed. **_Secondary storage_**, **simply called _storage_, refers to the devices and media that store data or information permanently.** A floppy disk (diskette) or hard disk is an example of this kind of storage. (Storage also holds the software—the computer programs.)

- **Output operation: _Output_ is whatever is output from ("put out of") the computer system—the results of processing, usually information.** Examples of output are numbers or pictures displayed on a screen, words printed out on paper by a printer, or music piped over some loudspeakers.

- **Communications operation:** These days, most (though not all) computers have communications ability, which offers an extension capability—in other words, it extends the power of the computer. With wired or wireless communications connections, data may be input from afar, processed in a remote area, stored in several different locations, and output in yet other places. However, you don't need communications ability to write letters, do calculations, or perform many other computer tasks.

These five operations are summarized in the illustration below. (● _See Panel 1.6._)

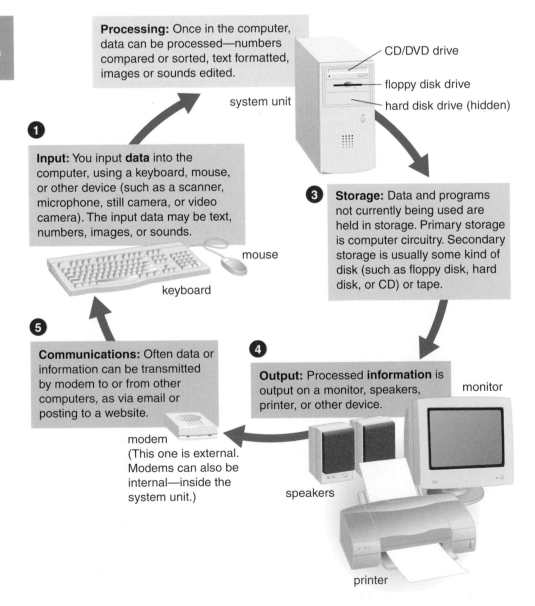

● PANEL 1.6
Basic operations of a computer

Processing: Once in the computer, data can be processed—numbers compared or sorted, text formatted, images or sounds edited.

CD/DVD drive

floppy disk drive

hard disk drive (hidden)

system unit

1

Input: You input **data** into the computer, using a keyboard, mouse, or other device (such as a scanner, microphone, still camera, or video camera). The input data may be text, numbers, images, or sounds.

mouse

keyboard

3 Storage: Data and programs not currently being used are held in storage. Primary storage is computer circuitry. Secondary storage is usually some kind of disk (such as floppy disk, hard disk, or CD) or tape.

5

Communications: Often data or information can be transmitted by modem to or from other computers, as via email or posting to a website.

modem
(This one is external. Modems can also be internal—inside the system unit.)

4

Output: Processed **information** is output on a monitor, speakers, printer, or other device.

monitor

speakers

printer

Pretending to Order (or Build) a Custom-Built Desktop Computer: Basic Knowledge of How a Computer Works

In what order would components be put together to build a custom desktop computer?

Now let's see how you would order a custom-built desktop PC, or even build one yourself. Remember, the purpose of this is to help you understand the internal workings of a computer so that you'll be knowledgeable about using one and buying one. (If you were going to build it yourself, you would pretend that someone had acquired the PC components for you from a catalog company and that you're now sitting at a table about to begin assembling them. All you would need is a combination Phillips/flathead screwdriver, perhaps a small wrench, and a static-electricity-arresting strap for your wrist, to keep static electricity from adversely affecting some computer components. You would also need the manuals that come with some of the components.) Although prices of components are always subject to change, we have indicated general ranges of prices for basic equipment current as of late 2005 so that you can get a sense of the relative importance of the various parts. ("Loaded" components—the most powerful and sophisticated equipment—cost more than the prices given here.)

Note: All the system components you or anyone else chooses *must be compatible*—in other words, each brand must work with other brands. If you work with one company—such as Dell, Apple, or Hewlett-Packard—to customize your system, you won't have to worry about compatibility. If you choose all the components yourself—for example, by going to a computer-parts seller such as Aberdeen (*www.aberdeeninc.com*)—you will have to check on compatibility as you choose each component. And you'll have to make sure each component comes with any necessary cables, instructions, and component-specific software (called a *driver*) that makes the component run.

This section of the chapter gives you a brief overview of the components, which are all covered in detail in Chapters 2–6. We describe them in the following order: (1) input hardware—keyboard and mouse; (2) processing and memory hardware; (3) storage hardware—disk drives; (4) output hardware—video and sound cards, monitor, speakers, and printer; (5) communication hardware—the modem; and (6) software—system and application.

Input Hardware: Keyboard & Mouse

What do the two principal input devices, keyboard and mouse, do?

Input hardware consists of devices that allow people to put data into the computer in a form that the computer can use. At minimum, you will need two things: a *keyboard* and a *mouse*.

Survival Tip
Hardware Info

Go to *www.bizrate.com/marketplace* for a listing of virtually all types of hardware, their descriptions, ratings, and prices, and the names of sellers.

Keyboard

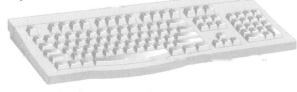

KEYBOARD (COST: $22–$70) On a microcomputer, a keyboard is the primary input device. **A _keyboard_ is an input device that converts letters, numbers, and other characters into electrical signals readable by the processor.** A microcomputer keyboard looks like a typewriter keyboard, but besides having keys for letters and numbers it has several keys (such as *F* keys and *Ctrl, Alt,* and *Del* keys) intended for computer-specific tasks. After other components are assembled, the keyboard will be plugged into the back of the computer in a socket intended for that purpose. (Cordless keyboards work differently.)

Mouse

MOUSE ($25–$100) **A _mouse_ is a nonkeyboard input device ("pointing device") that is used to manipulate objects viewed on the computer display screen.** The mouse cord is plugged into the back of the computer or into the back of the keyboard after the other components are assembled. (Cordless mice are also available.)

Processing & Memory Hardware: Inside the System Cabinet

How do I distinguish the processing and memory devices in a computer? What does the motherboard do?

Case

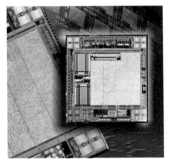

The brains of the computer are the *processing* and *memory* devices, which are installed in the case or system cabinet.

CASE & POWER SUPPLY (ABOUT $80–$160) **Also known as the *system unit*, the *case* or *system cabinet* is the box that houses the processor chip (CPU), the memory chips, and the motherboard with power supply, as well as some secondary storage devices**—floppy-disk drive, hard-disk drive, and CD or DVD drive, as we will explain. The case generally comes in desktop or tower models. It includes a power supply unit and a fan to keep the circuitry from overheating.

Processor chip

PROCESSOR CHIP ($65–$900 OR MORE) It may be small and not look like much, but it could be the most expensive hardware component of a build-it-yourself PC—and doubtless the most important. **A _processor chip (CPU, for central processing unit)_ is a tiny piece of silicon that contains millions of miniature electronic circuits.** The speed at which a chip processes information is expressed in *megahertz (MHz)*, millions of processing cycles per second, or *gigahertz (GHz)*, billions of processing cycles per second. The faster the processor, the more expensive it is. For $65, you might get a 1.33-GHz chip, which is adequate for most student purposes. For $350, you might get a 2.13-GHz chip, which you would want if you're running software with spectacular graphics and sound, such as those with some new videogames. Only older processors' speed is measured in megahertz now, but if you want a cheap processor—for instance, because you plan to work only with text documents—you could get a 233-MHz processor for $40.

MEMORY CHIPS ($20–$150) These chips are also small. **_Memory chips_, also known as _RAM (random access memory) chips_, represent _primary_ storage, or temporary storage; they hold data before processing and information after processing, before it is sent along to an output or storage device.** You'll want enough memory chips to hold at least 128 megabytes, or roughly 128 million characters, of data, which is adequate for most student purposes. If you work with large graphics files, you'll need more memory capacity, 256 megabytes or perhaps even 512 megabytes or more. (We explain the numbers used to measure storage capacities in a moment.)

Memory chip
(RAM chip)

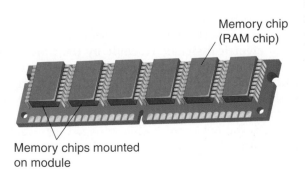

Memory chips mounted
on module

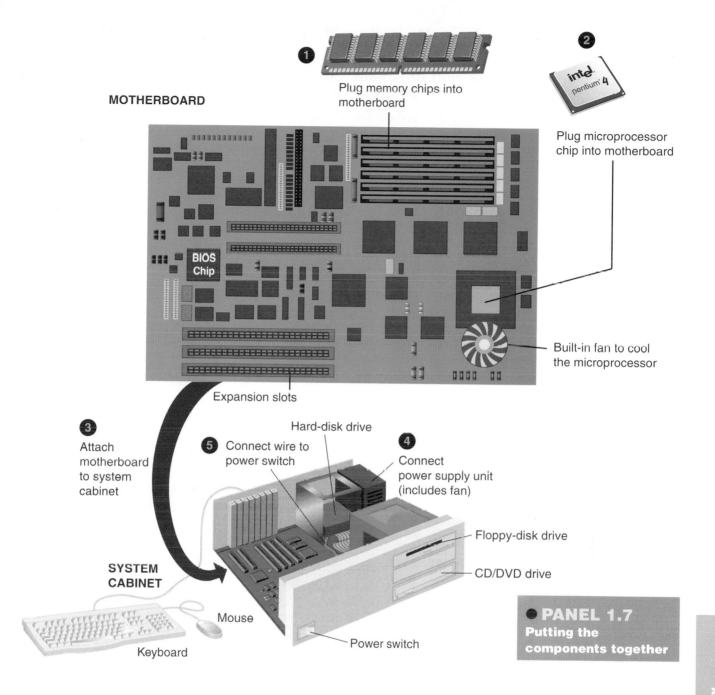

MOTHERBOARD

Plug memory chips into motherboard

Plug microprocessor chip into motherboard

intel pentium 4

Built-in fan to cool the microprocessor

Expansion slots

BIOS Chip

3 Attach motherboard to system cabinet

5 Connect wire to power switch

Hard-disk drive

4 Connect power supply unit (includes fan)

Floppy-disk drive

CD/DVD drive

SYSTEM CABINET

Mouse

Keyboard

Power switch

● **PANEL 1.7**
Putting the components together

MOTHERBOARD (ABOUT $100–$180) **Also called the *system board*, the *motherboard* is the main circuit board in the computer.** This is the big green circuit board to which everything else—such as the keyboard, mouse, and printer—attaches through connections (called *ports*) in the back of the computer. The processor chip and memory chips are also installed on the motherboard.

The motherboard has *expansion slots*—**for expanding the PC's capabilities—which give you places to plug in additional circuit boards,** such as those for video, sound, and communications (modem). (● See *Panel 1.7.*)

PUTTING THE COMPONENTS TOGETHER Now the components can be put together. As the illustration above shows, ❶ the memory chips are plugged into the motherboard. Then ❷ the processor chip is plugged into the motherboard. Now ❸ the motherboard is attached to the system cabinet. Then ❹ the power supply unit is connected to the system cabinet. Finally, ❺ the wire for the power switch, which turns the computer on and off, is connected to the motherboard.

Storage Hardware: Floppy Drive, Hard Drive, & CD/DVD Drive

What kind of storage devices would I as a student probably want in my computer?

With the motherboard in the system cabinet, the next step is installation of the storage hardware. Whereas memory chips deal only with temporary storage, *secondary storage*, or *permanent storage*, stores your data for as long as you want.

For today's student purposes, you'll need a CD/DVD drive, and it often helps to have a floppy disk drive. (Some users may also want a Zip-disk drive.) These storage devices slide into the system cabinet from the front and are secured with screws. Each drive is attached to the motherboard by a flat cable (called a *ribbon cable*). Also, each drive must be hooked up to a plug extending from the power supply.

A computer system's data/information storage capacity is represented by bytes, kilobytes, megabytes, gigabytes, and terabytes. Roughly speaking:

1 byte	*= 1 character of data* (A character can be alphabetic—A, B, or C—or numeric—1, 2, or 3—or a special character—!, ?, *, $, %.)
1 kilobyte	*= 1,024 characters*
1 megabyte	*= 1,048,576 characters*
1 gigabyte	*= more than 1 billion characters*
1 terabyte	*= more than 1 trillion characters*
1 petabye	*= about 1 quadrillion characters*

FLOPPY-DISK & ZIP DRIVES ($25 FLOPPY – $150 ZIP) **A *floppy-disk drive* is a storage device that stores data on removable 3.5-inch-diameter diskettes.** These diskettes don't seem to be "floppy," because they are encased in hard plastic, but the mylar disk inside is indeed flexible or floppy. Each can store 1.44 million bytes (characters) or more of data. With the floppy-disk drive installed, you'll later be able to insert a diskette through a slot in the front and remove it by pushing the eject button. **A *Zip-disk drive* is a storage device that stores data on removable floppy-disk cartridges with 70–500 times the capacity of the standard floppy.** These days, more people use CD/DVD drives to save data (see below).

(Left) Floppy disk; (right) Zip disk

Floppy disk

Floppy-disk drive

External Zip drive

HARD-DISK DRIVE ($100 FOR 120 GIGABYTES OF STORAGE—HIGHER-CAPACITY AND LOWER-CAPACITY DRIVES ARE AVAILABLE) **A *hard-disk drive* is a storage device that stores billions of characters of data on a nonremovable disk platter.** With 120 gigabytes of storage, you should be able to handle most student needs.

Hard-disk drive (goes inside the computer case)

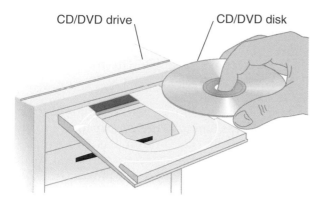

CD/DVD drive   CD/DVD disk

CD/DVD DRIVE ($50) **A** _**CD (compact-disk) drive**_**, or its more recent variant, a** _**DVD (digital video-disk) drive**_**, is a storage device that uses laser technology to read data from optical disks.** (Some companies call a DVD a "digital versatile disk.") These days new software is generally supplied on CDs.

The system cabinet has lights on the front that indicate when these drives are in use. (You must not remove a diskette or a cartridge from the drive until its light goes off, or else you risk damage to both disk and drive.) The wires for these lights need to be attached to the motherboard.

Output Hardware: Video & Sound Cards, Monitor, Speakers, & Printer

What kinds of output hardware are standard with a PC?

Output hardware consists of devices that translate information processed by the computer into a form that humans can understand—print, sound, graphics, or video, for example. Now a video card and a sound card need to be installed in the system cabinet. Next the monitor, speakers, and a printer are plugged in.

This is a good place to introduce the term _peripheral device_. **A** _**peripheral device**_ **is any component or piece of equipment that expands a computer's input, storage, and output capabilities.** In other words, a peripheral device is not part of the essential computer. Peripheral devices can be inside the computer or connected to it from the outside. Examples include printers and disk drives.

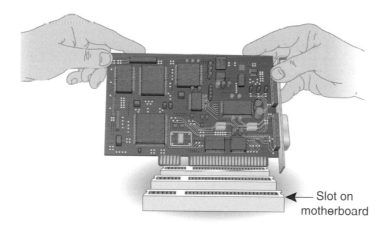

Slot on motherboard

VIDEO CARD ($50–$500) You doubtless want your monitor to display color (rather than just black-and-white) images. Your system cabinet will therefore need to have a device to make this possible. **A** _**video card**_ **converts the processor's output information into a video signal that can be sent through a cable to the monitor.** Remember the expansion slots we mentioned? Your video card is plugged into one of these on the motherboard. (You can also buy a motherboard with built-in video.)

SOUND CARD ($30–$200 AND HIGHER) You may wish to listen to music on your PC. If so, you'll need a _**sound card**_**, which enhances the computer's sound-generating capabilities by allowing sound to be output through speakers.** This, too, would be plugged into an expansion slot on the motherboard. (Once again, you can buy a motherboard with built-in sound.) With the CD drive connected to the card, you can listen to music CDs.

Monitor

MONITOR ($100–$200 OR HIGHER FOR A 17-INCH MODEL, $200–$1,000 FOR A 19-INCH MODEL) As with television sets, the inch dimension on monitors is measured diagonally corner to corner. **The** _**monitor**_ **is the display device that takes the electrical signals from the video card and forms an image using points of colored light on the screen.** Later, after the system cabinet has been closed up, the monitor will be connected by means of a cable to the back of the computer, using the clearly marked connector. The power cord for the monitor will be plugged into a wall plug.

Speakers

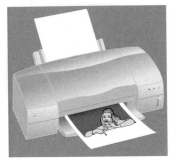

Printer

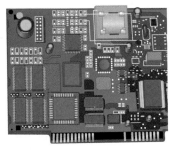

Modem card

System software—a version of Microsoft Windows XP

PAIR OF SPEAKERS ($25–$250) _**Speakers**_ **are the devices that play sounds transmitted as electrical signals from the sound card.** They may not be very sophisticated, but unless you're into high-fidelity recordings they're probably good enough. The two speakers are connected to a single wire that is plugged into the back of the computer once installation is completed.

PRINTER ($50–$700) Especially for student work, you certainly need a _**printer**_**, an output device that produces text and graphics on paper.** There are various types of printers, as we discuss later. The printer has two connections. One, which relays signals from the computer, goes to the back of the PC, where it connects with the motherboard. The other is a power cord that goes to a wall plug. Color printers are more expensive than black-and-white printers, and fast printers cost more than slow ones.

Communications Hardware: Modem

How is a modem installed?

Computers can be stand-alone machines, unconnected to anything else. If all you're doing is word processing to write term papers, you can do it with a stand-alone system. As we have seen, however, the communications component of the computer system vastly extends the range of a PC. Thus, while the system cabinet is still open, there is one more piece of hardware to install.

MODEM ($40–$100) **A standard _modem_ is a device that sends and receives data over telephone lines to and from computers.** The modem is mounted on an expansion card, which is fitted into an expansion slot on the motherboard. Later you can run a telephone line from the telephone wall plug to the back of the PC, where it will connect to the modem.

Other types of communications connections exist, which we cover in Chapters 2 and 6. However, standard modems are still often used.

Now the system cabinet is closed up. The person building the system will plug in all the input and output devices and turn on the power "on" button. Your microcomputer system will look similar to the one opposite. (● _See Panel 1.8._) Are you now ready to roll? Not quite.

Software

In what order are the two kinds of software installed?

With all the pieces put together, the person assembling the computer (you, if you're building it yourself) needs to check the motherboard manual for instructions on starting the system. One of the most important tasks is to install software, the electronically encoded instructions that tell the computer hardware what to do. Software is what makes the computer worthwhile. There are two types—_system software_ and _application software_.

SYSTEM SOFTWARE First, system software must be installed. _**System software**_ **helps the computer perform essential operating tasks and enables the application software to run.** System software consists of several electronically coded programs. The most important is the _operating system_, the master control program that runs the computer. Examples of operating system software for the PC are various Microsoft programs (such as Windows 95, 98, XP), Unix, and Linux. The Apple Macintosh microcomputer is another matter altogether. As we explain in Chapter 3, it has its own hardware components and software, which often aren't directly transferable to the PC.

System software comes most often on CDs. The person building your computer system will insert these into your CD drive and follow the on-screen

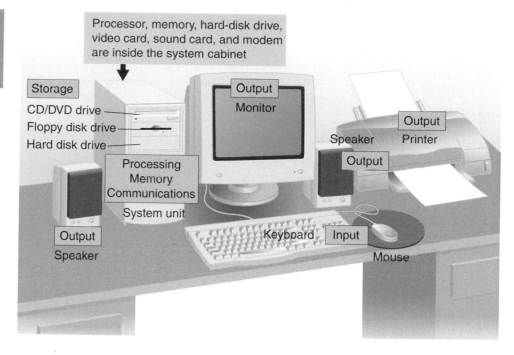

Processor, memory, hard-disk drive, video card, sound card, and modem are inside the system cabinet

Storage
CD/DVD drive
Floppy disk drive
Hard disk drive

Output
Monitor

Output
Speaker

Output
Printer

Processing
Memory
Communications
System unit

Output
Speaker

Keyboard

Input
Mouse

Application software for rendering art

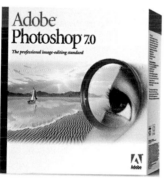

Application software for photo manipulation

directions for installation. (*Installation* is the process of copying software programs from secondary storage media—CDs, for example—onto your system's hard disk, so that you can have direct access to your hardware.)

After the system software is installed, setup software for the hard drive, the video and sound cards, and the modem must be installed. These setup programs (*drivers*, discussed in Chapter 3) will probably come on CDs (or maybe floppy disks). Once again, the installer inserts these into the appropriate drive and then follows the instructions that appear on the screen.

APPLICATION SOFTWARE   Now we're finally getting somewhere! After the application software has been installed, you can start using the PC. ***Application software* enables you to perform specific tasks—solve problems, perform work, or entertain yourself.** For example, when you prepare a term paper on your computer, you will use a word processing program. (Microsoft Word and Corel WordPerfect are two brands.) Application software is specific to the system software you use. If you want to run Microsoft Word, for instance, you'll need to first have Microsoft Windows system software on your system, not Unix or Linux.

Application software comes on CDs packaged in boxes that include instructions. You insert the CDs into your computer and then follow the instructions on the screen for installation. Later on you may obtain entire application programs by getting (downloading) them off the internet, using your modem or another type of communications connection.

We discuss software in more detail in Chapter 3.

Is Getting a Custom-Built PC Worth the Effort?

Why might I want to build a PC myself—and why not?

Does the foregoing description make you want to try putting together a PC yourself? If you add up the costs of all the components (not to mention the value of your time), and then start checking ads for PCs, you might wonder why anyone would bother going to the trouble of building one. And nowadays you would probably be right. "If you think you'd save money by putting together a computer from scratch," says David Einstein, "think again. You'd be lucky to match the price PC-makers are charging these days in their zeal to undercut the competition."[64]

But had you done this for real, it would not have been a wasted exercise: by knowing how to build a system yourself, not only would you be able to impress your friends but you'd also know how to upgrade any store-bought system to include components that are better than standard. For instance, as Einstein points out, if you're into videogames, knowing how to construct your own PC would enable you to make a system that's right for games. You could include the latest three-dimensional graphics video card and a state-of-the-art sound card, for example. More important, you'd also know how to order a custom-built system (as from Dell, Hewlett-Packard, or Gateway, some of the mail-order/online computer makers) that's right for you. In Chapters 4 and 5, we'll expand on this discussion so that you can really know what you're doing when you go shopping for a microcomputer system.

1.6

Where Is Information Technology Headed?

What are three directions of computer development and three directions of communications development?

How far we have come. At the beginning of the 19th century, most people thought they would live the same life their parents did. Today most people aren't surprised by the prediction that the Information Age will probably transform their lives beyond recognition. Let's consider the trends in the development of computers and communications and, most excitingly, the area where they intersect.

Three Directions of Computer Development: Miniaturization, Speed, & Affordability

What are other words I could use to describe the three ways computers have developed?

Since the days of ENIAC, computers have developed in three directions—and are continuing to do so.

MINIATURIZATION   Everything has become smaller. ENIAC's old-fashioned radio-style vacuum tubes gave way after 1947 to the smaller, faster, more reliable transistor. A *transistor* is a small device used as a gateway to transfer electrical signals along predetermined paths (circuits).

The next step was the development of tiny *integrated circuits*. Integrated circuits are entire collections of electrical circuits or pathways that are now etched on tiny squares (chips) of silicon half the size of your thumbnail. *Silicon* is a natural element found in sand. In pure form, it is the base material for computer processing devices.

The miniaturized processor, or microprocessor, in a personal desktop computer today can perform calculations that once required a computer filling an entire room.

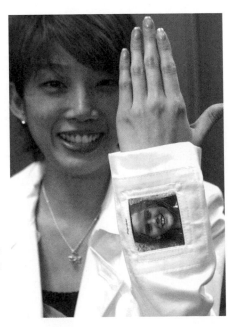

Miniaturization. A woman tries on a prototype of a wearable computer, a jacket with a built-in display in its sleeve. This type of computer, from Pioneer, is expected to aid medical, firefighting, and farm workers.

SPEED Thanks to miniaturization and new material used in making processors, computer makers can cram more hardware components into their machines, providing faster processing speeds and more data storage capacity.

AFFORDABILITY Processor costs today are only a fraction of what they were 15 years ago. A state-of-the-art processor costing less than $1,000 provides the same processing power as a huge 1980s computer costing more than $1 million.

These are the three major trends in computers. What about communications?

Three Directions of Communications Development: Connectivity, Interactivity, & Multimedia

What are three things I do that represent these three features—connectivity, interactivity, and multimedia?

Once upon a time, we had the voice telephone system—a one-to-one medium. You could talk to your Uncle Joe and he could talk to you, and with special arrangements (conference calls) more than two people could talk with one another. We also had radio and television systems—one-to-many media (or mass media). News announcers could talk to you on a single medium such as television, but you couldn't talk to them.

There have been three recent developments in communications:

CONNECTIVITY *Connectivity* **refers to the connection of computers to one another by a communications line in order to provide online information access and/or the sharing of peripheral devices.** The connectivity resulting from the expansion of computer networks has made possible email and online shopping, for example.

INTERACTIVITY *Interactivity* **refers to two-way communication; the user can respond to information he or she receives and modify what a computer is doing.** That is, there is an exchange or dialogue between the user and the computer, and the computer responds to user requests. A noninteractive program, once started, continues without requiring human contact, or interaction. The ability to interact means users can be active rather than passive participants in the technological process. On the television networks MSNBC or CNN, for example, you can immediately go on the internet and respond to news from broadcast anchors. Today, most application software is interactive. In the future, cars may respond to voice commands or feature computers built into the dashboard.

MULTIMEDIA Radio is a single-dimensional medium (sound), as is most email (mainly text). As mentioned earlier in this chapter, *multimedia* **refers to technology that presents information in more than one medium—such as text, pictures, video, sound, and animation—in a single integrated communication.** The development of the World Wide Web expanded the internet to include pictures, sound, music, and so on, as well as text.

Exciting as these developments are, truly mind-boggling possibilities emerge as computers and communications cross-pollinate.

When Computers & Communications Combine: Convergence, Portability, & Personalization

What is the meaning of convergence, portability, and personalization?

Interactivity. A dashboard computer allows drivers to request information about the car's operation, location, and nearby services.

Sometime in the 1990s, computers and communications started to fuse together, beginning a new era within the Information Age. The result has been four further developments, which have only just begun.

CONVERGENCE *Convergence* describes the combining of several industries through various devices that exchange data in the format used by computers. The industries are computers, communications, consumer electronics, entertainment, and mass media. Convergence has led to electronic products that perform multiple functions, such as TVs with internet access, cellphones that are also digital cameras, and a refrigerator that allows you to send email.

PORTABILITY In the 1980s, portability, or mobility, meant trading off computing power and convenience in return for smaller size and weight. Today, however, we are close to the point where we don't have to give up anything. As a result, experts have predicted that small, powerful, wireless personal electronic devices will transform our lives far more than the personal computer has done so far. "The new generation of machines will be truly personal computers, designed for our mobile lives," wrote one journalist back in 1992. "We will read office memos between strokes on the golf course and answer messages from our children in the middle of business meetings."[65] Today such activities are commonplace, and smartphones are taking on other functions. The risk they bring is that, unless we're careful, work will completely invade our personal time.

PERSONALIZATION Personalization is the creation of information tailored to your preferences—for instance, programs that will automatically cull recent news and information from the internet on just those topics you have designated. Companies involved in e-commerce can send you messages about forthcoming products based on your pattern of purchases, usage, and other criteria. Or they will build products (cars, computers, clothing) customized to your heart's desire.

COLLABORATION A more recent trend is mass collaboration. Says *New York Times* technology writer John Markoff, "A remarkable array of software systems makes it simple to share anything instantly, and sometimes enhance it along the way."[66] Adds *BusinessWeek* writer Robert Hof, "The nearly

SECURITY

1 billion people online worldwide—along with their shared knowledge, social contacts, online reputations, computing power, and more—are rapidly becoming a collective force of unprecedented power."[67] Examples are file-sharing, photo-sharing websites, calendar-sharing services, group-edited sites called *wikis*, social networking services, and so-called citizen-journalism sites, in which average people write their own news items on the internet and comment on what other people post—an interactive, democratic form of mass media.[68] Pooled ratings, for instance, enable people to create personalized net music radio stations or Amazon.com's millions of customer-generated product reviews.

"E" Also Stands for Ethics

What are the principal ethical concerns I should be conscious of in the use of information technology?

Every computer user will have to wrestle with ethical issues related to the use of information technology. **_Ethics_ is defined as a set of moral values or principles that govern the conduct of an individual or a group.** Because ethical questions arise so often in connection with information technology, we will note them, wherever they appear in this book, with the symbol shown at left. Below, for example, are some important ethical concerns pointed out by Tom Forester and Perry Morrison in their book *Computer Ethics*.[69] These considerations are only a few of many; we'll discuss others in subsequent chapters.

SPEED & SCALE Great amounts of information can be stored, retrieved, and transmitted at a speed and on a scale not possible before. Despite the benefits, this has serious implications "for data security and personal privacy," as well as employment, Forester and Morrison say, because information technology can never be considered totally secure against unauthorized access.

UNPREDICTABILITY Computers and communications are pervasive, touching nearly every aspect of our lives. However, at this point, compared to other pervasive technologies—such as electricity, television, and automobiles—information technology seems a lot less predictable and reliable.

COMPLEXITY Computer systems are often incredibly complex—some so complex that they are not always understood even by their creators. "This," say Forester and Morrison, "often makes them completely unmanageable," producing massive foul-ups or spectacularly out-of-control costs.

Ethics and security can often be talked about in the same breath, since secure computer systems obviously go a long way toward keeping people ethical and honest. We devote considerable effort to this discussion, as indicated by the icon at left, which you will see throughout the book

Experience Box
Better Organization & Time Management: Dealing with the Information Deluge in College—& in Life

An Experience Box appears at the end of each chapter. Each box offers you the opportunity to acquire useful experience that directly applies to the Digital Age. This first box illustrates skills that will benefit you in college, in this course and others. (Students reading the first six editions of our book have told us they received substantial benefit from these suggestions.)

"How on earth am I going to be able to keep up with what's required of me?" you may ask yourself. "How am I going to handle the information glut?" The answer is: *by learning how to learn.* By building your skills as a learner, you certainly help yourself do better in college, and you also train yourself to be an information manager in the future.

Using Your "Prime Study Time"

Each of us has a different energy cycle. The trick is to use it effectively. That way, your hours of best performance will co-incide with your heaviest academic demands. For example, if your energy level is high during the evenings, you should plan to do your studying then.

To capitalize on your prime study time, take the following steps: (1) Make a study schedule for the entire term, and in-dicate the times each day during which you plan to study. (2) Find some good places to study—places where you can avoid distractions. (3) Avoid time wasters, but give yourself frequent rewards for studying, such as a TV show, a favorite piece of music, or a conversation with a friend.

Improving Your Memory Ability

Memorizing is, of course, one of the principal requirements for succeeding in college. And it's a great help for success in life afterward. Some suggestions:

Get Rid of Distractions. Distractions are a major impedi-ment to remembering (as they are to other forms of learning). External distractions are those over which you have no control—hallway noises, instructor's accent, people talking. If you can't banish the distraction by moving, you might try to increase your interest in the subject you are studying. Internal distractions are daydreams, personal worries, hunger, and ill-ness. Small worries can be shunted aside by listing them on a page for future handling. Large worries may require talking with a friend or counselor.

Space Your Studying, Rather Than Cramming. Cramming—making a frantic, last-minute attempt to memorize massive amounts of material—is probably the least effective means of absorbing information. Research shows that it's best to space out your studying of a subject over successive days. A series of study sessions over several days is prefer-able to trying to do it all during the same number of hours on

one day. It is repetition that helps move information into your long-term memory bank.

Review Information Repeatedly—Even "Overlearn It." By repeatedly reviewing information—known as "rehearsing"—you can improve both your retention and your understanding of it. Overlearning is continuing to review material even after you appear to have absorbed it.

Use Memorizing Tricks. There are several ways to organize information so that you can retain it better. For example, you can make drawings or diagrams (as of the parts of a com-puter system). Some methods of establishing associations between items you want to remember are given in the box. (● *See Panel 1.9.*)

How to Improve Your Reading Ability: The SQ3R Method

SQ3R stands for "survey, question, read, recite, and re-view."[70] The strategy behind it is to break down a reading as-signment into small segments and master each before mov-ing on. The five steps of the SQ3R method are as follows:

1. *Survey the chapter before you read it:* Get an overview of the chapter before you begin reading it. If you have a sense of what the material is about before you begin reading it, you can predict where it is going. In this text, we offer on the first page of every chapter a list of the main heads and accompanying key questions. At the end of each chapter we offer a Summary, which explains what the chapter's terms and concepts mean and why they are important.

2. *Question the segment in the chapter before you read it:* This step is easy to do, and the point, again, is to get you involved in the material. After surveying the entire chapter, go to the first segment—section, sub-section, or even paragraph, depending on the level of difficulty and density of information. Look at the topic heading of that segment. In your mind, restate the heading as a question. In this book, following each section head we present a Key Question. An example in this chapter was "What are three directions of com-puter development and three directions of communica-tions development?"

   After you have formulated the question, go to steps 3 and 4 (read and recite). Then proceed to the next segment and restate the heading there as a question.

3. *Read the segment about which you asked the ques-tion:* Now read the segment you asked the question about. Read with purpose, to answer the question you

- **Mental and physical imagery:** Use your visual and other senses to construct a personal image of what you want to remember. Indeed, it helps to make the image humorous, action-filled, or outrageous in order to establish a personal connection. Example: To remember the name of the 21st president of the United States, Chester Arthur, you might visualize an author writing the number "21" on a wooden chest. This mental image helps you associate chest (Chester), author (Arthur), and 21 (21st president).

- **Acronyms and acrostics:** An acronym is a word created from the first letters of items in a list. For instance, *Roy G. Biv* helps you remember the colors of the rainbow in order: red, orange, yellow, green, blue, indigo, violet. An acrostic is a phrase or sentence created from the first letters of items on a list. For example, *Every Good Boy Does Fine* helps you remember that the order of musical notes on the treble staff is *E-G-B-D-F*.

- **Location:** Location memory occurs when you associate a concept with a place or imaginary place. For example, you could learn the parts of a computer system by imagining a walk across campus. Each building you pass could be associated with a part of the computer system.

- **Word games:** Jingles and rhymes are devices frequently used by advertisers to get people to remember their products. You may recall the spelling rule "I before E except after C or when sounded like A as in *neighbor* or *weigh*." You can also use narrative methods, such as making up a story.

● PANEL 1.9

Some memorizing tricks

Mental and physical imagery: Use your visual and other senses to construct a personal image of something you want to remember. Indeed, you might find it helpful to make the image humorous, action-filled, sexual, bizarre, or outrageous in order to establish a personal connection. Example: To remember the name of the 21st president of the United States, Chester Arthur, you might visualize an author writing the number "21" on a wooden chest. This mental image helps you associate chest (Chester), author (Arthur), and 21 (21st president).

formulated. Underline or color-mark sentences that you think are important, if they help you answer the question. Read this portion of the text more than once, if necessary, until you can answer the question. In addition, determine whether the segment covers any other significant questions, and formulate answers to these, too. After you have read the segment, proceed to step 4. (Perhaps you can see where this is all leading. If you read in terms of questions and answers, you will be better prepared when you see exam questions about the material later.)

4. *Recite the main points of the segment:* Recite means "say aloud." Thus, you should speak out loud (or softly) the answer to the principal question or questions about the segment and any other main points.

5. *Review the entire chapter by repeating questions:* After you have read the chapter, go back through it and review the main points. Then, without looking at the book, test your memory by repeating the questions.

Clearly the SQ3R method takes longer than simply reading with a rapidly moving color marker or underlining pencil. However, the technique is far more effective because it requires your involvement and understanding. This is the key to all effective learning.

Learning from Lectures

Does attending lectures really make a difference? Research shows that students with grades of B or above were more apt to have better class attendance than students with grades of C- or below.[71]

Some tips for getting the most out of lectures:

Take Effective Notes by Listening Actively. Research shows that good test performance is related to good note taking.[72] And good note taking requires that you listen actively—that is, participate in the lecture process. Here are some ways to take good lecture notes:

- *Read ahead and anticipate the lecturer:* Try to anticipate what the instructor is going to say, based on your previous reading. Having background knowledge makes learning more efficient.

- *Listen for signal words:* Instructors use key phrases such as "The most important point is . . .," "There are four reasons for . . .," "The chief reason . . .," "Of special importance . . .," "Consequently . . ." When you hear such signal phrases, mark your notes with a ! or *.

- *Take notes in your own words:* Instead of just being a stenographer, try to restate the lecturer's thoughts in your own words, which will make you pay attention more.

- *Ask questions:* By asking questions during the lecture, you necessarily participate in it and increase your understanding.

Review Your Notes Regularly. Make it a point to review your notes regularly—perhaps on the afternoon after the lecture, or once or twice a week. We cannot emphasize enough the importance of this kind of reviewing.

Summary

application software (p. 33) Software that has been developed to solve a particular problem, perform useful work on general-purpose tasks, or provide entertainment. Why it's important: *Application software such as word processing, spreadsheet, database management, graphics, and communications packages are commonly used tools for increasing people's productivity.*

avatar (p. 6) Computer depiction of a human, often found in online videogames. Why it's important: *Avatars can be helpful in training, such as by representing imaginary customers.*

case (p. 28) Also known as the *system unit* or *system cabinet;* the box that houses the processor chip (CPU), the memory chips, and the motherboard with power supply, as well as storage devices—floppy-disk drive, hard-disk drive, and CD or DVD drive. Why it's important: *The case protects many important processing and storage components.*

CD/DVD drive   CD/DVD disk

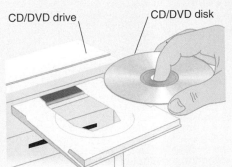

CD (compact-disk) drive (p. 31) Storage device that uses laser technology to read data from optical disks. Why it's important: *New software is generally supplied on CDs rather than diskettes. And even if you can get a program on floppies, you'll find it easier to install a new program from one CD than to repeatedly insert and remove many diskettes. The newest version is called DVD (digital video disk). The DVD format stores even more data than the CD format.*

central processing unit (CPU) *See* **processor chip.**

chip *See* **processor chip.**

clients (p. 24) Computers and other devices connected to a server, a central computer. Why it's important: *Client/server networks are used in many organizations for sharing databases, devices, and programs.*

communications technology (p. 5) Also called *telecommunications technology;* consists of electromagnetic devices and systems for communicating over long distances. Why it's important: *Communications systems using electronic connections have helped to expand human communication beyond face-to-face meetings.*

computer (p. 4) Programmable, multiuse machine that accepts data—raw facts and figures—and processes (manipulates) it into useful information, such as summaries and totals. Why it's important: *Computers greatly speed up problem solving and other tasks, increasing users' productivity.*

computer savvy (p. 3) Knowing what computers can do and what they can't, knowing how they can benefit you and how they can harm you, and knowing when you can solve computer problems and when you have to call for help. Why it's important: *You will know how to make better buying decisions, how to fix ordinary computer problems, how to upgrade your equipment and integrate it with new products, how to use the internet most effectively, how to protect yourself against online villains, and what kinds of computer uses can advance your career.*

connectivity (p. 35) Ability to connect computers to one another by communications lines, so as to provide online information access and/or the sharing of peripheral devices. Why it's important: *Connectivity is the foundation of the advances in the Information Age. It provides online access to countless types of information and services. The connectivity resulting from the expansion of computer networks has made possible email and online shopping, for example.*

course-management software (p. 5) Software for administering online assignments, schedules, examinations, and grades. Why it's important: *It helps to make administrative "housekeeping" more efficient.*

cyberspace (p. 17) Term used to refer to not only the online world and the internet in particular but also the whole wired and wireless world of communications in general. Why it's important: *More and more human activities take place in cyberspace.*

data (p. 25) Raw facts and figures that are processed into information. Why it's important: *Users need data to create useful information.*

desktop PC (p. 22) Microcomputer unit that sits on a desk, with the keyboard in front and the monitor often on top. Why it's important: *Desktop PCs and tower PCs are the most commonly used types of microcomputer.*

distance learning (p. 5) Also known as *e-learning*; name given to online education programs. Why it's important: *Provides students increased flexibility because they do not have to be in an actual classroom.*

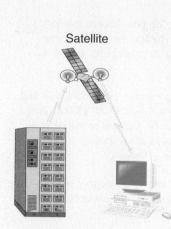
Satellite

download (p. 9) To transfer data from a remote computer to one's own computer. Why it's important: *Allows text, music, and images to be transferred quickly by telecommunications.*

DVD (digital video-disk) drive *See* **CD drive.**

email (electronic mail) (p. 5) Messages transmitted over a computer network, most often the internet. Why it's important: *Email has become universal; one of the first things new computer users learn is how to send and receive email.*

ethics (p. 37) Set of moral values or principles that govern the conduct of an individual or a group. Why it's important: *Ethical questions arise often in connection with information technology.*

expansion slots (p. 29) Internal "plugs" used to expand the PC's capabilities. Why it's important: *Expansion slots give you places to plug in additional circuit boards, such as those for video, sound, and communications (modem).*

floppy-disk drive (p. 30) Storage device that stores data on removable 3.5-inch-diameter flexible diskettes encased in hard plastic. Why it's important: *In the past, floppy-disk drives were included on almost all microcomputers and made many types of files portable.*

hard-disk drive (p. 30) Storage device that stores billions of characters of data on a nonremovable disk platter usually inside the computer case. Why it's important: *Hard disks hold much more data than diskettes do. Nearly all microcomputers use hard disks as their principal secondary-storage medium.*

hardware (p. 25) All machinery and equipment in a computer system. Why it's important: *Hardware runs under the control of software and is useless without it. However, hardware contains the circuitry that allows processing.*

information (p. 25) Data that has been summarized or otherwise manipulated for use in decision making. Why it's important: *The whole purpose of a computer (and communications) system is to produce (and transmit) usable information.*

information technology (IT) (p. 4) Technology that helps to produce, manipulate, store, communicate, and/or disseminate information. Why it's important: *Information technology is bringing about the fusion of several important industries dealing with computers, telephones, televisions, and various handheld devices.*

input (p. 25) Whatever is put in ("input") to a computer system. Input devices include the keyboard and the mouse. Why it's important: *Useful information cannot be produced without input data.*

interactivity (p. 35) Two-way communication; a user can respond to information he or she receives and modify the process. Why it's important: *Interactive devices allow the user to actively participate in a technological process instead of just reacting to it.*

internet (the "net") (p. 17) Worldwide computer network that connects hundreds of thousands of smaller networks linking computers at academic, scientific, and commercial institutions, as well as individuals. Why it's important: *Thanks to the internet, millions of people around the world can share all types of information and services*

keyboard (p. 27) Input device that converts letters, numbers, and other characters into electrical signals readable by the processor. Why it's important: *Keyboards are the most common kind of input device.*

local area network (LAN) (p. 22) Network that connects, usually by special cable, a group of desktop PCs and other devices, such as printers, in an office or a building. Why it's important: *LANs have replaced mainframes for many functions and are considerably less expensive.*

mainframe (p. 22) Second-largest computer available, after the supercomputer; capable of great processing speeds and data storage. Costs $5,000–$5 million. Small mainframes are often called *midsize computers.* Why it's important: *Mainframes are used by large organizations (banks, airlines, insurance companies, universities) that need to process millions of transactions.*

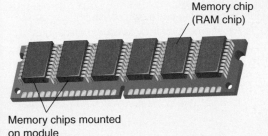

Memory chip (RAM chip)

Memory chips mounted on module

memory chip (p. 28) Also known as *RAM* (for "random access memory") *chip;* represents primary storage or temporary storage. Why it's important: *Holds data before processing and information after processing, before it is sent along to an output or storage device.*

microcomputer (p. 22) Also called *personal computer;* small computer that fits on or next to a desk or can be carried around. Costs $500–$5,000. Why it's important: *The microcomputer has lessened the reliance on mainframes and has provided more ordinary users with access to computers. It can be used as a stand-alone machine or connected to a network.*

microcontroller (p. 23) Also called an *embedded computer;* the smallest category of computer. Why it's important: *Microcontrollers are the tiny, specialized microprocessors built into "smart" electronic devices, such as appliances and automobiles.*

micropayments (p. 8) Electronic payments of as little as 25 cents in transactions for which it is uneconomical to use a credit card. Why it's important: *Allows products to be sold that previously weren't worth the effort of merchandising.*

modem (p. 32) Device that sends and receives data over telephone lines to and from computers. Why it's important: *A modem enables users to transmit data from one computer to another by using standard telephone lines instead of special communications equipment.*

monitor (p. 31) Display device that takes the electrical signals from the video card and forms an image using points of colored light on the screen. Why it's important: *Monitors enable users to view output without printing it out.*

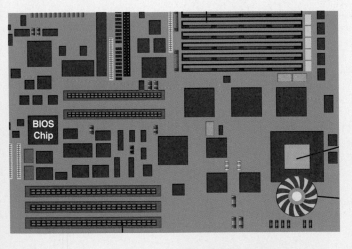

BIOS Chip

motherboard (p. 29) Also called the *system board;* main circuit board in the computer. Why it's important: *This is the big green circuit board to which everything else—such as the keyboard, mouse, and printer—is attached. The processor chip and memory chips are also installed on the motherboard.*

mouse (p. 28) Nonkeyboard input device, called a "pointing device," used to manipulate objects viewed on the computer display screen. Why it's important: *For many purposes, a mouse is easier to use than a keyboard for inputting commands. Also, the mouse is used extensively in many graphics programs.*

multimedia (p. 35) From "multiple media"; technology that presents information in more than one medium—including text graphics, animation, video, and sound—in a single integrated communication. Why it's important: *Multimedia is used increasingly in business, the professions, and education to improve the way information is communicated.*

nanotechnology (p. 22) Technology whereby molecule-size nanostructures are used to create tiny machines for holding data or performing tasks. Why it's important: *Could result in tremendous computer power in molecular-size devices.*

network (p. 5) Communications system connecting two or more computers. Why it's important: *Networks allow users to share applications and data and to use email. The internet is the largest network.*

notebook computer (p. 23) Also called *laptop computer;* lightweight portable computer with a built-in monitor, keyboard, hard-disk drive, battery, and adapter; weighs 1.8–9 pounds. Why it's important: *Notebook and other small computers have provided users with computing capabilities in the field and on the road.*

online (p. 5) Using a computer or some other information device, connected through a network, to access information and services from another computer or information device. Why it's important: *Online communication is widely used by businesses, services, individuals, and educational institutions.*

output (p. 26) Whatever is output from ("put out of") the computer system; the results of processing. Why it's important: *People use output to help them make decisions. Without output devices, computer users would not be able to view or use the results of processing.*

peripheral device (p. 31) Any component or piece of equipment that expands a computer's input, storage, and output capabilities. Examples include printers and disk drives. Why it's important: *Most computer input and output functions are performed by peripheral devices.*

personal digital assistant (PDA) (p. 23) Also known as *handheld computer* or *palmtop;* used as a schedule planner and address book and to prepare to-do lists and send email and faxes. Why it's important: *PDAs make it easier for people to do business and communicate while traveling.*

primary storage (p. 25) Also called *memory;* internal computer circuitry that temporarily holds data waiting to be processed. Why it's important: *By holding data, primary storage enables the processor to process.*

printer (p. 32) Output device that produces text and graphics on paper. Why it's important: *Printers provide one of the principal forms of computer output.*

processing (p. 25) The manipulation a computer does to transform data into information. Why it's important: *Processing is the essence of the computer, and the processor is the computer's "brain."*

processor chip (p. 28) Also called the *processor* or the *CPU (central processing unit);* tiny piece of silicon that contains millions of miniature electronic circuits used to process data. Why it's important: *Chips have made possible the development of small computers.*

robot (p. 7) Automatic device that performs functions ordinarily performed by human beings. Why it's important: *Robots help perform tasks that humans find difficult or impossible to do.*

secondary storage (p. 26) Also called *storage;* devices and media that store data and programs permanently—such as disks and disk drives, tape and tape drives, CDs and CD drives. Why it's important: *Without secondary storage, users would not be able to save their work. Storage also holds the computer's software.*

server (p. 24) Central computer in a network that holds collections of data (databases) and programs for connecting PCs, workstations, and other devices, which are called *clients.* Why it's important: *Servers enable many users to share equipment, programs, and data.*

software (p. 25) Also called *programs;* step-by-step electronically encoded instructions that tell the computer hardware how to perform a task. Why it's important: *Without software, hardware is useless.*

sound card (p. 31) Special circuit board that enhances the computer's sound-generating capabilities by allowing sound to be output through speakers. Why it's important: *Sound is used in multimedia applications. Also, many users like to listen to music CDs on their computers.*

speakers (p. 32) Devices that play sounds transmitted as electrical signals from the sound card. Speakers are connected to a single wire plugged into the back of the computer. Why it's important: *See* **sound card.**

supercomputer (p. 21) High-capacity computer with thousands of processors that is the fastest calculating device ever invented. Costs up to $350 million or more. Why it's important: *Supercomputers are used primarily for research purposes, airplane design, oil exploration, weather forecasting, and other activities that cannot be handled by mainframes and other less powerful machines.*

system software (p. 32) Software that helps the computer perform essential operating tasks. Why it's important: *Application software cannot run without system software. System software consists of several programs. The most important is the operating system, the master control program that runs the computer. Examples of operating system software for the PC are various Microsoft programs (such as Windows 95, 98, NT, Me, and XP), Unix, Linux, and the Macintosh operating system.*

system unit *See* **case.**

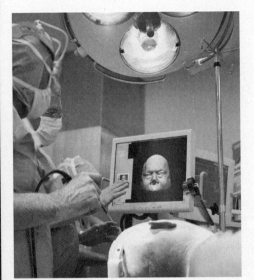

telemedicine (p. 6) Medical care delivered via telecommunications. Why it's important: *Allows physicians in remote areas to consult over a distance.*

terminal (p. 22) Input and output device that uses a keyboard for input and a monitor for output; it cannot process data. Why it's important: *Terminals are generally used to input data to and receive data from a mainframe computer system.*

tower PC (p. 23) Microcomputer unit that sits as a "tower," often on the floor, freeing up desk space. Why it's important: *Tower PCs and desktop PCs are the most commonly used types of microcomputer.*

video card (p. 31) Circuit board that converts the processor's output information into a video signal for transmission through a cable to the monitor. Why it's important: *Virtually all computer users need to be able to view video output on the monitor.*

virtual (p. 8) Something that is created, simulated, or carried on by means of a computer or a computer network. Why it's important: *Allows actual objects to be represented in computer-based form.*

workstation (p. 22) Smaller than a mainframe; expensive, powerful computer generally used for complex scientific, mathematical, and engineering calculations and for computer-aided design and computer-aided manufacturing. Why it's important: *The power of workstations is needed for specialized applications too large and complex to be handled by PCs.*

World Wide Web (the "web") (p. 18) The interconnected system of internet servers that support specially formatted documents in multimedia form—sounds, photos, and video as well as text. Why it's important: *The web is the most widely known part of the internet.*

Zip-disk drive (p. 30) Storage device that stores data on removable floppy-disk cartridges with at least 70 times the capacity of the standard floppy. Why it's important: *Zip drives are used to store large files.*

Chapter Review

More and more educators are favoring an approach to learning (presented by Benjamin Bloom and his colleagues in *Taxonomy of Educational Objectives*) that follows a hierarchy of six critical-thinking skills: (a) two lower-order skills—memorization and comprehension; and (b) four higher-order skills—application, analysis, synthesis, and evaluation. While you may be able to get through many introductory college courses by simply memorizing facts and comprehending the basic ideas, to advance further you will probably need to employ the four higher-order thinking skills.

In the Chapter Review at the end of each chapter, we have implemented this hierarchy in a three-stage approach, as follows:

- *Stage 1 learning—memorization:* "I can recognize and recall information." Self-test questions, multiple-choice questions, and true/false questions enable you to test how well you recall basic terms and concepts.

- *Stage 2 learning—comprehension:* "I can recall information in my own terms and explain it to a friend." Using open-ended short-answer questions, we ask you to reexpress terms and concepts in your own words.

- *Stage 3 learning—applying, analyzing, synthesizing, evaluating:* "I can apply what I've learned, relate these ideas to other concepts, build on other knowledge, and use all these thinking skills to form a judgment." In this part of the Chapter Review, we ask you to put the ideas into effect using the activities described, some of which include internet activities. The purpose is to help you take possession of the ideas, make them your own, and apply them realistically to your life.

stage 1 LEARNING MEMORIZATION

"I can recognize and recall information."

Self-Test Questions

1. The _____ _____ _____ refers to the part of the internet that presents information in multimedia form.

2. The two main types of microcomputers are the _____ _____, which sits on the desktop, and the _____ _____, which usually is placed on the floor.

3. "_____ technology" merges computing with high-speed communications.

4. A _____ is an electronic machine that accepts data and processes it into information.

5. The _____ is a worldwide network that connects hundreds of thousands of smaller networks.

6. _____ refers to information presented in nontextual forms such as video, sound, and graphics.

7. _____ are high-capacity machines with thousands of processors.

8. Embedded computers, or _____, are installed in "smart" appliances and automobiles.

9. The kind of software that enables users to perform specific tasks is called _____ software.

10. RAM is an example of _____ storage, and a hard drive is an example of _____ storage.

11. A _____ is a communications system connecting two or more computers.

12. The four basic operations of all computers are _____ _____ _____, and _____.

13. The first programmable computer in the USA, which appeared in 1946, was called the _____.

14. The _____ is the display device that takes the electrical signals from the video card and forms an image using points of colored light on the screen.

15. The base material for computer processing devices is _____, a natural element found in sand.

16. The general term for all the machinery and equipment in a computer system is _____.

17. _____ and _____ are the two most common input devices.

18. The processor chip, commonly called the _____ or a _____, is a tiny piece of silicon that contains millions of miniature electronic circuits.

19. One gigabyte is approximately _____ characters.

Multiple-Choice Questions

1. Which of the following devices converts computer output into displayed images?
   a. printer
   b. monitor
   c. floppy-disk drive
   d. processor
   e. hard-disk drive

2. Which of the following computer types is the smallest?

   a. mainframe

   b. microcomputer

   c. microcontroller

   d. supercomputer

   e. workstation

3. Which of the following is a secondary storage device?

   a. processor

   b. memory chip

   c. floppy-disk drive

   d. printer

   e. monitor

4. Since the days when computers were first made available, computers have developed in three directions. What are they?

   a. increased expense

   b. miniaturization

   c. increased size

   d. affordability

   e. increased speed

5. Which of the following operations constitute the four basic operations followed by all computers?

   a. input

   b. storage

   c. programming

   d. output

   e. processing

6. Supercomputers are used for

   a. breaking codes.

   b. simulations for explosions of nuclear bombs.

   c. forecasting weather.

   d. keeping planets in orbit.

   e. all of these

   f. only a, b, and c.

7. What is the leading use of computers?

   a. web surfing

   b. email

   c. e-shopping

   d. word processing

8. Which is the main circuit board in the computer?

   a. RAM chip (random access memory)

   b. CPU processor chip (central processing unit)

   c. motherboard (system board)

   d. hard drive

9. A terabyte is approximately

   a. one million characters.

   b. one billion characters.

   c. one trillion characters.

   d. one quadrillion characters.

10. Speakers are an example of

   a. an input device.

   b. an output device.

   c. a processor.

   d. a storage device.

True/False Questions

T F   1. Mainframe computers process faster than microcomputers.

T F   2. Main memory is a software component.

T F   3. The operating system is part of the system software.

T F   4. Processing is the manipulation by which a computer transforms data into information.

T F   5. Primary storage is the area in the computer where data or information is held permanently.

T F   6. The keyboard and the mouse are examples of input devices.

T F   7. Movies are a form of multimedia.

T F   8. Computers are becoming larger, slower, and more expensive.

T F   9. Modems store information.

T F   10. A microcomputer is used to view very small objects.

T F   11. A floppy disk is an example of software.

T F   12. Computers continue to get smaller and smaller.

T F   13. A CD-ROM can store more data than a floppy disk can.

T F   14. Supercomputers are particularly inexpensive.

 stage **2** **LEARNING**   COMPREHENSION

"I can recall information in my own terms and explain it to a friend."

Short-Answer Questions

1. What does *online* mean?

2. What is the difference between system software and application software?

3. Briefly define *cyberspace*.

4. What is the difference between software and hardware?

5. What is a local area network?

6. What is multimedia?

7. What is the difference between microcomputers and supercomputers?

8. What is the function of RAM?

9. What does *downloading* mean?

10. What is meant by *connectivity*?

11. Describe some ways that information technology can be used to help people find jobs and to help jobs find people.

12. Compare the use of email to the use of the telephone and of conventional letters sent via the postal system. Which kinds of communications are best suited for which medium?

 stage **LEARNING** APPLYING, ANALYZING, SYNTHESIZING, EVALUATING

"I can apply what I've learned, relate these ideas to other concepts, build on other knowledge, and use all these thinking skills to form a judgment."

Knowledge in Action

1. Do you wish there was an invention to make your life easier or better? Describe it. What would it do for you? Come up with ideas on how that device may be constructed.

2. Determine what types of computers are being used where you work or go to school. In which departments are the different types of computer used? Make a list of the input devices, output devices, and storage devices. What are they used for? How are they connected to other computers?

3. Imagine a business you could start or run at home. What type of business is it? What type of computer(s) do you think you'll need? Describe the computer system in as much detail as possible, including hardware components in the areas we have discussed so far. Keep your notes, and then refine your answers after you have completed the course.

4. Has reality become science fiction? Or has science fiction become science fact? First, watch an old futuristic movie, such as *2001– A Space Odyssey*, and take note of the then-futuristic technology displayed. Classify what you see according to input, output, processing, storage, and communications. Then watch a recent science fiction movie, and also list all the futuristic technology used according to the given categories. What was futuristic in the old movie that is now reality? What in the new movie is not yet reality but seems already feasible?

5. From what you've read and what you have experienced and/or observed in your life, do you have a positive, negative, or impartial view of our rapidly converging technological society? Why? Reevaluate your answers at the end of the course.

6. Computer prices are constantly falling. Whatever you pay for a computer today, you can be certain that you will be able to buy a more powerful computer for less money a year from now, and quite possibly even just a month from now. So how can you decide when it's a good time to upgrade to a better computer? Paradoxically, it seems that no matter how you time it, you'll always lose, because prices will go down again soon, and yet you will also always gain, because, since you were going to upgrade sooner or later anyway, you will reap the benefits of having the more powerful equipment that much longer.

Discuss the benefits and costs, both material and psychological, of "waiting until prices drop." Gather more information on this topic by asking friends and colleagues what choices they have made about upgrading equipment over the years and whether they feel satisfaction or regret about the timing when they finally did upgrade.

7. Computers are almost everywhere, and they affect most walks of life—business, education, government, military, hobbies, shopping, research, and so on. What aspects of your life can you think of that still seem relatively unaffected by computers and technology? Is this a good thing or a bad thing, and is it likely to last? What aspects of your life have been the most conspicuously affected by technology? Has anything been made worse or harder in your life by the advance of computers? What about things that have been made better or easier?

8. Have you become extremely dependent on some technologies? Some people no longer write down telephone numbers anywhere; instead, they simply program them into their cellphones. Some people feel helpless in a foreign country unless they have a calculator in hand to compute currency conversions. Many people rely on their email archive to hold essential information, such as addresses and appointments. When any of these technologies fails us, we can feel lost.

Make a list of technologies that have become indispensable to your life. Imagine the consequences if any of these technologies should fail you. What can you do to protect yourself against such failure?

9. It has been said that the computer is a "meta medium" because it can simulate (behave as) any other medium. Thus a computer can present text that can be read from virtual "pages" as if it were a book; it can let you compose and print text as if it were a typewriter; it can play music as if it were a boombox; it can display video as if it were a television set; it can make telephone calls as if it were a telephone; it can let you "draw" and "paint"; it can be programmed to serve as an answering machine; and so forth.

Imagine a future in which computers have replaced all the things they can emulate: instead of books and magazines and newspapers, we would have text only on computers. Telephones, PDAs, television sets, VCRs, DVD players, stereo sets, and other electronic devices would all be gone or, rather, subsumed by computers. What benefits to your life can you see in

Connecting to the Internet: Narrowband, Broadband, & Access Providers

What are the means of connecting to the internet, and how fast are they? What are the three kinds of internet access provider?

In general terms, **_bandwidth_, or *channel capacity*, is an expression of how much data—text, voice, video, and so on—can be sent through a communications channel in a given amount of time.** The type of data transmission that allows only one signal at a time is called *baseband transmission.* When several signals can be transmitted at once, it's called *broadband transmission.* **_Broadband_—very high speed—connections** include various kinds of high-speed wired connections (such as coaxial and fiber-optic, described in Chapter 6), as well as DSL, cable, and satellite and other wireless connections, discussed shortly. Today an estimated 55% of adult internet users—about a third of all adult Americans—have broadband at home or at work.[4]

THE PHYSICAL CONNECTION: WIRED OR WIRELESS?   What are your choices of a *physical connection*—the wired or wireless means of connecting to the internet? A lot depends on where you live. As you might expect, urban and many suburban areas offer more broadband connections than rural areas do. Among the principal means of connection are (1) telephone (dial-up) modem; (2) several high-speed phone lines—ISDN, DSL, and T1; (3) cable modem; and (4) wireless—satellite and other through-the-air links.

DATA TRANSMISSION SPEEDS   Data is transmitted in characters or collections of bits. A *bit,* as we will discuss later, is the smallest unit of information used by computers. Today's data transmission speeds are measured in *bits, kilobits, megabits,* and *gigabits* per second:

- **bps:** A computer with an older modem might have a speed of 28,800 bps, which is considered the minimum speed for visiting websites with graphics. The **_bps_ stands for _bits per second_.** (Eight bits equals one character, such as A, 3, or #.)
- **Kbps:** This is the most frequently used measure; **_kilobits per second_, or _Kbps_, are 1 thousand bits per second.** The speed of a modem that is 28,800 bps might be expressed as 28.8 Kbps.

1969	1970	1971	1973	1974	1975	1976
ARPANET established at 4 U.S. universities (4 computers linked by leased lines); led to internet (4 hosts)	Microprocessor chips come into use; 15 ARPANET sites established (universities/research), each with own address; 13 hosts on internet	Email invented by computer engineer Ray Tomlinson; 23 hosts on internet	ARPANet becomes international; 35 hosts on internet	TCP (Transmission Control Protocol) specification developed by U.S. Dept. of Defense; 62 hosts on internet	First micro-computer (MITS Altair 8800)	Queen Elizabeth sends the first royal email

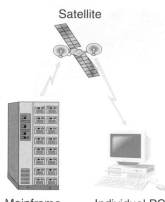

Satellite

Mainframe   Individual PC

Download
(reverse the direction of
data transmission to **upload**)

- **Mbps:** Faster means of connection are measured in ***megabits per second***, or ***Mbps***—1 million bits per second.
- **Gbps:** At the extreme are ***gigabits per second***, ***Gbps***—1 billion bits per second.

UPLOADING & DOWNLOADING   Why is it important to know these terms? Because the number of bits affects how fast you can upload and download information from a remote computer. As we've said, ***download* is the transmission of data from a remote computer to a local computer,** as from a website to your own PC. ***Upload* is the transmission of data from a local computer to a remote computer,** as from your PC to a website you are constructing.

Narrowband (Dial-Up Modem): Low Speed but Inexpensive & Widely Available

Why would I want to use dial-up to connect my computer to the internet?

The telephone line that you use for voice calls is still the cheapest means of online connection and is available everywhere. Although the majority of U.S. households, 51% in 2004, favor broadband internet connections, many home users still use what are called ***narrowband*, or low-bandwith,** connections.[5] This mainly consists of ***dial-up connections*—use of telephone modems to connect computers to the internet.**

CONNECTING THE MODEM   As we mentioned in Chapter 1, **a *modem* is a device that sends and receives data over telephone lines to and from computers.** These days, the modem is generally installed inside your computer, but there are also external modems. The modem is attached to the telephone wall outlet. (● *See Panel 2.3 on the next page.*) (We discuss modems in a bit more detail in Chapter 6.)

Most standard modems today have a maximum speed of 56 Kbps. That doesn't mean that you'll be sending and receiving data at that rate. The modem in your computer must negotiate with the modems used by your ***internet access provider*, the regional, national, or wireless organization or business that connects you to the internet.** Your provider may have modems operating at slower speeds, such as 28.8 Kbps. In addition, lower-quality phone lines or heavy traffic during peak hours—such as 5 P.M. to 11 P.M. in residential areas—can slow down your rate of transmission.

1976	1978	1979	1981	1984	1986	1987
Apple I computer (first personal computer sold in assembled form)	TCP/IP developed (released in 1983) as standard internet transmission protocol; 111 hosts on internet	First Usenet newsgroups; 188 hosts on internet	IBM introduces Personal Computer; 213 hosts on internet	Apple Macintosh; first personal laser printer; William Gibson coins term "cyberspace"; Domain Name System (DNS) introduced	NSFNET (National Science Foundation Network) backbone established	Digital cellular phones invented; first email message sent from China

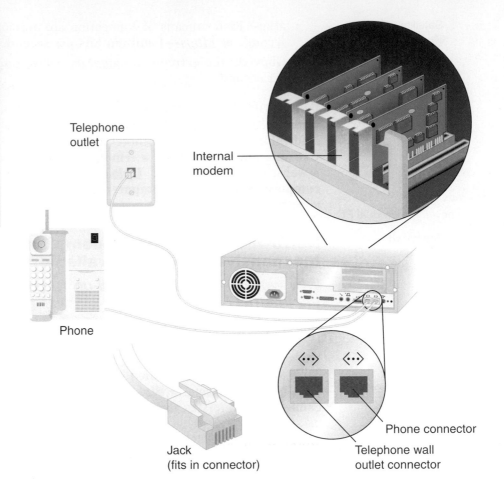

Telephone outlet

Internal modem

Phone

Jack (fits in connector)

Phone connector

Telephone wall outlet connector

MODEMS & CALL WAITING One disadvantage of a telephone modem is that while you're online you can't use that phone line to make voice calls unless you've installed special equipment. In addition, people who try to call you while you're using the modem will get a busy signal. (Call waiting may interrupt an online connection, so you need to talk to your phone company about disabling it or purchase a new modem that can handle call waiting. The Windows operating system also has a feature for disabling call waiting.)

You generally won't need to pay long-distance phone rates, since most access providers offer local access numbers. The cost of a dial-up modem connection to the internet is $10–$50 per month, plus a possible setup charge of $10–$25.

1989	1990	1984	1989–1991	1992	1993	1994
World Wide Web established by Tim Berners-Lee while working at the European Particle Physics Laboratory in Geneva, Switzerland; first home trials of fiber communications network; number of internet hosts breaks 100,000	ARPANET decommissioned; first ISP comes on-line (dial-up access); Berners-Lee develops first web browser, World Wide Web; 313,000 hosts on internet (9,300 domains)	9.6 K modem	14.4 K modem	"Surfing the internet" coined by Jean Armour Polly; 1,136,000 hosts on internet (18,1000 domains)	Multimedia desktop computers; NAPs replace NSFNET; first graphical web browser, Mosaic, developed by Marc Andreessen; the U.S. White House goes online; internet talk radio begins broadcasting; 2,056,000 hosts on internet (28,000 domains)	Apple and IBM introduce PCs with full-motion video built in; wireless data transmission for small portable computers; Netscape Navigator released; 28.8 K modem; the Japanese Prime Minister goes online; 3,864,000 hosts on internet (56,000 domains)

Check online connections

If you're using Windows on your computer, you can check your online connection speed by going to the taskbar and double-clicking the connection icon (bottom right of screen):

The result (dial-up):

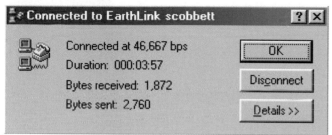

High-Speed Phone Lines: More Expensive but Available in Most Cities

What are my choices in high-speed phone lines?

Waiting while your computer's modem takes 25 minutes to transmit a 1-minute low-quality video from a website may have you pummeling the desk in frustration. To get some relief, you could enhance your ***POTS*—"plain old telephone system"**—connection with a high-speed adaptation. Among the choices are ISDN, DSL, and T1, available in most major cities, though not in rural and many suburban areas.

ISDN LINE **_ISDN (Integrated Services Digital Network)_ consists of hardware and software that allow voice, video, and data to be communicated over traditional copper-wire telephone lines.** Capable of transmitting 64 to 128 Kbps, ISDN is able to send digital signals over POTS lines. If you were trying to download an approximately 6-minute-long music video from the World Wide Web, it would take you about 4 hours and 45 minutes with a 28.8-Kbps modem. A 128-Kbps ISDN connection would reduce this to an hour.

Basically, ISDN is a viable solution for single users of small business networks when other high-speed options are not available. ISDN is not as fast as DSL, cable, or T1 and is expensive.

DSL LINE **_DSL (digital subscriber line)_ uses regular phone lines, a DSL modem, and special technology to transmit data in megabits per second.** Incoming data is significantly faster than outgoing data. That is, your computer can *receive* data at the rate of 1.5–9 Mbps, but it can *send* data at only 128 Kbps–1.5 Mbps. This arrangement may be fine, however, if you're principally interested in obtaining very large amounts of data (video, music) rather than in sending such data to others. With DSL, you could download that 6-minute music video perhaps in only 11 minutes (compared to an hour with ISDN). A big advantage of DSL is that it is always on (so you don't have to make a dial-up connection) and, unlike cable (discussed shortly), its transmission rate is relatively consistent. Also, you can talk on the phone and send data at the same time.

1995	1996	2000	2001	2002	2003	2004
NSFNET reverts to research project; internet now in commercial hands; the Vatican goes online	Microsoft releases Internet Explorer; 56 K modem invented; cable modem introduced, 12,881,000 hosts on internet (488,000 domains)	Web surpasses 1 billion indexable pages: 93,047,785 hosts on internet	AOL membership surpasses 28 million; Napster goes to court	Blogs become popular	First official Swiss online election; flash mobs start in New York City	More than 285,000,000 hosts on internet

Unlike dial-up services, broadband services, because they
are always switched on, make
your computer vulnerable to
over-the-internet security
breaches. Solution: Install firewall software (Chapter 6).

There is one big drawback to DSL: You have to live within 3.3 miles of a phone company central switching office, because the access speed and reliability degrade with distance. However, DSL is becoming more popular, and phone companies are building thousands of remote switching facilities to enhance service throughout their regions. Another drawback is that you have to choose from a list of internet service providers that are under contract to the phone company you use, although other DSL providers exist.

T1 LINE How important is high speed to you? Is it worth $1,500 a month? Then consider getting a **_T1 line_, essentially a traditional trunk line that carries 24 normal telephone circuits and has a transmission rate of 1.5 Mbps.** Generally, T1 lines are leased by corporate, government, and academic sites. Another high-speed line, the T3 line, transmits at 44.7 Mbps (the equivalent of 672 simultaneous voice calls) and costs about $4,000 or more a month. Telephone companies and other types of companies are making even faster connections available: An STS-1 connection runs at 51 Mbps, and an STS-48 connection speeds data along at 2.5 Gbps (2.5 billion bits per second). T1 and T3 lines are commonly used by businesses connecting to the internet, by internet access providers, and in the internet high-speed transmission lines.

Problem for Telephone Internet Connections: The Last Mile

Why does the "last mile" of a wired connection often slow down the data rate?

The distance from your home to your telephone's switching office, the local loop, is often called the "last mile." As we mentioned earlier, if you are using POTS for your initial internet connection—even if you use ISDN or DSL—data must pass back and forth between you and your telephone switching station. (This distance is usually more than a mile; it is shorter than 20 miles and averages about 3 miles in metropolitan areas.) The "last mile" of copper wire is what really slows things down. This problem can be solved by installing newer transmission media, but communications companies are slow to incur this cost. There are about 130 million phone lines in the United States that use 650 million miles of copper wire. Considering that our planet is only about 93 million miles from the sun, 650 million miles of wire represents a huge challenge to replace!

Cable modem

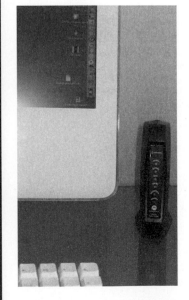

Cable Modem: Close Competitor to DSL

What are the advantages and disadvantages of a cable modem connection?

If DSL's 11 minutes to move a 6-minute video sounds good, 2 minutes sounds even better. That's the rate of transmission for cable modems, which can transmit outgoing data at about 1.4 Mbps and incoming data at up to 30 Mbps. (The common residential transmission rate is 3 Mbps.) **A _cable modem_ connects a personal computer to a cable-TV system that offers an internet connection.**

The advantage of a cable modem is that, like a DSL connection, it is always on. However, unlike DSL, you don't need to live near a telephone switching station. (● _See Panel 2.4._)

A disadvantage, however, is that you and your cable-TV-viewing neighbors are sharing the system and consequently, during peak-load times, your service may be slowed to the speed of a regular dial-up modem. Also, using an internet connection that is always on—and that, in the case of cable, you share with other people—invites outside interference with your computer, a risk that we discuss later in the book.

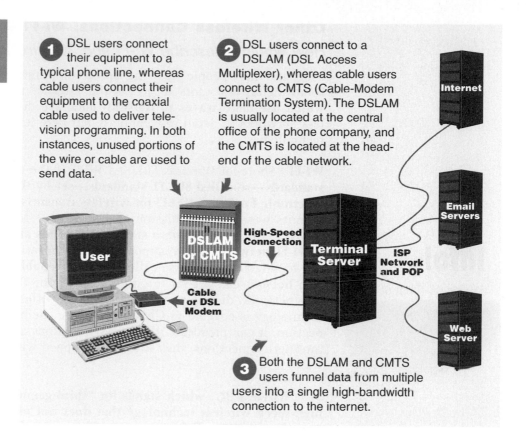

1 DSL users connect their equipment to a typical phone line, whereas cable users connect their equipment to the coaxial cable used to deliver television programming. In both instances, unused portions of the wire or cable are used to send data.

2 DSL users connect to a DSLAM (DSL Access Multiplexer), whereas cable users connect to CMTS (Cable-Modem Termination System). The DSLAM is usually located at the central office of the phone company, and the CMTS is located at the head-end of the cable network.

User

DSLAM or CMTS

High-Speed Connection

Cable or DSL Modem

Terminal Server

ISP Network and POP

Internet

Email Servers

Web Server

3 Both the DSLAM and CMTS users funnel data from multiple users into a single high-bandwidth connection to the internet.

Survival Tip
Connection Speeds

To test the speed of your internet connection, go to *www.groundcontrol.com/tools/speedtest.htm* and click on the *Speed Test* button.

Cable companies may contract you to use their own internet access provider, but more commonly you may choose your own. (Note that cable modems are for internet connections. They do not by themselves take the place of your voice phone system.)

Satellite Wireless Connections

Why might I consider having a satellite connection?

If you live in a rural area and are tired of the molasses-like speed of your cranky local phone system, you might—if you have an unobstructed view of the southern sky—consider taking to the air. With a pizza-size satellite dish on your roof, you can send data at the rate of 56–500 Kbps and receive data at about 1.5 Mbps from a ***communications satellite***, **a space station that transmits radio waves called *microwaves* from earth-based stations.**

Satellite internet connections are always on. To surf the internet using this kind of connection, you need an internet access provider that supports two-way satellite transmission. You will also have to lease or purchase satellite-access hardware, such as a dish. (We cover satellites in more detail in Chapter 6.)

Sky connection. Setting up a home satellite dish.

Other Wireless Connections: Wi-Fi & 3G

How would I describe Wi-Fi and 3G wireless connections?

More and more people are using laptop computers, smart cellphones, and personal digital assistants to access the internet through **_wireless networks_, which use radio waves to transmit data.** We discuss various types of wireless networks in detail in Chapter 6, but here let us mention just two of the technologies:

WI-FI Short for *Wireless Fidelity*, **_Wi-Fi_ is the name given to any of several standards—so-called 802.11 standards—set by the Institute of Electrical and Electronic Engineers (IEEE) for wireless transmission.** One standard, 802.11b, permits wireless transmission of data at 1–11 Mbps up to 300 feet from an **_access point_, or *hot spot*, a station that sends and receives data to and from a Wi-Fi network.** Many airports, hotels, libraries, convention centers, and fast-food facilities offer so-called **_hotspots_—public access to Wi-Fi networks.** The hotspot can get its internet access from DSL, cable modem, T1 local area network, dial-up phone service, or any other method. (Communications technology is covered in Chapter 6.) Once the hotspot has the internet connection, it can broadcast it wirelessly. Laptops are commonly used for Wi-Fi internet connections; they must be equipped with the necessary Wi-Fi hardware, however.

info!

Finding Wi-Fi Hot Spots

www.wififreespot.com/
www.wifihotspotlist.com/
www.wifinder.com/

3G Wireless **_3G_, which stands for "third generation," is loosely defined as high-speed wireless technology that does not need access points because it uses the existing cellphone system.** This technology, which is found in many new smartphones and PDAs that are capable of delivering downloadable video clips and high-resolution games, is being provided by Cingular, Sprint, Verizon, and T-Mobile.

The table at right shows the transmission rates for various connections, as well as their approximate costs (always subject to change, of course) and their pros and cons. (● *See Panel 2.5.*)

Internet Access Providers: Three Kinds

How do I distinguish among the three types of providers?

As we mentioned, in addition to having an access device and a means of connection, to get on the internet you need to go through an *internet access provider*. There are three types of such providers: *internet service providers, commercial online services,* and *wireless internet service providers.*

INTERNET SERVICE PROVIDERS (ISPs) **An _internet service provider (ISP)_ is a local, regional, or national organization that provides access to the internet.** Examples of well-known national providers are AT&T Worldnet and EarthLink. There are also some free ISPs.

info!

Finding ISPs

For comparison shopping, go online to www.thelist.com, which lists ISPs from all over the world and guides you through the process of finding one that's best for you. Which ISPs are available in your area? Which ones supply both dial-up and wireless services?

COMMERCIAL ONLINE SERVICES A *commercial online service* is a members-only company that provides not only internet access but other specialized content as well, such as news, games, and financial data. The two best-known subscriber-only commercial online services are AOL (America Online) and MSN (Microsoft Network).

WIRELESS INTERNET SERVICE PROVIDERS A *wireless internet service provider* enables users with computers containing wireless modems—mostly laptops/notebooks—and web-enabled mobile smartphones and personal digital assistants to gain access to the internet. Examples are AT&T Wireless (acquired by Cingular) and Verizon Wireless. Some ISPs, such as Earthlink and Net Zero, also provide wireless access services in addition to dial-up services.

S u r v i v a l   T i p
Some Free ISPs

www.freei.com
www.freelane.excite.com
www.juno.com

Service	Cost per Month (plus installation, equipment)	Maximum Speed (download only)	Pluses	Minuses
Telephone (dial-up) modem	$0–$45	56 Kbps	Inexpensive, available everywhere	Slow; connection supports only a single user
ISDN	$40–$110 (+ $350–$700 installation cost)	64–128 Kbps (1.5 Mbps with special wiring)	Faster than dial-up; uses conventional phone lines	More expensive than dial-up; no longer extensively supported by telephone companies for individuals; used mostly by small businesses
DSL	$50–$300, depending on speed	1.5–9 Mbps	Fast download, always on, higher security; uploads faster than cable; users can talk and transmit data at the same time	Needs to be close to phone company switching station; limited choice of service providers; supports only a single user
T1 line	$1,500 (+ $1,000 installation cost)	1.5 Mbps	Can support many users: 24 separate circuits of 64 Kbps each; reliable high-speed downloading and uploading; users can talk and transmit data at the same time	Expensive, best for businesses
Cable modem	$40–$100 (+ $5 monthly for leased cable modem)	Up to 30 Mbps (3 Mbps common)	Fast, always on, most popular broadband type of connection; can support many users; downloads faster than DSL; users can talk on phone and transmit cable data at the same time	Slower service during high-traffic times, vulnerability to outside intrusion, limited choice of service providers; not always available to businesses
Satellite	Up to $70 (+ about $600 installation)	1.5 Mbps	Wireless, fast, reliable, always on; goes where DSL and cable can't; users can talk on phone and transmit satellite data at the same time	High setup and monthly costs; users must have unobstructed view of the southern sky
Wi-Fi	Nothing for users accessing hotspots supplied by others; about $300 for access-point hardware and up to $25 per month for subscription to a Wi-Fi access service	2.5–4.5 Mbps (occasionally up to 11 Mbps)	Uses a beefed-up version of the current cellphone network; ultimately should be available everywhere	Range of access is usually only 50–300 feet (to access point)
3G	$250 and up	300 Kbps	Functions like Wi-Fi but without the need for hot spots; uses existing cellphone network	Low battery power on some models; phones are relatively large

PRACTICAL ACTION
Looking for an Internet Access Provider: Questions to Ask at the Beginning

If you belong to a college or company, you may get internet access free. Many public libraries also offer free net connections. If these options are not available to you, here are some questions to ask in your first phone call to an internet access provider:

- Is there a contract, and for what length of time? Is there a trial period, or are you obligated to stick with the provider for a while even if you're unhappy with it?

- Is there a setup fee? What kind of help do you get in setting up your connection?

- How much is unlimited access per month? Is there a discount for long-term commitments?

- Is the access number a local phone call so that you don't have long-distance phone tolls?

- Is there an alternative dial-up number if the main number is out of service?

- Is access available when you're traveling, through either local numbers or toll-free 800 numbers?

- Can you gain access to your email through the provider's website?

- Will the provider help you establish your personal website, if you want one, and is there sufficient space on the provider's server?

- Is there free, 24-hour technical support? Is it reachable through a toll-free number?

- How long does it take to get tech support? Ask for the tech-support number before you sign up, and then call it to see how long a response takes. Also try connecting through the web.

- Will the provider keep up with technology? For instance, is it planning to offer broadband access? Wireless access?

- Will the provider sell your name to marketers? What kind of service does it offer to block unwanted junk messages (spam)?

Once you have contacted an internet access provider and paid the required fee (charged to your credit card), you will be provided with information about local connections and necessary communications software for setting up your computer and modem to make a connection. For this you use your user name ("user ID") and your **_password_, a secret word or string of characters that enables you to _log on_, or make a connection to the remote computer.** The access provider will also help you establish your email address, which must be unique—that is, not the same as anyone else's.

2.2

How Does the Internet Work?

What is the basic structure of the internet, and who controls it?

The *international network* known as the *internet* consists of hundreds of thousands of smaller networks linking educational, commercial, nonprofit, and military organizations, as well as individuals. Central to this arrangement is the client/server network. **A _client_ computer is a computer requesting data or services. A _server_, or _host computer_, is a central computer supplying data or services requested of it.** When the client computer's request—for example, for information on various airline flights and prices—gets to a server computer, that computer sends the information back to the client computer.

Internet Connections: POPs, NAPs, Backbone, & Internet2

How do I distinguish a point of presence from a network access point, and what is the internet backbone? What does Internet2 do?

Your journey onto the internet starts, in the case of dial-up, with your client computer's modem connecting via a local phone call to your internet service provider (ISP). (● *See Panel 2.6.*) This is the slowest part of the internet connection. An ISP's headquarters and network servers may be located almost anywhere.

POINT OF PRESENCE   To avoid making its customers pay long-distance phone charges to connect, the ISP provides each customer with a ❶ *point of presence (POP)***—a local access point to the internet—a collection of modems and other equipment in a local area.** The POP acts as a local gateway to the ISP's network.

NETWORK ACCESS POINT   The ISP in turn connects to a ❷ *network access point (NAP)***, a routing computer at a point on the internet where several connections come together.** NAPs are owned by a *network service provider (NSP)*, a large communications company, such as AGIS or MCI. The four original NAPs in the United States are in San Francisco, Washington, D.C., Chicago, and New Jersey. These NAPs were established in 1993, when the original network that became the internet was privatized.

Much of the internet congestion occurs at NAPs. The four main NAPs quickly became overloaded, so Private/Peer NAPs, called *PNAPs* ("miniNAPs")

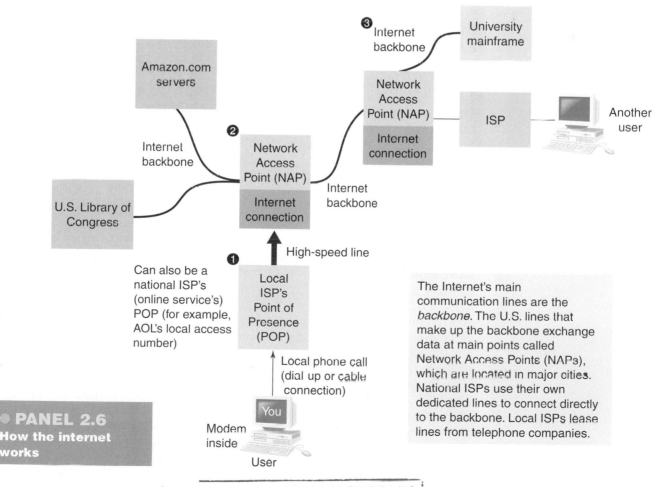

❸ Internet backbone

University mainframe

Network Access Point (NAP)

Internet connection

ISP

Another user

Amazon.com servers

Internet backbone

❷ Network Access Point (NAP)

Internet connection

Internet backbone

Internet backbone

U.S. Library of Congress

High-speed line

Can also be a national ISP's (online service's) POP (for example, AOL's local access number)

❶ Local ISP's Point of Presence (POP)

Local phone call (dial up or cable connection)

You

Modem inside

User

The Internet's main communication lines are the *backbone*. The U.S. lines that make up the backbone exchange data at main points called Network Access Points (NAPs), which are located in major cities. National ISPs use their own dedicated lines to connect directly to the backbone. Local ISPs lease lines from telephone companies.

● **PANEL 2.6**
How the internet works

The Internet & the World Wide Web

were established in the late 1990s. PNAPs facilitate more efficient routing (passing) of data back and forth on the internet by providing more backbone access locations. At present, there are more than 100 PNAPs in the United States, and the number is growing. PNAPs are provided by commercial companies, such as Savvis Communications.

INTERNET BACKBONE   Each NAP has at least one computer, whose task is simply to direct internet traffic from one NAP to the next. NAPs are connected by the equivalent of interstate highways, known collectively as the ❸ _internet backbone_, **high-speed, high-capacity transmission lines that use the newest communications technology to transmit data across the internet.** Backbone connections are supplied by internet backbone providers such as AT&T, Cable & Wireless, GTE, Sprint, Teleglobe, and UUNET.

INTERNET2   _Internet2_ **is a cooperative university/business education and research project that enables high-end users to quickly and reliably move huge amounts of data over high-speed networks.** In effect, Internet2 adds "toll lanes" to the older internet to speed things up. The purpose is to advance videoconferencing, research, and academic collaboration—to enable a kind of "virtual university." Presently Internet2 links more than 206 research universities. Requiring state-of-the-art infrastructure, Internet2 operates a high-speed backbone network called Abiline, which many colleges use.

A rival nonprofit organization is National LamdaRail, a fiber-optic network assembled for colleges' use. It and Internet2 are planning to merge in an experiment that may foreshadow something called Hybrid Optical and Packet Infrastructure (HOPI) as the blueprint for the next internet.[6] The National Science Foundation has also announced plans for a new network called the Global Environment for Networking Investigations, designed to reengineer the internet and overcome its shortcomings.

Internet Communications: Protocols, Packets, & Addresses

How do computers understand the data being transmitted over the internet?

When your modem connects to a modem at your ISP's POP location, the two modems go through a process called _handshaking,_ whereby the fastest-available transmission speed is established. Then _authentication_ occurs: your ISP needs to know you are who you say you are, so you will need to provide a user name and a password. These two items will have been established when you opened your account with your ISP.

PROTOCOLS   How do computers understand the data being transmitted? The key lies in data following the same _protocol_, **or set of rules, that computers must follow to transmit data electronically. The protocol that enables all computers to use data transmitted on the internet is called _Transmission Control Protocol/Internet Protocol_, or _TCP/IP_,** which was developed in 1978 by ARPA. TCP/IP is used for all internet transactions, from sending email to downloading pictures off a friend's website. Among other things, TCP/IP determines how the sending device will indicate that it has finished sending a message and how the receiving device will indicate that it has received the message.

PACKETS   Most importantly, perhaps, TCP/IP breaks the data in a message into separate _packets_, **fixed-length blocks of data for transmission.** This allows a message to be split up and its parts sent by separate routes yet still all wind up in the same place. IP is used to send the packets across the internet to their final destination, and TCP is used to reassemble the packets in the correct order. The packets do not have to follow the same network routes to reach their destination because all the packets have the same *IP address*, as we explain.

IP ADDRESSES **An *Internet Protocol (IP) address* uniquely identifies every computer and device connected to the internet.** An IP address consists of four sets of numbers between 0 and 255 separated by decimals (called a *dotted quad*)—for example, 1.160.10.240. An IP address is similar to a street address, but street addresses rarely change, whereas IP addresses often do. Each time you connect to your internet access provider, it assigns your computer a new IP address, called a *dynamic IP address,* for your online session. When you request data from the internet, it is transmitted to your computer's IP address. When you disconnect, your provider frees up the IP address you were using and reassigns it to another user.

A dynamic IP address changes each time you connect to the internet. A *static IP address* remains constant each time a person logs on to the internet. Established organizational websites—such as your ISP's—have static IP addresses.

It would be simple if every computer that connects to the internet had its own static IP number, but when the internet was first conceived, the architects didn't foresee the need for an unlimited number of IP addresses. Consequently, there are not enough IP numbers to go around. To get around that problem, many internet access providers limit the number of static IP addresses they allocate and economize on the remaining number of IP addresses they possess by temporarily assigning an IP address from a pool of IP addresses.

If your computer is constantly connected to the internet, through a local network at work or school, most likely you have a static IP address. If you have a dial-up connection to the internet or are using a computer that gets connected to the internet intermittently, you're most likely picking up a dynamic IP address from a pool of possible IP addresses at your internet access provider's network during each log-in.

After your connection has been made, your internet access provider functions as an interface between you and the rest of the internet. If you're exchanging data with someone who uses the same provider, the data may stay within that organization's network. Large national internet access providers operate their own backbones that connect their POPs throughout the country. Usually, however, data travels over many different networks before it reaches your computer.

Who Runs the Internet?

What does ICANN do?

Although no one owns the internet, everyone on the net adheres to standards overseen by the international Board of Trustees of ISOC, the *Internet Society (www.isoc.org)*. ISOC is a professional, nonprofit society with more than 100 organizational and 20,000 individual members in more than 180 countries. The organizations include companies, governments, and foundations. ISOC provides leadership in addressing issues that confront the future of the internet and is the organizational home for groups responsible for internet infrastructure standards.

In June 1998, the U.S. government proposed the creation of a series of nonprofit corporations to manage such complex issues as fraud prevention, privacy, and intellectual-property protection. The first such group, the ***Internet Corporation for Assigned Names and Numbers (ICANN),*** **was established to regulate human-friendly internet domain names—those addresses ending with .com, .org, .net, and so on, that overlie IP addresses and identify the website type.**

ICANN (which can be accessed at *www.icann..org*) is a global, private-sector, nonprofit corporation that has no statutory authority and imposes policies through contracts with its world members. Criticized for inefficiency, in 2003 it outlined what it called ICANN 2.0, intended to be a more responsive and agile agency that would consult better with the world internet community about the adoption of standards.[7] One of its improvements is the ICANN Whosis Database, which returns the name and address of any domain name entered (entering *microsoft.com*, for instance, returns the name and address of Microsoft Corp.). However, various groups, including the International Telecommunication Union, a United Nations agency, have suggested that the United States through ICANN has too much control over the internet, and several countries have proposed that an international body take over ICANN. The United States in July 2005 asserted that it intended to retain its role in internet management.[8]

At present, IP addresses in North America, South America, the Caribbean, and sub-Saharan Africa are administered by the American Registry for Internet Numbers (ARIN; *www.arin.net*). Two other organizations administer IP addresses in the Asia-Pacific region and in Europe and surrounding areas. These three groups, called *Regional Internet Registries*, work in close coordination with one another and with ISOC and ICANN.

2.3

The World Wide Web

How do the following work: websites, web pages, browsers, URLs, web portals, search tools, and search engines? What are HTML and hyperlinks?

The internet and the World Wide Web, we said, are not the same. The internet is a massive network of networks, connecting millions of computers via protocols, hardware, and communications channels. It is the infrastructure that supports not only the web but also other communications systems such as email, instant messaging, newsgroups, and other activities that we'll discuss. The part of the internet called the *web* is a *multimedia-based* technology that enables you to access more than just text. That is, you can also download art, audio, video, and animation and engage in interactive games.

**■ more
info!**

Whence the Word *Web?*

Why did people perceive the need for a "web," and how did Berners-Lee develop what they needed? Go to:
www.ibiblio.org/pioneers/lee.html
http://en.wikipedia.org/wiki/Tim_Berners-Lee

The Face of the Web: Browsers, Websites, & Web Pages

How would I explain to someone the difference between browsers, websites, and web pages?

If a Rip Van Winkle fell asleep in 1989—the year computer scientist Tim Berners-Lee developed the web software—and awoke today, he would be completely baffled by the new vocabulary that we now encounter on a daily basis: *browser, website, web page, www.* Let's see how we would explain to him what these and similar web terms mean.

BROWSERS: SOFTWARE FOR SURFING THE WEB   A *browser*, or *web browser*, is software that enables you to find and access the various parts of the web. The two best-known browsers are *Microsoft Internet Explorer*, more commonly used, and *Netscape Navigator*, once the leader but now used by fewer people. (● *See Panel 2.7.*) These and other browsers—examples are Mozilla Firefox, Opera, and Apple Macintosh Browser—allow you, like riding a wave with a surfboard, to surf the web. *Surf* means to explore the web by using your mouse to move via a series of connected paths, or links, from one location, or website, to another.

Tim Berners-Lee was born in London, England; his parents, both mathematicians, were employed together on the team that built the Manchester Mark I, one of the earliest computers. Berners-Lee graduated from the Queen's College of Oxford University, where he built a computer with a soldering iron. In 1980, while an independent contractor at CERN (European Organization for Nuclear Research), Berners-Lee proposed a project based on the concept of hypertext, to facilitate sharing and updating information among researchers. With other researchers, he built a prototype system named Enquire.

After leaving CERN, he used ideas similar to those used in Enquire to create the World Wide Web, for which he designed and built the first browser (called WorldWideWeb and developed into NeXTSTEP). Berners-Lee built the first website at *http://info.cern.ch/,* and it was first put online on August 6, 1991. It provided an explanation about what the World Wide Web was, how one could own a browser, how to set up a web server, and so on.

In 1994, Berners-Lee founded the World Wide Web Consortium (W3C) at the Massachusetts Institute of Technology. It comprised various companies willing to create standards and recommendations to improve the quality of the internet. It was not until 2000 and 2001 that popular browsers began to support this standard.

● **PANEL 2.7**
Internet Explorer (top) and Netscape

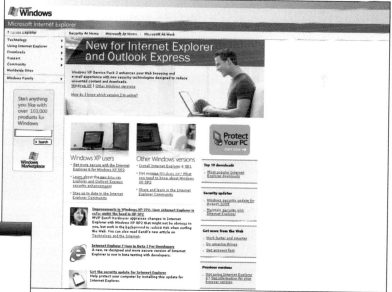

The Internet & the World Wide Web

WEBSITE: THE LOCATION ON THE COMPUTER A _website_, or simply _site_, **is a location on a particular computer on the web that has a unique address** (called a _URL,_ as we'll explain). When you decide to buy books at the online site of bookseller Barnes & Noble, you would visit its website, _www.barnesandnoble.com;_ the website is the location of a computer somewhere on the internet. That computer might be located in Barnes & Noble's offices, but it might be located somewhere else entirely.

WEB PAGES: THE DOCUMENTS ON A WEBSITE A website is composed of a web page or collection of related web pages. **A _web page_ is a document on the World Wide Web that can include text, pictures, sound, and video.** The first page you see at a website is like the title page of a book. This is the **_home page_, or welcome page, which identifies the website and contains links to other pages at the site.** (● _See Panel. 2.8.)_ If you have your own personal website, it might consist of just one page—the home page. Large websites have scores or even hundreds of pages.

How the Browser Finds Things: URLs

How would I describe how a browser connects?

Now let's look at the details of how the browser finds a particular web page.

URLs: ADDRESSES FOR WEB PAGES Before your browser can connect with a website, it needs to know the site's address, the URL. **The _URL (Uniform Resource Locator)_ is a string of characters that points to a specific piece of information anywhere on the web.** In other words, the URL is the website's unique address.

A URL consists of (1) the web _protocol,_ (2) the _domain name_ or web server, (3) the _directory_ (or folder) on that server, and (4) the _file_ within that directory

S u r v i v a l   T i p
Do Home Pages Endure?

The contents of home pages often change. Or they may disappear, and so the connecting links to them in other web pages become links to nowhere. To find out how to view "dead" pages, go to:
http://web.ticino.com/ multilingual/Search. htm#Find%20web%20pages

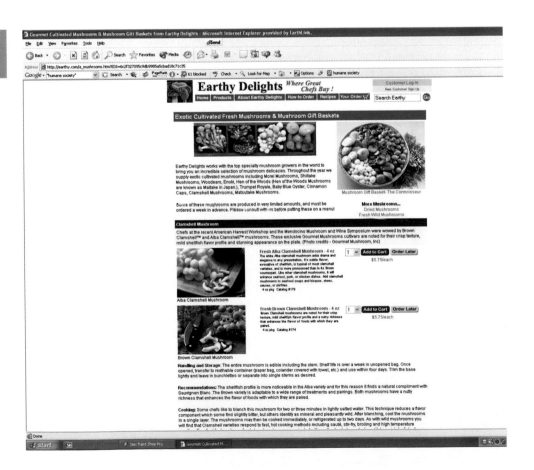

● **PANEL 2.8**
Home page

Marc Andreessen is best known as a cofounder of Netscape Communications Corporation and coauthor of Mosaic, an early web browser. Andreessen received his bachelor's degree in computer science from the University of Illinois, Urbana-Champaign. As an undergraduate, he interned one summer at IBM in Austin, Texas. He also worked at the university's National Center for Supercomputing Applications, where he became familiar with ViolaWWW, created by Pei-Yuan Wei, which was based on Tim Berners-Lee's open standards for the World Wide Web. These early browsers had been created to work only on expensive Unix workstations, so Andreessen created an improved and more user-friendly version with integrated graphics that would work on personal computers. The resulting code was the Mosaic web browser.

Soon after this development, Mosaic Communications Corporation was in business in Mountain View, California, with Andreessen as vice president. Mosaic Communications changed its name to Netscape Communications, and its flagship web browser became Netscape Navigator.

(perhaps with an extension such as *html* or *htm*). Consider the following example of a URL for a website offered by the National Park Service for Yosemite National Park:

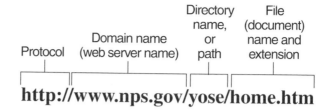

Let's look at these elements.

- **The protocol: http://** As mentioned, a protocol is a set of communication rules for exchanging information. The web protocol, HTTP, was developed by Tim Berners-Lee, and it appears at the beginning of some web addresses (as in *http://www.mcgraw-hill.com*). It stands for **<u>*HyperText Transfer Protocol (HTTP)*</u>, the communications rules that allow browsers to connect with web servers.** (Note: Most browsers assume that all web addresses begin with *http://*, and so you don't need to type this part; just start with whatever follows, such as *www.*)

- **The domain name (web server name): www.nps.gov/** A **<u>*domain*</u> is simply a location on the internet, the particular web server.** Domain names tell the location and the type of address. Domain-name components are separated by periods (called "dots"). The last part of the domain, called the *top-level domain*, is a three-letter extension that describes the domain type: *.gov, .com, .net, .edu, .org, .mil, .int*—government, commercial, network, educational, nonprofit, military, or international organization. In our example, the *www* stands for "World Wide Web," of course; the *.nps* stands for "National Park Service," and *.gov* is the top-level domain name indicating that this is a government website.

    The meanings of other internet top-level domain abbreviations appear in the box on the next page. (● *See Panel 2.9.*) Some top-level domain names also include a two-letter code extension for the country—for example, *.us* for United States, *.ca* for Canada, *.mx* for Mexico, *.uk* for United Kingdom, *.jp* for Japan, *.in* for India, *.cn* for China. These country codes are optional.

- **The directory name: yose/** The *directory* name is the name on the server for the directory, or folder, from which your browser needs to pull the file. Here it is *yose* for "Yosemite." For Yellowstone National Park, it is *yell*.

- **The file name and extension: home.htm** The *file* is the particular page or document that you are seeking. Here it is *home.htm*, because you have gone to the home page, or welcome page, for Yosemite National Park. The *.htm* is an extension to the file name, and this extension informs the browser that the file is an HTML file. Let us consider what HTML means.

URLs & EMAIL ADDRESSES: NOT THE SAME A URL, you may have observed, is *not* the same thing as an email address. The website for the White House (which includes presidential information, history, a tour, and a guide to federal services) is *www.whitehouse.gov*. Some people might type *president@whitehouse.gov* and expect to get a website, but that won't happen. We explain email addresses in another few pages.

Domain Name	Authorized Users	Example
.aero	air-transport industry	director@bigwings.aero
.biz	businesses	ceo@company.biz
.com	originally commercial; now anyone can use	editor@mcgraw-hill.com
.coop	cooperatives	buyer@greatgroceries.coop
.edu	postsecondary accredited educational and research institutions	professor@harvard.edu
.gov	U.S. government agencies and bureaus	president@whitehouse.gov
.info	information service providers	contact@research.info
.int	organizations established by international treaties between governments	sectretary_general@ unitednations.int
.jobs	human resources managers	AnnChu@Personnel.jobs
.mil	U.S. military organizations	chief_of_staff@pentagon.mil
.mobi	mobile devices	user@phonecompany.mobi
.museum	museums	curator@modernart.mus
.name	individuals	joe@smith.name
.net	networking organizations	contact@earthlink.net
.org	organizations, often nonprofit and professional (non-commercial)	director@redcross.org
.post	Universal Postal Union*	manager@UPS.post
.pro	credentialed professionals & related entities	auditor@accountant.pro
.travel	travel industry	JoeAgent@flyright.travel
.xxx	adults-only website	proprietor@badtaste.xxx

Note: The number of domain names will likely be expanded in the future; for a list of current domain names, go to *www.iana.org/gtld/gtld.htm.*

*Some groups pay $45,000 to ICANN for a particular domain name.

The Nuts & Bolts of the Web: HTML & Hyperlinks

What are HTML and hyperlinks?

The basic communications protocol that makes the internet work, as we described, is *TCP/IP*. The communications protocol used to access that part of the internet called the World Wide Web, we pointed out, is called *Hyper-Text Transfer Protocol (HTTP)*. A hypertext document uses *hypertext markup language (HTML)*, which uses *hypertext links*, to connect with other documents. The foundations of the World Wide Web, then, are HTML and hypertext links.

HYPERTEXT MARKUP LANGUAGE (HTML) **_Hypertext markup language (HTML)_ is the set of special instructions (called "tags" or "markups") that are used to specify document structure, formatting, and links to other multimedia documents on the web.** Extensible hypertext markup language (XHTML) is the successor to and the current version of HTML. The need for a more strict version of HTML was perceived primarily because World Wide Web content now needs to be delivered to many devices (such as mobile devices) that have fewer resources than traditional computers have.

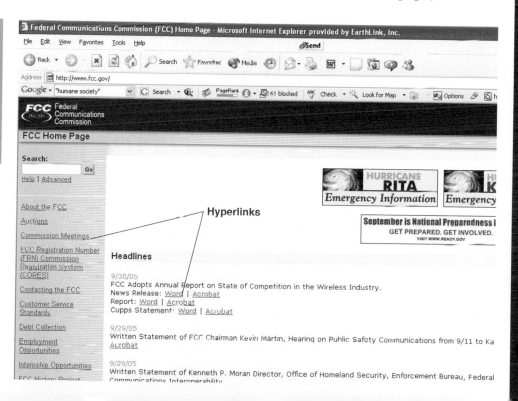

info!

HYPERTEXT LINKS *Hypertext links*—also called *hyperlinks, hotlinks,* or just *links*—are connections to other documents or web pages that contain related information; a word or phrase in one document becomes a connection to a document in a different place. Hyperlinks appear as underlined or color words and phrases. On a home page, for instance, the hyperlinks serve to connect the main page with other pages throughout the website. Other hyperlinks will connect to pages on other websites, whether located on a computer next door or one on the other side of the world.

An example of an HTML document with hyperlinks is shown below. (● *See Panel 2.10.*)

Using Your Browser to Get around the Web

What do I need to know to operate my browser?

You can find almost anything you want on the more than 1 billion web pages available around the world. In one month alone, one writer reported, he went on the web "to find, among other things, hotel reservations in California, a rose that will grow in the shade, information about volunteer fire departments, a used copy of an obscure novel, two ZIP codes, the complete text of *A Midsummer Night's Dream,* and the news that someone I went to high school with took early retirement. I also heard, out of the blue, from two people I haven't seen in years."[9] Among the droplets of what amounts to a Niagara Falls of information: Weather maps and forecasts. Guitar chords. Recipe archives. Sports schedules. Daily newspapers in all languages. Nielsen television ratings. The Alcoholism Research Data Base. U.S. government phone numbers. The Central Intelligence Agency world map. The daily White House press releases. And on and on. But it takes a browser and various kinds of search tools to find and make any kind of sense of this enormous amount of data.

FIVE BASIC ELEMENTS OF THE BROWSER If you buy a new computer, it will come with a browser already installed. Most browsers have a similar look and feel and similar navigational tools. Note that the web browser screen has five basic elements: *menu bar, toolbar, URL bar, workspace,* and *status bar.* To execute menu-bar and toolbar commands, you use the mouse to move the pointer over the word, known as a *menu selection,* and click the left button of the mouse. This will result in a pull-down menu of other commands for other options. (● *See Panel 2.11 on the next page.*)

● **PANEL 2.10**
An HTML document and hyperlinks. Using the mouse to click on the hyperlinked text connects to another location in the same website or at a different site.

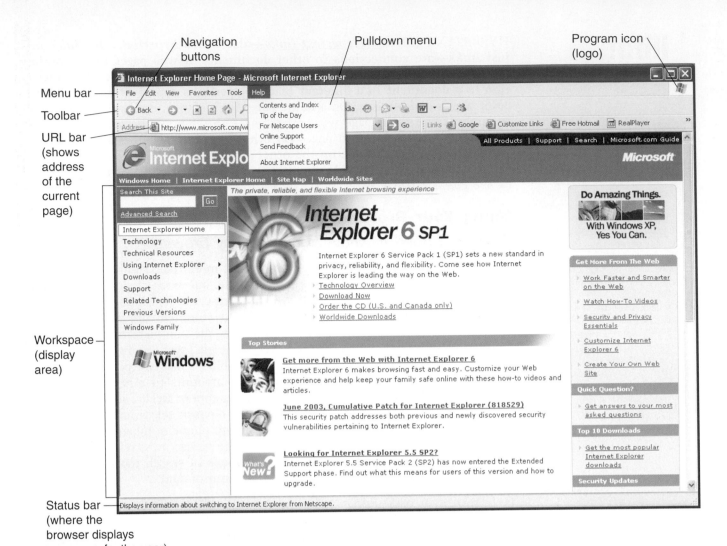

Navigation buttons

Pulldown menu

Program icon (logo)

Menu bar

Toolbar

URL bar (shows address of the current page)

Workspace (display area)

Status bar (where the browser displays messages for the user)

● **PANEL 2.11**
The commands on a browser screen (Internet Explorer)

After you've been using a mouse for a while, you may find moving the pointer around somewhat time consuming. As a shortcut, if you click on the right mouse button, you can reach many of the commands on the toolbar (*Back, Forward,* and so on) via a pop-up menu.

STARTING OUT FROM HOME: THE HOME PAGE   The first page you see when you start up your browser is the *home page* or *start page.* (You can also start up from just a blank page, if you don't want to wait for the time it takes to connect with a home page.) You can choose any page on the web as your start page, but a good start page offers links to sites you want to visit frequently. Often you may find that the internet access provider with which you arrange your internet connection will provide its own start page. However, you'll no doubt be able to customize it to make it your own personal home page.

PERSONALIZING YOUR HOME PAGE   Want to see the weather forecast for your college and/or hometown areas when you first log on? Or the day's news (general, sports, financial, health, and so on)? Or the websites you visit most frequently? Or a reminder page (as for deadlines or people's birthdays)? You can probably personalize your home page following the directions provided with the first start page you encounter. A customized start page is also provided by Yahoo, Google, AltaVista, and similar services.

GETTING AROUND: BACK, FORWARD, HOME, & SEARCH FEATURES Driving in a foreign city (or even Boston or San Francisco) can be an interesting experience in which street names change, turns lead into unknown

Menu bar

Back: Moves you to a previous page or site

Forward: Lets you revisit a page you have just returned from

Stop: You can halt any ongoing transfer of page information

Refresh: If page you are loading is garbled or stalled in transmission, this will retrieve it again

Home: To return to your start page

Search: Displays page containing a directory of search engine sites

Favorites: List of sites can be created so you can quickly jump to the ones used frequently (also called bookmarks)

History: Names and descriptions of sites most recently visited

Print: To print a page, click on this button

Logo: Technical support and free copies of the web browser

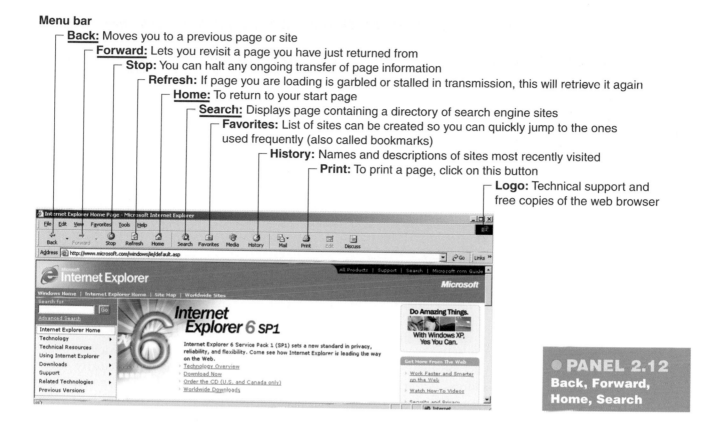

● **PANEL 2.12**
Back, Forward, Home, Search

neighborhoods, and signs aren't always evident, so that soon you have no idea where you are. That's what the internet is like, although on a far more massive scale. Fortunately, unlike being lost in Rome, here your browser toolbar provides navigational aids. (● *See Panel 2.12.) Back* takes you back to the previous page. *Forward* lets you look again at a page you returned from. If you really get lost, you can start over by clicking on *Home,* which returns you to your home page. *Search* lists various other search tools, as we will describe. Other navigational aides are *history lists* and *favorites* or *bookmarks.*

HISTORY LISTS If you are browsing through many web pages, it can be difficult to keep track of the locations of the pages you've already visited. The *history list* allows you to quickly return to the pages you have recently visited. (● *See Panel 2.13.)*

● **PANEL 2.13**
History list
If you want to return to a previously viewed site in Internet Explorer, click on *History.* In Netscape, click on *Communicator,* then *History.*

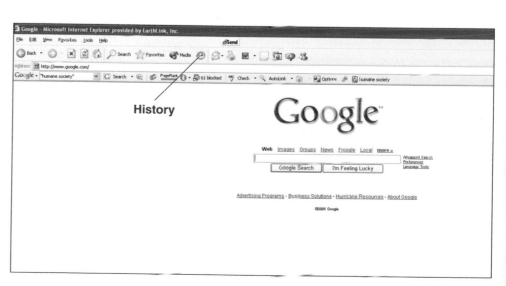

History

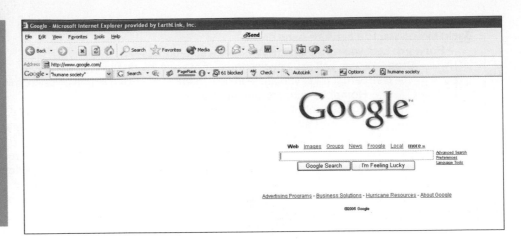

FAVORITES OR BOOKMARKS One great helper for finding your way is the Favorites or Bookmark feature, which lets you store the URLs of web pages you frequently visit so that you don't have to remember and retype your favorite addresses. (● *See Panel 2.14.*) Say you're visiting a site that you really like and that you know you'd like to come back to. You click on *Favorites* (in Internet Explorer) or *Bookmark* (in Netscape), which displays the URL on your screen, and then click on *Add*, which automatically stores the address. Later you can locate the site name on your Favorites menu and click on it, and the site will reappear. (When you want to delete it, you can use the right mouse button and select the *Delete* command.)

INTERACTIVITY: HYPERLINKS, RADIO BUTTONS, & SEARCH BOXES For any given web page that you happen to find yourself on, there may be one of three possible ways to interact with it—or sometimes even all three on the same page. (● *See Panel 2.15.*)

1. By using your mouse to click on the hyperlinks, which will transfer you to another web page (p. 69).
2. By using your mouse to click on a *radio button* and then clicking on a *Submit* command or pressing the *Enter* key. **Radio buttons are little circles located in front of various options; selecting an option with the mouse places a dot in the corresponding circle.**
3. By typing text in a **search box, a fill-in text box,** and then hitting the *Enter* key or clicking on a *Go* or *Continue* command, which will transfer you to another web page.

SCROLLING & FRAMES To the bottom and side of your screen display you will note **scroll arrows, small up/down and left/right arrows. Clicking on scroll arrows with your mouse pointer moves the screen so that you can see the rest of the web page, a movement known as _scrolling_.** You can also use the arrow keys on your keyboard for scrolling.

Some web pages are divided into different rectangles known as *frames*, each with its own scroll arrows. **A _frame_ is an independently controllable section of a web page.** A web page designer can divide a page into separate frames, each with different features or options.

LOOKING AT TWO PAGES SIMULTANEOUSLY If you want to look at more than one web page at the same time, you can position them side by side on your display screen. Select *New* from your File menu to open more than one browser window.

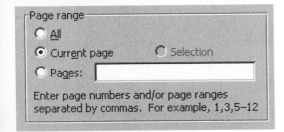

Web Portals: Starting Points for Finding Information

How can I benefit from understanding how portals work?

Using a browser is sort of like exploring an enormous cave with flashlight and string. You point your flashlight at something, go there, and at that location you can see another cave chamber to go to; meanwhile, you're unrolling the ball of string behind you so that you can find your way back.

But what if you want to visit only the most spectacular rock formations in the cave and skip the rest? For that you need a guidebook. There are many such "guidebooks" for finding information on the web, sort of internet superstations known as *web portals.*

TYPES OF WEB PORTALS **A *web portal*, or simply *portal*, is a type of gateway website that functions as an "anchor site" and offers a broad array of resources and services, online shopping malls, email support, community forums, current news and weather, stock quotes, travel information, and links to other popular subject categories.**

In addition, there are *wireless portals*, designed for web-enabled portable devices. An example is Yahoo! Everywhere, which offers the Yahoo! Wireless Directory, Yahoo! Auction to Go, and Yahoo! Movies on the Move.[10] Yahoo! users can access not only email, calendar, news, and stock quotes but also Yahoo's directory of wireless sites, movies, and auctions.

Portals may be general public portals (mega-portals), such as Yahoo!, AOL, Microsoft Network (MSN), Lycos, or (more recently) Google. (● *See Panel 2.16.*) There are also specialized portals—called *vertical portals*, or *vortals*, which focus on specific narrow audiences or communities—such as iVillage.com for women, Fool.com for investors, Burpee.com for gardeners, and SearchNetworking.com for network administrators.

LOGGING ON TO A PORTAL When you log on to a portal, you can do three things: (1) check the home page for general information, (2) use the subject guide to find a topic you want, and (3) use a keyword to search for a topic. (● *See Panel 2.17.*)

● **PANEL 2.16**
Portals: MSN and Fool.com

PANEL 2.17
Yahoo home page
You can read headlines for news and weather, use the directory to find a topic, or use keywords in the Search text box to research specific topics.

- **Check the home page for general information:** You can treat a portal's home or start page as you would a newspaper—for example, to get news headings, weather forecasts, sports scores, and stock-market prices.

- **Use the subject guide to find a topic:** Before they acquired their other features, many of these portals began as a type of search tool known as a *subject guide*, providing lists of several categories of websites classified by topic, such as (with Yahoo) "Business" or "Entertainment." Such a category is also called a *hypertext index*, and its purpose is to allow you to access information in specific categories by clicking on a hypertext link.

- **Use search box and keywords to search for a topic:** At the top of each portal's home page is a search box, a blank space into which you can type a **_keyword_, the subject word or words of the topic you wish to find.** If you want a biography on Apple Computer founder Steve Jobs, then *Steve Jobs* is the keyword. This way you don't have to plow through menu after menu of subject categories. The results of your keyword search will be displayed in a summary of documents containing the keyword you typed.

Some popular portal sites are listed below. (● *See Panel 2.18.*)

Search Services & Search Engines, & How They Work

How do search services build search engines, and what are the implications for me when I make a search?

Portal Name	Site
AOL	www.aol.com
Google	www.google.com
HotBot	www.hotbot.com
LookSmart	www.looksmart.com
Lycos	www.lycos.com
MSN	www.msn.com
Netscape	www.netscape.com
Yahoo	www.yahoo.com

Search services are organizations that maintain databases accessible through websites to help you find information on the internet. Examples are not only parts of portals such as Yahoo (Yahoo! Search) and MSN (MSN Search) but also Google, Ask Jeeves, and Gigablast, to name just a few. Search services also maintain **search engines, programs that enable you to ask questions or use keywords to help locate information on the web.**

Search services compile their databases by using special programs called **spiders**—also known as *crawlers, bots* (for "robots"), or *agents*—that crawl through the World Wide Web, following links from one web page to another and indexing the words on that site. This method of gathering information has two important implications:

A SEARCH NEVER COVERS THE ENTIRE WEB Whenever you are doing a search with a search engine, you are never searching the entire web. As one writer points out, "You are actually searching a portion of the web, captured on a fixed index created at an earlier date."[11] In addition, you should realize that there are a lot of databases whose material is not publicly available. Finally, a lot of published material from the 1970s and earlier has never been scanned into databases and made available. (An exception: Some news databases, such as Yahoo! News or Google Breaking News, offer up-to-the-minute reports on a number of subjects.)

SEARCH ENGINES DIFFER IN WHAT THEY COVER Search engines list their results according to some kind of relevance ranking, and different search engines use different ranking schemes. Some search engines, for instance, rank web pages according to popularity (frequency of visits by people looking for a particular keyword), but others don't.

Four Web Search Tools: Keyword Indexes, Subject Directories, Metasearch Engines, & Specialized Search Engines

What are the differences between the four different web search tools?

There are many types of search tools, but the most popular versions can be categorized as (1) *keyword indexes,* (2) *subject directories,* and (3) *metasearch engines.*[12] We also discuss (4) *specialized search engines.* The most popular search sites, measured in share of visitors, are Google, Yahoo! Search, and MSN Search.

KEYWORD INDEXES A **keyword index allows you to search for information by typing one or more keywords, and the search engine then displays a list of web pages, or "hits," that contain those key words,** ordered from most likely to least likely to contain the information you want. **Hits are defined as the sites that a search engine returns after running a keyword search.**

Examples of this kind of search engine are Google, Gigablast, HotBot, MSN Search, and Teoma. The search engine known as Ask Jeeves allows users to ask questions in a natural way, such as "What is the population of the United States?" GuruNet's Answers.com is a site-and-software combination providing instant "one-click" reference answers rather than lists of search engine links.[13] (● *See Panel 2.19 on the next page.*)

SUBJECT DIRECTORIES A **subject directory allows you to search for information by selecting lists of categories or topics,** such as "Business and Commerce" or "Arts and Humanities."

Examples of subject directories are Beaucoup, Galaxy, LookSmart, MSN directory, Netscape, Yahoo!, and Open Directory Project. (Some portals such as Yahoo! are subject directories.) Some subject directories, such as MSN Search and Yahoo! Search, also offer keyword indexes. (● *See Panel 2.20.*)

Search Tool	Site
Answers.com	www.answers.com
Ask Jeeves	www.ask.com
Gigablast	www.gigablast.com
Google	www.google.com
HotBot	www.hotbot.com
MSN Search	search.msn.com
Teoma	www.teoma.com

Search Tool	Site
Beaucoup	www.beaucoup.com
Galaxy	www.galaxy.com
LookSmart	www.looksmart.com
MSN directory	www.msn.com
Netscape	www.netscape.com
Open Directory Project	http://dmoz.org
MSN	www.msn.com
Yahoo	www.yahoo.com

METASEARCH ENGINES A _metasearch engine_ allows you to search several search engines simultaneously.

Examples are Dogpile, Ixquick, Mamma, MetaCrawler, ProFusion, Search, and Vivisimo. (● *See Panel 2.21.*) The Vivisimo Clustering Engine, or Clusty, organizes search results into folders grouping similar results together. Thus, for example, if you do a search on the word "pearl," the top 250–500 results will be grouped into folders such as Jewelry, Pearl Harbor, Pearl Jam, Steinbeck Novel, and (late journalist) Daniel Pearl.[14] Grokker, which operates as plug-in for standard web browsers, also displays categories in a circular map instead of ranked lists.[15] Searching on the word "pearl" with Grokker produces the categories American, Cultured, Daniel, Drum, Jam, Jewelry, Quality plus General and More.

SPECIALIZED SEARCH ENGINES There are also *specialized search engines*, which help locate specialized subject matter, such as material about movies, health, and jobs. These overlap with the specialized portals, or vortals, we discussed above. (● *See Panel 2.22.*)

Smart Searching: Three General Strategies

How do I find what I want on the web?

The phrase "trying to find a needle in a haystack" will come vividly to mind the first time you type a word into a search engine and back comes a response

Search Tool	Site
Clusty	http://clusty.com
Dogpile	www.dogpile.com
Grokker	www.grokker.com
Ixquick	www.ixquick.com
Mamma	www.mamma.com
MetaCrawler	www.metacrawler.com
ProFusion	www.profusion.com
Search	www.search.com
Vivisimo	www.vivisimo.com

Search Tool	Site
Career.com (jobs)	www.career.com
Expedia (travel)	www.expedia.com
Internet Movie Database (movies)	www.imdb.com
Monster Board (jobs)	www.monster.com
Motley Fool (personal investments)	www.fool.com
U.S. Census Bureau (statistics)	www.census.gov
WebMD (health)	www.webmd.com

on the order of "63,173 listings found." Clearly, it becomes mandatory that you have a strategy for narrowing your search. The following are some tips.[16]

IF YOU'RE JUST BROWSING—TWO STEPS If you're just trying to figure out what's available in your subject area, do as follows:

- **Try a subject directory:** First try using a subject directory, such as Yahoo or Open Directory Project.
- **Try a metasearch engine:** Next enter your search keywords into a metasearch engine, such as Dogpile or Mamma—just to see what else is out there.

Example: You could type *"search engine tutorial"* first into Yahoo and then into ProFusion.

IF YOU'RE LOOKING FOR SPECIFIC INFORMATION If you're looking for specific information, you can try GuruNet's Answers.com "one-click" search *(www.answers.com).* Or you can go to a major search engine such as Google or MSN and then go to a specialized search engine.

Example: You could type *"Life expectancy in U.S."* first into Google and then into the Centers for Disease Control and Prevention's search engine *(www.cdc.gov/search).*

IF YOU'RE LOOKING FOR EVERYTHING YOU CAN FIND ON A SUBJECT If you want to gather up everything you can find on a certain subject, try the same search on several search engines.

Example: You could type *pogonip* (a type of dense winter fog) into more than one search tool. (Of course, you will probably get some irrelevant responses, so it helps to know how to narrow your search, as explained in the box on page 80.)

Multimedia Search Tools: Image, Audio, & Video Searching

What kind of nontext search engines are available?

Most web searches involve text, but there are many nontext kinds of resources as well. (● *See Panel 2.23.)*

Search Tool	Site
A9 (Amazon.com)	http://a9.com
Blinkx	www.blinkx.com
Google	www.google.com/video
ShadowTV	www.shadowtv.com
StreamSage	www.streamsage.com
Virage	www.virage.com
Yahoo!	http://video.search.yahoo.com

The Internet & the World Wide Web

>>

PRACTICAL ACTION
Evaluating & Sourcing Information Found on the Web

Want to know what a term means? You could try the immensely popular Wikipedia (*http://en.wikipedia.org*), a free online encyclopedia to which anyone around the world can contribute, add, or edit. It has 1.5 million entries in 76 languages.

A wiki, which founding programmer Ward Cunningham got from the Hawaiian term for "quick" ("wiki wiki") when he created the WikiWiki Web in 1995, is a simple piece of software that can be downloaded for free and used to make a website that can be edited by anyone you like. Thus, for example, corporations such as Kodak use business wikis for cross-company collaboration, such as a Word document memo that is worked on by several coworkers simultaneously. That use of wikis is valuable. But the Wikipedia is not considered reliable or authoritative by academics and librarians. As Larry Sanger, Wikipedia's former editor in chief who now lectures at Ohio State University, says, "The wide-open nature of the internet encourages people to disregard the importance of expertise." As a result, Sanger does not allow his students to use Wikipedia for their papers.[17]

"You can expect to find everything on the web," points out one library director: "silly sites, hoaxes, frivolous and serious personal pages, commercials, reviews, articles, full-text documents, academic courses, scholarly papers, reference sources, and scientific reports."[18] It is "easy to post information on the internet, usually with no editorial oversight whatsoever, and that means it is often of questionable quality," adds a Columbia University instructor. "Few students are able to separate the good research from the bad, which is less of a problem with printed texts."[19]

If you're relying on web sources for research for a term paper, how do you determine what's useful and what's not? And what is the form for citing web-based research?

Guidelines for Evaluating Web Resources

Anyone (including you) can publish anything on the World Wide Web—and all kinds of people do. Here are some ways to assess credibility of the information you find there:[20]

- **On what kind of website does the information appear?** Websites may be *professional sites,* maintained by recognized organizations and institutions. They may be *news and journalistic sites,* which may be anything from *The New York Times* to e-zines (electronic magazines, or small web-based publications) such as *Network Audio Bits.* They may be *commercial sites,* sponsored by companies ranging from the Disney Company to The Happy House Painter. They may be *special-interest sites,* maintained by activists ranging from those of the major political parties to proponents of legalization of marijuana. They may be *personal home pages,* maintained by individuals of all sorts, from professors to children to struggling musicians.

- **Does the website author appear to be a legitimate authority?** What kind of qualifications and credentials does the author have, and what kind of organization is he or she associated with? Does a web search show that the author published in other scholarly and professional publications?

- **Is the website objective, complete, and current?** Is the website trying to sell you on a product, service, or point of view? Is the language balanced and objective, or is it one-sided and argumentative? Does the author cite sources, and do they seem to come from responsible publications?

A variant on these guidelines has been framed by Butler University librarian Brad Matthies as CRITIC. (● *See Panel 2.24.)*

Citing Web Sources in College Papers

The four principal kinds of styles for citing sources—books, articles, and so on—are (1) the Modern Language Association style (for humanities), (2) the American Psychological Association style (for social science subjects), (3) the University of Chicago Press style, and (4) the Turabian style. To learn the format of these styles—including those for internet sources—go to *http://library.wustl.edu/research/citesource.html* and to *www.lib.berkeley.edu/TeachingLib/Guides/Internet/Style.html.*

- **C—Claim:** Is the source's claim clear and reasonable, timely and relevant? Or is there evidence of motivationally based language?
- **R—Role of the claimant:** Is the author of the information clearly identifiable? Are there reasons to suspect political, religious, philosophical, cultural, or financial biases?
- **I—Information backing the claim:** Is evidence for the claim presented? Can it be verified, or is the evidence anecdotal or based on testimony? Does the author cite credible references?
- **T—Testing:** Can you test the claim, as by conducting your own quantitative research?
- **I—Independent verification:** Have reputable experts evaluated and verified the claim?
- **C—Conclusion:** After taking the preceding five steps, can you reach a conclusion about the claim?

Source: Adapted from Brad Matthies, "The Psychologist, the Philosopher, and the Librarian: The Information-Literacy Version of CRITIC," *Skeptical Inquirer*, May/June 2005, pp. 49–52.

AUDIO & VIDEO Other multimedia search engines offer audio as well as image and video searching. If you go to CampaignSearch.com, for instance, and type in key words of your choosing, the search engine StreamSage will search audio and video broadcasts by analyzing speech. ShadowTV can provide continuous access to live and archived television content via the web. Yahoo! allows users to search for closed captioning associated with a broadcast and then to click for full-motion video of the words being spoken. Google announced plans to offer searches of selected televised content of PBS, the NBA, Fox News, and C-SPAN, as well as other broadcasters.[21]

MULTIPLE SOURCES, INCLUDING MUSIC A9 culls information from multiple sources, including the web, images from Google, inside-the-book text from Amazon.com, movies (Internet Movie Database), and reference materials from GuruNet.[22] A small software developer called Rocket Mobile has a music search program for cellphones called Song Identity. "You can hold your cellphone up to a music source for 10 seconds," says one account, "and it will identify the singer, the song, and the album."[23]

SCHOLARLY Google offers Google Scholar (*www.scholar.google.com*), described as "a one-stop shop of scholarly abstracts, books, peer-reviewed papers, and technical papers intended for academics and scientists."[24]

Google has also launched an ambitious project in which it is scanning page by page more than 50 million books (at a cost of about $10 for each book scanned) from several libraries—those at Harvard, Stanford, Oxford, and the University of Michigan, plus the New York Public Library. "This is one of the most transformative events in the history of information distribution since Gutenberg," says New York Public Library CEO Paul LeClerc.[25] The work is expected to take 5–10 years.

Desktop Search: Tools for Searching Your Computer's Hard Disk

How can I find things on my hard disk?

These days an inexpensive desktop computer, points out *BusinessWeek* technology writer Stephen Wildstrom, comes with enough hard-drive storage capacity to store the text of 13,000 copies of the long novel *War and Peace.* "All that space," he observes, "means that anything you save, from Grandma's email messages to a web page for a quaint bed-and-breakfast, is likely to stay there forever. Good luck trying to find it."[26]

The solution: a *desktop search engine,* a tool that extends searching beyond the web to the contents of your personal computer's hard disk. Desktop search allows users to quickly and easily find words and concepts stored on the hard-disk drive, using technology similar to that in web search engines. Desktop tools must be downloaded from the internet, often as part of a toolbar (a bar across the top of the display window on your computer screen, offering frequently executed options or commands). The tools remain

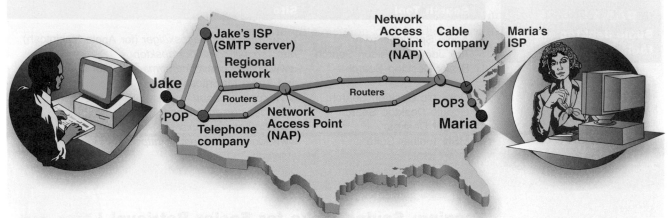

1 Jake on the West Coast logs into his dial-up ISP and sends an email message to Maria on the East Coast.

2 The message travels from Jake's computer over a standard telephone line to his ISP's local POP (point of presence). From there, the message is sent to an SMTP (Simple Mail Transport) server used to send outgoing mail.

3 The SMTP server breaks down the message into multiple packets that take different routes to their eventual destination. The packets

are forwarded by routers and may travel over faster regional networks before eventually ending up on one of several high-speed internet backbones.

4 The packets are eventually routed off the backbone and delivered to the proper email server at Maria's ISP. The packets are reassembled to re-create the message.

5 Maria uses her cable modem to connect to the internet. When she checks her email, she is sending a request from her PC to the CMTS (Cable-Modem Termination System),

which connects her to her ISP's network.

6 Maria's ISP routes the "check email" request to its POP3 (Post Office Protocol 3) email server where messages are temporarily stored. The POP3 server sends Jake's email message to Maria's PC, along with other waiting messages.

● **PANEL 2.26**
Sending and receiving emails

Adapted from *How Computers Work, www.SmartComputing.com.*

Key to Connection types		
Regular phone lines DSL Cable Satellite	T1, T2, T3 Lines	Fiber optic cable

are sent to your software's *inbox*, where they are ready to be opened and read. Examples of such programs are Microsoft's Outlook Express, Netscape's Mail, Apple Computer's Apple Mail, or QualComm's Eudora.

The advantage of standard email programs is that you can easily integrate your email with other applications, such as calendar, task list, and contact list.

WEB-BASED EMAIL With *web-based email*, you send and receive messages by interacting via a browser with a website, such as Yahoo! Mail, Hotmail, Bluebottle, or Sacmail.

The advantage of web-based email is that you can easily send and receive messages while traveling anywhere in the world. Moreover, because all your outgoing and incoming messages and folders for storing them (explained below) are stored on the mail server, you can use any personal computer and browser to keep up with your email.

Many users will rely mostly on an email program in their personal computer, but when traveling without their PCs, they will switch over to web-based email (using computers belonging to friends or available—for a fee—in airports and hotels) to check messages. Or they use portable devices such as a BlackBerry to do text messaging. (● *See Panel 2.27.*)

How to Use Email

What are some tips for being effective with email?

You'll need an email address, of course, a sort of electronic mailbox used to send and receive messages. All such addresses follow the same approach: *username@domain.* These are somewhat different from web URLs, which do not use the "@" (called "at") symbol. You can check with your internet access provider to see if a certain name is available.

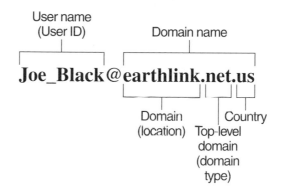

- **The user name: Joe_Black** The *user name*, or *user ID*, identifies who is at the address—in this case, *Joe_Black* (note the underscore). There are many ways that Joe Black's user name might be designated, with and without capital letters: *Joe_Black, joe_black, joe.black, joeblack, jblack, joeb,* and so on.

- **Domain name: @earthlink** The *domain name*, which is located after the @ ("at") symbol, tells the location and type of address. Domain-name components are separated by periods (called "dots"). The domain portion of the address (such as *Earthlink,* an internet service provider) provides specific information about the location—where the message should be delivered.

- **Top-level domain: .net** The *top-level domain*, or *domain code*, is a three-letter extension that describes the domain type: *.net, .com, .gov, .edu, .org, .mil, .int*—network, commercial, government, educational, nonprofit, military, or international organization.

- **Country: .us** Some domain names also include a two-letter extension for the country—*.us* for United States, *.ca* for Canada, *.mx* for Mexico.

The illustration on the next page shows how to send, receive, and reply to email. (● *See Panel 2.28.*) Some tips about using email are as follows:

TYPE ADDRESSES CAREFULLY You need to type the address exactly as it appears, including capitalization and all underscores and periods. If you type an email address incorrectly (putting in spaces, for example), your message will be returned to you labeled "undeliverable."

USE THE REPLY COMMAND When responding to an e-message someone has sent you, the easiest way to avoid making address mistakes is to use the Reply command, which will automatically fill in the correct address in the "To" line. Be careful not to use the Reply All command unless you want your reply to be sent to *all* the recipients of the original email.

USE THE ADDRESS-BOOK FEATURE You can store the email addresses of people sending you messages in your program's "address book." This feature also allows you to organize your email addresses according to a nickname or the person's real name so that, for instance, you can look up your friend Joe Black under his real name, instead of under his user name,

● **PANEL 2.27**
Mobile mail.
Portable devices such as this BlackBerry Wireless Handheld allow you to send and receive email messages from many locations.

info!
Country Abbreviations

What do you think the country abbreviations are for Micronesia? Botswana? Saint Lucia? Go to *www.eubank-web.com/ William/Webmaster/c-codes htm* or *www.thrall.org/ domains.htm* and find out.

The Internet & the World Wide Web

83

Sending email

Send: Command for sending messages

cc: For copying ("carbon/courtesy copy") message to others

bcc: For copying others ("blind carbon copy") without the primary recipient knowing it

Message area

You can conclude every message with a custom "signature"

Address Book: Lists email addresses you use most; can be attached automatically to messages

Subject line: Preview incoming email by reviewing the subject lines to see if you really need to read the messages

Receiving email

Reply, Reply All, Forward, Delete: For helping you handle incoming email

Inbox lists messages waiting in email box. (Unopened envelope icon shows unread mail.)

Selected message displayed here

Replying to email

Use the **Reply** command icon, and the email program automatically fills in the To, From, and Subject lines in your reply.

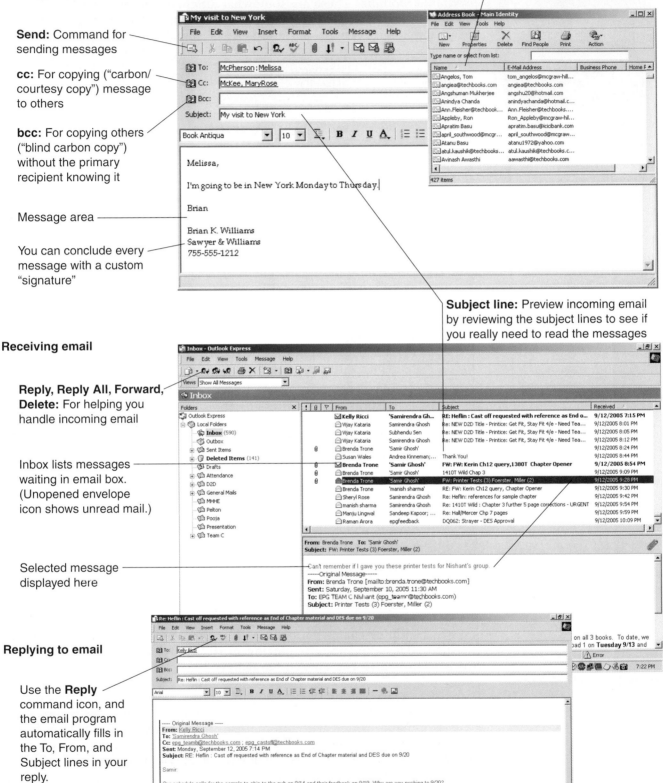

History of the @ Sign

In 1972, Ray Tomlinson sent the first electronic message using the @ symbol to indicate the location or institution of the recipient. Tomlinson knew that he had to use a symbol that would not appear in anyone's name. Before the @ sign became a character on the keyboard, where was it used? Some linguists believe that the symbol dates back to the 6th or 7th centuries, when Latin scribes adapted the Latin word *ad*, meaning "at," "to," or "toward." Other linguists say that @ dates to its use in the 18th century as a symbol in commerce to indicate price per unit, as in 4 CDs @ $5 [each]. In 2000, a professor of history in Italy discovered some original 14th-century documents clearly marked with the @ sign to indicate a measure of quantity, based on the word *amphora*, meaning "jar." The amphora was a standard-size earthenware jar used by merchants for wine and grain. In Florence, the capital "A" was written with a flourish and later became @ ("at the price of").

Other countries have different names for this symbol. For example:

South Africa—*aapstet,* or "monkey's tail"

Czech Republic—*zavinac,* or "pickled herring"

Denmark—*snable-a,* or "elephant's trunk"

France—*petit escargot,* "little snail"

Greece—*papaki,* "little duck"

Hungary—*kukac,* "worm"

Taiwan—*xiao lao-shu,* "mouse sign"

Norway—*grischale,* "pig's tail"

Russia—*sobachka,* "little dog"

Turkey—*kulak,* "oar"

Source: *www.pcwebopedia.com,* January 28, 2003, and *www.webopedia.com,* August 2005.

bugsme2, which you might not remember. The address book also allows you to organize addresses into various groups—such as your friends, your relatives, club members—so that you can easily send all members of a group the same message with a single command.

DEAL WITH EACH EMAIL ONLY ONCE   When a message comes in, delete it, respond to it, or file it away in a folder. Don't use your inbox for storage.

DON'T "BLOAT" YOUR EMAIL   Email messages loaded with fancy typestyles, logos, and background graphics take longer to download. Keep messages simple.

Sorting Your Email

How can I keep my emails organized?

On an average day, billions of business and personal emails are sent in North America. If, as many people do, you receive 50–150 emails per day, you'll have to keep them organized so that you don't lose control.

One way to stay organized is by using instant organizers, also called *filters,* which sort mail on the basis of the name of the sender or the mailing list and put particular emails into one folder. (● *See Panel 2.29.)* Then you can read emails sent to this folder later when you have time, freeing up your inbox for mail that needs your more immediate attention. Instructions on how to set up such organizers are in your email program's Help section.

Attachments

What are the benefits of being able to do email attachments?

You have written a great research paper and you immediately want to show it off to someone. If you were sending it via the Postal Service, you would

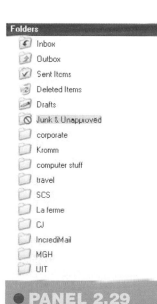

● **PANEL 2.29**
Sorting email.
Email folders

write a cover note—"Folks, look at this great paper I wrote about globalization! See attached"—then attach it to the paper, and stick it in an envelope. Email has its own version of this. If the file of your paper exists in the computer from which you are sending email, you can write your email message (your cover note) and then use the Attach File command to attach the document. (● *See Panel 2.30.*) (Note: It's important that the person receiving the email attachment have the exact same software that created the attached file, such as Microsoft Word, or have software that can read and convert the attached file.) Downloading attachments from the mail server can take a lot of time—so you may want to discourage friends from sending you many attachments. You may also want to use compression software to reduce the size of your attachments. (Compression is covered in detail in Chapter 8.)

While you could also copy your document into the main message and send it that way, some email software loses formatting options such as **bold** or *italic* text or special symbols. And if you're sending song lyrics or poetry, the lines of text may break differently on someone else's display screen than they do on yours. Thus, the benefit of the attachment feature is that it preserves all such formatting, provided the recipient is using the same word processing software that you used. You can also attach pictures, sounds, videos, and other files to your email message.

Note: Many *viruses*—those rogue programs that can seriously damage your PC or programs—ride along with email as attached files. Thus, you should *never open an attached file from an unknown source.* (We describe viruses in Chapter 6.)

Instant Messaging

What are the benefits and drawbacks of my using instant messaging?

By 2007, it's predicted that businesses will support 182 million instant-messaging users.[30] Instant messages are like a cross between email and phone, allowing communication that is far speedier than conventional email. With **instant messaging (IM), any user on a given email system can send a message and have it pop up instantly on the screen of anyone else logged onto that system.** As soon as you use your computer or portable device to connect to the internet and log on to your IM account, you see a *buddy* list (or *contacts* list), a list you have created that consists of other IM users you want to communicate with. If all parties agree, they can initiate online typed conversations in real time (in a "chat room"). The messages appear on the display screen in a small **window—a rectangular area containing a document or activity—**so that users can exchange messages almost instantaneously while operating other programs.

GETTING INSTANT-MESSAGING CAPABILITY Examples of present instant-message systems are AOL Instant Messenger, MSN Messenger, ICQ ("I Seek You," also from AOL), AT&T IM Anywhere, and Yahoo Messenger. Some of these, such as Yahoo's, allow voice chats among users if their PCs have microphones and speakers. Yahoo Messenger also allows users to make free worldwide PC-to-PC voice calls without a phone.[31]

To get instant messaging, which is often free, you download software and register with the service, providing it with a user name and password. When your computer or portable device is connected to the internet, the software checks in with a central server, which verifies your identity and looks to see if any of your "buddies" are also online. You can then start a conversation by sending a message to any buddy currently online.

Sending an email attachment

3 Third, use your email
software's toolbar buttons
or menus to attach the file
that contains the attachment.

4 Fourth, click on *Send*
to send the email
message and attachment.

1 First, address the
person who will
receive the attachment.

2 Second, write a "cover
letter" email advising
the recipient of the
attachment.

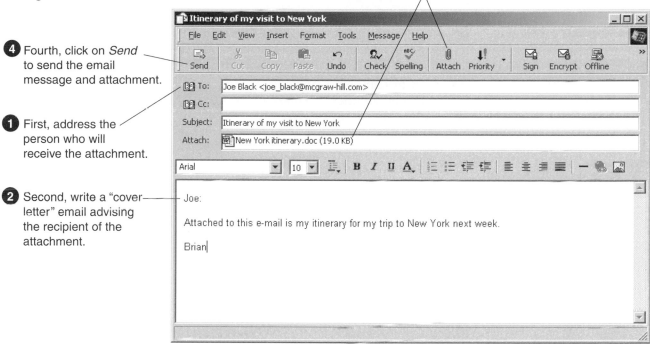

Receiving an email attachment

When you receive a
file containing an
attachment, you'll see
an icon indicating the
message contains more
than just text. You can
click on the icon to
see the attachment. If
you have the software
the attached file was
created in, you can
open the attachment
immediately to read or
print, or you can save
the attachment in a
location of your choice
(on your computer). You
can also forward the
attachment to another
person.

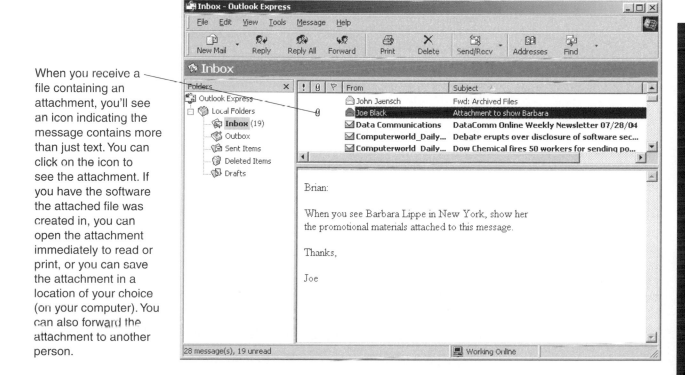

THE DOWNSIDE OF IM IM has become a hit with many users. Instant messaging is especially useful in the workplace as a way of reducing long-distance telephone bills when you have to communicate with colleagues who are geographically remote but with whom you must work closely. However, you need to be aware of a few drawbacks:

- **Lack of privacy:** Most IM services lack the basic privacy that most other forms of communication provide. Efforts are under way to develop IM security software and procedures, but for now IM users should be aware that they have virtually no privacy.

- **Lack of common standards:** Many IM products do not communicate with one another. If you're using AOL's IM, you cannot communicate with a buddy on Yahoo.

- **Time wasters when you have to get work done:** An instant message "is the equivalent of a ringing phone because it pops up on the recipient's screen right away," says one writer.[32] Some analysts suggest that, because of its speed, intrusiveness, and ability to show who else is online, IM can destroy workers' concentration in some offices. You can put off acknowledging email, voice mail, or faxes. But instant messaging is "the cyber-equivalent of someone walking into your office and starting up a conversation as if you had nothing better to do," says one critic. "It violates the basic courtesy of not shoving yourself into other people's faces."[33]

You can turn off your instant messages, but that is like turning off the ringer on your phone; after a while people will wonder why you're never available. Buddy lists or other contact lists can also become very in-groupish. When that happens, people are distracted from their work as they worry about staying current with their circle (or being shut out of one). Some companies have reportedly put an end to instant messaging, sending everyone back to the use of conventional email.

FTP—for Copying All the Free Files You Want
What does FTP allow me to do?

Many net users enjoy "FTPing"—cruising the system and checking into some of the tens of thousands of FTP sites, which predate the web and offer interesting free or inexpensive files to download. ___FTP (File Transfer Protocol)___ **is a software standard for transferring files between computers with different operating systems. You can connect to a remote computer called an** *FTP site* **and transfer files to your own microcomputer's hard disk via TCI/IP over the internet.** Free files offered cover nearly anything that can be stored on a computer: software, games, photos, maps, art, music, books, statistics. (● *See Panel 2.31.*)

Some FTP files are open to the public (at *anonymous FTP sites*); some are not. For instance, a university might maintain an FTP site with private files (such as lecture transcripts) available only to professors and students with assigned user names and passwords. It might also have public FTP files open to anyone with an email address. You can download FTP files using either your web browser or special software (called an *FTP client program*), such as Fetch and Cute.

Newsgroups—for Online Typed Discussions on Specific Topics
What would probably be my favorite kind of newsgroup?

A ___newsgroup___ (or *forum*) **is a giant electronic bulletin board on which users conduct written discussions about a specific subject.** (● *See Panel 2.32.*)

info!

FTP Clients and Servers

FTP software programs, called *FTP clients*, come with many operating systems, including Windows, DOS, Unix, and Linux. Web browsers also come with FTP clients. For lists of FTP clients and FTP servers, go to www.answers.com/main/ntquery?method+4&dsis+2222&dekey=FTP+client&gwp+11&curtab+2222

The concise
Tech Encyclopedia

A **free**
community
resource

File Transfer Protocol (FTP)

Free FTP Clients
Free, easy, and secure File
Transfer Protocol Transfers

Free and Easy FTP Clients
File Transfer Protocol Quick &
Easy File Transfers

IPSec White Papers
Free White Papers and in-
depth Reports on IP Security.

SSL FTP Server
Secure Windows FTP Server.
128-bit SSL and Remote
Admin.

A common method of moving files from system to system using TCP/IP. To work properly, it requires an FTP client to contact an FTP server in order to transmit data back and forth.

Selected File Transfer Protocol (FTP) links:

© 2004 tech-encyclopedia.com. If you have comments or additions that you wish to make, please email us. If you found this site useful, feel free to tell others or link to it from your site! tech-encyclopedia.com is a purely informational website, and should not be used as a substitute for professional legal, medical or technical advice.

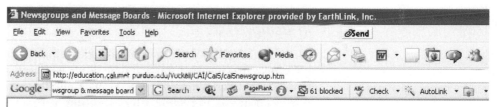

Newsgroups and Message Boards - Microsoft Internet Explorer provided by EarthLink, Inc.

File   Edit   View   Favorites   Tools   Help   ⊘Send

Back ▾   ✕ 🗘 🏠   🔍 Search ⭐ Favorites 🎬 Media   ✉ 📰 ▾ 🖨 ₩ ▾ 🗖 🗔 💬 🐟

Address http://education.calumet.purdue.edu/Vockell/CAI/Cai5/cai5newsgroup.htm

Google ▾ wsgroup & message board ▾ | G Search ▾ 🔍 | 📰 PageRank 🟢 ▾ 🔲 61 blocked | ᴬᴮ Check ▾ 🔍 AutoLink ▾ 📋 ▾

Newsgroups and Message Boards

Newsgroups and message boards are forums in which people post messages and share information with other people intere
world). The following is a segment of the Message Board from a recent session of Ed Vockell's graduate course on Compu

[Post Message] [Archive]
(View by Thread) (View by Author) (View by Date)

- ● Contributing to Existing Web Page - **Mary Ann Sytsma** *19:40:45 2/22/2001* (0)
- ● Feb. 22 Class - **Mary Ann Sytsma** *14:50:59 2/22/2001* (0)
- ● chapter 1 - **Jennifer Christine Harwood** *12:01:38 2/22/2001* (0)
- ● How to Use the Internet Tutorial Evaluation - **Julie Appel Parker** *22:12:00 2/21/2001* (1)
  - ○ Re:How to Use the Internet Tutorial Evaluation - **David R Coyle** *13:58:32 2/22/2001* (0)
- ● not able to log onto the 'online book' site - **Deborah Lynn Cox** *20:47:07 2/21/2001* (1)
  - ○ Re:not able to log onto the 'online book' site - **Thomas John Kekelik** *03:11:02 2/22/2001* (0)
- ● Internet Log-Other Activities - **Michelle M Westlund** *18:39:54 2/21/2001* (1)
  - ○ Re:Internet Log-Other Activities - **Thomas John Kekelik** *02:55:51 2/22/2001* (0)

There are thousands of internet newsgroups—which charge no fee—and they cover an amazing array of topics. In addition, for a small fee, services such as Meganetnews.com and Corenews.com will get you access to more than 100,000 newsgroups all over the world. Newsgroups take place on a special network of computers called **_Usenet_**, **a worldwide public network of servers that can be accessed through the internet** *(www.usenet.com)*. To participate, you need a **_newsreader_**, **a program included with most browsers that allows you to access a newsgroup and read or type messages.** (Messages, incidentally, are known as *articles*.)

One way to find a newsgroup of interest to you is to use a portal such as Yahoo or MSN to search for specific topics. Or you can use Google's Groups *(http://groups.google.com/grphp?hl+en&ie=UTF-8)*, which will present the newsgroups matching the topic you specify. About a dozen major topics, identified by abbreviations ranging from *alt* (alternative topics) to *talk* (opinion and discussion), are divided into hierarchies of subtopics.

Listservs: Email-Based Discussion Groups

Why would I want to sign up for a listserv?

Want to receive email from people all over the world who share your interests? You can try finding a mailing list and then "subscribing"—signing up, just as you would for a free newsletter or magazine. **A _listserv_ is an automatic mailing-list server that sends email to subscribers who regularly participate in discussion topics.** (● *See Panel 2.33.*) Listserv companies include L-Soft's Listserv *(www.lsoft.com)*, Majordomo *(www.majordomo.com)*, and *http://listserve.com/*. To subscribe, you send an email to the list-server moderator and ask to become a member, after which you will automatically receive email messages from anyone who responds to the server.

Mailing lists are one-way or two-way. A one-way list either accepts or sends information, but the user interacts only with the list server and not other users. Most one-way mailing lists are used for announcements, newsletters, and advertising (and "spam," discussed shortly). Two-way lists, which are limited to subscribers, let users interact with other subscribers to the mailing list; this is the discussion type of mailing list.

Real-Time Chat—Typed Discussions among Online Participants

What discussion subjects might I like to participate in using RTC?

With mailing lists and newsgroups, participants may contribute to a discussion, go away, and return hours or days later to catch up on others' typed contributions. With **_real-time chat (RTC)_, participants have a typed discussion ("chat") while online at the same time,** just like a telephone conversation except that messages are typed rather than spoken. Otherwise, the format is much like a newsgroup, with a message board to which participants may send ("post") their contributions. To start a chat, you use a service available on your browser such as IRC (Internet Relay Chat) that will connect you to a chat server.

Unlike instant messaging, which tends to involve one-on-one conversation, real-time chat usually involves several participants. As a result, RTC is often like being at a party, with many people and many threads of conversation occurring at once.

● **PANEL 2.33**
Listserv mailing list

Netiquette: Appropriate Online Behavior

What are the rules of courtesy for using email?

You may think etiquette is about knowing which fork to use at a formal dinner. Basically, though, etiquette has to do with politeness and civility—with rules for getting along so that people don't get upset or suffer hurt feelings.

New internet users, known as "newbies," may accidentally offend other people in a discussion group or in an email simply because they are unaware of _netiquette_, or "network etiquette"—appropriate online behavior. In general, netiquette has two basic rules: (1) Don't waste people's time, and (2) don't say anything to a person online that you wouldn't say to his or her face.

Some more specific rules of netiquette are shown below:

- **Consult FAQs:** Most online groups post _FAQs (frequently asked questions)_ **that explain expected norms of online behavior for a particular group.** Always read these first—before someone in the group tells you you've made a mistake.

- **Avoid flaming:** A form of speech unique to online communication, _flaming_ **is writing an online message that uses derogatory, obscene, or inappropriate language.** Flaming is a form of public humiliation inflicted on people who have failed to read FAQs or have otherwise not observed netiquette (although it can happen just because the sender has poor impulse control and needs a course in anger management). Something that smoothes communication online is the use of _emoticons_, keyboard-produced pictorial representations of expressions. (● _See Panel 2.34._)

- **Don't SHOUT:** Use of all-capital letters is considered the equivalent of SHOUTING. Avoid, except when they are required for emphasis of a word or two (as when you can't use italics in your e-messages).

- **Be careful with jokes:** In email, subtleties are often lost, so jokes may be taken as insults or criticism.

- **Avoid sloppiness, but avoid criticizing others' sloppiness:** Avoid spelling and grammatical errors. But don't criticize those same errors in others' messages. (After all, they may not speak English as a native language.) Most email software comes with spell-checking capability, which is easy to use.

- **Don't send huge file attachments, unless requested:** Your cousin living in the country may find it takes minutes rather than seconds for his or her computer to download a massive file (as of a video that you want to share). Better to query in advance before sending large files as attachments. Also, whenever you send an attachment, be sure the recipient has the appropriate software to open your attachment.

- **When replying, quote only the relevant portion:** If you're replying to just a couple of matters in a long email posting, don't send back the entire message. This forces your recipient to wade through lots of text to find the reference. Instead, edit his or her original text down to the relevant paragraph and then put in your response immediately following.

● **PANEL 2.34**
Some emoticons

.-)	Happy face	<g>	Grin
:-(	Sorrow or frown	BTW	By the way
:-O	Shock	IMHO	In my humble opinion
:-/	Sarcasm	FYI	For your information
;-)	Wink		

- **Don't "overforward":** Don't automatically forward emails to your friends without checking if the contents are true and appropriate.

The Online Gold Mine: Telephony, Multimedia, Webcasting, Blogs, E-Commerce, & Other Resources

What are internet telephony, various kinds of multimedia, and RSS and different web feeds—webcasting, blogging, podcasting—and types of e-commerce?

"For vivid reporting from the enormous zone of tsunami disaster," says a newspaper account, "it was hard to beat the blogs."[34]

Blogs, as technology writer Lee Gomes points out, used to be regarded as web-based "daily diaries of people with no real lives to chronicle in the first place."[35] But the December 2004 Indian Ocean calamity that resulted in over 143,000 people killed and more than 146,000 missing also showed how quickly and effectively this form of web technology could be in spreading instant news, often beating out the mainstream news media. In particular, the tsunami spurred the distribution of *video blogs,* or *vblogs,* consisting of video footage mostly shot by vacationing foreign tourists during and after the disaster. One video, for example, showed an elderly couple overpowered by a wave, filmed at a beach hotel by a factory worker from Sweden.[36]

The opportunities offered by the internet and the web seem inexhaustible. Here we'll examine several resources available to you.

Telephony: The Internet Telephone & Videophone

Why would I want to make phone calls via the internet?

As we stated earlier, the internet breaks up conversations (as it does any other transmitted data) into "information packets" that can be sent over separate lines and then regrouped at the destination, whereas conventional voice phone lines carry a conversation over a single path. Thus, the internet can move a lot more traffic over a network than the traditional telephone link can.

With **internet telephony,** or **VoIP phoning (short for *Voice over Internet Protocol*)—using the net to make phone calls, either one to one or for audio-conferencing—**you can make long-distance phone calls that are surprisingly inexpensive or even free. (● *See Panel 2.35.)* Indeed, it's possible to do this without owning a computer, simply by picking up your standard telephone and dialing a number that will "packetize" your conversation. However, people also can use a PC with a sound card and a microphone, a modem linked

● **PANEL 2.35**
Internet telephony

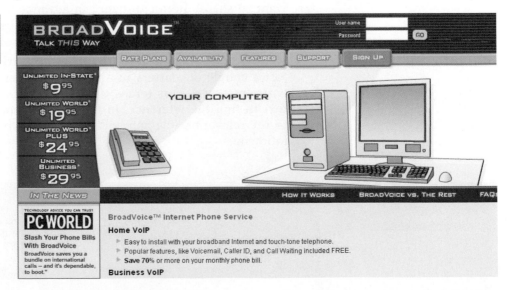

to a standard internet service provider, and internet telephone software such as Netscape Conference (part of Netscape's browser) or Microsoft NetMeeting (part of Microsoft Internet Explorer). VoIP is being offered by Comcast Corp., Vonage Holdings Co., and Verizon Communications, among others. It needs to be pointed out, however, that a 2005 study found internet-based phone services still very inferior to traditional phone connections in reliability and sound quality.[37]

Besides carrying voice signals, internet telephone software also allows videoconferencing, in which participants are linked by a videophone that will transmit their pictures, thanks to a video camera attached to their PCs.

Multimedia on the Web

How can I get images, sound, video, and animation as well as text?

Many websites (especially those trying to sell you something) employ complicated multimedia effects, using a combination of text, images, sound, video, and animation. While most web browsers can handle basic multimedia elements on a web page, eventually you'll probably want more dramatic capabilities.

PLUG-INS   In the 1990s, as the web was evolving from text to multimedia, browsers were unable to handle many kinds of graphic, sound, and video files. To do so, external application files called *plug-ins* had to be loaded into the system. **A _plug-in_—also called a *player* or a *viewer*—is a program that adds a specific feature to a browser, allowing it to play or view certain files.**

For example, to view certain documents, you may need to download Adobe Acrobat Reader. (● *See Panel 2.36.*) To view high-quality video and hear radio, you may need to download RealPlayer. (● *See Panel 2.37 on the next page.*) QuickTime is an audio/video plug-in for the Apple Macintosh. Plug-ins are required by many websites if you want to fully experience their content.

Recent versions of Microsoft Internet Explorer and Netscape can handle a lot of multimedia. Now if you come across a file for which you need a plug-in, the browser will ask whether you want it and then tell you how to go about downloading it, usually at no charge.

DEVELOPING MULTIMEDIA: APPLETS, JAVA, & VISUAL STUDIO.NET
How do website developers get all those nifty special multimedia effects? Often web pages contain links to **_applets_, small programs (software) that can**

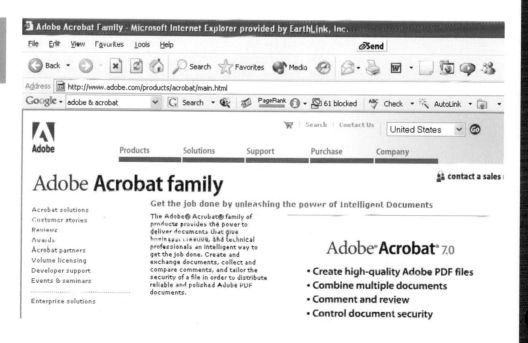

● PANEL 2.36
Adobe Acrobat
Reader

be quickly downloaded and run by most browsers. Applets are written in *Java*, **a complex programming language that enables programmers to create animated and interactive web pages.** Java applets enhance web pages by playing music, displaying graphics and animation, and providing interactive games. Java-compatible browsers such as Internet Explorer and Netscape automatically download applets from the website and run them on your computer so that you can experience the multimedia effects. Microsoft offers Visual Studio.NET to compete with Java.

TEXT & IMAGES Of course, you can call up all kinds of text documents on the web, such as newspapers, magazines, famous speeches, and works of literature. You can also view images, such as scenery, famous paintings, and photographs. Most web pages combine both text and images.

One interesting innovation is that of aerial maps. Google Earth (*www.google.earth.com*) is a satellite imaging program that Google describes as "part flight simulator, part search tool."[38] You type in your ZIP code or street address and it feels like you're looking down on a high-resolution aerial view of your house from a plane at 30,000 feet. Using Google Local Search, you can search for a business or other attraction in some city, and you'll get an indicator on your satellite image; if you click again, the establishment's web page opens. "Google sightseers can zoom in close enough to see airplanes parked in the desert, the baseball diamond at Wrigley Field, and cars in the Mall of America parking lot," says one writer.[39] Similar services are offered by Yahoo, Microsoft (MSN Virtual Earth, *www.virtualearth.msn.com*), and Amazon's A9 search engine, which incorporates street-level photographs into its results.

ANIMATION *Animation* **is the rapid sequencing of still images to create the appearance of motion,** as in most video games as well as in moving banners displaying sports scores or stock prices.

VIDEO Video can be transmitted in two ways. (1) A file, such as a movie or video clip, may have to be completely downloaded before you can view it. This may take several minutes in some cases. (2) A file may be displayed as streaming video and viewed while it is still being downloaded to your computer. *Streaming video* **is the process of transferring data in a continuous flow so that you can begin viewing a file even before the end of the file is sent.** For instance, RealPlayer offers live, television-style broadcasts over the internet as streaming video for viewing on your PC screen. You download and install this software and then point your browser to a site featuring RealVideo. That will produce a streaming-video television image in a window a few inches wide.

A milestone in streaming video occurred in July 2005 when America Online broadcast seven separate live feeds of the Live 8 concerts from London, Philadelphia, Paris, Berlin, Rome, and Toronto, as well as a separate global feed that included footage from four other venues. Although more than half of American households have the broadband connections that permit streaming video, relatively few have computers attached to their TVs, which have bigger screens.[40] Even so, the four most visited internet portals—Yahoo, America Online, MSN, and Google—are all moving aggressively into video programming, just like the traditional TV networks.[41]

AUDIO   Audio, such as sound or music files, may also be transmitted in two ways:

- **Downloaded completely before the file can be played:** Many online music services, such as Yahoo's Music Unlimited, Apple's iTunes Music Store, RealNetworks, and Napster, offer music for a fee, either by subscription or by the song.[42] Generally, music must be downloaded completely before the file can be played on a computer or portable music player.

- **Downloaded as streaming audio:** Music that is downloaded as **_streaming audio_ allows you to listen to the file while the data is still being downloaded to your computer.** A popular standard for transmitting audio is RealAudio. Supported by most web browsers, it compresses sound so that it can be played in real time, even though sent over telephone lines. You can, for instance, listen to 24-hour-a-day net radio, which features "vintage rock," or English-language services of 19 shortwave outlets from World Radio Network in London. Many large radio stations outside the United States have net radio, allowing people around the world to listen in. (We explain a form of web audio known as *podcasting* on the next page.)

The Web Automatically Comes to You: Webcasting, Blogging, & Podcasting

How would I explain webcasting, blogging, and podcasting to someone?

It used to be that you had to do the searching on the web. Now, if you wish, the web will come to you. Let's consider three variations—webcasting, blogging, and podcasting.

PUSH TECHNOLOGY & WEBCASTING   The trend began in the late 1990s with **_push technology_, software that automatically downloads information to personal computers** (as opposed to *pull technology*, in which you go to a website and pull down the information you want—in other words, the web page isn't delivered until a browser requests it). One result of push technology was **_webcasting_, in which customized text, video, and audio are sent to you automatically on a regular basis.**

The idea here is that you choose the categories, or (in Microsoft Internet Explorer) the channels, of websites that will automatically send you updated information. Thus, webcasting saves you time because you don't have to go out searching for the information. Webcasting companies are also called *subscription services,* because they sell on-demand services. However, a lot of push technology fell out of favor because it clogged networks with information that readers often didn't want.[43] Then along came RSS.

BLOGGING—RSS, XML, & THE RISE OF THE BLOGOSPHERE RSS was built to be simpler than push technology. **_RSS newsreaders_, or _RSS aggregators_, are programs that scour the web, sometimes hourly or more frequently, and pull together in one place web "feeds" from several websites.** The developers of RSS technology don't agree on what the abbreviation stands for, but some say it means "Really Simple Syndication" or "Rich Site Summary," although there are other variations as well.[44] "RSS allows you to play news editor and zero in on the information you really need," says one account, "even as you expand the number of sites you sample."[45] This is because the information is so specifically targeted.

RSS is based on **_XML_, or _extensible markup language_, a web-document tagging and formatting language that is an advance over HTML and that two computers can use to exchange information.** XML, in the form of RSS, has allowed people to have, in Gomes's phrase, "access to a whole new universe of content." One of the earliest adopters, for instance, was the Mormon Church, which used the system to keep in touch with members. Now, however, it has morphed into something called the **_blogosphere_, the total universe of blogs—_blog_ being short for _web log_, a diary-style web page.** (● *See Panel 2.38.*)

"Blogs can be anything their creators want them to be, from newsy to deeply personal, argumentative to poetic," says one article. "Some are written by individuals. Some are group projects. Some have readerships in the thousands and influence world media. Others are read by a handful of people—or not read at all, because they're all pictures. Bloggers are a new breed of homegrown journal writers and diarists who chronicle life as it happens, with words, photos, sound, and art."[46] Many bloggers use online communities such as Xanga (*xanga.com*) for posting their online diaries and journals. One 2005 study found that 16% of the U.S. population reads blogs and that 6% of adults have created a blog.[47]

Among other variations are *video blogs*, or *vblogs* or *vogs*, which seem to be simply video versions of blogs, and *moblogs*, or *mobile blogs*, in which picture blogs are posted to websites directly from a camera-enabled cellphone.

PODCASTING **_Podcasting_ involves the recording of internet radio or similar internet audio programs.** The term derives not only from *webcasting* but also from the Apple *iPod* portable music player and other mobile listening devices, such as MP3 players.

● **PANEL 2.38**
The blogosphere

Blogging and podcasting seem to represent the frontier of personalized media, a subject to which we will return throughout the book, particularly in Chapter 7.

E-Commerce: B2B Commerce, Online Finance, Auctions, & Job Hunting

What are ways I might personally benefit from online e-commerce possibilities?

The explosion in <u>*e-commerce (electronic commerce)*</u>—**conducting business activities online**—is not only widening consumers' choice of products and services but also creating new businesses and compelling established businesses to develop internet strategies. Many so-called brick-and-mortar retailers—those operating out of physical buildings—have been surprised at the success of such online companies as Amazon.com, seller of books, CDs, and other products. As a result, traditional retailers from giant Wal-Mart to very small one-person businesses now offer their products online.

Retail goods can be classified into two categories—hard and soft. *Hard goods* are those that can be viewed and priced online, such as computers, clothes, groceries, and furniture, but are then sent to buyers by mail or truck. *Soft goods* are those that can be downloaded directly from the retailer's site, such as music, software, travel tickets, and greeting cards.

Some specific forms of e-commerce are as follows:

B2B COMMERCE Of course, every kind of commerce has taken to the web, ranging from travel bookings to real estate. One of the most important variations is <u>***B2B (business-to-business) commerce***</u>, **the electronic sale or exchange of goods and services directly between companies, cutting out traditional intermediaries.** Expected to grow even more rapidly than other forms of e-commerce, B2B commerce covers an extremely broad range of activities, such as supplier-to-buyer display of inventories, provision of wholesale price lists, and sales of closed-out items and used materials—usually without agents, brokers, or other third parties.

ONLINE FINANCE: TRADING, BANKING, & E-MONEY The internet has changed the nature of stock trading. Anyone with a computer, a connection to the global network, and the information, tools, and access to transaction systems required to play the stock market can do so online. Companies such as E*Trade have built one-stop financial supermarkets offering a variety of money-related services, including home mortgage loans and insurance. More than 1,000 banks have websites, offering services that include account access, funds transfer, bill payment, loan and credit-card applications, and investments.

AUCTIONS: LINKING INDIVIDUAL BUYERS & SELLERS Today millions of buyers and sellers are linking up at online auctions, where everything is available from comic books to wines. The internet is also changing the tradition-bound art and antiques business (dominated by such venerable names as Sotheby's, Christie's, and Butterfield & Butterfield). There are generally two types of auction sites:

- **Person-to-person auctions:** Person-to-person auctions, such as eBay, connect buyers and sellers for a listing fee and a commission on sold items. (● *See Panel 2.39.*)

- **Vendor-based auctions:** Vendor-based auctions, such as OnSale, buy merchandise and sell it at discount. Some auctions are specialized, such as Priceline, an auction site for airline tickets and other items.

more info!

Some Auction Websites

eBay, *www.ebay.com*
OnSale, *www.onsale.com*
Priceline, *www.priceline.com*

The Internet & the World Wide Web

97

ONLINE JOB HUNTING There are more than 2,000 websites that promise to match job hunters with an employer. Some are specialty "boutique" sites looking for, say, scientists or executives. Some are general sites, the leaders being Monster.com, CareerPath.com, CareerMosaic.com, USAJOBS (U.S. government/federal jobs; *www.usajobs.opm.gov*), and CareerBuilder .com. Job sites can help you keep track of job openings and applications by downloading them to your own computer. Résumé sites such as Employment911.com help you prepare professional-quality résumés.

Relationships: Online Matchmaking & Social Relationships

Would I use a relationship or social-networking website?

It's like walking into "a football stadium full of single people of the gender of your choice," says Trish McDermott, an expert for Match.com, a San Francisco online dating service. People who connect online before meeting in the real world, she points out, have the chance to base their relationship on personality, intelligence, and sense of humor rather than purely physical attributes. "Online dating allows people to take some risks in an anonymous capacity," she adds. "When older people look back at their lives, it's the risks that they didn't take that they most regret."[48] (Still, serious risks exist in trying to establish intimacy through online means because people may pretend to be quite different from who they really are.)

People can also use search sites such as Infospace.com and Switchboard.com to try to track down old friends and relatives. Others find common bonds by joining online communities such as The WELL, the women's site Ivillage.com, the older people's site Third Age, and the gardening site GardenWeb. The LinkedIn website allows business professionals to make contacts for sales leads, as well as for job leads.[49]

info!

Some Online Communities

Friendster, *www.friendster.com*
GardenWeb,
www.gardenweb.com
Ivillage.com, *www.ivillage.com*
LinkedIn, *www.linkedin.com*
MySpace, *www.myspace.com*
Third Age, *www.thirdage.com*
Tribe, *www.tribe.com*
The Well, *www.well.com*

2.6

The Intrusive Internet: Snooping, Spamming, Spoofing, Phishing, Pharming, Cookies, & Spyware

How can I protect myself against snoopers, spam, spoofing, phishing, pharming, cookies, and spyware—adware, browser and search hijackers, and key loggers?

"The current internet model is just too wide open," says Orion E. Hill, president of the Napa Valley (California) Personal Computer Users Group, which educates users about PCs. "The internet is just too accessible, and it's too easy for people to make anything they want out of it."[50] Thus, although the internet may be affected to a degree by governing and regulatory bodies, just like society in general that doesn't mean that it doesn't have its pitfalls or share of users who can do you real harm.

We consider some of the other serious internet issues (such as viruses, worms, crackers, pornography, and gambling) in Chapters 6 and 9, but here let us touch on a few of immediate concern that you should be aware of: snooping, spam, spoofing, phishing, pharming, cookies, and spyware.

Snooping on Your Email: Your Messages Are Open to Anyone

Who's able to look at my private emails?

The single best piece of advice that can be given about sending email is this: *Pretend every electronic message is a postcard that can be read by anyone.* Because the chances are high that it could be. (And this includes email on college campus systems as well.)

Think the boss can't snoop on your email at work? The law allows employers to "intercept" employee communications if one of the parties involved agrees to the "interception." The party "involved" is the employer.[51] And in the workplace, email is typically saved on a server, at least for a while. Indeed, federal laws require that employers keep some email messages for years.

Think you can keep your email address a secret among your friends? You have no control over whether they might send your e-messages on to someone else—who might in turn forward them again. (One thing you can do for your friends, however, is delete their names and addresses before sending one of their messages on to someone.)

Think your internet access provider will protect your privacy? Often service providers post your address publicly or even sell their customer lists.

If you're really concerned about preserving your privacy, you can try certain technical solutions—for instance, installing software that encodes and decodes messages (discussed in Chapter 6). But the simplest solution is the easiest: Don't put any sensitive or embarrassing information in your email. Even deleted email removed from trash can still be traced on your hard disk. To guard against this, you can use software such as Spytech Eradicator and Webroot's Window Washer to completely eliminate deleted files. (Be aware, however: your email may already have been backed up on the company—or campus—server.)

SECURITY

Spam: Electronic Junk Mail

Is it really possible to manage spam?

Survival Tip
Look for the Padlock Icon

To avoid having people spying on you when you are sending information over the web, use a secure connection. This is indicated at the bottom of your browser window by an icon that looks like a padlock or key.

Several years ago, Monty Python, the British comedy group, did a sketch in which restaurant customers were unable to converse because people in the background (a group of Vikings, actually) kept chanting "Spam, spam, eggs and spam . . ." The term *spam* was picked up by the computer world to describe another kind of "noise" that interferes with communication. Now **spam refers to unsolicited email, or junk mail, in the form of advertising or chain letters.** But the problem of spam has metastasized well beyond the stage of annoyance.

Spam has become so pestiferous that *Smart Computing* magazine refers to it as a "cockroach infestation."[52] One recent survey found that spam accounts for 5 minutes of every hour spent online, which amounts to 10 eight-hour workdays a year.[53] So annoying is spam that the members of the European Union voted to outlaw bulk email, which includes spam.

Usually, of course, you don't recognize the spam sender on your list of incoming mail, and often the subject line will give no hint, stating something such as "The status of your application" or "It's up to you now." The solicitations can range from money-making schemes to online pornography. To better manage spam, some users get two email boxes. One is used for online shopping, business, research, and the like—which will continue to attract spam. The other is used (like an unlisted phone number) only for personal friends and family—and will probably not receive much spam.

The box below gives some other tips for fighting spam.

>>

PRACTICAL ACTION
Tips for Fighting Spam

Following are some tips for fighting spam:[54]

- **Delete without opening the message:** Opening the spam message can actually send a signal to the spammer that someone has looked at the on-screen message and therefore that the email address is valid—which means you'll probably get more spams in the future. If you don't recognize the name on your inbox directory or the topic on the inbox subject line, you can simply delete the message without reading it. Or you can use a preview feature in your email program to look at the message without actually opening it; then delete it. (Hint: Be sure to get rid of all the deleted messages from time to time; otherwise, they will build up in your "trash" area.)

- **Never reply to a spam message:** The following advice needs to be taken seriously: *Never reply in any way to a spam message!* Replying confirms to the spammer that yours is an active email address. Some spam senders will tell you that if you want to be removed from their mailing list, you should type the word *remove* or *unsubscribe* in the subject line and use the Reply command to send the message back. Invariably, however, all this does is confirm to the spammer that your address is valid, setting you up to receive more unsolicited messages.

- **Opt out:** When you sign up for or buy something online and are asked for an email address, remember to opt out of everything you're sure you don't want to receive. When you register for a website, for example, read its privacy policy to find out how it uses email addresses—and don't give the site permission to pass along yours.

- **Enlist the help of your internet access provider, or use spam filters:** Your IAP may offer a free spam filter (for example, Earthlink's Spaminator and Hotmail's Brightmail) to stop the stuff before you even see it. If it doesn't, you can sign up for a filtering service, such as ImagiNet (*www.imagin.net*) for a small monthly charge. Or there are do-it-yourself spam-stopping programs. Examples: Choicemail (*www.digiportal.com*), Spam-Assassin (*http://spamassassin.apache.org/index.html*), Junk Spy (*www.junkspy.com*), and McAfee SpamKiller (*http://us.mcafee.com/root/product.asp?productid=msk*). More complicated spam-blocker packages exist for businesses. Finally, you can subscribe to email forwarding services, such as Sneakemail (*www.sneakemail. com*), which forward only those messages you say in advance you will accept.

  Be warned, however: Even so-called spam killers don't always work. "Nothing will work 100%, short of changing your email address," says the operator of an online service called SpamCop. "No matter how well you try to filter a spammer, they're always working to defeat the filter."[55]

- **Fight back:** If you want to get back at spammers—and other internet abusers—check with abuse.net (*www.abuse.net*) or Ed Falk's Spam Tracking Page (*www.rahul.net/falk*). Spamhaus (*www.spamhaus.org*) tracks the internet's worst spammers and works with ISPs and law enforcement agencies to identify and remove persistent spammers from the internet. It also provides a free database of IP addresses of verified spammers. These groups will tell you where to report spammers, the appropriate people to complain to, and other spam-fighting tips.

Spoofing, Phishing, & Pharming: Phony Email Senders & Websites

How would I know if I were being spoofed, phished, or pharmed?

A message shows up in your email supposedly from someone named "Sonia Saunders." The subject line reads "Re: Hey cutie." It could have been from someone you know, but it's actually a pitch for porn. Or you receive what appears to be an "Urgent notice from eBay," the online auction company, stating that "failure to update billing information will result in cancellation of service" and asking you to go to the web address indicated and update your credit-card information. In the first instance you've been *spoofed*, in the second you've been *phished*.[56]

info!

Deciphering Fake Email

For more about spoofing and how to identify origins of fake emails, go to *www .mailsbroadcast.com/email .broadcast.faq/46.email .spoofing.htm*.

SPOOFING—USING FAKE EMAIL SENDER NAMES *Spoofing* **is the forgery of an email sender name so that the message appears to have originated from someone or somewhere other than the actual source.** Spoofing is one of the main tactics used by spammers (and virus writers) to induce or trick recipients into opening and perhaps responding to their solicitations. Spoofing is generally not illegal and might even serve a legitimate purpose under some circumstances—say, a "whistle-blowing" employee fearful of retaliation who reports a company's illegalities to a government agency. It is illegal, however, if it involves a direct threat of violence or death.[57]

PHISHING—USING TRUSTED INSTITUTIONAL NAMES TO ELICIT CONFIDENTIAL INFORMATION *Phishing* **(pronounced "fishing" and short for *password harvesting fishing*) is (1) the sending of a forged email that (2) directs recipients to a replica of an existing web page, both of which pretend to belong to a legitimate company. The purpose of the fraudulent sender is to "phish" for, or entice people to share, their personal, financial, or password data.** The names may be trusted names such as Citibank, eBay, or Best Buy. A variant is *spear-phishing*, in which a message seems to originate within your company, as when a "Security Department Assistant" asks you to update your name and password or risk suspension.[58] Thus, you should be suspicious of *any* email that directs you to a website that requests confidential information, such as credit-card or Social Security number.[59]

info!

Verifying Valid Websites

For more on verifying if you're dealing with a legitimate company website, go to *www .trustwatch.com*.

PHARMING—REDIRECTING YOU TO AN IMPOSTOR WEB PAGE Pharming is a relatively new kind of phishing that is harder to detect. **In *pharming*, thieves implant malicious software on a victim's computer that redirects the user to an impostor web page even when the individual types the correct address into his or her browser.** One way to protect yourself is to make sure you go to special secure web pages, such as any financial website, which begin with *https* rather than the standard *http* and which use encryption to protect data transfer.[60]

Cookies: Convenience or Hindrance?

Do I really want to leave cookies on my computer?

Cookies **are little text files—such as your log-in name, password, and preferences—left on your hard disk by some websites you visit. The websites retrieve the data when you visit again.**

THE BENEFITS OF COOKIES Cookies can be a convenience. If you visit an online merchant—such as BarnesandNoble.com for a book—and enter all your address and other information, the merchant will assign you an identification number, store your information with that number on its server, and send the number to your browser as a cookie, which stores the ID number

You can use your browser's
Help function to accept or reject
cookies. For instance, in Internet Explorer, go to the Tools
menu and click *Internet
Options;* on the General tab,
click *Settings;* then click *View
files,* select the cookie you want
to delete, and on the File menu
click *Delete.* Software such as
Cookie Pal (*www.kburra.com*)
will also help block and
control cookies, as will Window
Washer 5.

info!

Fighting Spyware

More information about ways to
combat spyware may be found
at:
*www.microsoft.com/athome/
security/spyware/default.mspx
www.pcpitstop.com
www.spywarewarrior.com/
rogue_anti-spyware.htm
www.webroot.com
www.cleansoftware.org
www.lavasoftusa.com/software/
adaware/*

on your hard disk. The next time you go to that merchant, the number is
sent to the server, which looks you up and sends you a customized web page
welcoming you. "Cookies actually perform valuable services," says technology writer and computer talk-radio-show host Kim Komando. "For instance,
they can shoot you right into a site so you don't have to enter your password."[61] Says another writer, "They can also fill in a username on a site that
requires logging in, or helping a weather site remember a Zip code so that it
can show a local forecast on return visits."[62]

THE DRAWBACKS OF COOKIES Cookies are not necessarily dangerous—
they can't transmit computer viruses, for example. However, some websites
sell the information associated with your ID number on their servers to marketers, who might use it to target customers for their products. "Unsatis-
factory cookies," in Microsoft's understated term, are those that might allow
someone access to personally identifiable information that could be used
without your consent for some secondary purpose. This can lead to *spyware,*
as we describe next.

Spyware—Adware, Browser & Search Hijackers, & Key Loggers: Intruders to Track Your Habits & Steal Your Data

What should I be afraid of about spyware?

You visit a search site such as Yahoo or Google and click on a text ad that
appears next to your search results. Or you download a free version of some
software, such as Kazaa, the popular file-sharing program. Or you simply visit
some web merchant to place an order.

The next thing you know, you seem to be getting ***pop-up ads***, **a form of
online advertising in which, when you visit certain websites, a new window
opens, or "pops up," to display advertisements.** You have just had an
encounter with *spyware,* of which pop-up ads are only one form. ***Spyware*
may be defined as deceptive software that is surreptitiously installed on a
computer via the web; once installed on your hard disk, it allows an outsider
to harvest confidential information,** such as keystrokes, passwords, your
email address, or your history of website visits. Spyware was found on the
personal computers of 80% of the 329 homes participating in a 2004 study
conducted by America Online Inc. and the National Cyber Security
Alliance.[63] Ways to avoid getting spyware are shown in the box opposite.

The most common forms of spyware are the following:[64]

ADWARE OR POP-UP GENERATORS ***Adware*, or *pop-up generators,* is a
kind of spyware that tracks web surfing or online buying so that marketers
can send you targeted and unsolicited pop-up and other ads.** This is the most
common, and benign, type of spyware. Adware can be developed by legitimate companies such as Verizon and Panasonic but also by all kinds of fly-
by-night purveyors of pornography and gambling operating from computer
servers in Russia, Spain, and the Virgin Islands.

BROWSER HIJACKERS & SEARCH HIJACKERS More damaging kinds of
spyware are ***browser hijackers*, which change settings in your browser without your knowledge, often changing your browser's home page and replacing
it with another web page,** and ***search hijackers*, which intercept your legitimate search requests made to real search engines and return results from
phony search services designed to send you to sites they run.** "A better name
for these programs is 'scumware,'" says one writer. "One of them might reset
your home page to a porn site or an obscure search engine and then refuse
to let you change the page back. Or it might hijack your search requests and
then direct them to its own page."[65]

KEY LOGGERS *Key loggers*, or *keystroke loggers,* **can record each charac-ter you type and transmit that information to someone else on the internet, making it possible for strangers to learn your passwords and other information.** For instance, some may secretly record the keystrokes you use to log in to online bank accounts and then send the information off to who knows where.

Almost all spyware is written to run on Microsoft Windows and won't run on Apple Macintoshes. In the past spyware writers have relied on flaws in Microsoft's Internet Explorer. On the PC, you may be protected if you keep your security features updated. Some people believe that getting rid of Inter-net Explorer and going to Mozilla Firefox, the most popular alternative, is a better solution.[66] Another option is Opera. Internet access providers such as AOL and Earthlink offer scan-and-removal tools, but you can also employ specialized antispyware software. Some of the good ones appear below. (● *See Panel 2.40.)*

One big problem with spyware is that overburdened PCs begin to run more slowly as the hard drives cope with random and uncontrollable processes. If none of the antispyware works, you will need to wipe your hard drive clean of programs and data and start from scratch—a complicated matter that we discuss in Chapter 4.

● **PANEL 2.40**
Some antispyware programs

Program Name	Site
Ad Aware	www.lavasoftusa.com/software/adaware
AntiSpyware	http://us.mcafee.com
Pest Patrol	www.ca.com/products/pestpatrol
Spybot Search & Destroy	http://spywaresoftware.net
SpyCather	www.tenebril.com
SpyCop	www.spywareinfo.com/downloads.php
Spy Sweeper	www.webroot.com
Yahoo Toolbar with Anti-Spy	www.download.com/Yahoo-Toolbar-with-Anti-Spy/3000-2379_4-10310983.html

>>

PRACTICAL ACTION
Tips for Avoiding Spyware

You may not be able to completely avoid spyware, but doing the following may help:

● *Be careful about free and illegal downloads:* Be choosy about free downloadings, as from Grokster and Kazaa, or illegal downloads of songs, movies, or TV shows. Often they use a form of spyware. File-sharing programs, which are popular with students, often con-tain spyware. Pornographic websites also are common carriers of spyware.

● *Don't just say "I agree"; read the fine print:* Sites that offer games, music-sharing videos, screen savers, and weather data often are paid to distribute spyware.

When you install their software, you might be asked to agree to certain conditions. If you simply click "I agree" without reading the fine print, you may be authorizing installation of spyware. "People have gotten in the habit of clicking next, next, next, without reading" when they install software, says a manager at McAfee Inc., which tracks spyware and viruses.[67]

● *Beware of unsolicited downloads:* If while you're surfing the net your browser warns you a file is being downloaded and you're asked if you choose to accept, keep clicking *no* until the messages stop.

Experience Box
Web Research, Term Papers, & Plagiarism

No matter how much students may be able to rationalize cheating in college—for example, trying to pass off someone else's term paper as their own (plagiarism)—ignorance of the consequences is not an excuse. Most instructors announce the penalties for cheating at the beginning of the course—usually a failing grade in the course and possible suspension or expulsion from school.

Even so, probably every student becomes aware before long that the World Wide Web contains sites that offer term papers, either for free or for a price. Some dishonest students may download papers and just change the author's name to their own. Others are more likely just to use the papers for ideas. Perhaps, suggests one article, "the fear of getting caught makes the online papers more a diversion than an invitation to wide-scale plagiarism."[68]

How the Web Can Lead to Plagiarism

Two types of term-paper websites are as follows:

- **Sites offering papers for free:** Such a site requires that users fill out a membership form and then provides at least one free student term paper. (Good quality is not guaranteed, since free-paper mills often subsist on the submissions of poor students, whose contributions may be subliterate.)

- **Sites offering papers for sale:** Commercial sites may charge $6–$10 a page, which users may charge to their credit card. (Expense is no guarantee of quality. Moreover, the term-paper factory may turn around and make your $350 custom paper available to others—even fellow classmates working on the same assignment—for half the price.)

How Instructors Catch Cheaters

How do instructors detect and defend against student plagiarism? Professors are unlikely to be fooled if they tailor term-paper assignments to work done in class, monitor students' progress—from outline to completion—and are alert to papers that seem radically different from a student's past work.[69]

Eugene Dwyer, a professor of art history at Kenyon College, requires that papers in his classes be submitted electronically, along with a list of World Wide Web site references. "This way I can click along as I read the paper. This format is more efficient than running around the college library, checking each footnote."[70]

Just as the internet is the source of cheating, it is also a tool for detecting cheaters. Search programs make it possible for instructors to locate texts containing identified strings of words from the millions of pages found on the web. Thus, a professor can input passages from a student's paper into a search program that scans the web for identical blocks of text. Indeed, some websites favored by instructors build a database of papers over time so that students can't recycle work previously handed in by others. One system can lock on to a stolen phrase as short as eight words. It can also identify copied material even if it has been changed slightly from the original. (More than 1,000 educational institutions have turned to Oakland, California-based Turnitin.com—*www.turnitin.com*—a service that searches documents for unoriginality.)[71] Another program professors use is the Self-Plagiarism Detection Tool, or SplaT (*http://splat.cs.arizona.edu*).[72]

How the Web Can Lead to Low-Quality Papers

William Rukeyser, coordinator for Learning in the Real World, a nonprofit information clearinghouse, points out another problem: The web enables students "to cut and paste together reports or presentations that appear to have taken hours or days to write but have really been assembled in minutes with no actual mastery or understanding by the student."[73]

Philosophy professor David Rothenberg, of New Jersey Institute of Technology, reports that as a result of students' doing more of their research on the web, he has seen "a disturbing decline in both the quality of the writing and the originality of the thoughts expressed."[74] How does an instructor spot a term paper based primarily on web research? Rothenberg offers four clues:

- **No books cited:** The student's bibliography cites no books, just articles or references to websites. Sadly, says Rothenberg, "one finds few references to careful, in-depth commentaries on the subject of the paper, the kind of analysis that requires a book, rather than an article, for its full development."

- **Outdated material:** A lot of the material in the bibliography is strangely out of date, says Rothenberg. "A lot of stuff on the web that is advertised as timely is actually at least a few years old."

- **Unrelated pictures and graphs:** Students may intersperse the text with a lot of impressive-looking pictures and graphs that actually bear little relation to the precise subject of the paper. "Cut and pasted from the vast realm of what's out there for the taking, they masquerade as original work."

- **Superficial references:** "Too much of what passes for information [online] these days is simply advertising for information," points out Rothenberg. "Screen after screen shows you where you can find out more, how you can connect to this place or that." Other kinds of information are detailed but often superficial: "pages and pages of federal documents, corporate propaganda, snippets of commentary by people whose credibility is difficult to assess."

Summary

E-shopping
Price anything from plane tickets to cars; order anything from books to sofas.

access point (p. 58) Station that sends and receives data to and from a Wi-Fi network. Why it's important: *Many public areas, such as airports and hotels, offer hot spots, or access points, that enable Wi-Fi-equipped users to go online wirelessly.*

adware (p. 102) Also called *pop-up generators:* kind of spyware that tracks web surfing or buying online. Why it's important: *Adware enables marketers to send you targeted and unsolicited pop-up and other ads.*

animation (p. 94) The rapid sequencing of still images to create the appearance of motion, as in a cartoon. Why it's important: *Animation is a component of multimedia; it is used in online video games as well as in moving banners displaying sports scores or stock prices.*

applets (p. 93) Small programs that can be quickly downloaded and run by most browsers. Why it's important: *Web pages contain links to applets, which add multimedia capabilities.*

B2B (business-to-business) commerce (p. 97) Electronic sale or exchange of goods and services directly between companies, cutting out traditional intermediaries. Why it's important: *Expected to grow even more rapidly than other forms of e-commerce, B2B commerce covers an extremely broad range of activities, such as supplier-to-buyer display of inventories, provision of wholesale price lists, and sales of closed-out items and used materials—usually without agents, brokers, or other third parties.*

backbones *See* **internet backbone.**

bandwidth (p. 52) Also known as *channel capacity:* expression of how much data—text, voice, video, and so on—can be sent through a communications channel in a given amount of time. Why it's important: *Different communications systems use different bandwidths for different purposes. The wider the bandwidth, the faster the data can be transmitted.*

bits per second (bps) (p. 52) Eight bits make up a character. Why it's important: *Data transfer speeds are measured in bits per second.*

blog (p. 96) Short for *web log,* an internet journal. Blogs are usually updated daily; they reflect the personality and views of the blogger. Why it's important: *Blogs are becoming important sources of current information.*

blogosphere (p. 96) The total universe of blogs. Why it's important: *The blogosphere has allowed the rise of a new breed of homegrown journal writers and diarists to chronicle life as it happens.*

broadband (p. 52) Very high speed connection. Why it's important: *Access to information is much faster than access with traditional phone lines.*

browser *See* **web browser.**

browser hijacker (p. 102) A damaging kind of spyware that changes settings in your browser without your knowledge. Why it's important: *This spyware can reset your home page to a porn site or obscure search engine or change your home page and replace it with another web page.*

cable modem (p. 56) Device connecting a personal computer to a cable-TV system that offers an internet connection. Why it's important: *Cable modems transmit data faster than do standard modems.*

client (p. 60) Computer requesting data or services. Why it's important: *Part of the client/server network, in which the server is a central computer supplying data or services requested of it to the client computer.*

search hijacker (p. 102) A damaging kind of spyware that can intercept your legitimate search requests made to real search engines and return results from phony search services. Why it's important: *Phony search services may send you to sites they run.*

search service (p. 75) Organization that maintains databases accessible through websites. Why it's important: *A search service helps you find information on the internet.*

server (p. 60) Central computer supplying data or services. Why it's important: *Part of the client/server network, in which the central computer supplies data or services requested of it to the client computer.*

site (p. 66) *See* **website.**

spam (p. 99) Unsolicited email in the form of advertising or chain letters. Why it's important: *Spam filters are available that can spare users the annoyance of receiving junk mail, ads, and other unwanted email.*

spider (p. 75) Also known as *crawler, bot,* or *agent;* special program that crawls through the World Wide Web, following links from one web page to another. Why it's important: *A spider indexes the words on each site it encounters and is used to compile the databases of a search service.*

spoofing (p. 101) The forgery of an email sender name so that the message appears to have originated from someone or somewhere other than the actual source. Why it's important: *Spoofing is one of the main tactics used by spammers to induce or trick recipients into responding to their solicitations.*

spyware (p. 102) Deceptive software that is surreptitiously installed on a computer via the web. Why it's important: *Once spyware is installed on your hard disk, it allows an outsider to harvest confidential information, such as keystrokes, passwords, or your email address.*

streaming audio (p. 95) Process of downloading audio in which you can listen to the file while the data is still being downloaded to your computer. Why it's important: *Users don't have to wait until the entire audio is downloaded to the hard disk before listening to it.*

streaming video (p. 94) Process of downloading video in which the data is transferred in a continuous flow so that you can begin viewing a file even before the end of the file is sent. Why it's important: *Users don't have to wait until the entire video is downloaded to the hard disk before watching it.*

subject directory (p. 75) Type of search engine that allows you to search for information by selecting lists of categories or subjects. Why it's important: *Subject directories allow you to look for information by categories such as "Business and Commerce" or "Arts and Humanities."*

surf (p. 64) To explore the web by using your mouse to move via a series of connected paths, or links, from one location, or website, to another. Surfing requires a browser. Why it's important: *Surfing enables you to easily find information on the web that's of interest to you.*

3G (third generation) (p. 58) High-speed wireless technology that does not need access points because it uses the existing cellphone system. Why it's important: *The technology is found in many new smartphones and PDAs that are capable of delivering downloadable video clips and high-resolution games.*

T1 line (p. 56) Traditional trunk line that carries 24 normal telephone circuits and has a transmission rate of 1.5 Mbps. Why it's important: *High-capacity T1 lines are used at many corporate, government, and academic sites; these lines provide greater data transmission speeds than do regular modem connections.*

tags (p. 81). Do-it-yourself labels that people can put on anything found on the internet, from articles to photos to videos. Why it's important: *Tagging is more powerful than a bookmark because they can be shared easily with other people.*

Transmission Control Protocol/Internet Protocol (TCP/IP) (p. 62) Protocol that enables all computers to use data transmitted on the internet by determining (1) the type of error checking to be used, (2) the data compression method, if any, (3) how the sending device will indicate that it has finished sending a message, and (4) how the receiving device will indicate that it has received a message. TCP/IP breaks data into *packets,* which are the largest blocks of data that can be sent across the internet (less than 1,500 characters, or 128 kilobytes). IP is used to send the packets across the internet to their final destination, and TCP is used to reassemble the packets in the correct order. Why it's important: *Internet computers use TCP/IP for all internet transactions, from sending email to downloading stock quotes or pictures off a friend's website.*

upload (p. 53) To transmit data from a local computer to a remote computer. Why it's important: *Uploading allows users to easily exchange files over networks.*

URL (Uniform Resource Locator) (p. 66) String of characters that points to a specific piece of information anywhere on the web. A URL consists of (1) the web protocol, (2) the name of the web server, (3) the directory (or folder) on that server, and (4) the file within that directory (perhaps with an extension such as *html* or *htm*). Why it's important: *URLs are necessary to distinguish among websites.*

```
                        Directory   File
                        name,       (document)
        Domain name     or          name and
Protocol (web server name) path     extension
┌──┐  ┌──────────┐  ┌──┐  ┌────────┐
http://www.nps.gov/yose/home.htm
```

Usenet (p. 89) Worldwide network of servers that can be accessed through the internet. Why it's important: *Newsgroups take place on Usenet.*

VoIP phoning. *See* **internet telephony.**

web-based email (p. 82) Type of email in which you send and receive messages by interacting via a browser with a website. Why it's important: *Unlike standard email, web-based email allows you to easily send and receive messages while traveling anywhere in the world and to use any personal computer and browser to access your email.*

web browser (browser) (p. 64) Software that enables users to locate and view web pages and to jump from one page to another. Why it's important: *Users can't surf the web without a browser. Examples of browsers are Microsoft Internet Explorer, Netscape Navigator, Mozilla Firefox, Opera, and Apple Macintosh Browser.*

webcasting (p. 95) Service, based on push technology, in which customized text, video, and audio are sent to the user automatically on a regular basis. Why it's important: *Users choose the categories, or the channels, of websites that will automatically send updated information. Thus, webcasting saves time because users don't have to go out searching for the information.*

web page (p. 66) Document on the World Wide Web that can include text, pictures, sound, and video. Why it's important: *A website's content is provided on web pages. The starting page is the home page.*

web portal (p. 73) Type of gateway website that functions as an "anchor site" and offers a broad array of resources and services, online shopping malls, email support, community forums, current news and weather, stock quotes, travel information, and links to other popular subject categories. The most popular portals are America Online, Yahoo, Google, Microsoft Network, Netscape, Lycos. Why it's important: *Web portals provide an easy way to access the web.*

website (site) (p. 66) Location of a web domain name in a computer somewhere on the internet. Why it's important: *Websites provide multimedia content to users.*

Wi-Fi (p. 58) Short for "wireless fidelity." The name given to any of several standards—so-called 802.11 standards—set by the Institute of Electrical and Electronic Engineers for wireless transmission. Why it's important: *Wi-Fi enables people to use their Wi-Fi-equipped laptops to go online wirelessly in certain areas such as airports that have public access to Wi-Fi networks.*

window (p. 86) A rectangular area on a computer display screen that contains a document or activity. Why it's important: *In instant messaging, a window allows a user to exchange IM messages with others almost simultaneously while operating other programs.*

wireless network (p. 58) Network that uses radio waves to transmit data, such as Wi-Fi. Why it's important: *Wireless networks enable people to access the internet without having a cabled or wired connection, using wireless equipped laptops and smart cellphones.*

XML (extensible markup language) (p. 96) A Web-document tagging and formatting language that two computers can use to exchange information. Why it's important: *XML is an improvement over HTML and enables the creation of RSS newsreaders.*

Chapter Review

stage **LEARNING** MEMORIZATION

"I can recognize and recall information."

Self-Test Questions

1. Today's data transmission speeds are measured in _____, *Kbps*, _____, and _____.

2. A(n) _____ _____ connects a personal computer to a cable-TV system that offers an internet connection.

3. A space station that transmits data as microwaves is a _____.

4. A company that connects you through your communications line to its server, which connects you to the internet, is a(n) _____.

5. A rectangular area on the computer screen that contains a document or displays an activity is called a(n) _____.

6. _____ is writing an online message that uses derogatory, obscene, or inappropriate language.

7. A(n) _____ is software that enables users to view web pages and to jump from one page to another.

8. A computer with a domain name is called a(n) _____.

9. _____ comprises the communications rules that allow browsers to connect with web servers.

10. A(n) _____ is a program that adds a specific feature to a browser, allowing it to play or view certain files.

11. Unsolicited email in the form of advertising or chain letters is known as _____.

12. The expression of how much data—text, voice, video, and so on—can be sent through a communications channel in a given amount of time is known as _____.

13. A(n) _____ is a string of characters that points to a specific piece of information somewhere on the web.

14. Some websites may leave files left on your hard disk that contain information such as your name, password, and preferences; they are called _____.

15. Using trusted institutional names to elicit confidential information is called _____.

16. The kind of spyware that can record each character you type and transmit that information to someone else on the internet, making it possible for strangers to learn your passwords and other information, is called _____.

Multiple-Choice Questions

1. Kbps means _____ bits per second.
   a. 1 billion
   b. 1 thousand
   c. 1 million
   d. 1 hundred
   e. 1 trillion

2. A location on the internet is called a(n)
   a. network.
   b. user ID.
   c. domain.
   d. browser.
   e. web.

3. In the email address *Kim_Lee@earthlink.net.us,* Kim_Lee is the
   a. domain.
   b. URL.
   c. site.
   d. user ID.
   e. location.

4. Which of the following is *not* one of the four components of a URL?
   a. web protocol
   b. name of the web server
   c. name of the browser
   d. name of the directory on the web server
   e. name of the file within the directory

5. Which of the following is the fastest method of data transmission?
   a. ISDN
   b. DSL
   c. modem
   d. T1 line
   e. cable modem

6. Which of the following is *not* a netiquette rule?
   a. Consult FAQs.
   b. Flame only when necessary.
   c. Don't shout.
   d. Avoid huge file attachments.
   e. Avoid sloppiness and errors.

7. Which protocol is used to retrieve email messages from the server to your computer?

   a. HTTP (HyperText Transfer Protocol)

   b. SMTP (Simple Mail Transfer Protocol)

   c. POP3 (Post Office Protocol version 3)

   d. POP (point of presence)

8. Who owns the internet?

   a. Microsoft

   b. IBM

   c. Apple

   d. U.S. government

   e. No one owns the internet; the components that make up the Internet are owned and shared by thousands of public and private entities.

9. Each time you connect to your ISP, it will assign your computer a new address called a(n)

   a. domain.

   b. IP address.

   c. plug-in.

   d. POP.

   e. URL (Universal Resource Locator).

10. ISPs that don't run their own backbones connect to an internet backbone through a

    a. NAP network access point.

    b. web portal.

    c. web browser.

    d. URL.

    e. TCP/IP.

11. Which of the following is *not* a protocol?

    a. TCP/IP

    b. IE

    c. HTTP

    d. SMTP

12. The sending of phony email that pretends to be from a credit-card company or bank, luring you to a website that attempts to obtain confidential information from you, is called

    a. spoofing.

    b. phishing.

    c. spamming.

    d. keylogging.

    e. cookies.

True or False

T F   1. POP3 is used for sending email, and SMTP is used for retrieving email.

T F   2. A dial-up modem is an ISP (internet service provider).

T F   3. Replying to spam email messages with the statement "remove" will always get spammers to stop sending you unsolicited email.

T F   4. All computer communications use the same bandwidth.

T F   5. A T1 line is the slowest but cheapest form of internet connection.

T F   6. A dynamic IP address gives you faster internet access than a static IP address does.

T F   7. A bookmark lets you return to a favorite website quickly.

T F   8. Radio buttons are used for listening to radio stations on the internet.

T F   9. Spoofing means using fake email sender names.

T F   10. Hypertext refers to text presented with very large letters.

stage **LEARNING** COMPREHENSION

"I can recall information in my own terms and explain them to a friend."

Short-Answer Questions

1. Name three methods of data transmission that are faster than a regular modem connection.

2. What does *log on* mean?

3. What is netiquette, and why is it important?

4. Briefly define *bandwidth*.

5. Many web documents are "linked." What does that mean?

6. Compare and contrast a cable modem service to a DSL service.

7. Explain the basics of how the internet works.

8. What expanded functions does IMAP (Internet Message Access Protocol) have?

9. Briefly explain what TCP/IP does.

10. Why was ICANN established?

11. What's the difference between a dynamic IP address and a static IP address?

12. Explain what a blog is.

13. How would you answer a person who asked you the question "Who owns the internet?"

14. Explain the difference between a "mega-portal" and a "vortal."

15. What is B2B commerce?

16. List and briefly describe three kinds of spyware.

"I can apply what I've learned, relate these ideas to other concepts, build on other knowledge, and use all these thinking skills to form a judgment."

Knowledge in Action

1. Distance learning uses electronic links to extend college campuses to people who otherwise would not be able to take college courses. Are you or is someone you know involved in distance learning? If so, research the system's components and uses. What hardware and software do students need in order to communicate with the instructor and classmates? What courses are offered? Discuss the pros and cons of distance learning compared to classroom-based learning.

2. It's difficult to conceive how much information is available on the internet and the web. One method you can use to find information among the millions of documents is to use a search engine, which helps you find web pages on the basis of typed keywords or phrases. Use your browser to go to the Google home page, and click in the *Search* box. Type the keywords *"personal computers";* then click on *Google Search,* or press the *Enter* key. How many results did you get?

3. As more and more homes get high-speed broadband internet connections, the flow of data will become exponentially faster and will open up many new possibilities for sharing large files such as video. What types of interactive services can you envision for the future?

4. Draw a diagram of what happens when you log onto your ISP; include all the connections you think possible for your situation.

5. How do the latest cellphones incorporate the internet into their functions? What functions could be improved? Have any of these extra functions affected your daily life?

6. How has the internet affected your life? Start keeping a list.

7. Email and instant messaging (IM) are ways of sending text messages back and forth to other people on the internet. They seem very similar: in both, you compose a message, and when it's ready, you send it; and when someone else sends something to you, you receive it on your device and can read it.

   But in practice, email and IM can be surprisingly different; each has its own rhythm, its own strengths and weaknesses, its own sociology, its own etiquette. Instant messaging is like using the telephone, whereas email is more like corresponding by letter.

   As you use email and IM during the course of the term, watch for differences between them. Which medium is more appropriate for which kinds of relationships and communications? Which medium is more stressful to use? Which takes more time? Which is more convenient for you? Which one is more useful for getting real work done? Which medium would you use if you knew that whatever you wrote was eventually going to be published in a book? If you were restricted to using only one of these communications methods, which would it be?

8. Internet service providers (ISPs) often place limits on upload speeds, thus making it take much longer to send (upload) a large file than it would take to receive (download) a file of the same size from someone else. Do a comparison between upload and download speeds on your internet connection, perhaps by emailing yourself a file large enough to allow you to notice the difference. Why do you think there is a difference? (Consider both technological and economic factors.)

9. Imagine that an elderly relative wants to start using the internet for the first time. You want to help the person get started, but you need to be careful not to overwhelm your relative with more information than he or she can use. What three or four things would you tell and show your relative first? What things do you think will be hardest for him or her to master? How do you think using the internet is likely to change this person's life? If possible, seek out such a relative or neighbor and actually introduce him or her to the internet.

10. As we have discussed in this chapter, the internet is both a goldmine and a minefield. There are vast riches of information, entertainment, education, and communication to be found, but there are also snoopers, spam, spoofing, phishing, spyware, adware, browser hijackers, and key loggers. What should you do to avoid these threats?

11. Some websites require you to register before you are allowed to use them. Others require that you have a paid membership. Others allow limited free access to everyone but require payment for further content. Why do you think different sites adopt these different attitudes toward use of their material?

Web Exercises

1. Some websites go overboard with multimedia effects, while others don't include enough. Locate a website that you think makes effective use of multimedia. What is the purpose of the site? Why is the site's use of multimedia effective? Take notes, and repeat the exercise for a site with too much multimedia and one with too little.

2. If you have never done a search before, try this: Find out how much car dealers pay for your favorite cars, what they charge consumers for them, and what you should know about buying a new car. A company called Edmunds publishes a magazine with all that information, but you can get the same information on its website for free.

   Using the Google search engine (*www.google.com*), type *"automobile buyer's guide"* and *Edmunds* in the search box, and hit the *Google Search* button. How many entries did you get? Click on a link to the Edmunds website. Explore the site, and answer the questions at the beginning of this exercise.

3. Ever wanted your own dot-com in your name? Visit these sites to see if your name is still available:

    www.register.com
    www.namezero.com
    www.domainname.com
    www.checkdomain.com/
    www.domaindirect.com/

4. Interested in PC-to-phone calls through your internet connection? Visit these sites and check out their services:

    www.dialpad.com
    www.net2phonedirect.com
    www.iconnecthere.com
    www.skype.com

5. HTTP (HyperText Transfer Protocol) on the World Wide Web isn't the only method of browsing and transferring data. FTP is the original method and is still a useful internet function. To use FTP, you'll need an FTP client software program, just as you need a web browser to surf the web. Download one of these shareware clients and visit its default FTP sites, which come preloaded:

CuteFtp	www.globalscape.com
WS_FTP	www.ipswitch.com
FTP Voyager	www.ftpvoyager.com

    You will need an FTP client program to upload files to a server if you ever decide to build a website. Some on-line website builders have browser uploaders, but the conventional method has always been FTP. After you download an FTP client, visit www.oth.net to search for FTP servers that house files you would like to download.

    On some Macintosh computers, an FTP function is built into the Mac OS Finder (for downloading only—you have to use an FTP client to upload).

6. Video chat extensions to chat software are now generally available, and they really constitute at least two-party videoconferencing. These extensions are available through AIM, Yahoo Messenger, MSN Messenger, and Apple's iChat. All you need is a camera, a microphone, and an internet connection. Check out these sites and decide if this type of "video chat" interests you.

7. Make a simple website using HTML. If you're using Windows, open up Notepad and type the following:

    ```
    <html>
    <head>
    <title> A very simple demonstration of how to use HTML.
    </head>
    <body>
    Hey look at me. I made a web page.
    </body>
    </html>
    ```

A very simple demonstration of how to use HTML.

Hey look at me. I made a web page.

After you've typed that in, choose *File, Save As,* and *Save As File Type;* switch from text document to *All Files,* and type the filename *lookatme.html.* Then double-click the file you just created to see how easily a website can be created. After you are done, visit these websites to learn more about creating websites with HTML:

www.lissaexplains.com
www.make-a-web-site.com
http://archive.ncsa.uiuc.edu/General/Internet/WWW/
    HTMLPrimer.html
Or do an internet search for *"html primer"* and for *"learn html."*

8. To learn more about internet conventions, go to:

    complete smiley emoticon dictionary, *http://members .tripod.com/~paandaa/smiley.htm*
    complete instant messaging acronym dictionary,

    www.gaarde.org/acronyms/?lookup=A-Z
    http://paul.merton.ox.ac.uk/ascii/smileys.html
    http://research.microsoft.com/~mbj/Smiley/Smiley.html
    http://piology.org/smiley.txt
    www.cygwin.com/acronyms/

9. Some hobbies have been dramatically changed by the advent of the World Wide Web. Particularly affected are the "collecting" hobbies, such as stamp collecting, coin collecting, antique collecting, memorabilia collecting, plate collecting, and so forth. Choose some such hobby that you know something about or have some interest in. Run a web search about the hobby and see if you can find:

    a. a mailing list about the hobby.

    b. an auction site that lists rare items and allows you to bid on them.

    c. a chat room or other discussion forum allowing enthusiasts to gather and discuss the hobby.

    d. a site on which someone's formidable collection is beautifully and proudly displayed.

10. When the web first came into widespread use, the most popular search engine was AltaVista. For several years there were a variety of search engines available, but in recent years one search engine, Google, has become predominant and the word *googling* has entered the language as a term that means "to use a search engine to find information."

    Visit *http://searchenginewatch.com/links/* for a list of many alternative search engines, as well as explanations about how they work and how to get your site listed on them.

11. E-commerce is booming. For any given product you may wish to buy on the web, there may be hundreds or thousands of possible suppliers, with different prices and terms—and not all of them will provide equally reputable and reliable service. The choices can be so numerous that it may sometimes seem difficult to know how to go about choosing a vendor.

    Websites that do comparison shopping for you can be a great help. Such a service communicates with many individual vendors' websites, gathering information as it proceeds; it then presents its findings to you in a convenient form. Often, ratings of the various vendors are provided as well, and sales tax and shipping charges are calculated for you.

Here are a few sites that can assist with comparison shopping:

www.epinions.com/
www.bizrate.com/
www.pricescan.com/
www.addall.com/
http://shopper.cnet.com/

Practice "catch-and-release e-commerce" by researching the best deals you can find for:

a rare or at least out-of-print book that you'd like to have.
a high-end DVD recorder.
a replacement ink or toner cartridge for your laser or inkjet printer.
a pair of athletic shoes exactly like the shoes you currently have.

Pursue each transaction right up to the last step before you would have to enter your credit-card number and actually buy the item, and then quit. (Don't buy it. You can always do that another time.)

12. WebCams, or web cameras, are used by some websites to show pictures of their locations—either live video or still shots. To run your own WebCam site requires a suitable camera and a continuous internet connection. But to look at other people's WebCam sites requires only a web browser.

For example, try searching for "WebCam Antarctica." Or go to *www.webcam-index.com*.

Find and bookmark at least one interesting WebCam site in each of the following places: Africa, Asia, South America, Europe, Australia, Antarctica, Hawaii.

More HTML information

In HTML, every command is surrounded by < and >. And in most commands, you need to tell the web browser when to end this comand. You do this by putting a back slash (/) in front of the ending command, as shown below. Since HTML isn't case sensitive, <title> is the same as <TITLE>, which is the same as <TiTLe>. Next, you need to decide what you want to put on your page. Text, links, graphics, and text fields are just a few ideas. Here are some HTML-tagged text lines:

TITLE

Let's say that your title is going to be "John Doe's Web Page." Type:

< title>John Doe's Web Page</title>

HEADINGS

HTML has six levels of headings, numbered 1 through 6, with 1 being the largest. Headings are displayed in larger or smaller typestyles, and usually bolder. If you want to type "Hello," this is what you would type for each heading and what the outcome would be.

<H1>Hello</H1>   <H2>Hello</H2>

Hello   ## Hello

<H3>Hello</H3>   <H4>Hello</H4>

Hello   #### Hello

<H5>Hello</H5>   <H6>Hello</H6>

Hello   **Hello**

PARAGRAPHS

To make a paragraph of "This is a web page. How do you like what I've done? Please email me with any suggestions at a@a.com," type:

<P>This is a web page. How do you like what I've done? Please email me with any suggestions at a@a.com</P>

The outcome is

This is a web page. How do you like what I've done? Please email me with any suggestions at a@a.com

Software

Tools for Productivity & Creativity

Chapter Topics & Key Questions

3.1 System Software: The Power behind the Power What does the operating system (OS) do, and what is a user interface?

3.2 The Operating System: What It Does What are the principal functions of the operating system?

3.3 Other System Software: Device Drivers & Utility Programs What are the characteristics of device drivers and utility programs?

3.4 Common Features of the User Interface What are some common features of the graphical software environment, and how do they relate to the keyboard and the mouse?

3.5 Common Operating Systems What are some common desktop, network, and portable OSs?

3.6 Application Software: Getting Started What are five ways of obtaining application software, tools available to help me learn to use software, three common types of files, and the types of software?

3.7 Word Processing What can I do with word processing software that I can't do with pencil and paper?

3.8 Spreadsheets What can I do with an electronic spreadsheet that I can't do with pencil and paper and a standard calculator?

3.9 Database Software What is database software, and what is personal information management software?

3.10 Specialty Software What are the principal uses of specialty software?

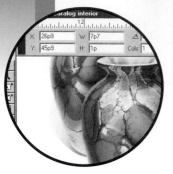

W

hat we need is a science called *practology*, a way of thinking about machines that focuses on how things will actually be used."

So says Alan Robbins, a professor of visual communications, on the subject of *machine interfaces*—the parts of a machine that people actually manipulate.[1] An *interface* is a machine's "control panel," ranging from the volume and tuner knobs on an old radio to all the switches and dials on the flight deck of a jetliner. You may have found, as Robbins thinks, that on too many of today's machines—digital watches, VCRs, even stoves—the interface is often designed to accommodate the machine or some engineering ideas rather than the people actually using it. Good interfaces are intuitive—that is, based on prior knowledge and experience—like the twin knobs on a 1950s radio, immediately usable by both novices and sophisticates. Bad interfaces, such as a software program with a bewildering array of menus and icons, force us to relearn the required behaviors every time.

So how well are computer hardware and software makers doing at giving us useful, helpful interfaces? The answer is, They're getting better all the time, but they still have some leftovers from the past. For instance, some interface screens have so many icons that they are confusing instead of helpful. And some microcomputer keyboards still come with a *SysRq* (for "System Request") key, which was once used to get the attention of the central computer but now is rarely used. (The Scroll Lock key is also seldom used.)

Improving interfaces is the province of *human-computer interaction (HCI)*, which is concerned with the study, design, construction, and implementation of human-centric interactive computer systems. HCI goes beyond improving screens and menus into the realm of adapting interfaces to human reasoning and studying the long-term effects that computer systems have on humans. HCI encompasses the disciplines of information technology, psychology, sociology, anthropology, linguistics, and others. As computers become more pervasive in our culture, HCI designers are increasingly looking for ways to make interfaces easier, safer, and more efficient.

In time, as interfaces are refined, computers may become no more difficult to use than a car. Until then, however, for smoother computing you need to know something about how software works. Today people communicate one way, computers another. People speak words and phrases; computers process bits and bytes. For us to communicate with these machines, we need an intermediary, an interpreter. This is the function of software, particularly system software.

3.1

System Software: The Power behind the Power

What are three components of system software, what does the operating system (OS) do, and what is a user interface?

As we mentioned in Chapter 1, *software,* or *programs,* consists of all the electronic instructions that tell the computer how to perform a task. These instructions come from a software developer in a form (such as a CD, or compact disk) that will be accepted by the computer. **<u>Application software</u> is software that has been developed to solve a particular problem for users— to perform useful work on specific tasks or to provide entertainment. <u>System software</u> enables the application software to interact with the computer and helps the computer manage its internal and external resources.** We interact mainly with the application software, which interacts with the system software, which controls the hardware.

● PANEL 3.1
Three components of system software
System software is the interface between the user and the application software and the computer hardware.

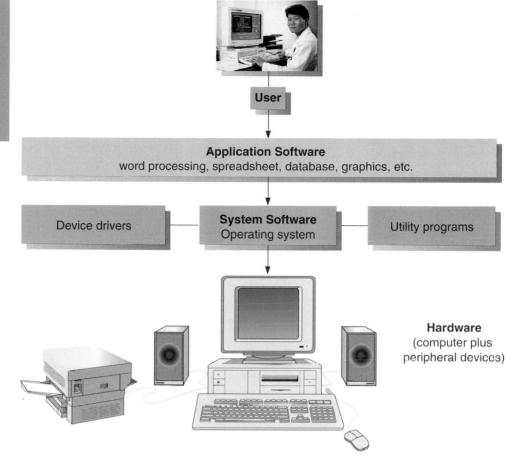

New microcomputers are usually equipped not only with system software but also with some application software.

There are three basic components of system software that you need to know about. (● *See Panel 3.1.*)

- **Operating systems:** An operating system is the principal component of system software in any computing system.
- **Device drivers:** Device drivers help the computer control peripheral devices.
- **Utility programs:** Utility programs are generally used to support, enhance, or expand existing programs in a computer system.

A fourth type of system software, *language translators*, is described in Chapter 10.

3.2

The Operating System: What It Does

What are the principal functions of the operating system?

The _operating system (OS)_, **also called the** *software platform*, **consists of the low-level, master system of programs that manage the basic operations of the computer.** These programs provide resource management services of many kinds. In particular, they handle the control and use of hardware resources, including disk space, memory, CPU time allocation, and peripheral devices. Every general-purpose computer must have an operating system to run other programs. The operating system allows you to concentrate on your own tasks or applications rather than on the complexities of managing the computer. Each application program is written to run on top of a particular operating system.

Different sizes and makes of computers have their own operating systems. For example, Cray supercomputers use UNICOS and COS; IBM mainframes use MVS and VM; PCs run Windows or Unix and Apple Macintoshes run the Macintosh OS. Some pen-based computers have their own operating systems—PenRight, for instance—that enable users to write scribbles and notes on the screen. In general, an operating system written for one kind of hardware will not be able to run on another kind of machine. In other words, *different operating systems are mutually incompatible.*

Microcomputer users may readily experience the aggravation of such incompatibility when they acquire a new or used microcomputer. Do they get an Apple Macintosh with Macintosh system software, which won't always run PC programs? Or do they get a PC (such as Dell or Hewlett-Packard), which won't run Macintosh programs?

Before we try to sort out these perplexities, we should have an idea of what operating systems do. We will consider:

- Booting
- CPU management
- File management
- Task management
- Security management

Booting

What is the boot process?

The work of the operating system begins as soon as you turn on, or "boot," the computer. **Booting is the process of loading an operating system into a computer's main memory.** This loading is accomplished by programs stored permanently in the computer's electronic circuitry (called *read-only memory*, or *ROM*, described in Chapter 4). When you turn on the machine, programs called *diagnostic routines* test the main memory, the central processing unit, and other parts of the system to make sure they are running properly. Next, BIOS (for "basic input/output system") programs are copied to main memory and help the computer interpret keyboard characters or transmit characters to the display screen or to a diskette. Then the boot program obtains the operating system, usually from the hard disk, and loads it into the computer's main memory, where it remains until you turn the computer off. (● *See Panel 3.2.)*

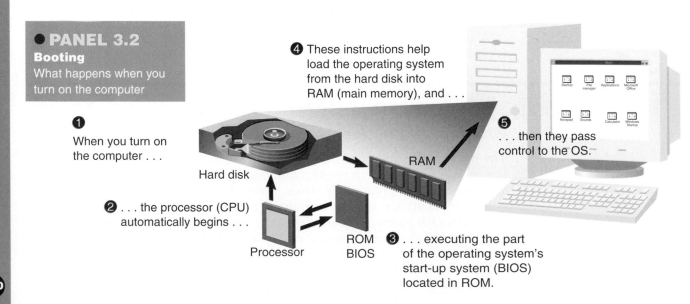

● **PANEL 3.2**
Booting
What happens when you turn on the computer

❶ When you turn on the computer . . .

❷ . . . the processor (CPU) automatically begins . . .

Hard disk

Processor

ROM BIOS

❸ . . . executing the part of the operating system's start-up system (BIOS) located in ROM.

❹ These instructions help load the operating system from the hard disk into RAM (main memory), and . . .

RAM

❺ . . . then they pass control to the OS.

StartUp File manager Applications Microsoft Office
Notepad Sounds Calculator Windows StartUp

COLD BOOTS & WARM BOOTS   When you power up a computer by turning on the power "on" switch, this is called a *cold boot*. If your computer is already on and you restart it, this is called a *warm boot* or a *warm start* (done by simultaneously pressing the *Ctrl + Alt + Del* keys or pressing the *Reset* button on your computer).

THE BOOT DISK   Normally, your computer would boot from the hard drive, but if that drive is damaged, you can use a disk called a *boot disk* to start up your computer. A boot disk is a floppy disk or a CD that contains all the files needed to launch the OS. When you insert the boot disk into your computer's floppy or CD drive, you force-feed the OS files to the BIOS, thereby enabling it to launch the OS and complete the start-up routine. After the OS loads completely, you then can access the contents of the Windows drive, run basic drive maintenance utilities, and perform troubleshooting tasks that will help you resolve the problem with the drive.

CPU Management

How does CPU management work?

The central component of the operating system is the supervisor. Like a police officer directing traffic, the **_supervisor_, or *kernel*, manages the CPU** (the central processing unit or processor, as mentioned in Chapter 1). **It remains in memory (main memory or primary storage) while the computer is running and directs other "nonresident" programs (programs that are not in memory) to perform tasks that support application programs.**

MEMORY MANAGEMENT   The operating system also manages memory—it keeps track of the locations within main memory where the programs and data are stored. It can swap portions of data and programs between main memory and secondary storage, such as your computer's hard disk, as so-called *virtual memory*. This capability allows a computer to hold only the most immediately needed data and programs within main memory. Yet it has ready access to programs and data on the hard disk, thereby greatly expanding memory capacity.

GETTING IN LINE: QUEUES, BUFFERS, & SPOOLING   Programs and data that are to be executed or processed wait on disk in *queues* (pronounced "Qs"). A queue is a first-in, first-out sequence of data and/or programs that "wait in line" in a temporary holding place to be processed. The disk area where the programs or documents wait is called a *buffer*. Print jobs are usually *spooled*—that is, placed—into a buffer, where they wait in a queue to be printed. This happens because the computer can send print jobs to the printer faster than the printer can print them, so the jobs must be stored and then passed to the printer at a rate it can handle. Once the CPU has passed a print job to the buffer, it can take on the next processing task. (The term *spooling* dates back to the days when print jobs were reeled, or copied, onto spools of magnetic tape, on which they went to the printer.)

File Management

What should I know about file management?

A **_file_ is (1) a named collection of data (data file) or (2) a program that exists in a computer's secondary storage (program file),** such as floppy disk, hard disk, or CD/DVD. Examples of data files are a word processing document, a spreadsheet, images, songs, and the like. Examples of program files are a word processing program or spreadsheet program. (We cover files in more detail in Chapters 4 and 8.)

Escape Key

You can press **Esc** to quit a task you are performing.

Caps Lock and Shift Keys

These keys let you enter text in uppercase (ABC) and lowercase (abc) letters.

Press **Caps Lock** to change the case of all letters you type. Press the key again to return to the original case.

Press **Shift** in combination with another key to type an uppercase letter.

Function Keys

These keys let you quickly perform specific tasks. For example, in many programs you can press **F1** to display help information.

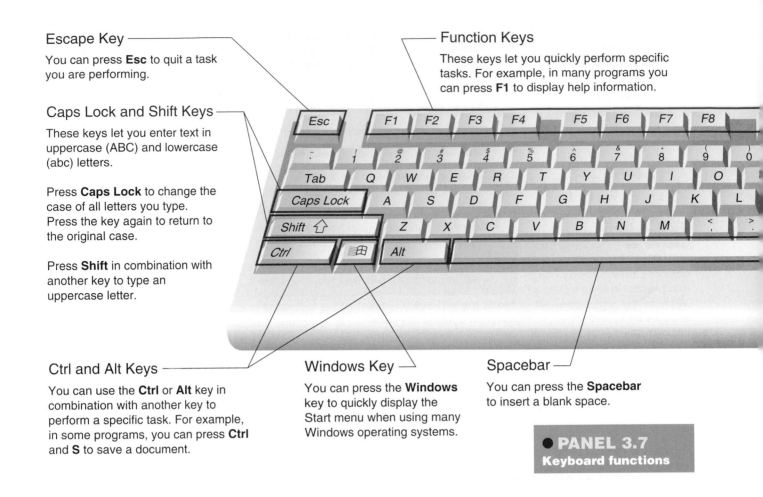

Ctrl and Alt Keys

You can use the **Ctrl** or **Alt** key in combination with another key to perform a specific task. For example, in some programs, you can press **Ctrl** and **S** to save a document.

Windows Key

You can press the **Windows** key to quickly display the Start menu when using many Windows operating systems.

Spacebar

You can press the **Spacebar** to insert a blank space.

● **PANEL 3.7**
Keyboard functions

FUNCTION KEYS *Function keys*, labeled "F1," "F2," and so on, are positioned along the top or left side of the keyboard. They are used to execute commands specific to the software being used. For example, one application software package may use F6 to exit a file, whereas another may use F6 to underline a word.

MACROS Sometimes you may wish to reduce the number of keystrokes required to execute a command. To do this, you use a macro. **A** *macro*, **also called a** *keyboard shortcut*, **is a single keystroke or command—or a series of keystrokes or commands—used to automatically issue a longer, predetermined series of keystrokes or commands.** Thus, you can consolidate several activities into only one or two keystrokes. The user names the macro and stores the corresponding command sequence; once this is done, the macro can be used repeatedly. (To set up a macro, pull down the Help menu and type in *macro*.)

Although many people have no need for macros, individuals who find themselves continually repeating complicated patterns of keystrokes say they are quite useful.

THE MOUSE & POINTER You will also frequently use your mouse to interact with the user interface. The mouse allows you to direct an on-screen pointer to perform any number of activities. **The** *pointer* **usually appears as an arrow, although it changes shape depending on the application. The mouse is used to move the pointer to a particular place on the display screen or to point to little symbols, or icons.** You can activate the function corresponding to the symbol by pressing ("clicking") buttons on the mouse. Using the mouse, you can pick up and slide ("drag") an image from one side of the screen to the other or change its size. (● *See Panel 3.8.*)

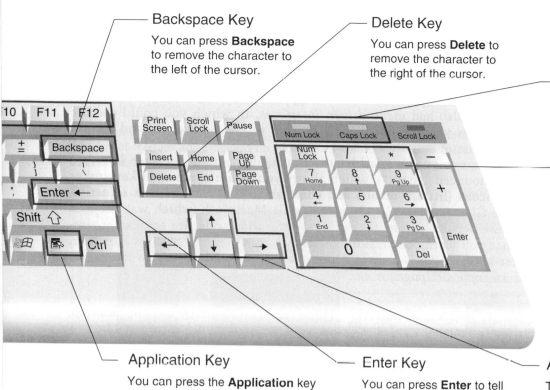

Backspace Key

You can press **Backspace** to remove the character to the left of the cursor.

Delete Key

You can press **Delete** to remove the character to the right of the cursor.

Status Lights

These lights indicate whether the **Num Lock** or **Caps Lock** features are on or off.

Numeric Keypad

When the **Num Lock** light is on, you can use the number keys (0 through 9) to enter numbers. When the **Num Lock** light is off, you can use these keys to move the cursor around the screen. To turn the light on or off, press **Num Lock**.

Application Key

You can press the **Application** key to quickly display the shortcut menu for an item on your screen. Shortcut menus display a list of commands commonly used to complete a task related to the current activity.

Enter Key

You can press **Enter** to tell the computer to carry out a task. In a word processing program, press this key to start a new paragraph.

Arrow Keys

These keys let you move the cursor around the screen.

Term	Action	Purpose
Point	Move mouse across desk to guide pointer to desired spot on screen. The pointer assumes different shapes, such as arrow, hand, or I-beam, depending on the task you're performing.	To execute commands, move objects, insert data, or similar actions on screen
Click	Press and quickly release left mouse button.	To select an item on the screen
Double-click	Quickly press and release left mouse button twice.	To open a document or start a program
Drag and drop	Position pointer over item on screen, press and hold down left mouse button while moving pointer to location in which you want to place item, then release.	To move an item on the screen
Right-click	Press and release right mouse button.	To display a shortcut list of commands, such as a pop-up menu of options

● PANEL 3.8
Mouse language

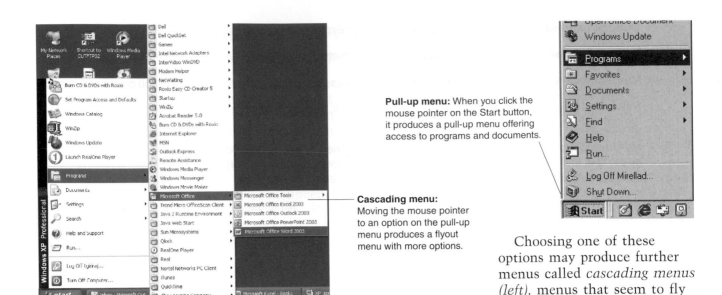

Pull-up menu: When you click the mouse pointer on the Start button, it produces a pull-up menu offering access to programs and documents.

Cascading menu: Moving the mouse pointer to an option on the pull-up menu produces a flyout menu with more options.

Choosing one of these options may produce further menus called *cascading menus (left)*, menus that seem to fly back to the left or explode out to the right, wherever there is space.

A *pull-up menu* is a list of options that pulls up from the menu bar at the bottom of the screen. In Windows XP, a pull-up menu appears in the lower left-hand corner when you click on the Start button.

A *pop-up menu* is a list of command options that can "pop up" anywhere on the screen when you click the right mouse button. In contrast to pull-down or pull-up menus, pop-up menus are not connected to a menu bar.

DOCUMENTS, TITLE BARS, MENU BARS, TOOLBARS, TASKBARS, & WINDOWS (SMALL "W") If you want to go to a document, there are three general ways to begin working from a typical Microsoft Windows GUI desktop: (1) You can click on the *Start* button at lower left corner and then make a selection from the pull-up menu that appears. Or (2) you can click on the *My Computer* icon on the desktop and pursue the choices offered there. (3) Or click on the *My Documents* icon and then on the folder that contains the document you want. In each case, the result is the same: the document is displayed in the window. (● *See Panel 3.10.*)

Once past the desktop—which is the GUI's opening screen—if you click on the *My Computer* icon, you will encounter various "bars" and window functions. (● *See Panel 3.11 on page 134.*)

- **Title bar: The _title bar_ runs across the very top of the display window and shows the name of the folder you are in**—for example, "My Computer."

- **Menu bar:** Below the title bar is the **_menu bar_, which shows the names of the various pull-down menus available.** Examples of menus are File, Edit, View, Favorites, Tools, and Help.

- **Toolbar: The _toolbar_, below the menu bar, displays menus and icons representing frequently used options or commands.** An example of an icon is the picture of two pages in an open folder with a superimposed arrow, which issues a *Copy to* command.

- **Taskbar:** In Windows, the **_taskbar_ is the bar across the bottom of the desktop screen that contains the Start button and that appears by default.** Small boxes appear here that show the names of open files. You can switch among the files by clicking on the boxes.

- **Windows:** When spelled with a capital "W," Windows is the name of Microsoft's system software (Windows 95, 98, Me, XP, Vista, and so on). When spelled with a lowercase "w," a **_window_ is a rectangular frame on the computer display screen. Through this frame you can view a file of data—such as a document, spreadsheet, or database—or an application program.**

● PANEL 3.10
Three ways to go to
a document in
Windows XP

From Start menu

Click on Start button to produce
Start menu, then go to *Documents* option,
then to *My Documents*. Click on the item you want.

From My Computer

Click on *My Computer*
icon, which opens a window
that provides access to
information on your computer.

Click on C, which opens a window that provides
access to information stored on your hard disk.

Click on
My Documents icon,
which opens a window
providing access to
document files
and folders.

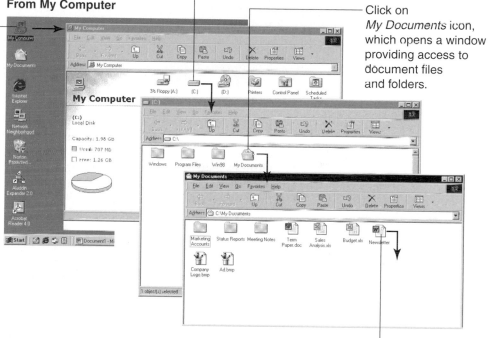

Click on
the Document
you want.

From My Documents icon

Click on
My Documents,
which opens
a window
that shows
the names
of your
documents/
document
folders.

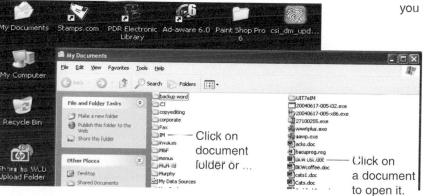

Click on
document
folder or ...

Click on
a document
to open it.

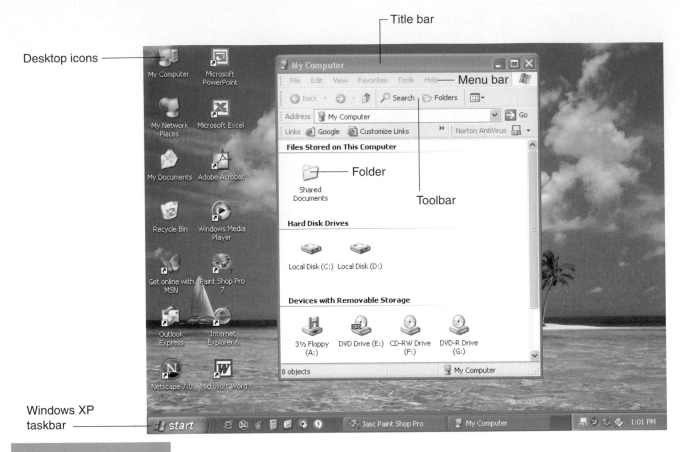

Title bar

Desktop icons

Menu bar

Folder

Toolbar

Windows XP taskbar

● PANEL 3.11
"Bars" and windows functions

Survival Tip
Don't Trash Those Icons

Don't delete unwanted software programs by using the mouse to drag their icons to the recycle bin. This might leave behind system files that could cause problems. Instead, use an "uninstall" utility. In Windows, go to *Start, Settings, Control Panel;* double-click *Add/ Remove Programs.* Find the program you want to delete, and click the *Add/Remove* button.

In the upper right-hand corner of the Windows title bar are some window controls—three icons that represent *Minimize, Maximize and Restore,* and *Close.* By clicking on these icons, you can *minimize* the window (shrink it down to an icon at the bottom of the screen), *maximize* it (enlarge it or restore it to its original size), or *close* it (exit the file and make the window disappear). You can also use the mouse to move the window around the desktop, by clicking on and dragging the title bar.

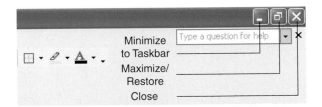

Minimize to Taskbar

Maximize/ Restore

Close

Finally, you can create *multiple windows* to show programs running concurrently. For example, one window might show the text of a paper you're working on, another might show the reference section for the paper, and a third might show something you're downloading from the internet. If you have more than one window open, click on the Maximize button of the window you want to be the main window to *restore* it.

The Help Command

What can the Help command do for me?

Don't understand how to do something? Forgotten a command? Accidentally pressed some keys that messed up your screen layout and you want to undo it? Most toolbars contain a **Help command**—a command generating a table

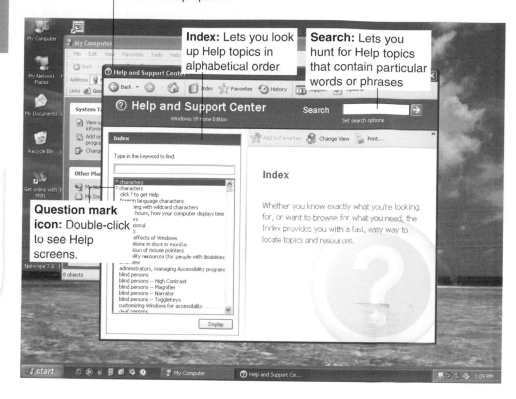

┌─ The *Help* menu provides
  a list of help options.

Index: Lets you look
up Help topics in
alphabetical order

Search: Lets you
hunt for Help topics
that contain particular
words or phrases

**Question mark
icon:** Double-click
to see Help
screens.

of contents, an index, and a search feature that can help you locate answers. In addition, many applications have *context-sensitive help,* which leads you to information about the task you're performing. (● *See Panel 3.12.*)

3.5

Common Operating Systems

What are some common desktop, network, and portable OSs?

The **_platform_** is the particular processor model and operating system on which a computer system is based. For example, there are "Mac platforms" (Apple Macintosh) and "Windows platforms" or "PC platforms" (for personal computers such as Dell, Compaq, Gateway, Hewlett-Packard, or IBM that run Microsoft Windows). Sometimes the latter are called *Wintel platforms,* for "Windows + Intel," because they often combine the Windows operating system with the Intel processor chip. (We discuss processors in Chapter 4.)

Despite the dominance of the Windows platform, some so-called *legacy systems* are still in use. A legacy system is an older, outdated, yet still functional technology, such as the **_DOS operating system. DOS_** (rhymes with "boss")—for *Disk Operating System*—was the original operating system produced by Microsoft and had a hard-to-use command-driven user interface. (● *See Panel 3.13.*) Its initial 1982 version was designed to run on the IBM PC as PC-DOS. Later Microsoft licensed the same system to other computer makers as MS-DOS.

```
C:\WINDOWS\System32\cmd.exe
(C) Copyright 1985-2001 Microsoft Corp.

C:\Documents and Settings\Stacey C. Sawyer>
C:\Documents and Settings\Stacey C. Sawyer>DIR
 Volume in drive C has no label
 Volume Serial Number is B0ED 7091

 Directory of C:\Documents and Settings\Stacey C. Sawyer

04/25/2003  07:14 PM    <DIR>          .
04/25/2003  07:14 PM    <DIR>          ..
03/01/2003  12:58 PM    <DIR>          .java
03/01/2003  12:58 PM    <DIR>          .jpi_cache
04/25/2003  07:17 PM             1,106 .plugin140_01.trace
08/23/2003  03:31 PM    <DIR>          Desktop
02/11/2003  11:35 AM               831 Eudora.lnk
08/23/2003  03:29 PM    <DIR>          Favorites
08/25/2003  11:48 AM    <DIR>          My Documents
08/25/2003  10:45 AM         3,145,728 ntuser.dat
```

Here we briefly describe the principal platforms used on single-user computers today, both desktops and laptops: the Macintosh OS and Windows. We discuss operating systems for servers and for handheld computers in a few pages.

Macintosh Operating System

What significant contribution has the Mac OS made to personal computing?

The *Macintosh operating system (Mac OS)*, **which runs only on Apple Macintosh computers, set the standard for icon-oriented, easy-to-use graphical user interfaces.** Apple based its new interface on work done at Xerox, which in turn had based its work on early research at Stanford Research Institute (now SRI International). (See the timeline starting on page 162.) The software generated a strong legion of fans shortly after its launch in 1984 and inspired rival Microsoft to upgrade DOS to the more user-friendly Windows operating systems. Much later, in 1998, Apple introduced its iMac computer (the "i" stands for "internet"), which added capabilities such as small-scale networking.

MAC OS X   The Mac OS is *proprietary*, meaning that it is privately owned and controlled by a company. Many Mac users still use System 9, introduced in October 1999, which added an integrated search engine, updated the GUI, and improved networking services. The next version of the operating system, Mac OS X, broke with 15 years of Mac software to use Unix (discussed shortly) to offer a dramatic new look and feel. (● *See Panel 3.14.)* Many Apple users claim that OS X (pronounced "ten") won't allow software conflicts, a frequent headache with Microsoft's Windows operating systems. (For example, you might install a game and find that it interferes with the device driver for a sound card. Then, when you uninstall the game, the problem persists. With Mac OS X, when you try to install an application program that conflicts with any other program, the Mac simply won't allow you to run it.) Mac OS X also offers free universal email services, improved graphics and printing, CD burning capability, a DVD player, easier ways to find files, and support for building and storing web pages.

The latest version of OS X, called Tiger, released in mid-2005, has several notable features:

- **Spotlight—for desktop search:** A desktop search engine, you'll recall from Chapter 2, is a tool that extends searching beyond the web to the contents of your personal computer's hard disk. The Macintosh version, called Spotlight, helps you find information stored in the

● **PANEL 3.14**
Mac OS X
The latest version is called Tiger.

thousands of files on your hard drive. Spotlight also offers Smart Folders, permanent lists of search results that are automatically updated as files are added or deleted.

- **Dashboard—for creating desktop "widgets":** Tiger's Dashboard is a collection of little applications that Apple calls "widgets," such as clock, calculator, weather report, or stock ticker, that can be available on your desktop at any time. Dashboard enables programmers to create customized widgets that provide notification on the desktop whenever there's a newsworthy development, such as a change in stock price, that's important to users—something that "could tame the flood of real-time data that threatens to overwhelm us," says *BusinessWeek's* Stephen Wildstrom.[2]

- **Automator—for handling repetitive tasks:** Automator is a personal robotic assistant that lets you streamline repetitive tasks into a script that will perform a sequence of actions with a single click, such as checking for new messages from a particular sender, then sending you an alert.

WHERE IS MAC KING?   Macintosh is still considered king in areas such as desktop publishing, and Macs are still favored in many educational settings. For very specialized applications, most programs are written for the Windows platform. However, programs for games and for common business uses such as word processing and spreadsheets are also widely available for the Mac.

Note that many Macs can accept documents created with basic Microsoft PC applications, such as Word—if the Mac version of these applications has been installed on the Mac. You can even use a PC-formatted diskette in many Macs. (The reverse is not true: PCs will not accept Mac-formatted disks.)

Microsoft Windows

What has been the evolution of Windows?

In the 1980s, taking its cue from the popularity of Mac's easy-to-use GUI, Microsoft began working on Windows—to make DOS more user-friendly. Also a proprietary system, **_Microsoft Windows_ is the most common operating system for desktop and portable PCs.** Early attempts (Windows 1.0, 2.0, 3.0) did not catch on. However, in 1992, *Windows 3.X* emerged as the preferred system among PC users. (Technically, Windows 3.X wasn't a full operating system; it was simply a layer or shell over DOS.)

Bill Gates. The founder of Microsoft has been identified with every PC operating system since the 1980 DOS.

EARLY VERSIONS OF WINDOWS: WINDOWS 95, 98, & ME Windows 3.X evolved into the *Windows 95* operating system, which was succeeded by *Windows 98* and *Windows Me*. Among other improvements over their predecessors, Windows 95, 98, and Me (for *Millennium Edition*), which are still used in many homes and businesses, adhere to a standard called Universal Plug and Play, which is supposed to let a variety of electronics seamlessly network with each other. *Plug and Play* is defined as the ability of a computer to automatically configure a new hardware component that is added to it.

WINDOWS XP   **_Microsoft Windows XP_, introduced in 2001, combines elements of Windows networking software and Windows Me with a new GUI.** It has improved stability and increased driver and hardware support. It also features built-in instant messaging; centralized shopping managers to help you keep track of your favorite online stores and products; and music, video, and photography managers.

PANEL 3.15
Windows XP (right) and Windows Tablet PC

Start menu (Harold):
Internet — Internet Explorer
E-mail — Outlook Express
Windows Media Player
MSN Explorer
Windows Movie Maker
Tour Windows XP
Windows Product Catalog
Files and Settings Transfer Wizard
All Programs

My Documents
My Pictures
My Music
My Computer
Control Panel
Help and Support
Search
Run...

Survival Tip
Service Packs 1 & 2

In late 2002, Microsoft updated XP to a "second edition" by supplying a downloadable service pack called *Windows XP Service Pack 1 (SP1)*. In August 2004 Microsoft released *Service Pack 2 (SP2)*. These service pack updates improve the OS's performance and strengthen security. However, they can cause compatibility problems with existing applications on a user's computer. So, if you are considering updating your version of XP, research compatibility and backup issues on the internet before going to Microsoft's website to download these updates.

Windows XP Home Edition is for typical home users. Windows XP Professional Edition is for businesses of all sizes and for home users who need to do more than get email, browse the internet, and do word processing. Windows XP Tablet PC Edition is for business notebook computers that support data entry via a special pen used to write on the display screen. (● *See Panel 3.15.*)

WINDOWS XP MEDIA CENTER EDITION In late 2004, Microsoft released *Windows XP Media Center Edition 2005,* a media-oriented operating system that supports DVD burning, high-definition television (HDTV), two TV tuners, cable TV, and satellite TV and that provides a refreshed user interface. A key feature is support for wireless technology and in particular Media Center Extender, a technology that lets users wirelessly connect up to five TVs to the Media Center PC.

Windows XP Media Center Edition is a premium version of Windows XP, designed to make the PC the media and entertainment hub of the home. In addition to performing traditional PC tasks, the system can serve music, pictures,

Portable media center

video, and live television to portable devices, stereos, and TVs while also enforcing digital rights set by content owners. Users can access a Windows XP Media Center PC—a combination of special Windows OS and hardware that includes a TV tuner, remote control, and other multimedia equipment—with a remote control through a special user interface on their TV. Users will be able to exchange MSN instant messages through their TV and remotely program their Media Center 2005 system through a service offered by MSN. You can't purchase or upgrade to the Windows Media Center OS; you must buy it on a new PC.

Desktop media center

WINDOWS VISTA   Scheduled to be rolled out in late 2006, *Windows Vista* (codenamed Longhorn during development) is the first big upgrade since Windows XP in 2001. An early (beta) version was delivered in August 2005 for technical users. Promised improvements included the following:[3]

- **Improved reliability:** Improved security, privacy, performance, reliability, and ease of deployment is supposed to be at the top of the list. One reason that development had been delayed was that Microsoft learned it needed to focus acutely on the growing problem of viruses and spyware, which had plagued Windows XP. Something called User Account Protection will limit users to certain protected modes, which will help prevent the installation of spyware.

- **New ways of organizing information:** Vista is supposed to bring clarity to the organization and use of information. For instance, icons used to represent files, such as a photo or a letter, will consist of tiny snapshots of the actual files, not generic graphics.

- **Better connections:** Microsoft is putting more emphasis on search capability, so that users can find files more intuitively.

Network Operating Systems: NetWare, Windows NT/2000/2003, Unix, & Linux

How would I distinguish among the several network OSs?

The operating systems described so far were principally designed for use with stand-alone desktop and laptop machines. Now let's consider the important operating systems designed to work with networks—NetWare, Windows NT/2000/2003, Unix/Solaris, and Linux:

NOVELL'S NETWARE—PC NETWORKING SOFTWARE   *NetWare* **has long been a popular network operating system for coordinating microcomputer-based local area networks (LANs) throughout a company or a campus.** LANs allow PCs to share programs, data files, and printers and other devices. Novell, the maker of NetWare, thrived as corporate data managers realized that networks of PCs could exchange information more cheaply than the previous generation of mainframes and midrange computers. The biggest challenge to NetWare has been Windows NT, Windows 2000/2003, and Windows XP. However, Novell is continuing to improve its networking OS with its new version, Netware 6.5. It is also moving forward with its strategy to combine Netware with the benefits of a more recently popular operating system, Linux (to be discussed shortly); this combined product, Novell Open Enterprise, was introduced in early 2005.

WINDOWS NT & 2000/2003—THE CHALLENGE FROM MICROSOFT Windows desktop operating systems (95/98/Me/XP) can be used to link PCs in small networks in homes and offices. However, something more powerful was needed to run the huge networks linking a variety of computers—PCs, workstations, mainframes—used by many companies, universities, and other organizations, which previously were served principally by Unix and NetWare operating systems. *Microsoft Windows NT* **(the *NT* stands for "New Technology"), later upgraded to** *Windows 2000*, **is the company's multitasking operating system designed to run on network servers in businesses of all sizes.** It allows multiple users to share resources such as data, programs, and printers and to build web applications and connect to the internet.

- **Two versions—NT Server and 2000:** When it first appeared, in 1993, the system came in two versions. The *Windows NT Workstation* version enabled graphic artists, engineers, and others using stand-alone workstations to do intensive computing at their desks. The *Windows*

Y: 45p9   H: 1p   Cols: 1

PRACTICAL ACTION
Get a PC or Get a Mac?
Dealing with Security Issues

SECURITY Nervous about spam, spyware, phishing, viruses, and other threats to your computer? That's certainly been the case for many Windows PC users.

"Microsoft has paid so little attention to security over the years," says distinguished technology writer Walter Mossberg, "that consumers who use Windows have been forced to spend more and more of their time and money fending off" these invasive demons.[4]

In late 2004, the software giant rolled out a major, free operating system upgrade called *Service Pack 2,* or *SP2.* Among other things, this was supposed to reduce the risk to Microsoft's Internet Explorer web browser of online attacks that had frustrated users and slowed businesses. However, the company conceded that SP2 interfered with about 50 known programs, including corporate products and a few games. And a couple of months later, Microsoft issued several "security advisories" urging consumers and businesses to patch 21 new flaws in Windows software products.[5] In February 2005 it said that it was upgrading its Internet Explorer against malicious software and expected to include security defenses in its new operating system, Vista, scheduled for release in 2006.[6]

Microsoft's Service Packs versus Patches

A *service pack* (sometimes called a "service release") is a collection of files for various updates, software bug fixes, and security improvements. A *patch* (sometimes called a "hotfix") is an update that occurs between service packs; most patches are built to correct security vulnerabilities. Service packs are planned, or strategic, releases. Patches are unplanned, interim solutions.

A new service pack is supposed to "roll up" all previous service packs and patches. Microsoft recommends installing both packs and patches as they become available but checking the online bulletin accompanying every new patch to see what risks vulnerability poses to your particular hardware and software configuration.

It would appear, however, that the pack-and-patch approach to security for Microsoft products will be with us for a

long time. "We don't feel like we've ever crossed the finish line," says Windows' lead product manager. "We have to keep outrunning the bad guys."[7]

Switch to Mac?

Perhaps because Apple Macintosh has only around 5% of the market share for microcomputers, it seems to have eluded most of the attention of the hackers and virus writers. Similarly, for whatever reasons, Linux-based computers hardly get infected or invaded at all.

Big-business users, it's suggested, have too much money invested in Windows machines to think of switching to Macintoshes, and the costs of making the change would be astronomical.[8] Individual users, however, might wish to give it some thought.

What if you already have Microsoft Office files, such as Microsoft Word or Microsoft Excel, on a PC and want to move them to a Mac? In that case, you'll need the current version of Microsoft Office for the Mac, which will handle the PC version. Apple's OS X also comes with a program called *Mail* and a word processor called *Text Edit* that will deal with Microsoft Word documents that come to you online. You can also get a Macintosh version of Internet Explorer, which is, in fact, considered safer than the Windows versions.[9]

Could Xen Be a Threat to Windows?

A free software known as Xen, being developed through an open-source project at the University of Cambridge in England, allows a single computer to act as several "virtual machines"—that is, to run multiple operating systems, simultaneously performing multiple tasks on different operating systems. A computer running Xen, in one description, "could operate an accounting application on one virtual machine, while managing the data on another. If the accounting application crashes, the data remain stable. Meanwhile, the user saves time and money by running both tasks on a single physical machine."[10] Could Xen threaten the dominance of Windows in the personal computer world?

NT Server version was designed to benefit multiple users tied together in networks. In early 2000, Microsoft rolled out its updated version, *Windows 2000,* to replace Windows NT.

The choice of the name Windows 2000 is somewhat confusing. This is not the successor to Windows 95 and 98 for home and non-network use. If you're interested in arcade-style games, for instance, 2000 won't work, since it was specifically designed for businesses. (Windows Me was designed for home users.)

- **Windows for servers:** Windows 2000 includes the Windows 2000 server family, for various levels of network servers: Windows 2000 Professional, Windows 2000 Server, Windows 2000 Advanced Server, and Windows 2000 Datacenter Server. In 2003, Microsoft introduced a new version of Windows 2000 called the *Windows Server 2003* family. This version enhances security, reliability, and manageability, among other things. A server version of Windows Vista was expected to be announced in late 2006.

UNIX, SOLARIS, AND BSD—FIRST TO EXPLOIT THE INTERNET Unix (pronounced "*you*-nicks") was developed at AT&T's Bell Laboratories in 1969 as an operating system for minicomputers. By the 1980s, AT&T entered into partnership with Sun Microsystems to develop a standardized version of Unix for sale to industry. Today __Unix__ **is a proprietary multitasking operating system for multiple users that has built-in networking capability and versions that can run on all kinds of computers.** (● *See Panel 3.16.*) It is used mostly on mainframes, workstations, and servers, rather than on PCs. Government agencies, universities, research institutions, large corporations, and banks all use Unix for everything from designing airplane parts to currency trading. Unix is also used for website management and runs the backbone of the internet. The developers of the internet built their communications system around Unix because it has the ability to keep large systems (with hundreds of processors) churning out transactions day in and day out for years without fail.

● **PANEL 3.16**
Unix screen

- **Versions of Unix:** Sun Microsystems' *Solaris* is a version of Unix that is popular for handling large e-commerce servers and large websites. Another interesting variant is *BSD*, free software derived from Unix. BSD began in the 1970s in the computer science department of the University of California, Berkeley, when students and staff began to develop their own derivative of Unix, known as the Berkeley Software Distribution, or BSD. There are now three variations, which are distributed online and on CD. Other Unix variations are made by Hewlett-Packard (HP-UX) and IBM (AIX).

- **Unix interface—command or shell?** Like MS-DOS, Unix uses a command-line interface (but the commands are different for each system). Some companies market Unix systems with graphical interface *shells* that make Unix easier to use.

Some Common Unix Commands

^h, [backspace]	erase previously typed character
^u	erase entire line of input typed so far
cp	copy files
whoami	who is logged on to this terminal
mkdir	make new directory
mv	change name of directory
mail	read/send email
gzip, gunzip	compress, recompress a file
lpr	send file to printer
wc	count characters, words, and lines in a file
head	show first few lines of a file
tail	show last few lines of a file
find	find files that match certain criteria

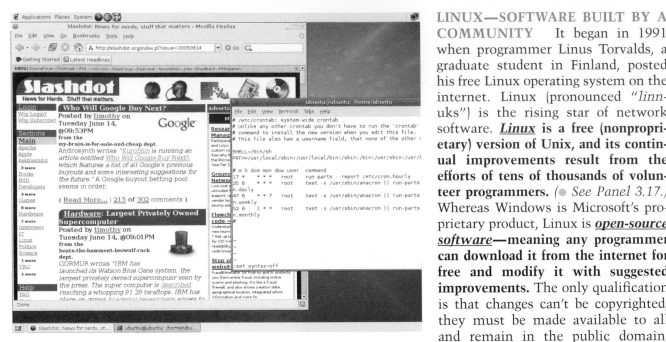

LINUX—SOFTWARE BUILT BY A COMMUNITY It began in 1991 when programmer Linus Torvalds, a graduate student in Finland, posted his free Linux operating system on the internet. Linux (pronounced "*linn-uks*") is the rising star of network software. **_Linux_ is a free (nonproprietary) version of Unix, and its continual improvements result from the efforts of tens of thousands of volunteer programmers.** (● *See Panel 3.17.*) Whereas Windows is Microsoft's proprietary product, Linux is **_open-source software_—meaning any programmer can download it from the internet for free and modify it with suggested improvements.** The only qualification is that changes can't be copyrighted; they must be made available to all and remain in the public domain.

● **PANEL 3.17**
Linux screen

From these beginnings, Linux has attained cultlike status. "What makes Linux different is that it's part of the internet culture," says an IBM general manager. "It's essentially being built by a community."[11]

- **Linux and China:** In 2000, the People's Republic of China announced that it was adopting Linux as a national standard for operating systems because it feared being dominated by the OS of a company of a foreign power—namely, Microsoft. In 2005, Red Flag Software Company, Ltd., the leading developer of Linux software in China, joined the Open Source Development Labs (OSDL, *www.osdl.org*), a global consortium dedicated to accelerating the adoption of Linux in the business world.

- **The permutations of Linux:** If Linux belongs to everyone, how do companies like Red Hat Software—a company that bases its business on Linux—make money? Their strategy is to give away the software but then sell services and support. Red Hat, for example, makes available an inexpensive application software package that offers word processing, spreadsheets, email support, and the like for users of its PC OS version. It also offers more powerful versions of its Linux OS for small and medium-size businesses, along with applications, networking capabilities, and support services.

  Wal-Mart offers a Microtel PC with a Linux operating system called *Linspire* (formerly Lindows). The Linspire operating system delivers the stability of Linux with the ease of Windows. And it includes a trial membership to a library of more than 1,000 software programs for business, home, and entertainment needs. (The Linux-based operating system in these PCs may not be compatible with some dial-up internet services.)

  Lycoris offers Linux-based *DesktopLX* for desktop PCs, and some companies, such as Ibex, make dual-boot PCs that can switch back and forth between Windows and Linux by rebooting. Win4Lin, Inc., a leading supplier of specialized operating systems that run on Linux, has released Win4Lin Pro, which runs Windows 2000 and Windows XP applications on Linux.

- **Linux in the future:** Because it was originally built for use on the internet, Linux is more reliable than Windows for online applications. Hence, it is better suited to run websites and e-commerce software.

more info!

Open-Source Search

Open-source utilities—some command-line interfaces, some GUIs—are available for Linux and Mac systems. What keywords could you use to find these utilities?

Its real growth, however, may come as it reaches outward to other applications and, possibly, replaces Windows in many situations. IBM, Red Hat, Motorola Computing, Panasonic, Sony, and many other companies have formed the nonprofit, vendor-neutral Embedded Linux Consortium, which is working to make Linux a top operating system of choice for developers designing embedded systems, as we discuss next.

The three major microcomputer operating systems are compared in the box below. (● *See Panel 3.18.*)

Operating Systems for Handhelds: Palm OS & Windows CE/Pocket PC/Windows CE .NET

What are some different embedded OSs for handheld computers and PDAs?

An *embedded system* is any electronic system that uses a CPU chip but that is not a general-purpose workstation, desktop, or laptop computer. It is a specialized computer system that is part of a larger system or a machine. Embedded systems are used in automobiles, planes, trains, space vehicles, machine tools, watches, appliances, cellphones, PDAs, and robots. Handheld computers and personal digital assistants rely on specialized operating systems, embedded operating systems, that reside on a single chip. Such operating systems include the Palm OS, Windows CE/Pocket PC/Windows CE .NET, and Symbian OS.

more info!

Palm OS

Ron Nicholson's Palm OS Computing General Information Page: *www.nicholson.com/rhn/palm.*

PALM OS—THE DOMINANT OS FOR HANDHELDS The *Palm OS*, produced by PalmSource, is the dominant operating system for handhelds. In 1994, Jeff Hawkins took blocks of mahogany and plywood into his garage and emerged with a prototype for the PalmPilot, which led to the revolution in handheld computing. Today the Japanese-owned company PalmSource, Inc., sells Palm OS 5, the enhanced Palm OS Garnet and Palm OS Cobalt, and

Windows XP	Linux	Mac OS X
Pros:	*Pros:*	*Pros:*
Runs on a wide range of hardware	Runs on a wide range of hardware	Easy to install
Has largest market share	Has largest number of user interface types	Best GUI
Has many built-in utilities		Secure and stable
	Can be used as server or desktop PC OS	
Cons:	*Cons:*	*Cons:*
Security problems	Limited support for games	Supports only Apple computers
Not efficient used as a server OS	Limited commercial applications available	Base hardware more expensive than other platforms
Have to reboot every time a network configuration is changed	Can be difficult to learn	Fewer utilities available
		Fewer games than for Windows
		Some applications still being updated to run with OS X

● PANEL 3.18
OS comparison

Software

143

palmOne Treo smartphone

info!

Windows CE

The Windows CE Web Community at *www.windowscewebring.com* offers links to several websites that focus on using Windows CE.

info!

For More on OSs

Things undoubtedly have changed a bit on the OS scene between the time we wrote this chapter and the time you are reading it. Go to *www.osnews.com* and use some keywords from this chapter to search for updates on, for example, Linux versus Windows for the PC user, Linux in embedded systems, Unix market share, responses to any updates to Windows, and availability of new applications for Mac OS X.

mFone for smartphones. The company known as Palm sells the popular Palm Tungsten T5 PDA, the Treo smartphone, and Zire PDAs. (Smartphones are cellular phones that have built-in music players and video recorders and that run computerlike applications and support internet activities beyond just email. We discuss them in detail in Chapter 7.)

Early versions of handhelds cannot be upgraded to run recent versions of Palm OS. If you buy a PDA or a smartphone, therefore, you need to make sure that the OS is a current version and that the manufacturer plans to continue to support it.

Microsoft Pocket PC for handhelds.

WINDOWS CE/POCKET PC/CE .NET— MICROSOFT WINDOWS FOR HANDHELDS In 1996, Microsoft released **_Microsoft Windows CE_**, now known as **_Microsoft Pocket PC_** or **_Windows Mobile_**—or, in its newest form, *Windows CE .NET*. **This OS is a slimmed-down version of Windows for handhelds** such as those made by Symbol Technologies, Garmin, ASUS, Dell, and Hewlett-Packard. The Windows handheld OS has the familiar Windows look and feel and includes mobile versions of word processing, spreadsheet, email, web browsing, text messaging, and other software. Microsoft also makes a Pocket PC Phone Edition. O2's Xda II mini, the smallest PDA-phone in the market so far that includes a camera, a video recorder, and an MP3 player, is powered by the second edition of the Microsoft Windows Mobile 2003 Pocket PC Phone operating system.

WINDOWS MEDIA PLAYER FOR CELLPHONES & PDAS Windows Media Player 10 Mobile is designed for Windows-based cellphones and personal digital assistants (PDAs). The software turns these devices into portable media players that will work with Media Center PCs, allowing users to take content with them.

SYMBIAN OS—THE NEWEST MOBILE OS *Symbian* is the new name for the old EPOC OS for Psion handhelds, which were very popular in Europe but less so in the United States. Britain-based Symbian was set up with financial backing from the world's biggest mobile phone vendors, including Psion, Nokia, Matsushita, Motorola, and Sony Ericsson, which wanted a strong alternative supplier to Microsoft. While Microsoft aims for corporate clients looking to supply smartphones that work well with other Microsoft software on office computers, Symbian is looking for big-volume sales in the mass consumer markets where multimedia and video phones are becoming more popular with the advent of faster networks.

Symbian is currently the world's biggest producer of software for smartphones; Symbian phones have their own office applications and multitasking, open-source software. The first phones with Symbian Operating System version 9 were introduced in the United States in 2005.

Symbian OS

Application Software: Getting Started

What are five ways of obtaining application software, tools available to help me learn to use software, three common types of files, and the types of software?

At one time, just about everyone paid for microcomputer application software. You bought it as part of the computer or in a software store, or you downloaded it online with a credit card charge. Now, other ways exist to obtain software.

Application Software: For Sale, for Free, or for Rent?

What are the various ways I can obtain software?

Although most people pay for software, usually popular brands that they can use with similar programs owned by their friends and coworkers, it's possible to rent programs or get them free. (● *See Panel 3.19.*) Let's consider these categories.

COMMERCIAL SOFTWARE *Commercial software,* also called *proprietary software* or *packaged software,* is software that's offered for sale, such as Microsoft Word, Microsoft Office XP, or Adobe PhotoShop. Although such software may not show up on the bill of sale when you buy a new PC, you've paid for some of it as part of the purchase. And, most likely, whenever you order a new game or other commercial program, you'll have to pay for it. This software is copyrighted. A *copyright* is the exclusive legal right that prohibits copying of intellectual property without the permission of the copyright holder.

Software manufacturers don't sell you their software; rather, they sell you a license to become an authorized user of it. What's the difference? In paying for a ***software license*, you sign a contract in which you agree not to make copies of the software to give away or resell.** That is, you have bought only the company's permission to use the software and not the software itself. This legal nicety allows the company to retain its rights to the program and limits the way its customers can use it. The small print in the licensing agreement usually allows you to make one copy *(backup copy* or *archival copy)* for your own use. (Each software company has a different license; there is no industry standard.)

Several types of software licenses exist:

- *Site licenses* allow the software to be used on all computers at a specific location.

● PANEL 3.19
Choices among application software

Types	Definition
Commercial software	Copyrighted. If you don't pay for it, you can be prosecuted.
Public-domain software	Not copyrighted. You can copy it for free without fear of prosecution.
Shareware	Copyrighted. Available free, but you should pay to continue using it.
Freeware	Copyrighted. Available free.
Rentalware	Copyrighted. Lease for a fee.

- *Concurrent-use licenses* allow a certain number of copies of the software to be used at the same time.
- A *multiple-user license* specifies the number of people who may use the software.
- A *single-user license* limits software use to one user at a time.

Most personal computer software licenses allow you to run the program on only one machine and make copies of the software only for personal backup purposes. Personal computer users often buy their software in shrink-wrapped packages; once you have opened the shrink-wrap, you have accepted the terms of the software license.

Every year or so, software developers find ways to enhance their products and put forth new versions or new releases. A *version* is a major upgrade in a software product, traditionally indicated by numbers such as 1.0, 2.0, 3.0. More recently, other notations have been used. After 1995, for a while Microsoft labeled its Windows and Office software versions by year instead of by number, as in Microsoft's Office 97, Office 2000, and so forth. However, its latest software version is Office XP. A *release,* which now may be called an "add" or "addition," is a minor upgrade. Often this is indicated by a change in number after the decimal point. (For instance, 3.0 may become 3.1, 3.11, 3.2, and so on.) Some releases are now also indicated by the year in which they are marketed. And, unfortunately, some releases are not clearly indicated at all. (These are "patches," which may be downloaded from the software maker's website, as can version updates.)

PUBLIC-DOMAIN SOFTWARE *Public-domain software* **is not protected by copyright and thus may be duplicated by anyone at will.** Public-domain programs—sometimes developed at taxpayer expense by government agencies—have been donated to the public by their creators. They are often available through sites on the internet. You can download and duplicate public-domain software without fear of legal prosecution.

SHAREWARE *Shareware* **is copyrighted software that is distributed free of charge but requires that users make a monetary contribution, or pay a registration fee, to continue using it**—in other words, you can try it before you buy it. Once you pay the fee, you usually get supporting documentation, access to updated versions, and perhaps some technical support. Shareware is distributed primarily through the internet, but because it is copyrighted, you cannot use it to develop your own program that would compete with the original product. If you copy shareware and pass it along to friends, they are expected to pay the registration fee also, if they choose to use the software.

FREEWARE *Freeware* **is copyrighted software that is distributed free of charge,** today most often over the internet. Why would any software creator let his or her product go for free? Sometimes developers want to see how users respond, so that they can make improvements in a later version. Sometimes they want to further some scholarly or humanitarian purpose—for instance, to create a standard for software on which people are apt to agree. In its most recent form, freeware is made available by companies trying to make money some other way—actually, by attracting viewers to their advertising. (The web browsers Internet Explorer and Netscape Navigator are of this type.) Freeware developers generally retain all rights to their programs; technically, you are not supposed to duplicate and redistribute the programs. (Freeware is different from free software, or public-domain software, which has no restrictions on use, modification, or redistribution.)

info!

Shareware & Freeware

What kinds of shareware and freeware are available? To find out, go to *www.downloadalot.com, www.shareware.com, www.tucows.com,* and *www.sharewareking.com.*

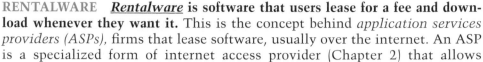

info!

More about ASPs

If you want to learn more about ASPs, go to *www.aspnews.com.*

RENTALWARE   *Rentalware* **is software that users lease for a fee and download whenever they want it.** This is the concept behind *application services providers (ASPs),* firms that lease software, usually over the internet. An ASP is a specialized form of internet access provider (Chapter 2) that allows users—usually companies—to have access to software stored on the ASP's servers and supplies support and other services. Sometimes leased applications and services can be furnished on-site. This information technology service reduces companies' expenditures and need to constantly update.

PIRATED SOFTWARE   *Pirated software* **is software obtained illegally,** as when you get a CD-ROM from a friend who has made an illicit copy of, say, a commercial video game. Sometimes pirated software can be downloaded off the internet. Sometimes it is sold in retail outlets in foreign countries. If you buy such software, not only do the original copyright owners not get paid for their creative work but you risk getting inferior goods and, worse, picking up a virus. To discourage software piracy, many software manufacturers, such as Microsoft, require that users register their software when they install it on their computers. If the software is not registered, it will not work properly.

ABANDONWARE   "Abandonware" does not refer to a way to obtain software. It refers to software that is no longer being sold or supported by its publisher. U.S. copyright laws state that copyrights owned by corporations are valid for up to 95 years from the date the software was first published. Copyrights are not considered abandoned even if they are no longer enforced. Therefore, abandoned software does not enter the public domain just because it is no longer supported. Don't copy it.

CUSTOM SOFTWARE   Occasionally, companies or individuals need software written specifically for them, to meet unique needs. This software is called *custom software,* and it's created by software engineers and programmers. We discuss programming in Chapter 10.

Tutorials & Documentation

How could software tutorials and documentation be helpful to me?

How are you going to learn a given software program? Most commercial packages come with tutorials and documentation.

TUTORIALS   A *tutorial* is an instruction book or program that helps you learn to use the product by taking you through a prescribed series of steps. For instance, our publisher offers several how-to books that enable you to learn different kinds of software. Tutorials may also form part of the software package.

DOCUMENTATION   *Documentation* is all information that describes a product to users, including a user guide or reference manual that provides a narrative and graphical description of a program. While documentation may be print-based, today it is usually available on CD, as well as via the internet. Documentation may be instructional, but features and functions are usually grouped by category for reference purposes. For example, in word processing documentation, all features related to printing are grouped together so that you can easily look them up.

A Few Facts about Files & the Usefulness of Importing & Exporting

What are three types of files, and what do importing and exporting mean?

There is only one reason for having application software: to take raw data and manipulate it into useful files of information. A *file,* as we said earlier, is (1) a named collection of data (data file) or (2) a program that exists in a computer's secondary storage (program file), such as floppy disk, hard disk, or CD/DVD.

THREE TYPES OF DATA FILES   Three well-known types of data files are as follows:

- **Document files:** Document files are created by word processing programs and consist of documents such as reports, letters, memos, and term papers.

- **Worksheet files:** Worksheet files are created by electronic spreadsheets and usually consist of collections of numerical data such as budgets, sales forecasts, and schedules.

- **Database files:** Database files are created by database management programs and consist of organized data that can be analyzed and displayed in various useful ways. Examples are student names and addresses that can be displayed according to age, grade-point average, or home state.

EXCHANGING FILES: IMPORTING & EXPORTING   It's useful to know that often files can be exchanged—that is, imported and exported—between programs.

- **Importing: *Importing* is defined as getting data from another source and then converting it into a format compatible with the program in which you are currently working.** For example, you might write a letter in your word processing program and include in it—that is, import—a column of numbers from your spreadsheet program. The ability to import data is very important in software applications because it means that one application can complement another.

- **Exporting: *Exporting* is defined as transforming data into a format that can be used in another program and then transmitting it.** For example, you might work up a list of names and addresses in your database program and then send it—export it—to a document you wrote in your word processing program. Exporting implies that the sending application reformats the data for the receiving application; importing implies that the receiving application does the reformatting.

Types of Application Software

What is productivity software?

Application software can be classified in many ways—for entertainment, personal, education/reference, productivity, and specialized uses. (● *See Panel 3.20.)*

In the rest of this chapter we will discuss types of ***productivity software—such as word processing programs, spreadsheets, and database managers—whose purpose is to make users more productive at particular tasks.*** Some productivity software comes in the form of an *office suite,* which bundles several applications together into a single large package. Microsoft Office, for example, includes (among other things) Word, Excel, and Access—word

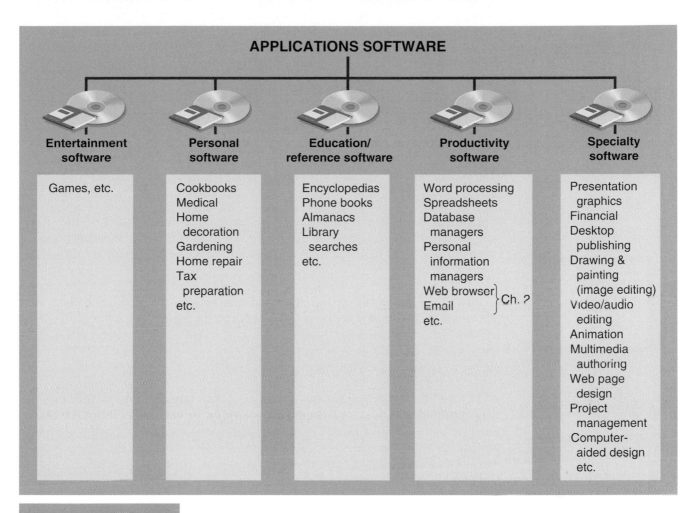

APPLICATIONS SOFTWARE

Entertainment software	Personal software	Education/ reference software	Productivity software	Specialty software

Games, etc.	Cookbooks
	Medical
	Home
	  decoration
	Gardening
	Home repair
	Tax
	  preparation
	etc.
		Encyclopedias
		Phone books
		Almanacs
		Library
		  searches
		etc.
			Word processing
			Spreadsheets
			Database
			  managers
			Personal
			  information
			  managers
			Web browser ⎱
			Email ⎰ Ch. ?
			etc.
				Presentation
				  graphics
				Financial
				Desktop
				  publishing
				Drawing &
				  painting
				  (image editing)
				Video/audio
				  editing
				Animation
				Multimedia
				  authoring
				Web page
				  design
				Project
				  management
				Computer-
				  aided design
				etc.

● PANEL 3.20
Types of application software

processing, spreadsheet, and database programs, respectively. Corel Corp. offers similar programs, such as the WordPerfect word processing program. Other productivity software, such as Lotus Notes, is sold as *groupware*—online software that allows several people to collaborate on the same project and share some resources.

We will now consider the three most important types of productivity software: word processing, spreadsheet, and database software (including personal information managers). We will then discuss more specialized software: presentation graphics, financial, desktop-publishing, drawing and painting, project management, computer-aided design, web page design, image/video/audio editing, and animation software.

3.7

Word Processing

What can I do with word processing software that I can't do with pencil and paper?

After a long and productive life, the typewriter has gone to its reward. Indeed, it is practically as difficult today to get a manual typewriter repaired as to find a blacksmith. Word processing software offers a much-improved way of dealing with documents.

**Word processing software** allows you to use computers to create, edit, format, print, and store text material, among other things. Word processing is the most common software application. The best-known word processing program is probably Microsoft Word, but there are others, such as Corel WordPerfect and the word processing components of Lotus Smart Suite 9.8 and Sun Microsystems' StarOffice 7. Word processing software allows users to work through a document and *delete, insert,* and *replace* text, the principal edit/correction activities. It also offers such additional features as *creating, formatting, printing,* and *saving.*

Creating Documents

What are word processing features I would use when creating a document?

Creating a document means entering text using the keyboard or the dictation function associated with speech-recognition software. Word processing software has three features that affect this process—the *cursor, scrolling,* and *word wrap.*

Cursor

Scrolling

CURSOR The __cursor__ **is the movable symbol on the display screen that shows you where you may next enter data or commands.** The symbol is often a blinking rectangle or an I-beam. You can move the cursor on the screen using the keyboard's directional arrow keys or a mouse. The point where the cursor is located is called the *insertion point.*

SCROLLING __Scrolling__ **means moving quickly upward, downward, or sideways through the text or other screen display.** A standard computer screen displays only 20–22 lines of standard-size text. Of course, most documents are longer than that. Using the directional arrow keys, or the mouse and a scroll bar located at the side of the screen, you can move ("scroll") through the display screen and into the text above and below it.

WORD WRAP __Word wrap__ **automatically continues text to the next line when you reach the right margin.** That is, the text "wraps around" to the next line. You don't have to hit a "carriage-return" key or Enter key, as was necessary with a typewriter.

SOME OTHER FEATURES To help you organize term papers and reports, the *Outline View* feature puts tags on various headings to show the hierarchy of heads—for example, main head, subhead, and sub-subhead. Word processing software also allows you to insert footnotes that are automatically numbered and renumbered when changes are made. The basics of word processing are shown in the accompanying illustration. (● *See Panel 3.21.*)

Editing Documents

What are various kinds of editing I can do in a word processing document?

Editing is the act of making alterations in the content of your document. Some features of editing are *insert* and *delete, undelete, find and replace, cut/copy and paste, spelling checker, grammar checker,* and *thesaurus.* Some of these commands are in the Edit pull-down menu and icons on the toolbar.

INSERT & DELETE *Inserting* is the act of adding to the document. Simply place the cursor wherever you want to add text and start typing; the existing characters will be pushed along. If you want to write over (replace) text as you write, press the *Insert* key before typing. When you're finished typing, press the *Insert* key again to exit Insert mode.

Deleting is the act of removing text, usually using the *Delete* or *Backspace* key.

Toolbar:
Allows quick access to frequently used commands

Menu bar:
Allows access to all commands

Title bar:
Shows name of document you're working on

Spelling and Grammar button:
Click on to check for misspelled words and incorrect grammar.

Text alignment buttons:
Click on to align text to be left, center, right, or full justified.

Style button:
Click on to access variety of format styles.

Ruler:
Shows tabs and margins

Insertion point:
Blinking symbol shows where the next character you type will appear

Status bar:
Shows details about the document you're working on

Window controls:
Let you enlarge a window, restore its previous position, or hide it from view

Mouse pointer:
Use the mouse to move the insertion point, to click on icons, or to select text for editing.

Scroll bars:
Let you scroll the document to reveal hidden portions

Taskbar

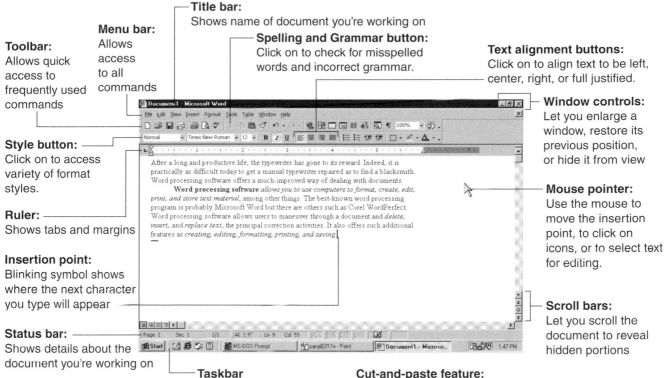

Outline feature:
Enables you to view headings in your document

Cut-and-paste feature:
Enables you to move blocks of text. First highlight the text. On Edit menu, select Cut option. Then, on Edit menu, select Paste option. (You can also use the icons in the toolbar.)

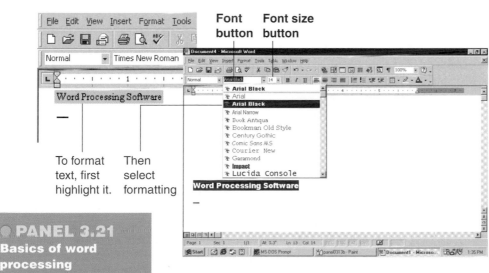

Font button

Font size button

To format text, first highlight it.

Then select formatting

Formatting feature:
Enables you to change type font and size. First highlight the text. Then click next to Font button for pull-down menu of fonts. Click next to Font Size button for menu of type sizes.

● **PANEL 3.21**
Basics of word processing

A character, word, or image that appears faintly in the background of a printed document is called a *watermark*. To add one to your Word document, select *Format, Background* and *Printed Watermark*. Then select *Picture Watermark* or *Text Watermark* and provide the requested information.

The *Undo command* allows you to change your mind and restore text that you have deleted. Some word processing programs offer as many as 100 layers of "undo," so that users who delete several paragraphs of text, but then change their minds, can reinstate the material.

FIND & REPLACE The *Find*, or *Search, command* allows you to find any word, phrase, or number that exists in your document. The *Replace command* allows you to automatically replace it with something else.

CUT/COPY & PASTE Typewriter users who wanted to move a paragraph or block of text from one place to another in a manuscript used scissors and glue to "cut and paste." With word processing, moving text takes only a few keystrokes. You select (highlight with the mouse) the portion of text you want to copy or move. Then you use the *Copy* or *Cut command* to move it to the *clipboard,* a special holding area in the computer's memory. From there, you use *Paste* to transfer the material to any point (indicated with the cursor) in the existing document or in a new document. The clipboard retains its material, so repeated pastes of the same item will work without your having to recopy each time.

SPELLING CHECKER Most word processors have a *spelling checker,* which tests for incorrectly spelled words. As you type, the spelling checker indicates (perhaps with a squiggly line) words that aren't in its dictionary and thus may be misspelled. (● *See Panel 3.22.*) Special add-on dictionaries are available for medical, engineering, and legal terms.

In addition, programs such as Microsoft Word have an Auto Correct function that automatically fixes such common mistakes as transposed letters—replacing "teh" with "the," for instance.

GRAMMAR CHECKER A *grammar checker* highlights poor grammar, wordiness, incomplete sentences, and awkward phrases. The grammar checker won't fix things automatically, but it will flag (perhaps with a different-color squiggly line) possible incorrect word usage and sentence structure. (● *See Panel 3.23.*)

● **PANEL 3.22**
Spelling checker
How a word processing program checks for misspelled words and offers alternatives

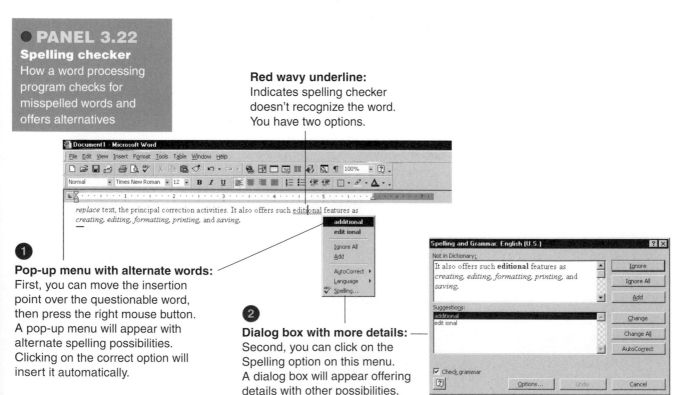

Red wavy underline:
Indicates spelling checker doesn't recognize the word. You have two options.

❶ **Pop-up menu with alternate words:**
First, you can move the insertion point over the questionable word, then press the right mouse button. A pop-up menu will appear with alternate spelling possibilities. Clicking on the correct option will insert it automatically.

❷ **Dialog box with more details:**
Second, you can click on the Spelling option on this menu. A dialog box will appear offering details with other possibilities.

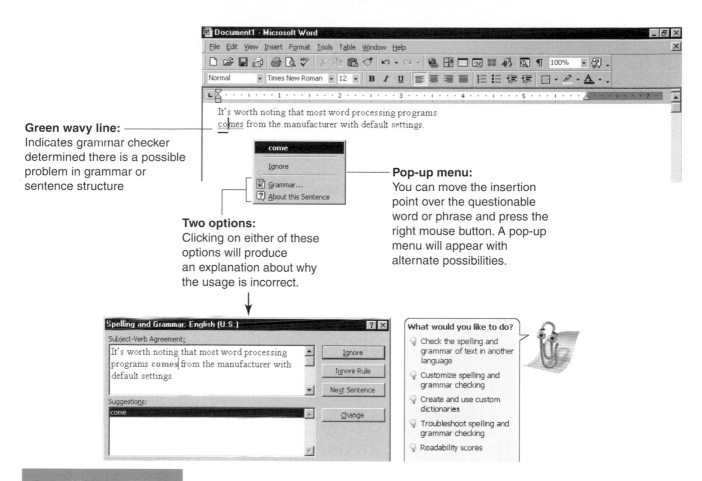

Green wavy line: Indicates grammar checker determined there is a possible problem in grammar or sentence structure

Pop-up menu: You can move the insertion point over the questionable word or phrase and press the right mouse button. A pop-up menu will appear with alternate possibilities.

Two options: Clicking on either of these options will produce an explanation about why the usage is incorrect.

● PANEL 3.23
Grammar checker
This program points out possible errors in sentence structure and word usage and suggests alternatives.

THESAURUS If you find yourself stuck for the right word while you're writing, you can call up an on-screen *thesaurus,* which will present you with the appropriate word or alternative words.

Formatting Documents with the Help of Templates & Wizards

What are some types of formatting I can do?

In the context of word processing, *formatting* means determining the appearance of a document. To this end, word processing programs provide two helpful devices—templates and wizards. **A _template_ is a preformatted document that provides basic tools for shaping a final document**—the text, layout, and style for a letter, for example. Simply put, it is a style guide for documents. Because most documents are fairly standard in format, every word processing program comes with at least a few standard templates. When you use a template, you're actually opening a copy of the template. In this way you'll always have a fresh copy of the original template when you need it. After you open a copy of the template and add your text, you save this version of the template under the filename of your choice. In this way, for example, in a letterhead template, your project's name, address, phone number, and web address are included every time you open your letterhead template file.

A _wizard_ is an interactive computer utility program that acts as an interface to lead the user through a task by using dialog steps to ask questions. For example, a "letter wizard" within a word processing application can lead you through the steps of producing different types of correspondence. In Word, you can use the Memo wizard to create preformatted, professional-looking memos or the Résumé wizard to create a résumé.

Among the many aspects of formatting are the following:

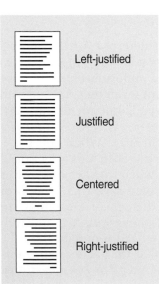

10 point
Times Roman

**14 point
Arial Black**

16 point
Courier

60

(60 point Arial)

Left-justified

Justified

Centered

Right-justified

FONT You can decide what *font*—typeface and type size—you wish to use. For instance, you can specify whether it should be Arial, Courier, or *Freestyle Script*. You can indicate whether the text should be, say, 10 points or 12 points in size and the headings should be 14 points or 16 points. (There are 72 points in an inch.) You can specify what parts should be underlined, *italic*, or **boldface.**

SPACING & COLUMNS You can choose whether you want the lines to be *single-spaced* or *double-spaced* (or something else). You can specify whether you want text to be *one column* (like this page), *two columns* (like many magazines and books), or *several columns* (like newspapers).

MARGINS & JUSTIFICATION You can indicate the dimensions of the margins—left, right, top, and bottom—around the text. You can specify the text *justification*—how the letters and words are spaced in each line. To *justify* means to align text evenly between left and right margins, as in most newspaper columns and the preceding two paragraphs. To *left-justify* means to align text evenly on the left. (Left-justified text has a "ragged-right" margin, as do many business letters and this paragraph.) *Centering* centers each text line in the available white space between the left and right margins.

HEADERS, FOOTERS, & PAGE NUMBERS You can indicate headers or footers and include page numbers. A *header* is common text (such as a date or document name) printed at the top of every page. A *footer* is the same thing printed at the bottom of every page. If you want page numbers, you can determine what number to start with, among other things.

OTHER FORMATTING You can specify *borders* or other decorative lines, *shading, tables,* and *footnotes.* You can even import *graphics* or drawings from files in other software programs, including *clip art*—collections of ready-made pictures and illustrations available online or on CDs/DVDs.

DEFAULT SETTINGS It's worth noting that word processing programs (and indeed most forms of application software) come from the manufacturer with

When Several Word
Documents Are Open

You can write with several Word
documents open simultaneously.
To go ("toggle") back and forth,
hold down *Ctrl* and press *F6*. To
go backward, press *Ctrl, Shift*,
and press *F6*. To display several
documents at once, go to the
Window menu and select
Arrange All. You can cut and
paste text from one document
to another.

default settings. **_Default settings_ are the settings automatically used by a program unless the user specifies otherwise, thereby overriding them.** Thus, for example, a word processing program may automatically prepare a document single-spaced, left-justified, with 1-inch right and left margins, unless you alter these default settings.

Output Options: Printing, Faxing, or Emailing Documents

What are some of my output options?

Most word processing software gives you several options for printing. For example, you can print *several copies* of a document. You can print *individual pages* or a *range of pages*. You can even preview a document before printing it out. *Previewing (print previewing)* means viewing a document on-screen to see what it will look like in printed form before it's printed. Whole pages are displayed in reduced size.

You can also send your document off to someone else by fax or email attachment if your computer has the appropriate communications link.

Saving Documents

How can I save a document?

Saving means storing, or preserving, a document as an electronic file permanently—on floppy disk, hard disk, or CD, for example. Saving is a feature of nearly all application software. Having the document stored in electronic form spares you the tiresome chore of retyping it from scratch whenever you want to make changes. You need only retrieve it from the storage medium and make the changes you want. Then you can print it out again. (You should save your documents often while you are working; don't wait until the document is finished.)

Tracking Changes & Inserting Comments

How do I and any cowriters make changes visible in a document?

Four score and Eighty-seven
years ago, our fathers and
mothers brought forth on this
continent a new nation

What if you have written an important document and have asked other people to edit it? Word processing software allows editing changes to be tracked by highlighting them, underlining additions, and crossing out deletions. Each person working on the document can choose a different color so that you can tell who's done what. And anyone can insert hidden questions or comments that become visible when you pass the mouse pointer over yellow-highlighted words or punctuation. An edited document can be printed out showing all the changes, as well as a list of comments keyed to the text by numbers. Or it can be printed out "clean," showing the edited text in its new form, without the changes.

Web Document Creation

How do I format a document to put it on the web?

Most word processing programs allow you to automatically format your documents into HTML so that they can be used on the web. To do this in Microsoft Word, open *File, Save As, Save As Type: Web page (*.htm, *.html)*.

Spreadsheets

What can I do with an electronic spreadsheet that I can't do with pencil and paper and a standard calculator?

What is a spreadsheet? Traditionally, it was simply a grid of rows and columns, printed on special light-green paper, that was used to produce financial projections and reports. A person making up a spreadsheet spent long days and weekends at the office penciling tiny numbers into countless tiny rectangles. When one figure changed, all other numbers on the spreadsheet had to be erased and recomputed. Ultimately, there might be wastebaskets full of jettisoned worksheets.

In 1978, Daniel Bricklin was a student at the Harvard Business School. One day he was staring at columns of numbers on a blackboard when he got the idea for computerizing the spreadsheet. He created the first electronic spreadsheet, now called simply a spreadsheet. **The _spreadsheet_ allows users to create tables and financial schedules by entering data and formulas into rows and columns arranged as a grid on a display screen.** Before long, the electronic spreadsheet was the most popular small business program. Unfortunately for Bricklin, his version (called VisiCalc) was quickly surpassed by others. Today the principal spreadsheets are Microsoft Excel, Corel Quattro Pro, and Lotus 1-2-3. Spreadsheets are used for maintaining student grade books, tracking investments, creating and tracking budgets, calculating loan payments, estimating project costs, and creating other types of financial reports.

The Basics: How Spreadsheets Work

What are the basic principles involved with manipulating a spreadsheet?

A spreadsheet is arranged as follows. (● *See Panel 3.24.*)

HOW A SPREADSHEET IS ORGANIZED   A spreadsheet's arrangement of columns, rows, and labels is called a *worksheet.*

- **Column headings:** In the worksheet's frame area (work area), lettered *column headings* appear across the top ("A" is the name of the first column, "B" the second, and so on).
- **Row headings:** Numbered *row headings* appear down the left side ("1" is the name of the first row, "2" the second, and so forth).
- **Labels:** *Labels* are any descriptive text that identifies categories, such as APRIL, RENT, or GROSS SALES.

You use your computer's keyboard to type in the various headings and labels. Each worksheet has 256 columns and 65,536 rows, and each spreadsheet file holds up to 255 related worksheets.

CELLS: WHERE COLUMNS & ROWS MEET   Each worksheet has more than 16 million cells.

- **Cells & cell addresses: A _cell_ is the place where a row and a column intersect; its position is called a _cell address._** For example, "A1" is the cell address for the top left cell, where column A and row 1 intersect.
- **Ranges: A _range_ is a group of adjacent cells**—for example, A1 to A5.

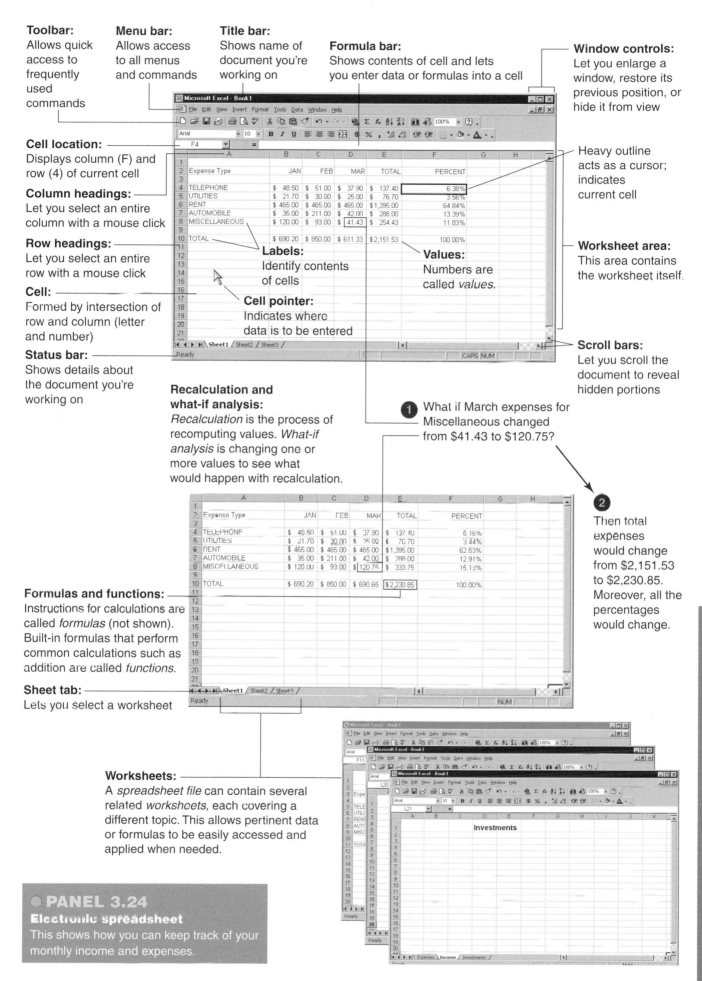

Toolbar: Allows quick access to frequently used commands

Menu bar: Allows access to all menus and commands

Title bar: Shows name of document you're working on

Formula bar: Shows contents of cell and lets you enter data or formulas into a cell

Window controls: Let you enlarge a window, restore its previous position, or hide it from view

Cell location: Displays column (F) and row (4) of current cell

Column headings: Let you select an entire column with a mouse click

Row headings: Let you select an entire row with a mouse click

Cell: Formed by intersection of row and column (letter and number)

Status bar: Shows details about the document you're working on

Labels: Identify contents of cells

Cell pointer: Indicates where data is to be entered

Values: Numbers are called *values.*

Heavy outline acts as a cursor; indicates current cell

Worksheet area: This area contains the worksheet itself.

Scroll bars: Let you scroll the document to reveal hidden portions

Recalculation and what-if analysis: *Recalculation* is the process of recomputing values. *What-if analysis* is changing one or more values to see what would happen with recalculation.

① What if March expenses for Miscellaneous changed from $41.43 to $120.75?

② Then total expenses would change from $2,151.53 to $2,230.85. Moreover, all the percentages would change.

Formulas and functions: Instructions for calculations are called *formulas* (not shown). Built-in formulas that perform common calculations such as addition are called *functions.*

Sheet tab: Lets you select a worksheet

Worksheets: A *spreadsheet file* can contain several related *worksheets,* each covering a different topic. This allows pertinent data or formulas to be easily accessed and applied when needed.

● PANEL 3.24

Electronic spreadsheet

This shows how you can keep track of your monthly income and expenses.

Software

- **Values: A number or date entered in a cell is called a _value_.** The values are the actual numbers used in the spreadsheet—dollars, percentages, grade points, temperatures, or whatever. Headings, labels, and formulas also go into cells.

- **Cell pointer:** A *cell pointer*, or *spreadsheet cursor*, indicates where data is to be entered. The cell pointer can be moved around like a cursor in a word processing program.

FORMULAS, FUNCTIONS, RECALCULATION, & WHAT-IF ANALYSIS

Why has the spreadsheet become so popular? The reasons lie in the features known as formulas, functions, recalculation, and what-if analysis.

- **Formulas: _Formulas_ are instructions for calculations; they define how one cell relates to other cells.** For example, a formula might be =SUM(A5:A15) or @SUM(A5:A15), meaning "Sum (that is, add) all the numbers in the cells with cell addresses A5 through A15."

- **Functions: _Functions_ are built-in formulas that perform common calculations.** For instance, a function might average a range of numbers or round off a number to two decimal places.

- **Recalculation:** After the values have been entered into the worksheet, the formulas and functions can be used to calculate outcomes. However, what was revolutionary about the electronic spreadsheet was its ability to easily do recalculation. **_Recalculation_ is the process of recomputing values,** either as an ongoing process as data is entered or afterward, with the press of a key. With this simple feature, the hours of mind-numbing work required to manually rework paper spreadsheets has become a thing of the past.

- **What-if analysis:** The recalculation feature has opened up whole new possibilities for decision making. In particular, **_what-if analysis_ allows the user to see how changing one or more numbers changes the outcome of the calculation.** That is, you can create a worksheet, putting in formulas and numbers, and then ask, "What would happen if we change that detail?"—and immediately see the effect on the bottom line.

Mooving data. A dairy farmer enters feeding data into a spreadsheet.

WORKSHEET TEMPLATES You may find that your spreadsheet software makes worksheet templates available for specific tasks. *Worksheet templates* are forms containing formats and formulas custom-designed for particular kinds of work. Examples are templates for calculating loan payments, tracking travel expenses, monitoring personal budgets, and keeping track of time worked on projects. Templates are also available for a variety of business needs—providing sales quotations, invoicing customers, creating purchase orders, and writing a business plan.

MULTIDIMENSIONAL SPREADSHEETS Most spreadsheet applications are *multidimensional*, meaning that you can link one spreadsheet to another. A three-dimensional spreadsheet, for example, is like a stack of spreadsheets all connected by formulas. A change made in one spreadsheet automatically affects the other spreadsheets.

Analytical Graphics: Creating Charts

What are analytical graphics?

You can use spreadsheet packages to create analytical graphics, or charts. **_Analytical graphics_, or business graphics, are graphical forms**

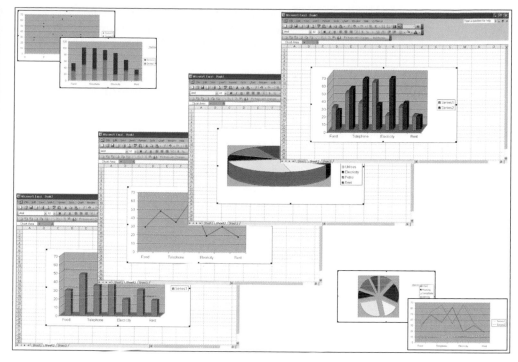

that make numeric data easier to analyze than it is when organized as rows and columns of numbers. Whether viewed on a monitor or printed out, analytical graphics help make sales figures, economic trends, and the like easier to comprehend and visualize. In Excel, you enter your data to the worksheet, select the data, and use the Chart wizard to step through the process of choosing the chart type and various options.

Examples of analytical graphics are *column charts, bar charts, line graphs, pie charts,* and *scatter charts.* (● *See Panel 3.25.)* If you have a color printer, these charts can appear in color. In addition, they can be displayed or printed out so that they look three-dimensional. Spreadsheets can even be linked to more exciting graphics, such as digitized maps.

3.9

Database Software

What is database software, and what is personal information management software?

In its most general sense, a database is any electronically stored collection of data in a computer system. In its more specific sense, a ___database___ **is a collection of interrelated files in a computer system.** These computer-based files are organized according to their common elements, so that they can be retrieved easily. (Databases are covered in detail in Chapter 8.) Sometimes called a *database manager* or *database management system (DBMS),* ___database software___ **is a program that sets up and controls the structure of a database and access to the data.**

The Benefits of Database Software

What are two advantages of database software over the old ways of organizing files?

When data is stored in separate files, the same data will be repeated in many files. In the old days, each college administrative office—registrar, financial aid, housing, and so on—might have a separate file on you. Thus, there was *redundancy*—your address, for example, was repeated over and over. This means that when you changed addresses, all the college's files on you had to be updated separately. Thus, database software has two advantages.

Software

159

INTEGRATION With database software, the data is not in separate files. Rather, it is *integrated*. Thus, your address need only be listed once, and all the separate administrative offices will have access to the same information.

INTEGRITY For that reason, information in databases is considered to have more *integrity*. That is, the information is more likely to be accurate and up to date.

Databases are a lot more interesting than they used to be. Once they included only text. Now they can also include pictures, sound, and animation. It's likely, for instance, that your personnel record in a future company database will include a picture of you and perhaps even a clip of your voice. If you go looking for a house to buy, you will be able to view a real estate agent's database of video clips of homes and properties without leaving the realtor's office.

Today the principal microcomputer database programs are Microsoft Access and FileMaker Pro. (In larger systems, Oracle is a major player.)

The Basics: How Databases Work

What are the basic principles involved with manipulating a database?

Let's consider some basic features of databases:

HOW A RELATIONAL DATABASE IS ORGANIZED: TABLES, RECORDS, & FIELDS The most widely used form of database, especially on PCs, is the **<u>*relational database*</u>, in which data is organized into related tables.** Each table contains rows and columns; the rows are called *records*, and the columns are called *fields*. An example of a record is a person's address—name, street address, city, and so on. An example of a field is that person's last name; another field would be that person's first name; a third field would be that person's street address; and so on. (● *See Panel 3.26.*)

Just as a spreadsheet may include a workbook with several worksheets, so a relational database might include a database with several tables. For instance, if you're running a small company, you might have one database headed *Employees*, containing three tables—*Addresses*, *Payroll*, and *Benefits*. You might have another database headed *Customers*, with *Addresses*, *Orders*, and *Invoices* tables.

LINKING RECORDS, USING A KEY In relational databases a **<u>*key*</u>—also called *key field, sort key, index,* or *keyword*—is a field used to sort data.** For example, if you sort records by age, then the age field is a key. The most frequent key field used in the United States is the Social Security number, but any unique identifier, such as employee number or student number, can be used. Most database management systems allow you to have more than one key so that you can sort records in different ways. One of the keys is designated the *primary key* and must hold a unique value for each record. A key field that identifies records in different tables is called a *foreign key*. Foreign keys are used to cross-reference data among relational tables.

FINDING WHAT YOU WANT: QUERYING & DISPLAYING RECORDS The beauty of database software is that you can locate records quickly. For example, several offices at your college may need access to your records but for different reasons: registrar, financial aid, student housing, and so on. Any of these offices can *query records*—locate and display records—by calling them up on a computer screen for viewing and updating. Thus, if you move, your address field will need to be corrected for all relevant offices of the college. A person making a search might make the query, *"Display the address*

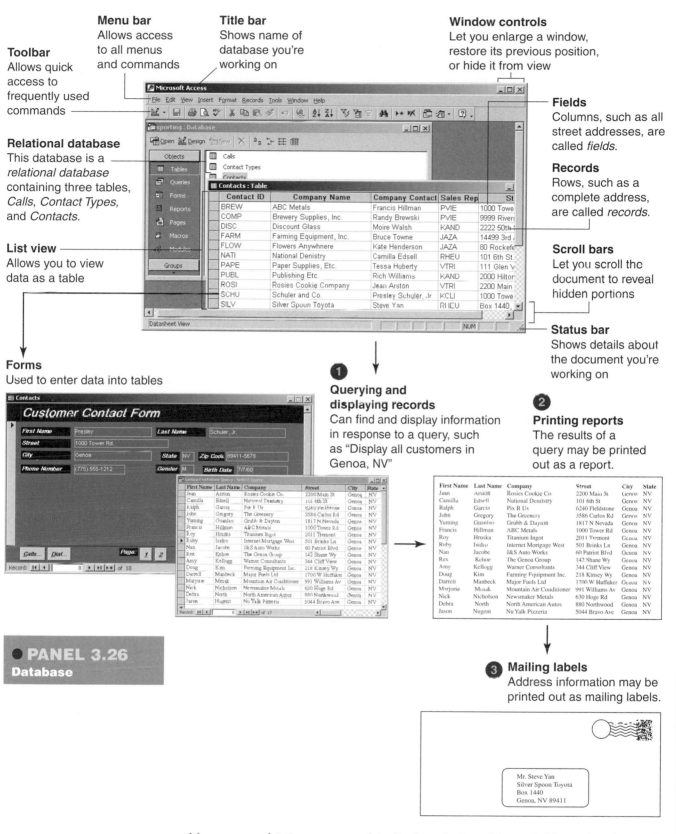

Toolbar
Allows quick access to frequently used commands

Menu bar
Allows access to all menus and commands

Title bar
Shows name of database you're working on

Window controls
Let you enlarge a window, restore its previous position, or hide it from view

Relational database
This database is a *relational database* containing three tables, *Calls*, *Contact Types*, and *Contacts*.

List view
Allows you to view data as a table

Fields
Columns, such as all street addresses, are called *fields*.

Records
Rows, such as a complete address, are called *records*.

Scroll bars
Let you scroll the document to reveal hidden portions

Status bar
Shows details about the document you're working on

Forms
Used to enter data into tables

① Querying and displaying records
Can find and display information in response to a query, such as "Display all customers in Genoa, NV"

② Printing reports
The results of a query may be printed out as a report.

③ Mailing labels
Address information may be printed out as mailing labels.

● **PANEL 3.26**
Database

of *[your name]*." Once a record is displayed, the address field can be changed. Thereafter, any office calling up your file will see the new address.

SORTING & ANALYZING RECORDS & APPLYING FORMULAS With database software you can easily find and change the order of records in a table—in other words, they can be *sorted* in different ways—arranged alphabetically, numerically, geographically, or in some other order. For example, they can be rearranged by state, by age, or by Social Security number.

Software

In addition, database programs contain built-in mathematical *formulas* so that you can analyze data. This feature can be used, for example, to find the grade-point averages for students in different majors or in different classes.

PUTTING SEARCH RESULTS TO USE: SAVING, FORMATTING, PRINTING, COPYING, OR TRANSMITTING Once you've queried, sorted, and analyzed the records and fields, you can simply save them to your hard disk, floppy disk, or CD. You can format them in different ways, altering headings and typestyles. You can print them out on paper as reports, such as an employee list with up-to-date addresses and phone numbers. A common use is to print out the results as names and addresses on *mailing labels*—adhesive-backed stickers that can be run through your printer and then stuck on envelopes. You can use the Copy command to copy your search results and then paste them into a paper produced on your word processor. You can also cut and paste data into an email message or make the data an attachment file to an email, so that it can be transmitted to someone else.

Personal Information Managers

How could a personal information manager be valuable to me?

Pretend you are sitting at a desk in an old-fashioned office. You have a calendar, a Rolodex-type address file, and a notepad. Most of these items could also be found on a student's desk. How would a computer and software improve on this arrangement?

Many people find ready uses for specialized types of database software known as personal information managers. **A *personal information manager (PIM)* is software that helps you keep track of and manage information you use on a daily basis, such as addresses, telephone numbers, appointments, to-do lists, and miscellaneous notes.** Some programs feature phone dialers, outliners (for roughing out ideas in outline form), and ticklers (or reminders). With a PIM, you can key in notes in any way you like and then retrieve them later based on any of the words you typed.

Popular PIMs are Microsoft Outlook, Lotus SmartSuite Organizer, and Microsoft Scheduler. Microsoft Outlook, for example, has sections such as Inbox, Calendar, Contacts, Tasks (to-do list), Journal (to record interactions with people), Notes (scratchpad), and Files. (● *See Panel 3.27.*) Other PIM programs are AZZ Cardfile, Day-Timer, and iSBiSTER 5.5.0 International Time & Chaos.

Timeline
Developments in software

3000 BCE	1621 CE	1642	1801	1820	1833
Abacus is invented in Babylonia	Slide rule invented (Edmund Gunther)	First mechanical adding machine (Blaise Pascal)	A linked sequence of punched cards controls the weaving patterns in Jacquard's loom	The first mass-produced calculator, the Thomas Arithnometer	Babbage's difference engine (automatic calculator)

● PANEL 3.27
Personal information manager
This shows the calendar available with Microsoft Outlook.

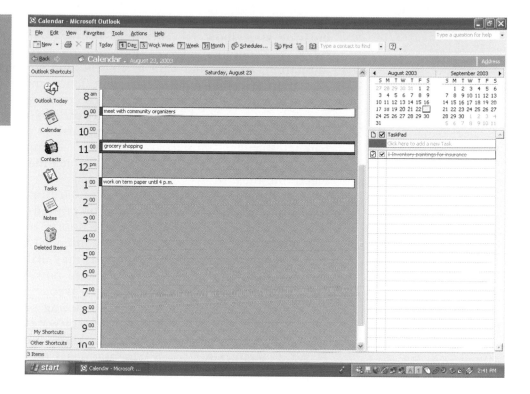

3.10

Specialty Software

What are the principal uses of specialty software?

After learning some of the productivity software just described, you may wish to become familiar with more specialized programs. For example, you might first learn word processing and then move on to desktop publishing, or first learn spreadsheets and then learn personal-finance software. We will consider the following kinds of software, although they are but a handful of the thousands of specialized programs available: *presentation graphics, financial, desktop-publishing, drawing and painting, project management, computer-aided design, video/audio editing, animation,* and *web page design software.*

Presentation Graphics Software

How could I do a visual presentation, using presentation graphics software?

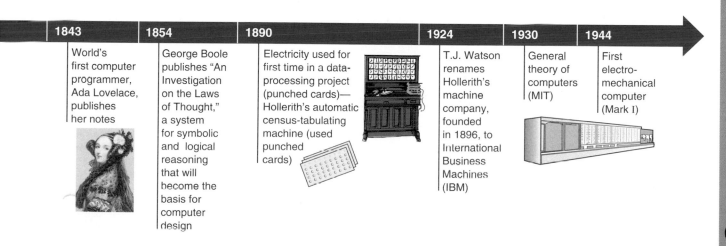

1843	1854	1890	1924	1930	1944
World's first computer programmer, Ada Lovelace, publishes her notes	George Boole publishes "An Investigation on the Laws of Thought," a system for symbolic and logical reasoning that will become the basis for computer design	Electricity used for first time in a data-processing project (punched cards)—Hollerith's automatic census-tabulating machine (used punched cards)	T.J. Watson renames Hollerith's machine company, founded in 1896, to International Business Machines (IBM)	General theory of computers (MIT)	First electro-mechanical computer (Mark I)

Presentation graphics software is intended primarily for the business user, for creating slide-show presentations, overhead transparencies, reports, portfolios, and training materials. ***Presentation graphics software* uses graphics, animation, sound, and data or information to make visual presentations.** Presentation graphics are much more fancy and complicated than are analytical graphics. Pages in presentation software are often referred to as *slides*, and visual presentations are commonly called *slide shows*. They can consist, however, not only of 35-mm slides but also of paper copies, overhead transparencies, video, animation, and sound. Completed presentations are frequently published in multiple formats which may include print, the web, and electronic files.

Most often, presentation projects are used in live sessions. They are commonly projected onto large screens or printed on overhead transparencies. The slides may be distributed in printed form as handouts to accompany the live presentation. Slides are generally intended to be followed in an ordered sequence, although some presentations may utilize interactive forms of navigation. More and more of this software now has the ability to export to HTML for posting presentations on the web.

You may already be accustomed to seeing presentation graphics because many college instructors now use such software to accompany their lectures. Well-known presentation graphics packages include Microsoft PowerPoint, Harvard Graphics Advanced Presentations, Corel Presentations (as part of the WordPerfect Office Suite), Open Office IMPRESS, and Serious Magic's Visual Communicator Pro. (● *See Panel 3.28.*)

Presentation graphics packages often come with slide sorters, which group together a dozen or so slides in miniature. The person making the presentation can use a mouse or a keyboard to bring the slides up for viewing or even start a self-running electronic slide show. You can also use a projection system from the computer itself.

Let's examine the process of using presentation software:

USING TEMPLATES TO GET STARTED Just as word processing programs offer templates for faxes, business letters, and the like, presentation graphics programs offer templates to help you organize your presentation, whether it's for a roomful of people or over the internet. Templates are of two types: design and content.

- **Design templates:** These offer formats, layouts, background patterns, and color schemes that can apply to general forms of content material.
- **Content templates:** These offer formats for specific subjects. For instance, PowerPoint offers templates for "Selling Your Ideas," "Facilitating a Meeting," and "Motivating a Team."

1946	1967	1969–1971	1970
First programmable electronic computer in United States (ENIAC)	A graphical user interface (GUI) is a main theme of Jeff Raskin, who later became an Apple Macintosh team leader; handheld calculator	Unix is developed and released by Bell Laboratories	Microprocessor chips come into use; floppy disk introduced for storing data

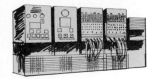

1 Outline View
This view helps you organize the content of your material in standard outline form.

2 Dressing up your presentation
PowerPoint offers professional design templates of text format, background, and borders. You place your text for each slide into one of these templates. You can also import a graphic from the clip art that comes with the program.

3 Slide View
This view allows you to see what a single slide will look like. You can use this view to edit the content and looks of each slide.

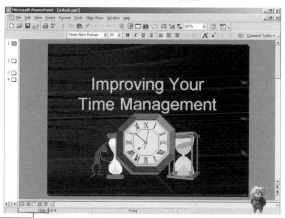

4 Notes Page View
This view displays a small version of the slide plus the notes you will be using as speaker notes.

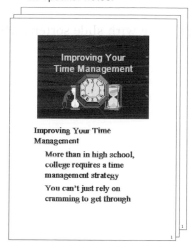

View icons
Clicking on these offers different views: *Slide, Outline, Slide Sorter, Notes Page,* and *Slide Show*

5 Slide Sorter View
This view displays miniatures of each slide, enabling you to adjust the order of your presentation.

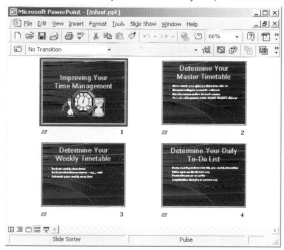

● **PANEL 3.28**
Presentation graphics
Microsoft PowerPoint helps you prepare and make visual presentations.

1973	1975	1976	1977
Xerox PARC develops an experimental PC that uses a mouse and a GUI	Bill Gates and Paul Allen start Microsoft in Albuquerque, N.M. (move to Seattle in 1979); first microcomputer (MITS Altair 0000)	Apple I computer (first personal computer sold in assembled form)	Apple II's floppy disk drive leads to writing of many software programs

Software

The software offers wizards that walk you through the process of filling in the template.

GETTING ASSISTANCE ON CONTENT DEVELOPMENT & ORGANIZATION To provide assistance as you're building your presentation, PowerPoint displays three windows on your screen at the same time—*Outline View, Slide View,* and *Notes Page View.* This enables you to add new slides, create and edit the text on the slides, and create notes (to use as lecture or speech notes) while developing your presentation.

- **Outline View:** This helps you organize the content of your material in standard outline form. The text you enter into the outline is automatically formatted into slides according to the template you selected. If you wish, you can pull in (import) your outline from a word processing document.
- **Slide View:** This helps you see what a single slide will look like. The outline text appears as slide titles and subtitles in subordinate order.
- **Notes Page View:** This displays the notes you will be using as speaker notes. It includes a small version of the slide.

Two other views are helpful in organizing and practicing. *Slide Sorter View* allows you to view a number of slides (4 to 12 or more) at once, so you can see how to order and reorder them. *Slide Show View* presents the slides in the order in which your audience will view them, so you can practice your presentation.

DRESSING UP YOUR PRESENTATION Presentation software makes it easy to dress up each visual page ("slide") with artwork by pulling in ("dragging and dropping") clip art from other sources. Although presentations may make use of some basic analytical graphics—bar, line, and pie charts—they generally use much more sophisticated elements. For instance, they may display different textures (speckled, solid, cross-hatched), color, and three-dimensionality. In addition, you can add sound clips, special visual effects (such as blinking text), animation, and video clips. (You can, in fact, drag and drop art and other enhancements into desktop-publishing, word processing or other standard PC applications.)

Financial Software

What are the features of financial software that could be useful to me?

Financial software is a growing category that ranges from personal-finance managers to entry-level accounting programs to business financial-management packages.

more info!

Adding Clip Art

For information on adding clip art to presentation slides, check out *www. communicateusingtechnology .com/articles/using_clip_art_ photo.htm.*
    For information on obtaining clip art, see the following three websites: *www.clipartinc.com, www.bizart.com/links-sources .html, http://dir.yahoo.com/ Business_and_Economy/ Business_to_Business/ Computers/Software/Graphics/ Clip_Art.*

1978	1980	1981	1982	1983	1984
The first electronic spreadsheet, VisiCalc, is introduced; WordStar, the first commercial word processor for consumers, is introduced	Microsoft obtains DOS version that becomes PC-DOS for IBM PC.	Xerox introduces mouse-operated icons, buttons, and menus on the Star computer; IBM introduces personal computer (IBM PC)	Portable computers	Bill Gates announces the first version of the Windows operating system (and releases it two years later)	Apple Macintosh; first personal laser printer; the Apple Macintosh introduces the first widely used GUI; Mac System 1.0 is introduced

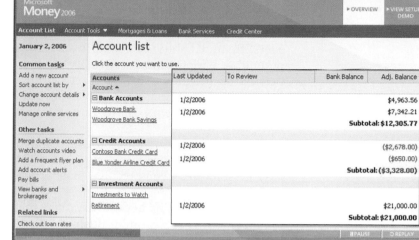

Consider the first of these, which you may find particularly useful. ***Personal-finance managers* let you keep track of income and expenses, write checks, do online banking, and plan financial goals.** (● *See Panel 3.29.)* Such programs don't promise to make you rich, but they can help you manage your money. They may even get you out of trouble. Many personal-finance programs, such as Quicken and Microsoft Money, include a calendar and a calculator.

FEATURES OF FINANCIAL SOFTWARE   The principal features are the following:

- **Tracking of income and expenses:** The programs allow you to set up various account categories for recording income and expenses, including credit card expenses.

- **Checkbook management:** All programs feature checkbook management, with an on-screen check writing form and check register that look like the ones in your checkbook. Checks can be purchased to use with your computer printer.

- **Reporting:** All programs compare your actual expenses with your budgeted expenses. Some will compare this year's expenses to last year's.

- **Income tax:** All programs offer tax categories, for indicating types of income and expenses that are important when you're filing your tax return.

1985	1986	1987	1988	1990	1991
Aldus PageMaker becomes the first integrated desktop publishing program; Microsoft Windows 1.0 is released; Mac. System 2.0	Mac System 3.0	Microsoft's Excel program introduced; Mac system 4.0, then 5.0	Mac System 6.0	Microsoft introduces Windows 3.0 in May, intensifying its legal dispute with Apple over the software's "look and feel" resemblance to the Macintosh operating system	Linus Torvalds introduces Linux; Mac System 7.0

Software

- **Other:** Some of the more versatile personal-finance programs also offer financial-planning and portfolio-management features.

GOING BEYOND PERSONAL FINANCE   Besides personal-finance managers, financial software includes small business accounting and tax software programs, which provide virtually all the forms you need for filing income taxes. Tax programs such as TaxCut and TurboTax make complex calculations, check for mistakes, and even unearth deductions you didn't know existed. Tax programs can be linked to personal-finance software to form an integrated tool.

Many financial software programs may be used in all kinds of enterprises. For instance, accounting software automates bookkeeping tasks, while payroll software keeps records of employee hours and produces reports for tax purposes.

Some programs go beyond financial management and tax and accounting management. For example, Business Plan Pro and Small Business Management Pro can help you set up your own business from scratch.

Finally, there are investment software packages, such as StreetSmart Pro from Charles Schwab and Online Xpress from Fidelity, as well as various retirement-planning programs.

Desktop Publishing

What are the principal features of desktop-publishing software?

Adobe Systems was founded in 1982, when John Warnock and Charles Geschke began to work on solving some of the long-standing problems that plagued the relationship between microcomputers and printers. Collaboration with Apple Computers produced the first desktop-publishing package, using Adobe PostScript, a printer language that can handle many fonts and graphics, in 1984. By 1987, Adobe had agreements with IBM, Digital, AST Research, Hewlett-Packard, and Texas Instruments for them to use PostScript in their printers.

Desktop publishing (DTP) involves mixing text and graphics to produce high-quality output for commercial printing, using a microcomputer and mouse, scanner, laser or ink-jet printer, and DTP software. Often the printer is used primarily to get an advance look before the completed job is sent to a typesetter service bureau for even higher-quality output. Service bureaus have special machines that convert the DTP files to film, which can then be used to make plates for offset printing. Offset printing produces higher-quality documents, especially if color is used, but is generally more expensive than laser printing.

1992	1993	1994	1995	1996	1997
Microsoft's Access database program released	Multimedia desktop computers	Apple and IBM introduce PCs with full-motion video built in; wireless data transmission for small portable computers; Netscape's first web browser is introduced (based on Mosaic, introduced in 1993)	Windows 95 is released	First version of Windows CE is released	Mac OS 8 sells 1.25 million copies in its first two weeks; Mac System 8.0

Desktop-publishing software. Named after the subatomic particle proposed as the building block for all matter, the Quark company was founded in Colorado in 1981. QuarkXPress was released in 1987 and made an immediate impact on the fledgling desktop publishing business. QuarkXPress introduced precision typography, layout, and color control to the desktop computer.

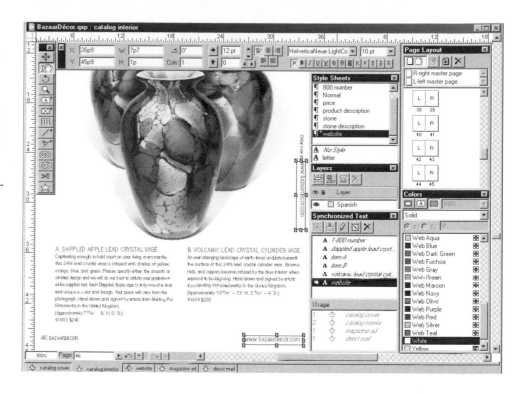

FEATURES OF DESKTOP PUBLISHING Desktop publishing has the following characteristics:

- **Mix of text with graphics:** Desktop-publishing software allows you to precisely manage and merge text with graphics. As you lay out a page on-screen, you can make the text "flow," liquidlike, around graphics such as photographs. You can resize art, silhouette it, change the colors, change the texture, flip it upside down, and make it look like a photo negative.

- **Varied type and layout styles:** As do word processing programs, DTP programs support a variety of fonts, or typestyles, from readable Times Roman to staid Tribune to wild Jester and Scribble. Additional fonts can be purchased on disk or downloaded online. You can also create all kinds of rules, borders, columns, and page-numbering styles.

- **Use of files from other programs:** It's usually not efficient to do word processing, drawing, and painting with the DTP software. As a rule, text is composed on a word processor, artwork is created with drawing and painting software, and photographs are input using a scanner and then modified and stored using image-editing software. Prefabricated art to illustrate DTP documents may be obtained from clip-art

1998	2000	2001	2003	2005
Windows 98 is released	Windows 2000 (ME) is released; Mac System 9.0	Windows XP becomes available; Mac OS X ships	Microsoft Longhorn OS (Prebeta) first introduced; Windows Mobile released	Microsoft's new Vista OS (previously named Longhorn) released in beta version (to be commercially available late 2006)

sources. The DTP program is used to integrate all these files. You can look at your work on the display screen as one page, as two facing pages (in reduced size), or as "thumbnails." Then you can see it again after it has been printed out. (● *See Panel 3.30.*)

BECOMING A DTP PROFESSIONAL Not everyone can be successful at desktop publishing, because many complex layouts require experience, skill, and knowledge of graphic design. Indeed, use of these programs by nonprofessional users can lead to rather unprofessional-looking results. Nevertheless, the availability of microcomputers and reasonably inexpensive software has opened up a career area formerly reserved for professional typographers and printers.

Professional DTP programs are QuarkXPress, Adobe InDesign, and Adobe PageMaker. Microsoft Publisher is a "low-end," consumer-oriented DTP package. Some word processing programs, such as Word and WordPerfect, also have many DTP features, although still not at the sophisticated level of the specialized DTP software. DTP packages, for example, give you more control over typographical characteristics and provide more support for full-color output.

Drawing & Painting Programs

How do drawing and painting programs differ?

It may be no surprise to learn that commercial artists and fine artists have largely abandoned the paintbox and pen and ink for software versions of

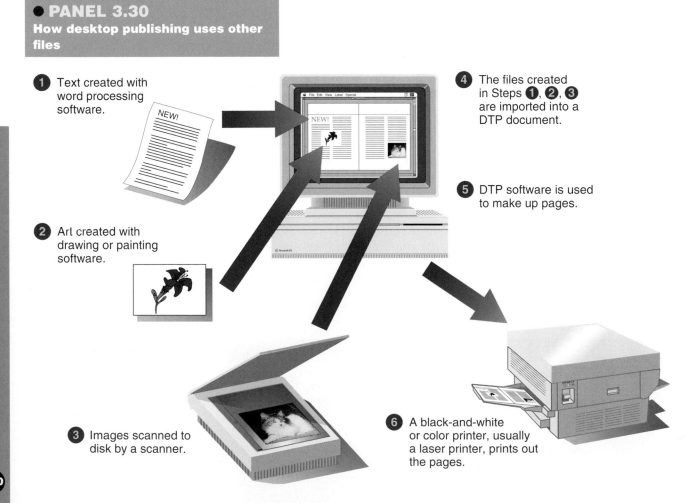

● PANEL 3.30
How desktop publishing uses other files

❶ Text created with word processing software.

❷ Art created with drawing or painting software.

❸ Images scanned to disk by a scanner.

❹ The files created in Steps ❶, ❷, ❸ are imported into a DTP document.

❺ DTP software is used to make up pages.

❻ A black-and-white or color printer, usually a laser printer, prints out the pages.

palettes, brushes, and pens. However, even nonartists can produce good-looking work with these programs.

There are two types of computer art programs, also called *illustration software*—drawing and painting.

DRAWING PROGRAMS A *drawing program* is graphics software that allows users to design and illustrate objects and products. Some drawing programs are CorelDRAW, Adobe Illustrator, and Macromedia Freehand.

Drawing programs create *vector images*—images created from geometrical formulas. Almost all sophisticated graphics programs use vector graphics.

PAINTING PROGRAMS *Painting programs* are graphics programs that allow users to simulate painting on-screen. A mouse or a tablet stylus is used to simulate a paintbrush. The program allows you to select "brush" sizes, as well as colors from a color palette. Examples of painting programs are Adobe PhotoShop, Microsoft Photo Editor, Corel Photopaint, and JASC's PaintShop Pro.

Painting programs produce *bit-mapped images*, or *raster images*, made up of little dots.

Painting software is also called *image-editing software* because it allows you to retouch photographs, adjust the contrast and the colors, and add special effects, such as shadows.

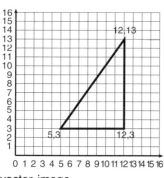

vector image

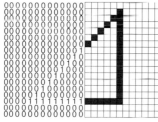

bit-mapped image

SOME GRAPHICS FILE FORMATS When you create an image, it's important to choose the most appropriate graphics file format, which specifies the method of organizing information in a file. Among the most important graphics formats you are apt to encounter are the following:

- **.bmp (BitMaP):** This bitmap graphic file format is native to Microsoft Windows and is used on PCs. Microsoft Paint creates .bmp file formats.
- **.gif (Graphic Interchange Format):** This format is used in Web pages and for downloadable online images.
- **.jpeg (Joint Photographic Experts Group):** Pronounced "*jay-peg*," this bitmap format is used for websites and for photos and other high-resolution images.
- **.tiff (Tagged Image File Format):** This bitmap format is used on both PCs and Macs for high-resolution files that will be printed.
- **.png (Portable network Graphics):** This file format was specifically created for web page images and can be used as a public domain alternative to .gif for compression.

Video/Audio Editing Software

Could I benefit from using video/audio editing software?

The popularity of digital camcorders ("camera recorders") has caused an increase in sales of video-editing software. This software allows you to import video footage to your PC and edit it, for example, deleting parts you don't want, reordering sequences, and adding special effects. Popular video-editing software packages include Adobe Premiere, Sony Pictures Digital Vegas, Apple Final Cut Express, Pinnacle Studio DV, and Ulead VideoStudio.

Audio-editing software provides similar capabilities for working with sound tracks, and you can also clean up background noise (called *artifacts*) and emphasize certain sound qualities. Sound-editing software includes Windows Sound Recorder, Sony Pictures Digital Sound Forge, Audacity (freeware), Felt Tip Software's Sound Studio (shareware), GoldWave, and WavePad.

Video and audio are covered in more detail in Chapter 5.

S u r v i v a l   T i p
Compressing Web
& Audio Files

Video and audio files tend to be very large, so they need to be edited down and compressed to be as short as possible, especially if they are to be used on web pages. Your software documentation will explain how to do this.

Sophisticated application software. Animation artist at work at the Studio Ghibli, Mitaka, Japan.

more info!

About Animation

For sources about animation, go to *http://webreference.com/3d/*. For schools offering training in computer-based graphics, including animation, check out *www.computertrainingschools .com/?googleanimation=y&got= 3d_animation_training&t=30.*

Animation Software

How does animation software differ from video software?

Animation is the simulation (illusion) of movement created by displaying a series of still pictures, or frames, very quickly in sequence. *Computer animation* refers to the creation of moving images by means of a computer. Whereas video devices record continuous motion and break it up into discrete frames, animation starts with independent pictures and puts them together to form the illusion of continuous motion. Animation is one of the chief ingredients of multimedia presentations and is commonly used on web pages. There are many software applications that enable you to create animations that you can display on a computer monitor.

The first type of animation to catch on for web use was called *GIF* (for Graphics Interchange Format) animation, and it is still very popular today. GIF files contain a group of images that display very quickly to simulate movement when a web page viewer clicks on the file icon. Animated GIF Construction Professional enables users to easily create animation via the use of a wizard. It allows the creation of many special effects and supports compression, as well as offering tutorials. Among the many other GIF animation software packages are 3D GIF Designer and The Complete Animator.

Screen from a GIF animation program

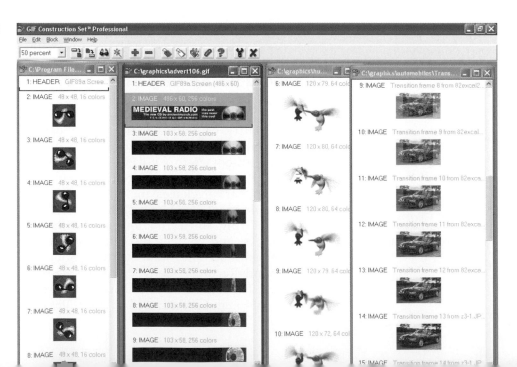

Macromedia Dreamweaver is a leading web-development tool.

Multimedia Authoring Software

What is multimedia authoring software?

Multimedia authoring software combines text, graphics, video, animation, and sound in an integrated way to create stand-alone multimedia applications. Content can be burned to CDs/DVDs or delivered via the web. Until the mid-'90s, multimedia applications were relatively uncommon, owing to the expensive hardware required. With increases in performance and decreases in price, however, multimedia is now commonplace. Nearly all PCs are capable of displaying video, though the resolution available depends on the power of the computer's video adapter and CPU. Macromedia Director and Macromedia Authorware are two popular multimedia authoring packages.

Many websites, as well as business training centers and educational institutions, use multimedia to develop interactive applications.

Web Page Design Software

How could using web page design software be to my benefit?

Web page design software is used to create web pages with sophisticated multimedia features. A few of these packages are easy enough for even beginners to use. A few of the best-known are Macromedia Dreamweaver, Macromedia Flash, Adobe GoLive, and Microsoft FrontPage.

Internet access providers also offer some free, easy-to-use web-authoring tools for building simple websites. They help you create web pages using icons and menus to automate the process; you don't need to know hypertext markup language (HTML, Chapter 2) to get the job done. These automated tools let you select a prepared, template web page design, type a few words, add a picture or two—and you're done. To save the pages and make them accessible on the internet, the provider grants you a certain amount of space on its web servers.

PRACTICAL ACTION
Help in Building Your Web Page

Netscape Navigator comes with a web-building program, as does Microsoft Explorer. Local and national internet access providers also offer tools, as well as space on their servers for storing your web page. Other sources of information for designing and building web pages are as follows.

For Novices

- **Yahoo:** Yahoo offers web page–building tools and templates under the name PageWizards. For advanced users, it also offers a service called PageBuilder, which enables you to add music to your web pages and have components that track how many people visit your site. Yahoo offers 15 megabytes of storage for PageWizard or PageBuilder pages in its GeoCities area (*http://geocities.yahoo.com*). Advertising banners will appear on your pages unless you pay a small monthly fee not to have them.

- **Lycos:** Lycos offers templates and tools, but it also offers tutorials to help you get started if you want to build your own pages from scratch. Lycos also offers free space in its Tripod area (*www.tripod.lycos.com*). As with Yahoo, you will have advertisements on your pages unless you pay a small fee to get rid of them.

- **America Online:** For its subscribers, AOL offers more than 100 templates, 5,000 free images, and 2 megabytes per screen name.

- **Microsoft:** Microsoft offers Internet Assistant, which can be used with Microsoft Word. You can create a web page from a template, from an existing Word document by choosing *Save as Web page,* or by using the Web Page wizard.

Once you've created your website, you'll need to "publish" it—upload it to a web server for viewing on the internet. You can get upload instructions from your online service or internet access provider, which may also provide space (for free or for a fee) on its servers.

For Advanced Training

If you've taken up desktop publishing, you should know that Adobe PageMaker and Microsoft Publisher come with simple web page editors. Macromedia's web-authoring software, Dreamweaver, and Microsoft's FrontPage allow you to create complicated web pages from scratch, but you will need to learn some HTML to use these packages.

Big businesses treat web authoring differently, because it's much more complicated than individually authored web design. However, Homestead Technologies (*www.homestead.com*) is making it easier for small businesses whose employees have little or no technical expertise to create smart-looking websites.

Handzon offers web page–building services, page storage, answers to questions, and training videos. Go to *www.handzon.com.*

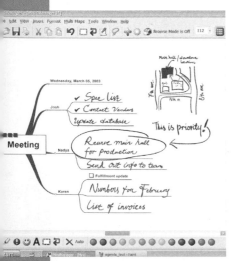

Project Management Software

Should I learn to use project management software?

As we have seen, a personal information manager (PIM) can help you schedule your appointments and do some planning. That is, it can help you manage your own life. But what if you need to manage the lives of others in order to accomplish a full-blown project, such as steering a political campaign or handling a nationwide road tour for a band? Strictly defined, a *project* is a one-time operation involving several tasks and multiple resources that must be organized toward completing a specific goal within a given period of time. The project can be small, such as an advertising campaign for an in-house advertising department, or large, such as construction of an office tower or a jetliner.

Project management software is a program used to plan and schedule the people, costs, and resources required to complete a project on time. For instance, the associate producer on a feature film might use such software to keep track of the locations, cast and crew, materials, dollars, and schedules needed to complete the picture on time and within budget. The software would show the scheduled beginning and ending dates for a particular task—such as shooting all scenes on a certain set—and then the date that task was actually completed. Examples of project management software are Mindjet MindManager X5 Pro, Microsoft Project, and Primavera Suretrack Project Manager.

Computer-Aided Design

What could I do with a CAD program?

Computers have long been used in engineering design. **_Computer-aided design (CAD)_ programs are intended for the design of products, structures, civil engineering drawings, and maps.** CAD programs, which are available for microcomputers, help architects design buildings and workspaces and help engineers design cars, planes, electronic devices, roadways, bridges, and subdivisions. CAD and drawing programs are similar. However, CAD programs provide precise dimensioning and positioning of the elements being drawn, so they can be transferred later to computer-aided manufacturing (CAM) programs. Also, CAD programs lack the special effects for illustrations that come with drawing programs. One advantage of CAD software is that the product can be drawn in three dimensions and then rotated on the screen, so the designer can see all sides. (● *See Panel 3.31.*) Examples of CAD programs for beginners are Autosketch, AutoCAD, Turbocad, and CorelCAD.

Computer-aided design/computer-aided manufacturing (CAD/CAM) software allows products designed with CAD to be input into an automated manufacturing system that makes the products. For example, CAD/CAM systems brought a whirlwind of enhanced creativity and efficiency to the fashion industry. Some CAD systems, says one writer, "allow designers to electronically drape digital-generated mannequins in flowing gowns or tailored suits that don't exist, or twist imaginary threads into yarns, yarns into weaves, weaves into sweaters without once touching needle to garment."[12] The designs and specifications are then input into CAM systems that enable robot pattern-cutters to automatically cut thousands of patterns from fabric with only minimal waste. Whereas previously the fashion industry worked about a year in advance of delivery, CAD/CAM has cut that time to 8 months—a competitive edge for a field that feeds on fads.

● PANEL 3.31

CAD

CAD software is used for nearly all three-dimensional designing.

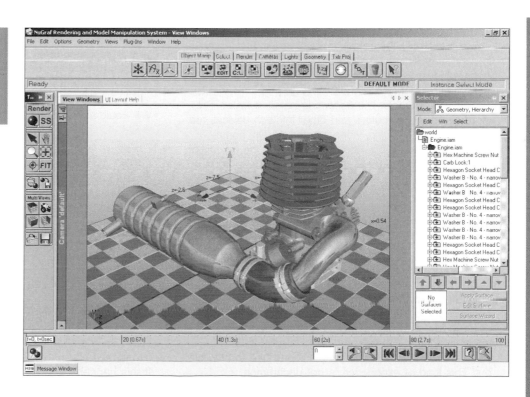

use of hardware and software. The three types of user interface are command-driven, menu-driven, and graphical (GUI), which is now most common. Without user interfaces, no one could operate a computer system.

utility programs (p. 124) Also known as *service programs;* system software components that perform tasks related to the control and allocation of computer resources. Why it's important: *Utility programs enhance existing functions or provide services not supplied by other system software programs. Most computers come with built-in utilities as part of the system software; they usually include backup, data recovery, virus protection, data compression, and file defragmentation, along with check (scan) disk and disk cleanup.*

value (p. 158) A number or date entered in a spreadsheet cell. Why it's important: *Values are the actual numbers used in the spreadsheet—dollars, percentages, grade points, temperatures, or whatever.*

web page design software (p. 173) Software used to create web pages with sophisticated multimedia features. Why it's important: *Allows even beginners to create web pages.*

what-if analysis (p. 158) Spreadsheet feature that employs the recalculation feature to investigate how changing one or more numbers changes the outcome of the calculation. Why it's important: *Users can create a worksheet, putting in formulas and numbers, and then ask, "What would happen if we change that detail?"—and immediately see the effect.*

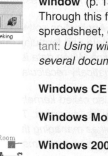

window (p. 132) Rectangular frame on the computer display screen. Through this frame users can view a file of data—such as a document, spreadsheet, or database—or an application program. Why it's important: *Using windows, users can display at the same time portions of several documents and/or programs on the screen.*

Windows CE *See* **Microsoft Pocket PC.**

Windows Mobile *See* **Microsoft Pocket PC.**

Windows 2000 *See* **Microsoft Windows NT/2000.**

wizard (p. 153) Interactive computer utility program that acts as an interface to lead the user through a task by using dialog steps to ask questions. Why it's important: *Wizards make it relatively easy for users to learn new computer functions.*

word processing software (p. 149) Application software that allows users to use computers to format, create, edit, print, and store text material, among other things. Why it's important: *Word processing software allows users to maneuver through a document and delete, insert, and replace text, the principal correction activities. It also offers such additional features as creating, editing, formatting, printing, and saving.*

word wrap (p. 150) Special feature that automatically continues text to the next line by "wrapping around" when the user reaches the right margin. Why it's important: *You don't have to hit a "carriage-return" key or Enter key to move to the next line.*

Chapter Review

"I can recognize and recall information."

Self-Test Questions

1. _____ software enables the computer to perform essential operating tasks.

2. _____ _____ is the term for programs designed to perform specific tasks for the user.

3. _____ is the activity in which a computer works on more than one process at a time.

4. _____ is the scattering of portions of files about the disk in nonadjacent areas, thus greatly slowing access to the files.

5. Windows and Mac OS are generally used on _____ computers.

6. _____ is the process of loading an operating system into a computer's main memory.

7. A(n) _____ is a utility that will find all the scattered files on your hard disk and reorganize them as contiguous files.

8. The _____ is the component of system software that comprises the master system of programs that manage the basic operations of the computer.

9. The _____ is the user-controllable display screen that allows you to communicate, or interact, with your computer.

10. Disk scanner and disk cleanup utilities detect and correct certain types of common problems on hard disks, such as removing unnecessary files called _____ files that are created by Windows only for short tasks and system restore after system problems.

11. OSs allow users to control access to their computers via use of a _____ and a _____.

12. Software or hardware that is _____ means that it is privately owned and controlled by a company.

13. Linux is _____ - _____ software—meaning any programmer can download it from the internet for free and modify it with suggested improvements.

14. When you power up a computer by turning on the power "on" switch, this is called a _____ boot. If your computer is already on and you restart it, this is called a _____ boot.

15. _____ software allows you to create and edit documents.

16. _____ is the activity of moving upward or downward through the text or other screen display.

17. Name four editing features offered by word processing programs: _____, _____, _____, _____.

18. In a spreadsheet, the place where a row and a column intersect is called a _____.

19. The _____ is the movable symbol on the display screen that shows you where you may next enter data or commands.

20. When you buy software, you pay for a _____, a contract by which you agree not to make copies of the software to give away or resell.

21. Records in a database are sorted according to a _____.

22. _____ involves mixing text and graphics to produce high-quality output for commercial printing.

23. A _____ allows users to create tables and do "what-if" financial analyses by entering data and formulas into rows and columns arranged as a grid on a display screen.

24. _____ automatically continues text to the next line when you reach the right margin.

25. Settings that are automatically used by a program unless the user specifies otherwise are called _____.

26. _____-_____ software is not protected by copyright and may be copied by anyone.

27. _____ _____ are specialized software programs that allow input and output devices to communicate with the rest of the computer system.

Multiple-Choice Questions

1. Which of the following are functions of the operating system?
   a. file management
   b. CPU management
   c. task management
   d. booting
   e. all of these

2. Which of the following was the first major microcomputer OS?
   a. Mac OS
   b. Windows
   c. DOS
   d. Unix
   e. Linux

3. Which of the following is a prominent network operating system?
   a. Linux
   b. Unix
   c. Windows NT/2000
   d. DOS
   e. Mac OS

4. Which of the following is the newest Microsoft Windows operating system?
   a. Windows Me
   b. Windows XP
   c. Windows 2000
   d. Windows NT
   e. Windows CE

5. Which of the following refers to the execution of two or more programs by one user almost at the same time on the same computer with one central processor?
   a. multitasking
   b. multiprocessing
   c. time-sharing
   d. multiprogramming
   e. coprocessing

6. Which of the following are specialized software programs that allow input and output devices to communicate with the rest of the computer system?
   a. multitasking
   b. boot-disks
   c. utility programs
   d. device drivers
   e. service packs

7. Which of the following is *not* an advantage of using database software?
   a. integrated data
   b. improved data integrity
   c. lack of structure
   d. elimination of data redundancy

8. Which of the following is (are) *not* a type of menu?
   a. cascading menu
   b. pop-in menu
   c. pop-out menu
   d. pull-down menu
   e. pull-out menu

9. Which of the following is *not* a feature of word processing software?
   a. spelling checker
   b. cell address
   c. formatting
   d. cut and paste
   e. find and replace

10. What is the common consumer computer interface used today?
    a. command-driven interface
    b. graphical user interface
    c. menu-driven interface
    d. electronic user interface
    e. biometric user interface

11. Which type of software can you download and duplicate without any restrictions whatsoever and without fear of legal prosecution?
    a. commercial software
    b. shareware
    c. public-domain software
    d. pirated software
    e. rentalware

True/False Questions

T F   1. The supervisor manages the CPU.

T F   2. The first graphical user interface was provided by Microsoft Windows.

T F   3. Formatting a floppy disk will not affect any data already written on it.

T F   4. All operating systems are mutually compatible.

T F   5. *Font* refers to a preformatted document that provides basic tools for shaping the final document.

T F   6. Unix crashes often and thus is not normally used for running important large systems.

T F   7. Windows NT is the most recent Microsoft OS.

T F   8. Spreadsheet software enables you to perform what-if calculations.

T F   9. Rentalware is software that users lease for a fee.

T F 10. Public-domain software is protected by copyright and so is offered for sale by license only.

T F 11. The records within the various tables in a database are linked by a key field.

T F 12. QuarkXPress, Adobe InDesign, and Adobe PageMaker are professional desktop-publishing programs.

T F 13. The best-known graphical user interface is the command-driven one.

T F 14. Microsoft PowerPoint and Corel Presentations are examples of financial software.

T F 15. Drawing programs create vector images, and painting programs produce bit-mapped images.

T F 16. General computer users can design their own web pages using Macromedia Dreamweaver, Macromedia Flash, and Microsoft FrontPage.

"I can recall information in my own terms and explain it to a friend."

Short-Answer Questions

1. Briefly define *booting*.
2. What is the difference between a command-driven interface and a graphical user interface (GUI)?
3. Why can't you run your computer without system software?
4. Why is multitasking useful?
5. What is a device driver?
6. What is a utility program?
7. What is a platform?
8. What are the three components of system software? What is the basic function of each?
9. What is open-source software?
10. What does defragmenting do?
11. What is an embedded system?
12. What are the following types of application software used for?

    a. project management software
    b. desktop-publishing software
    c. database software
    d. spreadsheet software
    e. word processing software

13. Which program is more sophisticated, analytical graphics or presentation graphics? Why?
14. How are the following different? Pop-up menu; pull-down menu; cascading menu.
15. What is importing? Exporting?
16. Briefly compare drawing programs and painting programs.
17. Explain what computer-aided design (CAD) programs do.
18. Discuss the various software licenses: site licenses, concurrent-use licenses, multiple-user licenses, single-user license.

"I can apply what I've learned, relate these ideas to other concepts, build on other knowledge, and use all these thinking skills to form a judgment."

Knowledge in Action

1. Here's a Windows exercise in defragmenting your hard-disk drive. Defragmenting is a housekeeping procedure that will speed up your system and often free up hard-disk space.

   Double-click on *My Computer* on your Windows desktop (opening screen). Now use your right mouse button to click on *C drive*, then right-click on *Properties*, then left-click on the *General* tab, and you will see how much free space there is on your hard disk. Next left-click on the *Tools* tab; to clear out any errors, click the *Check Now* button; this will run a scan.

   Once the scan is complete, return to the *Tools* window and click the *Defragment Now* button. Click on *Show Details*. This will visually display on the screen the process of your files being reorganized into a contiguous order.

   Many times when your PC isn't performing well, such as when it's sluggish, running both ScanDisk (Check Now) and Defragment will solve the problem.

2. Go to the box "Various task management operations," page 123 in this chapter. Create noncomputer analogies for multitasking, multiprogramming, time-sharing, and multiprocessing. For instance, for time-sharing, you could imagine a waiter taking menu requests, because he or she spends a fixed amount of time with each customer (program) before going on to the next one.

3. What do you think is the future of operating systems? Look up Yale computer scientist David Gelernter's paper "The Second Coming—A Manifesto" on the technology forum *www.edge.org*. (Type the article's title into the Edge search box.) Do you agree with him that data and computer processing will be increasingly spread across thousands, if not millions, of interconnected computers so that it is less likely that Microsoft or any other company will dominate the field?

4. What do you think is the future of Linux? Experts currently disagree about whether Linux will become a serious competitor to Windows. Research Linux on the web. Which companies are creating application software to run on Linux? Which businesses are adopting Linux as an OS? What are the predictions about Linux use?

5. How do you think you will obtain software for your computer in the future? Explain your answer.

6. Design your own handheld. Draw what your ideal handheld would look like, and draw screens of what your user interface would look like. Describe the key features of your handheld.

7. What sorts of tasks do operating systems *not* do that you would like them to do?

8. If you were in the market for a new microcomputer today, what application software would you want to use

on it? Why? What are some "dream" applications that you would like that have not yet been developed?

9. Several websites include libraries of shareware programs. Visit the *www.5star-shareware.com* site and identify three shareware programs that interest you. State the name of each program, the operating system it runs on, and its capabilities. Also, describe the contribution you must make to receive technical support. What about freeware? Check out *www.freewarehome.com*.

10. What is your opinion of downloading free music from the web to play on your own PC and/or CDs? Much attention has been given lately to music downloading and copyright infringement. Research this topic in library magazines and newspapers or on the internet, and take a position in a short report.

11. How do you think you could use desktop publishing at home? For personal items? Family occasions? Holidays? What else? What hardware and software would you have to buy?

12. Think of three new ways that software companies could prevent people from pirating their software.

13. What is your favorite application software program of all? Why?

14. Did your computer come with a Windows Startup disk, and have you misplaced it? If your computer crashes, you'll need this disk to reinstall the operating system.

    To learn the benefits of having a Startup disk, visit *www.microsoft.com*. Type *startup* in the "search for" box; then click on the links that interest you.

Web Exercises

1. Go to *http://list.driverguide.com/list/company243/* and identify the drivers that correspond to equipment you use. How does this website let you know which devices the drivers are for and which operating systems are compatible with them? If you own your own computer, go to the manufacturer's website and locate its resource for updating drivers. Does the manufacturer recommend any driver updates that you could use?

2. Use a web search tool such as Google or Yahoo! to find some online antivirus sites—sites where users can regularly download updates for their antivirus software. Do you know what kind of antivirus software is installed on your computer?

3. Microsoft offers "patches," or updates, for its Windows OS. Go to *www.microsoft.com* and search for the list of updates. What kinds of problems do these updates fix? Do you need any?

4. The History of Operating Systems: Visit the following websites to get an overview of the evolution and history of the theory and function of operating systems:

   www.microsoft.com/windows/winhistoryintro.mspx
   www.computinghistorymuseum.org/teaching/papers/ research/history_of_operating_system_Moumina.pdf
   www.osdata.com/kind/history.htm
   www.answers.com/operating%20systems?gwp=11

5. Write Your Own Operating System: Visit the following websites to get a better picture of the groundwork that underlies designing an operating system and how to build your own:

   http://cdsmith.twu.net/professional/osdesign.html

   www.openbg.net/sto/os/
   www.groovyweb.uklinux.net/index.php?page_ name=how%20to%20write%20your%20own%20os

6. Security Issue: Read about some security flaws and limitations of Microsoft Windows operating systems:

   www.samag.com/documents/s=1152/sam0104o/ 0104o.htm
   www.newsfactor.com/perl/story/15458.html
   www.sans.org/top20/
   www.zdnetasia.com/news/security/ 0,39044215,39242898,00.htm

7. Some people are fascinated by the error message commonly referred to as the "Blue Screen of Death" (BSOD) or "Doom." Run a search on the internet and find websites that sell T-shirts with the BSOD image on it, photo galleries of public terminals displaying the BSOD, fictional stories of BSOD attacks, and various other forms of entertainment based on the infamous error message.

   Do a search on the web to find users' hypotheses of why the BSOD occurs, and find methods to avoid it. Here are a few sites:

   www.answers.com/topic/blue-screen-of-death?method= 8www.bbspot.com/News/2002/10/bsod_ads.html
   http://bsod.org/
   http://bbspot.com/News/2000/9/bsod_death.html
   www.ntbrad.com/bsod.htm

8. Many productivity programs designed after 1997 have features built into them for converting files into web pages. If you have Microsoft Word 97 or later, or Microsoft PowerPoint 97 or later, try saving a document as a website. Under the File menu, select *Save As, Save As Type: Web Page HTML,* and then view the file in your web browser. (Open your browser, then go to *File, Open, Browse.*) What possibilities does this open up for you?

9. Using Microsoft Excel or another spreadsheet program, make a food shopping list incorporating the estimated price for each item and any coupon discounts you have, and then have Excel calculate the overall cost. Then go buy your groceries and compare Excel's price with the supermarket's price. What else could Excel help you with?

10. The Windows operating system comes with a basic word processing program called *Wordpad*. Go to the Microsoft home page and find out how Wordpad differs from Microsoft Word. Then use a keyword search in a search engine to get more information about these programs. Which one is right for you?

11. Curriculum Data Wales (CDW) is a public/private partnership that has been charged by the Welsh Assembly Government with the task of designing, building, and maintaining the National Grid for Learning Cymru as a bilingual service to schools and colleges in Wales. CDW's website includes some short tutorials on desktop-publishing (DTP), spreadsheet, word processing, and database management software:

   www.ngfl-cymru.org.uk/vtc-home/vtc-ks4-home/ vtc-ks4-ict/vtc-ks4-ict-application_software.htm

   Work through the tutorials. Did they expand your knowledge of these applications?

   Do a search for *"applications software" & tutorials.* What other useful tutorials did you find?

Hardware: The CPU & Storage

How to Choose a Multimedia Computer System

Chapter Topics & Key Questions

4.1   **Microchips, Miniaturization, & Mobility** What are the differences between transistors, integrated circuits, chips, and microprocessors?

4.2   **The System Unit: The Basics** How is data represented in a computer, what are the components of the system cabinet, and what are processing speeds?

4.3   **More on the System Unit** How do the processor and memory work, and what are some important ports, buses, and cards?

4.4   **Secondary Storage** What are the features of floppy disks, hard disks, optical disks, magnetic tape, smart cards, flash memory, and online secondary storage?

4.5   **Future Developments in Processing & Storage** What are some forthcoming developments that could affect processing power and storage capacity?

The microprocessor was "the most important invention of the 20th century," says Michael Malone, author of *The Microprocessor: A Biography.*[1]

Quite a bold claim, considering the incredible products that have issued forth during those 100 years. More important than the airplane? More than television? More than atomic energy?

According to Malone, the case for the exalted status of this thumbnail-size information-processing device is demonstrated, first, by its pervasiveness in the important machines in our lives, from computers to transportation. Second, "The microprocessor is, intrinsically, something special," he says. "Just as [the human being] is an animal, yet transcends that state, so too the microprocessor is a silicon chip, but more." Why? Because it can be programmed to recognize and respond to patterns in the environment, as humans do. Malone writes:

> Implant [a microprocessor] into a traditional machine—say an automobile engine or refrigerator—and suddenly that machine for the first time can learn, it can adapt to its environment, respond to changing conditions, become more efficient, more responsive to the unique needs of its user.[2]

4.1

Microchips, Miniaturization, & Mobility

What are the differences between transistors, integrated circuits, chips, and microprocessors?

The microprocessor has presented us with gifts that we may only barely appreciate—*portability* and *mobility* in electronic devices.

In 1955, for instance, portability was exemplified by the ads showing a young woman holding a Zenith television set over the caption: IT DOESN'T TAKE A MUSCLE MAN TO MOVE THIS LIGHTWEIGHT TV. That "lightweight" TV weighed a hefty 45 pounds. Today, by contrast, there is a handheld Casio 2.3-inch color TV weighing a mere 10.2 ounces.

Had the transistor not arrived, as it did in 1947, the Age of Portability and consequent mobility would never have happened. To us a "portable" telephone might have meant the 40-pound backpack radio-phones carried by some American GIs during World War II, rather than the 6-ounce shirt-pocket cellular models available today.

From Vacuum Tubes to Transistors to Microchips

How do transistors and integrated circuits differ, and what does a semiconductor do?

A *circuit* is a closed path followed or capable of being followed by an electric current. Without circuits, electricity would not be controllable, and so we would not have electric or electronic appliances. Old-time radios used vacuum tubes—small lightbulb-size electronic tubes with glowing filaments, or wire circuits, inside them—to facilitate the transmission (flow) of electrons.

One computer with these tubes, the ENIAC, was switched on in 1946 at the University of Pennsylvania and employed about 18,000 of them. Unfortunately, a tube failure occurred on average once every 7 minutes. Since it took more than 15 minutes to find and replace the faulty tube, it was difficult to get any useful computing work done—during a typical week, ENIAC was down for about one-third of the time. Moreover, the ENIAC was enormous, occupying 1,800 square feet and weighing more than 30 tons. ENIAC could perform about 5,000 calculations per second—more than 10,000 times *slower* than modern PCs. Yet even at that relatively slow speed, ENIAC took about 20 seconds to complete a problem that had taken experts 1 or 2 days to complete manually.

THE TRANSISTOR ARRIVES   The transistor changed all that. **A _transistor_ is essentially a tiny electrically operated switch, or gate, that can alternate between "on" and "off" many millions of times per second.** The transistor was developed by Bell Labs in 1947. The first transistors were one-hundredth the size of a vacuum tube, needed no warm-up time, consumed less energy, and were faster and more reliable. (● *See Panel 4.1.*) Moreover, they marked the beginning of a process of miniaturization that has not ended yet. In 1960 one transistor fit into an area about a half-centimeter square. This was sufficient to permit Zenith, for instance, to market a transistor radio weighing about 1 pound (convenient, the company advertised, for "pocket or purse"). Today more than 3 million transistors can be squeezed into a half centimeter, and a Sony headset radio, for example, weighs only 6.2 ounces. Hewlett-Packard is working on a transistor about 0.1 nanometer square. One nanometer is 1 billionth of a meter; a human hair is about 80,000 nanometers thick.

In the old days, transistors were made individually and then formed into an electronic circuit with the use of wires and solder. Today transistors are part of an **_integrated circuit_—an entire electronic circuit, including wires, formed on a single "chip," or piece, of special material, usually silicon,** as part of a single manufacturing process. Integrated circuits were developed by Jack Kilby at Texas Instruments, who demonstrated the first one in 1958. (● *See the timeline, Panel 4.2, starting on the next page.*)

An integrated circuit embodies what is called solid-state technology. **In a _solid-state device_, the electrons travel through solid material**—in this case, silicon. They do not travel through a vacuum, as was the case with the old radio vacuum tubes.

SILICON & SEMICONDUCTORS   What is silicon, and why use it? **_Silicon_ is an element that is widely found in clay and sand. It is used not only because its abundance makes it cheap but also because it is a semiconductor.**

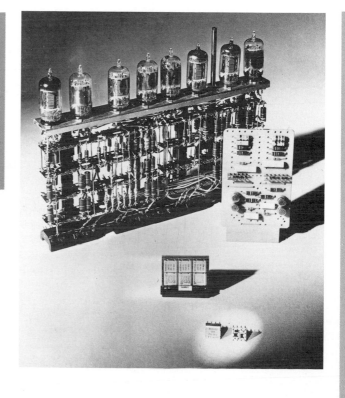

● PANEL 4.1
Shrinking components
The lightbulb-size 1940s vacuum tube was replaced in the 1950s by a transistor one-hundredth its size. Today's transistors are much smaller, being microscopic in size.

A _semiconductor_ is material whose electrical properties are intermediate between a good conductor of electricity and a nonconductor of electricity. (An example of a good conductor of electricity is the copper in household wiring; an example of a nonconductor is the plastic sheath around that wiring.) Because it is only a semiconductor, silicon has partial resistance to electricity. As a result, highly conducting materials can be overlaid on the silicon to create the electronic circuitry of the integrated circuit. (● *See Panel 4.3, opposite page.*)

Silicon alone has no processing power. A _chip_, or _microchip_, **is a tiny piece of silicon that contains millions of microminiature integrated electronic circuits.** Chip manufacture requires very clean environments, which is why chip manufacturing workers appear to be dressed for a surgical operation. Such workers must also be highly skilled, which is why chip makers are not found everywhere in the world.

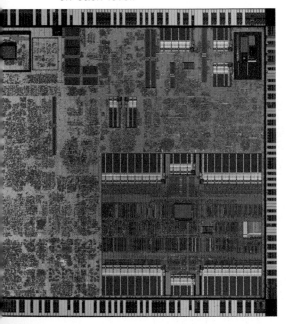

Modern chip with etched transistors. This chip would be about 1/2 inch by 1/2 inch and be several layers deep, with transistors etched on each level.

Miniaturization Miracles: Microchips, Microprocessors, & Micromachines

What is a microprocessor?

Microchips—"industrial rice," as the Japanese call them—are responsible for the miniaturization that has revolutionized consumer electronics, computers, and communications. They store and process data in all the electronic gadgetry we've become accustomed to—from microwave ovens to videogame controllers to music synthesizers to cameras to automobile fuel-injection systems to pagers to satellites.

There are different kinds of microchips—for example, microprocessor, memory, logic, communications, graphics, and math coprocessor chips. We discuss some of these later in this chapter. Perhaps the most important is the microprocessor chip. A _microprocessor_ ("microscopic processor" or "processor on a chip") **is the miniaturized circuitry of a computer processor—the CPU, the part that processes, or manipulates, data into information.** When modified for use in machines other than computers, microprocessors are called *microcontrollers,* or *embedded computers.*

Mobility

How have microprocessors helped make information technology more mobile?

Smallness in TVs, phones, radios, camcorders, CD players, and computers is now largely taken for granted. In the 1980s portability, or mobility, meant

● **PANEL 4.2**

Timeline: Developments in processing and storage

3000 BCE	1621 CE	1642	1666	1801	1820	1833
Abacus is invented in Babylonia	Slide rule invented (Edmund Gunther)	First mechanical adding machine (Blaise Pascal)	First mechanical calculator that can add and subtract (Samuel Morland)	A linked sequence of punched cards controls the weaving patterns in Jacquard's loom	The first mass-produced calculator, the Thomas Arithmometer	Babbage's difference engine (automatic calculator)

1. A large drawing of the electrical circuitry is made; it looks something like the map of a train yard. The drawing is photographically reduced hundreds of times, to microscopic size.

2. That reduced photograph is then duplicated many times so that, like a sheet of postage stamps, there are multiple copies of the same image or circuit.

Chip designers checking out an enlarged drawing of chip circuits

3. That sheet of multiple copies of the circuit is then printed (in a printing process called *photolithography*) and etched onto a round slice of silicon called a *wafer.* Wafers have gone from 4 inches in diameter to 6 inches to 8 inches, and now are moving toward 12 inches; this allows semiconductor manufacturers to produce more chips at lower cost.

4. Subsequent printings of layer after layer of additional circuits produce multilayered and interconnected electronic circuitry built above and below the original silicon surface.

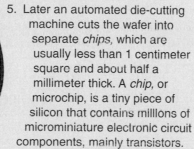
A wafer imprinted with many microprocessors.

5. Later an automated die-cutting machine cuts the wafer into separate *chips,* which are usually less than 1 centimeter square and about half a millimeter thick. A *chip,* or microchip, is a tiny piece of silicon that contains millions of microminiature electronic circuit components, mainly transistors. An 8-inch silicon wafer will have a grid of nearly 300 chips, each with as many as 5.5 million transistors.

6. After testing, each chip is mounted in a protective frame with protruding metallic pins that provide electrical connections through wires to a computer or other electronic device.

(above) Pentium 4 microprocessor chip mounted in protective frame with pins that can be connected to an electronic device such as a microcomputer.

● **PANEL 4.3**
Making of a chip
How microscopic circuitry is put onto silicon.

1843	1854	1877	1890	1915
World's first computer programmer, Ada Lovelace, publishes her notes	George Boole publishes "An Investigation on the Laws of Thought," a system for symbolic and logical reasoning that will become the basis for computer design	Thomas Edison invents the phonograph	Electricity used for first time in a data-processing project (punched cards)—Hollerith's automatic census-tabulating machine (used punched cards) 	78 rpm record platters are introduced

Try the magazine *Smart
Computing in Plain English*
(*www.smartcomputing.com*).
A more technical publica-
tion is *PC Magazine*
(*www.pcmag.com*). See also
*www.webopedia, http://
computer.howstuffworks.com,*
and *www.geek.com*. At
www.pitstop.com you can get
all sorts of information about
testing your system's performance and improving it.

trading off computing power and convenience in return for smaller size and
less weight. Today, however, we are getting close to the point where we don't
have to give up anything. As a result, experts have predicted that small, powerful, wireless personal electronic devices will transform our lives far more
than the personal computer has done so far. "The new generation of
machines will be truly personal computers, designed for our mobile lives,"
wrote one reporter back in 1992. "We will read office memos between strokes
on the golf course, and answer messages from our children in the middle of
business meetings."[3] Today such activities are commonplace.

Choosing an Inexpensive Personal Computer: Understanding Computer Ads

*What kind of features does a multimedia personal computer
have?*

You're in the market for a new PC and are studying the ads. What does "256
MB DDR SDRAM" mean? How about "40GB (7200 RPM) SATA Hard
Drive"? Let's see how to interpret a typical computer ad. (● *See Panel 4.4.*)

Most desktop computers are *multimedia computers,* with sound and
graphics capability. As we explained in Chapter 1, the word *multimedia*
means "combination of media"—the combination of pictures, video, animation, and sound in addition to text. A multimedia computer features such
equipment as a fast processor, DVD drive, sound card, graphics card, and
speakers, and you may also wish to have headphones and a microphone.
(Common peripherals are printer, scanner, sound recorder, and digital
camera.)

Let us now go through the parts of a computer system so that you can
understand what you're doing when you buy a new computer. First we look
at how the system processes data. In the remainder of this chapter, we will
consider the *system unit* and *storage devices.* In Chapter 5, we look at *input
devices* and *output devices.*

4.2

The System Unit: The Basics

*How is data represented in a computer, what are the
components of the system cabinet, and what are processing
speeds?*

Computers run on electricity. What is the most fundamental thing you can
say about electricity? Electricity is either *on* or *off.* This two-state situation
allows computers to use the binary system to represent data and programs.

1924	1930	1936	1944	1945	1946	1947
T.J. Watson renames Hollerith's machine company, founded in 1896, to International Business Machines (IBM)	General theory of computers (MIT)	Konrad Zuse, German, develops the concept of a computer memory to hold binary information	First electro-mechanical computer (Mark I)	John von Neumann introduces the concept of a stored program	First programmable electronic computer in United States (ENIAC)	Magnetic tape enters the U.S. market; the first transistor is developed

Great PC Buy!

- 7-Bay Mid-Tower Case
- Intel Pentium 4 Processor 2.80 GHz
- 512 MB 533 MHz DDR2 SDRAM
- 1 MB L2 Cache
- 6 USB 2.0 Ports
- 56 Kbps Internal Modem
- 3D AGP Graphics Card (64 MB)
- Sound Blaster Digital Sound Card
- 160 GB SATA 7200 RPM Hard Drive
- 24X DVD/CD-RW Combo Drive
- 104-Keyboard
- Microsoft IntelliMouse
- 17" Flat Panel Display
- HP Business Inkjet 1000 Printer

Details of this ad are explained throughout this chapter and the next. See the little magnifying glass:

The Binary System: Using On/Off Electrical States to Represent Data & Instructions

What does a computer's binary system do, and what are some binary coding schemes?

The decimal system that we are accustomed to has 10 digits (0, 1, 2, 3, 4, 5, 6, 7, 8, 9). By contrast, the **_binary system_ has only two digits: 0 and 1.** Thus, in the computer, the 0 can be represented by the electrical current being off and the 1 by the current being on. Although the use of binary systems is not restricted to computers, *all data and program instructions that go into the computer are represented in terms of these binary numbers.* (● *See Panel 4.5 on the next page.*)

For example, the letter "G" is a translation of the electronic signal 01000111, or off-on-off-off-off-on-on-on. When you press the key for "G" on the computer keyboard, the character is automatically converted into the series of electronic impulses that the computer can recognize. Inside the

1947–1948	1949	1952	1954	1956	1958	1962	1963
Magnetic drum memory is introduced as a data storage device for computers	45 rpm record platters are introduced	UNIVAC computer correctly predicts election of Eisenhower as U.S. President	Texas Instruments introduces the silicon transistor	First computer hard disk is used	Stereo records are produced	Integrated circuit is nicknamed the "chip"; timesharing becomes common	The American National Standards Institute accepts ASCII-7 code for information exchange

Hardware: The CPU & Storage

195

● **PANEL 4.5**
Binary data representation
How the letters "G-R-O-W" are represented in one
type of on/off, 1/0 binary code.

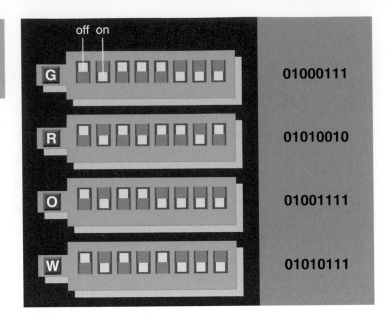

off   on

G   01000111

R   01010010

O   01001111

W   01010111

computer, the character "G" is represented by a combination of eight *transistors* (as we will describe). Some are off, or closed (representing the 0s), and some are on, or open (representing the 1s).

MEASURING CAPACITY   How many representations of 0s and 1s can be held in a computer or a storage device such as a hard disk? Capacity is denoted by *bits* and *bytes* and multiples thereof:

- **Bit:** In the binary system, **each 0 or 1 is called a _bit_, which is short for "binary digit."**

- **Byte:** To represent letters, numbers, or special characters (such as ! or *), bits are combined into groups. **A group of 8 bits is called a _byte_, and a byte represents one character, digit, or other value.** (As we mentioned, in one scheme, 01000111 represents the letter "G.") The capacity of a computer's memory or of a floppy disk is expressed in numbers of bytes or multiples such as kilobytes and megabytes. (There are 256 combinations of 8 bits available: $2^8 = 256$.)

- **Kilobyte:** A _kilobyte (K, KB)_ is about 1,000 bytes. (Actually, it's precisely 1,024 bytes, but the figure is commonly rounded.) The kilobyte was a common unit of measure for memory or secondary storage capacity on older computers. 1 KB equals about one-half page of text.

- **Megabyte:** A _megabyte (M, MB)_ is about 1 million bytes (1,048,576 bytes). Measures of microcomputer primary storage capacity today are expressed in megabytes. 1 MB equals about 500 pages of text.

1964	1965	1968	1969	1970
IBM introduces 360 line of computers; IBM's seven-year long Sabre project, allowing travel agents anywhere to make airline reservations via terminals, is fully implemented; Control Data Corp.'s CDC 6600, designed by Seymour Cray, becomes the first commercially successful supercomputer	Audio cassette tape introduced; Gordon Moore pronounces "Moore's Law"	Robert Noyce, Andy Grove, and Gordon Moore establish Intel, Inc.	Klass Compaan conceives idea for CD	Microprocessor chips come into use; floppy disk introduced for storing data; a chip ⅒ inches square contains 1,000 transistors; the first and only patent on the smart card was filed

more info!

Bigger Than an Exabyte?

How big is a zettabyte? A yottabyte? Do an online search to find out.

- **Gigabyte:** A *gigabyte* *(G, GB)* **is about 1 billion bytes** (1,073,741,824 bytes). This measure was formerly used mainly with "big iron" (mainframe) computers, but it is typical of the secondary storage (hard-disk) capacity of today's microcomputers. One gigabyte equals about 500,000 pages of text.

- **Terabyte:** A *terabyte* *(T, TB)* **represents about 1 trillion bytes** (1,009,511,627,776 bytes). 1 TB equals about 500,000,000 pages of text. Some high-capacity disk storage is expressed in terabytes.

- **Petabyte:** A *petabyte* *(P, PB)* **represents about 1 quadrillion bytes** (1,048,576 gigabytes). The huge storage capacities of modern databases are now expressed in petabytes.

- **Exabyte:** An *exabyte* *(EB)* **represents about 1 quintillion bytes**—that's *1 billion billion* bytes (1,024 petabytes—or 1,152,921,504,606,846,976 bytes). This number is seldom used. It is estimated that all the printed material in the world represents about 5 exabytes.[4]

BINARY CODING SCHEMES Letters, numbers, and special characters are represented within a computer system by means of binary coding schemes. (● *See Panel 4.6 on the next page.*) That is, the off/on 0s and 1s are arranged in such a way that they can be made to represent characters, digits, or other values.

Survival Tip
Decimal to Binary Conversion

Use the Windows Calculator for quick decimal to binary conversions. Open Windows Calculator by choosing *Start, All programs, Accessories, Calculator.* Or you could just go to *Start, Run, Calc.* When the calculator opens, choose *View, Scientific.* This will change your calculator interface from a standard calculator to a scientific calculator. Type in your number, and click the circle that says *BIN.* Also, while having the calculator on the Bin setting, you can type in any combination of 0s and 1s and convert them to decimal form by clicking on the *DEC* circle.

- **EBCDIC:** Pronounced *"eb-see-dick," **EBCDIC (Extended Binary Coded Decimal Interchange Code)* is a binary code used with large computers, such as mainframes.** It was deveoped in 1963–1964 by IBM and uses 8 bits (1 byte) for each character.

- **ASCII:** Pronounced *"ask-ee," **ASCII (American Standard Code for Information Interchange)* is the binary code most widely used with microcomputers.** Depending on the version, ASCII uses 7 or 8 bits (1 byte) for each character. Besides having the more conventional characters, the version known as Extended ASCII includes such characters as math symbols and Greek letters. ASCII's 256 characters, however, are not enough to handle such languages as Chinese and Japanese, with their thousands of characters.

- **Unicode:** Developed in the early 1990s, *Unicode* **uses 2 bytes (16 bits) for each character, rather than 1 byte (8 bits).** Instead of having the 256 character combinations of ASCII, Unicode can handle 65,536 character combinations. Thus, it allows almost all the written languages of the world to be represented using a single character set.

1971	1972	1973	1974	1975	1976	1978
First pocket calculator; the Intel 4004 microprocessor is developed—a "computer on a chip"	Intel 8008 8-bit microprocessor	Large-scale integration: 10,000 components are placed on a 1-sq.-cm. chip	A DRAM chip becomes available	First microcomputer (MITS Altair 8800)	Apple I computer (first personal computer sold in assembled form); has 512 KB RAM	5¼" floppy disk; Atari home videogame; Intel's first 16-bit microprocessor, the 8086, debuts

Character	EBCDIC	ASCII-8	Character	EBCDIC	ASCII-8
A	1100 0001	0100 0001	N	1101 0101	0100 1110
B	1100 0010	0100 0010	O	1101 0110	0100 1111
C	1100 0011	0100 0011	P	1101 0111	0101 0000
D	1100 0100	0100 0100	Q	1101 1000	0101 0001
E	1100 0101	0100 0101	R	1101 1001	0101 0010
F	1100 0110	0100 0110	S	1110 0010	0101 0011
G	1100 0111	0100 0111	T	1110 0011	0101 0100
H	1100 1000	0100 1000	U	1110 0100	0101 0101
I	1100 1001	0100 1001	V	1110 0101	0101 0110
J	1101 0001	0100 1010	W	1110 0110	0101 0111
K	1101 0010	0100 1011	X	1110 0111	0101 1000
L	1101 0011	0100 1100	Y	1110 1000	0101 1001
M	1101 0100	0100 1101	Z	1110 1001	0101 1010
0	1111 0000	0011 0000	5	1111 0101	0011 0101
1	1111 0001	0011 0001	6	1111 0110	0011 0110
2	1111 0010	0011 0010	7	1111 0111	0011 0111
3	1111 0011	0011 0011	8	1111 1000	0011 1000
4	1111 0100	0011 0100	9	1111 1001	0011 1001
!	0101 1010	0010 0001	;	0101 1110	0011 1011

The Parity Bit

Why does computer code need parity bits?

Dust, electrical disturbance, weather conditions, and other factors can cause interference in a circuit or communications line that is transmitting a byte. How does the computer know if an error has occurred? Detection is accomplished by use of a parity bit. **A _parity bit_, also called a *check bit*, is an extra bit attached to the end of a byte for purposes of checking for accuracy.**

Parity schemes may be *even parity* or *odd parity*. In an even-parity scheme, for example, the ASCII letter "H" (01001000) contains two 1s. Thus, the ninth bit, the parity bit, would be 0 in order to make the sum of the bits come out even. With the letter "O" (01001111), which has five 1s, the ninth bit would be 1 to make the byte come out even. (● *See Panel 4.7.*) The system software in the computer automatically and continually checks the parity scheme for accuracy. (If the message "Parity Error" appears on your screen, you need a technician to look at the computer to see what is causing the problem.)

1979	1981	1982	1983	1984	1985
Motorola introduces the 68000 chip, which later will support the Mac	IBM introduces personal computer (with 8088 CPU and 16 KB RAM)	Portable computers	The capacity of floppy disks is expanded to 360 KB; CDs are introduced to U.S. market	Apple Macintosh; first personal laser printer; Sony and Phillips introduce the CD-ROM; Intel's 80286 chip is released	Intel's 80386 32-bit microprocessor is introduced

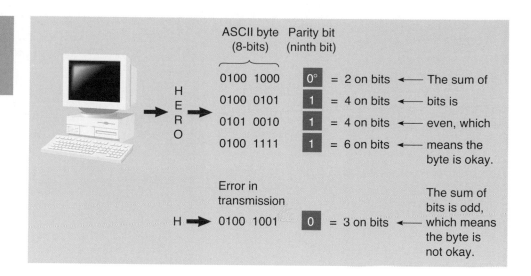

	ASCII byte (8-bits)	Parity bit (ninth bit)		
H	0100 1000	0°	= 2 on bits ←	The sum of
E	0100 0101	1	= 4 on bits ←	bits is
R	0101 0010	1	= 4 on bits ←	even, which
O	0100 1111	1	= 6 on bits ←	means the byte is okay.

Error in transmission

H →	0100 1001	0	= 3 on bits ←	The sum of bits is odd, which means the byte is not okay.

Machine Language

How would I define machine language?

Every brand of computer has its own binary language, called *machine language*. **Machine language is a binary-type programming language built into the CPU that the computer can run directly.** The machine language is specific to the particular CPU model; this is why, for example, software written for a Macintosh will not run on a Dell PC. To most people, an instruction written in machine language, consisting only of 0s and 1s, is incomprehensible. To the computer, however, the 0s and 1s represent precise storage locations and operations.

How do people-comprehensible program instructions become computer-comprehensible machine language? Special system programs called *language translators* rapidly convert the instructions into machine language. This translating occurs virtually instantaneously, so you are not aware it is happening. Machine language is discussed in more detail in Chapter 10.

Because the type of computer you will most likely be working with is the microcomputer, we'll now take a look at what's inside the microcomputer's system unit.

The Computer Case: Bays, Buttons, & Boards

What is a bay used for?

The *system unit* houses the motherboard (including the processor chip and memory chips), the power supply, and storage devices. (● *See Panel 4.8.*) In

1986	1988	1989	1990
The 3½" diskette is introduced for the Mac and becomes popular for the PC as well	Motorola's 32-bit 88000 series of RISC microprocessors is introduced	Double-sided, double-density floppy disks come on the market, increasing the 5¼" diskette to 1.2 MB and the 3½" diskette to 1.4 MB; Intel's 80486 chip with 1.2 million transistors is introduced; first portable Mac	Motorola's 68040 and Intel's 1486 chips are released

The system unit
Overhead view of the box, or case. It includes the motherboard, power supply, and storage devices. (The arrangement of the components varies among models.)

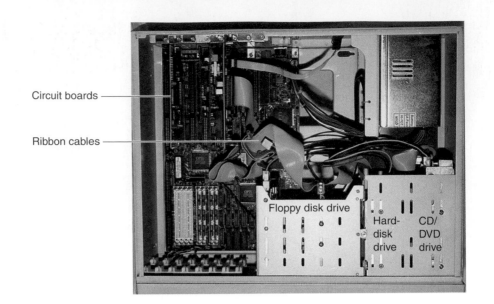

Circuit boards

Ribbon cables

Floppy disk drive

Hard-disk drive

CD/DVD drive

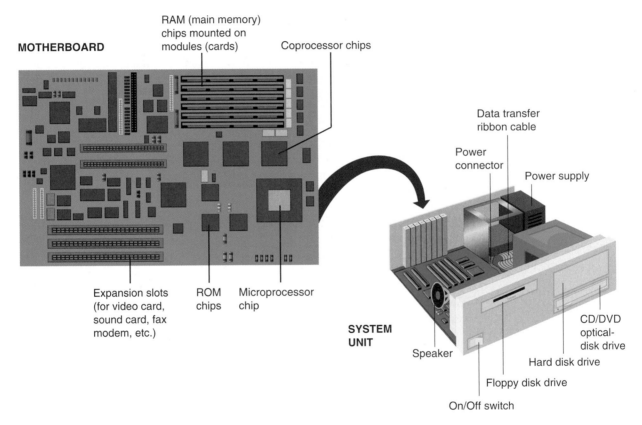

MOTHERBOARD

RAM (main memory) chips mounted on modules (cards)

Coprocessor chips

Data transfer ribbon cable

Power connector

Power supply

Expansion slots (for video card, sound card, fax modem, etc.)

ROM chips

Microprocessor chip

SYSTEM UNIT

Speaker

On/Off switch

Floppy disk drive

Hard disk drive

CD/DVD optical-disk drive

1993	1994	1995	1997	1998	1999	2000
Multimedia desktop computers; Intel introduces its first Pentium chip; Motorola releases the Power PC CPU	Apple and IBM introduce PCs with full-motion video built in; wireless data transmission for small portable computers; Power Macintosh based on Motorola's Power PC 601 microprocessor; DNA computing proof of concept released	Intel's Pentium Pro	Intel's Pentium II and Pentium MMX for games and multimedia	Apple iMac	Intel's Pentium III; AMD's Athlon CPU (800 MHz); Power Mac G4 available; end of the floppy disk predicted	Intel's Pentium 4; AMD's Athlon CPU reaches 1 GHz

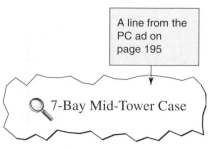

A line from the PC ad on page 195

7-Bay Mid-Tower Case

Survival Tip
Bay Access

Drive bays are the openings in your computer's case into which drives are installed. If the bay is "accessible," it's open to the outside of the PC (tape, floppy, CD/DVD drives). If it's "hidden," it's closed inside the PC case (hard drive).

computer ads, the part of the system unit that is the empty box with just the power supply is called the *case* or *system cabinet*.

For today's desktop PC, the system unit may be advertised as something like a "4-bay mini-tower case" or an "8-bay mid-tower case." **A _bay_ is a shelf or an opening used for the installation of electronic equipment,** generally storage devices such as a hard drive or DVD drive. A computer may come equipped with four to eight bays. Empty bays are covered by a panel.

A *tower* is a cabinet that is tall, narrow, and deep (so that it can sit on the floor beside or under a table) rather than short, wide, and deep. Originally a tower (full tower) was considered to be 24 inches high. Mini- (micro) and mid-towers may be less than half that size. At 6.5 inches square and 2 inches high, the Mac mini is the smallest desktop microcomputer.

The number of buttons on the outside of the computer case will vary, but the on/off power switch will appear somewhere, probably on the front or back. There may also be a "sleep" switch; this allows you to suspend operations without terminating them, so that you can conserve electrical power without the need for subsequently "rebooting," or restarting, the computer.

Inside the case—not visible unless you remove the cabinet—are various electrical circuit boards, chief of which is the motherboard, as we'll discuss.

Power Supply

What does a computer's power supply do?

The electricity available from a standard wall outlet is alternating current (AC), but a microcomputer runs on direct current (DC). **The _power supply_ is a device that converts AC to DC to run the computer.** The on/off switch in your computer turns on or shuts off the electricity to the power supply. Because electricity can generate a lot of heat, a fan inside the computer keeps the power supply and other components from becoming too hot.

Electrical power drawn from a standard AC outlet can be quite uneven. For example, a sudden surge, or "spike," in AC voltage can burn out the low-voltage DC circuitry in your computer ("fry the motherboard"). Instead of plugging your computer directly into a wall electrical outlet, it's a good idea to plug it into a power protection device. The three principal types are surge protectors, voltage regulators, and UPS units:

SURGE PROTECTOR A *surge protector*, or *surge suppressor*, is a device that protects a computer from being damaged by surges (spikes) of high voltage. The computer is plugged into the surge protector, which in turn is plugged into a standard electrical outlet. (*See the Practical Action box on page 203.*)

2001	2002	2003	2004	2005
Pentium 4 reaches 2 GHz; USB 2.0 is introduced	Pentium 4 reaches 3.06 GHz; Power Mac has 2 1-GHz Power PC CPUs; about 1 billion PCs have been shipped worldwide since the mid-70s	Intel's Pentium M/Centrino for mobile computing; 64-bit processors	Intel Express chipsets for built-in sound and video capabilities (no cards needed); IBM sells its PC computing division to Lenovo Group	Intel and AMD introduce dual-core processors; 64-bit processors enter the market

Surge protector   UPS

VOLTAGE REGULATOR A *voltage regulator*, or line conditioner, is a device that protects a computer from being damaged by insufficient power—"brownouts" or "sags" in voltage. Brownouts can occur when a large machine such as a power tool starts up and causes the lights in your house to dim. They also may occur on very hot summer days when the power company has to lower the voltage in an area because too many people are running their air conditioners all at once.

UPS A *UPS (uninterruptible power supply)* is a battery-operated device that provides a computer with electricity if there is a power failure. The UPS will keep a computer going for 5–30 minutes or more. It goes into operation as soon as the power to your computer fails.

Power supply units are usually rated in *joules*, named after a 19th-century English physicist. The higher the number of joules, the better the power protection. (One hundred joules of energy keep a 100-watt light going for 1 second.)

The Motherboard & the Microprocessor Chip

How is the motherboard important, and what are types of processor chips?

As we mentioned in Chapter 1, the *motherboard*, or *system board*, is the main circuit board in the system unit. The motherboard consists of a flat board that fills one side of the case. It contains both soldered, nonremovable components and sockets or slots for components that can bc removed—microprocessor chip, RAM chips, and various expansion cards, as we explain later. (● *See Panel 4.9.)*

Making some components removable allows you to expand or upgrade your system. **Expansion is a way of increasing a computer's capabilities by adding hardware to perform tasks that are beyond the scope of the basic system.** For example, you might want to add video and sound cards. **Upgrading means changing to newer, usually more powerful or sophisticated versions,** such as a more powerful microprocessor or more memory chips.

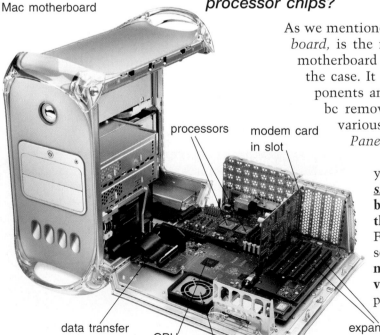

Mac motherboard

processors

modem card in slot

data transfer ribbon

CPU

fan

expansion slots

PRACTICAL ACTION
Preventing Problems from Too Much or Too Little Electrical Power to Your Computer

"When the power disappears, so can your data," writes *San Jose Mercury News* computer columnist Philip Robinson. "I say this with authority, sitting here in the dark in the wake of severe storms that have hit my part of California."[5] (Deprived of use of his computer, Robinson dictated his column by phone.)

Too little electricity can devastate your data. Too much electricity can devastate your computer hardware.

Here are a few things you can do to keep both safe:

- *Back up data regularly:* You should faithfully make backup (duplicate) copies of your data every few minutes as you're working. Then, if your computer has power problems, you'll be able to get back in business fairly quickly once the machine is running again.

- *Use a surge protector to protect against too much electricity:* Plug all your hardware into a surge protector (suppressor), which will prevent damage to your equipment if there is a power surge. (You'll know you've experienced a power surge when the lights in the room suddenly get very bright.) Surge protectors cost $20–$25.

- *Use a voltage regulator to protect against too little electricity:* Plug your computer into a voltage regulator (line conditioner) to adjust for power sags or brownouts. If power is too low for too long, it's as though the computer were turned off.

- *Consider using a UPS to protect against complete absence of electricity:* Consider plugging your computer into a UPS, or uninterruptible power supply. (A low-cost one, available at electronics stores, sells for about $100–$150.) The UPS is kind of a short-term battery that, when the power fails, will keep your computer running long enough (5–30 minutes) for you to save your data before you turn off the machine. It also acts as a surge protector.

- *Turn ON highest-power-consuming hardware first:* When you turn on your computer system, you should turn on the devices that use the most power first. This will avoid causing a power drain on smaller devices. The most common advice is to turn on (1) printer, (2) other external peripherals, (3) system unit, (4) monitor—in that order.

- *Turn OFF lowest-power-consuming hardware first:* When you turn off your system, follow the reverse order. This avoids a power surge to the smaller devices.

- *Unplug your computer system during lightning storms:* Unplug all your system's components—including phone lines—during thunder and lightning storms. If lightning strikes your house or the power lines, it can ruin your equipment.

RAM (main memory) chips mounted on modules (cards)

Microprocessor chip (with CPU)

THE MICROPROCESSOR CHIP The motherboard may be thought of as your computer's central nervous system. The brain of your computer is the microprocessor chip. As Chapter 1 described, a *microprocessor* is the miniaturized circuitry of a computer processor, contained on a small silicon chip. It stores program instructions that process, or manipulate, data into information.

- **Transistors—key parts of a chip:** The key parts of the microprocessor are transistors. *Transistors,* we said, are tiny electronic devices that act as on/off switches, which process the on/off (1/0) bits used to represent data. According to *Moore's law,* named for legendary Intel cofounder Gordon Moore, the number of transistors that can be packed onto a chip doubles about every 18 months, while the price stays the same, which has enabled the industry to shrink the size and cost of things such as computers and cellphones while improving their performance. In 1961 a chip had only 4 transistors. In 1971 it had 2,300; in 1979 it had 30,000; and in 1997 it had 7.5 million. In 2005, Intel announced a new Itanium chip with 1.7 billion transistors.[6]

- **The chipset—chips for controlling information among system components: The _chipset_ consists of groups of interconnected chips on the motherboard that control the flow of information between the microprocessor and other system components connected to the motherboard.** The chipset determines what types of processors, memory, and video card ports will work on the same motherboard. It also establishes the types of multimedia, storage, network, and other hardware the motherboard supports.

Intel Pentium 4 Processor 2.80 GHz

TRADITIONAL MICROCOMPUTER MICROPROCESSORS Most personal computers in use today have one of two types of microprocessors—one for the PC, one for the Macintosh. The leading chip makers are Intel, AMD, IBM, and Motorola/Freescale.

- **Intel-type processors for PCs—Intel and AMD chips:** Most PCs use CPUs manufactured by Intel Corporation or Advanced Micro Devices (AMD). Indeed, the Microsoft Windows operating system was designed to run on Intel chips. **_Intel-type chips_ have a similar internal design and are made to run PCs.** They are used by manufacturers such as Dell, Gateway, and Hewlett-Packard in their PC microcomputers.

  Since 1993, Intel has marketed its chips under such names as "Pentium," "Pentium Pro," "Pentium II," and "Pentium III." New computers usually have a "Pentium 4" (P4), "Celeron," "Xeon," or "Itanium" processor. Many ads for PCs contain the logo "Intel inside" to show that the systems run an Intel microprocessor. (AMD has competed with Intel in its K6 and Athlon processor series.)

  All future Intel desktop and mobile processors will fall into broader categories, using different number series—300, 400, 500, and so on. Thus, for its dual-core Pentium Extreme Edition 840, the name conspicuously does not include the words "Pentium" and "4" arranged together, although, as one observer points out, "the Extreme Edition packs a thoroughly Pentium 4 heritage."[7] Using the name switch to simplify consumer choices, Intel will place desktop-based Celeron chips in the 300 series, Pentium 4 chips in the 400 series, and Pentium 4 Extreme Editions in the 700 and 800 series. The notebook series will see mobile Celerons in the 400 series, Pentium 4 Mobile chips in the 500 series, and Pentium M chips in the 700 series.

- **Motorola/Freescale and IBM processors for Macintoshes:** Apple computers have used a design different from Intel-type processors. For

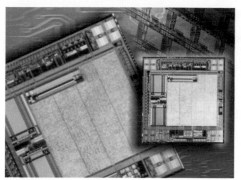

Motorola processor

many years, _**Motorola-type chips**_ were made by Motorola and later its subsidiary Freescale Semiconductor for Apple Macintosh computers up to and including the Apple Mac G4. PowerMac G5 Macintoshes use the PowerPC family of processors, which were originally developed in a cooperative effort by Apple, Freestone, and IBM. **The advantage of _PowerPC chips_ is that, with certain hardware or software configurations, a Macintosh can run PC as well as Mac applications software.**

NEW MICROCOMPUTER MICROPROCESSORS Two new developments in chips are certain to have significant effects:

- **Intel processors for Macintoshes:** In mid-2005, Apple announced that it would end its current line of products based on PowerPC chips from IBM and Freescale and would have all its machines running on Intel chips by the end of June 2007. To help existing programs run on Intel-powered Macs, Apple planned to introduce a utility program called _Rosetta_ that would automatically adapt PowerPC programs.

  Although risky for Apple, the advantage of this major switch is that an Intel processor gives Mac users access to the vast array of Windows-based games and programs (at least those users who are willing to run a Windows OS as well as Mac OS on their systems). This should appeal to Windows users who are attracted to Apple's designs but have stayed away from Macs because they don't run Windows programs.[8]

  Even Windows users will benefit, suggests well-known technology journalist Walter Mossberg, because, as the most innovative major computer maker, Apple's impact on the industry is vastly greater than its market share. "Almost everything it does is later copied by the Windows PC makers," he points out, "so keeping Apple strong and innovating is good for Windows users, too."[9] (However, it's doubtful that people will be able to run the Mac operating system on non-Apple machines.)

- **Multicore processors for PCs:** "You'll be typing along on an email when suddenly the PC stops responding to your keystrokes, then catches up a few seconds later," writes _BusinessWeek_'s Stephen Wildstrom. "Or a program that used to load in a few seconds inexplicably takes three times as long. Processors are faster than ever, but the demands of even routine computing are overwhelming them."[10]

  The reason computers are bogging down is that, as Wildstrom explains, "no matter how fast a processor runs, it can do only one thing at a time." Adding more transistors won't help because they generate too much heat. Enter a new kind of microcomputer chip— the _**multicore processor**_, **which is designed to let the operating system divide the work over more than one processor, with two or more processor "cores" on a single piece of silicon.**

  The concept is not new; large computer systems, such as IBM's Blue Gene/L supercomputer, have featured as many as 65,000 processors working in unison.[11] But the beauty of having two or more cores is that chips can take on several tasks at once, eliminating the annoying pauses that Wildstrom mentioned. AMD calls its "dual-core" version the _Athlon 64 X2_; Intel's entry is the _Pentium D_ chip. (For videogamers, AMD offers the even faster _64 FX-57_; Intel announced a dual-core Montecito Itanium to appear in servers.[12]) In 2005, IBM, Sony, and Toshiba announced a multicore chip called _Cell_ that will have nine separate cores, which could tackle nine tasks at once; it was anticipated that Cell would be used in the PlayStation 3.[13] (The multicore technology takes advantage of a technology called _hyperthreading_, explained in a few pages.)

Processing Speeds: From Megahertz to Picoseconds

How does the system clock work in my computer, and how is its speed measured?

Often a PC ad will say something like "Intel Celeron processor 2.40 GHz," "Intel Pentium 4 processor 2.80 GHz," or "AMD Athlon 64 FX-57 2.8 GHz." *GHz* stands for "gigahertz." These figures indicate how fast the microprocessor can process data and execute program instructions.

Every microprocessor contains a ***system clock***, **which controls how fast all the operations within a computer take place.** The system clock uses fixed vibrations from a quartz crystal to deliver a steady stream of digital pulses or "ticks" to the CPU. These ticks are called *cycles*. Faster *clock speeds* will result in faster processing of data and execution of program instructions, as long as the computer's internal circuits can handle the increased speed.

There are four main ways in which processing speeds are measured, as follows.

more info!

Which Intel Processor Is Better?

Which is the better Intel processor, Pentium or Celeron, for basic word processing and web surfing versus video games and complex applications? Find out by doing a web search.

FOR MICROCOMPUTERS: MEGAHERTZ & GIGAHERTZ Older microcomputer microprocessor speeds are expressed in ***megahertz (MHz)*****, a measure of frequency equivalent to 1 million cycles (ticks of the system clock) per second.** The original IBM PC had a clock speed of 4.77 megahertz, which equaled 4.77 million cycles per second. The latest-generation processors from AMD and Intel operate in ***gigahertz (GHz)***—**a billion cycles per second.** Intel's Pentium 4 operates at up to 3.60 gigahertz, or 3.60 billion cycles per second. Some experts predict that advances in microprocessor technology will produce a 50-gigahertz processor by 2010, the kind of power that will be required to support such functions as true speech interfaces and real-time speech translation. However, unfortunately, the faster a CPU runs, the more power it consumes and the more waste heat it produces. Thus, rather than increasing clock speeds, which requires smaller transistors and creates tricky engineering problems, chip makers such as Intel and AMD are now concentrating on employing a second CPU core and running it in parallel—dual core or multicore technology, as we've described.[14]

As for you, since a new high-speed processor can cost many hundred dollars more than a previous-generation chip, experts often recommend that buyers fret less about the speed of the processor (since the work most people do on their PCs doesn't even tax the limits of the current hardware) and more about spending money on extra memory. (However, game playing *does* tax the system. Thus, if you're an avid computer game player, you may want to purchase the fastest processor.)

Comparison of Some Popular Recent Microcomputer Processors

Year	Processor Name	Clock Speed	Transistors
2006	Intel Pentium EE 840 dual core	3.2 GHZ (each core)	230 million
2005	Intel Pentium 4 660	3.6–3.7 GHz	169 million
2005	AMD Athlon 64 X2 dual core	2 GHz (each core)	105.9 million
2005	Intel Itanium 2 Montecito dual core	2 GHz (each core)	1.7 billion
2004	IBM PowerPC 970FX (G5)	2.2 GHz	58 million
2003	AMD Opteron	2–2.4 GHz	37.5 million
2002	Intel Itanium 2	1 GHz and up	221 million
2002	AMD Athlon MP	1.53–1.6 GHz	37.5 million
2001	Intel Xeon	1.4–2.8 GHz	140 million
2001	Intel Mobile Pentium 4	1.4–3.06 GHz	55 million
2001	AMD Athlon XP	1.33–1.73 GHz	37.5 million
2001	Intel Itanium	733–800 MHz	25.4–60 million
2000	Intel Pentium 4	1.4–3.06 GHz	42–55 million
1999	Motorola PowerPC 7400 (G4)	400–500 MHz	10.5 million

FOR WORKSTATIONS, MINICOMPUTERS, & MAINFRAMES: MIPS Processing speed can also be measured according to the number of instructions per second that a computer can process. **_MIPS_ stands for "millions of instructions per second."** MIPS is used to measure processing speeds of mainframes, minicomputers, and workstations. A workstation might perform at 100 MIPS or more, a mainframe at 200–1,200 MIPS.

FOR SUPERCOMPUTERS: FLOPS The abbreviation **_flops_ stands for "floating-point operations per second."** A _floating-point operation_ is a special kind of mathematical calculation. This measure, used mainly with supercomputers, is expressed as _megaflops_ (_mflops_, or millions of floating-point operations per second), _gigaflops_ (_gflops_, or billions), and _teraflops_ (_tflops_, or trillions). IBM's Blue Gene/L (for "Lite") supercomputer cranks out 280.6 teraflops, or 280.6 trillion calculations per second. (With the previous speed of 70.72 teraflops, a person able to complete one arithmetic calculation every second would take more than a million years to do what Blue Gene/L does in a single second.) New supercomputer speeds will be measured in _petaflops_ (1 quadrillion operations per second). IBM is working on an improved Blue Gene supercomputer that is supposed to run at 360 teraflops.

FOR ALL COMPUTERS: FRACTIONS OF A SECOND Another way to measure cycle times is in fractions of a second. A microcomputer operates in microseconds, a supercomputer in nanoseconds or picoseconds—thousands or millions of times faster. A _millisecond_ is one-thousandth of a second. A _microsecond_ is one-millionth of a second. A _nanosecond_ is one billionth of a second. A _picosecond_ is one-trillionth of a second.

4.3

More on the System Unit

How do the processor and memory work, and what are some important ports, buses, and cards?

Technology moves on—toward simplifying on the one hand, toward mastering more complexity on the other. An example of simplifying: How about a $100 laptop that could be put into the hands of billions of schoolchildren in the world's poorest nations? Nicholas Negroponte of the Media Lab at the Massachusetts Institute of Technology has proposed just such a "world computer." It wouldn't have a hard drive or expensive display screen, it would run on Linux, and it would come with less capable batteries and a hand crank for recharging them.[15] An example of mastering more complexity: As mentioned, in 2005, IBM, Sony, and Toshiba announced a processor chip called Cell that has nine processing units, or cores, each of which can work independently, allowing the processor to tackle nine tasks at once. Cell's processing power is such that it would have been among the top 500 supercomputers in 2002.

To understand the different ways computers are diverging, let's take a deeper look at how the system unit works.

How the Processor or CPU Works: Control Unit, ALU, Registers, & Buses

How does the CPU and its parts work?

Once upon a time, the processor in a computer was measured in feet. A processing unit in the 1946 ENIAC (which had 20 such processors) was about 2 feet wide and 8 feet high. Today, computers are based on _micro_processors, less than 1 centimeter square. It may be difficult to visualize components so tiny. Yet it is necessary to understand how microprocessors work if you are

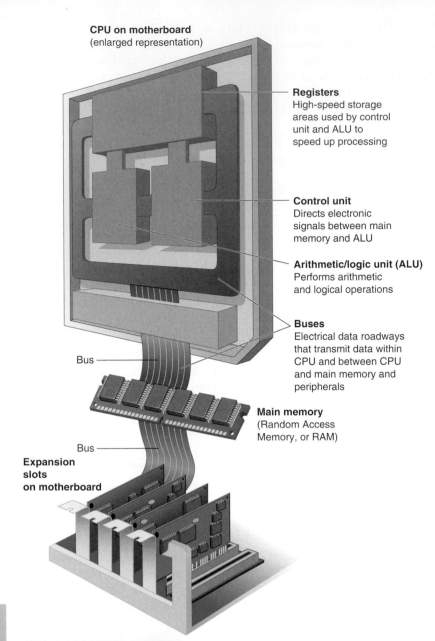

CPU on motherboard
(enlarged representation)

Registers
High-speed storage areas used by control unit and ALU to speed up processing

Control unit
Directs electronic signals between main memory and ALU

Arithmetic/logic unit (ALU)
Performs arithmetic and logical operations

Buses
Electrical data roadways that transmit data within CPU and between CPU and main memory and peripherals

Main memory
(Random Access Memory, or RAM)

Bus

Bus

Expansion slots on motherboard

● **PANEL 4.10**
The CPU and main memory
The two main CPU components on a microprocessor are the control unit and the ALU, which contain working storage areas called *registers* and are linked by a kind of electronic roadway called a *bus*.

to grasp what PC advertisers mean when they throw out terms such as "256 MB DDR-SDRAM" or "512 K Level 2 Advanced Transfer Cache."

WORD SIZE   Computer professionals often discuss a computer's word size. **_Word size_ is the number of bits that the processor may process at any one time.** The more bits in a word, the faster the computer. A 32-bit computer—that is, one with a 32-bit-word processor—will transfer data within each microprocessor chip in 32-bit chunks, or 4 bytes at a time. (Recall there are 8 bits in a byte.) A 64-bit-word computer is faster; it transfers data in 64-bit chunks, or 8 bytes at a time.

THE PARTS OF THE CPU   A processor is also called the *CPU,* and it works hand in hand with other circuits known as *main memory* to carry out processing. The **_CPU (central processing unit)_ is the "brain" of the computer; it follows the instructions of the software (program) to manipulate data into information. The CPU consists of two parts—(1) the control unit and (2) the arithmetic/logic unit (ALU), both of which contain registers, or high-speed storage areas** (as we discuss shortly). All are linked by a kind of electronic "roadway" called a *bus.* (● *See Panel 4.10.*)

- **The control unit—for directing electronic signals: The _control unit_ deciphers each instruction stored in the CPU and then carries out the instruction.** It directs the movement of electronic signals between main memory and the arithmetic/logic unit. It also directs these electronic signals between main memory and the input and output devices.
  For every instruction, the control unit carries out four basic operations, known as the *machine cycle.* In the **_machine cycle_, the CPU (1) fetches an instruction, (2) decodes the instruction, (3) executes the instruction, and (4) stores the result.** (● *See Panel 4.11.*)

- **The arithmetic/logic unit—for arithmetic and logical operations: The _arithmetic/logic unit (ALU)_ performs arithmetic operations and logical operations and controls the speed of those operations.**
  As you might guess, *arithmetic operations* are the fundamental math operations: addition, subtraction, multiplication, and division.
  Logical operations are comparisons. That is, the ALU compares two pieces of data to see whether one is equal to (=), greater than (>), greater than or equal to (>=), less than (<), less than or equal to (<=), or not equal to ( ) the other.

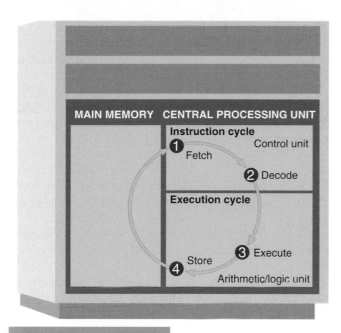
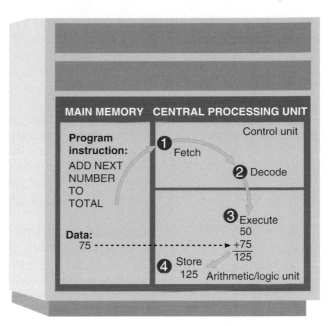

MAIN MEMORY CENTRAL PROCESSING UNIT

Instruction cycle
❶ Fetch
Control unit
❷ Decode

Execution cycle
❸ Execute
❹ Store
Arithmetic/logic unit

MAIN MEMORY CENTRAL PROCESSING UNIT

Program instruction:
ADD NEXT NUMBER TO TOTAL

Data:
75

❶ Fetch
Control unit
❷ Decode
❸ Execute
50
+75
125
❹ Store
125 Arithmetic/logic unit

● **PANEL 4.11**

The machine cycle

(Left) The machine cycle executes instructions one at a time during the instruction cycle and the execution cycle. *(Right)* Example of how the addition of two numbers, 50 and 75, is processed and stored in a single cycle.

- **Registers—special high-speed storage areas:** The control unit and the ALU also use registers, special CPU areas that enhance the computer's performance. _**Registers**_ **are high-speed storage areas that temporarily store data during processing.** They may store a program instruction while it is being decoded, store data while it is being processed by the ALU, or store the results of a calculation.

  All data must be represented in a register before it can be processed. For example, if two numbers are to be multiplied, both numbers must be in registers, and the result is also placed in a register. (The register can contain the address of a memory location where data is stored rather than the actual data itself.)

  The number of registers that a CPU has and the size of each (number of bits) help determine the power and speed of a CPU. For example, a 32-bit CPU is one in which each register is 32 bits wide. Therefore, each CPU instruction can manipulate 32 bits of data. (There are several types of registers, including *instruction register, address register, storage register, and accumulator register.*)

- **Buses—data roadways:** _**Buses**_**,** or *bus lines,* **are electrical data roadways through which bits are transmitted within the CPU and between the CPU and other components of the motherboard.** A bus resembles a multilane highway: The more lanes it has, the faster the bits can be transferred. The old-fashioned 8-bit-word bus of early microprocessors had only eight pathways. Data is transmitted four times faster in a computer with a 32-bit bus, which has 32 pathways, than in a computer with an 8-bit bus. Intel's Pentium chip is a 64-bit processor, as are the Macintosh G5 processors (some models have two). Some supercomputers have 128-bit processors.

We return to a discussion of buses in a few pages.

How Memory Works: RAM, ROM, CMOS, & Flash

How do I distinguish among the four principal types of memory chips?

So far we have described only the kinds of chips known as microprocessors. But other silicon chips called *memory chips* are attached to the motherboard. The four principal types of memory chips are *RAM, ROM, CMOS,* and *flash.*

RAM CHIPS—TO TEMPORARILY STORE PROGRAM INSTRUCTIONS & DATA Recall from Chapter 1 that there are two types of storage, primary and secondary. Primary storage is temporary or working storage and is often called *memory* or *main memory*; secondary storage, usually called just *storage*, is relatively permanent storage. *Memory* refers to storage media in the form of chips, and *storage* refers to media such as disks and tape. **_RAM (random access memory) chips_ temporarily hold (1) software instructions and (2) data before and after it is processed by the CPU.** Think of RAM as the primary workspace inside your computer. When you open a file, a copy of the file transfers from the hard disk to RAM, and this copy in RAM is the one that changes as you work with the file. When you activate the Save command, the changed copy transfers from RAM back to permanent storage on the hard drive.

Because its contents are temporary, RAM is said to be **_volatile_—the contents are lost when the power goes off or is turned off.** This is why you should *frequently*—every 5–10 minutes, say—transfer (save) your work to a secondary storage medium such as your hard disk, in case the electricity goes off while you're working. (However, there is one kind of RAM, called *flash RAM*, that is not temporary, as we'll discuss shortly.)

Several types of RAM chips are used in personal computers—*DRAM, SDRAM, SRAM,* and *DDR-SDRAM:*

- **DRAM:** The first type (pronounced "dee-ram"), *DRAM (dynamic RAM)*, must be constantly refreshed by the CPU or it will lose its contents.

- **SDRAM:** The second type of RAM is *SDRAM (synchronous dynamic RAM)*, which is synchronized by the system clock and is much faster than DRAM. Often in computer ads, the speed of SDRAM is expressed in megahertz.

- **SRAM:** The third type, *static RAM,* or *SRAM* (pronounced "ess-ram"), is faster than DRAM and retains its contents without having to be refreshed by the CPU.

- **DDR-SDRAM:** The fourth type, *DDR-SDRAM (double-data rate synchronous dynamic RAM)*, is the newest type of RAM chip and is the one most commonly used in PCs and in Apple computers today. Its speed is also measured in megahertz.

Microcomputers come with different amounts of RAM, which is usually measured in megabytes or gigabytes. An ad may list "256 MB SDRAM," but you can get more. The Macintosh G5, for instance, can provide up to 8 gigabytes of RAM. The more RAM you have, the more efficiently the computer operates and the better your software performs. *Having enough RAM is a critical matter.* Before you buy a software package, look at the outside of the box or check the manufacturer's website to see how much RAM is required. Microsoft Office XP, for instance, states that a minimum of 24–64 megabytes of RAM is required, depending on the operating system, plus 8 megabytes of RAM for *each* application the user plans to run simultaneously. Current recommendations suggest 512 megabytes as the minimum system RAM requirement.

512 MB 533 MHz DDR2 SDRAM

If you're short on memory capacity, you can usually add more RAM chips by plugging a RAM *memory module* into the motherboard. A memory module is a small fiberglass circuit board that can be plugged into an expansion slot on the motherboard. There are two types of such modules: SIMMs and the newer DIMMs.

SIMM

- **SIMM:** A *SIMM (single inline memory module)* has RAM chips on only one

DIMM

more info!

RAM-ifications

To find out more about how RAM works, go to http://computer.howstuffworks.com/ram.htm.

side. SIMMs are available in either FPM (fast page mode) or EDO (extended data output) speeds, with EDO being the faster of the two. SIMMs are more expensive than modern memory modules because they are no longer in demand, which also makes them difficult to obtain.

- **DIMM:** A *DIMM (dual inline memory module)* has RAM chips on both sides. DIMMs are the most popular and common type of RAM used today.

ROM CHIPS—TO STORE FIXED START-UP INSTRUCTIONS   Unlike RAM, to which data is constantly being added and removed, ___ROM (read-only memory)___ **cannot be written on or erased by the computer user without special equipment. ROM chips contain fixed start-up instructions.** That is, ROM chips are loaded, at the factory, with programs containing special instructions for basic computer operations, such as those that start the computer (BIOS) or put characters on the screen. These chips are nonvolatile; their contents are not lost when power to the computer is turned off.

In computer terminology, ___read___ **means to transfer data from an input source into the computer's memory or CPU. The opposite is** ___write___**—to transfer data from the computer's CPU or memory to an output device.** Thus, with a ROM chip, *read-only* means that the CPU can retrieve programs from the ROM chip but cannot modify or add to those programs. A variation is *PROM (programmable read-only memory)*, which is a ROM chip that allows you, the user, to load read-only programs and data. However, this can be done only once.

CMOS CHIPS—TO STORE FLEXIBLE START-UP INSTRUCTIONS   Pronounced "*see*-moss," ___CMOS (complementary metal-oxide semiconductor)___ ___chips___ **are powered by a battery and thus don't lose their contents when the power is turned off.** CMOS chips contain flexible start-up instructions—such as time, date, and calendar—that must be kept current even when the computer is turned off. Unlike ROM chips, CMOS chips can be reprogrammed, as when you need to change the time for daylight savings time. (Your system software may prompt you to do this; newer systems do it automatically.)

FLASH MEMORY CHIPS—TO STORE FLEXIBLE PROGRAMS   Also a nonvolatile form of memory, ___flash memory chips___ **can be erased and reprogrammed more than once** (unlike PROM chips, which can be programmed only once). Flash memory, which doesn't require a battery and which can range from 32 to 128 megabytes in capacity, is used to store programs not only in personal computers but also in pagers, cellphones, MP3 players, Palm organizers, printers, and digital cameras. Flash memory is also used in newer PCs for BIOS instructions; they can be updated electronically on flash memory—the chip does not need to be replaced, as a ROM chip would.

How Cache Works

How do cache and virtual memory differ?

Because the CPU runs so much faster than the main system RAM, it ends up waiting for information, which is inefficient. To reduce this effect, we have cache. Pronounced "cash," ___cache___ **temporarily stores instructions and data that the processor is likely to use frequently. Thus, cache speeds up processing.** SRAM chips are commonly used for cache.

THREE KINDS OF CACHE   There are three kinds of cache, as follows:

- **Level 1 (L1) cache—part of the microprocessor chip:** *Level 1 (L1) cache*, also called *internal cache*, is built into the processor chip.

1 MB L2 Cache

Ranging from 8 to 256 kilobytes, its capacity is less than that of Level 2 cache, although it operates faster.

- **Level 2 (L2) cache—not part of the microprocessor chip:** This is the kind of cache usually referred to in computer ads. *Level 2 (L2) cache,* also called *external cache,* resides outside the processor chip and consists of SRAM chips. Capacities range from 64 kilobytes to 2 megabytes. (In Intel ads, L2 is called *Advanced Transfer Cache.*) L2 cache is generally quite a bit larger than L1 cache (most new systems have at least 512 kilobytes of L2 cache) and is the most commonly cited type of cache when measuring PC performance.

- **Level 3 (L3) cache—on the motherboard:** *Level 3 (L3) cache* is a cache separate from the processor chip on the motherboard. It is found only on very high-end computers, only those that use L2 Advanced Transfer Cache.

Cache is not upgradable; it is set by the type of processor purchased with the system.

VIRTUAL MEMORY   In addition to including cache, most current computer operating systems allow for the use of **_virtual memory_—that is, some free hard-disk space is used to extend the capacity of RAM.** The processor searches for data or program instructions in the following order: first L1, then L2, then RAM, then hard disk (or CD). In this progression, each kind of memory or storage is slower than its predecessor.

Other Methods of Speeding Up Processing

What are interleaving, bursting, pipelining, superscalar architecture, and hyperthreading?

The placement of memory chips on the motherboard has a direct effect on system performance. Because RAM must hold all the information the CPU needs to process, the speed at which the data can travel between memory and the CPU is critical to performance. And because the exchanges of data between the CPU and RAM are so intricately timed, the distance between them becomes another critical performance factor. Ways to speed up data traveling between memory and CPU are *interleaving, bursting, pipelining, superscalar architecture,* and *hyperthreading.* When you see these terms in a computer ad, you'll know that the processor's speed is being boosted.

INTERLEAVING   The term *interleaving* refers to a process in which the CPU alternates communication between two or more memory banks. Interleaving is generally used in large systems such as servers and workstations. For example, SDRAM chips are each divided into independent cell banks. Interleaving between the two cell banks produces a continuous flow of data.

BURSTING   The purpose of *bursting* is to provide the CPU with additional data from memory based on the likelihood that it will be needed. So, instead of the CPU retrieving data from memory one piece at a time, it grabs a block of information from several consecutive addresses in memory. This saves time because there's a statistical likelihood that the next data address the CPU will request will be sequential to the previous one.

PIPELINING   *Pipelining* divides a task into a series of stages, with some of the work completed at each stage. That is, large tasks are divided into smaller overlapping ones. The CPU does not wait for one instruction to complete the machine cycle before fetching the next instruction. Pipelining is available in most PCs; each processor can pipeline up to four instructions.

SUPERSCALAR ARCHITECTURE & HYPERTHREADING *Superscalar architecture* means the computer has the ability to execute more than one instruction per clock cycle (a 200-MHz processor executes 200 million clock cycles per second). One type of such architecture is hyperthreading. With *hyperthreading,* software and operating systems treat the microprocessor as though it's two microprocessors. This technology lets the microprocessor handle simultaneous requests from the OS or from software, initially improving performance by around 30%–40%. A processor using hyperthreading technology manages the incoming data instructions in parallel by switching between the instructions every few nanoseconds, essentially letting the processor handle two separate threads of code at once.

Ports & Cables

What ports will I probably use most?

A *port* is a connecting socket or jack on the outside of the system unit into which are plugged different kinds of cables. (● *See Panel 4.12 on the next page.*) A port allows you to plug in a cable to connect a peripheral device, such as a monitor, printer, or modem, so that it can communicate with the computer system.

Ports are of several types, as follows.

DEDICATED PORTS—FOR KEYBOARD, MOUSE, MONITOR, AUDIO, & MODEM/NETWORK CONNECTION *Dedicated ports* are ports for special purposes, such as the round ports for connecting the keyboard and the mouse (if they're not USB), the monitor port, the audio ports (green for speakers or headphones, pink for microphone, yellow for home stereo connection), the modem port to connect your computer to a phone line, and a network port for a high-speed internet connection. (There is also one connector that is not a port at all—the power plug socket, into which you insert the power cord that brings electricity from a wall plug.)

SERIAL, PARALLEL, & SCSI PORTS Most ports other than the dedicated ones just described are generally multipurpose. We consider serial, parallel, and SCSI ports:

- Serial ports—for transmitting slow data over long distances: **A line connected to a *serial port* will send bits one at a time, one after another,** like cars on a one-lane highway. Because individual bits must follow each other, a serial port is usually used to connect devices that do not require fast transmission of data, such as keyboard, mouse, monitors, and dial-up modems. It is also useful for sending data over a long distance.

- Parallel ports—for transmitting fast data over short distances: **A line connected to a *parallel port* allows 8 bits (1 byte) to be transmitted simultaneously,** like cars on an eight-lane highway. Parallel lines move information faster than serial lines do, but they can transmit information efficiently only up to 15 feet. Thus, parallel ports are used principally for connecting printers or external disk or magnetic-tape backup storage devices.

- SCSI ports—for transmitting fast data to up to seven devices in a daisy chain: Pronounced "scuzzy," a *SCSI (small computer system interface) port* **allows data to be transmitted in a "daisy chain" to up to seven devices connected to a single port at speeds (32 bits at a time) higher than those possible with serial and parallel ports.** Among the devices that may be connected are external hard-disk drives, CD drives, scanners, and magnetic-tape backup units. The term *daisy chain* means that several devices are connected in series to each

The Jhai computer, a low-wattage computer run by a car battery drawing power from a bicycle-powered generator. The computer is linked to a wireless network that supports small businesses in remote Laotian villages.

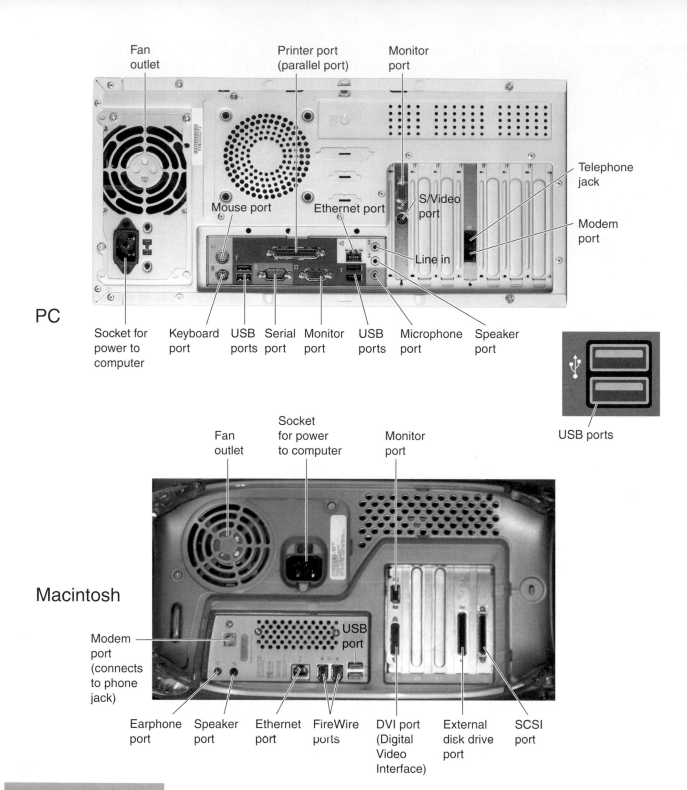

PC

Fan outlet

Printer port (parallel port)

Monitor port

Telephone jack

Modem port

S/Video port

Mouse port

Ethernet port

Line in

Socket for power to computer

Keyboard port

USB ports

Serial port

Monitor port

USB ports

Microphone port

Speaker port

USB ports

Macintosh

Fan outlet

Socket for power to computer

Monitor port

Modem port (connects to phone jack)

USB port

Earphone port

Speaker port

Ethernet port

FireWire ports

DVI port (Digital Video Interface)

External disk drive port

SCSI port

● **PANEL 4.12**
Ports
The backs of a PC and a Macintosh. Additional USB ports may be on the front or side of some system units.

🔍 6 USB 2.0 Ports

other, so that data for the seventh device, for example, has to go through the other six devices first. Sometimes the equipment on the chain is inside the computer, an internal daisy chain; sometimes it is outside the computer, an external daisy chain. (● *See Panel 4.13.*)

USB PORTS—FOR TRANSMITTING DATA TO UP TO 127 DEVICES IN A DAISY CHAIN A single *USB (universal serial bus) port* can theoretically **connect up to 127 peripheral devices in a daisy chain.** USB ports are multipurpose, useful for all kinds of peripherals, and are included on all new computers.

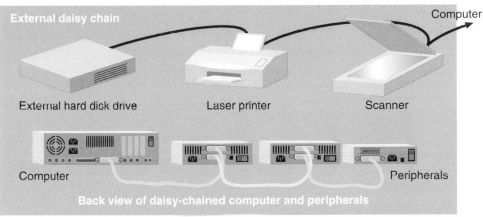

Internal daisy chain (inside computer)

SCSI controller in expansion slot

Hard disk drive

Tape drive

CD/DVD drive

External daisy chain

Computer

External hard disk drive

Laser printer

Scanner

Computer

Peripherals

Back view of daisy-chained computer and peripherals

USB hub connectors

- **The goals of USB:** The designers of the USB standard had several goals in mind. They wanted it to . . .

  1. Be low-cost so that it could be used in cheap peripherals such as mice and game controllers.

  2. Be able to connect lots of devices and have sufficient speed that it could replace all the different ports on computers with a single standard.

  3. Be "hot swappable" or "hot pluggable," meaning that it could allow USB devices to be connected or disconnected even while the PC is running.

  4. Permit _**plug and play**_—**to allow peripheral devices and expansion cards to be automatically configured while they are being installed**—to avoid the hassle of setting switches and creating special files, as was required of early users.

  USB has fulfilled these goals, so now just about every peripheral is available in a USB version, and it's expected that soon microcomputers will have nothing but USB ports.

- **USB connections—daisy chains, hubs, and wireless:** Can you really connect up to 127 devices on a single chain? An Intel engineer did set a world record at an industry trade show before a live audience by connecting 111 peripheral devices to a single USB port on a personal computer. But many USB peripherals do not support such long daisy chains. Thus, because some PCs contain only two USB ports, it's worth shopping around to find a model with more than two.

  You can also hook up a USB hub to one of the USB ports. A _USB hub_ typically has four USB ports. You plug the hub into your computer, and then plug your devices—or other hubs—into that. By chaining hubs together, you can build up dozens of available USB ports on a single computer. However, using a hub weakens data signals, so some peripherals, such as cable modems, work better when plugged directly into one of the computer's USB ports.

Hardware: The CPU & Storage

215

USB peripheral devices can
draw their power from their
USB connection. Printers and
scanners have their own power
supply. Mice and digital
cameras draw power from the
PC. If you have lots of self-
powered devices (like printers
and scanners), your hub need
not be powered. If you have
lots of unpowered devices (like
mice and cameras), you probably need a powered hub.

Find the USB port by looking
for the ⟶ icon. If there's a
tiny + (plus sign) next to the
icon, you have USB 2.0.

FireWire port

USB port

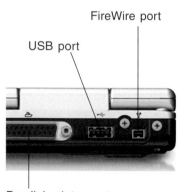

Parallel printer port on a
notebook computer

Infrared port

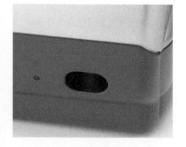

Individual USB cables can run as long as 5 meters (about 5.4 yards). With hubs, devices can be up to 30 meters (about 32.8 yards) away. Each USB cable has an A connector, which plugs into the computer, and a B connector, which plugs into the peripheral USB device.

Wireless USB technology was introduced in 2004, and wireless USB products became available in 2006. This connection technology will be useful in home and office networks in ranges of 9 to about 30 feet.

- **USB standards:** Just about every peripheral made now comes in a USB version. Common USB standards are *USB 1.1* and the more recent *USB 2.0*. Nearly all new PCs have USB 2.0 ports. Users with an older PC with USB 1.1 ports will have to buy a 2.0 add-on upgrade card to be able to hook up 2.0 peripherals. (USB 2.0 data "throughput"—the speed at which data passes through the cable—is 2–13 times faster than USB 1.1.) Windows 95 and Windows NT do not support USB.

  An even newer USB standard, *USB On-The-Go (USB OTG)* is being developed. USB OTG enables devices other than a personal computer to be connected together, such as PDAs, cellphones, and digital cameras.

- **USB connectors:** There are four types of USB connectors: A, B, Mini B, and Mini A. USB 1.1 specifies the Type A and Type B. USB 2.0 specifies the Type A, Type B, and Mini B. The Mini A connector was developed as part of the USB OTG specification and is used for smaller peripherals, such as cellphones.

SPECIALIZED EXPANSION PORTS—FIREWIRE, MIDI, IRDA, BLUE-TOOTH, & ETHERNET   Some specialized expansion ports are these:

- **FireWire ports—for camcorders, DVD players, and TVs:** FireWire was created by Apple Computer and later standardized as IEEE-1394 (IEEE is short for Institute of Electrical and Electronics Engineers). It actually preceded USB and had similar goals. The difference is that **_FireWire (IEEE-1394)_ is intended for devices working with lots of data—not just mice and keyboards but digital video recorders, DVD players, gaming consoles, and digital audio equipment.** Like USB, FireWire is a serial bus. However, whereas USB is limited to 12 megabits per second, FireWire currently handles up to 400 megabits per second. USB can handle 127 devices per bus, while FireWire handles 63. Both USB and FireWire allow you to plug and unplug devices at any time.

  FireWire doesn't always require the use of a PC; you can connect a FireWire camcorder directly to a digital TV, for example, without a PC in the middle. Like USB devices, FireWire devices can be powered or unpowered, and they are also hot pluggable. FireWire requires a special card in the PC.

- **MIDI ports—for connecting musical instruments:** A **_MIDI_ (pronounced "_mid_-dee" and short for _Musical Instrument Digital Interface_) _port_ is a specialized port used in creating, recording, editing, and performing music.** It is used for connecting amplifiers, electronic synthesizers, sound cards, drum machines, and the like.

- **IrDA ports—for cableless connections over a few feet:** When you use a handheld remote unit to change channels on a TV set, you're using invisible radio waves of the type known as infrared waves. **An _IrDA_, or _infrared_, _port_ allows a computer to make a cableless connection with infrared-capable devices,** such as some printers. (IrDA stands for "Infrared Data Association," which sets the standards.) This type of connection requires an unobstructed line of sight between transmitting and receiving ports, and they can be only a few feet apart.

Bluetooth logo

- **Bluetooth ports—for wireless connections up to 30 feet:** Bluetooth technology consists of short-range radio waves that transmit up to 30 feet. It is used to connect cellphones to computers but also to connect computers to printers, keyboards, headsets, and other appliances (including refrigerators).

- **Ethernet—for LANs:** Developed by the Xerox Corporation in the mid-1970s, Ethernet is a network standard for linking all devices in a local area network. (The name comes from the concept of "ether.") It's commonly used to connect microcomputers, cable modems, and printers. (To use Ethernet, the computer must have an Ethernet network interface card, and special Ethernet cables are required.) (Bluetooth and Ethernet are also discussed in Chapter 6.)

MULTIMEDIA PORTS "The trend toward multimedia notebooks has introduced a whole slew of port types you may not be familiar with, but that you might need," states a computer magazine.[16] These include ports for connecting your computer to the following: gaming consoles or camcorders (Composite In/RCA ports); cable boxes, TVs, or VCRs, so you can watch and record TV content on your computer (TV tuner/75-ohm coaxial port); speakers or stereo receivers to play digital audio (S/PDIF ports); digital projectors to LCD panels to display in large format video content or a PowerPoint presentation (DVI port); digital cameras to display video stored on the cameras (S-video In port); and plasma or newer TV so you can view video content streaming from the computer (S-Video Out).

Expandability: Buses & Cards

What is the purpose of expansion buses and cards?

Today many new microcomputer systems can be expanded. As mentioned earlier, *expansion* is a way of increasing a computer's capabilities by adding hardware to perform tasks that are not part of the basic system. *Upgrading* means changing to a newer, usually more powerful or sophisticated version. (Computer ads often make no distinction between expansion and upgrading. Their main interest is simply to sell you more hardware or software.)

CLOSED & OPEN ARCHITECTURE Whether a computer can be expanded depends on its "architecture"—closed or open. *Closed architecture* means a computer has no expansion slots; *open architecture* means it does have expansion slots. (An alternative definition is that closed architecture is a computer design whose specifications are not made freely available by the manufacturer. Thus, other companies cannot create ancillary devices to work with it. With open architecture, the manufacturer shares specifications with outsiders.)

Expansion slots **are sockets on the motherboard into which you can plug expansion cards.** _Expansion cards_**—also known as** _expansion boards,_ _adapter cards,_ _interface cards,_ _plug-in boards,_ _controller cards,_ _add-ins,_ **or** _add-ons_**—are circuit boards that provide more memory or that control peripheral devices.** (● *See Panel 4.14 on the next page.*)

COMMON EXPANSION CARDS & BUSES Common expansion cards connect to the monitor (graphics card), speakers and microphones (sound card), and network (network card), as we'll discuss. Most computers have four to eight expansion slots, some of which may already contain expansion cards included in your initial PC purchase.

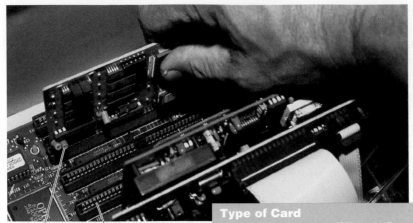

Expansion card   Expansion slot

Type of Card (Board)	What It Does
Accelerator board	Speeds up processing; also known as turbo board or upgrade board
Cache card	Improves disk performance
Coprocessor board	Contains specialized processor chips that increase processing speed of computer system
Disk controller card	Allows certain type of disk drive to be connected to computer system
Emulator board	Permits microcomputer to be used as a terminal for a larger computer system
Fax modem board	Enables computer to transmit and receive fax messages and data over telephone lines
Graphics (video) adapter board	Permits computer to have a particular graphics standard
Memory expansion board	Enables additional RAM to be added to computer system
Sound board	Enables certain types of systems to produce sound output

Expansion cards are made to connect with different types of buses on the motherboard. As we mentioned, *buses* are electrical data roadways through which bits are transmitted. The bus that connects the CPU within itself and to main memory is the *frontside bus,* also called the *memory bus* or the *local bus.* (● *See Panel 4.15.)* The buses that connect the CPU with expansion slots on the motherboard and thus with peripheral devices are *expansion buses.* We already described the universal serial bus (USB), whose purpose, in fact, is to *eliminate* the need for expansion slots and expansion cards, since you can just connect USB devices in a daisy chain outside the system unit.

Two expansion buses to be aware of are *PCI* and *AGP:*

- **PCI bus—for high-speed connections: At 32 or 64 bits wide, the _PCI (peripheral component interconnect) bus_ is a high-speed bus** that has been widely used to connect PC graphics cards, sound cards, modems, and high-speed network cards.

- **AGP bus—for even higher speeds and 3D graphics:** The PCI bus was adequate for many years, providing enough bandwidth for all the peripherals most users wanted to connect—except graphics cards. In the mid-1990s, however, graphics cards were becoming more powerful, and three-dimensional (3-D) games were demanding higher performance. Because the PCI bus couldn't handle all the information

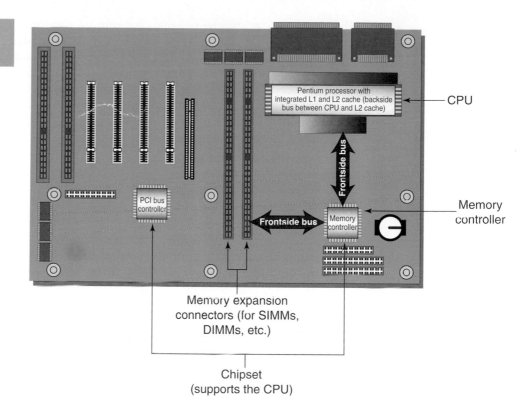

CPU

Pentium processor with integrated L1 and L2 cache (backside bus between CPU and L2 cache)

Frontside bus

Memory controller

Frontside bus

Memory controller

PCI bus controller

Memory expansion connectors (for SIMMs, DIMMs, etc.)

Chipset (supports the CPU)

passing between the main processor and the graphics processor, Intel developed the AGP bus. **The _AGP (accelerated graphics port) bus_, which transmits data at twice the speed of a PCI bus, is designed to support video and 3-D graphics.** Although PCI continues to be the bus of choice for most peripherals, AGP has taken over the specialized task of graphics processing.

TYPES OF EXPANSION CARDS   Among the types of expansion cards are graphics, sound, modem, and network interface cards. A special kind of card is the PC card.

- Graphics cards—for monitors: Graphics cards are included in all PCs. **Also called a _video card_ or _video adapter_, a _graphics card_ converts signals from the computer into video signals that can be displayed as images on a monitor.** Each graphics card has its own memory chips, a graphics BIOS ROM chip, and a dedicated processor. The processor is designed specifically to handle the intense computational requirements of displaying graphics. Graphics cards are usually connected to an AGP slot on the motherboard.

🔍 3D AGP Graphics Card (64 MB)

- Sound cards—for speakers and audio output: **A _sound card_ is used to convert and transmit digital sounds through analog speakers, microphones, and headsets.** Sound cards come installed on most new PCs. Cards such as PCI wavetable sound cards are used to add music and sound effects to computer videogames. _Wavetable synthesis_ is a method of creating music based on a wave table, which is a collection of digitized sound samples taken from recordings of actual instruments. The sound samples are then stored on a sound card and are edited and mixed together to produce music. Wavetable synthesis produces higher-quality audio output than other sound techniques.

🔍 Sound Blaster Digital Sound Card

56 Kbps Internal Modem

- **Modem cards—for remote communication via phone lines:** Occasionally you may still see a modem that is outside the computer. Most new PCs, however, come with internal modems—modems installed inside as circuit cards. The modem not only sends and receives digital data over telephone lines to and from other computers but can also transmit voice and fax signals. (Modems are discussed in more detail in later chapters.)

- **Network interface cards—for remote communication via cable:** A _network interface card (NIC)_ **allows the transmission of data over a cable network,** which connects various computers and other devices such as printers. (Various types of networks are covered in Chapter 6.)

- **PC cards—for laptop computers:** Originally called *PCMCIA cards* (for the Personal Computer Memory Card International Association), **_PC cards_ are thin, credit card–size (2.1- by 3.4-inch) devices used principally on laptop computers to expand capabilities.** (● *See Panel 4.16.*) Examples are extra memory (flash RAM), sound cards, modems, hard disks, and even pagers and cellular communicators. At present there are three sizes for PC cards—I (thin), II (thick), and III (thickest). Type I is used primarily for flash memory cards. Type II, the kind you'll find most often, is used for fax modems and network interface cards. Type III is for rotating disk devices, such as hard-disk drives, and for wireless communication devices.

4.4

Secondary Storage

What are the features of floppy disks, hard disks, optical disks, magnetic tape, smart cards, flash memory, and online secondary storage?

You're on a trip with your laptop, or maybe just a cellphone or a personal digital assistant, and you don't have a crucial file of data. Or maybe you need to look up a phone number that you can't get through the phone company's directory assistance. Fortunately, you backed up your data online, using any one of several storage services (for example, Driveway, *www.driveway.com*; MagicalDesk, *www.magicaldesk.com*; or X:Drive, *www.xdrive.com*), and are able to access it through your modem.

Here is yet another example of how the World Wide Web is offering alternatives to traditional computer functions that once resided within standalone machines. We are not, however, fully into the all-online era just yet. Let us consider more traditional forms of **_secondary storage hardware_**,

● **PANEL 4.16**
PC card
An example of a PC card used in a laptop.

devices that permanently hold data and information as well as programs. We will look at the following types of secondary storage devices:

- Floppy disks and Zip disks
- Hard disks
- Optical disks
- Magnetic tape
- Smart cards
- Flash memory
- Online secondary storage

Floppy disk

Floppy Disks & Zip Disks

Are floppy disks and Zip disks still being used?

A *floppy disk*, often called a *diskette* or simply a *disk*, is a removable flat piece of mylar plastic packaged in a 3.5-inch plastic case. Data and programs are stored on the disk's coating by means of magnetized spots, following standard on/off patterns of data representation (such as ASCII). The plastic case protects the mylar disk from being touched by human hands. Originally, when most disks were larger (5.25 inches) and covered in paper, the disks actually were "floppy"; now only the disk inside the rigid plastic case is flexible, or floppy. In some places, such as South Africa, 3.5-inch floppy disks are called *stiffies* or *stiffy disks* because of their stiff (rigid) cases.

Floppy disks each store about 1.44 megabytes, the equivalent of 400 typewritten pages. SuperDisks can store 120 megabytes. Today's floppies carry the label "2HD." The *2* stands for "double-sided" (the disk holds data on both sides), and *HD* stands for "high density" (it stores more data than the previous standard—*DD*, for "double density").

Front

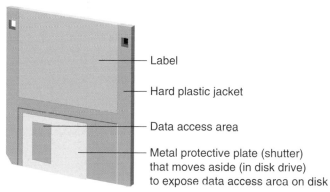

— Label

— Hard plastic jacket

— Data access area

— Metal protective plate (shutter) that moves aside (in disk drive) to expose data access area on disk

Back

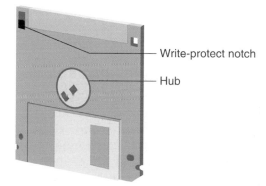

— Write-protect notch

— Hub

HOW FLOPPY DISKS WORK Floppy disks are inserted into a floppy-disk drive, a device that holds, spins, reads data from, and writes data to a floppy disk. *Read* means that the data in secondary storage is converted to electronic signals and a copy of that data is transmitted to the computer's memory (RAM). *Write* means that a copy of the electronic information processed by the computer is transferred to secondary storage.

Some details about floppy disks:

- **The write-protect notch:** Floppy disks have a *write-protect notch*, which allows you to prevent a diskette from being written to. In other words, it allows you to protect the data already on the disk. To write-protect, use your thumbnail or the tip of a pen to move the small sliding tab on the lower right side of the disk (viewed from the back), thereby uncovering the square hole.

- **Tracks, sectors, and clusters:** On the diskette, **data is recorded in concentric recording bands called *tracks*.** Unlike on a vinyl phonograph record, these tracks are neither visible grooves nor a single

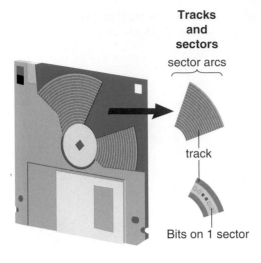

Tracks and sectors

sector arcs

track

Bits on 1 sector

spiral. Rather, they are closed concentric rings; each track forms a full circle on the disk. **When a disk is formatted, the disk's storage locations are divided into wedge-shaped sections, which break the tracks into small arcs called _sectors_.** When you save data from your computer to a diskette, the data is distributed by tracks and sectors on the disk. That is, the system software uses the point at which a sector intersects a track to reference the data location. The smallest unit of disk space that holds data is called a cluster. A *cluster* is a group of sectors on a storage device. Tracks, sectors, and clusters are also used in hard disks, as we'll discuss.

- **The read/write head:** When you insert a floppy disk into the slot (the *drive gate* or *drive door*) in the front of the disk drive, the disk is fixed in place over the spindle of the drive mechanism. **The _read/write head_ is used to transfer data between the computer and the disk.** When the disk spins inside its case, the read/write head moves back and forth over the *data access area* on the disk. When the disk is not in the drive, a metal or plastic shutter covers this access area. An access light goes on when the disk is in use. After using the disk, you can retrieve it by pressing an eject button beside the drive.

Zip disk

ZIP DISKS One alternative to floppy disks is **_floppy-disk cartridges_—higher-capacity removable disks**, such as Zip disks. Produced by Iomega Corp., **_Zip disks_ are disks with a special high-quality magnetic coating that have a capacity of 100, 250, or 750 megabytes.** Even at 100 megabytes, this is nearly 70 times the storage capacity of the standard floppy. Zip disks require their own Zip-disk drives, which you may request to have installed in a bay on a new computer, although external Zip drives (usually USB) are also available. Zip disks are used to store large spreadsheet files, database files, image files, multimedia presentation files, and websites.

Zip disk drive

acer

ARE FLOPPIES & ZIP DISKS OBSOLETE? Although one can usually get a floppy-disk drive or Zip drive as optional equipment with new personal computers, most manufacturers have dropped it as standard equipment. If you've bought a PC without a built-in floppy drive and later decide you need one, you can buy an external USB floppy drive, which can be connected with a USB cable. Most PC users now rely on CD-ROM, CD-RW, or DVD/CD-RW combo drives for interchangeable disks, as we'll explain.

160 GB SATA 7200 RPM Hard Drive

Hard Disks

How does the hard disk in my computer work?

Floppy disks use flexible plastic, but hard disks are rigid. **_Hard disks_ are thin but rigid metal, glass, or ceramic platters covered with a substance that allows data to be held in the form of magnetized spots.** Most hard-disk drives have at least two platters; the greater the number of platters, the larger the capacity of the drive. The

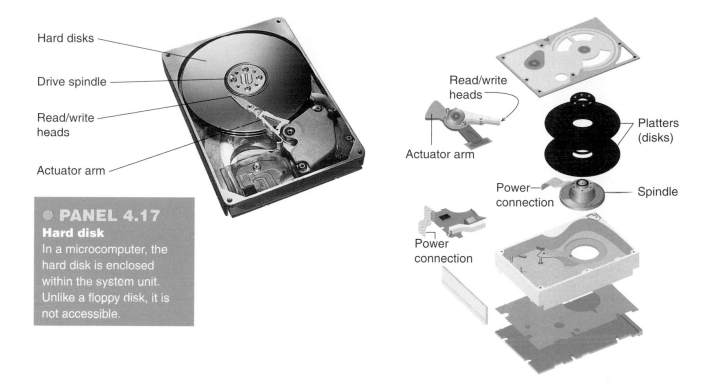

Hard disks

Drive spindle

Read/write heads

Actuator arm

Read/write heads

Actuator arm

Platters (disks)

Power connection

Spindle

Power connection

Relative size of a hard disk drive

info!

Storage in the Old Days

In 1978, a pioneering drive for home computers, the external Shugart SA4000, weighed 35 pounds, wholesaled for $2,550, and stored only 14.5 megabytes. Hard drives began to become popular with PC users in 1983, when Seagate made one that fit into the PC case. For more history, go to *www.computer-museum.org.*

platters in the drive are separated by spaces and are clamped to a rotating spindle that turns all the platters in unison. Hard disks are tightly sealed within an enclosed hard-disk-drive unit to prevent any foreign matter from getting inside. Data may be recorded on both sides of the disk platters. (● *See Panel 4.17.*)

KEEPING TRACK OF DATA: THE VIRTUAL FILE ALLOCATION TABLE
Like floppy disks, hard disks store data in *tracks, sectors,* and *clusters.* Computer operating systems keep track of hard-disk sectors according to clusters. The Windows operating system assigns a unique number to each cluster and then keeps track of files on a disk by using a kind of table, a *Virtual File Allocation Table (VFAT),* as a method for storing and keeping track of files according to which clusters they use. (The VFAT was called a FAT in earlier operating systems.) That is, the VFAT includes an entry for each cluster that describes where on the disk the cluster is located.

Occasionally, the operating system numbers a cluster as being used even though it is not assigned to any file. This is called a *lost cluster.* You can free up lost clusters and thus increase disk space in Windows by using the ScanDisk utility.

HEAD CRASHES Hard disks are sensitive devices. The read/write head does not actually touch the disk but rather rides on a cushion of air about 0.000001 inch thick. (● *See Panel 4.18 on the next page.*) The disk is sealed from impurities within a container, and the whole apparatus is manufactured under sterile conditions. Otherwise, all it would take is a human hair, a dust particle, a fingerprint smudge, or a smoke particle to cause what is called a *head crash.* A *head crash* happens when the surface of the read/write head or particles on its surface come into contact with the surface of the hard-disk platter, causing the loss of some or all of the data on the disk. A head crash can also happen when you bump a computer too hard or drop something heavy on the system cabinet. An incident of this sort could, of course, be a disaster if the data has not been backed up. There are firms that specialize in trying to retrieve data from crashed hard disks (for a hefty price), though this cannot always be done.

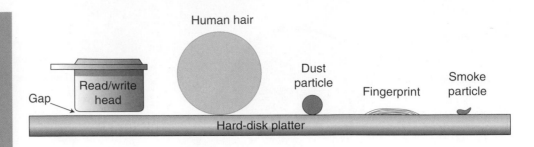

Bits on disk. Magnetic bits on a disk surface, caught by a magnetic force microscope. The dark stripes are 0 bits; the bright stripes are 1 bits.

NONREMOVABLE HARD DISKS   **An internal _nonremovable hard disk_, also known as a *fixed disk*, is housed in the microcomputer system unit and is used to store nearly all programs and most data files.** Usually it consists of several metallic or glass platters, 1–5.25 inches (most commonly 3.5 inches) in diameter, stacked on a spindle, with data stored on both sides. Read/write heads, one for each side of each platter, are mounted on an access arm that moves back and forth to the right location on the platter. The entire apparatus is sealed within an airtight case to protect it from contaminants such as dust.

The storage capacities of nonremovable hard drives range from 40 to over 300 gigabytes—or even 500 gigabytes, in the case of the Mac G5. A gigabyte holds around 20,000 pages of regular text, but video and sound files often run to 30 megabytes or more.

A hard disk spins many times faster than a floppy disk, allowing faster access. Computer ads frequently specify hard-disk speeds in *revolutions per minute (rpm)*, usually 5,400–7,200 rpm. Whereas a floppy-disk drive rotates at only 360 rpm, a 7,200-rpm hard drive is going about 300 miles per hour. (And some new hard-disk drives are even faster.)

PORTABLE HARD-DRIVE SYSTEMS: EXTERNAL & REMOVABLE HARD DISKS   Two types of portable hard-drive systems are available:

- **External hard disk:** An *external hard disk* is a freestanding hard-disk drive enclosed in an airtight case, which is connected by cable to the computer system unit through a FireWire, USB, or other port. Storage capacities run to 250 gigabytes or more.

  Western Digital and Iomega make portable hard-disk drives that hold 80–250 gigabytes.

- **Removable hard disk: A _removable hard disk_, or *hard-disk cartridge*, consists of one or two platters enclosed along with read/write heads in a hard plastic case, which is inserted into a cartridge drive built into the microcomputer's system unit.** Cartridges, which have storage capacities of 80 gigabytes and more, are frequently used to back up and transport huge files of data, such as those for large spreadsheets or desktop-publishing files.

HARD-DISK CONTROLLERS   When advertising hard-disk drives, computer ads may specify the type of *hard-disk controller* used, a special-purpose circuit board that positions the disk and read/write heads and manages the flow of data and instructions to and from the disk. The hard-disk controller may be part of the hard disk or an adapter card located inside the system unit. Two hard-disk controllers for connecting external hard-disk drives are FireWire and USB. Other standards are as follows:

- **EIDE:** *EIDE (Enhanced Integrated Drive Electronics)* can support up to four hard disks at 137 gigabytes per disk. EIDE controllers are marketed under such names as SATA (for "Serial Advanced Technology Attachment"), Fast ATA, Ultra ATA, Fast IDE, ATA-2, ATA/100, and Serial ATA. Nowadays consumers are advised to buy SATA or Serial ATA, which have faster data transfer and less magnetic interference.

- **SCSI:** *SCSI (small computer system interface),* pronounced "scuzzy," supports several disk drives as well as other peripheral devices by linking them in a daisy chain. SCSI controllers are faster and have more storage capacity than EIDE controllers; they are typically found in servers and workstations.

- **Fibre Channel:** *Fibre Channel* is an up-and-coming standard that is not expected to be used much with personal computers, though it may some day replace SCSI. This standard has the advantage of very fast data speeds and so will probably be used with host computers and high-end servers.

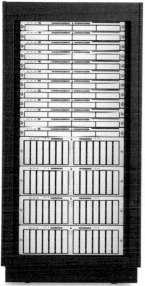

RAID unit

HARD-DISK TECHNOLOGY FOR LARGE COMPUTER SYSTEMS: RAID Large databases, such as those maintained by insurance companies or AOL, require far bigger storage systems than the fixed-disk drives we've been describing, which send data to a computer along a single path. **A _RAID (redundant array of independent [or inexpensive] disks) storage system_, which links any number of disk drives within a single cabinet or connected along a SCSI chain, sends data to the computer along several parallel paths simultaneously.** Response time is thereby significantly improved.

Optical Disks: CDs & DVDs

Why would I prefer an optical disk over other forms of secondary storage?

Everyone who has ever played an audio CD is familiar with optical disks. **An _optical disk_ is a removable disk, usually 4.75 inches in diameter and less than one-twentieth of an inch thick, on which data is written and read through the use of laser beams.** An audio CD holds up to 74 minutes (2 billion bits' worth) of high-fidelity stereo sound. Some optical disks are used strictly for digital data storage, but many are used to distribute multimedia programs that combine text, visuals, and sound.

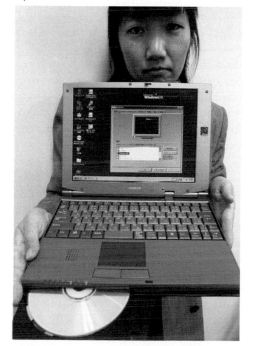

Optical disk

HOW OPTICAL-DISK STORAGE WORKS With an optical disk, there is no mechanical arm, as with floppy disks and hard disks. Instead, a high-power laser beam is used to write data by burning tiny pits or indentations into the surface of a hard plastic disk. To read the data, a low-power laser light scans the disk surface: Pitted areas are not reflected and are interpreted as 0 bits; smooth areas are reflected and are interpreted as 1 bits. (● *See Panel 4.19 on the next page.)* Because the pits are so tiny, a great deal more data can be represented than is possible in the same amount of space on a diskette and many hard disks. An optical disk can hold more than 4.7 gigabytes of data, the equivalent of 1 million typewritten pages.

Nearly every PC marketed today contains a CD/DVD drive, which can also read audio CDs. These, along with their recordable and rewritable variations, are the two principal types of optical-disk technology used with computers.

CD-ROM—FOR READING ONLY The first kind of optical disk for microcomputers was the CD-ROM. **_CD-ROM (compact disk read-only memory)_ is an optical-disk format that is used to**

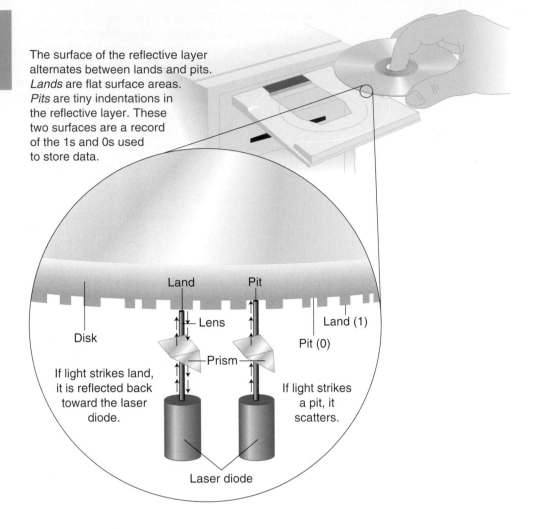

The surface of the reflective layer alternates between lands and pits. *Lands* are flat surface areas. *Pits* are tiny indentations in the reflective layer. These two surfaces are a record of the 1s and 0s used to store data.

Land   Pit

Disk

Lens

Land (1)
Pit (0)

Prism

If light strikes land, it is reflected back toward the laser diode.

If light strikes a pit, it scatters.

Laser diode

hold prerecorded text, graphics, and sound. Like music CDs, a CD-ROM is a read-only disk. *Read-only* means the disk's content is recorded at the time of manufacture and cannot be written on or erased by the user. As the user, you have access only to the data imprinted by the disk's manufacturer. A CD-ROM disk can hold up to 650 megabytes of data, equal to over 300,000 pages of text.

A CD-ROM drive's speed is important because with slower drives, images and sounds may appear choppy. In computer ads, drive speeds are indicated by the symbol "X," as in "56X," which is a high speed. *X* denotes the original data transfer rate of 150 kilobytes per second. The data transfer rate is the time the drive takes to transmit data to another device. A 56X drive runs at 56 times 150, or 8,400 kilobytes (8.4 megabytes) per second. If an ad carries the word *Max*, as in "56X Max," this indicates the device's maximum speed. Drives range in speed from 16X to 75X; the faster ones are more expensive.

CD-R—FOR RECORDING ON ONCE _**CD-R (compact disk–recordable) disks**_ **can be written to only once but can be read many times.** This allows users to make their own CD disks. Once recorded, the information cannot be erased. CD-R is often used by companies for archiving—that is, to store vast amounts of information. A variant is the Photo CD, an optical disk developed by Kodak that can digitally store film photos taken with an ordinary 35-millimeter camera. Once you've shot a roll of color photographs, you take it for processing to a photo shop, which produces a disk containing your images. You can view the disk on any personal computer with a CD-ROM drive and the right software.

24X DVD/CD-RW Combo Drive

CD-RW—FOR REWRITING MANY TIMES A _CD-RW (compact disk–rewritable) disk_, also known as an _erasable optical disk_, allows users to record and erase data, so the disk can be used over and over again. CD-RW drives are becoming more common on microcomputers. CD-RW disks are useful for archiving and backing up large amounts of data or work in multimedia production or desktop publishing. CD-RW disks cannot be read by CD-ROM drives. CD-RW disks have a capacity of 650–700 megabytes.

DVD-ROM—THE VERSATILE VIDEO DISK A _DVD-ROM (digital versatile disk_ or _digital video disk, with read-only memory)_ is a CD-style disk with extremely high capacity, able to store 4.7 or more gigabytes. How is this done? Like a CD or CD-ROM, the surface of a DVD contains microscopic pits, which represent the 0s and 1s of digital code that can be read by a laser. The pits on the DVD, however, are much smaller and grouped more closely together than those on a CD, allowing far more information to be represented. Also, the laser beam used focuses on pits roughly half the size of those on current audio CDs. In addition, the DVD format allows for two layers of data-defining pits, not just one. Finally, engineers have succeeded in squeezing more data into fewer pits, principally through data compression.

Most new computer systems now come with a DVD drive as standard equipment. A great advantage is that these drives can also take standard CD-ROM disks, so now you can watch DVD movies and play CD-ROMs using just one drive. DVDs are replacing CDs for archival storage, mass distribution of software, and entertainment. They not only store far more data but are different in quality from CDs. As one writer points out, "DVDs encompass much more: multiple dialogue tracks and screen formats, and best of all, smashing sound and video."[18] The theater-quality video and sound, of course, are what have made DVD a challenger to videotape as a vehicle for movie rentals.

Like CDs, DVDs have their recordable and rewritable variants:

- **DVD-R—recordable DVDs:** _DVD-R (DVD-recordable) disks_ allow **one-time recording by the user.** That is, they cannot be reused—written on more than once.

- **DVD-RW, DVD-RAM, DVD+RW—reusable DVDs:** Three types of reusable disks are _DVD-RW (DVD-rewritable), DVD-RAM (DVD–random access memory),_ and _DVD+RW (DVD+rewritable),_ all of which can be recorded on and erased (except for video) many times. DVD-R disks have a capacity of 4.7 (single-sided) to 9.4 (double-sided) gigabytes.

BLU-RAY: THE NEXT-GENERATION OPTICAL DISK _Blu-ray,_ also known as _Blu-ray Disc (BD),_ is the name of a next-generation optical-disk format jointly

CD/DVD, label side up

Slide-out tray for DVD/CD disk drive

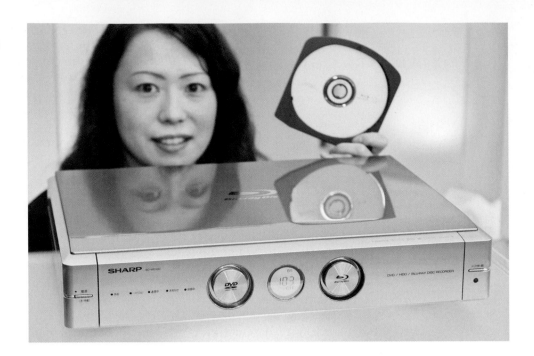

Sharp introduces the company's first Blu-ray disc recorder.

developed by the Blu-ray Disc Association (BDA), a group of consumer electronics and PC companies (including Dell, Hitachi, HP, JVC, LG, Mitsubishi, Panasonic, Pioneer, Philips, Samsung, Sharp, Sony, TDK, and Thomson). **The _Blu-ray_ optical format was developed to enable recording, rewriting, and playback of high-definition (HD) video, as well as storing large amounts of data.**

A single-layer Blu-ray Disc can hold 25 gigabytes, which can be used to record over 2 hours of HDTV or more than 13 hours of standard-definition TV. There are also dual-layer versions of the disks that can hold 50 gigabytes.

While current optical-disk technologies such as DVD, DVD-R, DVD-RW, and DVD-RAM use a red laser to read and write data, the new format uses a blue-violet laser instead, hence the name Blu-ray. Blu-ray products are backward-compatible and allow use of CDs and DVDs. The benefit of using a blue-violet laser is that it has a shorter wavelength than a red laser, which makes it possible to focus the laser spot with even greater precision. This allows data to be packed more tightly and stored in less space, so it's possible to fit more data on the disk even though it's the same size as a CD/DVD.

Not all the optical-disk drives and different types of optical media (read-only, rewritable, and so on) are mutually compatible. Thus, you need to check the product information before you buy to make sure you get what you want.

DVD Capacities

Type	Sides	Layers	Capacity		Type	Sides	Layers	Capacity
Read only					**Rewritable (100,000 cycles)**			
DVD-Video	1	1	4.7 GB (DVD-5)		DVD-RAM Ver. 1	1	1	2.6 GB
and	1	2	8.5 GB (DVD-9)		DVD-RAM Ver. 1	2	1	5.2 GB
DVD-ROM	2	1	9.4 GB (DVD-10)		DVD-RAM Ver. 2	1	1	4.7 GB
	2	2	17.0 GB (DVD-18)		DVD-RAM Ver. 2	2	1	9.4 GB
HD DVD	1	1	15.0 GB		DVD-RAM (80 mm)	1	1	1.46 GB
HD DVD	1	2	30.0 GB		DVD-RAM (80 mm)	2	1	2.92 GB
Write once					Blu-ray	1	1	27.0 GB
DVD-R (A)	1	1	3.95 GB		Blu-ray	1	2	54.0 GB
DVD-R (A)	1	1	4.7 GB		HD DVD	1	1	20.0 GB
DVD-R (G)	1	1	4.7 GB		HD DVD	1	2	32.0 GB
DVD-R (G)	2	1	9.4 GB		**Re-recordable (1,000 cycles)**			
DVD+R	1	1	4.7 GB		DVD-RW	1	1	4.7 GB
					DVD+RW	1	1	4.7 GB
					DVD+RW	2	1	9.4 GB

THE WORLD'S DVD ZONES As a way to maximize movie revenues, the film industry decided to split the world up into six DVD zones. This is to prevent the DVD version of a movie made in one country from being sold in another country in which the theater version has not yet opened. DVD disks with a particular region code will play only on DVD players with that region code. However, some of the newest DVD players are code-free—they are not "region locked." (Thus, if you plan to travel and use a DVD player for movies, check out the compatibility restrictions.)

Six DVD regions. The DVD world is basically divided into six regions. This means that DVD players and DVDs are encoded for operation in a specific geographical region in the world. For example, the U.S. is in region 1. This means that all DVD players sold in the U.S. are made to region 1 specifications. As a result, region 1 players can play only region 1 discs. On the back of each DVD package, you will find a region number (1 through 6).

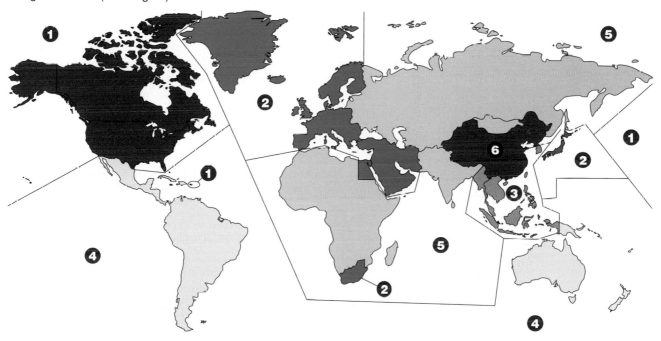

Magnetic Tape

Would I find magnetic tape storage useful?

Similar to the tape used on an audio tape recorder (but of higher density), **_magnetic tape_ is thin plastic tape coated with a substance that can be magnetized. Data is represented by magnetized spots (representing 1s) or non-magnetized spots (representing 0s).** Today, "mag tape" is used mainly for backup and archiving—that is, for maintaining historical records—where there is no need for quick access.

On large computers, tapes are used on magnetic-tape units or reels and in special cartridges. These tapes store 200 gigabytes and higher. On microcomputers, tape is used in the form of **_tape cartridges_, modules resembling audiocassettes that contain tape in rectangular, plastic housings.** The two most common types of tape drives are DAT (digital audio tape) from Hewlett-Packard, Sony, and Maxwell, for example, and Traven TR-3. (● *See Panel 4.20.*) An internal or external tape drive is required to use tape media.

Tape fell out of favor for a while, supplanted by such products as Iomega's Jaz (since discontinued) and Zip-disk cartridge drives. However, as hard drives swelled to multigigabyte size, using Zip disks for backup became less

convenient. Since a single tape cassette may well hold 200 gigabytes, tape is still an alternative. However, as we mentioned, CD-R and DVD-R are becoming the most popular backup methods.

Smart Cards

Would I ever have occasion to use smart cards?

Today in the United States, most credit cards are old-fashioned magnetic-strip cards. A *magnetic-strip card* has a strip of magnetically encoded data on its back and holds about 0.2–0.9 kilobytes of data. The encoded data might include your name, account number, and PIN (personal identification number). The strip contains information needed to use the card, but the strip also has drawbacks. First, it can degrade over time, making the data unreadable. Second, the magnetic strip doesn't hold much information. Third, such data as the magnetic strip does contain is easy to access and duplicate, raising the risk of fraud.

Two other kinds of cards, smart cards and optical cards, which hold far more information, are already popular in Europe. Manufacturers are betting they will become more popular in the United States.

SMART CARDS A *smart card* **looks like a credit card but has a microprocessor and memory chips embedded in it.** Smart cards hold more information than standard magnetic-strip credit cards—about 8–40 megabytes of data. Some can be reloaded for reuse. When inserted into a reader, a smart card transfers data to and from a central computer. It is more secure than a magnetic-strip card and can be programmed to self-destruct if the wrong password is entered too many times. As a financial transaction card, it can be loaded with digital money and used as a traveler's check, except that variable amounts of money can be spent until the balance is zero.

Smart cards are well suited for prepaid, disposable applications such as cash cards or telephone debit cards. For example, when you're using a phone card, which is programmed to contain a set number of available minutes, you insert the card into a slot in the phone, wait for a tone, and dial the number. The length of your call is automatically calculated on the card, and the corresponding charge is deducted from the balance. Other uses of smart cards are as student cards, building-entrance cards, bridge-toll cards, and (as in Germany) national-health-care cards.

Different forms of smart-card technology are available:

- **UltraCard:** A card from UltraCard, Inc., uses a magnetic-coated strip of metal (a "shim") that is drawn almost entirely out of the card by the drive and then is read, written to, and inserted back inside the card before the card is withdrawn from the drive. It provides 2 gigabytes of storage and protects security by allowing biometric information such as a fingerprint to be used.

- **Contact smart cards:** These kinds of cards, which must be swiped through card readers, are less prone to misalignment and being misread, but they tend to wear out from the contact.

A new smart card (*right*) compared to a regular credit card (*left*)

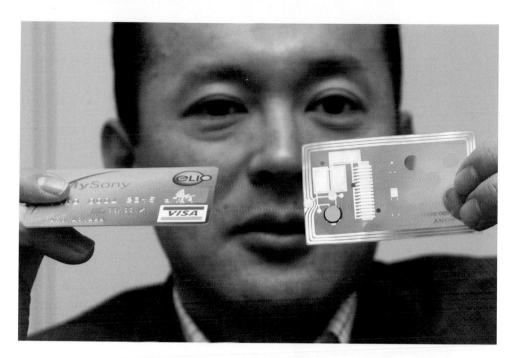

- **Contactless smart cards:** These cards, which are read when held in front of a low-powered laser, can be used in mobile applications, as by automated toll-collecting devices reading cards as drivers pass through toll booths without stopping.

OPTICAL CARDS   Optical cards use the same type of technology as music compact disks but look like silvery credit cards. ***Optical cards* are plastic, laser-recordable, wallet-type cards used with an optical-card reader.** Optical health cards have room not only for the individual's medical history and health-insurance information but also for digital images, such as electrocardiograms.

One form of optical card technology, the LaserCard, is used by the U.S. Immigration and Naturalization Service as new Permanent Resident and Border Crossing cards because of its highly secure, counterfeit-resistant features. In addition, LaserCards containing shipping manifest data are attached to shipping containers and sea vans, speeding up receipt processing considerably.

VISX Corporation uses a LaserCard to operate the STAR Excimer Laser System for each of its eye surgery patients. The high-capacity card stores parameter settings and patient data for each surgery performed. Honda dealers in the Philippines use the LaserCard as a comprehensive vehicle record database that Honda car owners keep. The cards are also used to obtain discounts at participating retail stores and restaurants.

Flash Memory

How useful is flash memory to me?

Disk drives, whether for diskettes, hard disks, or CD-ROMs, all involve moving parts—and moving parts can break. By contrast, *flash memory*, which is a variation on conventional computer memory chips, has no moving parts. Flash memory is also nonvolatile—it retains data even when the power is turned off. A drawback, however, is that flash memory circuits wear out after repeated use, limiting their life span.

Flash memory media are available in three forms: *flash memory cards, flash memory sticks,* and *flash memory drives.*

Survival Tip
What's the Life Span of Storage Media?

Hard drives: 3–6 years
CD/DVD-R: 2–15 years
CD/DVD-RW: 25–30 years
Flash drives: 10 years

FLASH MEMORY CARDS <u>*Flash memory cards*</u>, or *flash RAM cards*, are removable storage media that are inserted into a flash memory port in a digital camera, handheld PC, smartphone, or other mobile device. Unlike smart cards, flash memory cards have no processor; they are useful only for storage. Examples are Compact-Flash cards (up to 4 gigabytes), Secure Digital cards (up to 1 gigabyte), MiniSD cards (up to 512 megabytes), xD (or xD Picture) cards (up to 512 megabytes), Multimedia cards (up to 256 megabytes), and Smart-Media cards (up to 128 megabytes).

FLASH MEMORY STICKS Smaller than a stick of chewing gum, **a** <u>*flash memory stick*</u> **is a form of flash memory media that plugs into a memory stick port in a digital camera, camcorder, notebook PC, photo printer, and other devices.** It holds up to 1 gigabyte of data.

FLASH MEMORY DRIVES A <u>*flash memory drive*</u>, **also called a** *USB flash drive, keychain drive,* **or** *key drive,* **consists of a finger-size module of flash memory that plugs into the USB ports of nearly any PC or Macintosh.** Examples are M-Systems' DiskOnKey, Sony's Micro Vault, Lexar's Jump Drive, and SanDisk's Cruzer. They generally have storage capacities up to 2 gigabytes, making the device extremely useful if you're traveling from home to office, say, and don't want to carry a laptop. When you plug the device into your USB port, it shows up as an external drive on the computer.

(*Top*) A collection of flash drives; (*bottom*) Swiss Army Knife with a flash drive that folds out

Online Secondary Storage

Would I ever use an online storage service?

Online storage services, mentioned at the start of this section, allow you to use the internet to back up your data. Some services are free; others charge a small fee. When you sign up, you obtain software that lets you upload whatever files you wish to the company's server. For security, you are given a password, and the files are supposedly encrypted to guard against unwanted access.

From a practical standpoint, online backup should be used only for vital files. Tape, removable hard-disk (Zip) cartridges, CD/DVDs, and flash memory are all better media for backing up entire hard disks.

more info!

Online Storage

Examples of online storage services:
Backup, *www.atbackup.com*
Connected Online Backup, *www.connected.com*
Driveway, *www.driveway.com*
MagicalDesk.com, *www.magicaldesk.com*
X:Drive, *www.xdrive.com*

4.5

Future Developments in Processing & Storage

What are some forthcoming developments that could affect processing power and storage capacity?

Computer developers are obsessed with speed and power, constantly seeking ways to promote faster processing and more main memory in a smaller area. IBM, for instance, came up with a manufacturing process (called *silicon-on-insular,* or *SOI*) that had the effect of increasing a chip's speed and reducing its power consumption. This increasing power, says physicist Michio Kaku, is the reason you can get "a musical birthday card that contains more processing power than the combined computers of the Allied Forces in World War II."[19]

PRACTICAL ACTION
Starting Over with Your Hard Drive: Erasing, Reformatting, & Reloading

There may come a time when your hard drive is so compromised by spyware and other parasites and slows down your computer so much it's as though you lost two cylinders on your car engine. (This situation might have been avoided had you been running antispyware software, such as Ad-aware, AntiSpyware, CounterSpy, PestPatrol, Spybot Search & Destroy, SpyCatcher, or Spy Sweeper. But many people don't do this, which is why one study found that 80% of computers were infected with spyware.[20]) Or perhaps you've installed a new application, such as a speech-recognition program or new piece of hardware, whose drivers have the effect of causing such chaos in your system that you wish you'd never acquired the new item.

What should you do now? Give up your computer, buy a new one (if you can afford it), and transfer all your old data to it? (Some people actually do this.) Or erase everything on your hard drive and reinstall your software and data files? Here's what to do.[21]

Make a List of Everything in Your System & Tech Support Phone Numbers

The first thing you need to do is take paper and pencil and make a list of all (1) hardware components, (2) software registration codes, and (3) technical support phone numbers for your computer maker and your internet access provider, in case you run into problems while rebuilding your system.

Make Sure You Have Disks with Copies of Your Software

See if you have the original disks for all your software (your program files), including the Windows installation disks that came with your computer. If you don't, create CD/DVD copies. Or contact your computer maker for new ones (for which the company will probably charge you).

Make Backup Copies of Your Data Files

There are several different ways to back up your data files, those listed under My Documents, My Pictures, and My Music:

- *Copy to a server:* If you're on a network, you can save all your data files to a networked folder.
- *Copy to CDs or DVDs:* Using a CD/DVD burner, you can back up your data files onto writable CDs or DVDs. (A 5-gigabyte DVD may be sufficient. Or you may use several CDs, which hold about 700 megabytes each.)

- *Copy to Zip disks:* If you have a Zip drive, you can save your data files to Zip disks, though it'll take you longer than with CDs or DVDs because a Zip disk (at 250 megabytes) has less storage capacity.
- *Copy to a keychain drive or MP3 player:* You can save all your data files to a keychain drive, an MP3 player, or another external hard drive and then transfer them to another computer to burn them onto CD/DVD disks.

You might wish to make two copies of your backed-up files—just in case.

Reformat Your Hard Drive & Reinstall Your Operating System

If your operating system is Microsoft Windows XP, insert your XP CD-ROM installation disks into the CD drive. When your computer asks you if you want to reformat your hard dive, answer yes. When it advises that if you continue all files will be deleted, respond that this is okay. It will take perhaps 60 minutes for Windows to reformat the hard disk and reinstall itself. You will then be asked to fill in your name, time zone, country, and type of internet connection. Probably at this point you should also download any XP updates (patches) from *www.windowsupdate.microsoft.com.* Reboot your computer.

Reinstall Your Programs & Drivers

Reinstall all your programs, such as Microsoft Office, and all the drivers for your printer, camera, CD burner, and other peripherals. (Drivers may not be needed because XP can recognize almost anything and find the driver online.)

Install Security Software & Firewall

To be sure you're starting over with a secure system, now you should install updated security software (such as McAfee Virus Protection, Norton AntiVirus, or ZoneAlarm Antivirus), which contains antivirus software and a firewall that prevents unauthorized users from gaining access to your network. Reboot your computer.

Reinstall All Your Data Files

Copy your saved data files from the backup source. Run scans on everything, using the security software. Reboot. By now, it's hoped, you will have gotten rid of all your spyware and other nuisances.

Hardware: The CPU & Storage

Does this mean that Moore's law—Intel cofounder Gordon Moore's 1965 prediction that the number of circuits on a silicon chip would keep doubling every 18 months—will never be repealed? After all, the smaller circuits get, the more chip manufacturers bump up against material limits. "The fact that chips are made of atoms has increasingly become a problem for us," said Moore in 2000. "In the next two or three generations, it may slow down to doubling every five years. People are predicting we will run out of gas in about 2020, and I don't see how we get around that."[22] Later, in 2005, he said, "I think we've got quite a bit more to go . . . but then I can never see more than 10 years."[23] (● *See Panel 4.21.*) Scientists have assumed that silicon chips can't be made infinitely smaller because of leakage of electrons across boundaries that are supposed to serve as insulators.

Still, there have already been advances that will extend silicon's performance projections until at least 2016, according to experts. IBM has developed a transistor that is one-tenth the size of the most advanced transistors to date. It is 6 nanometers in length, or about 20,000 times smaller than the width of a human hair. The IBM development would allow 100 times more transistors to be put into a computer chip than is currently possible. By way of comparison, today Intel's 3.06 gigahertz Pentium 4 chip holds 55 million transistors.[24] AMD also announced technologies that would allow a 10-nanometer transistor, which could lead to 1-billion-transistor chips.[25] As mentioned, Intel unveiled its own strategy to cram additional transistors onto a chip, with a 1.7-billion-transistor microprocessor to appear in 2006. By the early 2010s, experts predict, hybrid chips containing elements of traditional silicon and some undetermined materials or structures will begin to appear.[26] In addition, engineers are creating three-dimensional chips that consist of a multistory silicon stack that can run faster and cooler than ordinary chips.[27]

For now, RAM is volatile, but researchers have developed new nonvolatile forms of RAM that may soon be available. One form is *M-RAM*—the *M* stands for "magnetic"—which uses minuscule magnets rather than electric charges to store the 0s and 1s of binary data. M-RAM uses much less power than current RAM, and whatever is in memory when the computer's power is turned off or lost will remain there. A second type, *OUM (ovonic unified memory)* or *phase-change memory*, stores bits by generating different levels of low and high resistance on a glossy material.

● **PANEL 4.21**
Moore's law: How much longer?
Miraculously, Gordon Moore's prediction that the number of transistors on a silicon chip will double every 18 months has held up since the 1960s. Eventually transistors will become so tiny that their components will approach the size of molecules, and the laws of physics will no longer allow this kind of doubling.

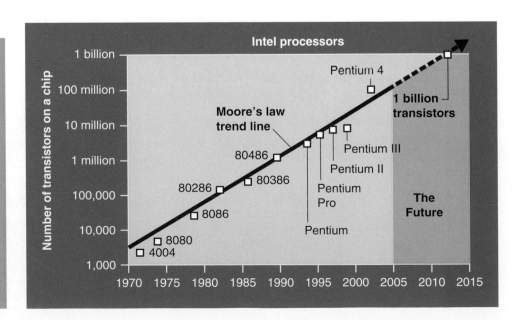

Dr. Milan Stojanovik of Columbia University developed the first game-playing DNA computer, called Maya.

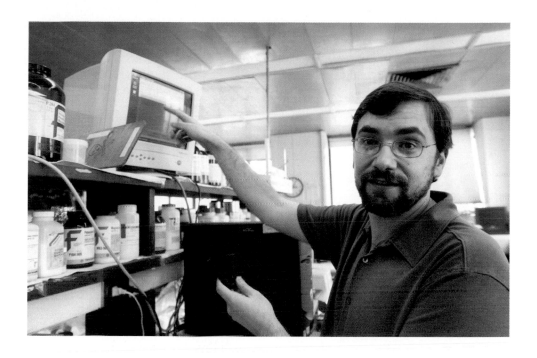

Future Developments in Processing

What are ways that processing power could be increased in the future?

What are the possible directions in which processors can go? Let us consider some of them:

An ant carrying a 1-millimeter-square microchip illustrates the kind of scale involved in nanotechnology.

info!

More about Nanotechnology

Is nanotechnology almost here? To begin to find out, check out nanotechnology's leading forum, the Foresight Institute *(www.foresight.org)*, and leading critic, ETC *(www.etcgroup.org)*.

SELLING PROCESSING POWER OVER THE INTERNET In 2002, IBM launched a service, Virtual Linux Service, selling computer processing power on an as-needed basis, in the same way that power companies sell electricity. When, for example, a company wants to update its email or other programs, instead of buying more servers or the latest hardware to provide processing power, it rents the processing time on Linux servers.[28]

NANOTECHNOLOGY Nanotechnology, nanoelectronics, nanostructures, nanofabrication—all start with a measurement known as a *nanometer*, a billionth of a meter, which means we are operating at the level of atoms and molecules. (● *See Panel 4.22.*)

In *nanotechnology*, molecules are used to create tiny machines for holding data or performing tasks. Experts attempt to do nanofabrication by building tiny nanostructures one atom or molecule at a time. When applied to chips and other electronic devices, the field is called *nanoelectronics*.

Today scientists are trying to simulate the on/off of traditional transistors by creating transistor switches that manipulate a single electron, the subatomic particle that is the fundamental unit of electricity. In theory, a trillion of these electrons could be put on a chip the size of a fingernail—a chip made not of silicon, which has its physical limits, but of cylindrical wisps of carbon atoms known as carbon nanotubes. IBM has built some working nanocircuits, but it will probably not be until 2015 that factory chipmakers will be able to make them in huge quantities.[29]

OPTICAL COMPUTING Today's computers are electronic; tomorrow's might be optical, or opto-electronic—using light, not electricity. With optical technology, a machine using lasers, lenses, and mirrors would represent the on/off codes of data with pulses of light.

Light is much faster than electricity. Indeed, fiber-optic networks, which consist of hair-thin glass fibers instead of copper wire, can move information

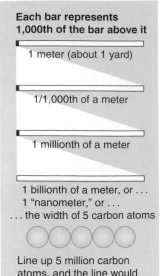

Each bar represents 1,000th of the bar above it

1 meter (about 1 yard)

1/1,000th of a meter

1 millionth of a meter

1 billionth of a meter, or . . .
1 "nanometer," or . . .
. . . the width of 5 carbon atoms

Line up 5 million carbon atoms, and the line would be about as long as the diameter of a grain of sand this big: •

Source: Ryan Randazzo, "Science of Small," *Reno-Gazette Journal*, June 16, 2002, pp. 1E, 8E; Knight-Ridder.

● **PANEL 4.22**
Nanotechnology: A matter of scale
The world of nanotechnology is so small it defies imagination.

at speeds 3,000 times faster than conventional networks. However, the signals get bogged down when they have to be processed by silicon chips. Optical chips would remove that bottleneck. It's suggested that mass-produced versions of optical chips could not only slash costs of voice and data networks but also become a new type of technology for delivering high-bandwidth movies, music, and games.[30]

DNA COMPUTING Potentially, biotechnology could be used to grow cultures of bacteria that, when exposed to light, emit a small electrical charge, for example. The properties of this "biochip" could be used to represent the on/off digital signals used in computing. Or a strand of synthetic DNA might represent information as a pattern of molecules, and the information might be manipulated by subjecting it to precisely designed chemical reactions that could mark or lengthen the strand. For instance, instead of using binary, it could manipulate the four nucleic acids (represented by *A, T, C, G*), which holds the promise of processing big numbers. This is an entirely *nondigital* way of thinking about computing.[31]

QUANTUM COMPUTING Sometimes called the "ultimate computer," the *quantum computer* is based on quantum mechanics, the theory of physics that explains the erratic world of the atom. Whereas an ordinary computer stores information as 0s and 1s represented by electrical currents or voltages that are either high or low, a quantum computer stores information by using states of elementary particles. Scientists envision using the energized and relaxed states of individual atoms to represent data. For example, hydrogen atoms could be made to switch off and on like a conventional computer's transistors by moving from low energy states (off) to high energy states (on).[32]

ETHICAL MATTERS Nanotechnology has probably received the most attention of these future developments, and indeed the U.S. government has launched the National Nanotechnology Initiative. In fact, however, nanotechnology and other important fields have been melding into a new field of science vital to U.S. security and economic development. This field is known by the acronym *NBIC* (pronounced either "*en*-bick" or "*nib*-bick"), which represents the convergence of nanotechnology, biotechnology, information technology, and cognitive (brain) science. "NBIC are the power tools of the 21st century," says James Canton, president of a technology trends research firm in San Francisco called Institute for Global Futures.[33]

But is there a possibility of "Gray Goo"? This is a scenario hypothesized in *Wired* magazine in which self-replicating molecule-size robots run amok and transform all earthly matter into nanobots. (It is also the plot for Michael Crichton's science-fiction thriller *Prey.*) People concerned about adverse effects of nanotechnology worry that the tiny particles might embed themselves in live tissue, with unknown harmful effects.[34]

Future Developments in Secondary Storage

What are ways that secondary storage could be increased in the future?

As for developments in secondary storage, when IBM introduced the world's first disk drive in 1956, it was capable of storing 2,000 bits per square inch. Today the company is shipping hard disks with densities of 14.3 billion bits, or 1.43 gigabits, per square inch. But that, too, is sure to change. Magnetic disk drives consisting of a single disk now hold 100 gigabytes of data.

HIGHER-DENSITY DISKS With 650 megabytes of storage space, and prices well under $1 each, blank CDs have been replacing floppy disks as the most

Chapter 4

236

Conventional (Longitudinal) Recording

Each bit of information is represented by a collection of magnetized particles, with their north and south poles oriented in one direction or the other. In longitudinal recording, the particles' north and south poles are lined up parallel to the disk's surface in a ring around its center. Magnetic repulsion limits how closely packed those bits can be and still maintain data integrity.

Perpendicular Recording

In this type of recording, the poles are arranged perpendicular to the disk's surface, which allows more bits to be packed onto a disk and reduces problems from magnetic interference.

cost-effective storage medium. Present DVD disks currently hold up to 4.7 gigabytes of data. We mentioned the Blu-ray, which uses a blue-violet laser to record information; Blu-ray disks will be able (at 27 gigabytes) to record more than 13 hours of standard TV programming and over 2 hours of digital high-definition video.

Hard drive makers have begun employing a technology known as *perpendicular recording technology*, which involves stacking magnetic bits vertically on the surface of a platter (instead of horizontally, as is usual). Based on this innovation, in mid-2005 Seagate Technology announced a disk drive for notebook computers that stores 160 gigabytes of data, 25% more than had been customary for other high-end hard drives for portable PCs. It also introduced a 1-inch drive that stores 8 gigabytes, which could be used in digital cameras, MP3 players, and handheld PCs, and a 500-gigabyte drive that could be used for storing video in home-entertainment centers.[35]

Also using perpendicular recording, Hitachi Ltd. announced in April 2005 that it had achieved record storage densities of 230 gigabits (230 billion bits) per square inch on a hard drive—an achievement, as one writer noted, "that would make possible a desktop computer drive capable of storing a trillion bytes of information, roughly twice the capacity of today's disk."[36]

MOLECULAR ELECTRONICS—STORAGE AT THE SUBATOMIC LEVEL
An emerging field, molecular electronics, may push secondary storage into another dimension entirely. Some possibilities include polymer memory, holograms, molecular magnets, subatomic lines, and bacteria.

Polymer memory involves developing an alternative to silicon to create chips that store data on polymer, or plastics, which are cheaper than silicon devices. A bonus with polymers is that, unlike conventional RAM memory, it is nonvolatile—it retains information even after the machine is shut off. Polymer memory involves storing data based on the polymer's electrical resistance. One manufacturer, Coatue, hoped to have memory chips available that would store 32 gigabits, outperforming flash memory. IBM is also developing so-called *probe storage* that uses tiny probes to burn pits in a Plexiglas-like polymer, in which each hole is about 10 nanometers wide, resulting in an experimental storage device of 200 gigabits per square inch.

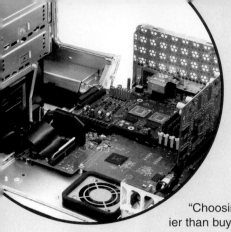

Experience Box
How to Buy a Laptop

"Choosing a laptop computer is trickier than buying a desktop PC," says *Wall Street Journal* technology writer Walter Mossberg.[37] The reason: laptops are more diverse and more personal and include "a mind-boggling array of computers, from featherweight models that are great for travel to bulky multimedia machines that double as TV sets." In addition, desktop "generic boxes" tend to be similar, at least within a price class.[38] Trying to choose among the many Windows-based laptops is a particularly brow-wrinkling experience. Macintosh laptops tend to be more straightforward.

Nevertheless, here are some suggestions:[39]

Purpose. What are you going to use your laptop for? You can get one that's essentially a desktop replacement and won't be moved much. If you expect to use the machine a lot in class, in libraries, or on airplanes, however, weight and battery life are important.

Whatever kind of computer you get (whether laptop or desktop), you want to be sure it works with your school's system. Some departments within a college may require a specific computer configuration or software, so check beforehand. Certainly it should be up to date and able to get internet access through Ethernet and wireless connections. "Don't bring clunkers [to college]," says one article. "Older computers tend to be buggy, and often lack protection against viruses and other mailicious software."[40]

Basics. Technology writer Walter Mossberg advises: "Get at least 512 megabytes of memory. Don't worry much about processor speed. Buy as much hard-drive capacity as you can. Make sure your computer has multiple USB 2.0 ports. . . . Also, on Windows laptops, security is crucial. Make sure you get the more secure SP2 version of Windows XP and that you immediately install antivirus, antispam, antispyware, antipopup, and firewall software. If you plan to use your laptop in public wireless hot spots, take the time to enable its wireless security features."[41]

Budget & Weight. Laptops range from $500 (with rebates) to $4,000, with high-end brands aimed mainly at businesspeople, hard-core videogamers, and people doing video production. Weight can range from less than a pound to over 10 pounds.

- *Low-budget:* Laptops costing in the $500–$1,000 range weigh from 5.4 to 7.5 pounds, have 14- or 15-inch screens, run on Intel's Celeron Mobile or AMD's Sempron chip with 256 megabytes of RAM, and have 30- to 40-gigabyte hard drives. Be sure the laptop comes with a CD burner or CD burner/DVD-ROM combination and Wi-Fi connection. Examples are the Compaq Presario V2000, the eMachines M5405, and the HP Pavillion ze4900.

- *Middle range:* Generally priced at $1,000–$1,300 and weighing 4 to 7 pounds, these laptops have power-conserving processors such as the Pentium M, 512 megabytes of RAM, 80-gigabyte hard drives, wider (14.1- to 15.4-inch) screens, and DVD/CD-RW combination drive. Examples are the Sony Vaio S series and the Apple 15-inch PowerBook.

- *Lightweight ultraportables:* Weighing under 4 pounds, ultraportable laptops have screen sizes ranging from 8.9 inches to 12.1 inches, 1.0- to 1.8-gigahertz processors, and 60-gigabyte hard drives. Some models lack internal CD or DVD drives. Battery life varies from 2 to 6 hours. Cost is generally $1,300 to $2,500. Examples are the ThinkPad X41 and the Sony Vaio T25.

- *High-end desktop replacements:* Weighing 7–10 pounds and costing $2,500–$4,000, high-end laptops are bulky machines designed to be desktop replacements. Their 15.1- to 17-inch display screens with wide viewing angles allow four or five people to watch at once. These machines feature graphics processors, 80-gigabyte hard drives, DVD-RW drives, and TV tuners for watching TV on your display screen. Examples are Toshiba's Qosimio models, the Dell Inspiron 9200, and the Gateway M675XL.

Batteries: The Life-Weight Trade-Off. You should strive to find a laptop that will run for at least 3 hours on its battery. A rechargeable lithium-ion battery lasts longer than the nickel–metal hydride battery. Even so, a battery in the less expensive machines will usually run continuously for only about $2\frac{1}{4}$ hours. (DVD players are particularly voracious consumers of battery power, so it's the rare laptop that will allow you to finish watching a 2-hour movie.)

The trade-off is that heavier machines usually have longer battery life. The lightweight machines tend to get less than 2 hours, and toting extra batteries offsets the weight savings. (A battery can weigh a pound or so.)

Software. Laptops might come with less software than you would get with a typical desktop, though what you get will probably be adequate for most student purposes. For instance, notebook PCs might offer Microsoft Works rather than the more powerful Microsoft Office to handle word processing, spreadsheets, databases, and the like.

Keyboards & Pointing Devices. The keys on a laptop keyboard are usually the same size as those on a desktop machine, although they can be smaller. However, the up-and-down action feels different, and the keys may feel wobbly. In addition, some keys may be omitted altogether, or keys may do double duty or appear in unaccustomed arrangements.

Most laptops have a small touch-sensitive pad in lieu of a mouse—you drag your finger across the touchpad to move the cursor. Others use a pencil-eraser-size pointing stick in the middle of the keyboard.

Screens. If you're not going to carry the notebook around much, go for a big, bright screen. Most people find they are comfortable with a 12- to 14-inch display, measured diagonally, though screens can be as small as 10.4 inches and as large as 15.4 inches (which is better for viewing movies).

The best screens are active-matrix display (TFT). However, some low-priced models have the cheaper passive-matrix screens (HPA, STN, or DSTN), which are harder to read, though you may find you can live with them. XGA screens (1,024 × 768 pixels) have a higher resolution than SVGA screens (800 × 600 pixels), but fine detail may not be important to you. A UXGA screen will cost about $100 more than XGA.

Memory, Speed, & Storage Capacity. If you're buying a laptop to complement your desktop, you may be able to get along with reduced memory, slow processor, small hard disk, and no CD-ROM. Otherwise, all these matters become important.

Memory (RAM) is the most important factor in computer performance, even though processor speed is more heavily hyped. Most laptops have at least 256 megabytes (MB) of memory, but 512 MB of high-speed DDR SDRAM is best. A microprocessor running 1.6 gigaherz should be adequate.

Sometimes notebooks are referred to as "three-spindle" or "two-spindle" machines. In a three-spindle machine, a hard drive, a floppy-disk drive, and a CD/DVD drive all reside internally (not as external peripherals). A two-spindle machine has a hard drive and space for either a CD/DVD drive or a second battery. A hard drive of 60 gigabytes or more is sufficient for most people.

Wireless. Most laptops come with Wi-Fi for wireless networking; 802.11b technology is fine for ordinary use, but try to buy a laptop with the "g" version.

For more information about buying computers, go to *www.zdnet.com/computershopper*.

Summary

AGP (accelerated graphics port) bus (p. 219) Bus that transmits data at very high speeds; designed to support video and three-dimensional (3-D) graphics. Why it's important: *An AGP bus is twice as fast as a PCI bus.*

arithmetic/logic unit (ALU) (p. 208) Part of the CPU that performs arithmetic operations and logical operations and controls the speed of those operations. Why it's important: *Arithmetic operations are the fundamental math operations: addition, subtraction, multiplication, and division. Logical operations are comparisons such as "equal to," "greater than," or "less than."*

ASCII (American Standard Code for Information Interchange) (p. 197) Binary code used with microcomputers. Besides having the more conventional characters, the Extended ASCII version includes such characters as math symbols and Greek letters. Why it's important: *ASCII is the binary code most widely used in microcomputers.*

bay (p. 201) Shelf or opening in the computer case used for the installation of electronic equipment, generally storage devices such as a hard drive or DVD drive. Why it's important: *Bays permit the expansion of system capabilities. A computer may come equipped with four or eight bays.*

binary system (p. 195) A two-state system used for data representation in computers; has only two digits—0 and 1. Why it's important: *In the computer, 0 can be represented by electrical current being off and 1 by the current being on. All data and program instructions that go into the computer are represented in terms of these binary numbers.*

EBCDIC	ASCII-8
1100 0001	0100 0001
1100 0010	0100 0010
1100 0011	0100 0011
1100 0100	0100 0100
1100 0101	0100 0101
1100 0110	0100 0110
1100 0111	0100 0111
1100 1000	0100 1000
1100 1001	0100 1001
1101 0001	0100 1010

bit (p. 196) Short for "binary digit," which is either a 0 or a 1 in the binary system of data representation in computer systems. Why it's important: *The bit is the fundamental element of all data and information processed and stored in a computer system.*

Blu-ray (p. 228) The Blu-ray optical format was developed to enable recording, rewriting, and playback of high-definition video, as well as storing of large amounts of data. Why it's important: *It's possible to fit more data on a Blu-ray disk even though it's the same size as a CD/DVD.*

bus (p. 209) Also called *bus line;* electrical data roadway through which bits are transmitted within the CPU and between the CPU and other components of the motherboard. Why it's important: *A bus resembles a multilane highway: The more lanes it has, the faster the bits can be transferred.*

byte (p. 196) Group of 8 bits. Why it's important: *A byte represents one character, digit, or other value. It is the basic unit used to measure the storage capacity of main memory and secondary storage devices (kilobytes and megabytes).*

cache (p. 211) Special high-speed memory area on a chip that the CPU can access quickly. It temporarily stores instructions and data that the processor is likely to use frequently. Why it's important: *Cache speeds up processing.*

CD-R (compact disk–recordable) disk (p. 226) Optical-disk form of secondary storage that can be written to only once but can be read many times. Why it's important: *This format allows consumers to make their own CD disks, though it's a slow process. Once recorded, the information cannot be erased. CD-R is often used by companies for archiving—that is, to store vast amounts of information. A variant is the Photo CD, an optical disk developed by Kodak that can digitally store photographs taken with an ordinary 35-millimeter camera.*

CD-ROM (compact disk read-only memory) (p. 225) Optical-disk form of secondary storage that is used to hold prerecorded text, graphics, and sound. Why it's important: *Like music CDs, a CD-ROM is a read-only disk. Read-only means the disk's content is recorded at the time of* manufacture and cannot be written on or erased by the user. A CD-ROM disk can hold up to 650 megabytes of data, equal to over 300,000 pages of text.

CD-RW (compact disk–rewritable) disk (p. 227) Also known as *erasable optical disk;* optical-disk form of secondary storage that allows users to record and erase data, so the disk can be used over and over again. Special CD-RW drives and software are required. Why it's important: *CD-RW disks are useful for archiving and backing up large amounts of data or work in multimedia production or desktop publishing; however, they are relatively slow.*

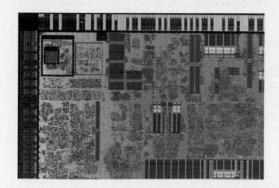

chip (p. 192) Also called a *microchip,* or *integrated circuit;* consists of millions of microminiature electronic circuits printed on a tiny piece of silicon. Silicon is an element widely found in sand that has desirable electrical (or "semiconducting") properties. Why it's important: *Chips have made possible the development of small computers.*

chipset (p. 204) Groups of interconnected chips on the motherboard that control the flow of information between the microprocessor and other system components connected to the motherboard. Why it's important: *The chipset determines what types of processors, memory, and video card ports will work on the same motherboard. It also establishes the types of multimedia, storage, network, and other hardware the motherboard supports.*

CMOS (complementary metal-oxide semiconductor) chips (p. 211) Battery-powered chips that don't lose their contents when the power is turned off. Why it's important: *CMOS chips contain flexible start-up instructions—such as time, date, and calendar—that must be kept current even when the computer is turned off. Unlike ROM chips, CMOS chips can be reprogrammed, as when you need to change the time for daylight savings time.*

control unit (p. 208) Part of the CPU that deciphers each instruction stored in it and then carries out the instruction. Why it's important: *The control unit directs the movement of electronic signals between main memory and the arithmetic/logic unit. It also directs these electronic signals between main memory and the input and output devices.*

CPU (central processing unit) (p. 208) The processor; it follows the instructions of the software (program) to manipulate data into information. The CPU consists of two parts—(1) the control unit and (2) the arithmetic/logic unit (ALU), both of which contain registers, or high-speed storage areas. All are linked by a kind of electronic "roadway" called a *bus.* Why it's important: *The CPU is the "brain" of the computer.*

DVD-R (DVD-recordable) disks (p. 227) DVD disks that allow one-time recording by the user. Why it's important: *Recordable DVDs offer the user yet another option for storing large amounts of data.*

DVD-ROM (digital versatile disk or digital video disk, with read-only memory) (p. 227) CD-type disk with extremely high capacity, able to store 4.7 or more gigabytes. Why it's important: *It is a powerful and versatile secondary storage medium.*

exabyte (EB) (p. 197) Approximately 1 quintillion bytes—1 billion billion bytes (1,024 petabytes—or 1,152,921,504,606,846,976 bytes). Why it's important: *Although this number is seldom used, it is estimated that all the printed material in the world represents about 5 exabytes.*

EBCDIC (Extended Binary Coded Decimal Interchange Code) (p. 197) Binary code used with large computers. Why it's important: *EBCDIC is commonly used in mainframes.*

expansion (p. 202) Way of increasing a computer's capabilities by adding hardware to perform tasks that are beyond the scope of the basic system. Why it's important: *Expansion allows users to customize and/or upgrade their computer systems.*

expansion card (p. 217) Also known as *expansion board, adapter card, interface card, plug-in board, controller card, add-in,* or *add-on;* circuit board that provides more memory or that controls peripheral devices. Why it's important: *Common expansion cards connect to the monitor (graphics card), speakers and microphones (sound card), and network (network card). Most computers have four to eight expansion slots, some of which may already contain expansion cards included in your initial PC purchase.*

expansion slot (p. 217) Socket on the motherboard into which the user can plug an expansion card. Why it's important: See *expansion card.*

FireWire (p. 216) A specialized serial-bus port intended to connect devices working with lots of data, such as digital video recorders, DVD players, gaming consoles, and digital audio equipment. Why it's important: *Whereas the USB port handles only 12 megabits per second, FireWire handles up to 400 megabits per second.*

flash memory card (p. 232) Also known as *flash RAM cards;* form of secondary storage consisting of circuitry on credit-card–size cards that can be inserted into slots connecting to the motherboard on notebook computers. Why it's important: *Flash memory is nonvolatile, so it retains data even when the power is turned off.*

flash memory chip (p. 211) Chip that can be erased and reprogrammed more than once (unlike PROM chips, which can be programmed only once). Why it's important: *Flash memory, which can range from 32 to 128 megabytes in capacity, is used to store programs not only in personal computers but also in pagers, cellphones, printers, and digital cameras. Unlike standard RAM chips, flash memory is nonvolatile—data is retained when the power is turned off.*

flash memory drive (p. 232) Also called a *USB flash drive, keychain drive,* or *key drive;* a finger-size module of flash memory that plugs into the USB ports of nearly any PC or Macintosh. Why it's important: *They generally have storage capacities up to 2 gigabytes, making the device extremely useful if you're traveling from home to office and don't want to carry a laptop. When you plug the device into your USB port, it shows up as an external drive on the computer.*

flash memory stick (p. 232) Smaller than a stick of chewing gum, a form of flash memory media that plugs into a memory stick port in a digital camera, camcorder, notebook PC, photo printer, and other devices. Why it's important: *It holds up to 1 gigabyte of data and is very convenient to transport and use.*

floppy disk (p. 221) Often called *diskette* or simply *disk;* removable flat piece of mylar plastic packaged in a 3.5-inch plastic case. Data and programs are stored on the disk's coating by means of magnetized spots, following standard on/off patterns of data representation (such as ASCII). The plastic case protects the mylar disk from being touched by human hands. Why it's important: *Floppy disks are still used on some older microcomputers.*

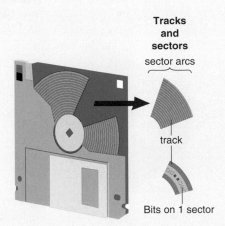

Tracks and sectors

sector arcs

track

Bits on 1 sector

floppy-disk cartridges (p. 222) High-capacity removable 3.5-inch disks, such as Zip disks. Why it's important: *These cartridges store more data than regular floppy disks and are just as portable.*

flops (p. 207) Stands for "floating-point operations per second." A *floating-point operation* is a special kind of mathematical calculation. This measure, used mainly with supercomputers, is expressed as *megaflops* (*mflops,* or millions of floating-point operations per second), *gigaflops* (*gflops,* or billions), and *teraflops* (*tflops,* or trillions). Why it's important: *The measure is used to express the processing speed of supercomputers.*

gigabyte (G, GB) (p. 197) Approximately 1 billion bytes (1,073,741,824 bytes); a measure of storage capacity. Why it's important: *This measure was formerly used mainly with "big iron" (mainframe) computers but is typical of the secondary storage (hard-disk) capacity of today's microcomputers.*

gigahertz (GHz) (p. 206) Measure of speed used for the latest generation of processors: 1 billion cycles per second. Why it's important: *Since a new high-speed processor can cost many hundred dollars more than a previous-generation chip, experts often recommend that buyers fret less about the speed of the processor (since the work most people do on their PCs doesn't even tax the limits of the current hardware) and more about spending money on extra memory.*

graphics card (p. 219) Also called a *video card* or *video adapter;* expansion card that converts signals from the computer into video signals that can be displayed as images on a monitor. Why it's important: *The power of a graphics card, often expressed in megabytes, as in 8, 16, or 32 MB, determines the clarity of the images on the monitor.*

hard disk (p. 222) Secondary storage medium; thin but rigid metal, glass, or ceramic platter covered with a substance that allows data to be stored in the form of magnetized spots. Hard disks are tightly sealed within an enclosed hard-disk-drive unit to prevent any foreign matter from getting inside. Data may be recorded on both sides of the disk platters. Why it's important: *Hard disks hold much more data than do floppy disks. All microcomputers use hard disks as their principal storage medium.*

integrated circuit (p. 191). An entire electronic circuit, including wires, formed on a single "chip," or piece, of special material, usually silicon. Why it's important: *In the old days, transistors were made individually and then formed into an electronic circuit with the use of wires and solder. An integrated circuit is formed as part of a single manufacturing process.*

Intel-type chip (p. 204) Processor chip for PCs; made principally by Intel Corp. and Advanced Micro Devices (AMD), but also by Cyrix, DEC, and others. Why it's important: *These chips are used by manufacturers such as Compaq, Dell, Gateway, Hewlett-Packard, and IBM. Since 1993, Intel has marketed its chips under the names "Pentium," "Pentium Pro," "Pentium MMX," "Pentium II," "Pentium III," "Pentium 4," "Xeon," "Itanium," and "Celeron." Many ads for PCs contain the logo "Intel inside" to show that the systems run an Intel microprocessor.*

IrDA port (p. 216) Port that allows a computer to make a cableless connection with infrared-capable devices, such as some printers. Why it's important: *Infrared ports eliminate the need for cabling.*

kilobyte (K, KB) (p. 196) Approximately 1,000 bytes (1,024 bytes); a measure of storage capacity. Why it's important: *The kilobyte was a common unit of measure for memory or secondary storage capacity on older computers.*

machine cycle (p. 208) Series of operations performed by the control unit to execute a single program instruction. It (1) fetches an instruction, (2) decodes the instruction, (3) executes the instruction, and (4) stores the result. Why it's important: *The machine cycle is the essence of computer-based processing.*

machine language (p. 199) Binary code (language) that the computer uses directly. The 0s and 1s represent precise storage locations and operations. Why it's important: *For a program to run, it must be in the machine language of the computer that is executing it.*

magnetic tape (p. 229) Thin plastic tape coated with a substance that can be magnetized. Data is represented by magnetized spots (representing 1s) or nonmagnetized spots (representing 0s). Why it's important: *Today, "mag tape" is used mainly for backup and archiving—that is, for maintaining historical records—where there is no need for quick access.*

megabyte (M, MB) (p. 196) Approximately 1 million bytes (1,048,576 bytes); measure of storage capacity. Why it's important: *Microcomputer primary storage capacity is expressed in megabytes.*

megahertz (MHz) (p. 206) Measure of microcomputer processing speed, controlled by the system clock; 1 million cycles per second. Why it's important: *Generally, the higher the megahertz rate, the faster the computer can process data. A 550-MHz Pentium III–based microcomputer, for example, processes 550 million cycles per second.*

microprocessor (p. 192) Miniaturized circuitry of a computer processor. It stores program instructions that process, or manipulate, data into information. The key parts of the microprocessor are transistors. Why it's important: *Microprocessors enabled the development of microcomputers.*

MIDI port (p. 216) Pronounced "*mid-*dee," and short for *Musical Instrument Digital Interface.* A specialized port used in creating, recording, editing, and performing music. Why it's important: *It is used for connecting amplifiers, electronic synthesizers, sound cards, drum machines, and the like.*

MIPS (p. 207) Stands for "millions of instructions per second"; a measure of processing speed. Why it's important: *MIPS is used to measure processing speeds of mainframes, minicomputers, and workstations. A workstation might perform at 100 MIPS or more, a mainframe at 200–1,200 MIPS.*

Motorola-type chips (p. 205) Microprocessors made by Motorola for Apple Macintosh computers. Why it's important: *From 1993 until the arrival of the PowerPC, Motorola provided an alternative to the Intel-style chips made for PC microcomputers.*

multicore processor (p. 205) Microcomputer chip such as AMD's Athlon 64 X2, Intel's Pentium D, and IBM/Sony/Toshiba's Cell, with two or more processor "cores" on a single piece of silicon. Why it's important: *Chips can take on several tasks at once because the operating system can divide its work over more than one processor.*

network interface card (NIC) (p. 220) Expansion card that allows the transmission of data over a cable network. Why it's important: *Installation of a network interface card in the computer enables the user to connect with various computers and other devices such as printers.*

nonremovable hard disk (p. 224) Also known as *fixed disk;* hard disk housed in a microcomputer system unit and used to store nearly all programs and most data files. Usually it consists of several 3.5-inch metallic platters sealed inside a drive case the size of a small sandwich, which contains disk platters on a drive spindle, read/write heads mounted on an access arm that moves back and forth, and power connections and circuitry. Operation is much the same as for a diskette drive: The read/write heads locate specific instructions or data files according to track or sector. Hard disks can also come in removable cartridges. Why it's important: See ***hard disk.***

optical card (p. 231) Plastic, laser-recordable, wallet-type card used with an optical-card reader. Why it's important: *Because optical cards can cram so much data (6.6 megabytes) into so little space, they may become popular in the future. For instance, a health card based on an optical card would have room not only for the individual's medical history and health-insurance information but also for digital images, such as electrocardiograms.*

optical disk (p. 225) Removable disk, usually 4.75 inches in diameter and less than one-twentieth of an inch thick, on which data is written and read through the use of laser beams. Why it's important: *An audio CD holds up to 74 minutes (2 billion bits' worth) of high-fidelity stereo sound. Some optical disks are used strictly for digital data storage, but many are used to distribute multimedia programs that combine text, visuals, and sound.*

parallel port (p. 213) A connector for a line that allows 8 bits (1 byte) to be transmitted simultaneously, like cars on an eight-lane highway. Why it's important: *Parallel lines move information faster than serial lines do. However, because they can transmit information efficiently only up to 15 feet, they are used principally for connecting printers or external disk or magnetic-tape backup storage devices.*

parity bit (p. 198) Also called a *check bit;* an extra bit attached to the end of a byte. Why it's important: *The parity bit enables a computer system to check for errors during transmission (the check bits are organized according to a particular coding scheme designed into the computer).*

PC card (p. 220) Thin, credit card–size (2.1- by 3.4-inch) hardware device. Why it's important: *PC cards are used principally on laptop computers to expand capabilities.*

PCI (peripheral component interconnect) bus (p. 218) High-speed bus; at 32 or 64 bits wide. Why it's important: *PCI has been widely used in microcomputers to connect graphics cards, sound cards, modems, and high-speed network cards.*

petabyte (P, PB) (p. 197) Approximately 1 quadrillion bytes (1,048,576 gigabytes); measure of storage capacity. Why it's important: *The huge storage capacities of modern databases are now expressed in petabytes.*

plug and play (p. 215) USB peripheral connection standard that allows peripheral devices and expansion cards to be automatically configured while they are being installed. Why it's important: *Plug and play avoids the hassle of setting switches and creating special files, which plagued earlier users.*

port (p. 213) A connecting socket or jack on the outside of the system unit into which are plugged different kinds of cables. Why it's important: *A port allows the user to plug in a cable to connect a peripheral device, such as a monitor, printer, or modem, so that it can communicate with the computer system.*

power supply (p. 201) Device that converts AC to DC to run the computer. Why it's important: *The electricity available from a standard wall outlet is alternating current (AC), but a microcomputer runs on direct current (DC).*

PowerPC chip (p. 205) Family of processor developed by Apple, Freestone, and IBM for Apple Macintoshes, such as the G5. Why it's important: *The advantage of PowerPC chips is that with certain hardware or software configurations a Macintosh can run PC as well as Macintosh applications software.*

RAID (redundant array of independent disks) storage system (p. 225) Secondary storage system consisting of two or more disk drives within a single cabinet or connected along a SCSI chain; sends data to the computer along parallel paths simultaneously. Why it's important: *The system not only holds more data than a fixed-disk drive within the same amount of space but is also more reliable.*

RAM (random access memory) chips (p. 210) Also called *primary storage* and *main memory;* chips that temporarily hold software instructions and data before and after it is processed by the CPU. RAM is a volatile form of storage. Why it's important: *RAM is the working memory of the computer. Having enough RAM is critical to users' ability to run many software programs.*

read (p. 211) To transfer data from an input source into the computer's memory or CPU. Why it's important: *Reading, along with writing, is an essential computer activity.*

read/write head (p. 222) Mechanism used to transfer data between the computer and the disk. When the disk spins inside its case, the read/write head moves back and forth over the data access area on the disk. Why it's important: *The read/write head enables the essential activities of reading and writing data.*

registers (p. 209) High-speed storage areas that temporarily store data during processing. Why it's important: *Registers may store a program instruction while it is being decoded, store data while it is being processed by the ALU, or store the results of a calculation.*

removable hard disk (p. 224) Also called *hard-disk cartridge;* one or two platters enclosed along with read/write heads in a hard plastic case, which is inserted into a microcomputer's cartridge drive. Typical capacity is 80 gigabytes and more. Why it's important: *These cartridges offer users greater storage capacity than do floppy disks but with the same portability.*

ROM (read-only memory) (p. 211) Memory chip that cannot be written on or erased by the computer user without special equipment. Why it's important: *ROM chips contain fixed start-up instructions. They are loaded, at the factory, with programs containing special instructions for basic computer operations, such as starting the computer or putting characters on the screen. These chips are nonvolatile; their contents are not lost when power to the computer is turned off.*

SCSI (small computer system interface) port (p. 213) Pronounced "scuzzy," a connector that allows data to be transmitted in a "daisy chain" to up to seven devices at speeds (32 bits at a time) higher than those possible with serial and parallel ports. The term *daisy chain* means that several devices are connected in series to each other, so that data for the seventh device, for example, has to go through the other six devices first. Why it's important: *A SCSI enables users to connect external hard-disk drives, CD-ROM drives, scanners, and magnetic-tape backup units.*

secondary storage hardware (p. 220) Devices that permanently hold data and information as well as programs. Why it's important: *Secondary storage—as opposed to primary storage—is nonvolatile; that is, saved data and programs are permanent, or remain intact, when the power is turned off.*

sectors (p. 222) The small arcs created in tracks when a disk's storage locations are divided into wedge-shaped sections. Why it's important: *The system software uses the point at which a sector intersects a track to reference the data location.*

semiconductor (p. 192) Material, such as silicon (in combination with other elements), whose electrical properties are intermediate between a good conductor and a nonconductor of electricity. When highly conducting materials are laid on the semiconducting material, an electronic circuit can be created. Why it's important: *Semiconductors are the materials from which integrated circuits (chips) are made.*

serial port (p. 213) A connector for a line that sends bits one after another, like cars on a one-lane highway. Why it's important: *Because individual bits must follow each other, a serial port is usually used to connect devices that do not require fast transmission of data, such as keyboard, mouse, monitors, and modems. It is also useful for sending data over a long distance.*

silicon (p. 191) An element that is widely found in clay and sand and is used in the making of solid-state integrated circuits. Why it's important: *It is used not only because its abundance makes it cheap but also because it is a good semiconductor. As a result, highly conducting materials can be overlaid on the silicon to create the electronic circuitry of the integrated circuit.*

smart card (p. 230) Wallet-type card that looks like a credit card but has a microprocessor and memory chips embedded in it. When inserted into a reader, it transfers data to and from a central computer. Why it's important: *Unlike conventional credit cards, smart cards can hold a fair amount of data and can store some basic financial records. Thus, they are used as telephone debit cards, health cards, and student cards.*

solid-state device (p. 191) Electronic component made of solid materials with no moving parts, such as an integrated circuit. Why it's important: *Solid-state integrated circuits are far more reliable, smaller, and less expensive than electronic circuits made from several components.*

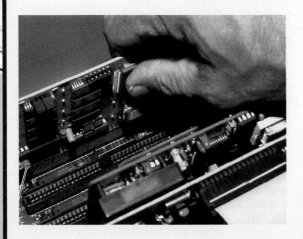

sound card (p. 219) Expansion card used to convert and transmit digital sounds through analog speakers, microphones, and headsets. Why it's important: *Cards such as PCI wavetable sound cards are used to add music and sound effects to computer video games.*

system clock (p. 206) Internal timing device that uses fixed vibrations from a quartz crystal to deliver a steady stream of digital pulses or "ticks" to the CPU. These ticks are called *cycles.* Why it's important: *Faster clock speeds will result in faster processing of data and execution of program instructions, as long as the computer's internal circuits can handle the increased speed.*

tape cartridge (p. 229) Module resembling an audiocassette that contains tape in a rectangular plastic housing. The two most common types of tape drives are DAT and Traven TR-3. Why it's important: *Tape cartridges are used for secondary storage on microcomputers and also on some large computers. Tape is used mainly for archiving purposes and backup.*

terabyte (T, TB) (p. 197) Approximately 1 trillion bytes (1,009,511,627,776 bytes); measure of storage capacity. Why it's important: *Some high-capacity disk storage is expressed in terabytes.*

tracks (p. 221) The rings on a diskette along which data is recorded. Why it's important: See *sectors.*

transistor (p. 191) Tiny electronic device that acts as an on/off switch, switching between "on" and "off" millions of times per second. Why it's important: *Transistors are part of the microprocessor.*

Unicode (p. 197) Binary coding scheme that uses 2 bytes (16 bits) for each character, rather than 1 byte (8 bits). Why it's important: *Instead of the 256 character combinations of ASCII, Unicode can handle 65,536 character combinations. Thus, it allows almost all the written languages of the world to be represented using a single character set.*

upgrading (p. 202) Changing to newer, usually more powerful or sophisticated versions, such as a more powerful microprocessor or more memory chips. Why it's important: *Through upgrading, users can improve their computer systems without buying completely new ones.*

USB (universal serial bus) port (p. 214) Port that can theoretically connect up to 127 peripheral devices daisy-chained to one general-purpose port. Why it's important: *USB ports are useful for peripherals such as digital cameras, digital speakers, scanners, high-speed modems, and joysticks. Being "hot pluggable" or "hot swappable" means that USB allows such devices to be connected or disconnected even while the PC is running.*

virtual memory (p. 212) Type of hard-disk space that mimics primary storage (RAM). Why it's important: *When RAM space is limited, virtual memory allows users to run more software at once, provided the computer's CPU and operating system are equipped to use it. The system allocates some free disk space as an extension of RAM; that is, the computer swaps parts of the software program between the hard disk and RAM as needed.*

volatile (p. 210) Temporary; the contents of volatile storage media, such as RAM, are lost when the power is turned off. Why it's important: *To avoid data loss, save your work to a secondary storage medium, such as a hard disk, in case the electricity goes off while you're working.*

word size (p. 208) Number of bits that the processor may process at any one time. Why it's important: *The more bits in a word, the faster the computer. A 32-bit computer—that is, one with a 32-bit-word processor—will transfer data within each microprocessor chip in 32-bit chunks, or 4 bytes at a time. A 64-bit computer transfers data in 64-bit chunks, or 8 bytes at a time.*

write (p. 211) To transfer data from the computer's CPU or memory to an output device. Why it's important: See **read.**

Zip disk (p. 222) Floppy-disk cartridge with a capacity of 100, 250, or 750 megabytes. At 100 megabytes, this is nearly 70 times the storage capacity of the standard floppy. Why it's important: *Among their uses, Zip disks are used to store large spreadsheet files, database files, image files, multimedia presentation files, and websites. Zip disks require their own Zip-disk drives, which may come installed on new computers, although external Zip drives are also available.* See also **floppy-disk cartridges.**

Chapter Review

stage **1** **LEARNING** MEMORIZATION

"I can recognize and recall information."

Self-Test Questions

1. A(n) _____ is about 1,000 bytes; a(n) _____ is about 1 million bytes; a(n) _____ is about 1 billion bytes.

2. The _____ is the part of the microprocessor that tells the rest of the computer how to carry out a program's instructions.

3. The process of retrieving data from a storage device is referred to as _____; the process of copying data to a storage device is called _____.

4. To avoid losing data, users should always _____ their files.

5. Formatted disks have _____ and _____ that the system software uses to reference data locations.

6. The _____ is often referred to as the "brain" of a computer.

7. The electrical data roadways through which bits are transmitted are called _____.

8. A cable connected to a _____ port sends bits one at a time, one after the other; a cable connected to a _____ port sends 8 bits simultaneously.

9. Part of the disk-drive mechanism, the _____ transfers data between the computer and the disk.

10. _____ chips, also called *main memory,* and are critical to computer performance.

11. _____ operations are the fundamental math operations: addition, subtraction, multiplication, and division. _____ operations are comparisons such as "equal to," "greater than," or "less than."

12. A group of 8 bits is a _____.

13. The extra bit attached to the end of a byte for error checking is a _____ bit.

14. A tiny electronic device that acts as an on/off switch, switching between "on" and "off" millions of times per second, is called a _____.

15. _____ is a form of flash memory media that plugs into a special port in a digital camera, camcorder, notebook PC, or photo printer and holds up to 1 gigabyte of data.

Multiple-Choice Questions

1. Which of the following is another term for primary storage?
   a. ROM
   b. ALU
   c. CPU
   d. RAM
   e. CD-R

2. Which of the following is *not* included on a computer's motherboard?
   a. RAM chips
   b. ROM chips
   c. keyboard
   d. microprocessor
   e. expansion slots

3. Which of the following is used to hold data and instructions that will be used shortly by the CPU?
   a. ROM chips
   b. peripheral devices
   c. RAM chips
   d. CD-R
   e. hard disk

4. Which of the following coding schemes is widely used on microcomputers?
   a. EBCDIC
   b. Unicode
   c. ASCII
   d. Microcode
   e. Unix

5. Which of the following is used to measure processing speed in microcomputers?
   a. MIPS
   b. flops
   c. picoseconds
   d. megahertz
   e. millihertz

6. Which expansion bus specializes in graphics processing?
   a. PCI
   b. ROM
   c. CMOS
   d. AGP
   e. AMR

7. Which company is the main manufacturer of microprocessors for non-Apple computers?
   a. Intel
   b. CMOS

c. Motorola

d. Pentium

e. Cyrix

8. Which element is commonly used in the making of solid-state integrated circuits?

a. pentium

b. lithium

c. copper

d. iron

e. silicon

9. DVD-Rs allow

a. repeated rewriting.

b. one-time recording by the user; they cannot be written on more than once.

c. no writing; they are read-only.

True/False Questions

T F   1. A bus connects a computer's control unit and ALU.

T F   2. The machine cycle comprises the instruction cycle and the execution cycle.

T F   3. Magnetic tape is the most common secondary storage medium used with microcomputers.

T F   4. Main memory is nonvolatile.

T F   5. Pipelining is a method of speeding up processing.

T F   6. Today's laptop computers can perform more calculations per second than the ENIAC, an enormous machine occupying more than 1,800 square feet and weighing more than 30 tons.

T F   7. USB can theoretically connect up to 127 peripheral devices.

T F   8. A petabyte is approximately 1 quadrillion bytes.

T F   9. Online secondary storage services test your computer's RAM capacity.

T F   10. Keychain memory is a pipelining device.

 stage **LEARNING**   COMPREHENSION

"I can recall information in my own terms and explain it to a friend."

Short-Answer Questions

1. What is ASCII, and what do the letters stand for?

2. Why should measures of capacity matter to computer users?

3. What's the difference between RAM and ROM?

4. What is the significance of the term *megahertz*?

5. What is a motherboard? Name at least four components of a motherboard.

6. What are the most convenient forms of back-up storage? Why?

7. Why is it important for your computer to be expandable?

8. What are three uses of a smart card?

9. What is nanotechnology?

10. What are the uses of a surge protector, voltage regulator, and UPS, and why are these devices important?

11. Explain the binary system.

12. What are Unicode and ASCII?

13. Why is silicon used in the manufacture of microprocessors?

14. What is Blu-ray used for?

stage **LEARNING**   APPLYING, ANALYZING, SYNTHESIZING, EVALUATING

"I can apply what I've learned, relate these ideas to other concepts, build on other knowledge, and use all these thinking skills to form a judgment."

Knowledge in Action

1. If you're using Windows XP, you can easily determine what microprocessor is in your computer and how much RAM it has. To begin, click the *Start* button in the Windows desktop pull-up menu bar and then choose *Control Panel*. Then locate the System icon in the Control Panel window and double-click on the icon.

     The System Properties dialog box will open. It contains several tabs: General, Computer Name, Hardware, System Restore, Automatic Updates, and

Advanced. The name of your computer's microprocessor will display on the General tab, as well as its speed and your computer's RAM capacity.

2. Visit a local computer store and note the system requirements listed on five software packages. What are the requirements for processor? RAM? Operating system? Available hard-disk space? CD/DVD speed? Audio/video cards? Are there any output hardware requirements?

3. Develop a binary system of your own. Use any two objects, states, or conditions, and encode the following statement: "I am a rocket scientist."

4. The floppy drive no longer comes with a standard Dell PC (other companies are following suit). What do you think will be the next "legacy" device to be abandoned?

5. Storing humans: If the human genome is 800 million bytes (according to Raymond Kurzweil), how many humans could you fit on a 120-GB hard drive?

6. What are the predictions about how long Moore's Law will continue to apply? Do a web search for four opinions; list the website sources and their predictions, and state how reliable you believe the sites are, and why.

Web Exercises

1. The objective of this project is to introduce you to an online encyclopedia that's dedicated to computer technology. The *www.webopedia* website is a good resource for deciphering computer ads and clearing up difficult concepts. For practice, visit the site and type *processor* into the Search text box and then press the *Enter* key or click on the *Go!* button. Click on some of the links that are displayed. Search for information on other topics of interest to you.

2. You can customize your own PC through a brand-name company such as Dell or Gateway, or you can create your own personal model by choosing each component on your own. Decide which method is best for you. Go to the following sites and customize your ideal PCs:

   www.dell.com
   www.gateway.com
   www.compaq.com
   www.hpshopping.com
   Write down the prices for your ideal customized PCs. Then go to

   www.pricewatch.com
   www.dealsdepot.com/Shop/Control/fp/SFV/14930/view_page/custom/RID/118778?engine=adwords+1&keyword=custom+1
   http://store.sysbuilder.com/desktop.html
   and see if you could save money by putting your own PC together piece by piece. (This includes purchasing each component separately and verifying compatibility of all components.)

For a tutorial on building your own computer, go to *www.pcmech.com/build.htm*.

3. DVD formats: DVD+R, DVD-R, DVD+RW, DVD-RW, so many formats! Are they all the same? Visit these websites to get current information on the issues surrounding recordable DVD media:

   www.manifest-tech.com/media_dvd/dvd_compat.htm
   www.dvddemystified.com/dvdfaq.html#4.3
   www.plextor.com/english/support/faqs/G00015.htm
   www.idvd.ca/dvd-format-guide.htm

4. What is a Qubit? You've learned about binary digits in this chapter; now learn about the Qubit, the basic unit of information in a quantum computer. Beware: When you step into the realm of quantum theory, things become bizarre.

   http://whatis.techtarget.com/definition/0,,sid9_gci341232,00.html
   www.thefreedictionary.com/quantum+bit
   www.yourdictionary.com/ahd/q/q0015350.html
   www.answers.com/topic/qubit?method=6

5. DNA computing: Visit the following websites to learn more about DNA software and computing:

   www.arstechnica.com/reviews/2q00/dna/dna-1.html
   www.cis.udel.edu/~dna3/DNA/dnacomp.html
   www.hypography.com/topics/dnacomputers.cfm
   http://computer.howstuffworks.com/dna-computer.htm

6. You mean the Matrix was real? Visit this website for analysis of real-world applications of concepts from the movie *Matrix: www.kurzweilai.net/index.html?flash=2.*

SECURITY 7. Security issue—credit card fraud: When buying parts or making any kind of purchase over the internet, always make sure that the web address says HTTPS to let you know it is an encrypted SSL (Secured Socket Layer) website. Visit the site below for safety tips when using your credit card online:

http://familyinternet.about.com/od/onlineshoppingsafety/a/onlineshopping.htm

Hardware: Input & Output

Taking Charge of Computing & Communications

Chapter Topics & Key Questions

5.1   Input & Output How is input and output hardware used by a computer system?

5.2   Input Hardware What are the three categories of input hardware, what devices do they include, and what are their features?

5.3   Output Hardware What are the two categories of output hardware, what devices do they include, and what are their features?

5.4   Input & Output Technology & Quality of Life: Health & Ergonomics What are the principal health and ergonomic issues relating to computer use?

5.5   The Future of Input & Output What are some examples of future input and output technology?

Automated teller machines have become so common, it now seems there are almost as many places to get cash as to spend it," says one account.[1]

Not only are *automated teller machines (ATMs)*, or cash machines, in office buildings, convenience stores, nightclubs, and even the lobbies of some big apartment buildings; they are also becoming something quite different from simple devices for people who need fast cash. They are migrating into different kinds of *kiosks* (pronounced "*key*-osks"), computerized booths or small standing structures providing any number of services, from electronic banking options to corporate job benefits, from tourism advice to garage-sale permits. (● *See Panel 5.1.*)

In Tampa, Florida, Kimberly Ruggiero, 22, a hospital inventory tracker, avoids a 30-minute bank line on Fridays by stopping at a 7-Eleven and using an ATM-like kiosk to cash her paycheck. "It's extremely convenient and fast," she says, "and I don't have to wait in line."[2] In New York City, kiosks can be used by citizens to pay parking tickets and check for building-code violations. In San Antonio, they provide information on animals available for adoption. In Seattle, commuters at car-ferry terminals view images of traffic conditions on major highways. Many colleges and universities use kiosks to provide students with information about classes, schedules, activity locations, maps, and so on.[3]

Kiosks also sell stamps, print out checks, and issue movie and plane tickets. Alamo car rental offices have kiosks at which travelers can print out directions, get descriptions of hotel services, and obtain restaurant menus and reviews—in four languages. At MainStay Suites, guests may not even find front-desk clerks; kiosks have replaced them at many locations. Many kiosks have been transformed into full-blown multimedia centers, offering publicized corporate and governmental activities, job listings, and benefits, as well as ads, coupons, and movie previews. One company has even experimented with launching an 18-foot-wide vending machine described as looking like "a 7-Eleven in a box," to dispense everything from olive oil and milk to towels and pantyhose.[4]

The kiosk presents the two faces of the computer that are important to humans: It allows them to input data and to output information. For example, many kiosks use touch screens (and sometimes also keyboards) for input and thermal printers for output. In this chapter, we discuss what the principal input and output devices are and how you can make use of them.

Input & Output

How is input/output hardware used by a computer system?

Recall from Chapter 1 that *input* refers to data entered into a computer for processing—for example, from a keyboard or from a file stored on disk. Input includes program instructions that the CPU receives after commands are issued by the user. Commands can be issued by typing keywords, defined by the application program, or pressing certain keyboard keys. Commands can also be issued by choosing menu options or clicking on icons. Finally, input includes user responses—for example, when you reply to a question posed by the application or the operating system, such as "Are you sure you want to put this file in the Recycle Bin?" *Output* refers to the results of processing—that is, information sent to the screen or the printer or to be stored on disk or sent to another computer in a network. Some devices combine input and output functions, examples being not only ATMs and kiosks, as we just mentioned, but also combination scanner-printer-fax devices.

In this chapter we focus on the common input and output devices used with a computer. (● *See Panel 5.2.*) **Input hardware consists of devices that translate data into a form the computer can process.** The people-readable form of the data may be words like those on this page, but the computer-readable form consists of binary 0s and 1s, or off and on electrical signals.

Output hardware consists of devices that translate information processed by the computer into a form that humans can understand. The computer-processed information consists of 0s and 1s, which need to be translated into words, numbers, sounds, and pictures.

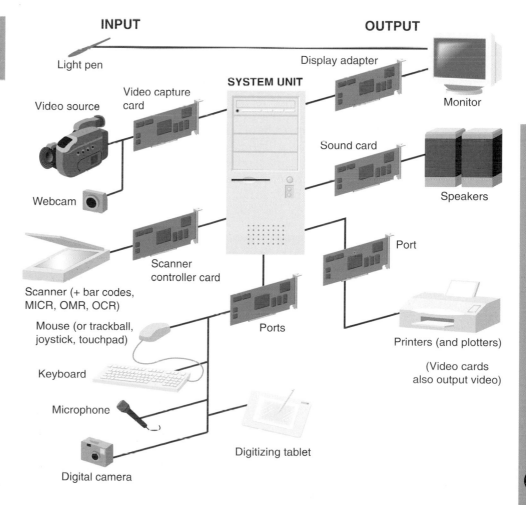

INPUT

OUTPUT

Light pen

Display adapter

SYSTEM UNIT

Video source

Video capture card

Monitor

Sound card

Webcam

Speakers

Scanner controller card

Port

Scanner (+ bar codes, MICR, OMR, OCR)

Mouse (or trackball, joystick, touchpad)

Ports

Printers (and plotters)

Keyboard

(Video cards also output video)

Microphone

Digitizing tablet

Digital camera

Input Hardware

What are the three categories of input hardware, what devices do they include, and what are their features?

Consider the *joystick,* a vertical lever mounted on a base that is used as a pointing device in videogames or flying or driving simulators. The joystick's translation of human movement into machine movement elegantly satisfies the three virtues of an input device—simplicity, efficiency, and control—according to Ben Bederson, director of the Human-Computer Interaction Lab at the University of Maryland. "It's pretty successful," Bederson says.[5]

The pointing device, however, is only one of three types of input hardware devices: *keyboards, pointing devices,* and *source data-entry devices.* (● *See Panel 5.3.)* Quite often a computer system will combine all three.

Keyboards

How do traditional and specialty keyboards differ?

A **_keyboard_ is a device that converts letters, numbers, and other characters into electrical signals that can be read by the computer's processor.** The keyboard does this with its own processor and a grid of circuits underneath the keys. When you press a key or combination of keys, the current flowing through the circuits is interrupted. The processor determines where the break occurs and compares the location information with a character map (organized by x,y coordinates) located on the keyboard's ROM chip. The character information is briefly stored in the keyboard's memory buffer, which usually holds about 16 bytes. The keyboard then sends the data in a stream to the PC via a wired or wireless connection. (Laptops use an internal wired connection.) The computer has a keyboard controller, an integrated circuit whose job it is to process all the data that comes from the keyboard and forward it to the operating system. The operating system checks to see if the keyboard data is operating-system-specific or application-specific. If, for example, you press *Ctrl+Alt+Delete*—the keyboard command for rebooting your computer—the OS will recognize the data as operating-system-specific and react accordingly (reboot). If, in contrast, you are doing word processing in Microsoft Word and press *Alt+F,* the OS will recognize the data as application-specific and send it along to the current application, Word, to be executed. All this happens so quickly that you notice no time lapse between pressing keys and seeing results.

● PANEL 5.3

Three types of input devices

Keyboards	Pointing Devices	Source Data-Entry Devices
Traditional computer keyboards	Mice, trackballs, pointing sticks, touchpads	Scanner devices: imaging systems, bar-code readers, mark- and character-recognition devices (MICR, OMR, OCR), fax machines
Specialty keyboards and terminals: dumb terminals, intelligent terminals (ATMs, POS terminals), internet terminals	Touch screens	Audio-input devices
	Pen-based computer systems, light pens, digitizers (digitizing tablets)	Webcams and video-input devices
		Digital cameras
		Speech-recognition systems
		Sensors
		Radio-frequency identification
		Human-biology input devices

Inside a computer keyboard. In most keyboards, each key sits over a small, flexible rubber dome with a hard carbon center. When the key is pressed, a plunger on the bottom of the key pushes down against the dome. This pushes the carbon center down, which presses against the keyboard circuitry to complete an electrical circuit and send a signal to the computer.

The keyboard may look like a typewriter keyboard to which some special keys have been added. Alternatively, it may look like the keys on a bank ATM or the keypad of a pocket computer. It may even be a Touch-Tone phone or cable-TV set-top box.

Let's look at *traditional computer keyboards* and various kinds of *specialty keyboards and terminals.*

TRADITIONAL COMPUTER KEYBOARDS Picking up where we left off with the PC ad presented in Chapter 4 (Panel 4.4 on p. 195), we see that the seller lists a "104-Key Keyboard." Conventional computer keyboards have all the keys on typewriter keyboards, plus other keys unique to computers. This totals 104–108 keys for desktop computers and 85 keys for laptops. Newer keyboards include extra keys for special activities such as instant web access, CD/DVD controls, and Windows shortcut keys.

Wired keyboards connect a cable to the computer via a serial port or a USB port. Wireless keyboards use either infrared-light (IR) technology or radio frequency (RF) technology to transmit signals to a receiver device plugged into the computer, usually via a USB port. Infrared wireless keyboards have a transmission range of 6–10 feet and cannot have any obstacles in the transmission path (called the *line of sight*). (Some laptop/notebook computers and printers come with built-in IR ports.) Radio-based keyboards have a range of up to 100 feet and have no line-of-sight problems. A final interesting variation is the VKB Virtual Keyboard. The size of a writing pen, it uses light to project a full-size computer keyboard onto almost any surface; the image disappears when not in use. The Virtual Keyboard can be used with PDAs and smartphones, allowing users a practical way to do email and word processing without having to take along a laptop computer.

Wireless keyboard

The keyboard illustration in Chapter 3 shows keyboard functions. *(Refer back to Panel 3.7, p. 128.)*

Seeing the light. The Virtual Keyboard uses light to project a full-size computer keyboard onto almost any surface. Used with smartphones and PDAs, this technology provides a way to do email, word processing, and other basic tasks without having to carry a notebook computer.

Survival Tip

How Do I Use the Print Scrn key?

This PC key is a holdover from DOS days, before Windows. Then it sent the contents of the screen to the printer. It still can, but in a roundabout way. When you press the key, an image of the screen is placed on the Windows clipboard. Open a blank document in any program that accepts graphics and choose *Edit, Paste*. The image is inserted, and you can print it.

SPECIALTY KEYBOARDS & TERMINALS Specialty keyboards range from Touch-Tone telephone keypads to keyboards featuring pictures of food for use in fast-food restaurants. Here we will consider dumb terminals, intelligent terminals, and internet terminals:

- **Dumb terminals: A _dumb terminal_, also called a *video display terminal (VDT)*, has a display screen and a keyboard and can input and output but cannot process data.** Usually the output is text only. For instance, airline reservations clerks use these terminals to access a mainframe computer containing flight information. Dumb terminals cannot perform functions independent of the mainframe to which they are linked.[6]

- **Intelligent terminals: An _intelligent terminal_ has its own memory and processor, as well as a display screen and a keyboard.** Such a terminal can perform some functions independent of any mainframe to which it is linked. One example is the familiar *automated teller machine (ATM),* the self-service banking machine that is connected through a telephone network to a central computer. Another example is the *point-of-sale (POS) terminal,* used to record purchases at a store's checkout counter. Intelligent terminals can be connected to the main (usually mainframe) computer system wirelessly or via cables.

  A third example of an intelligent terminal is the *mobile data terminal (MDT),* a rugged notebook PC found in police cruisers, which must endure high-speed chases and nonstop use. MDTs may run Windows operating systems but use specialized software that lets officers track dispatch information, search for local and national warrants, verify license plate and vehicle registration, check criminal records, and more. Some systems, such as that used by the Tulsa, Oklahoma, police department, let officers complete reports online to avoid time-consuming in-office paperwork.

- **Internet terminals: An _internet terminal_ provides access to the internet**—that is, it powers up directly into a browser. There are

Terminals. (*Left*) A dumb terminal at an airline check-in counter. (*Right*) A point-of-sale (POS) terminal at a retail store. It records purchases and processes the buyer's credit card.

Internet terminals—PDAs

several variants: (1) the *set-top box* or *web terminal*, which displays web pages on a TV set; (2) the *network computer*, a cheap, stripped-down computer that connects people to networks; (3) the *online game player*, which not only lets the user play games but also connects to the internet; (4) the full-blown *PC/TV* (or *TV/PC*), which merges the personal computer with the television set; and (5) the *wireless pocket PC* or *personal digital assistant (PDA)*, a handheld computer with a tiny keyboard that can do two-way wireless messaging; some smartphones also act as internet terminals.

SPECIAL KEYBOARDS FOR HANDHELDS Users of PDAs generally use a penlike device (a stylus, as we'll describe) to enter data and commands. Some PDAs do include a small keyboard, but it is usually limited in function and hard to use. However, there are alternatives:

- **Foldable PDA keyboards:** Many manufacturers offer foldable keyboards for PDAs. One example is Think Outside's Stowaway Portable Keyboard, which folds up to roughly the size of a PDA and then unfolds into a full-size keyboard. Once the keyboard driver is installed, you can simply clip the PDA onto the keyboard, which runs off the PDA's power, thus requiring no external power source. Wireless foldable keyboards are also available.

- **One-hand PDA keyboards:** The FrogPad is a 20-key gadget designed to be used with just the five fingers of one hand. A clamshell allows people to plug in a smartphone and use the whole kit like a mini-laptop.

Stowaway Portable Keyboard for PDAs. This keyboard folds up to roughly the size of a PDA (3.8 inches wide × 5.1 inches deep × 0.8 inch thick) and unfolds into a full-size keyboard measuring 13.8 inches wide × 5.1 inches deep × 0.4 inch thick. To use the keyboard, you simply unfold it and slide the sections together.

FrogPad one-hand, 20-key keyboard for PDAs. Keys are the same size as those on a regular computer keyboard. Fifteen keys represent letters, numbers, and punctuation marks; four keys alternate the symbols for each key; a Shift key is for capital letters. The keyboard can also handle page navigation.

Pointing Devices

How do pointing devices work, such as the mouse and its variants, touch screens, and pen-based systems?

One of the most natural of all human gestures, the act of pointing is incorporated in several kinds of input devices. **_Pointing devices_ control the position of the cursor or pointer on the screen and allow the user to select options displayed on the screen.** Pointing devices include the *mouse* and its variants, the *touch screen*, and various forms of *pen input*. We also describe recent innovations in *handwriting input*.

Microsoft IntelliMouse

THE MOUSE The principal pointing tool used with microcomputers is the **_mouse_, a device that is rolled about on a desktop mouse pad and directs a pointer on the computer's display screen.** The name is derived from the device's shape, which is a bit like a mouse, with the cord to the computer being the tail. The mouse went public in 1984 with the introduction of the Apple Macintosh. (● *See the timeline, Panel 5.4.*) Once Microsoft Windows 3.1 made the GUI the PC standard, the mouse also became a standard input device. In 2005, Apple introduced its Mighty Mouse, capable of working on both Macs and PCs. [7]

Mouse on a mouse pad—the cord looks a bit like a tail.

info!

Hands-Free Mice

If your mouse is causing arm and wrist pain, you could try the hands-free voice mouse models at *www.fentek-ind .com/ergmouse.htm*.

- **How the mouse works:** When the mouse is moved, a ball inside the mouse touches the desktop surface and rolls with the mouse. Two rollers inside the mouse touch the ball. One of the rollers detects motions in the *x* direction, and the other roller detects motions in the *y* direction (90 degrees opposite from the *x* direction). The rollers connect to a shaft that spins a disk with holes in it. On either side of the disk there is an infrared-light-emitting diode (LED) and an infrared sensor. The holes in the disk break the beam of light coming from the LED so that the infrared sensor sees pulses of light. A processor chip in the mouse reads the pulses and turns them into binary data that the computer can understand. The chip sends the data to the computer through the mouse cord (serial, USB, or PS/2) or through a wireless connection (infrared or radio transmission).

- **Mechanical versus optical mouse:** A mouse pad—a rectangular rubber/foam pad—provides traction for the traditional mouse, often called a *mechanical mouse* or a *wheeled mouse*. Newer mice are *optical*; that is, they use laser beams and special chips to encode data for the computer. Optical mice have no moving parts, have a smoother response, and don't require a mouse pad.

- **Mouse pointers and buttons:** The *mouse pointer*—an arrow, a rectangle, a pointing finger—is the symbol that indicates the position of the

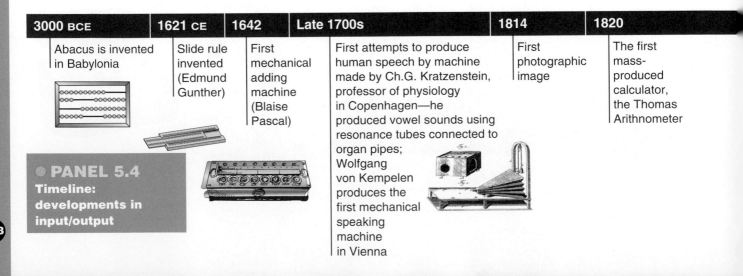

3000 BCE	1621 CE	1642	Late 1700s	1814	1820
Abacus is invented in Babylonia	Slide rule invented (Edmund Gunther)	First mechanical adding machine (Blaise Pascal)	First attempts to produce human speech by machine made by Ch.G. Kratzenstein, professor of physiology in Copenhagen—he produced vowel sounds using resonance tubes connected to organ pipes; Wolfgang von Kempelen produces the first mechanical speaking machine in Vienna	First photographic image	The first mass-produced calculator, the Thomas Arithnometer

● **PANEL 5.4**
Timeline: developments in input/output

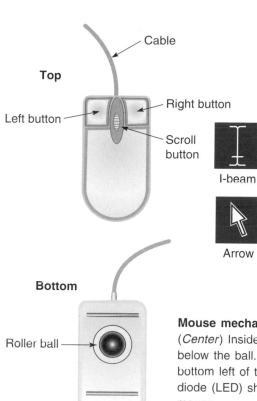

Top

Left button

Right button

Scroll button

Cable

I-beam

Arrow

Bottom

Roller ball

Mouse mechanics. (*Left*) Basic parts of a mouse. (*Center*) Inside a mouse; the processor chip is just below the ball. The rollers are to the top right and bottom left of the ball. (*Right*) The light-emitting diode (LED) shines through the bottom of the mouse.

mouse on the display screen or that activates icons. When the mouse pointer changes to the shape of an I-beam, it shows the place where text may be inserted or selected for special treatment.

On the top side of the mouse are one to five buttons. The first button is used for common functions, such as clicking and dragging. The functions of the other buttons are determined by the software you're using. Some mice have a scroll wheel on top to make it easier for you to scroll down the screen.

- **Specialty mice:** Many specialty mice versions are available. For example, the Siemens ID Mouse has a sensor that reads the user's index fingerprint and keeps unauthorized people from using the system.

Interlinks Electronics makes the DuraPoint Mouse, with stainless-steel casing that can survive almost anything, including being dropped five stories, being used as a hockey puck, and being run over by an 18-wheel truck. Immune to liquids, gases, dirt and dust, the DuraPoint is ideal for industrial, manufacturing, and field-computing environments and any home with especially curious kids.

1821	1829	1843	1844	1876	1877
First microphone	William Austin patents the first workable typewriter in America	World's first computer programmer, Ada Lovelace, publishes her notes; facsimile transmission (faxing) over wires invented by Alexander Bain, Scottish mechanic (via telegraph wires)	Samuel Morse sends a telegraph message from Washington to Baltimore	A. G. Bell patents the electric telephone	Thomas Edison patents the phonograph

Pros		Cons
• Relatively inexpensive • Very little finger movement needed to reach buttons	**Mouse**	• When gripped too tightly can cause muscle strain • Uses more desk space than other pointing devices • Must be cleaned occasionally
• Uses less desk space than mouse • Requires less arm and hand movement than mouse	**Trackball**	• Wrist is bent during use • More finger movement needed to reach buttons than with other pointing devices • Requires frequent cleaning because of finger oils
• Small footprint • Least prone to dust • Needs little cleaning	**Touchpad**	• Places more stress on index finger than other pointing devices do • Small active area makes precise cursor control difficult

● PANEL 5.5
Mouse, trackball, touchpad

Trackball

Boost Technology's Tracer Mouse can be used by people who have trouble using their hands. It's worn on and controlled by the head via a tiny gyroscope and radio waves.

VARIATIONS ON THE MOUSE: TRACKBALL, POINTING STICK, & TOUCHPAD There are three main variations on the mouse. (● *See Panel 5.5, above.*)

- Trackball: The *trackball* **is a movable ball, mounted on top of a stationary device, that can be rotated using your fingers or palm.** In fact, the trackball looks like the mouse turned upside down. Instead of moving the mouse around on the desktop, you move the trackball with the tips of your fingers. A trackball is not as accurate as a mouse, and it requires more frequent cleaning, but it's a good alternative when desktop space is limited. Trackballs come in wired and wireless versions, and newer optical trackballs use laser technology.

1897	1898	1912	1924	1927
Karl Ferdinand Braun, German physicist, invents the first cathode-ray tube (CRT), the basis of all TV and computer monitors	First telephone answering machine	Motorized movie camera replaces hand-cranked movie camera	T.J. Watson renames Hollerith's machine company, founded in 1896, to International Business Machines (IBM); first political convention photos faxed via AT&T telephone fax technology	The first electronic TV picture is transmitted

- **Pointing stick: A _pointing stick_ looks like a pencil eraser protruding from the keyboard between the G, H, and B keys. When you move the pointing stick with your finger, the screen pointer moves accordingly.** IBM developed the pointing stick for use with its notebook computers. (A forerunner of the pointing stick is the joystick, which consists of a vertical handle like a gearshift lever mounted on a base with one or two buttons.)

- **Touchpad: A _touchpad_ is a small, flat surface over which you slide your finger, using the same movements as you would with a mouse.** The cursor follows the movement of your finger. You "click" by tapping your finger on the pad's surface or by pressing buttons positioned close by the pad. Touchpads are most often found on laptop computers, but freestanding touchpads are available for use with PCs.

Pointing stick

Touchpad

TOUCH SCREEN A _touch screen_ is a video display screen that has been sensitized to receive input from the touch of a finger. (● _See Panel 5.6._) The specially coated screen layers are covered with a plastic layer. Depending on the type of touch screen, the pressure of the user's finger creates a connection of electrical current between the layers, decreases the electrical charge at the touched point, or otherwise disturbs the electrical field. The change in electrical current creates a signal that is sent to the computer. You can input requests for information by pressing on displayed buttons or menus. (● _See Panel 5.7 on the next page._) The answers to your requests are then output as displayed words or pictures on the screen. (There may also be sound.)

● PANEL 5.6
Touch screens
(_Left_) Touch screen menus at a Holiday Inn; they can keep track of the charges, as well as the carbs and the calories. (_Below_) Touch screen used to operate a sewing machine.

1931	1936	1939	1944	1946	1959
Reynold B. Johnson, a Michigan high-school science teacher, invents a test-scoring machine that senses conductive pencil marks on answer sheets	Bell Labs invents the voice-recognition machine	First electrical speech-producing machine—New York World's Fair; computer technology takes over speech synthesis in about 1970	First electro-mechanical computer (Mark I)	First programmable electronic computer in United States (ENIAC)	General Electric produces the first system to process checks in a banking application via magnetic-ink character recognition (MICR)

● PANEL 5.7
Touch-screen menus
(Left) Touch-screen menu on a self-scanning check-out system. *(Right)* Touch-screen voter's ballot.

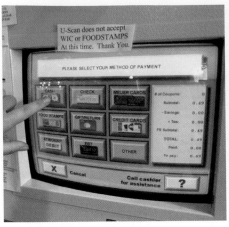

You find touch screens in kiosks, ATMs, airport tourist directories, hotel TV screens (for guest checkout), department store bridal registries, and campus information kiosks making available everything from lists of coming events to (with proper ID and personal code) student financial-aid records and grades.

PEN INPUT   Some input devices use variations on an electronic pen. Examples are *pen-based systems, light pens,* and *digitizers:*

- **Pen-based computer systems: _Pen-based computer systems_ allow users to enter handwriting and marks onto a computer screen by means of a penlike stylus rather than by typing on a keyboard.** A *stylus* is a penlike device that is used to write text or draw lines on a touch-sensitive surface as input to a computer. Pen computers use *handwriting-recognition* software that translates handwritten characters made by the stylus into data that is usable by the computer. **_Handwriting recognition_ refers to the ability of a computer to receive intelligible written input.** The system requires special software that interprets the movements of the stylus across the writing surface and translates the resulting cursive writing into digital information. (Alternatively, written text may be scanned from a piece of paper, using optical character recognition, as we describe in the next section.)

  Handwriting recognition is commonly used as an input method for PDAs and some handheld videogame systems, such as the Nintendo DS.[8] Earlier attempts to allow written input—the Apple Newton, IBM's Thinkpad tablet computer, and Microsoft Windows for Pen—were not commercially successful. Then Palm launched a successful

1960	1962	1963	1967	1968	1970
DEC introduces the PDP-1, the first commercial computer with a monitor for output and a keyboard for input	Bell Laboratories develops software to design, store, and edit synthesized music	Ivan Sutherland uses the first interactive computer graphics in his Ph.D. thesis, which used a light pen to create engineering graphics	Hand-held calculator	World debut of the computer mouse, developed since 1965 by Doug Engelbart at Stanford Research Institute (SRI)	Microprocessor chips come into use; floppy disk introduced for storing data; barcodes come into use; the daisy wheel printer makes its debut

series of PDAs based on the Graffiti recognition system. A currently successful handwriting-recognition system is Microsoft's version of the Windows XP operating system for the Tablet PC. The Tablet PC is Microsoft's version of a *tablet PC*, a special notebook computer outfitted with a digitizer tablet and a stylus that allows a user to handwrite text on the unit's screen. (● *See Panel 5.8.*) The stylus can take the place of a keyboard when users use an on-screen input panel or tap letters and numbers directly on an on-screen keyboard. Some tablets (such as ViewSonic and Motion Computing) also include a standard keyboard so that the computer can function as a laptop when the screen is repositioned.[9] We discuss tablet PCs further in Chapter 7.

Although handwriting recognition has become a popular input form, it is still generally accepted that keyboard input is both faster and more reliable.

- **Light pen:** The **_light pen_** **is a light-sensitive penlike device that uses a wired connection to a computer terminal.** The user brings the pen to a desired point on the display screen and presses the pen button, which identifies that screen location to the computer. Light pens are used by engineers, graphic designers, and illustrators. They also are used in the health, food service, and chemical fields in situations in which users' hands need to be covered. (● *See Panel 5.9.*)

● **PANEL 5.8**
Pen-based computer systems
Handheld tablet PC

● **PANEL 5.9**
Light pen
This person is using a light pen to input to the computer.

1973	1974	1976	1984	1988	1990
The Alto, an experimental PC that uses a mouse and a GUI, is developed at Xerox PARC	Simple version of optical character recognition (OCR) developed	IBM develops the ink-jet printer	Apple Macintosh; first personal laser printer	Hewlett-Packard markets the first inkjet printer for $1,000	Dragon speech-recognition program recognizes 30,000 words; SVGA video standard

- Digitizer: A _digitizer_ uses an electronic pen or a mouselike copying device called a _puck_ that can convert drawings and photos to digital data. One form of digitizer is the _digitizing tablet_, used in engineering and architecture applications, in which a specific location on an electronic plastic board corresponds to a location on the screen. (● _See Panel 5.10._)

Several electronic pens have come to market that capture handwritten notes as digital data. (● _See Panel 5.11._) In Logitech's Io2 Personal Digital Pen, a pinhole video camera next to the tip of the pen records your scrawl of words and graphics as the pen rides across a special paper, which has been preprinted with faintly visible dots, and the pen captures and stores up to 40 handwritten pages, which later can be transferred (through a USB docking port) to your PC. Seiko's InkLink doesn't require special paper but rather uses a special clip, which attaches to a pad of regular paper, with two receivers that trace inaudible beeps of infrared ultrasound emitted by the pen and transfers those signals via cable to your PC or PDA. The talking computer pen called Fly (from LeapFrog, see right) can "translate a written English word into Spanish," says one account. Users can also "sketch a piano keyboard and play it, or draw a calculator to solve an arithmetic problem. . . . A child can draw icons (called Flycons) and have the pen reply orally."[10] Designed for 8- to 13-year-olds, the Fly pen also contains an alarm clock and calendar.[11]

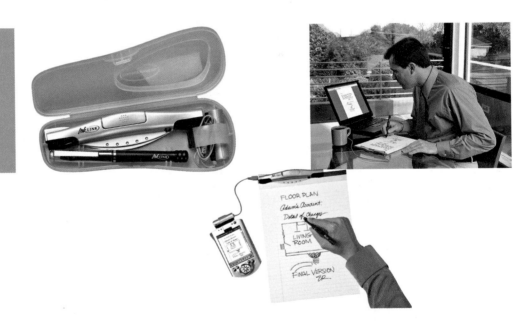

1991	1994	2004	2005
SoundBlaster Pro introduced	Apple and IBM introduce PCs with full-motion video built in; wireless data transmission for small portable computers; web browser first invented	Wireless monitors available for certain industrial applications	Wireless desktop printers commercially available

Scanning & Reading Devices

How is source data entry different from keyboard entry?

In old-fashioned grocery stores, checkout clerks read the price on every can and box and then enter those prices on the keyboard—a wasteful, duplicated effort. In most stores, of course, the clerks merely wave the products over a scanner, which automatically enters the price (from the bar code) in digital form. This is the difference between keyboard entry and source data entry.

Source data-input devices do not require keystrokes (or require only a few keystrokes) to input data to the computer. In most cases, data is entered directly from the source, without human intervention. **_Source data-entry devices_ create machine-readable data on magnetic media or paper or feed it directly into the computer's processor.** One type of source data-entry device includes scanning and reading devices—scanners, bar-code readers, mark- and character-recognition devices, and fax machines.

Talking pentop. The LeapFrog talking pentop computer called Fly; it uses special paper to enable the pen to track positions of characters.

SCANNERS _Scanners_, **or** _optical scanners_, **use light-sensing (optical) equipment to translate images of text, drawings, photos, and the like into digital form.** (● _See Panel 5.12._) The images can then be processed by a computer, displayed on a monitor, stored on a storage device, or transmitted to another computer.

Scanners have led to a whole new industry called _electronic imaging_, the software-controlled integration and manipulation of separate images, using scanners, digital cameras, and advanced graphic computers. This technology has become an important part of multimedia. PC users can get a decent scanner with good software for less than $200 or a fantastic one for about $1,000.

● PANEL 5.12
Image scanners
(Left) Desktop scanner used in desktop publishing. (Right) Handheld scanner used to scan information from a hospital patient's ID bracelet.

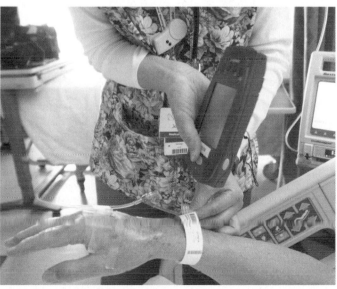

**Standard Color
Desktop Scanner
Resolutions, in dpi**

300 × 200
600 × 600
600 × 1,200
1,200 × 1,200
1,200 × 4,800
2,400 × 2,400

more info!

What Is TWAIN?

Maybe "Technology Without An
Interesting Name"? It's a driver
that's a go-between for a scanner and the applications on the
computer. Borrowing from a
Rudyard Kipling poem ("and
never the twain [that is, two]
shall meet," about Eastern and
Western cultures), someone
used the word *TWAIN* to
express the difficulty, at the
time, of connecting scanners
and personal computers,
although this is no longer the
case.

Do a web search for
TWAIN. When was it developed, and how does it work?

- **Dots and bitmaps:** Scanners are similar to photocopy machines except they create electronic files of scanned items instead of paper copies. The system scans each image—color or black and white—with light and breaks the image into rows and columns of light and dark dots or color dots. Dots are stored in computer memory as digital code called a *bitmap,* a grid of dots. A *dot* is the smallest identifiable part of an image, and each dot is represented by one or more bits. The more bits in each dot, the more shades of gray and the more colors that can be represented. The amount of information stored in a dot is referred to as *color depth,* or *bit depth.* Good scanners have a 48-bit color depth.

- **Resolution:** Scanners vary in resolution. <u>**Resolution**</u> **refers to the clarity and sharpness of an image and is measured in** <u>**dots per inch (dpi)**</u>**—the number of columns and rows of dots per inch.** The higher the number of dots, the clearer and sharper the image—that is, the higher the resolution. Popular color desktop scanners currently vary in dpi from 300 × 200 up to 2,400 × 2,400; some commercial scanners operate at 4,800 × 9,600 dpi. The quality of the scanner's optical equipment also affects the quality of the scanned images.

- **Types of scanners:** One of the most popular types of scanners is the <u>*flatbed scanner*</u>**, or** *desktop scanner,* **which works much like a photocopier—the image being scanned is placed on a glass surface, where it remains stationary, and the scanning beam moves across it.** Three other types of scanners are *sheet-fed, handheld,* and *drum.* The four types are compared below. (● *See Panel 5.13.*) Other single-purpose scanners are available, such as business-card, slide, and photo scanners. There are even scanner pens, such as the DocuPen, that can scan text from books and articles.[12]

BAR-CODE READERS On June 26, 1974, a customer at Marsh's Super-market in Troy, Ohio, made the first purchase of a product with a bar code—a pack of Wrigley's Juicy Fruit chewing gum (which pack is now in the Smithsonian National Museum of National History in Washington, D.C.). <u>**Bar codes**</u> **are the vertical, zebra-striped marks you see on most manufactured retail products**—everything from candy to cosmetics to comic books. (● *See Panel 5.14.*) In North America, supermarkets, food manufacturers, and others have agreed to use a bar-code system called the *Universal Product Code*

● **PANEL 5.13**
**Types of scanners
compared**

Flatbed: Costs $60–$400. The type most PC users have. Operates like a photocopier—image lies atop glass, which is scanned by scanning beam. Useful for single-sheet documents, books, photos. Some models scan transparencies, slides.

Sheet-fed: Costs $300–$800. Also popular for desktops because of compact size, which is smaller than flatbed. Operates similarly except sheet with image being scanned is fed into a slot and drawn past sensor. Useful only for single-page documents, photos; some accept slides.

Handheld: Costs $130–$160. Popular for use in factory and field settings and student and research use because of portability. Handheld scanners are physically dragged by hand across a document. Most models are pen-shaped and scan in swaths 5 inches wide or less, and special software "knits" strips of images together into a complete image. Some models translate single words into other languages. Images can be transmitted to a PC, PDA, or cellphone via serial, infrared, or USB connection.

Drum: Costs in tens of thousands of dollars. Used by publishing industry to capture extremely detailed images. Image to be scanned is mounted on a glass cylinder, inside of which are sensors (photomultiplier tubes) that convert light signals into digital images.

7 12345 12345

Conventional 1D bar code

2D bar code

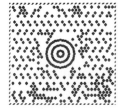

Maxicode is a 2D bar code
used by United Parcel
Service (UPS)

(UPC), established by the Uniform Code Council (UCC). Other kinds of bar-
code systems are used on everything from FedEx and Postal Service packages
to railroad cars, video-store videos, and the jerseys of long-distance runners.

- **Bar-code readers:** ***Bar-code readers* are photoelectric (optical) scanners
that translate the symbols in the bar code into digital code.** In this
system, the price of a particular item is set within the store's computer. Once the bar code has been scanned, the corresponding price
appears on the salesclerk's point-of-sale terminal and on your receipt.
Records of sales from the bar-code readers are input to the store's
computer and used for accounting, restocking store inventory, and
weeding out products that don't sell well.

  "Want to skip long lines?" asks one writer. "Be your own
cashier."[13] The reference is to self-scanning, which is becoming
increasingly available in stores throughout the United States. Do-it-
yourself checkout is an automated process that enables shoppers to
scan, bag, and pay for their purchases without human assistance. The
self-scanning checkout lane looks like a traditional checkout lane
except that the shopper interacts with a computer's user interface
instead of a store employee. (● *See Panel 5.15 on the next page.*) Self-
scanning has become popular in Wal-Mart, Home Depot, Kroger, and
many other name-brand stores.[14]

- **Types of bar codes:** Bar codes may be 1D, 2D, or 3D. *1D* codes,
today's ordinary vertical bar code, can hold up to 16 ASCII characters.
2D bar codes—composed of different-size rectangles, with data
recorded along both the height and the length of each rectangle, can
hold 1,000–2,000 ASCII characters. 2D codes are used on medication
containers and for other purposes in which there is limited space for
a bar-code label; shipping company UPS uses a special 2D code (based
on hexagons). *3D* bar codes, used on items such as automobile tires,
are called "bumpy" bar codes because they are read by a scanner that

The ModelMaker 3D scanning system can scan all types of surfaces in all sorts of lighting. It is used, among other things, to scan cars and motorcycles, or specific components, into a CAD/CAM system so that new models can be created.

differentiates by symbol height. 3D codes are used on metal, hard rubber, and other surfaces to which ordinary bar codes will not adhere.

Soon store 1D bar codes may yield to so-called *smart tags,* radio-frequency identification (RFID) tags already now employed for the tracking of cattle or of electronic components in a warehouse. We discuss RFID in another few pages.

MARK-RECOGNITION/CHARACTER-RECOGNITION DEVICES There are three types of scanning devices that sense marks or characters. They are usually referred to by their abbreviations—MICR, OMR, and OCR:

- **Magnetic-ink character recognition:** *__Magnetic-ink character recognition (MICR)__* **is a character-recognition system that uses magnetizable ink and special characters.** When an MICR document needs to be read, it passes through a special scanner that magnetizes the special

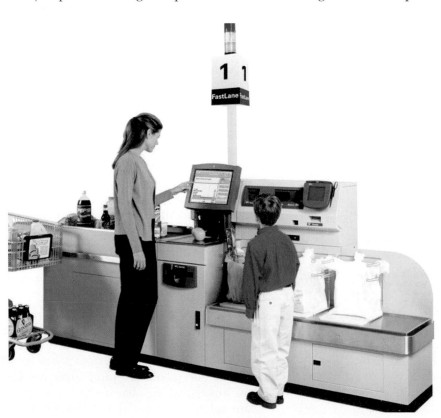

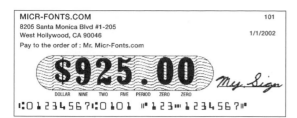

ink and then translates the magnetic information into characters. MICR technology is used by banks. Numbers and characters found on the bottom of checks (usually containing the check number, sort number, and account number) are printed with a laser printer that accepts MICR toner. MICR provides a secure, high-speed method of scanning and processing information.

- **Optical mark recognition:** ***Optical mark recognition (OMR)* uses a special scanner that reads "bubble" marks and converts them into computer-usable form.** The best-known example is the OMR technology used to read students' answers to the College Board Scholastic Aptitude Test (SAT) and the Graduate Record Examination (GRE). In these cases, the scanner reads pencil marks that fill in circles, or bubbles, on specially designed documents. OMR is also used in forms and surveys.

- **Optical character recognition:** These days almost all scanners come with OCR software. ***Optical character recognition (OCR)* software converts scanned text from images (pictures of the text) to an editable text format (usually ASCII) that can be imported into a word processing application and manipulated.** Special OCR characters appear on utility bills and price tags on department-store merchandise. The wand reader is a common OCR scanning device. (● *See Panel 5.16.*)

  OCR software can deal with nearly all printed characters, but script fonts and handwriting still present problems. In addition, OCR accuracy varies with the quality of the scanner—a text with 1,200 dpi will take longer to scan than one with 72 dpi, but the accuracy will be higher. Some OCR programs are better than others, with lesser versions unable to convert tables, boxes, or other extensive formatting; high-quality OCR software can read such complex material without difficulty. Users wanting to scan text with foreign-language (diacritical) marks, such as French accents, should check the OCR package to see if the program can handle this.

FAX MACHINES A *fax machine*—or *facsimile transmission machine*—**scans an image and sends it as electronic signals over telephone lines to a receiving fax machine, which prints out the image on paper.**

There are two types of fax machines—dedicated fax machines and fax modems:

● PANEL 5.16
Optical character recognition
OCR is often used in stores to encode and read price tags. A handheld wand is used as a reading device.

OCR-A	
NUMERIC	0123456789
ALPHA	ABCDEFGHIJ
SYMBOLS	KLMNOPQRST
	UVWXYZ
	>$/-+-#"

OCR-B	
NUMERIC	00123456789
ALPHA	ACENPSTVX
SYMBOLS	<+>-¥

Dedicated fax machine

Fax modem circuit board, which plugs into an expansion slot inside the computer

PC modem card, which plugs into a USB port on a notebook computer

- **Dedicated fax machines:** *Dedicated fax machines* are specialized devices that do nothing except send and receive fax documents. These are what we usually think of as fax machines. They are found not only in offices and homes but also alongside regular phones in public places such as airports.

- **Fax modems:** A *fax modem* is installed as a circuit board inside the computer's system cabinet. It is a modem with fax capability that enables you to send signals directly from your computer to someone else's fax machine or computer fax modem. With this device, you don't have to print out the material from your printer and then turn around and run it through the scanner on a fax machine. The fax modem allows you to send information more quickly than you could if you had to feed it page by page into a machine.

  The fax modem is another feature of mobile computing; it's especially powerful as a receiving device. Fax modems are installed inside portable computers, including pocket PCs and PDAs. If you link up a cellular phone to a fax modem in your portable computer, you can send and receive wireless fax messages no matter where you are in the world.

Why, you might wonder, does fax technology still exist in the era of email? Although email is certainly faster and cheaper (in terms of the cost of supplies), the fax machine is still more reliable, more secure, and cheaper in its equipment cost (a fax machine costs only $100 whereas the cost of a computer and monitor is more).[15]

Audio-Input Devices

What are ways of digitizing audio input?

An <u>*audio-input device*</u> **records analog sound and translates it for digital storage and processing.** An analog sound signal is a continuously variable wave within a certain frequency range. For the computer to process them, these variable waves must be converted to digital 0s and 1s. The principal use of audio-input devices is to produce digital input for multimedia computers.

TWO WAYS OF DIGITIZING AUDIO An audio signal can be digitized in two ways—by a *sound board* or a *MIDI board:*

- **Sound board:** Analog sound from a cassette player or a microphone goes through a special circuit board called a *sound board.* **A <u>sound board</u> is an add-on circuit board in a computer that converts analog sound to digital sound and stores it for further processing and/or plays it back, providing output directly to speakers or an external amplifier.**

- **MIDI board:** A <u>**MIDI board**</u>—**MIDI, pronounced "middie," stands for "Musical Instrument Digital Interface"—uses a standard for the interchange of musical information between musical instruments, synthesizers, and computers.**

MICROPHONES Also supporting audio input are *microphones,* devices that take varying air pressure waves created by voice or other sound sources and convert them into varying electric signals. Many new microcomputers and notebooks come with built-in microphones; stand-alone microphones can be connected via USB or some other connection. Microphones are used in voice-recognition input (discussed on page 274).

● PANEL 5.17
Webcam in use
The cameras are mounted on the top of the screens.

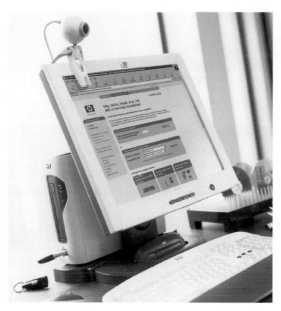

Webcams & Video-Input Cards

What are my options in video input?

Arc you the type who likes to show off for the camera? Maybe, then, you'd like to acquire a **_webcam_**, **a video camera attached to a computer to record live moving images that can then be posted on a website in rcal time.** (● _See Panel 5.17.)_ Webcam connections require special software, usually included with the camera, and a USB or video cable or a wireless radio-frequency connection.

Initially intended for personal videoconferencing, the webcam has become popular with web users. You can join thousands of other webcam users out there who are hosting such riveting material as a live 24-hour view of the aquarium of a turtle named Pixel. Or you might show your living quarters or messy desk for all to see. Samsung's Sports Cam Xtreme is a digital video camera with a lens that straps onto your helmet—perfect for capturing your moves as you ski down a black-diamond trail.

As with sound, most film and videotape traditionally has been in analog form; the signal is a continuously variable wave. For computer use, the signals that come from a VCR or camcorder must be converted to digital form through a special digitizing card—a _video-capture card_ or simply _video card_—that is installed in the computer. There are two types of video cards—frame-grabber and full-motion.

FRAME-GRABBER VIDEO CARD The _frame-grabber video card_ can capture and digitize only a single frame at a time.

FULL-MOTION VIDEO CARD The _full-motion video card_ can convert analog to digital signals at ratcs up to 30 frames per second, giving the effect of a continuously flowing motion picture. (● _See Panel 5.18 on the next page.)_

Digital Cameras

How does a digital camera differ from a film camera?

Digital cameras, which now outscll film cameras in the United States, are particularly interesting because they have changed the entire industry of photography. The environmentally undesirable stage of chemical development required for conventional film is completely eliminated.

271

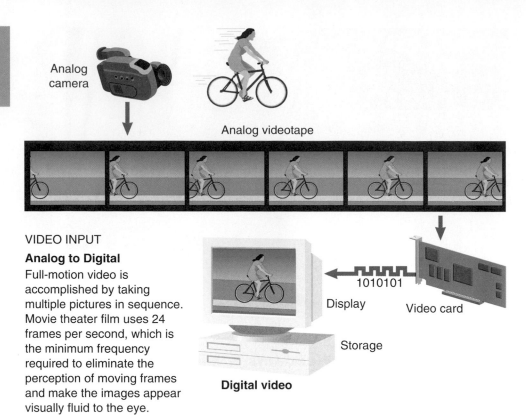

Analog camera

Analog videotape

VIDEO INPUT

Analog to Digital

Full-motion video is accomplished by taking multiple pictures in sequence. Movie theater film uses 24 frames per second, which is the minimum frequency required to eliminate the perception of moving frames and make the images appear visually fluid to the eye.

1010101

Display

Video card

Storage

Digital video

TV video generates 30 interlaced frames per second, which is actually transmitted as 60 half frames ("fields" in TV lingo) per second.

Video that has been digitized and stored in the computer can be displayed at varying frame rates, depending on the speed of the computer. The slower the computer, the jerkier the movement.

Digital photo in an HP Photosmart digital camera; the dock sends pictures to the computer.

Instead of using traditional (chemical) film, **a _digital camera_ uses a light-sensitive processor chip to capture photographic images in digital form and store them on a small diskette inserted into the camera or on flash memory cards**. (● *See Panel 5.19.*) The bits of digital information can then be copied right into a computer's hard disk for manipulation, emailing, posting on websites, and printing out.

Many digital cameras can be connected to a computer by a USB or FireWire connection (discussed in Chapter 6), so your computer's operating system must support these connections to recognize the camera's driver. Some cameras store picture data on 3.5-inch diskettes that can simply be inserted into your PC's floppy drive. Others use flash memory cards and memory sticks, from which later you can transmit photo data to your computer through a USB cable. Popular software applications for sophisticated digital photo manipulation include Adobe Photoshop, Jasc Paint Shop Pro, and Microsoft Picture It! Digital Image Pro.

CAMERA CAVEATS Three facts that all digital-camera users should be aware of:

- **Don't use the "delete" function:** Instead of using a camera's "delete" function to remove images you don't want from the camera's memory card each night, you should download all the images to your computer and then reformat the card. This will prevent possible fragmentation and loss of your pictures.

- **Use a photo shop for printing:** Photos printed out on standard computer printers do not last nearly as long as photos printed in the usual manner—that is, by a photo shop on good-quality paper.

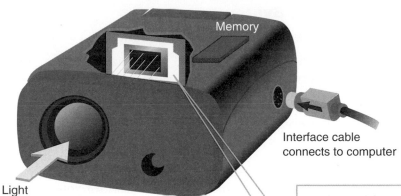

Memory

Light

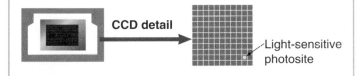

Interface cable connects to computer

3. The digital information is stored in the camera's electronic memory, either built-in or removable.

4. Using an interface cable, the digital photo can be downloaded onto a computer, where it can be manipulated, printed, placed on a web page, or emailed.

1. Light enters the camera through the lens.

2. The light is focused on the charge-coupled device (CCD), a solid-state chip made up of tiny, light-sensitive photosites. When light hits the CCD, it records the image electronically, just like film records images in a standard camera. The photosites convert light into electrons, which are then converted into digital information.

A look at CCDs
The smallest CCDs are 1/8 the size of a frame of 35mm film. The largest are the same size as a 35mm frame.

Smallest CCD

- Lower-end cameras start with 180,000 photosites.
- Professional cameras can have up to 6 million photosites.

CCD detail

Light-sensitive photosite

● **PANEL 5.19**
Digital cameras and how they work
(*Right*) This digital camera is attached to a desktop computer in order to download its images to the computer.

info!

Digital-Camera Resource

For more information on digital cameras, go to *www.dcresource.com*.

- **Be aware of limitations of storage media:** If you store your photos on such secondary storage media as CDs, be aware that such media may very well not be usable by information technology equipment 20 or 30 years from now.

CAMERA PHONES Digital-camera technology has, of course, migrated to cellphones (which can also be used for web surfing, playing games, and downloading music), enabling you to visually share your vacation experiences in real time. You compose the shot on your phone's color LCD screen, point and shoot, then wait a minute or so for the picture to "develop," and then send it. We discuss camera phones further in Chapter 7.

Speech-Recognition Systems

How might I use a speech-recognition system?

Can your computer tell whether you want it to "recognize speech" or "wreck a nice beach"? **A _speech-recognition system_, using a microphone (or a telephone) as an input device, converts a person's speech into digital signals by comparing the electrical patterns produced by the speaker's voice with a set of prerecorded patterns stored in the computer. (● _See Panel 5.20._)** Most of today's speech-recognition packages have a database of about 200,000 words from which they try to match the words you say. These programs let you accomplish two tasks: turn spoken dictation into typed text and issue oral commands (such as "Print file" or "Change font") to control your computer.

● **PANEL 5.20**
How a speech-recognition system works

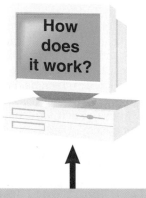

Speech
A person who's going to use speech recognition software must first go through an _enrollment_. This consists of the person dictating text that is already known to the software for 10 minutes to an hour. From this sampling, the software creates a table of _vocal references_, which are the ways in which the speaker's pronunciation of phonemes varies from models of speech based on a sampling of hundreds to thousands of people. _Phonemes_ are the smallest sound units that combine into words, such as "duh," "aw" and "guh" in "dog." There are 48 phonemes in English. After enrollment, the speaker dictates the text he wants the software to transcribe into a microphone, preferably one that uses _noise-cancellation_ to eliminate background sounds. The quality of the microphone and the computer's processing power are the most important hardware factors in speech recognition. The speaker can use continuous speech, which is normal speech without pauses between words.

Output
Computer recognizes word string and prints it on the screen.

Recognition Search
Using the data from 1, 2, and 3, the computer tries to find the best matching sequence of words as learned from a variety of examples.

1. Phonetic Models
Describe what codes may occur for a given speech sound.
> In the word _how_, what is the probability of the "ow" sound appearing between an H and a D?

2. Dictionary
Defines the phonetic pronunciation (sequence of sounds) of each word.
> **How does it work**
> _haw   daz   it werk_

3. Grammar
Defines what words may follow each other, using parts of speech.
> **How does <it> [work]**
> _adv   vt   pron   v_

Signal Processing
The sound wave is transformed into a sequence of codes that represent speech sounds.

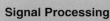

Speech-recognition systems have had to overcome many difficulties, such as different voices, pronunciations, and accents. Recently, however, the systems have measurably improved, at least up to a point.[16] Three major recognition systems are ScanSoft's Open Speech Dialog, Apple Speech Recognition, and ScanSoft's Dragon Naturally Speaking. Windows XP has some basic built-in speech-recognition features. ScanSoft's NAVIGON MobileNavigator 5 adds speech recognition to a variety of automotive, mobile, and PC-based applications. Special speech-recognition software is available for specific professions, such as law, medicine, and public safety.

Speech-recognition systems are finding many uses. Warehouse workers are able to speed inventory taking by recording inventory counts verbally. Traders on stock exchanges can communicate their trades by speaking to computers. Radiologists can dictate their interpretations of X-rays directly into transcription machines. Nurses can fill out patient charts by talking to a computer. Drivers can talk to their car radios to change stations. Speakers of Chinese can speak to machines that will print out Chinese characters. And for many individuals with disabilities, a computer isn't so much a luxury or a productivity tool as a necessity, providing freedom of expression, independence, and empowerment.

Sensors

What are various kinds of uses for sensors?

A _sensor_ is an input device that collects specific data directly from the environment and transmits it to a computer. Although you are unlikely to see such input devices connected to a PC in an office, they exist all around us, often in nearly invisible form and as part of some larger electronic system. Sensors can be used to detect all kinds of things: speed, movement, weight, pressure, temperature, humidity, wind, current, fog, gas, smoke, light, shapes, images, and so on. An electronic device is used to measure a physical quantity such as temperature, pressure, or volume and convert it into an electronic signal of some kind (for instance, a voltage).

Adidas markets a running shoe with a sensor that adjusts cushioning for different surfaces. (● *See Panel 5.21.*) Building security systems use sensors to detect movement. Sensors are used to detect the speed and volume of traffic and adjust traffic lights. They are used on highways in wintertime in Iowa as weather-sensing devices to tell workers when to roll out snowplows. In aviation, sensors are used to detect ice buildup on airplane wings or to alert pilots to sudden changes in wind direction. In California, sensors have been planted along major earthquake fault lines in an experiment to see whether scientists can predict major earth movements. Also in California, environmental scientists have hooked up a network of sensors, robots, cameras, and computers to produce an ecological picture of a lush world that is home to more than 30 rare and endangered species.[17]

● **PANEL 5.21**
Sensors
(*Top*) On October 9, 2003, the 2,080-pound Liberty Bell made a 963-foot journey to a new home on Independence Mall in Philadelphia. High-tech sensors monitored the famous crack, to make sure it did not get larger. (*Bottom*) The Adidas-1 running shoe is billed as the world's first "intelligent shoe." A microprocessor beneath the arch sends instructions to a small battery-powered motor that adjusts the cushioning level of the shoe.

Radio-Frequency Identification Tags

Is there any way I could benefit from RFID technology?

<u>Radio-frequency identification (RFID) tags</u> are based on an identifying tag bearing a microchip that contains specific code numbers. These code numbers are read by the radio waves of a scanner linked to a database. Drivers with RFID tags breeze through tollbooths without having to even roll down their windows; the toll is automatically charged to their accounts. Radio-wave-readable ID tags are also used by the Postal Service to monitor the flow of mail, by stores for inventory control and warehousing, and in the railroad industry to keep track of rail cars. Delta Airlines is installing a system to use RFID to track lost baggage. The Food and Drug Administration (FDA) and drug companies are working on tagging popular drugs with RFID devices to guard against counterfeit medicines. The technology is used to make supposedly "thiefproof" keys in millions of Fords, Toyotas, and Nissans (although researchers have succeeded in cracking the security).[18] Gambling casinos have put RFID tags inside gambling chips to help them track the betting habits of high rollers.[19] Visa and MasterCard "no-swipe" or "wave-and-pay" credit cards have been developed with embedded RFID technology that means cards don't have to be swiped across a magnetic-strip reader but can just be waved near a scanner, which could help speed up lines in stores.[20] Wal-Mart has required that all its vendors replace bar codes with RFID technology in order to improve inventory control. In the long run, the tags could end up on every product we buy, so that in the future we will no longer wait for a checkout counter but will breeze past readers that will scan our purchases in milliseconds.[21] All U.S. passports now must have RFID tags.

Representing a significant next step, RFID tags are now being implanted under the skin of animals and even people. When injected into dogs and cats, for example, they allow veterinarians with the right scanning equipment to identify the animals if they become separated from their owners, although there are still some questions about standards.[22] (Still, a California man was reunited with his cat, Ted, in this manner after 10 years of separation.[23]) The technology could also be used to identify cattle that share a certain feed, so their meat could be tracked in the event of an outbreak of mad cow disease. In Mexico, several police officers and officials reportedly had the rice-grain-size chips implanted in their arms as a security precaution: Only people with the chips could get past the scanners into a new anticrime information center in Mexico City.[24] In the United States, the FDA approved implantation of an implantable chip, the VeriChip, for medical purposes, on the premise that patient-specific information stored in the chip could speed vital information about a person's medical history. However, privacy advocates fear that the technology could be used to track people's movements and put sensitive personal information at risk to hackers. "If privacy protections aren't built in at the outset," says a policy analyst with the Health Privacy Project, "there could be harmful consequences for patients."[25]

RFID. (*Left, center*) Drivers can buy RFID tags to drive through tollbooths without having to stop; the tolls are automatically charged to their account, usually established with a credit card. FastTrack is a common tollbooth RFID program. (*Right*) RFID tags are starting to replace bar codes.

● **PANEL 5.22**
Some types of biometric devices
(*Left*) Palm print recognition. (*Right*) Screen from a face recognition program.

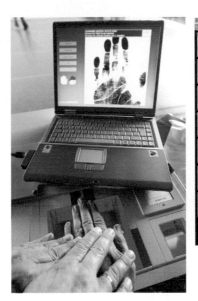

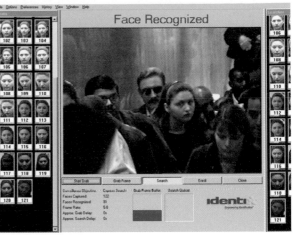

Human-Biology-Input Devices

Have I ever seen biometrics in use?

SECURITY Security concerns following the terrorist attacks of September 11, 2001, on the New York World Trade Center and the Pentagon made more people aware of _**biometrics**_, **the science of measuring individual body characteristics.** Biometric security devices identify a person through a fingerprint; hand, eye, or facial characteristics; voice intonation; or some other biological trait. (● *See Panel 5.22.*) Hewlett-Packard, Toshiba, and Lenovo have produced notebook computers equipped with biometric sensors that read fingerprints, instead of passwords, before allowing access to networks.[26] We return to this subject in more detail in Chapter 9.

5.3

Output Hardware

What are the two categories of output hardware, what devices do they include, and what are their features?

Hardcopy

Softcopy

Are we back to old-time radio? Almost. Except that you can call up local programs by downloading them from the internet. The sound quality often isn't even as good as that of AM radio, but no doubt that will improve eventually. Computer output is taking more and more innovative forms and getting better and better.

As mentioned, output hardware consists of devices that convert machine-readable information, obtained as the result of processing, into people-readable form. The principal kinds of output are softcopy and hardcopy. (● *See Panel 5.23 on the next page.*)

- **Softcopy:** _**Softcopy**_ **is data that is shown on a display screen or is in audio or voice form; it exists only electronically.** This kind of output is not tangible; it cannot be touched. It's like music: You can see musical scores and touch CDs and tapes, but the music itself is intangible. Similarly, you can touch floppy disks on which programs are stored, but the software itself is intangible. *Soft* is also used to describe things that are easily changed or impermanent. In contrast, *hard* is used to describe things that are relatively permanent.

- **Hardcopy:** _**Hardcopy**_ **is printed output**. The principal examples are printouts, whether text or graphics, from printers. Film, including microfilm and microfiche, is also considered hardcopy output.

Softcopy Devices	Hardcopy Devices	Other Devices
CRT display screens	Impact printers: dot-matrix printer	Sound output
Flat-panel display screen (e.g., liquid-crystal display)	Nonimpact printers: laser, ink-jet, thermal	Voice output
		Video output

There are several types of softcopy and hardcopy output devices. In the following three sections, we discuss, first, traditional *softcopy* output—*display screens*; second, traditional *hardcopy* output—*printers*; and, third, *mixed* output—including *sound, voice,* and *video.*

Traditional Softcopy Output: Display Screens

What are the principal things I need to know to make a good choice of display screen?

<u>Display screens</u>—**also variously called *monitors* or simply *screens*—are output devices that show programming instructions and data as they are being input and information after it is processed.** The monitor is the component that displays the visual output from your computer as generated by the video card. It is responsible not for doing any real computing but rather for showing the results of computing.

As with TV screens, the size of a computer screen is measured diagonally from corner to corner in inches. For desktop microcomputers, the most common sizes are 13, 15, 17, 19, 21, and 24 inches. For laptop computers, they are 12.1, 13.3, 14.1, and 15.1 inches. Increasingly, computer ads state the actual display area, called the *viewable image size (vis),* which may be an inch or so less. A 15-inch monitor may have a 13.8-inch vis; a 17-inch monitor may have a 16-inch vis.

monitor screen size	viewable image size
15 inches	14 inches
17 inches	16 inches
21 inches	20 inches

In deciding which display screen to buy, you will need to consider issues of screen clarity (dot pitch, resolution, color depth, and refresh rate), type of display technology (CRT versus flat panel, active-matrix flat panel versus passive-matrix flat panel), and color and resolution standards (SVGA and XGA).

Pixels

SCREEN CLARITY: DOT PITCH, RESOLUTION, COLOR DEPTH, & REFRESH RATE Among the factors affecting screen clarity (often mentioned in ads) are *dot pitch, resolution, color depth,* and *refresh rate.* These relate to the individual dots on the screen known as *pixels,* which represent the images on the screen. **A <u>pixel</u>, for "*pic*ture *el*ement," is the smallest unit on the screen that can be turned on and off or made different shades.** Pixels are tiny squares, not circles.

- **Dot pitch: <u>Dot pitch (dp)</u> is the amount of space between the centers of adjacent pixels; the closer the pixels, the crisper the image.** For a .25-dp monitor, for instance, the dots (pixels) are 25/100ths of a millimeter apart. Generally, a dot pitch of .25 dp will provide clear images.

- **Resolution:** Here, <u>resolution</u> **refers to the image sharpness of the display screen; the more pixels, or dots, there are per square inch, the**

Larger
dot
pitch

Smaller
dot
pitch

The Letter " i "

Standard monitor resolutions, in pixels

640 × 480

800 × 600

1,024 × 768

1,280 × 1,024

1,600 × 1,200

1,920 × 1,400

Standard bit depths for color

4-bit—16 colors

8-bit—256 colors

16-bit—65,536 colors

24-bit—10 million colors

finer the level of detail. As with scanners, resolution is expressed in *dots per inch (dpi)*, the number of columns and rows of dots per inch. The higher the number of dots, the clearer and sharper the image. Resolution clarity is measured by the formula *horizontal-row pixels × vertical-row pixels*. For example, a 640 × 480 screen displays 640 pixels on each of 480 lines, for a total of 307,200 pixels. On color monitors, each pixel is assigned some red, some green, some blue, or particular shades of gray.

● **Color depth:** As we said about scanners, *__color depth__*, or *bit depth*, **is the amount of information, expressed in bits, that is stored in a dot.** The more bits in a dot or pixel, the more shades of gray and colors can be represented. With 24-bit color depth, for example, 8 bits are dedicated to each primary color—red, green, and blue. Eight-bit color is standard for most of computing; 24-bit, called *true color*, requires more resources, such as video memory.

● **Refresh rate:** *__Refresh rate__* **is the number of times per second that the pixels are recharged so that their glow remains bright.** That is, refresh rate refers to the number of times that the image on the screen is redrawn each second. The higher the refresh rate, the more solid the image looks on the screen—that is, the less it flickers. In general, displays are refreshed 56–120 times per second, or *hertz (Hz)*, with speeds of 70–87 hertz being common. A high-quality monitor has a refresh rate of 90 hertz—the screen is redrawn 90 times per second. A low-quality monitor will be under 72 hertz, which will cause noticeable flicker and lead to headaches and eyestrain. (The measurement *hertz* was named after the German professor of physics Heinrich Rudolf Hertz [1847–1894], who was the first to broadcast and receive radio waves.)

Televisions and older computer monitors have a lower refresh rate than most new computer monitors. To help adjust for the lower rate, they use a method called *interlacing*. This means that the electron gun in the television's CRT will scan through all the odd rows from top to bottom and then start again with the even rows. The phosphors hold the light long enough that your eyes are tricked into thinking that all the lines are being drawn together. Modern monitors are *noninterlaced*, because that is what modern video systems typically use.

● PANEL 5.24
CRT *(left)* versus flat-panel displays

17" Flat-Panel Display

TWO TYPES OF MONITORS: CRT & FLAT PANEL Display screens are of two types: CRT and flat panel. (● *See Panel 5.24.*)

● **CRT: A *__CRT (cathode-ray tube)__* is a vacuum tube used as a display screen in a computer or video display terminal.** (● *See Panel 5.25 on the next page.*) The same kind of technology is found not only in the screens of desktop computers but also in television sets and flight-information monitors in airports. *Note:* Advertisements for desktop computers often *do not* include a monitor as part of the system. You need to be prepared to spend a few hundred dollars extra for the monitor.

● **Flat panel:** Compared to CRTs, flat-panel displays are much thinner, weigh less, and consume less power, which is why they are used in portable computers. *__Flat-panel displays__* **are made up of two plates of glass separated by a layer of a substance in which light is manipulated. One flat-panel technology is *__liquid crystal display (LCD)__*, in which molecules of liquid crystal line up in a way that alters their optical properties, creating images on the screen by transmitting or blocking out light.**

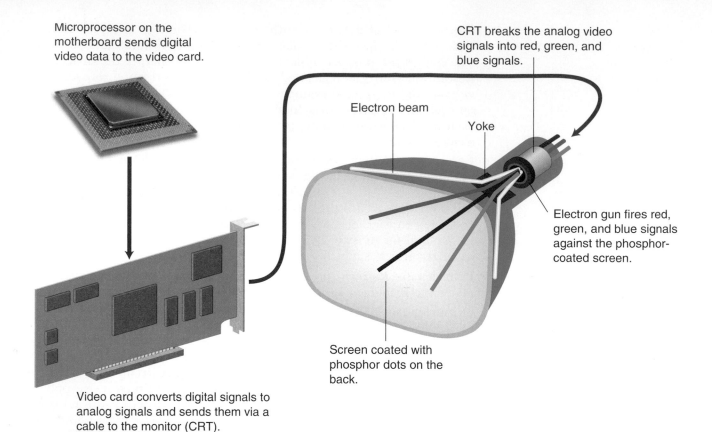

Microprocessor on the motherboard sends digital video data to the video card.

Video card converts digital signals to analog signals and sends them via a cable to the monitor (CRT).

CRT breaks the analog video signals into red, green, and blue signals.

Electron beam

Yoke

Electron gun fires red, green, and blue signals against the phosphor-coated screen.

Screen coated with phosphor dots on the back.

● **PANEL 5.25**
How a CRT works
A stream of bits from the computer's CPU is sent to the electron gun, which converts the bits into electrons. The gun then shoots a beam of electrons through the yoke, which deflects the beam. When the beam hits the phosphor coating on the inside of the CRT screen, a number of pixels light up, making the image on the screen.

Flat-panel monitors are available for desktop computers as well, and because they are smaller than CRTs, they fit more easily onto a crowded desk. CRTs are still slightly cheaper, but flat-panel prices have come down, and they have been matched by technical advances that make flat panels a better choice.[27] The best flat-panel displays can also now do double duty as desktop television sets.[28] Still, because flat panels don't redraw the images on the screen as fast as CRTs do—a problem called *latency*—they're not advisable for game players.

In conjunction with Microsoft, several manufacturers have developed wireless flat-panel screens to be used with Microsoft Windows XP Smart Display for desktop computers and Microsoft Windows CE for handheld computers (PDAs).

The Smart Display wireless flat-panel 15-inch touch-sensitive screen has a wireless connection to a Windows computer so it can be used in any room in the home or office, similar to a cordless home telephone. Thus, you can surf the web from your couch, check the news in bed, or balance your budget in the den, wirelessly, with the touch of a stylus. The Smart Display also supports wireless mice and keyboards, as well as handwriting recognition.

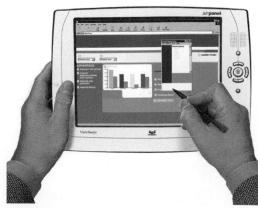

Smart Display wireless panel

ACTIVE-MATRIX VERSUS PASSIVE-MATRIX PANEL DISPLAYS Flat-panel screens are either active-matrix or passive-matrix displays, according to where their transistors are located.

- **Active matrix: In an _active-matrix display_, also known as _TFT (thin-film transistor) display_, each pixel on the flat-panel screen is controlled by its own transistor.** Active-matrix screens are much brighter and sharper than passive-matrix screens, but they are more complicated and thus more expensive. They also require more power, affecting the battery life in laptop computers. A newer type of TFT, called _organic TFT_, or _organic light-emitting diode (OLED)_, uses films of organic molecules sandwiched between two charged electrodes to produce even brighter and more readable displays than traditional TFT displays.

- **Passive matrix: In a _passive-matrix display_, a transistor controls a whole row or column of pixels on the flat-screen display.** Passive matrix provides a sharp image for one-color (monochrome) screens but is more subdued for color. The advantage is that passive-matrix displays are less expensive and use less power than active-matrix displays, but they aren't as clear and bright and can leave "ghosts" when the display changes quickly. Passive-matrix displays go by the abbreviations _HPA, STN_, or _DSTN_.

COLOR & RESOLUTION STANDARDS FOR MONITORS: SVGA & XGA
As mentioned earlier, PCs come with _graphics cards_ (also known as _video cards_ or _video adapters_) that convert signals from the computer into video signals that can be displayed as images on a monitor. The monitor then separates the video signal into three colors: red, green, and blue signals. Inside the monitor, these three colors combine to make up each individual pixel. Video cards have their own memory, video RAM, or VRAM, which stores the information about each pixel. The more VRAM you have, which can range from 2 to 64 megabytes, the higher the resolution you can use. Video gamers and desktop publishers (Photoshop users) will want a video card with lots of VRAM.

The common color and resolution standards for monitors are _SVGA, XGA, SXGA, UXGA_, and, most recently, _QXGA_. (● _See Panel 5.26._)

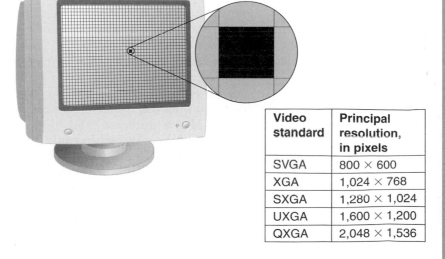

● **PANEL 5.26**
Video graphics standards
compared for pixels

A single pixel

Video standard	Principal resolution, in pixels
SVGA	800 × 600
XGA	1,024 × 768
SXGA	1,280 × 1,024
UXGA	1,600 × 1,200
QXGA	2,048 × 1,536

Hardware: Input & Output

281

- **SVGA:** <u>*SVGA (super video graphics array)*</u> **supports a resolution of 800 × 600 pixels, or variations, producing up to 16 million possible simultaneous colors** (depending on the amount of VRAM). SVGA is the most common standard used today with 15-inch monitors. This standard is best for simpler applications: word processing, email, and so on.

- **XGA:** <u>*XGA (extended graphics array)*</u> **has a resolution of up to 1,024 × 768 pixels, with 65,536 possible colors.** It is used mainly with 17-, 19-, and 21-inch monitors. It's useful for simple applications plus spreadsheets and graphics software.

- **SXGA:** <u>*SXGA (super extended graphics array)*</u> **has a resolution of up to 1,280 × 1,024 pixels.** It is often used with 19- and 21-inch monitors by graphic designers, engineers, and programmers.

- **UXGA:** <u>*UXGA (ultra extended graphics array)*</u> **has a resolution of up to 1,600 × 1,200 pixels and supports up to 16.8 million colors.** Common applications for it include CAD (computer-aided design) and business presentations, such as trade show displays.

- **QXGA:** <u>*QXGA (quantum extended graphics array)*</u> **is a new display standard with a resolution of up to 2,048 × 1,536 pixels.** Presently costing upward of $10,000, QXGA is being used for large LCD screens for computer users needing to view extreme detail, for businesspeople needing to enlarge images that will still be crisp in big-screen presentations, for high-density television (discussed in Chapter 7), and for special applications involving viewing multiple images on a single screen.

Traditional Hardcopy Output: Printers

What kind of options do I have in printers?

HP Business Inkjet 1000 Printer

The prices in ads for computer systems often do not include a printer. Thus, you will need to budget an additional $100–$1,000 or more for a printer. **A <u>*printer*</u> is an output device that prints characters, symbols, and perhaps graphics on paper or another hardcopy medium.** As with scanners, the resolution, or quality of sharpness, of the printed image is indicated by *dots per inch (dpi)*, a measure of the number of rows and columns of dots that are printed in a square inch. For microcomputer printers, the resolution is in the range of 60–1,500 dpi.

Printers can be separated into two categories, according to whether or not the image produced is formed by physical contact of the print mechanism with the paper. *Impact printers* do have contact with paper; *nonimpact printers* do not. We will also consider plotters and multifunction printers.

IMPACT PRINTERS An <u>*impact printer*</u> **forms characters or images by striking a mechanism such as a print hammer or wheel against an inked ribbon, leaving an image on paper.** A *dot-matrix printer* contains a print head of small pins that strike an inked ribbon against paper, to form characters or images. Print heads are available with 9, 18, or 24 pins; the 24-pin head offers the best quality. Dot-matrix printers can print *draft quality*, a coarser-looking 72 dpi, or *near-letter-quality (NLQ)*, a crisper-looking 144 dpi. The machines print 40–300 characters per second and can handle graphics as well as text.

Impact printers are the only desktop printers that can use multilayered forms to print "carbon copies." A disadvantage, however, is the noise they produce, because of the print head striking the paper. Nowadays such printers are more commonly used with mainframes than with personal computers.

NONIMPACT PRINTERS Nonimpact printers are faster and quieter than impact printers because no print head strikes paper. ***Nonimpact printers* form characters and images without direct physical contact between the printing mechanism and paper.** Two types of nonimpact printers often used with microcomputers are *laser printers* and *inkjet printers*. A third kind, the *thermal printer,* is seen less frequently.

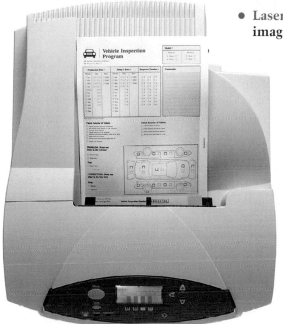

- **Laser printers:** Like a dot-matrix printer, a ***laser printer* creates images with dots. However, as in a photocopying machine, these images are produced on a drum, treated with a magnetically charged inklike toner (powder), and then transferred from drum to paper.** (● *See Panel 5.27.*) (Laser printers are still sometimes called *page printers*, because they print one page at a time.)

  Laser printers run with software called a *page description language (PDL)*. This software tells the printer how to lay out the printed page, and it supports various fonts. A laser printer comes with one of two types of PDL: PostScript (developed by Adobe) or PCL (Printer Control Language, developed by Hewlett-Packard). In desktop publishing, PostScript is the preferred PDL. Laser printers have their own CPU, ROM, and memory (RAM), usually 8 megabytes (expandable generally up to 32 megabytes for low- to medium-cost printers). When you need to print out graphics-heavy color documents, your printer will need more memory—up to 416 megabytes.

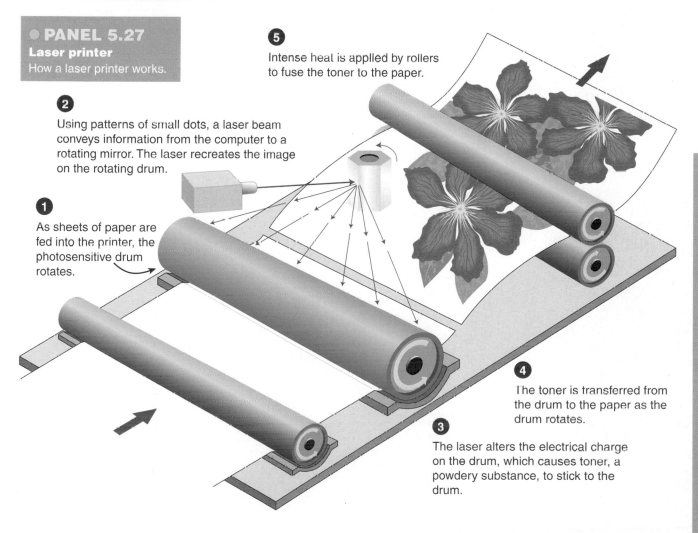

● PANEL 5.27
Laser printer
How a laser printer works.

❺ Intense heat is applied by rollers to fuse the toner to the paper.

❷ Using patterns of small dots, a laser beam conveys information from the computer to a rotating mirror. The laser recreates the image on the rotating drum.

❶ As sheets of paper are fed into the printer, the photosensitive drum rotates.

❹ The toner is transferred from the drum to the paper as the drum rotates.

❸ The laser alters the electrical charge on the drum, which causes toner, a powdery substance, to stick to the drum.

There are good reasons that laser printers are among the most common types of nonimpact printer. They produce sharp, crisp images of both text and graphics. They are quiet and fast—able to print 4–12 pages per minute (ppm) in color and 16–32 black-and-white pages per minute for individual microcomputers and up to 200 pages per minute for mainframes. They can print in different *fonts*—that is, sets of typestyles and type sizes. The more expensive models can print in different colors. Laser printers usually have a dpi of 600–2,400. You can buy a laser printer from Dell, the Laser Printer 1100, for just $99.[29]

- **Inkjet printers:** <u>*Inkjet printers*</u> **spray onto paper small, electrically charged droplets of ink from four nozzles through holes in a matrix at high speed.** (● *See Panel 5.28.*) Like laser and dot-matrix printers, inkjet printers form images with little dots. Inkjet printers have a dpi of 300 to 2,400 and spray ink onto the page a line at a time, in both high-quality black-and-white text and high-quality color graphics. (To achieve impressive color images, you should use high-quality, high-gloss paper, which prevents inkjet-sprayed dots from *feathering*, or spreading.) Inkjet cartridges come in various combinations: a single cartridge for black and all color inks, two separate black and color cartridges, or separate cartridges for black and each color. Some cartridges also include the print head, which is apt to wear out before the rest of the machine.

  The advantages of inkjet printers are that they can print in color, are quiet, and are generally less expensive than color laser printers. The disadvantages have been that they print a bit less precisely than laser printers do and traditionally they have been slower. Until

S u r v i v a l   T i p
Do I Have to Print the Whole Thing?

If you click on the shortcut icon on the Windows toolbar (the one that looks like a printer), your entire document will be printed. To print only a portion of your document, go to *File, Print* and check the option you want in the dialog box that appears.

● **PANEL 5.28**
Inkjet printer
How an inkjet printer works.

1 Four removable ink cartridges are attached to print heads with 64 firing chambers and nozzles apiece.

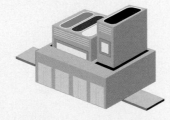

2 As the print heads move back and forth across the page, software instructs them where to apply dots of ink, what colors to use, and in what quantity.

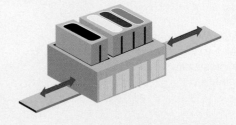

3 To follow those instructions, the printer sends electrical pulses to thin resistors at the base of the firing chambers behind each nozzle.

Ink — Resistor — Vapor bubble

4 The resistor heats a thin layer of ink, which in turn forms a vapor bubble. That expansion forces ink through the nozzle and onto the paper at a rate of about 6,000 dots per second.

5 A matrix of dots forms characters and pictures. Colors are created by layering multiple color dots in varying densities.

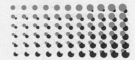

Homaro Cantu, the 28-year-old chef at Chicago's Moto Restaurant, prints sushi on a Canon i560 inkjet printer using edible paper made of soybeans and cornstarch and food-based inks of his own concoction. He then flavors the back of the paper with powdered soy sauce and seaweed. Even Mr. Cantu's menu is edible; diners crunch it up into soups.

Thermal printed cash-register receipt

recently, inkjet-printing a document with high-resolution color graphics could take 10 minutes or more for a single page. That may be changing, however, since the Hewlett-Packard Photosmart 8250, an inkjet photo printer announced in 2005, can print a shiny 4- by 6-inch print in a mere 14 seconds.[30]

Another disadvantage is that inkjet cartridges have to be replaced more often than laser-toner cartridges do and so may cost more in the long run. Moreover, a freshly inkjet-printed page is apt to smear unless handled carefully. For users who print infrequently, laser printers (which use toner, a dry powder) have the advantage of not drying out. Laser owners don't have to deal with dried-out cartridges clogging nozzles and wasting expensive ink and time—a common problem for inkjet users.

Still, experts maintain that users who want the best-quality photo output should get a photo inkjet printer. Laser printers in general are known for their mediocre photo output, with cheap personal laser printers often doing an especially poor job.

Laser printers are known for handling a much higher volume of printouts than inkjet printers. Buyers who do a lot of printing should pay close attention to a printer's monthly duty cycle, which determines how many printouts a printer can comfortably handle in a month. Going over this can often shorten the life of a printer. In general, cheaper printers have much lower duty cycles than the higher-end printers.

- **Thermal printers:** _**Thermal printers**_ **are low- to medium-resolution printers that use a type of coated paper that darkens when heat is applied to it.** The paper is moved past a line of heating elements that burn dots onto the paper. This technology is typically used in business for bar-code label applications and is replacing dot-matrix printers for printing cash register receipts because it is faster and quieter. Until about 2000, most fax machines used direct thermal printing, though now only the cheapest models use it, the rest having switched to either thermal wax-transfer, laser, or inkjet printing.

- **Thermal wax-transfer printers:** *Thermal wax-transfer printers* print a wax-based ink onto paper. As the paper and ribbon travel in unison beneath the thermal print head, the wax-based ink from the transfer ribbon melts onto the paper. After it becomes cool, the wax adheres permanently to the paper. Although such printers are highly reliable, because of the small number of moving parts, they don't compare with modern inkjet printers and color laser printers. However, because of their waterfastness, they find uses in industrial label printing.

Wax-transfer thermal printed label

- **Photo printers:** *Photo printers* are specialized machines for printing continuous-tone photo prints (typically 3 × 5 or 4 × 6 inches), using special dye-receptive paper and ribbons with special transparent color dyes. Paper and ribbon pass together over the printhead, which contains thousands of heating elements producing varying amounts of heat. The hotter the element, the more dye is released, and as the temperature is varied, shades of each color can be overlaid on top of one another. The dyes are transparent and blend (sublimate) into continuous-tone color.

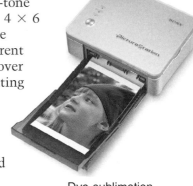

Dye-sublimation photo printer

>>

PRACTICAL ACTION
Buying a Printer

Some questions to consider when you're buying a printer:

- *Do I need color, or will black-only do?* Are you mainly printing text, or will you need to produce color charts and illustrations (and, if so, how often)? If you print lots of black text, consider getting a laser printer. If you might occasionally print color, get an inkjet that will accept cartridges for both black and color. Unless you are in the publishing or design business, you will probably not need an expensive color laser printer.

- *Do I have other special output requirements?* Do you need to print envelopes or labels? Special fonts (type styles)? Multiple copies? Transparencies or on heavy paper? Unusual paper size? Find out if the printer comes with envelope feeders, sheet feeders holding at least 100 sheets, or whatever will meet your requirements.

- *Is the printer easy to set up?* Can you easily pull the unit together, plug in the hardware, and adjust the software (the driver programs) to make the printer work with your computer?

- *Is the printer easy to operate?* Can you add paper, replace ink/toner cartridges or ribbons, and otherwise operate the printer without much difficulty?

- *Does the printer provide the speed and quality I want?* A laser printer prints 4–32 pages per minute (ppm); a color inkjet prints 10–14 ppm. Colors and graphics take longer to print. Are the blacks dark enough and the colors vivid enough?

- *Will I get a reasonable cost per page?* Special paper, ink or toner cartridges (especially color), and ribbons are all ongoing costs. Inkjet color cartridges, for example, may last 100–500 pages and cost $2–$30 new. Laser toner cartridges can cost up to $100 each but last much longer. Ribbons for dot-matrix printers are inexpensive. Ask the seller what the cost per page works out to.

- *Does the manufacturer offer a good warranty and good telephone technical support?* Find out if the warranty for a printer lasts at least 2 years. See if the printer's manufacturer offers telephone support in case you have technical problems. The best support systems offer toll-free numbers and operate evenings and weekends as well as weekdays. (If you can, try calling tech support before you buy and see what happens.)

Wax-transfer printers are faster and less expensive to use than photo printers, but the dye-sublimation photo printers produce photo-realistic quality.

PLOTTERS A **_plotter_** is a specialized output device designed to produce large, high-quality graphics in a variety of colors. (● *See Panel 5.29.*) Plotter lines are not made up of dots; they are actually drawn.

The plotter was the first computer output device that could not only print graphics but also accommodate full-size engineering, three-dimensional, and architectural drawings, as well as maps. Using different colored pens, it was also able to print in color long before inkjet printers became an alternative. Plotters are still the most affordable printing device for computer-aided design (CAD) and offer much higher resolutions than desktop printers do. Plotters are controlled by PCL and HPGL (Hewlett-Packard Control Language).

The three principal kinds of plotters are pen, electrostatic, and large-format:

- **Pen:** A *pen plotter* uses one or more colored pens to draw on paper or transparencies.
- **Electrostatic:** In an *electrostatic plotter,* paper lies partially flat on a tablelike surface, and toner is used in a photocopier-like manner.
- **Large-format:** *Large-format plotters* operate somewhat like an inkjet printer but on a much larger scale. This type of plotter is often used by graphic artists.

MULTIFUNCTION PRINTERS: PRINTERS THAT DO MORE THAN PRINT **_Multifunction printers_** combine several capabilities, such as printing, scanning, copying, and faxing. (● *See Panel 5.30.*) Brother, Canon, Epson, and Hewlett-Packard make machines in a price range of $89–$266 that combine a photocopier, fax machine, scanner, and inkjet printer.[31] Multifunction printers take up less space and cost less than the four separate office machines that they replace. The drawback is that if one component breaks, nothing works.

SPECIALTY PRINTERS
Specialty printers exist for such purposes as printing certain types of labels, tickets, and text in Braille.

Proofreader Ed Kochanowski proofreads a Braille edition of *Beowulf* at the National Braille Press in Boston.

Mixed Output: Sound, Voice, & Video

How would I explain sound, voice, and video output?

Most PCs are now multimedia computers, capable of displaying and printing not only traditional softcopy and hardcopy text and graphics but also sound, voice, and video, as we consider next.

Survival Tip

Customize Your Sounds

In Windows, open the Control Panel and the Sounds section. Once there you can change all the sounds Windows outputs. Click on the event you want to alter the sound for, click the *Browse* button, and then select a new sound file from the folders. Don't assign long sound sequences to common events; they'll become annoying. Many websites offer "themes" for your computer—you can download sound bites from your favorite movies and TV shows.

SOUND OUTPUT _Sound-output devices_ **produce digitized sounds, ranging from beeps and chirps to music.** To use sound output, you need appropriate software and a sound card. The sound card could be Sound Blaster or, since that brand has become a de facto standard, one that is "Sound Blaster–compatible." Well-known brands include Creative Labs, Diamond, and Turtle Beach. The sound card plugs into an expansion slot in your computer; on newer computers, it is integrated with the motherboard.

A dozen years ago, the audio emerging from a computer had the crackly sound of an old vacuum-tube radio. Then, in 1997, audio in personal computers began to shift to three-dimensional sound. Now, a PC with two speakers can sound more like a "surround sound" movie house. Unlike conventional stereo sound, 3-D audio describes an expanded field of sound—a broad arc starting at the right ear and curving around to the left. Thus, in a video game, you might hear a rocket approach, go by, and explode off your right shoulder. The effect is achieved by boosting certain frequencies that provide clues to a sound's location in the room or by varying the timing of sounds from different speakers. You can augment your computer's internal speakers with external speakers for high-quality sound.

more info!

AT&T's TTS Program

To find out more about this program and listen to audio demonstrations, go to *www .naturalvoices.att.com/demos.*

VOICE OUTPUT _Voice-output devices_ **convert digital data into speechlike sounds.** You hear such forms of voice output on telephones ("Please hang up and dial your call again"), in soft-drink machines, in cars, in toys and games, and in mapping software for vehicle-navigation devices. Voice portals read news and other information to users on the go.

One form of voice output that is becoming popular is *text-to-speech (TTS) systems,* which convert computer text into audible speech. TTS benefits not only the visually impaired but also anyone with a computer system sound card and speakers who wants to reduce reading chores (and eyestrain) and do other tasks at the same time. Windows XP offers a TTS program called Narrator. Others are CoolSpeech, CrazyTalk, and Digalo (which can read computer text in eight languages). AOL, Yahoo!, and MapQuest have licensed a high-quality TTS program developed by the AT&T Natural Voices Lab to read email, give driving directions, provide stock quotes, and more.

VIDEO OUTPUT _Video_ **consists of photographic images, which are played at 15–29 frames per second to give the appearance of full motion.** Video is input into a multimedia system using a video camera or VCR and, after editing, is output on a computer's display screen. Because video files can require a great deal of storage—a 3-minute video may require 1 gigabyte of storage—video is often compressed. Good video output requires a powerful processor as well as a video card.

Another form of video output is _videoconferencing_, in which people in different geographic locations can have a meeting—can see and hear one another—using computers and communications. Videoconferencing systems range from videophones to group conference rooms with cameras and multimedia equipment to desktop systems with small video cameras, microphones, and speakers. (We discuss videoconferencing further in Chapter 6.)

Input & Output Technology & Quality of Life: Health & Ergonomics

What are the principal health and ergonomic issues relating to computer use?

Danielle Weatherbee, 29, of Seattle, a medical supplies saleswoman who's on the road constantly, spends much of her time hunched over the keyboard of her notebook computer on planes, in coffee shops, in bed, and even in taxicabs. Her neck and wrists constantly ache. Her doctor has said she has the skeletal health of a 50-year-old. "But what can I do?" Weatherbee says. "My laptop is the only way to go for my work. I couldn't live without it."[32]

Health Matters

How could information technology adversely affect my health?

The computer clearly has negative health consequences for some people. College students, for example, may be susceptible to back, shoulder, wrist, and neck aches because they often use laptops, which—unlike desktop computers—have keyboard and screen too close to each other. "When you use a laptop, you can make your head and neck comfortable, or you can make your hands and arms comfortable, but it's impossible to do both," says Tom Albin of the Human Factors and Ergonomics Society, an organization that issues standards on the use of computers.[33] Observes another expert about students, "They sit in lecture halls with built-in tables, hunched over their laptops eight hours a day, and you can see it's very uncomfortable with them. Even if they could move their chairs, that would be a help."[34]

Let's consider some of the adverse health effects of computers. These may include repetitive strain injuries, eyestrain and headache, and back and neck pains. We will also consider the effects of electromagnetic fields and noise.

info!

RSIs

The United Food and Commercial Workers International Union provides detailed information about RSIs at www.ufcw.org/workplace_ connections/retail/safety_ health_news_and_facts/ rep_stress_overview.cfm.

REPETITIVE STRESS INJURIES *Repetitive stress* (or *strain*) *injuries (RSIs)* **are wrist, hand, arm, and neck injuries resulting when muscle groups are forced through fast, repetitive motions.** Most victims of RSI are in dentistry, meatpacking, automobile manufacturing, poultry slaughtering, and clothing manufacturing, which require awkward wrist positions Musicians, too, are often troubled by RSI (because of long hours of practice).

People who use computer keyboards—some superstar data-entry operators reportedly regularly average 15,000 keystrokes an hour—account for some RSI cases that result in lost work time. Before computers came along, typists would stop to make corrections or change paper. These motions had the effect of providing many small rest breaks. Today keyboard users must devise their own mini-breaks to prevent excessive use of hands and wrists (or install a computer program that uses animated characters to remind users to take breaks and provide suggestions for exercises). People who use a mouse for more than a few hours a day—graphic designers, desktop-publishing professionals, and the like—are also showing up with increased RSI injuries. The best advice is to find a mouse large enough so that your hand fits comfortably over it, and don't leave your hand on the mouse when you are not using it. Some cordless mice, such as the Logitech MX1000 and the MediaPlay cordless mouse, are said to offer relief to aching fingers.[35] You might also try using function keys instead of the mouse whenever possible.[36]

Among the various RSIs, some, such as muscle strain and tendinitis, are painful but usually not crippling. These injuries may be cured by rest, anti-inflammatory medication, and change in typing technique. One type of RSI, carpal tunnel syndrome, is disabling and often requires surgery. ___Carpal tunnel syndrome (CTS)___ **is a debilitating condition caused by pressure on the median nerve in the wrist, producing damage and pain to nerves and tendons in the hands.** It is caused by short repetitive movement, such as typing, knitting, and using vibrating tools for hours on end. The lack of rest in between these motions irritates and inflames the flexor tendons that travel with the median nerve to the hand through an area in the wrist called the "carpal tunnel," which is surrounded by bones and a transverse ligament. The inflamed tendons squeeze the nerve against the ligament. It must be pointed out, however, that there is some dispute as to whether continuous typing is necessarily a culprit in CTS. A 2003 study by researchers in Denmark found no association between computer use and CTS, and a 2001 Mayo Clinic study found that even 7 hours a day of computer use didn't increase the disorder.[37] Other researchers have found limitations in both studies.[38]

EYESTRAIN & HEADACHES Vision problems are actually more common than RSI problems among computer users. Computers compel people to use their eyes at close range for a long time. However, our eyes were made to see most efficiently at a distance. It's not surprising, then, that people develop what's called *computer vision syndrome.*

   ___Computer vision syndrome (CVS)___ **consists of eyestrain, headaches, double vision, and other problems caused by improper use of computer display screens.** By "improper use," we mean not only staring at the screen for too long but also failing to correct faulty lighting and screen glare and using screens with poor resolution.

BACK & NECK PAINS Improper chairs or improper positioning of keyboards and display screens can lead to back and neck pains. All kinds of adjustable, special-purpose furniture and equipment are available to avoid or diminish such maladies.

ELECTROMAGNETIC FIELDS Like kitchen appliances, hairdryers, and television sets, many devices related to computers and communications generate low-level electromagnetic field emissions. ___Electromagnetic fields (EMFs)___ **are waves of electrical energy and magnetic energy.**

   In recent years, stories have appeared in the mass media reflecting concerns that high-voltage power lines, cellphones, and CRT-type computer monitors might be harmful. There have been worries that monitors might be linked to miscarriages and birth defects and that cellphones and power lines might lead to some types of cancers.

   Is there anything to this? The answer, so far, is that no one is sure. The evidence seems scant that weak electromagnetic fields, such as those used for cellphones and found near high-voltage lines, cause cancer. Still, handheld cellphones do put the radio transmitter next to the user's head. This causes some health professionals concern about the effects of radio waves entering the brain as they seek out the nearest cellular transmitter. Thus, in the United Kingdom, the Independent Expert Study Group on Mobile Phones recommended that children should use cellphones only when necessary, and Norway's Ombudsman for Children has said that children under the age of 13 should not have their own mobile phones. Dr. Lief Salford of Lund University in Sweden, who has called the evolution of wireless phones "the largest biological experiment in the history of the world," reported that cellphone radiation damaged neurons in the brains of young rats.[39] On the other hand, an August 2005 study found no increased risk of brain tumors associated with using a cellphone for at least 10 years.[40]

   The Federal Communications Commission (FCC) has created a measurement called *Specific Absorption Rate (SAR)* to give consumers data on the radiation levels their phones produce. Currently, the Ericsson T28 World cellphone has the highest SAR level (1.49 SAR), and the Motorola StarTac 7860 has the lowest level (0.24 SAR). The World Health Organization has said that further research is needed for long-term health assessment.

As for CRT monitors, those made since the early 1980s produce very low emissions. Even so, users are advised to work no closer than arm's length to a CRT monitor. The strongest fields are emitted from the sides and backs of terminals. Alternatively, you can use laptop computers, because their liquid crystal display (LCD) screens emit negligible radiation.

The current advice from the Environmental Protection Agency is to exercise *prudent avoidance*. That is, we should take precautions that are relatively easy. However, we should not feel compelled to change our whole lives or to spend a fortune minimizing exposure to electromagnetic fields. Thus, we can take steps to put some distance between ourselves and a CRT monitor. However, it is probably not necessary to change residences because we happen to be living near some high-tension power lines.

NOISE   The chatter of impact printers or hum of fans in computer power units can be psychologically stressful to many people. Sound-muffling covers are available for impact printers. Some system units may be placed on the floor under the desk to minimize noise from fans.

Ergonomics: Design with People in Mind

Why is ergonomics important?

Previously, workers had to fit themselves to the job environment. However, health and productivity issues have spurred the development of a relatively new field, called *ergonomics*, that is concerned with fitting the job environment to the worker.

The purpose of _ergonomics_ is to make working conditions and equipment safer and more efficient. It is concerned with designing hardware and software that is less stressful and more comfortable to use, that blends more smoothly with a person's body or actions. Examples of ergonomic hardware are tilting display screens, detachable keyboards, and keyboards hinged in the middle to allow the users' wrists to rest in a more natural position.

We address some further ergonomic issues in the Experience Box at the end of this chapter.

info!

Ergonomics for Kids

To learn how to set up an ergonomic workstation for children, go to *www .businessweek.com/magazine/ content/02_51/b3813121.htm.* For more general government ergonomic guidelines, go to *www.osha.gov* and then search for *computer workstations.*

5.5

The Future of Input & Output

What are some examples of future input and output technology?

Biologists studying wildlife at the James San Jacinto Mountains Reserve in southern California found out something interesting about squirrels. It seems the little rodents chew on moss for the moisture. No one had known that before, but researchers learned it when the animals tripped motion sensors spread around a patch of wilderness, which activated tiny wireless cameras, which in turn were linked to a speedy wireless network.[41] Other scientists are using "smart dust," grids of simple, low-powered devices called "motes" to monitor the environment and wirelessly pass along collected data.[42]

The uses of such "intelligent sensors"—digital sensors placed in all kinds of places and linked by wireless networks—"could be bigger than the internet," suggests one expert.[43] Not only do they allow squirrels and other inhabitants of the natural world to communicate with the digital realm; they are also used for many other purposes—security, for example. Sensors that can detect biological agents are being placed in 4,000 locations across the United States to provide warnings of bioterror attacks.[44] Networks of sensors have been used to detect undersea intruders and accidents, as when the Russian submarine *Kursk* sank in August 2000.[45] Experiments are also being done in

info!

More on EMFs

Do you have CVS? What are computer eyeglasses? Find out at *www.allaboutvision.com/cvs/ faqs.htm.*

SECURITY

info!

Science Sensors

Want to see how biologists study 30 acres of nature with wireless-network-linked sensors? Go to *www.jamesreserve.edu*.

using sensors for a kind of personal security, as in using wireless biosensors worn inside the underwear of people with health problems to alert doctors when there's an emergency.[46]

As these examples show, input technology seems headed in two directions: (1) toward more input devices in remote locations and (2) toward more refinements in source data automation. Output is distinguished by (1) more output in remote locations and (2) increasingly realistic—even lifelike—forms, as we discuss in a few pages.

Toward More Input from Remote Locations

How can data be input from anywhere?

The linkage of computers and telecommunications means that data may be input from nearly anywhere. For instance, X-ray machines are now going digital, which means that a medical technician in the jungles of South America can take an X-ray of a patient and then transmit a perfect copy of it by satellite uplink to a hospital in Boston. Visa and MasterCard are moving closer to using "smart cards," or stored-value cards, for internet transactions.

info!

Mouse-Pointer Systems for the Disabled

To learn more about this subject, use the following keywords to search the web: *Quick Glance Eye-Gaze Tracking System, CameraMouse, Quadjoy, Cordless Gyro-HeadMouse.*

Toward More Source Data Automation

How might source data automation be of help to me?

Increasingly, input technology is being designed to capture data at its source, which will reduce the costs and mistakes associated with copying or otherwise preparing data in a form suitable for processing. We mentioned some possible innovations in high-capacity bar codes, more sophisticated scanners, smarter smart cards, and widespread use of sensors. Some reports from elsewhere on the input-technology front:

INPUT HELP FOR THE DISABLED   Some devices now available for people with physical disabilities, such as paraplegics, may portend new ways of entering and manipulating data input. For example, in one system a camera and special software enable users to operate the on-screen pointer with their eye movements instead of their hands. In another hands-free system a camera tracks the user's body movements and the system converts them into mouse-pointer movements on the screen. In yet another system, the user's breathing controls the screen pointer. In a fourth, the nose is used to direct the cursor on a computer screen. Finally, there is a system that can be attached to a baseball hat enabling head movements to control the pointer.

MORE SOPHISTICATED TOUCH DEVICES   Touch screens are becoming commonplace. Sometime in the near future, futurists have suggested, your car may have a dashboard touch screen linked to mobile electronic "yellow pages," which would enable you to reserve a motel room for the night or find the nearest Chinese restaurant.

Researchers in what is known as *hepatic*—active touch—systems are exploring how to create devices that will allow people to feel what isn't there. With this kind of "virtual touch," a dental student could train in drilling down into the decay of a simulated tooth without fear of destroying a real healthy tooth. Doctors would have new surgical tools, videogames would be made more realistic, and drivers would be able to manipulate dials and knobs without taking their eyes off the road.[47]

BETTER SPEECH RECOGNITION   It's possible that speech recognition may some day fulfill world travelers' fondest dream: You'll be able to speak in English, and a speech-recognition device will instantly translate your remarks into another language, whether French, Swahili, or Japanese. At the moment, translation programs such as Easy Translator can translate text on

(*Top*) One-handed keyboard. (*Bottom*) Special mouse/digitizer and pointing stick to press keys.

web pages (English to Spanish, French, and German and the reverse), although they do so imperfectly. Research is also going forward on speech-recognition software that can decode slight differences in pitch, timing, and amplitude, so that computers can recognize anger and pain, for example.[48] And already voice-recognition techniques have been applied to animal sounds, so that, for instance, farmers can be alerted to unrest in the pigpen or cows in heat.[49]

IMPROVED DIGITAL CAMERAS   Digital still cameras and video cameras are now commonplace. Now manufacturers are also bringing out compact digital camcorders, some that are coat-pocket size. Some will also produce both still photos and respectable video. Olympus has a camera with a 20-gigabyte hard drive that can store both photos and music.[50]

Microsoft has invented a wide-angle, fish-eye lens digital camera that can be worn like a badge, recording all the still and video images of the wearer's daily life.[51] A boon to bird-watchers, sports spectators, and opera goers is the appearance of binoculars with built-in digital cameras, so that one can bring home photos or digital movies.[52] The newest development is the camera-on-a-chip, which contains all the control components necessary to take a photograph or make a movie. Such a device, called an *active pixel sensor*, based on NASA space technology, is now being made by Micron and by Photobit.

GESTURE RECOGNITION   In the movie *Minority Report*, Tom Cruise, playing a police officer who fights crimes occurring in the future, gestures into the air and calls up computer images. Researchers have been trying to apply gesturing to computing input devices for decades, and the technology is still in its infancy. For now, developers of gesture-recognition systems are experimenting with cameras, special gloves, head gear, and even body suits. A computer keyboard developed by FingerWorks, for instance, allows you to do such tasks as clicking, scrolling, and dragging by gesturing or moving your fingers across a motion-sensitive keyboard.[53]

Gesture recognition also is a key component in the idea of *pervasive computing*, which refers to having access to computing tools any place at any time. For example, a display could appear on a wall or wherever you want. To make pervasive computing work, however, users need to have an input device available wherever they are, a process often referred to as *human-centric computing*. Through human-centric computing, the user is always connected to computing tools. Gesture-recognition technology could provide the needed input capabilities for both types of computing.

PATTERN-RECOGNITION & BIOMETRIC DEVICES   Would you believe a computer could read people's emotions from changes in their facial patterns, like surprise and sadness? Such devices are being worked on at Georgia Institute of Technology and elsewhere. (● *See Panel 5.31.*) Indeed, you can buy a face-recognition program so that you can have your computer respond only to your smile.

Camera on a chip. Active pixel sensor, a digital camera (exclusive of the optics) on a single 1 cm^2 chip. It can set the desired exposure time and control any number of imaging modes.

info!

Gesture Recognition

For an update on the progress in gesture-recognition technology, go to *www.cybernet.com/~ccohen/#Commercial*.

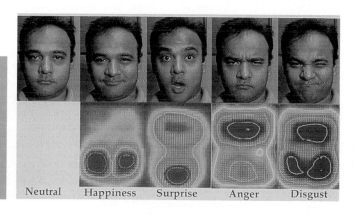

● **PANEL 5.31**
Computers tuned to emotions
At Georgia Institute of Technology, researchers have devised a computer system that can read people's emotions from changes in their faces. Expressions of happiness, surprise, and other movements of facial muscles are converted by a special camera into digitized renderings of energy patterns.

Neutral   Happiness   Surprise   Anger   Disgust

Columbia University scientists have also developed a computer-based system that can determine from the photograph of a person's eye what that person was looking at, including panoramic details to the left, to the right, and even slightly behind the person. Such images could be used in human-computer interactions.[54]

BRAINWAVE DEVICES Perhaps the ultimate input device analyzes the electrical signals of the brain and translates them into computer commands. In one experiment, 100 tiny sensors were implanted in the brain of a 25-year-old quadriplegic who, using just

Suitor (Simple User Internet Tracker) eye-gaze technology is an *attentive system* that pays attention to what users do so it can attend to what users need. Suitor tracks computer users according to what they look at and how often they look at it.

his thoughts, was able to control a computer well enough to operate a TV, open email, and play Pong with 70% accuracy.[55] At Duke University, scientists induced a monkey with electrodes implanted in its brain to move a robot arm, much as if it were moving its own arm.[56] Users have successfully moved a cursor on the screen through sheer power of thought.[57] Individuals have even typed words by simply thinking them.[58]

Although there is a very long way to go before brainwave-input technology becomes practical, the consequences not only for people with disabilities but for all of us could be tremendous.

Toward More Output in Remote Locations

How could information be output in far-flung places?

Clearly, output in remote locations is the wave of the future. As TV and the personal computer converge, we can expect more scenarios like this: Your PC continually receives any websites covering topics of interest to you—CNN's home page for news on Iraq, for example—which are stored on your hard disk for later viewing. The information, on up to 5,000 websites a day, is transmitted from a television broadcast satellite to an 18-inch satellite dish on your roof. Interested? The technology is available now with direct-broadcast services such as DirecTV Inc.

Toward More Realistic Output

In what way is output being made more realistic?

Once upon a time, having a "home theater" probably meant you were a wealthy film-industry figure with a large room containing a wall-size screen and movie-house seats. Now, says technology writer Phillip Robinson, "home theater has become a middle-income commodity with new audio and video gear promising to improve your television viewing so much you'll practically think you're in a theater instead of your living room or den."[59]

The enhanced qualities of home theater won't be limited to television viewing, of course. As analog and digital technologies merge, we will no doubt see this type of increased realism appearing in all forms of output. Let's consider what's coming into view.

E Ink's flexible screen on 0.3-millimeter thick electronic "paper" with millions of tiny capsules with black and white pigment chips and transmitting electrodes.

DISPLAY SCREENS: BETTER & CHEAPER Computer screens are becoming crisper, brighter, bigger, and cheaper. Newer LCD monitors, for instance, have as high a resolution as previous larger CRT monitors, and the prices have dropped significantly.

New gas-plasma technology is being employed to build flat-panel hang-on-the-wall screens as large as 50 inches from corner to corner. Using a technique known as *microreplication*, researchers have constructed a thin

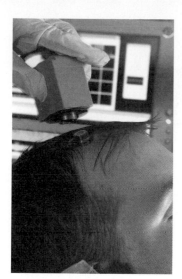

Quadriplegic Matt Nagle is connected to a computer by a cable that is screwed into his head. The 4-millimeter square chip, which is placed on the surface of the motor cortex area of the brain, contains 100 electrodes, each thinner than a hair, that detect neural electrical activity. The sensor is then connected to a computer via a small wire attached to a pedestal mounted on the skull. Nagle has been able to check email and play computer games simply by using thoughts. He can also turn lights on and off and control a television, all while talking and moving his head.

transparent sheet of plastic prisms that allows builders of portable computer screens to halve the amount of battery power required.

VIDEO: MOVIE QUALITY FOR PCs   Today the movement of many video images displayed on a microcomputer comes across as fuzzy and jerky, with a person's lip movements out of sync with his or her voice. This is because currently available equipment often is capable of running only about eight frames a second.

New technology based on digital wavelet theory, a complicated mathematical theory, has led to software that can compress digitized pictures into fewer bytes and do it more quickly than current standards. Indeed, the technology can display 30–38 frames a second—"real-time video." Images have the look and feel of a movie.

In addition, advanced graphics chips from firms such as Nvidia and NTI are increasing the realism of animation, making possible lifelike imagery. Nvidia's computer-generated mermaid named Nalu, for instance, is described as having "a cloud of golden tresses that realistically seem to reflect dappled light and flow with the water [and] rosy, unusually lifelike skin."[60] Such images use geometric building blocks called *polygons*, and Nalu is composed of 300,000 of them—far more than most video images today have.

THREE-DIMENSIONAL OUTPUT   In the 1930s, radiologists tried to create three-dimensional images by holding up two slightly offset X-rays of the same object and crossing their eyes. Now the same effects can be achieved by computers. With 3-D technology, flat, cartoonlike images give way to rounded objects with shadows and textures. Artists can even add "radiosity," so that a dog standing next to a red car, for instance, will pick up a red glow.

In the 1990s, Sanyo demonstrated an experimental system, which did not involve special glasses, that employed what appeared to be a normal 40-inch TV set with a screen using hundreds of tiny prisms/lenses. "The 3-D effect was stunning," wrote one reporter who saw it. "In one scene, water was sprayed from a hose directly at the camera. When watching the replay, I had to control the urge to jump aside."[61] Sharp now offers a flat-panel monitor that is mainly two-dimensional but can be made 3-D at the press of a button; it is intended for engineers, architects, and serious gamers.[62]

Three-dimensional printers using inkjet printer heads are able to output layer after layer of images printed on starch or plaster, producing 3-D objects. In the future, this technology may be used to print on plastics and metals to produce electronic components, such as transistors. Indeed, military engineers have developed something called a Mobile Parts Hospital in which technicians use workstations and robotic machine tools to fabricate replacement parts, such as bolts or machine-gun mounts, that lasers then will "print" as powdered metal, layer by layer.[63] Medical researchers have even used 3-D printing technology to print layers of cells, building intricate tissue structures.[64] Indeed, researchers have used older-model inkjet printers to spray cells onto a gauze scaffolding to create living tissue—artificial skin that may be of help to burn victims.[65]

more info!

Using the Smithsonian

What kind of technology does the Smithsonian Museum in Washington, D.C., have on display? Find the museum's website and enter keywords of computer terms.

Experience Box
Good Habits: Protecting Your Computer System, Your Data, & Your Health

Whether you set up a desktop computer and never move it or tote a portable PC from place to place, you need to be concerned about protecting not only your computer but yourself. You don't want your computer to get stolen or zapped by a power surge. You don't want to lose your data. And you certainly don't want to lose your health for computer-related reasons. Here are some tips for taking care of these vital areas.

Guarding against Hardware Theft & Loss

Portable computers are easy targets for thieves. Obviously, anything conveniently small enough to be slipped into your briefcase or backpack can be slipped into someone else's. Never leave a portable computer unattended in a public place.

It's also possible to simply lose a portable—for example, forgetting it's in the overhead-luggage bin in an airplane. To help in its return, use a wide piece of clear tape to tape a card with your name and address to the outside of the machine. You should tape a similar card to the inside also. In addition, scatter a few such cards in the pockets of the carrying case.

Desktop computers are also easily stolen. However, for under $25, you can buy a cable and lock, like those used for bicycles, and secure the computer, monitor, and printer to a work area. If your hardware does get stolen, its recovery may be helped if you have inscribed your driver's license number or home address on each piece. Some campus and city police departments lend inscribing tools for such purposes. Finally, insurance to cover computer theft or damage is surprisingly cheap. Look for advertisements in computer magazines. (If you have standard tenants' or homeowners' insurance, it may not cover your computer. Ask your insurance agent.)

Guarding against Heat, Cold, Spills, & Drops

"We dropped 'em, baked 'em, we even froze 'em," proclaimed the *PC Computing* cover, ballyhooing a story about its notebook "torture test."[66]

The magazine put eight notebook computers through durability trials a few years ago. One approximated putting these machines in a car trunk in the desert heat; another, leaving them outdoors in a Buffalo, New York, winter. A third test simulated sloshing coffee on a keyboard, and a fourth dropped computers from desktop height to a carpeted floor. All passed the bake test, but one failed the freeze test. Three completely flunked the coffee-spill test, one other revived, and the rest passed. One that was dropped lost the right side of its display; the others were unharmed. Of the eight, half passed all tests unscathed. In a more recent torture test, nine notebooks survived the heat, cold, and spill tests, but three failed the drop test.

This gives you an idea of how durable computers are. Designed for portability, notebooks may be hardier than desktop machines. Even so, you really don't want to tempt fate by dropping your computer, which could cause your hard-disk drive to fail. (Special "ruggedized" laptops exist, such as Panasonic's Toughbooks and Hewlett-Packard's nr3600, that meet military specifications for shock resistance and sealing against the elements for people in rough construction sites, deserts, and combat.[67])

Guarding against Damage to Software

Systems software and applications software generally come on CD-ROM disks. The unbreakable rule is simply this: copy the original disk, either onto your hard-disk drive or onto another disk. Then store the original disk in a safe place. If your computer gets stolen or your software destroyed, you can retrieve the original and make another copy.

Protecting Your Data

Computer hardware and commercial software are nearly always replaceable, although perhaps with some expense and difficulty. Data, however, may be major trouble to replace or even be irreplaceable. If your hard-disk drive crashes, do you have the same data on a backup disk? Almost every microcomputer user sooner or later has the experience of accidentally wiping out or losing material and having no copy. This is what makes people true believers in backing up their data—making a duplicate in some form. If you're working on a research paper, for example, it's fairly easy to copy your work onto a CD or keychain memory at the end of your work session. Then store the copy in a safe place.

Protecting Your Health

More important than any computer system and (probably) any data is your health. What adverse effects might computers cause? As we discussed earlier in the chapter, the most serious are painful hand and wrist injuries, eyestrain and headache, and back and neck pains.

Many people set up their computers in the same way as they would a typewriter. However, the two machines are ergonomically different for various reasons. With a computer, it's important to sit with both feet on the floor, thighs at right angles to your body. The chair should be adjustable and support your lower back. Your forearms should be parallel to the floor. You should look down slightly at the screen. (● *See Panel 5.32.*) This setup is particularly important if you are going to be sitting at a computer for hours at a stretch.

To avoid wrist and forearm injuries, keep your wrists straight and hands relaxed as you type. Instead of putting the keyboard on top of a desk, put it on a low table or in a keyboard drawer under the desk. Otherwise the nerves in your wrists will rub against the sheaths surrounding them, possibly leading to RSI pains. Or try setting an hourly alarm on your watch or on an alarm clock; when the alarm goes off, take a short break and rotate your wrists and hands a bit.

Eyestrain and headaches usually arise because of improper lighting, screen glare, and long shifts staring at the screen. Make sure that your windows and lights don't throw a glare on the screen and that your computer is not framed by an uncovered window.

Back and neck pains occur because furniture is not adjusted correctly or because of heavy computer use, especially on laptops. Adjustable furniture and frequent breaks should provide relief.

HEAD Directly over shoulders, without straining forward or backward, about an arm's length from screen.

NECK Elongated and relaxed.

SHOULDERS Kept down, with the chest open and wide.

BACK Upright or inclined slightly forward from the hips. Maintain the slight natural curve of the lower back.

ELBOWS Relaxed, at about a right angle, try to keep forearms parallel to floor.

WRISTS Relaxed, and in a neutral position, without flexing up or down.

KNEES Slightly lower than the hips.

CHAIR Sloped slightly forward to facilitate proper knee position.

LIGHT SOURCE Should come from behind the head.

SCREEN At eye level or slightly lower. Use an anti-glare screen.

FINGERS Gently curved.

KEYBOARD Best when kept flat (for proper wrist positioning) and at or just below elbow level. Computer keys that are far away should be reached by moving the entire arm, starting from the shoulders, rather than by twisting the wrists or straining the fingers. Take frequent rest breaks.

FEET Firmly planted on the floor. Shorter people may need a footrest.

● **PANEL 5.32**
How to set up your computer work area

Keep wrists above the pad, and tilt the keyboard downward.

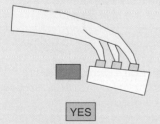

YES

Don't bend your hand in awkward angles to type key combinations.

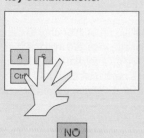

NO

Use both hands to type combination key strokes.

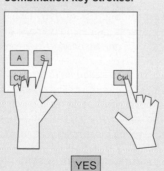

YES

Don't rest on the wrist pad.

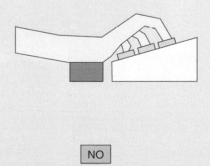

NO

Twisting your hands puts strain on them. Resting on a wrist rest, the table, or arm rests while typing forces you to twist your hand to reach some keys. Instead, keep your hands moving freely above the keyboard, letting the strong muscles of your arms move your hands.

It is also a bad idea to contort your hands in other ways. Your hand should be flat and parallel to the keyboard, without twisting. there should not be any pressure on your wrist or forearms while you type. You should NOT rest your wrists on a wrist rest except while taking a very short break from typing. A wrist rest of the proper height (level with the space bar) can serve as a reminder to keep your wrists straight. If you feel your wrist touching the rest, you know that your wrists are starting to dip.

Summary

active-matrix display (p. 281) Also known as *TFT (thin-film transistor) display;* flat-panel display in which each pixel on the screen is controlled by its own transistor. Why it's important: *Active-matrix screens are much brighter and sharper than passive-matrix screens, but they are more complicated and thus more expensive. They also require more power, affecting the battery life in laptop computers.*

audio-input device (p. 270) Hardware that records analog sound and translates it for digital storage and processing. Why it's important: *Analog sound signals are continuous variable waves within a certain frequency range. For the computer to process them, these variable waves must be converted to digital 0s and 1s. The principal use of audio-input devices is to produce digital input for multimedia computers. An audio signal can be digitized in two ways—by an audio board or a MIDI board.*

bar-code reader (p. 267) Photoelectric (optical) scanner that translates bar codes into digital codes. Why it's important: *With bar-code readers and the appropriate software system, store clerks can total purchases and produce invoices with increased speed and accuracy, and stores and other businesses can monitor inventory and services with increased efficiency.*

bar codes (p. 266) Vertical, zebra-striped marks imprinted on most manufactured retail products. Why it's important: *Bar codes provide a convenient means of identifying and tracking items. In North America, supermarkets, food manufacturers, and others have agreed to use a bar-code system called the* Universal Product Code (UPC). *Other kinds of bar-code systems are used on everything from FedEx packages to railroad cars to the jerseys of long-distance runners.*

biometrics (p. 277) Science of measuring individual body characteristics. Why it's important: *Biometric security devices identify a person through a fingerprint, voice intonation, or some other biological characteristic. For example, retinal-identification devices use a ray of light to identify the distinctive network of blood vessels at the back of the eyeball.*

carpal tunnel syndrome (CTS) (p. 290) Debilitating condition caused by pressure on the median nerve in the wrist, producing damage and pain to nerves and tendons in the hands. Why it's important: *CTS can be caused by overuse or misuse of computer keyboards.*

color depth (p. 279) Also called *bit depth;* the amount of information, expressed in bits, that is stored in a dot. Why it's important: *The more bits in a dot or pixel, the more shades of gray and colors can be represented. With 24-bit color depth, for example, 8 bits are dedicated to each primary color—red, green, and blue. Eight-bit color is standard for most of computing; 24-bit, called* true color, *requires more resources, such as video memory.*

computer vision syndrome (CVS) (p. 290) Eyestrain, headaches, double vision, and other problems caused by improper use of computer display screens. Why it's important: *CVS can be prevented by not staring at the display screen for too long, by correcting faulty lighting, by avoiding screen glare, and by not using screens with poor resolution.*

CRT (cathode-ray tube) (p. 279) Vacuum tube used as a display screen in a computer or video display terminal. Why it's important: *This technology is found not only in the screens of desktop computers but also in television sets and flight-information monitors in airports.*

digital camera (p. 272) Electronic camera that uses a light-sensitive processor chip to capture photographic images in digital form and store them on a small diskette inserted into the camera or on flash memory chips (cards). Why it's important: *The bits of digital information—the snapshots you have taken, say—can be copied right onto a computer's hard disk for manipulation and printing out. The environmentally undesirable stage of chemical development required for conventional film is completely eliminated.*

digitizer (p. 264) Input unit based on an electronic pen or a mouselike copying device called a *puck* that converts drawings and photos to digital data. Why it's important: See *digitizing tablet.*

digitizing tablet (p. 264) One form of digitizer; an electronic plastic board on which each specific location corresponds to a location on the screen. When the user uses a puck, the tablet converts his or her movements into digital signals that are input to the computer. Why it's important: *Digitizing tablets are often used to make maps and engineering drawings, as well as to trace drawings.*

display screen (p. 278) Also called *monitor, CRT,* or simply *screen;* output device that shows programming instructions and data as they are being input and information after it is processed. Why it's important: *Screens are needed to display softcopy output.*

dot pitch (dp) (p. 278) Amount of space between the centers of adjacent pixels; the closer the pixels (dots), the crisper the image. Why it's important: *Dot pitch is one of the measures of display-screen crispness. For a .25dp monitor, for instance, the dots are 25/100ths of a millimeter apart. Generally, a dot pitch of .25dp will provide clear images.*

dots per inch (dpi) (p. 266) Measure of the number of columns and rows of dots per inch. For microcomputer printers, resolution is in the range 60–1,500 dpi. Why it's important: *The higher the dpi, the better the resolution.* (See also *resolution.*)

dumb terminal (p. 256) Also called *video display terminal (VDT);* display screen and a keyboard hooked up to a computer system. It can input and output but not process data. Why it's important: *Dumb terminals are used, for example, by airline reservations clerks to access a mainframe computer containing flight information.*

electromagnetic fields (EMFs) (p. 290) Waves of electrical energy and magnetic energy. Why it's important: *Some people have worried that CRT monitors might be linked to miscarriages and birth defects and that cellphones and power lines might lead to some types of cancers. However, the evidence is unclear.*

ergonomics (p. 291) Study, or science, of working conditions and equipment with the goal of improving worker safety and efficiency. Why it's important: *On the basis of ergonomic principles, stress, illness, and injuries associated with computer use may be minimized.*

fax machine (p. 269). Also called a *facsimile transmission machine;* input device that scans an image and sends it as electronic signals over telephone lines to a receiving fax machine, which prints the image on paper. Two types of fax machines are dedicated fax machines and fax modems. Why it's important: *Fax machines permit the transmission of text or graphic data over telephone lines quickly and inexpensively. They are found not only in offices and homes but also alongside regular phones in some public places such as airports.*

flat-panel display (p. 279) Display screen that is much thinner, weighs less, and consumes less power than a CRT. Flat-panel displays are made up of two plates of glass separated by a layer of a substance in which light is manipulated. Why it's important: *Flat-panel displays are essential to portable computers, although they are available for desktop computers as well.*

flatbed scanner (p. 266) Also called *desktop scanner;* the image being scanned is placed on a glass surface, where it remains stationary, and the scanning beam moves across it. Three other types of scanners are *sheet-fed, handheld,* and *drum.* Why it's important: *Flatbed scanners are one of the most popular types of scanner.*

handwriting recognition (p. 262) System in which a computer receives intelligible written input, using special software to interpret the movement of a stylus across a writing service and translating the resulting cursive writing into digital information. Why it's important: *Handwriting recognition is a commonly used input method for PDAs, some handheld videogames, and tablet PCs.*

hardcopy (p. 277) Printed output. The principal examples are printouts, whether text or graphics, from printers. Film, including microfilm and microfiche, is also considered hardcopy output. Why it's important: *Hardcopy is an essential form of computer output.*

impact printer (p. 282) Printer that forms characters or images by striking a mechanism such as a print hammer or wheel against an inked ribbon, leaving an image on paper. Why it's important: *Nonimpact printers are more commonly used than impact printers, but dot-matrix printers are still used in some businesses.*

inkjet printer (p. 284) Printer that sprays onto paper small, electrically charged droplets of ink from four nozzles through holes in a matrix at high speed. Like laser and dot-matrix printers, inkjet printers form images with little dots. Why it's important: *Because they produce high-quality images on special paper, inkjet printers are often used in graphic design and desktop publishing. However, traditionally inkjet printers have been slower than laser printers and they print at a lower resolution on regular paper.*

input hardware (p. 253) Devices that translate data into a form the computer can process. Why it's important: *Without input hardware, computers could not function. The computer-readable form consists of 0s and 1s, represented as off and on electrical signals. Input hardware devices are categorized as three types: keyboards, pointing devices, and source data-entry devices.*

intelligent terminal (p. 256) Hardware unit with its own memory and processor, as well as a display screen and keyboard, hooked up to a larger computer system. Why it's important: *Such a terminal can perform some functions independent of any mainframe to which it is linked. Examples include the automated teller machine (ATM), a self-service banking machine connected through a telephone network to a central computer, and the point-of-sale (POS) terminal, used to record purchases at a store's customer checkout counter. Recently, many intelligent terminals have been replaced by personal computers.*

internet terminal (p. 256) Terminal that provides access to the internet. There are several variants of internet terminal: (1) the set-top box or web terminal, which displays web pages on a TV set; (2) the network computer, a cheap, stripped-down computer that connects people to networks; (3) the online game player, which not only lets you play games but also connects to the Internet; (4) the full-blown PC/TV (or TV/PC), which merges the personal computer with the television set; and (5) the wireless pocket PC or personal digital assistant (PDA), a handheld computer with a tiny keyboard that can do two-way wireless messaging. Why it's important: *In the near future, most likely, internet terminals will be everywhere.*

keyboard (p. 254) Input device that converts letters, numbers, and other characters into electrical signals that can be read by the computer's processor. Why it's important: *Keyboards are the most popular kind of input device.*

laser printer (p. 283) Nonimpact printer that creates images with dots. As in a photocopying machine, images are produced on a drum, treated with a magnetically charged inklike toner (powder), and then transferred from drum to paper. Why it's important: *Laser printers produce much better image quality than do dot-matrix printers and can print in many more colors; they are also quieter. Laser printers, along with page description languages, enabled the development of desktop publishing.*

light pen (p. 263) Light-sensitive penlike device connected by a wire to the computer terminal. The user brings the pen to a desired point on the display screen and presses the pen button, which identifies that screen location to the computer. Why it's important: *Light pens are used by engineers, graphic designers, and illustrators.*

liquid crystal display (LCD) (p. 279) Flat-panel display in which molecules of liquid crystal line up in a way that alters their optical properties, creating images on the screen by transmitting or blocking out light. Why it's important: *LCD is useful not only for portable computers but also as a display for various electronic devices, such as watches and radios.*

magnetic-ink character recognition (MICR) (p. 268) Scanning technology that reads magnetized-ink characters printed at the bottom of checks and converts them to digital form. Why it's important: *MICR technology is used by banks to sort checks.*

MIDI board (p. 270) *MIDI,* pronounced "middie," stands for "Musical Instrument Digital Interface." MIDI sound boards use this standard. Why it's important: *MIDI provides a standard for the interchange of musical information between musical instruments, synthesizers, and computers.*

mouse (p. 258) A pointing device that is rolled about on a desktop mouse pad and directs a pointer on the computer's display screen. The name is derived from the device's shape, which is a bit like a mouse, with the cord to the computer being the tail. Why it's important: *The mouse is the principal pointing tool used with microcomputers.*

multifunction printer (p. 287) Hardware device that combines several capabilities, such as printing, scanning, copying, and faxing. Why it's important: *Multifunction printers take up less space and cost less than the four separate office machines that they replace. The downside, however, is that if one component breaks, nothing works.*

nonimpact printer (p. 283) Printer that forms characters and images without direct physical contact between the printing mechanism and paper. Two types of nonimpact printers often used with microcomputers are laser printers and inkjet printers. A third kind, the thermal printer, is seen less frequently. Why it's important: *Nonimpact printers are faster and quieter than impact printers.*

optical character recognition (OCR) (p. 269) Software technology that converts scanned text from images (pictures of the text) to an editable text format (usually ASCII) that can be imported into a word processing application and manipulated. Why it's important: *Special OCR characters appear on utility bills and price tags on department-store merchandise. The wand reader is a common OCR scanning device. These days almost all scanners come with OCR software.*

optical mark recognition (OMR) (p. 269) Scanning technology that reads "bubble" marks and converts them into computer-usable form. Why it's important: *OMR technology is used to read the College Board Scholastic Aptitude Test (SAT) and the Graduate Record Examination (GRE).*

output hardware (p. 253) Hardware devices that convert machine-readable information, obtained as the result of processing, into people-readable form. The principal kinds of output are softcopy and hardcopy. Why it's important: *Without output devices, people would have no access to processed data and information.*

page description language (p. 283) Software that describes the shape and position of characters and graphics to the printer. PostScript and PCL are common page description languages. Why it's important: *Page description languages are essential to desktop publishing.*

passive-matrix display (p. 281) Flat-panel display in which a transistor controls a whole row or column of pixels. Passive matrix provides a sharp image for one-color (monochrome) screens but is more subdued for color. Why it's important: *Passive-matrix displays are less expensive and use less power than active-matrix displays, but they aren't as clear and bright and can leave "ghosts" when the display changes quickly. Passive-matrix displays go by the abbreviations HPA, STN, or DSTN.*

pen-based computer system (p. 262) Input system that allows users to enter handwriting and marks onto a computer screen by means of a penlike stylus rather than by typing on a keyboard. Pen computers use handwriting-recognition software that translates handwritten characters made by the stylus into data that is usable by the computer. Why it's important: *Many handheld computers and PDAs have pen input, as do digital notebooks.*

pixel (p. 278) Short for "picture element"; the smallest unit on the screen that can be turned on and off or made different shades. Why it's important: *Pixels are the building blocks that allow text and graphical images to be displayed on a screen.*

plotter (p. 287) Specialized output device designed to produce high-quality graphics in a variety of colors. The inkjet plotter employs the same principle as an inkjet printer; the paper is output over a drum, enabling continuous output. In an electrostatic plotter, paper lies partially flat on a tablelike surface, and toner is used in a photocopier-like manner. Why it's important: *Plotters are used to create hardcopy items such as maps, architectural drawings, and three-dimensional illustrations, which are usually too large for regular printers.*

pointing device (p. 258) Hardware that controls the position of the cursor or pointer on the screen. It includes the mouse and its variants, the touch screen, and various forms of pen input. Why it's important: *In many contexts, pointing devices permit quick and convenient data input.*

pointing stick (p. 261) Pointing device that looks like a pencil eraser protruding from the keyboard between the G, H, and B keys. The user moves the pointing stick with a forefinger. Why it's important: *Pointing sticks are used principally in video games, in computer-aided design systems, and in robots.*

printer (p. 282) Output device that prints characters, symbols, and perhaps graphics on paper or another hardcopy medium. Why it's important: *Printers provide one of the principal forms of computer output.*

QXGA (quantum extended graphics array) (p. 282) New, expensive display standard with a resolution of up to 2,048 × 1,536 pixels. Why it's important: *QXGA is used for large LCD screens for computer users needing to view extreme detail, for businesspeople needing to enlarge images that will still be crisp in big-screen presentations, for high-density television, and for special applications involving viewing multiple images on a single screen.*

radio-frequency identification (RFID) tags (p. 276) Source data-entry technology based on an identifying tag bearing a microchip that contains specific code numbers. These code numbers are read by the radio waves of a scanner linked to a database. Why it's important: *Drivers with RFID tags can breeze through tollbooths without having to even roll down their windows; the toll is automatically charged to their accounts. Radio-wave-readable ID tags are also used by the Postal Service to monitor the flow of mail, by stores for inventory control and warehousing, and in the railroad industry to keep track of rail cars.*

refresh rate (p. 279) Number of times per second that screen pixels are recharged so that their glow remains bright. In general, displays are refreshed 56–120 times per second. Why it's important: *The higher the refresh rate, the more solid the image looks on the screen—that is, the less it flickers.*

repetitive stress (or strain) injuries (RSIs) (p. 289) Several wrist, hand, arm, and neck injuries resulting when muscle groups are forced through fast, repetitive motions. They include muscle strain and tendinitis, which are painful but usually not crippling, and carpal tunnel syndrome, which is disabling and often requires surgery. Why it's important: *People who use computer keyboards account for some of the RSI cases that result in lost work time.*

resolution (p. 266, 278) Clarity or sharpness of display-screen/scanned/printed images; the more pixels (dots) there are per square inch, the finer the level of detail attained. Resolution is expressed in terms of the formula horizontal pixels × vertical pixels. Each pixel can be assigned a color or a particular shade of gray. Standard screen resolutions are 640 × 480, 800 × 600, 1,024 × 768, 1,280 × 1,024, and 1,600 × 1,200 pixels. Scanner resolutions are 300 × 200, 600 × 600, 600 × 1,200, 1,200 × 1,200, and 1,200 × 2,400. Why it's important: *Users need to know what resolution is appropriate for their purposes.*

scanner (p. 265) Source data-input device that uses light-sensing (optical) equipment to translate images of text, drawings, photos, and the like into digital form. Why it's important: *Scanners simplify the input of complex data. The images can be processed by a computer, displayed on a monitor, stored on a storage device, or communicated to another computer.*

sensor (p. 275) Input device that collects specific data directly from the environment and transmits it to a computer. Why it's important: *Although you are unlikely to see such input devices connected to a PC in an office, they exist all around us, often in nearly invisible form. Sensors can be used to detect all kinds of things: speed, movement, weight, pressure, temperature, humidity, wind, current, fog, gas, smoke, light, shapes, images, and so on. In aviation, for example, sensors are used to detect ice buildup on airplane wings and to alert pilots to sudden changes in wind direction.*

softcopy (p. 277) Data on a display screen or in audio or voice form. This kind of output is not tangible; it cannot be touched. Why it's important: *This term is used to distinguish nonprinted output from printed (hardcopy) output.*

sound board (p. 270) An add-on circuit board in a computer that converts analog sound to digital sound and stores it for further processing and/or plays it back, providing output directly to speakers or an external amplifier. Why it's important: *The sound board enables users to work with audible sound.*

sound-output device (p. 288) Hardware that produces digitized sounds, ranging from beeps and chirps to music. Why it's important: *To use sound output, the user needs appropriate software and a sound card. Such devices are used to produce the sound effects when the user plays a CD-ROM, for example.*

source data-entry devices (p. 265) Data-entry devices that create machine-readable data on magnetic media or paper or feed it directly into the computer's processor, without the use of a keyboard. Categories include scanning devices (imaging systems, bar-code readers, mark- and character-recognition devices, and fax machines), audio-input devices, video input, photographic input (digital cameras), voice-recognition systems, sensors, radio-frequency identification devices, and human-biology-input devices. Why it's important: *Source data-entry devices lessen reliance on keyboards for data entry and can make data entry more accurate.*

speech-recognition system (p. 274) Input system that uses a microphone (or a telephone) as an input device and converts a person's speech into digital signals by comparing the electrical patterns produced by the speaker's voice with a set of prerecorded patterns stored in the computer. Why it's important: *Voice-recognition technology is useful in situations where people are unable to use their hands to input data or need their hands free for other purposes.*

SVGA (super video graphics array) (p. 282) Graphics board standard that supports a resolution of 800 × 600 pixels, or variations, producing 16 million possible simultaneous colors. Why it's important: *SVGA is the most common standard used today with 15-inch monitors.*

SXGA (super extended graphics array) (p. 282) Graphics board standard that supports a resolution of 1,280 × 1,024 pixels. Why it's important: *SXGA is often used with 19- and 21-inch monitors.*

thermal printer (p. 285) Printer that uses colored waxes and heat to produce images by burning dots onto special paper. The colored wax sheets are not required for black-and-white output. Thermal printers are expensive, and they require expensive paper. Why it's important: *For people who want the highest-quality color printing available with a desktop printer, thermal printers are the answer.*

touch screen (p. 261) Video display screen that has been sensitized to receive input from the touch of a finger. The screen is covered with a plastic layer, behind which are invisible beams of infrared light. Why it's important: *Users can input requests for information by pressing on buttons or menus displayed. The answers to requests are displayed as output in words or pictures on the screen. (There may also be sound.) Touch screens are found in kiosks, ATMs, airport tourist directories, hotel TV screens (for guest checkout), and campus information kiosks making available everything from lists of coming events to (with proper ID and personal code) student financial-aid records and grades.*

touchpad (p. 261) Input device; a small, flat surface over which the user slides a finger, using the same movements as those used with a mouse. The cursor follows the movement of the finger. The user "clicks" by tapping a finger on the pad's surface or by pressing buttons positioned close by the pad. Why it's important: *Touchpads let users control the cursor/pointer with a finger, and they require very little space to use. Most laptops have touchpads.*

trackball (p. 260) Movable ball, mounted on top of a stationary device, that can be rotated by the user's fingers or palm. It looks like the mouse turned upside down. Instead of moving the mouse around on the desktop, you move the trackball with the tips of your fingers. Why it's important: *Trackballs require less space to use than does a mouse.*

UXGA (ultra extended graphics array) (p. 282) Graphics board standard that supports a resolution of 1,600 × 1,200 pixels, producing up to 16.8 million colors. Why it's important: *UXGA is popular with graphic artists, engineering designers, and others using 21-inch monitors.*

video (p. 288) Output consisting of photographic images played at 15–29 frames per second to give the appearance of full motion. Why it's important: *Video is input into a multimedia system using a video camera or VCR and, after editing, is output on a computer's display screen. Because video files can require a great deal of storage—a 3-minute video may require 1 gigabyte of storage—video is often compressed. Digital video has revolutionized the movie industry, as in the use of special effects.*

videoconferencing (p. 288) Form of video output in which people in different geographic locations can have a meeting—can see and hear one another—using computers and communications. Why it's important: *Many organizations use videoconferencing to take the place of face-to-face meetings. Videoconferencing systems range from videophones to group conference rooms with cameras and multimedia equipment to desktop systems with small video cameras, microphones, and speakers.*

voice-output device (p. 288) Hardware that converts digital data into speechlike sounds. Why it's important: *We hear such voice output on telephones ("Please hang up and dial your call again"), in soft-drink machines, in cars, in toys and games, and recently in mapping software for vehicle-navigation devices. For people with physical challenges, computers with voice output help to level the playing field.*

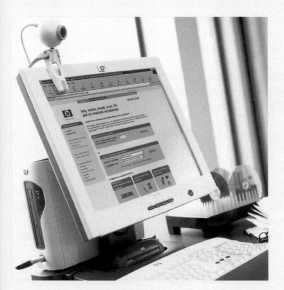

webcam (p. 271) A video camera attached to a computer to record live moving images that can then be posted on a website in real time. Why it's important: *The webcam is an affordable tool that enables users to have videoconferencing capabilities and may change the future of communications.*

XGA (extended graphics array) (p. 282) Graphics board display standard with a resolution of up to 1,024 × 768 pixels, corresponding to 65,536 possible colors. Why it's important: *XGA is used mainly for simple applications plus spreadsheets and graphics software.*

Chapter Review

"I can recognize and recall information."

Self-Test Questions

1. A(n) _____ terminal is entirely dependent for all its processing activities on the computer system to which it is connected.

2. The two main categories of printer are _____ and _____.

3. _____ is the study of the physical relationships between people and their work environment.

4. A(n) _____ is an input device that is rolled about on a desktop and directs a pointer on the computer's display screen.

5. _____ consists of devices that translate information processed by the computer into a form that humans can understand.

6. _____ is the science of measuring individual body characteristics.

7. CRT is short for _____.

8. LCD is short for _____.

9. A _____ is software that describes the shape and position of characters and graphics to the printer.

10. When people in different geographic locations can have a meeting using computers and communications, it is called _____.

11. _____-matrix screens are much brighter and sharper than _____-matrix screens, but they are more complicated and thus more expensive.

12. A debilitating condition caused by pressure on the median nerve in the wrist, producing damage and pain to nerves and tendons in the hands, is called _____.

13. The measure of the number of dots that are printed in a linear inch is called _____ or _____.

14. A printer that forms characters or images by striking a mechanism such as a print hammer or wheel against an inked ribbon, leaving images on a paper, is called a(n) _____ printer.

15. _____ printers enabled the development of desktop publishing.

Multiple-Choice Questions

1. Which of the following is *not* a pointing device?
   a. mouse
   b. touchpad
   c. keyboard
   d. joystick

2. Which of the following is not a source data-entry device?
   a. bar-code reader
   b. sensor
   c. digital camera
   d. scanner
   e. mouse

3. Which of the following display standards has the highest screen resolution?
   a. XGA
   b. UXGA
   c. VGA
   d. SVGA
   e. QXGA

4. Which of the following *isn't* considered hardcopy output?
   a. spreadsheet printout
   b. microfilm
   c. fax report
   d. Word document computer file
   e. printed invoice

5. Which of the following factors does not affect the quality of a screen display?
   a. refresh rate
   b. speed
   c. resolution
   d. pixels
   e. color depth

True/False Questions

T F 1. On a computer screen, the more pixels that appear per square inch, the higher the resolution.

T F 2. Photos taken with a digital camera can be downloaded to a computer's hard disk.

T F 3. Resolution is the amount of space between the centers of adjacent pixels.

T F 4. The abbreviation *dpi* stands for "dense pixel intervals."

T F 5. Pointing devices control the position of the cursor on the screen.

T F 6. Output hardware consists of devices that translate information processed by the computer into a form that humans can understand.

T F 7. Scanners use a driver called TWAIN.

T F 8. Optical character-recognition software reads "bubble" marks and converts them into computer-usable form.

T F 9. The lower the refresh rate, the more solid the image looks on the screen.

T F 10. CRTs consume more power than flat-panel displays do.

T F 11. Computer users have no need to be concerned about ergonomics.

T F 12. It has been proven that electromagnetic fields pose no danger to human beings.

T F 13. Plotters are used to print architectural drawings and in computer-aided design.

 stage **LEARNING** COMPREHENSION

"I can recall information in my own terms and explain it to a friend."

Short-Answer Questions

1. What is a common use of dumb terminals?

2. What characteristics determine the clarity of a computer screen?

3. Describe two situations in which scanning is useful.

4. What is source data entry?

5. What is *pixel* short for? What is a pixel?

6. Briefly describe RSI and CTS. Why are they problems?

7. What is a font?

8. Discuss the different types of printers and their features.

9. Explain the differences between CRT monitors and LCD monitors.

10. What is OCR used for?

 stage **LEARNING** APPLYING, ANALYZING, SYNTHESIZING, EVALUATING

"I can apply what I've learned, relate these ideas to other concepts, build on other knowledge, and use all these thinking skills to form a judgment."

Knowledge in Action

1. Cut out several advertisements from newspapers or magazines that feature new microcomputer systems. Circle all the terms that are familiar to you now that you have read the first five chapters of this text. Define these terms on a separate sheet of paper. Is this computer expandable? How much does it cost? Is a monitor included in the price? A printer?

2. *Paperless office* is a term that has been around for some time. However, the paperless office has not yet been achieved. Do you think the paperless office is a good idea? Do you think it's possible? Why do you think it has not yet been achieved?

3. Many PC warranties do not cover protection against lightning damage, which is thought to be an "act of God." Does your PC warranty provide coverage for "acts of God"? Read it to find out.

4. Compare and contrast the pros and cons of different types of monitors. Decide which one is best for you and explain why. Do some research on how each monitor type creates displayed images.

5. Do you have access to a computer with (a) speech-recognition software and (b) word processing software that determines writing level (such as eighth grade, ninth grade, and so on)? Dictate a few sentences about your day into the microphone. After your speech is encoded into text, use the word processing software to determine the grade level of your everyday speech.

6. A pixel is the smallest unit on the screen that can be turned on and off. In most high-quality digital photos, you can't see the pixilation unless you zoom in real close. However, even when you don't zoom in, you know that the pixilation is there, a series of different pixels all plotted on a grid. How can you relate this to our experience of reality? Via high-tech microscopes we see that everything is made of smaller particles not visible to the naked eye, such as atoms, subatomic particles, and quarks. How are the basic building blocks of computer imaging and the basic building blocks of physical matter alike, and how are they different?

7. Biometrics: Which form of biometric technology do you prefer for identification purposes: fingerprints, voice intonation, facial characteristics, or retinal identification? Which do you think will become most commonly used in the future?

Web Exercises

1. Visit an online shopping site such as *www.yahoo.com.* Click on *Shopping;* then type *Printers* in the Shopping search box. Investigate five different types of printers by clicking on the printer names and then on Full Specifications. Note (a) the type of printer, (b) its price, (c) its resolution, and (d) its speeds for black-and-white printing and for color printing, if applicable. Which operating system is each printer compatible with? Which printer would you choose? Why?

2. There is an abundance of information about electronic devices on the internet. People write all kinds of reviews either raving about the device that made their lives better or lamenting the device that became their evil nemesis. GO to *www.consumerreports.org/main/detailv3.jsp?CONTENT%3C%3Ecnt_id=29307&FOLDER%3C%3Efolder_id=333133&ASSORTMENT%3C%3East_id=333133&bmUID=1126478749294* and find out how people at ConsumerReports.org rate some of your favorite electronic devices.

3. Go to *www.infobeagle.com/computers/touchscreens.htm* and find a touch screen that appeals to you.

4. Concerned about electromagnetic radiation? For a question-and-answer session on electromagnetic frequencies, go to *http://vitatech.net/q_a.html*.

5. Do you see any ethical problems involved with self-scanning checkout? Can people cheat the system? Are store jobs being lost to automation? Are people without credit cards and/or computer experience being excluded? Do a keyword search for *self-scanning* and *self-checkout* and other terms related to these issues. Do you think self-scanning is a good idea?

6. Research the development of Smart Labels. Visit these websites for more information on radio-frequency tagging:

   www.idtechex.com/conference.html
   www.idtechex.com/sla.html

   http://electronics.howstuffworks.com/smart-label.htm
   www.copytag.com/
   www.ti.com/tiris/docs/news/news_releases/90s/rel02-25-99.shtml

7. The human cyborg: Visit Professor Kevin Warwick's website to learn about the implant microchips he has been creating and surgically implanting in his body to allow it to communicate with a computer. Investigate the many applications he has been working on. After visiting his site, run a search on *"Kevin Warwick"* to read what others have to say about him and his ideas.

   www.kevinwarwick.com/
   www.cyber.rdg.ac.uk/people/K.Warwick.htm
   www.wired.com/wired/archive/8.02/warwick.html
   (Kevin Warwick outlines his plan to become one with his computer)

8. PostScript is the most important page description language in desktop publishing. Read the information at Answer.com *www.answers.com/topic/postscript-1?hl=postscript&hl=printing* and find out why. You almost certainly will need some familiarity with PostScript.

9. What is the difference between a screen font and a printer font? Go to Answers.com and Wikipedia to find out. What does *WYSIWYG* mean?

10. How many images can the newest digital cameras hold? DO a web search and find out.

Communications, Networks, & Safeguards

The Wired & Wireless World

Chapter Topics & Key Questions

6.1   **From the Analog to the Digital Age** How do digital data and analog data differ, and what does a modem do?

6.2   **Networks** What are the benefits of networks, and what are their types, components, and variations?

6.3   **Wired Communications Media** What are types of wired communications media?

6.4   **Wireless Communications Media** What are types of wireless communications media, both long distance and short distance?

6.5   **Cyberthreats, Hackers, & Safeguards** What are areas I should be concerned about for keeping my computer system secure?

6.6   **The Future of Communications** What are the characteristics of the next generation of wired and wireless communications?

The essence of all revolution, stated philosopher Hannah Arendt, is the start of a *new story* in human experience.

Before the 1950s, computing devices processed data into information, and communications devices communicated information over distances. The two streams of technology developed pretty much independently, like rails on a railroad track that never merge. Now we have a new story, a revolution.

For us, the new story has been *digital convergence*—the gradual merger of computing and communications into a new information environment, in which the *same information is exchanged among many kinds of equipment, using the language of computers*. (● *See Panel 6.1.*) At the same time, there has been a convergence of several important industries—computers, telecommunications, consumer electronics, entertainment, mass media—producing new electronic products that perform multiple functions.

● PANEL 6.1

Digital convergence—the fusion of computer and communications technologies
Today's new information environment came about gradually from the merger of two separate streams of technological development—computers and communications.

Computer Technology

1621 CE	1642	1833	1843		1890
Slide rule invented (Edmund Gunther)	First mechanical adding machine (Blaise Pascal)	Babbage's difference engine (automatic calculator)	World's first computer programmer, Ada Lovelace, publishes her notes		Electricity used for first time in a data-processing project (punched cards); Hollerith's automatic census-tabulating machine (used punched cards)

Communications Technology

1562	1594	1639	1827	1835	1846	1857	1876	1888	1894
First monthly newspaper (Italy)	First magazine (Germany)	First printing press in North America	Photographs on metal plates	Telegraph (first long-distance digital communication system)	High-speed printing	Trans-atlantic telegraph cable laid	Telephone invented	Radio waves identified	Edison makes a movie

6.1

From the Analog to the Digital Age

How do digital data and analog data differ, and what does a modem do?

Why have the worlds of computers and of telecommunications been so long in coming together? Because *computers are digital, but most of the world has been analog.* Let's take a look at what this means. We will elaborate on two subjects we introduced earlier—digital signals and modems.

The Digital Basis of Computers: Electrical Signals as Discontinuous Bursts

What does "digital" mean?

Computers may seem like incredibly complicated devices but, as we've seen, their underlying principle is simple. Because they are based on on/off electrical states, they use the *binary system,* which consists of only two digits— 0 and 1. At their most basic level computers can distinguish between just these two values, 0 and 1, or off and on. There is no simple way to represent all the values in between, such as 0.25. All data that a computer processes must be encoded digitally, as a series of 0s and 1s.

In general, *digital* means "computer-based." Specifically, **_digital_ describes any system based on discontinuous data or events; in the case of computers, it refers to communications signals or information represented in a two-state (binary) way using electronic or electromagnetic signals. Each 0 and 1 signal represents a** *bit.*

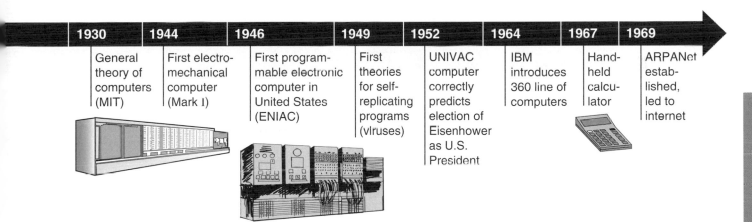

1930	1944	1946	1949	1952	1964	1967	1969
General theory of computers (MIT)	First electro-mechanical computer (Mark I)	First program-mable electronic computer in United States (ENIAC)	First theories for self-replicating programs (viruses)	UNIVAC computer correctly predicts election of Eisenhower as U.S. President	IBM introduces 360 line of computers	Hand-held calcu-lator	ARPANet estab-lished, led to internet

1895	1907	1912	1915	1928	1939	1946	1947	1948	1950
Marconi develops radio; motion-picture camera invented	First regular radio broadcast from New York	Motion pictures become a big business	AT&T long-distance service reaches San Francisco	First TV demonstrated; first sound movie	Commercial TV broad-casting	Color TV demon-strated	Transistor invented	Reel-to-reel tape recorder	Cable TV

Communications, Networks, & Safeguards

311

The Analog Basis of Life: Electrical Signals as Continuous Waves

What does "analog" mean?

"The shades of a sunset, the flight of a bird, or the voice of a singer would seem to defy the black or white simplicity of binary representation," points out one writer.[1] Indeed, these and most other phenomena of the world are **_analog_, continuously varying in strength and/or quality—fluctuating, evolving, or continually changing.** Sound, light, temperature, and pressure values, for instance, can be anywhere on a continuum or range. The highs, lows, and in-between states have historically been represented with analog devices rather than in digital form. Examples of analog devices are a speedometer, a thermometer, and a tire-pressure gauge, all of which can measure continuous fluctuations.

Humans experience most of the world in analog form—our vision, for instance, perceives shapes and colors as smooth gradations. But most analog events can be simulated digitally. A newspaper photograph, viewed through a magnifying glass, is made up of an array of dots—so small that most newspaper readers see the tones of the photograph as continuous (that is, analog).

Traditionally, electronic transmission of telephone, radio, television, and cable-TV signals has been analog. The electrical signals on a telephone line, for instance, have been analog-data representations of the original voices, transmitted in the shape of a wave (called a *carrier wave*). Why bother to change analog signals into digital ones, especially since the digital representations are only *approximations* of analog events? The reason is that *digital signals are easier to store and manipulate electronically.*

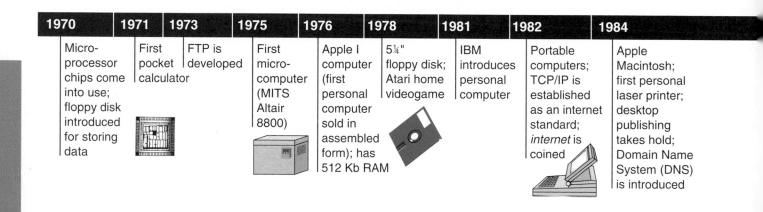

1970	1971	1973	1975	1976	1978	1981	1982	1984
Microprocessor chips come into use; floppy disk introduced for storing data	First pocket calculator	FTP is developed	First microcomputer (MITS Altair 8800)	Apple I computer (first personal computer sold in assembled form); has 512 Kb RAM	5¼" floppy disk; Atari home videogame	IBM introduces personal computer	Portable computers; TCP/IP is established as an internet standard; *internet* is coined	Apple Macintosh; first personal laser printer; desktop publishing takes hold; Domain Name System (DNS) is introduced

1952	1957	1961	1961–1968	1968	1975	1977	1979	1981	1982
Direct-distance dialing (no need to go through operator); transistor radio introduced	First satellite launched (Russia's Sputnik)	Push-button telephones	Packet-switching networks developed	Portable video recorders; video cassettes	Flat-screen TV; GPS	First inter-active cable TV	3-D TV demonstrated	First viruses appear "in the wild" (public domain)	Compact disks; European consortium launches multiple communications satellites

Purpose of the Dial-Up Modem: Converting Digital Signals to Analog Signals & Back

How does a telephone modem change analog to digital signals and the reverse?

Michelle Philips, an Indianapolis real-estate agent, used to drive to her office at odd hours just to check her email messages and search the web on her company's high-speed internet lines because her dial-up connection at home was too slow. "At home, I can do laundry, take a shower, and wash dishes while the computer is logging on to the internet," she said with a laugh.[2] No wonder more Americans are switching from the slower dial-up means of access to high-speed connections. Nevertheless, let's stick with the telephone modem for a moment to help illustrate the differences between digital and analog transmission.

Consider a graphic representation of an on/off digital signal emitted from a computer. Like a regular light switch, this signal has only two states—on and off. Compare this with a graphic representation of a wavy analog signal emitted as a signal. The changes in this signal are gradual, as in a dimmer switch, which gradually increases or decreases brightness.

Because telephone lines have traditionally been analog, you need to have a dial-up modem if your computer is to send communications signals over a telephone line. As we've seen, the modem translates the computer's digital signals into the telephone line's analog signals. The receiving computer also needs a modem to translate the analog signals back into digital signals. (● *See Panel 6.2 on the next page.*)

How, in fact, does a modem convert the continuous analog wave to a discontinuous digital pulse that can represent 0s and 1s? The modem can make adjustments to the frequency—the number of cycles per second, or the number of times a wave repeats during a specific time interval (the fastness/slowness). Or it can make adjustments to the analog signal's amplitude—the

Computer Technology

1993	1994	1997	2000	2001	2003	2005	2047?
Multimedia desktop computers; personal digital assistants	Apple and IBM introduce PCs with full-motion video built in; wireless data transmission for small portable computers; web browser Mosaic invented	Network computers; Pathfinder robot lands on Mars	Microsoft .NET announced; BlackBerry	Windows XP; Mac OS X; MP3	Mac G5; i Pod	Mac mini; Apple video iPod	By this date, some experts predict, all electronically encodable information will be in cyberspace

Communications Technology

1985	1990	1991	1994	1996	1997	2000	2001	2005
Cellular phone; Nintendo	IRS accepts electronically filed tax returns	CD-ROM games (Sega)	FCC selects HDTV standard	WebTV	Internet telephone-to-telephone service	Napster popular; 3.8% of music sales online	2.5 G wireless services; Wi-Fi and Bluetooth	3 G wireless services

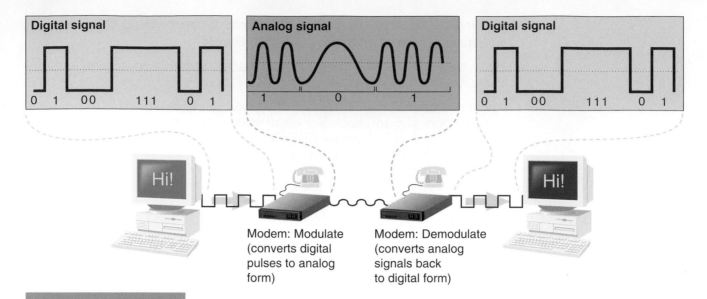

Digital signal	Analog signal	Digital signal
0 1 00 111 0 1	1 0 1	0 1 00 111 0 1

Modem: Modulate (converts digital pulses to analog form)

Modem: Demodulate (converts analog signals back to digital form)

● PANEL 6.2

Analog versus digital signals, and the modem

Note that an analog signal represents a continuous electrical signal in the form of a wave. A digital signal is discontinuous, expressed as discrete bursts in on/off electrical pulses.

height of the wave (the loudness/softness). Thus, in frequency, a slow wave might represent a 0 and a quick wave might represent a 1. In amplitude, a low wave might represent a 0 and a high wave might represent a 1. (● *See Panel 6.3, opposite.*)

Modem **is short for** *"modulate/demodulate"*; **a sending modem modulates digital signals into analog signals for transmission over phone lines. A receiving modem demodulates the analog signals back into digital signals.** The modem provides a means for computers to communicate with one another using the standard copper-wire telephone network, an analog system that was built to transmit the human voice but not computer signals.

Our concern, however, goes far beyond telephone transmission. How can the analog realities of the world be expressed in digital form? How can light, sounds, colors, temperatures, and other dynamic values be represented so that they can be manipulated by a computer? Let us consider this.

Converting Reality to Digital Form

How is sampling used to express analog reality in digital form?

Suppose you are using an analog tape recorder to record a singer during a performance. The analog wave from the microphone, which is recorded onto the tape as an analog wave as well, will produce a near duplicate of the sounds—including distortions, such as buzzings and clicks, or electronic hums if an amplified guitar is used.

The digital recording process is different. The way that music is captured for digital audio CDs, for example, does not provide a duplicate of a musical performance. Rather, the digital process uses a device (called an *analog-to-digital converter*) to record *representative selections,* or *samples,* of the sounds and convert the analog waves into a stream of numbers that the computer then uses to express the sounds. To play back the music, the stream of numbers is converted (by a *digital-to-analog converter*) back into an analog wave. The samples of sounds are taken at regular intervals— nearly 44,100 times a second—and the copy obtained is virtually exact and free from distortion and noise. The sampling rate of 44,100 times per second and the high precision fool our ears into hearing a smooth, continuous sound.

Digital photography also uses sampling: A computer takes samples of values such as brightness and color. The same is true of other aspects of real-life experience, such as pressure, temperature, and motion.

● PANEL 6.3
How analog waves
are modified to
resemble digital
pulses

The continuous, even
cycle of an analog wave...

Frequency

Amplitude

OR

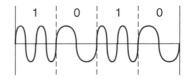

... is converted to digital form through *frequency modulation*—the frequency of the cycle increases to represent a 1 and stays the same to represent a 0.

1 0 1 0

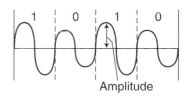

... or is converted to digital form through *amplitude modulation*—the height of the wave is increased to represent a 1 and stays the same to represent a 0.

1 0 1 0

Amplitude

info!

Sampling

For detailed information and diagrams on sampling, go to http://electronics.howstuffworks.com/analog-digital3.htm.

Does digital sampling cheat us out of our experience of "reality" by allowing computers to sample sounds, images, and so on? Actually, people willingly made this compromise years ago, before computers were invented. Movies, for instance, carve up reality into 24 frames a second. Television pictures are drawn at 30 frames per second. These processes happen so quickly that our eyes and brains easily jump the visual gaps. Digital processing of analog experience is just one more way of expressing or translating reality.

Turning analog reality into digital form provides tremendous opportunities. One of the most important is that *all kinds of multimedia can now be changed into digital form and transmitted as data to all kinds of devices.*

Now let us examine the digital world of telecommunications. We begin with the subject of networks and then discuss how networks are connected—first by wired means, then by wireless means.

6.2

Networks

What are the benefits of networks, and what are their types, components, and variations?

More and more people are now designing their homes to accommodate networks. For instance, when Lisa Guernsey and her husband, Rob Krupicka, were planning their new house, they wanted modern conveniences without the modern-day headaches. "We didn't want to face another nest of tangled cables or to see speaker wires snaking beneath the rugs," Guernsey said. "What Rob and I wanted was . . . a so-called networked home with 'digital plumbing' for internet and television connections hidden behind walls, and just enough equipment to make tapping into video and audio as easy as filling a glass of water."[3]

Whether wired or wireless or both, **a *network,* or *communications network,* is a system of interconnected computers, telephones, or other communications devices that can communicate with one another and share applications and data.** The tying together of so many communications devices in so many ways is changing the world we live in.

Internet cafe in Laos

The Benefits of Networks

What are five ways I might benefit from networks?

People and organizations use networks for the following reasons, the most important of which is the sharing of resources.

SHARING OF PERIPHERAL DEVICES Peripheral devices such as laser printers, disk drives, and scanners can be expensive. Consequently, to justify their purchase, management wants to maximize their use. Usually the best way to do this is to connect the peripheral to a network serving several computer users.

SHARING OF PROGRAMS & DATA In most organizations, people use the same software and need access to the same information. It is less expensive for a company to buy one word processing program that serves many employees than to buy a separate word processing program for each employee.

Moreover, if all employees have access to the same data on a shared storage device, the organization can save money and avoid serious problems. If each employee has a separate machine, some employees may update customer addresses while others remain ignorant of the changes. Updating information on a shared server is much easier than updating every user's individual system.

Finally, network-linked employees can more easily work together online on shared projects.

BETTER COMMUNICATIONS One of the greatest features of networks is electronic mail. With email, everyone on a network can easily keep others posted about important information.

SECURITY OF INFORMATION Before networks became commonplace, an individual employee might have been the only one with a particular piece of information, which was stored in his or her desktop computer. If the employee was dismissed—or if a fire or flood demolished the office—the company would lose that information. Today such data would be backed up or duplicated on a networked storage device shared by others.

ACCESS TO DATABASES Networks enable users to tap into numerous databases, whether private company databases or public databases available online through the internet.

Types of Networks: WANs, MANs, LANs, HANs, PANs, & Others

How do the sizes of networks differ?

Networks, which consist of various combinations of computers, storage devices, and communications devices, may be divided into several main categories, differing primarily in their geographic range and purposes.

WIDE AREA NETWORK A _wide area network (WAN)_ **is a communications network that covers a wide geographic area, such as a country or the world.** Most long-distance and regional telephone companies are WANs. A WAN may use a combination of satellites, fiber-optic cable, microwave, and copper-wire connections and link a variety of computers, from mainframes to terminals. (● _See Panel 6.4._)

WANs are used to connect local area networks (see below) together, so that users and computers in one location can communicate with users and computers in other locations. A wide area network may be privately owned or rented, but the term usually connotes the inclusion of public (shared-user) networks. The best example of a WAN is the internet.

METROPOLITAN AREA NETWORK A _metropolitan area network (MAN)_ **is a communications network covering a city or a suburb.** The purpose of a MAN is often to bypass local telephone companies when accessing long-distance services. Many cellphone systems are MANs.

LOCAL AREA NETWORK A _local area network (LAN),_ or _local net,_ **connects computers and devices in a limited geographic area, such as one office, one building, or a group of buildings close together.** LANs are the basis for most office networks. The LANs of different offices on a university campus may also be linked together into a so-called _campus-area network._

HOME AREA NETWORK A _home area network (HAN)_ **uses wired, cable, or wireless connections to link a household's digital devices**—not only multiple computers, printers, and storage devices but also VCRs, DVDs, televisions, fax machines, videogame machines, and home security systems. (A variant of the HAN is the GAN the _garden area network_—which can be used to link watering systems, outdoor lights, and alarm systems.)

PERSONAL AREA NETWORK Slightly different from a HAN because it doesn't use wires or cables, **a _personal area network (PAN)_ uses short-range wireless technology to connect an individual's personal electronics,** such as cellphone, PDA, MP3 player, notebook PC, and printer. PANs have been made possible with the arrival of such inexpensive, short-range wireless technologies as Bluetooth, ultra wideband, and wireless USB, which have a range of 30 feet or so, as we will describe.

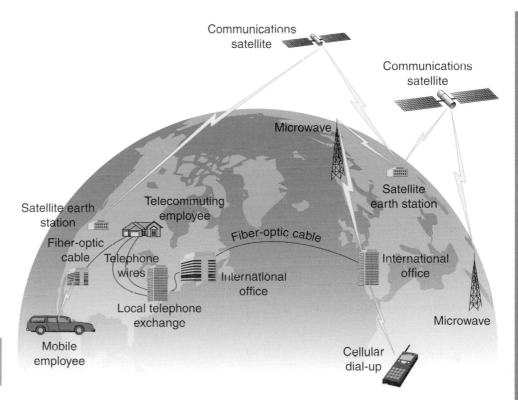

● **PANEL 6.4**
Wide area network

HOME AUTOMATION NETWORK A *home automation network* relies on very inexpensive, very short-range, low-power wireless technology in the under-200-Kbps range to link switches and sensors around the house. Such networks, which use wireless standards such as Insteon, ZigBee, and Z-Wave, as we will describe, run on inexpensive AA batteries and can control lights and switches, thermostats and furnaces, smoke alarms and outdoor floodlights.

How Networks Are Structured: Client/Server & Peer to Peer

What's the difference between client/server and peer-to-peer networks?

Two principal ways in which networks are structured are *client/server* and *peer to peer.* (● See Panel 6.5.)

CLIENT/SERVER NETWORKS A *client/server network* consists of *clients,* which are microcomputers that request data, and *servers,* which are computers used to supply data. The server is a powerful microcomputer that manages shared devices, such as laser printers. It runs server software for applications such as email and web browsing. Different servers may be used to manage different tasks. A *file server* is a computer that acts like a disk drive, storing the programs and data files shared by users on a LAN.

● PANEL 6.5
Two network structures: client/server and peer to peer

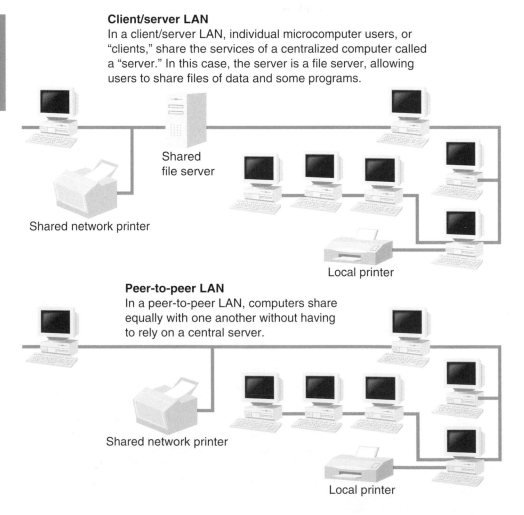

Client/server LAN
In a client/server LAN, individual microcomputer users, or "clients," share the services of a centralized computer called a "server." In this case, the server is a file server, allowing users to share files of data and some programs.

Shared file server

Shared network printer

Local printer

Peer-to-peer LAN
In a peer-to-peer LAN, computers share equally with one another without having to rely on a central server.

Shared network printer

Local printer

A *database server* is a computer in a LAN that stores data but doesn't store programs. A *print server* controls one or more printers and stores the print-image output from all the microcomputers on the system. *Web servers* contain web pages that can be viewed using a browser. *Mail servers* manage email.

PEER-TO-PEER NETWORKS The word *peer* denotes one who is equal in standing with another (as in the phrases "peer pressure" and "jury of one's peers"). **In a _peer-to-peer (P2P) network_, all microcomputers on the network communicate directly with one another without relying on a server.** Every computer can share files and peripherals with all other computers on the network, given that all are granted access privileges. Peer-to-peer networks are less expensive than client/server networks and work effectively for up to 25 computers. Beyond that, they slow down under heavy use. They are appropriate for small networks, such as *home networks.*

HOW THE DIFFERENCES AFFECT ILLEGAL MUSIC DOWNLOADING For record companies losing money to illegal file sharing—the illegal downloading of copyrighted songs by college students and others—the distinction between client/server and peer-to-peer networks has been crucial. Napster, one of the earliest of online file-sharing sites, used a central server to connect users to each other—a client/server network. The music industry sued Napster and shut it down by shutting down the service's central server. (It has since reinvented itself as another kind of business.) Other file-sharing sites—Kazaa, Limewire, Grokster, Gnutella, and Morpheus—bypass the bottleneck of the file server by enabling users to search among peers, scanning all the computers on a peer-to-peer network and then swapping bits. The software that lets users share music files for free over the internet resides on each user's computer, not on a central server, and goes out on the net to find other users to link up to. When the recording industry sued Kazaa, the company's founders sold it to new owners, who quickly broke it up in little pieces, parts of them located offshore of the United States. Now recording companies are going after individual users as well as colleges and other institutions whose networks are being used for illegal purposes. We return to the subject of file sharing in Chapter 9.

Intranets, Extranets, & VPNs

What are the differences among intranets, extranets, and VPNs?

Early in the Online Age, businesses discovered the benefits of using the World Wide Web to get information to customers, suppliers, or investors. For example, in the mid-1990s, FedEx found it could save millions by allowing customers to click through web pages to trace their parcels, instead of having FedEx customer-service agents do it. From there, it was a short step to the application of the same technology inside companies—in internal internet networks called *intranets.*

INTRANETS: FOR INTERNAL USE ONLY **An _intranet_ is an organization's internal private network that uses the infrastructure and standards of the internet and the web.** When a corporation develops a public website, it is making selected information available to consumers and other interested parties. When it creates an intranet, it enables employees to have quicker access to internal information and to share knowledge so that they can do their jobs better. Information exchanged on intranets may include employee email addresses and telephone numbers, product information, sales data, employee benefit information, and lists of jobs available within the organization.

EXTRANETS: FOR CERTAIN OUTSIDERS Taking intranet technology a few steps further, extranets offer security and controlled access. As we have seen, intranets are internal systems, designed to connect the members of a specific group or a single company. By contrast, _**extranets**_ **are private intranets that connect not only internal personnel but also selected suppliers and other strategic parties.** Extranets have become popular for standard transactions such as purchasing. Ford Motor Company, for instance, has an extranet that connects more than 15,000 Ford dealers worldwide. Called FocalPt, the extranet supports sales and servicing of cars, with the aim of improving service to Ford customers.

VIRTUAL PRIVATE NETWORKS Because wide area networks use leased lines, maintaining them can be expensive, especially as distances between offices increase. To decrease communications costs, some companies have established their own _**virtual private networks (VPNs)**_, **private networks that use a public network (usually the internet) to connect remote sites.** Company intranets, extranets, and LANs can all be parts of a VPN.

Components of a Network

What are the various parts of a network?

Regardless of size, networks all have several components in common.

WIRED AND/OR WIRELESS CONNECTIONS Networks use a wired or wireless connection system. Wired connections may be twisted-pair wiring, coaxial cable, or fiber-optic cable, and wireless connections may be infrared, microwave (such as Bluetooth), broadcast radio (such as Wi-Fi), or satellite, as we describe shortly.

HOSTS & NODES A client/server network has a _**host computer**_, **a mainframe or midsize central computer that controls the network.** The other devices on the network are called nodes. **A _node_ is any device that is attached to a network—for example, a microcomputer, terminal, storage device, or printer.**

PACKETS Electronic messages are sent as packets. **A _packet_ is a fixed-length block of data for transmission.** A sending computer breaks an electronic message apart into packets, each of which typically contains 1,000–1,500 bytes. The various packets are sent through a communications network—often using different (and most expedient) routes, at different speeds, and sandwiched in between packets from other messages. Once the packets arrive at their destination, the receiving computer reassembles them into proper sequence to complete the message.

PROTOCOLS A _protocol_, or _communications protocol_, **is a set of conventions governing the exchange of data between hardware and/or software components in a communications network.** Every device connected to a network has an internet protocol (IP) address so that other computers on the network can properly route data to that address. Sending and receiving devices must follow the same set of protocols.

Protocols are built into the hardware or software you are using. The protocol in your communications software, for example, will specify how receiver devices will acknowledge sending devices, a matter called _handshaking_. Handshaking establishes the fact that the circuit is available and operational. It also establishes the level of device compatibility and the speed of transmission. In addition, protocols specify the type of electrical connections used, the timing of message exchanges, and error-detection techniques.

A packet, or electronic message, carries four types of information that will help it get to its destination—namely, the sender's address (the IP), the intended receiver's address, how many packets the message has been broken into, and the number of the individual packet. The packets carry the data in the protocols that the internet uses—that is, TCP/IP.

NETWORK LINKING DEVICES: HUBS, SWITCHES, BRIDGES, GATE-WAYS, ROUTERS, & BACKBONES Networks are often linked together—LANs to MANs and MANs to WANs, for example. The means for connecting them are hubs, switches, bridges, routers, and gateways. (● *See Panel 6.6.*)

- **Hubs:** In general, a hub is the central part of a wheel where the spokes come together. In computer terminology, a **_hub_ is a common connection point for devices in a network—a place of convergence where data arrives from one or more directions and is forwarded out in one or more other directions.** With hubs, bandwidth is shared by

● **PANEL 6.6**
Components of a typical network

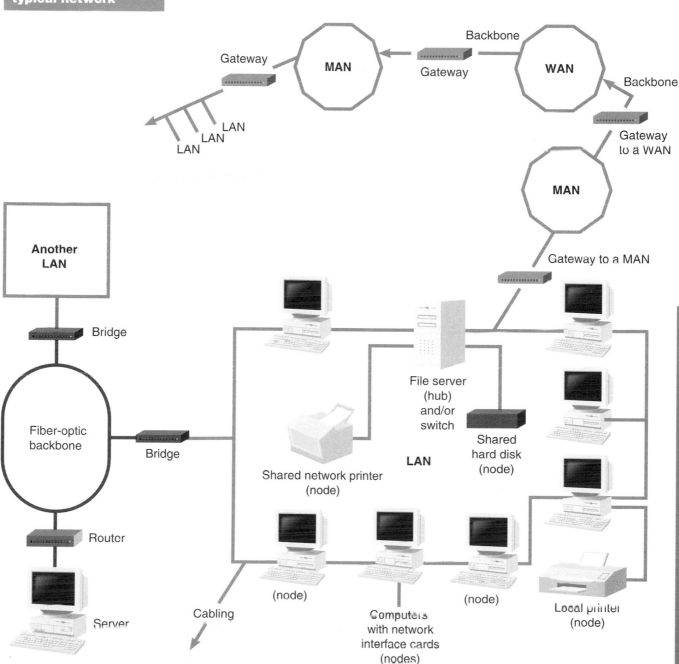

all components. Hubs are commonly used to connect segments of a LAN. A hub contains multiple ports. When a packet arrives at one port, it is copied to the other ports so that all segments of the LAN can see all packets. A hub is a *half-duplex* device, meaning it transmits data in both directions but only in one direction at a time.

- **Switches: A _switch_ is a device that connects computers to a network.** Unlike a hub, it sends messages only to a computer that is the intended recipient. A switch is a *full-duplex* device, meaning data is transmitted back and forth at the same time, which improves the performance of the network. Switches allow each component full use of the bandwidth. Switches are used only in certain configurations; hubs and switches can be used in combination.

- **Bridges: A _bridge_ is an interface used to connect the same types of networks.** For instance, similar local area networks can be joined together to create larger area networks.

- **Gateways: A _gateway_ is an interface permitting communication between dissimilar networks**—for instance, between a LAN and a WAN or between two LANs based on different network operating systems or different layouts. Gateways can be hardware, software, or a combination of the two.

- **Routers: A _router_ is a special computer that directs communicating messages when several networks are connected together.** High-speed routers can serve as part of the internet backbone, or transmission path, handling the major data traffic.

- **Backbones: The _backbone_ consists of the main highway—including gateways, routers, and other communications equipment—that connects all computer networks in an organization.** People frequently talk about the *internet backbone,* the central structure that connects all other elements of the internet. As we discussed in Chapter 2, several commercial companies provide these major high-speed links across the country; these backbones are connected at network access points (NAPs).

NETWORK INTERFACE CARDS   As we stated in Chapter 4, a *network interface card (NIC)* enables the computer to send and receive messages over a cable network. The network card can be inserted into an expansion slot in a microcomputer. Alternatively, a network card in a stand-alone box may serve a number of devices. New computers often come with network cards already installed.

NETWORK OPERATING SYSTEM   The *network operating system (NOS)* is the system software that manages the activity of a network. The NOS supports access by multiple users and provides for recognition of users based on passwords and terminal identifications. Depending on whether the LAN is client/server or peer-to-peer, the operating system may be stored on the file server, on each microcomputer on the network, or on a combination of both.

Examples of popular NOS software are Novell NetWare, Microsoft Windows NT/2000, Unix, and Linux. Peer-to-peer networking can also be accomplished with Microsoft Windows 95/98/Me/XP and Microsoft Windows for Workgroups.

Network Topologies: Bus, Ring, & Star

What are three popular configurations for networks?

Networks can be laid out in different ways. **The logical layout, or shape, of a network is called a _topology_.** The three basic topologies, or configurations, are *bus, ring,* and *star.*

more info!

Backbones

Some large companies that provide backbone connectivity are MCI, UUnet, British Telecom, AT&T, and Teleglobe. Scientists in the U.S. invented the internet, and the computers that oversee the network are still controlled by the U.S. Department of Commerce. Some people believe that the internet should be put under international control through the United Nations. What do you think?

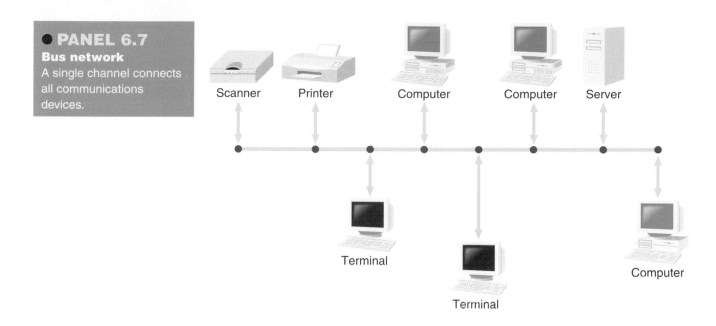

● PANEL 6.7
Bus network
A single channel connects all communications devices.

Scanner   Printer   Computer   Computer   Server

Terminal

Terminal

Computer

BUS NETWORK The bus network works like a bus system at rush hour, with various buses pausing in different bus zones to pick up passengers. In a bus network, all communications devices are connected to a common channel. (● *See Panel 6.7.*) That is, **in a _bus network_, all nodes are connected to a single wire or cable, the *bus*, which has two endpoints. Each communications device on the network transmits electronic messages to other devices.** If some of those messages collide, the sending device waits and tries to transmit again.

The advantage of a bus network is that it may be organized as a client/server or peer-to-peer network. The disadvantage is that extra circuitry and software are needed to avoid collisions between data. Also, if a connection in the bus is broken—as when someone moves a desk and knocks the connection out—the entire network may stop working.

● PANEL 6.8
Ring network
This arrangement connects the network's devices in a closed loop.

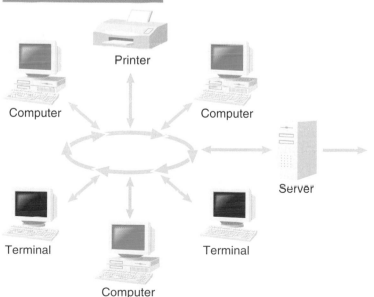

Computer   Printer   Computer

Terminal   Terminal   Server

Computer

RING NETWORK **A _ring network_ is one in which all microcomputers and other communications devices are connected in a continuous loop.** (● *See Panel 6.8.*) There are no endpoints.

Electronic messages are passed around the ring until they reach the right destination. There is no central server. An example of a ring network is IBM's Token Ring Network, in which a bit pattern (called a "token") determines which user on the network can send information, as we'll discuss.

The advantage of a ring network is that messages flow in only one direction. Thus, there is no danger of collisions. The disadvantage is that if a connection is broken, the entire network stops working.

Communications, Networks, & Safeguards

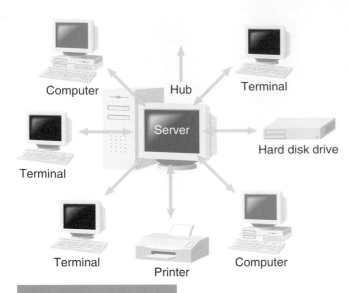

Computer　Hub　Terminal

Server

Hard disk drive

Terminal

Terminal　Printer　Computer

STAR NETWORK   A *star network* is one in which all microcomputers and other communications devices are directly connected to a central server. (● *See Panel 6.9.*) Electronic messages are routed through the central hub to their destinations. The central hub monitors the flow of traffic. A PBX system—a private telephone system, such as that found on a college campus, that connects telephone extensions to each other—is an example of a star network. Traditional star networks are designed to be easily expandable because hubs can be connected to additional hubs of other networks.

The advantage of a star network is that the hub prevents collisions between messages. Moreover, if a connection is broken between any communications device and the hub, the rest of the devices on the network will continue operating. However, if the hub goes down, the entire network stops.

Two Ways to Prevent Messages from Colliding: Ethernet & Token Ring

How do the two methods of keeping messages from colliding work?

When you deal with small LANs, especially when they use wired or cable (twisted-pair, coaxial, or fiber-optic) connections, you may hear about Ethernet and Token Ring, two LAN protocols, or technologies, used to keep messages from bumping into one another along the transmission line.

ETHERNET   In 1973, at Xerox Corporation's Palo Alto Research Center (more commonly known as PARC), researcher Bob Metcalfe designed and tested the first Ethernet network. While working on a way to link a particular computer to a printer, Metcalfe developed the physical method of cabling that connected devices on the Ethernet as well as the standards that governed communication on the cable. Ethernet has since become the most popular and most widely deployed network technology in the world.

Ethernet is a LAN technology that can be used with almost any kind of computer and that describes how data can be sent in packets in between computers and other networked devices usually in close proximity. When two nodes try to send data at the same time that might collide, Ethernet instructs the nodes to resend the data one packet at a time. It is frequently used in a star topology.

Ethernet devices used to be able to have only a few hundred meters of cable between them, making it impractical to connect geographically dispersed locations. Modern advancements have increased these distances considerably, allowing Ethernet networks to span tens of kilometers.

The most common version (called *10Base-T*) handles about 10 megabits per second. A newer version, *Fast Ethernet* (or *100Base-T*) transfers data at 100 megabits per second. The newest version, *Gigabit Ethernet* (or *1000Base-T*) transmits data at the rate of 1 gigabit (1,000 megabits) per second. Most new microcomputers come equipped with an Ethernet card and an Ethernet port (Chapter 4).

TOKEN RING   A technology developed by IBM, *Token Ring* is a LAN technology that transmits a special control message or message frame, called a "token," around a network to each node, signaling the node that it can then send a message. The node then sends the token on to the next node, thus guaranteeing that each computer or device on the network will transmit at regular intervals. The Token Ring standard is mainly used in star or ring topologies. The advantage of Token Ring is that broken cable connections are easily detected and thus easily fixed.

Wired Communications Media

What are types of wired communications media?

It used to be that two-way individual communications were accomplished mainly in two ways. They were carried by the medium of (1) a telephone wire or (2) a wireless method such as shortwave radio. Today there are many kinds of communications media, although they are still wired or wireless. ___Communications media___, or ___communications channels___, **carry signals over a communications path, the route between two or more communications media devices.** The speed, or data transfer rate, at which transmission occurs—and how much data can be carried by a signal—depends on the media and the type of signal.

Wired Communications Media: Wires & Cables

What is the difference between the three types of wired communications media?

Three types of wired communications media are _twisted-pair wire_ (conventional telephone lines), _coaxial cable_, and _fiber-optic cable_. The various kinds of wired internet connections discussed in Chapter 2—dial-up modem, DSL, ISDN, cable modem, T1 lines, as well as internet backbones—are created by using these wired communications media.

Twisted-pair wire

TWISTED-PAIR WIRE   The telephone line that runs from your house to the pole outside, or underground, is probably twisted-pair wire. ___Twisted-pair wire___ **consists of two strands of insulated copper wire, twisted around each other. This twisted-pair configuration (compared to straight wire) somewhat reduces interference (called "crosstalk") from electrical fields.** Twisted-pair is relatively slow, carrying data at the rate of 1–128 megabits per second. Moreover, it does not protect well against electrical interference. However, because so much of the world is already served by twisted-pair wire, it will no doubt be used for years to come, both for voice messages and for modem-transmitted computer data (dial-up connections).

The prevalence of twisted-pair wire gives rise to what experts call the "last-mile problem." That is, it is relatively easy for telecommunications companies to upgrade the physical connections between cities and even between neighborhoods. But it is expensive for them to replace the "last mile" of twisted-pair wire that connects to individual houses.

Coaxial cable

COAXIAL CABLE   ___Coaxial cable___, **commonly called "co-ax," is a high-frequency transmission cable that consists of insulated copper wire wrapped in a solid or braided metal shield and then in an external plastic cover.** Co-ax is widely used for cable television and cable internet connections. Thanks to the extra insulation, coaxial cable is much better than twisted-pair wiring at resisting noise. Moreover, it can carry voice and data at a faster rate (up to 200 megabits per second). Often many coaxial cables are bundled together.

FIBER-OPTIC CABLE   A ___fiber-optic cable___ **consists of dozens or hundreds of thin strands of glass or plastic that transmit pulsating beams of light rather than electricity.** These strands, each as thin as a human

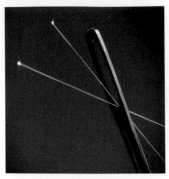

Fiber-optic strands

hair, can transmit up to about 2 billion pulses per second (2 gigabits); each "on" pulse represents 1 bit. When bundled together, fiber-optic strands in a cable 0.12 inch thick can support a quarter- to a half-million voice conversations at the same time. Moreover, unlike electrical signals, light pulses are not affected by random electromagnetic interference in the environment. Thus, fiber-optic cable has a much lower error rate than normal telephone wire and cable. In addition, fiber-optic cable is lighter and more durable than twisted-pair wire and co-ax cable, although it is more expensive. A final advantage is that it cannot easily be wiretapped, so transmissions are more secure.

At present, much of the world is awash in fiber-optic lines, stemming from the building boom of the early internet frenzy that preceded the telecommunications-industry crash of 2000.[4] "Despite a surge in internet usage since then," says one account, "the fiber glut is as bad as ever." Indeed, in the United States in mid-2005, less than 5% of the total transmission capacity of all fiber lines was being put to use—about the same amount as 4 years earlier.

Wired Communications Media for Homes: Ethernet, HomePNA, & HomePlug

What is the difference between the three types of wired communications media used for home networks?

Many households now have more than one computer, and many have taken steps to link their equipment in a home network. Indeed, some new high-tech homes include network technology that links as many as 12 televisions positioned around the house plus computers, telephones, lights, audio, and alarm systems.

Traditionally wired media have been used to connect equipment. Three wired network technologies are *Ethernet, HomePNA,* and *HomePlug.*

ETHERNET Most personal PCs come with Ethernet capability. Homes wanting to network with this technology use the kind of cabling (Cat5) that permits either regular Ethernet data speeds (10 megabits per second) or Fast Ethernet speeds (100 megabits per second). Besides cabling, which will have to be installed throughout the house (by you or by a professional installer), a home Ethernet network may require a router and a hub, which may be available from your internet access provider.

HOMEPNA: USING THE HOME'S EXISTING TELEPHONE WIRING Does your house have a phone jack in every room in which you have a computer? Then you might be interested in **_HomePNA (HPNA)_ technology, a standard that allows a household to use a home's existing telephone wiring for a home network,** transmitting data at speeds of 10–240 megabits per second. No hubs are needed, and network adapters are inexpensive. HomePNA was developed by the Home Phoneline Networking Alliance.

HOMEPLUG: USING THE HOME'S EXISTING ELECTRIC-POWER WIRING **_HomePlug_ technology is a standard that allows users to send data over a home's existing electrical (AC) power lines,** which can be transmitted at 14 megabits per second. This kind of communications medium has an advantage, of course, in that there is at least one power outlet in every room.

Some households have a combination of wired and wireless networks, but more are going over to all-wireless. "My clients now want wireless—they want to come home from work and be able to go anywhere in the house with their laptop," said the representative of a Wisconsin home builder at the 2005 International Builders' Show.[5]

PRACTICAL ACTION
Telecommuting & Telework: The Nontraditional Workplace

Christie Thomas, 33, of Morgan Hill, California, is a "mompreneur." A loan officer and independent contractor for Pacific West Financial, she does loan work and research while her 2-year-old daughter sleeps and when her husband, a teacher, gets home from work. "It's very lucrative," says Thomas, who also manages other loan officers. However, she adds, "You do have to manage your time very well. I can work in my pajamas."[6]

Home-based working moms, or mompreneurs, are the beneficiaries of technology such as high-speed internet access, the widespread availability of personal computers, and the willingness of budget-conscious companies to rely on home-based free agents. But if this kind of "home work" was once primarily the province of technical employees, it is also now moving into senior management ranks. The reasons driving the trend, according to The Wall Street Journal: "fast broadband internet connections between home and office; the September 11, 2001 terror attacks, which made companies recognize the value of placing executives apart geographically to help reduce disruptions; and managers' increasing comfort with the idea of working remotely."[7]

The rise of remote working arrangements may be part of a larger trend. "Powerful economic forces are turning the whole labor force into an army of freelancers—temps, contingents, and independent contractors and consultants," says business strategy consultant David Kline. The result, he believes, is that "computers, the net, and telecommuting systems will become as central to the conduct of 21st-century business as the automobile, freeways, and corporate parking lots were to the conduct of mid-20th-century business."[8] The transformation will really gather momentum, some observers believe, when more homes have broadband internet access.

Two offshoots of nontraditional office work are telecommuting and telework.

Telecommuting: Working from Home

Working at home while in telecommunication with the office is called telecommuting. In the United States, telecommuting has been gaining favor for several years. According to one research organization, 8.9 million Americans worked at home for a corporate job at least 3 days a month in 2004.[9] Other research of 936 large companies showed that 32% of them offered work-at-home/telecommuting arrangements in 2004.[10]

Telecommunication can have many benefits. The advantages to society are reduced traffic congestion, energy consumption, and air pollution. The advantages to employees are lower commuting and workplace-wardrobe costs and choices about how they handle their time. The advantages to employers, it's argued, are increased productivity, because telecommuters may experience fewer distractions at home than in the office and can work flexible hours. Absenteeism may be reduced, teamwork improved, and the labor pool expanded because hard-to-get employees don't have to uproot themselves from where they want to live. Costs for office space, parking, insurance, and other overhead are reduced.

Despite the advantages, however, telecommuting has some drawbacks. Employees sometimes feel isolated, even deserted, or they are afraid that working outside the office will hinder their career advancement. They also find the arrangements blur the line between office and home, straining family life. Employers may feel telecommuting causes resentments among office-bound employees, and they may find it difficult to measure employees' productivity. In addition, with teamwork now more a workplace requirement, managers may worry that telecommuters cannot keep up with the pace of change. Finally, some employers worry that telecommuters create more opportunities for security breaches by hackers or equipment thieves.[11]

Telework: Working from Anywhere

More recently, the term telework (or virtual office) has been adopted to replace the term "telecommuting" because it encompasses not just working from home but working from anywhere: "a client's office, a coffee shop, an airport lounge, a commuter train," in one description. "With cellphones, broadband at home, Wi-Fi, virtual private networks, and instant messaging becoming ubiquitous, telework has become easier than ever."[12]

Employees at big high-tech companies such as Sun Microsystems (which has a telework program called iWork), IBM, and Cisco work from their homes, cars, and other nontraditional work sites. Thus, the workplace exists more in virtual than physical space, and the actual office may be little more than a computer, a high-bandwidth connection, and a cellphone, with most communication with the outer world being through a voice mail system, an email address, a web page, and a post office box. (Indeed, you can live anywhere you want in the world but have a local business presence with a prestigious physical mailing address and local phone number by hiring a virtual office web service such as Officescape, whose staff can handle all your office information locally and forward your physical mail and packages.)

Despite the history of productivity, some managers fear losing control over employees who become teleworkers, picturing them at home watching TV instead of working. The way to avoid that, suggests Robert Smith, director of ITAC (formerly the International Telework Association & Council), is for companies to install ways of measuring productivity. "Good organizations put into place performance requirements," he says. "They have a process in which managers can set performance goals and properly evaluate whether those goals are met. If you have that in place, you'll have a more effective organization, whether an employee is 10 feet or 10 miles away."[13]

6.4

Wireless Communications Media

What are types of wireless communications media, both long distance and short distance?

His friends thought Hank Kahrs, an insurance auditor, was just being silly in taking along a cellphone back in 1997 when they hiked up California's Mount Whitney, the highest mountain in the continental United States. One reason for getting outdoors, they chided him, was to get away from civilization and its gadgets; anyway, they said, the phone probably wouldn't work at 14,494 feet. Kahrs, however, didn't want to break his long-standing custom of calling his wife every day.

At the top of the peak, he got a pleasant surprise. "It took 30 seconds before the phone started ringing" at his wife's number, he said. After he made his call, his hiking companions' attitudes changed. "When they saw the signal was fine, my friends all wanted to use the phone," said Kahrs.[14]

Very soon, it will be nearly impossible to *not* be able to make a phone call or at least page someone from anywhere on earth (except from a cave, perhaps). Already this has produced headaches for forest rangers, who have had to rescue too many hikers who embarked on wilderness treks without being properly equipped—except for a cellphone with which to call for help. (Don't try this yourself, since there are no guarantees that it will work in remote places.)

The cellphone, however, is only one kind of wireless communication. Let us take a closer look at this technology. We consider (1) the electromagnetic spectrum, (2) the four types of long-distance wireless communications media, (3) long-distance wireless, and (4) short-distance wireless.

The Electromagnetic Spectrum, the Radio-Frequency (RF) Spectrum, & Bandwidth

What is the electromagnetic spectrum, and how do electromagnetic waves differ?

Often it's inefficient or impossible to use wired media for data transmission, and wireless transmission is better. To understand wireless communication, we need to understand transmission signals and the electromagnetic spectrum.

THE ELECTROMAGNETIC SPECTRUM Telephone signals, radar waves, microwaves, and the invisible commands from a garage-door opener all represent different waves on what is called the electromagnetic spectrum of radiation. The *electromagnetic spectrum of radiation* **is the basis for *all* telecommunications signals, carried by both wired and wireless media.**

Part of the electromagnetic spectrum is the *radio-frequency (RF) spectrum*, **fields of electrical energy and magnetic energy that carry most communications signals.** (● *See Panel 6.10 on the next page.*) Internationally, the RF spectrum is allocated by the International Telecommunications Union (ITU) in Geneva, Switzerland. Within the United States, the RF spectrum is further allocated to nongovernment and government users.

The Federal Communications Commission (FCC), acting under the authority of Congress, allocates and assigns frequencies to nongovernment users. The National Telecommunications and Information Administration (NTIA) is responsible for departments and agencies of the U.S. government.

Electromagnetic waves vary according to *frequency*—the number of times a wave repeats, or makes a cycle, in a second. The radio-frequency spectrum ranges from low-frequency waves, such as those used for garage-door openers (40 megahertz), through the medium frequencies for certain cellphones (824–849 megahertz) and air-traffic control monitors (960–1,215 megahertz), to deep-space radio communications (2,290–2,300 megahertz). Frequencies at the very ends of the spectrum take the forms of infrared rays, visible light, ultraviolet light, X-rays, and gamma rays.

BANDWIDTH The *bandwidth* **is the range, or** *band*, **of frequencies that a transmission medium can carry in a given period of time.** For analog signals, bandwidth is expressed in *hertz (Hz)*, or *cycles per second*. For example, certain cellphones operate within the range 824–849 megahertz—that is, their bandwidth is 25 megahertz. *The wider a medium's bandwidth, the more frequencies it can use to transmit data and thus the faster the transmission.*

There are three general classes of bandwidth—narrow, medium, and broad—which can be expressed in hertz but also in *bits per second (bps):*

- **Narrowband: *Narrowband*, also known as *voiceband*, is used for regular telephone communications**—that is, for speech, faxes, and data. Transmission rates are usually 100 kilobits per second or less. Dial-up modems use this bandwidth.

- **Medium band: *Medium band* is used for transmitting data over long distances and for connecting mainframe and midrange computers.** It is also used to transmit pictures, video, and high-fidelity sound. Transmission rates are 100 kilobits to 1 megabit per second.

- **Broadband: *Broadband* is used to transmit high-speed data and high-quality audio and video.** Transmission speeds are 1 megabit per second to as high (for super-broadband and ultra-broadband) as 100 megabits per second.

Currently most residences in the United States have connections that top out at 4 or 5 megabits. That's why the United States ranks number 12 in broadband deployment, trailing such countries as South Korea and Japan, where blinding 100-megabit speeds are common.[15]

WAP: WIRELESS APPLICATION PROTOCOL Wireless handheld devices such as cellphones use the Wireless Application Protocol for connecting wireless users to the World Wide Web. Just as the protocol TCP/IP gives you a wired connection to your internet service provider, **the *Wireless Application Protocol (WAP)* is designed to link nearly all mobile devices to your telecommunications carrier's wireless network and content providers.**

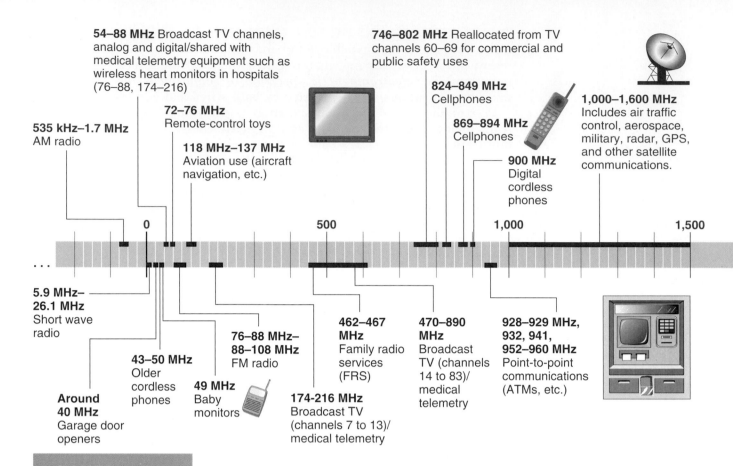

54–88 MHz Broadcast TV channels, analog and digital/shared with medical telemetry equipment such as wireless heart monitors in hospitals (76–88, 174–216)

72–76 MHz Remote-control toys

118 MHz–137 MHz Aviation use (aircraft navigation, etc.)

535 kHz–1.7 MHz AM radio

746–802 MHz Reallocated from TV channels 60–69 for commercial and public safety uses

824–849 MHz Cellphones

869–894 MHz Cellphones

900 MHz Digital cordless phones

1,000–1,600 MHz Includes air traffic control, aerospace, military, radar, GPS, and other satellite communications.

0   500   1,000   1,500

5.9 MHz–26.1 MHz Short wave radio

Around 40 MHz Garage door openers

43–50 MHz Older cordless phones

49 MHz Baby monitors

76–88 MHz–88–108 MHz FM radio

174-216 MHz Broadcast TV (channels 7 to 13)/ medical telemetry

462–467 MHz Family radio services (FRS)

470–890 MHz Broadcast TV (channels 14 to 83)/ medical telemetry

928–929 MHz, 932, 941, 952–960 MHz Point-to-point communications (ATMs, etc.)

● PANEL 6.10
The radio-frequency spectrum
The radio-frequency spectrum, which carries most communications signals, appears as part of the electromagnetic spectrum.

Four Types of Wireless Communications Media

What are the differences between the four types of wireless communications media?

Four types of wireless media are *infrared transmission, broadcast radio, microwave radio,* and *communications satellite.*

INFRARED TRANSMISSION **_Infrared wireless transmission_ sends data signals using infrared-light waves at a frequency too low (1–4 megabits per second) for human eyes to receive and interpret.** Infrared ports can be found on some laptop computers, PDAs, digital cameras, and printers, as well as wireless mice. TV remote-control units use infrared transmission. The drawbacks are that *line-of-sight* communication is required—there must be an unobstructed view between transmitter and receiver—and transmission is confined to short range.

BROADCAST RADIO When you tune in to an AM or FM radio station, you are using **_broadcast radio_, a wireless transmission medium that sends data over long distances at up to 2 megabits per second—between regions, states, or countries.** A transmitter is required to send messages and a receiver to receive them; sometimes both sending and receiving functions are combined in a *transceiver.*

In the lower frequencies of the radio spectrum, several broadcast radio bands are reserved not only for conventional AM/FM radio but also for broadcast television, CB (citizens band) radio, ham (amateur) radio, cellphones, and private radio-band mobile services (such as police, fire, and taxi dispatch). Some organizations use specific radio frequencies and networks to support wireless communications. For example, UPC (Universal Product Code) bar-code readers are used by grocery-store clerks restocking store shelves to

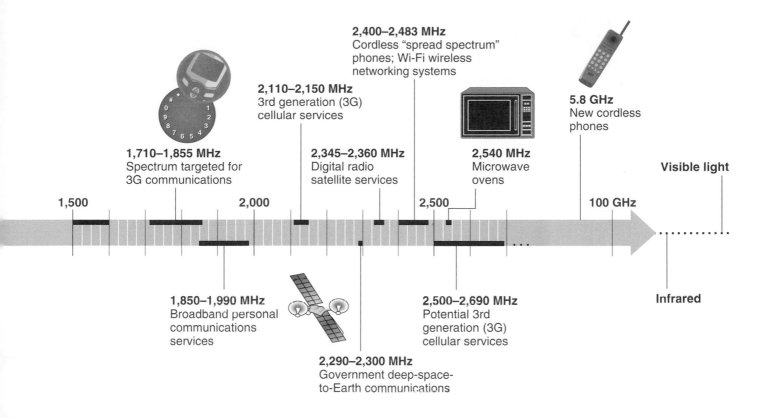

2,400–2,483 MHz
Cordless "spread spectrum" phones; Wi-Fi wireless networking systems

2,110–2,150 MHz
3rd generation (3G) cellular services

5.8 GHz
New cordless phones

1,710–1,855 MHz
Spectrum targeted for 3G communications

2,345–2,360 MHz
Digital radio satellite services

2,540 MHz
Microwave ovens

Visible light

1,500   2,000   2,500   100 GHz

1,850–1,990 MHz
Broadband personal communications services

2,500–2,690 MHz
Potential 3rd generation (3G) cellular services

Infrared

2,290–2,300 MHz
Government deep-space-to-Earth communications

communicate with a main computer so that the store can control inventory levels. In addition, there are certain web-enabled devices that follow standards such as *Wi-Fi (wireless fidelity),* as we discuss in a few pages.

MICROWAVE RADIO   ***Microwave radio*** **transmits voice and data at 45 megabits per second through the atmosphere as superhigh-frequency radio waves called** *microwaves,* **which vibrate at 1 gigahertz (1 billion hertz) per second or higher.** These frequencies are used not only to operate microwave ovens but also to transmit messages between ground-based stations and satellite communications systems. One short-range microwave standard used for communicating text is *Bluetooth,* as we shall discuss.

Nowadays horn-shaped microwave reflective dishes, which contain transceivers and antennas, are nearly everywhere—on towers, buildings, and hilltops. Why, you might wonder, do we have to interfere with nature by putting a microwave dish on top of a mountain? As with infrared waves, microwaves are line-of-sight; they cannot bend around corners or around the earth's curvature, so there must be an unobstructed view between transmitter and receiver. Thus, microwave stations need to be placed within 25–30 miles of each other, with no obstructions in between. The size of the dish

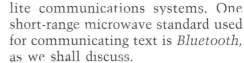

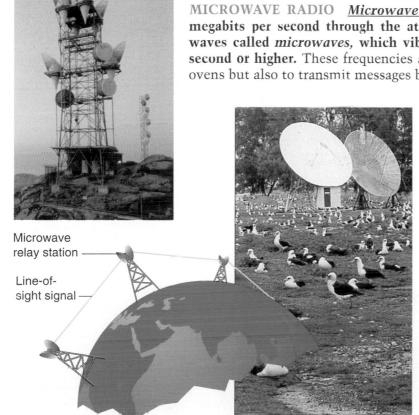

Microwave relay station

Line-of-sight signal

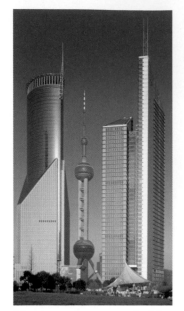

Microwave towers on buildings in Shanghai

varies with the distance (perhaps 2–4 feet in diameter for short distances, 10 feet or more for long distances). In a string of microwave relay stations, each station will receive incoming messages, boost the signal strength, and relay the signal to the next station.

More than half of today's telephone systems uses dish microwave transmission. However, the airwaves are becoming so saturated with microwave signals that future needs will have to be satisfied by other channels, such as satellite systems.

COMMUNICATIONS SATELLITES To avoid some of the limitations of microwave earth stations, communications companies have added microwave "sky stations"—communications satellites. **_Communications satellites_ are microwave relay stations in orbit around the earth.** Transmitting a signal from a ground station to a satellite is called *uplinking;* the reverse is called *downlinking.* The delivery process will be slowed if, as is often the case, more than one satellite is required to get the message delivered. Satellites cost from $300 million to $700 million each. A satellite launch costs between $50 million and $400 million. Communications satellites are the basis for the *Global Positioning System (GPS),* as we shall discuss.

Satellite systems may occupy one of three zones in space: *GEO, MEO,* and *LEO:*

Communications satellite

- **GEO:** The highest level, known as *geostationary earth orbit (GEO),* is 22,300 miles and up and is always directly above the equator. Because the satellites in this orbit travel at the same speed as the earth, they appear to an observer on the ground to be stationary in space—that is, they are geostationary. Consequently, microwave earth stations are always able to beam signals to a fixed location above. The orbiting satellite has solar-powered transceivers to receive the signals, amplify them, and retransmit them to another earth station. At this high orbit, fewer satellites are required for global coverage; however, their quarter-second delay makes two-way conversations difficult.

- **MEO:** The *medium-earth orbit (MEO)* is 5,000–10,000 miles up. It requires more satellites for global coverage than does GEO.
- **LEO:** The *low-earth orbit (LEO)* is 200–1,000 miles up and has no signal delay. LEO satellites may be smaller and are cheaper to launch.

Satellite launch in India

GEO

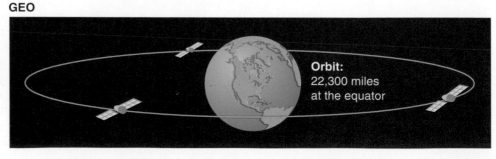

Orbit: 22,300 miles at the equator

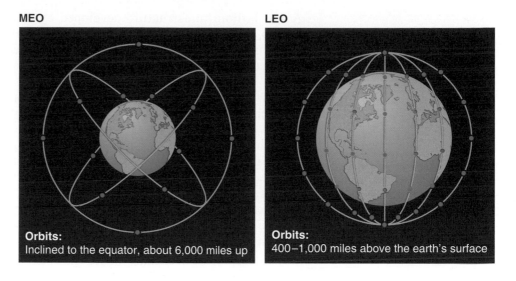

MEO   LEO

Orbits:
Inclined to the equator, about 6,000 miles up

Orbits:
400–1,000 miles above the earth's surface

Wouldn't it be great if you could get internet access anywhere by way of satellite-delivered service? "Surfing the web by bouncing signals off satellites thousands of miles above the Earth has been a dream since the early 1990s," says one report. "Users of early efforts were frustrated by balky equipment and transmission delays, while investor interest plunged after the massive implosions of many startups."[16] At present, only Direcway, part of DirectTV Group, offers nationally promoted satellite-based internet service, although a small company called WildBlue Communications is attempting to do the same. Certainly, the need is apparent: An estimated 20 million consumers, mostly in rural areas, can't get high-speed internet service from a phone or cable-TV company.

Long-Distance Wireless: One-Way Communication

How do the Global Positioning System and one-way pager systems work?

Mobile wireless communications have been around for some time. The Detroit Police Department started using two-way car radios in 1921. Mobile telephones were introduced in 1946. Today, however, we are in the midst of an explosion in mobile wireless use that is making worldwide changes.

There are essentially two ways to move information through the air long distance on radio frequencies—one way and two way. *One-way communications*, discussed below, is typified by the satellite navigation system known as the Global Positioning System and by most pagers. *Two-way communications*, described on page 337, is exemplified by cellphones.

THE GLOBAL POSITIONING SYSTEM   "There is a traffic jam about 800 meters ahead of you," says the electronic voice beamed over FM airwaves to your car radio. "It will take 10 minutes to get through it." You look at your computerized map on the dashboard, which shows your car's location. It confirms the bad news: The road ahead is blinking yellow, then red—serious congestion that means you should take an alternate route. This is an onboard car navigation system in actual use today—in Japan. Some day we may see the system, which relies on the Global Positioning System, in North America as well.[17]

A $10 billion infrastructure developed by the military in the mid-1980s, the **_Global Positioning System (GPS)_ consists of 24 earth-orbiting satellites continuously transmitting timed radio signals that can be used to identify earth locations.**

- **How GPS works:** The U.S. military developed and implemented this satellite network in the 1970s as a military navigation system, but on May 1, 2000, the federal government opened it up to everyone else. Each of these 3,000- to 4,000-pound, solar-powered satellites circles the earth twice a day at an altitude of 11,000 nautical miles. A GPS receiver—handheld or mounted in a vehicle, plane, or boat—can pick up transmissions from any four satellites, interpret the information from each, and pinpoint the receiver's longitude, latitude, and altitude. (● *See Panel 6.11.)* The system is accurate within 3–50 feet, with 10 feet being the norm.

- **The uses of GPS:** Besides being the technology behind car onboard navigation systems, the GPS is used for such activities as tracking trucks, buses, and taxis; locating stolen cars; orienting hikers; and aiding in surveying. An aerial camera connected to a GPS receiver can automatically tag photos with GPS coordinates, showing where the photograph was taken, which can be useful to surveyors, forestry managers, search-and-rescue teams, and archaeologists.[18] GPS has

This woman is using a GPS unit with a Braille keyboard and voice output to find her way around a park.

Brian Sniatkowski holds a GPS unit in the Kekeout Reservoir Forest while geocaching, engaging in a high-tech treasure hunt for GPS users. Participants set up caches all over the world and share the locations via the internet. GPS users can then use the location coordinates to find the caches. Each cache may provide the finder with a variety of rewards; each finder is asked to put something new in the cache.

Sunto's n3wrist watch, one of the first wristwatches to receive a steady stream of news, weather information, and other information via Microsoft's MSN Direct subscription service.

Many cars come with GPS units to guide users to their destinations.

● PANEL 6.11

GPS

The Global Positioning System uses 24 satellites, developed for military use, to pinpoint a location on the earth's surface.

been used by scientists to keep a satellite watch over a Hawaiian volcano, Mauna Loa, and to capture infinitesimal movements that may be used to predict eruptions. In one technology, cellular carriers have E-911 (for "Enhanced 911") capability that is able to locate, through tiny GPS receivers embedded in users' digital cellphones, the position of every person making an emergency 911 call.

"Geocaching" hobbyists practice an activity in which they hide a metal or plastic receptacle (a cache) filled with pens, pins, pliers, and

The ABCs of GPS

Developed by the U.S. military over the past three decades to aid ship and plane navigation, the Global Positioning System has evolved to serve a variety of purposes, with GPS receivers in everything from cars to handheld devices. The system has three main components—a satellite constellation, ground control, and receivers.

Space

There are about 24 satellites orbiting the Earth at an altitude of 11,000 nautical miles. Each is equipped with an atomic clock that keeps time to three-billionths of a second. The satellites send time-stamped radio signals to Earth at the speed of light. The signals include information about each satellite's exact position.

Ground control

Five stations around the world monitor the satellites and send them information about their orbital position. The main control center is in Colorado Springs, Colorado.

The receiver

The receiver must pick up signals from at least four satellites. It calculates its distance from each satellite by comparing the time stamp of the signal to the time it reached the receiver. The receiver's clock isn't nearly as accurate as the atomic clocks in the satellites, but mathematical adjustments are made to account for inaccuracies.

Connect four

The basic premise of GPS is a concept called triangulation. Using this concept, the exact location of a golf ball on a two-dimensional course can be calculated by determining its distance from three pins.

The distance from pin 1 reveals possible locations anywhere along the edge of an imaginary circle.

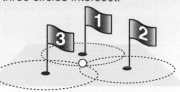

The distances from pins 1 and 2 reveal two possible ball locations   the point where the circles intersect.

The distances from pins 1, 2, and 3 reveal one possible location—the point where all three circles intersect.

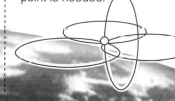

Triangulation works the same way in three-dimensional space, but with spheres instead of circles and a fourth reference point is needed.

other inexpensive trinkets; give the location a funny name ("Indiana Jones Fortress of Doom"); and post the GPS coordinates on the official geocachers' website *(www.geocaching.com)*. Cache hunters then punch the coordinates into their handheld GPS units ($90–$700 at sporting-goods and electronics stores) and follow electronic prompts to the site, where they sign a log and replace some trinkets with gewgaws of their own.[19]

- **The limitations of GPS:** Not all services based on GPS technology are reliable, as any frequent user of online mapping systems (such as those of MapQuest, MSN Maps and Directions, and Yahoo Maps) knows by now. Indeed, about 1 in 50 computer-generated directions is a dud, according to Doug Richardson, executive director of the Association of American Geographers, mostly because of inaccurate road information. "You have to have the latest data about road characteristics—things like one-way streets, turns, and exits in order for it to generate accurate directions," he says.[20] One woman who queried three mapping sites for directions to New York's La Guardia Airport received three different sets of instructions. One way to offset this is to put a GPS receiver in your car to constantly monitor your position and calculate the most efficient route. (Many GPS units sell to hiking buffs, but they don't replace knowing how to read a topographical map. GPS screens may show a good path from A to B, but they don't always show what might be on that path—such as a rocky cliff.[21])

    Emergency operators answering E-911 cellphone calls have also often found some unsettling results: The information giving the caller's location either didn't appear or was inaccurate. There are several reasons for this. (1) Some cellphone companies (Verizon Wireless, Nextel, and Sprint) use GPS to find callers, but the system works only if cellphones are equipped with a special GPS chip and if the local 911 emergency center has been upgraded to receive positioning data. (2) Other cellphone companies (Cingular and T-Mobile) don't use GPS but rather rely on triangulation—measuring the distance of a signal from three cellphone towers—to locate callers, which doesn't work well in certain urban and rural areas. (3) Internet-based phone companies (Vonage until recently) can't always connect callers with 911 operators because of the way the system is designed (such as allowing cellphone users to choose their area code or take their phone numbers with them when they travel). (4) Customers forget to register their addresses, or don't update them when they move. (5) E-911 operators aren't available at internet phone companies to answer calls, particularly after hours.[22]

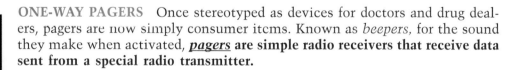

ONE-WAY PAGERS Once stereotyped as devices for doctors and drug dealers, pagers are now simply consumer items. Known as *beepers*, for the sound they make when activated, <u>**pagers**</u> **are simple radio receivers that receive data sent from a special radio transmitter.**

Pager

- **How a pager works:** The radio transmitter broadcasting to the pager sends signals over a specific frequency. All of the pagers for that particular network have a built-in receiver that is tuned to the same frequency broadcast from the transmitter. The pagers listen to the signal from the transmitter constantly as long as the pager is turned on. Often the pager has its own telephone number. When the number is dialed from a phone, the call goes by way of the transmitter straight to the designated pager.

    Some pagers do more than beep or vibrate, transmitting full-blown alphanumeric text (such as four-line, 80-character messages) and other data. Newer ones are mini-answering machines, capable of relaying digitized voice messages.

- **The uses of pagers:** Pagers are very efficient for transmitting one-way information—emergency messages, news, prices, stock quotations, delivery-route assignments, even sports news and scores—at low cost to single or multiple receivers. One Clearwater, Florida, child care center gives all parents a pager as part of its service. Then, if one child bites another or otherwise misbehaves, the head of the center can instantly alert the parents by paging them.

Long-Distance Wireless: Two-Way Communication

How do I distinguish among five types of long-distance wireless technologies?

There are five types of long-distance two-way wireless communications: (1) two-way pagers and wireless email devices such as BlackBerry, (2) analog cellphones, (3) 2G digital wireless, (4) 2.5G digital wireless, and (5) 3G digital wireless. (Other wireless methods, such as Wi-Fi, operate at short distances, as we discuss later.)

TWO-WAY PAGERS & WIRELESS EMAIL DEVICES: BLACKBERRY With two-way, or enhanced, paging, service is provided by carriers like SkyTel, and you can use gadgets such as the BlackBerry, which debuted in 1999 as a two-way pager. For instance, with one pager, users can send a preprogrammed message or acknowledgment that they have received a message. Another version allows consumers to compose and send email to anyone on the internet and to other pagers and wireless email gadgets. Typing a message on a tiny keyboard no larger than those on pocket calculators can pose a challenge for some people, however.

Today, with more than 3 million users, BlackBerry is the most popular of wireless email devices and two-way pagers, although it is being challenged by the Treo 650 (which doubles as a cellphone) from PalmOne and other devices from Intellisync, Seven Networks, Microsoft, and Motorola.[23] BlackBerry-like devices are used by delivery services and other users (including members of the U.S. Congress) who want to get simple messages to people spread over a wide geographic area. The subscription service, which is offered by such companies as Cingular Wireless, operates in three different frequencies (800, 900, or 1,900 megahertz). Today BlackBerry has expanded beyond the simple pager into more complicated devices, such as smartphones.

1G (FIRST-GENERATION) CELLULAR SERVICE: ANALOG CELLPHONES "In the fall of 1985," writes *USA Today* technology reporter Kevin Maney, "I went to Los Angeles to research a story about a new phenomenon called a cellphone. I got to use one for a day. The phones then cost $1,000 and looked like field radios from *M*A*S*H*. Calls were 45 cents a minute, and total cellphone users in the world totaled 200,000."[24] Now a research firm called Gartner Group estimates that in 2009 2.6 billion people will be using cellphones.[25]

Cellphones are essentially two-way radios that operate using either analog or digital signals. **_Analog cellphones_ are designed primarily for communicating by voice through a system of ground-area cells. Each cell is hexagonal in shape, usually 8 miles or less in diameter, and is served by a transmitter-receiving tower.** Communications are handled in the bandwidth of 824–894 megahertz. Calls are directed between cells by a mobile telephone switching office (MTSO). Movement between cells requires that calls be "handed off" by this switching office. (● *See Panel 6.12.*) This technology is known as 1G, for "first generation."

Handing off voice calls between cells poses only minimal problems. However, handing off data transmission (where every bit counts), with the inevitable gaps and pauses on moving from one cell to another, is much more difficult.

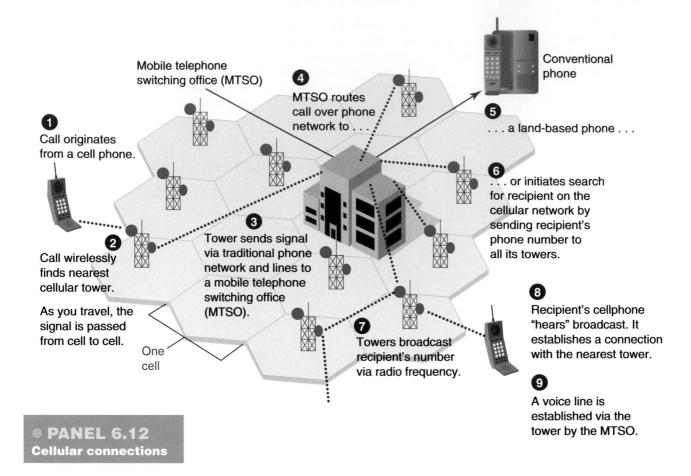

Mobile telephone switching office (MTSO)

1 Call originates from a cell phone.

2 Call wirelessly finds nearest cellular tower.

As you travel, the signal is passed from cell to cell.

One cell

3 Tower sends signal via traditional phone network and lines to a mobile telephone switching office (MTSO).

4 MTSO routes call over phone network to . . .

5 . . . a land-based phone . . .

Conventional phone

6 . . . or initiates search for recipient on the cellular network by sending recipient's phone number to all its towers.

7 Towers broadcast recipient's number via radio frequency.

8 Recipient's cellphone "hears" broadcast. It establishes a connection with the nearest tower.

9 A voice line is established via the tower by the MTSO.

> ● **PANEL 6.12**
> **Cellular connections**

S u r v i v a l   T i p
Cellphone Minutes

If you have a limited amount of daytime/anytime minutes on your cellphone service, you can conserve minutes by calling your cellphone voice mail from a land-line POTS telephone.

2G (SECOND-GENERATION) WIRELESS SERVICES: DIGITAL CELL-PHONES & PDAS *Digital wireless services*—**which support digital cell-phones and personal digital assistants—use a network of cell towers to send voice communications and data over the airwaves in digital form.** Known as *2G*, for "second-generation," technology, digital cellphones began replacing analog cellphones during the 1990s as telecommunications companies added digital transceivers to their cell towers. *2G (second generation)* technology was the first digital voice cellular network; data communication was added as an afterthought, with data speeds ranging from 9.6 to 14.4 kilobits per second. 2G technology not only dramatically increased voice clarity; it also allowed the telecommunications companies to cram many more voice calls into the same slice of bandwidth.

Other countries adhere to a single standard for 2G wireless, but in the United States there are different and incompatible digital technologies operating at different frequencies. For 2G, the two prevailing standards are *CDMA* and *GSM:*

- **CDMA:** *CDMA* (for "Code Division Multiple Access") has data transfer rates of about 14.4 kilobits per second. This is the wireless 2G network used by telecommunications companies Verizon and Sprint for both voice and data wireless transmissions.

- **GSM:** *GSM* (for "Global System for Mobile Communications") has data transfer rates of about 9.6 kilobits per second, slower than CDMA.

Inside a cellphone

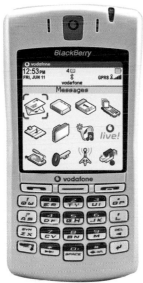

info!

The Ideal Cellphone
& Service

What's the ideal cellphone and
service for your area? Go to
www.cellmania.com or
*http://reviews.cnet.com/
4520-7609_7-5537615-1.html*
to compare plans.

This is the wireless 2G standard used by AT&T and T-Mobile, as well as by countries in Western Europe and many other regions, especially the Middle East and Asia. Unfortunately, in the United States, GSM is on a different frequency from the GSM phones in the rest of the world, which limits use of an American GSM cellphone in other countries. There are some phones on the market that include both U.S. and European-based GSM frequencies.

2.5G WIRELESS DIGITAL SERVICES   A step above 2G, *2.5G* networks offer data speeds of 30–100 kilobits per second, faster than 2G standards. An upgrade to GSM known as *GPRS* (for "General Packet Radio Service") offers data speeds of 30–50 kilobits per second. Another 2.5G upgrade, *EDGE* (for "Enhanced Data for Global Evolution"), offers speeds up to 236 kilobits per second. (● *See Panel 6.13.*)

3G (THIRD-GENERATION) WIRELESS DIGITAL SERVICES   3G cellphones look more like PDAs. *3G (third-generation) wireless digital services,* often called *broadband technology,* are based on the U.S. GSM standard and support devices that are "always on," carry data at high speeds (144 kilobits per second up to about 2 megabits per second), accept emails with attachments, provide internet and web access, are able to display color video and still pictures, and play music.

Two important upgrades to 3G wireless are the following. *(See Panel 6.13.)*

- **EV-DO:** *EV-DO* (for "Evolution Data Only") has average speeds of 400–700 kilobits per second, with peaks of 2 megabits. EV-DO uses radio waves similar to Wi-Fi (discussed shortly) but provides internet speeds comparable to DSL or cable modems. (Sprint and Verizon offer EV-DO.)
- **UMTS:** *UMTS* (for "Universal Mobile Telecommunications Systems") has average speeds of 220–320 kilobits per second.

Both EV-DO and UMTS have increasingly been deployed across Asia and Europe and are now being rolled out in the United States.[26] Indeed, you can now get special cellular EV-DO cards that can slide into the PC-card slot on your laptop, enabling you to communicate wirelessly without having to hunt down a public access zone or "hot spot" (as is required with Wi-Fi), since you have an entire metropolitan area as a hot spot. Moreover, EV-DO is very fast.[27]

Vodafone's BlackBerry 7100v smartphone: always-on, email everywhere. You don't have to retrieve your mail; it finds you.

WIMAX: PROMISING NEW LONG-DISTANCE WIRELESS STANDARD
To the long-distance wireless technologies of EDGE, EV-DO, and UMTS—which have a range of 1–5 miles—let us add another standard that promises eventually to be competitive for 3G technologies. This is *WiMax* (for "Worldwide Interoperability for Microwave Access"), a wireless standard capable of transmitting at a typical range of 6–10 miles (and a maximum of 20–30 miles)

● PANEL 6.13
Wireless upgrades

Technology		Standard	Speed
Upgrades to CDMA			
1xRTT	(Single Carrier Radio Transmission Technology)	First upgrade	1G	50–70 Kbps
EV-DO	(Evolution Data Only)	Second upgrade	3G	300–500 Kbps
EV-DV	(Evolution for Data and Voice)	Under development	3G	Up to 3.1 Mbps
Upgrades to GSM			
GPRS	(General Packet Radio Service)	First upgrade	2.5G	30–50 Kbps
EDGE	(Enhanced Data for Global Evolution)	Second upgrade	2.5G	Up to 236 Kbps
UMTS	(Universal Mobile Telecommunications System)	Third upgrade	3G	220–320 Kbps

Communications, Networks, & Safeguards

339

at up to 20 megabits per second. (Wi-Fi, the wildly popular short-range standard, is similar to WiMax but doesn't transmit as far.) There are two varieties of WiMax, one fixed and one mobile. First uses of WiMax will probably be by business customers, but eventually it's hoped the standard can provide broadband to homes in rural areas that can't get cable or DSL.[28]

Short-Range Wireless: Two-Way Communication

How do I distinguish among the three types of short-range wireless technologies?

We have discussed the standards for high-powered wireless digital communications in the 800–1,900 megahertz part of the radio-frequency spectrum, which are considered long-range waves. Now let us consider low-powered wireless communications in the 2.4–7.5 gigahertz part of the radio spectrum, which are short-range and effective only within several feet of a wireless access point—generally between 30 and 250 feet. This band is available globally for unlicensed, low-power uses and is set aside as an innovation zone where new devices can be tested without the need for a government license; it's also used for industrial, scientific, and medical devices.

There are three kinds of networks covered by this range:[29]

- **Local area networks—range 50–150 feet:** These include the popular Wi-Fi standard.
- **Personal area networks—range 30–32 feet:** These use Bluetooth, ultra wideband, and wireless USB.
- **Home automation networks—range 100–250 feet:** These use the Insteon, ZigBee, and Z-Wave standards.

Let's consider these.

SHORT-RANGE WIRELESS FOR LOCAL AREA NETWORKS: WI-FI B, A, G, & N Wi-Fi is known formally as an *802.11 network*, named for the wireless technical standard specified by the Institute of Electrical and Electronics Engineers (IEEE). As we mentioned in Chapter 2, *Wi-Fi*—short for *"wireless fidelity"*—is a short-range wireless digital standard aimed at helping portable computers and handheld wireless devices to communicate at high speeds and share internet connections at distances of 50–150 feet. You can find Wi-Fi connections, which operate at 2.4–5 gigahertz, inside offices, airports, and internet cafés, and some enthusiasts have set up transmitters on rooftops, distributing wireless connections throughout their neighborhoods. (● *See Panel 6.14.*)

Wi-Fi camera

There are four varieties of Wi-Fi standards—Wi-Fi a, Wi-Fi b, Wi-Fi g, and Wi-Fi n:

- **Wi-Fi b, a, and g:** Named for variations on the IEEE 802.11 standard (802.11b, 802.11a, 802.11g), these all typically transmit at data rates ranging from 11 megabits per second up to about 150 feet (for Wi-Fi b, the older and slower version) to 54 megabits per second up to 50 feet (for Wi-Fi a and g).
- **Wi-Fi n with MIMO:** Wi-Fi n is a promising standard that—when used with a technology called *MIMO* (pronounced "my-moh" and short for "multiple input multiple output"), which extends the range of Wi-Fi by using multiple transmitting and receiving antennas (instead of the one transmit and two receive antennas typical for most Wi-Fi)—will be able to transmit at 200 megabits per second up to about 150 feet.[30]

Wi-Fi

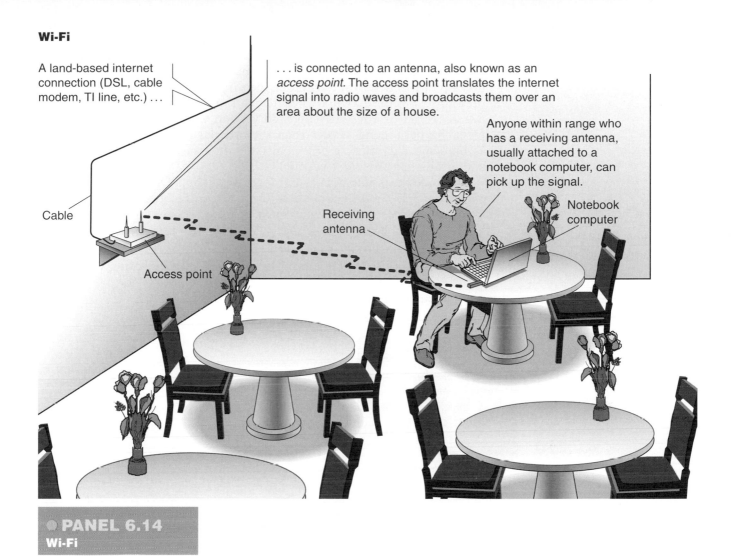

A land-based internet connection (DSL, cable modem, TI line, etc.) ...

... is connected to an antenna, also known as an *access point*. The access point translates the internet signal into radio waves and broadcasts them over an area about the size of a house.

Anyone within range who has a receiving antenna, usually attached to a notebook computer, can pick up the signal.

Cable

Access point

Receiving antenna

Notebook computer

● PANEL 6.14
Wi-Fi

SECURITY Increasingly, people are installing Wi-Fi networks in their homes, as well as going online through wireless hot spots. Either way, it's extremely important to make sure the Wi-Fi connection is secure against cyberspying. "If you're using a mobile connection, you may not even be aware that you're connected to an unverified access point because the connection can be made without your knowledge," says one writer. "Disable your Wi-Fi software, instead of leaving it on to autoconnect, whenever you're not using it. This will keep you from unknowingly connecting to a fraudulent network and wasting your laptop's battery."[31] We give some more security tips in another few pages.

SHORT-RANGE WIRELESS FOR PERSONAL AREA NETWORKS: BLUETOOTH, ULTRA WIDEBAND, & WIRELESS USB As we stated, personal area networks (PANs) use short-range wireless technology to connect personal electronics, such as cellphones, PDAs, MP3 players, and printers in a range of 30–32 feet. The principal wireless technology used so far has been Bluetooth, which is now being joined by ultra wideband (UWB) and wireless USB.[32]

more info!

Lastest Bluetooth Devices

What are the newest Bluetooth devices? See www.bluetooth.com

- Bluetooth: **_Bluetooth_ is a short-range wireless digital standard aimed at linking cellphones, PDAs, computers, and peripherals up to distances of 30 feet.** (The name comes from Harald Bluetooth, the 10th-century Danish king who unified Denmark and Norway.) Transmitting in the range of 720 kilobits per second, the original version of Bluetooth was designed to replace cables connecting PCs to printers and PDAs or wireless phones and to overcome line-of-sight problems with infrared transmission. When Bluetooth-capable devices come within range of one another, an automatic electronic "conversation" takes place to determine whether they have data to share, and then they form a mini-network to exchange that data. *(See ● Panel 6.15.)* *Bluetooth 2.0* is a recently updated version with a data rate of 3 megabits per seconds, faster than the original Bluetooth, though not as high-speed as Wi-Fi g or n. This version also has lower battery consumption than the older version.

Bluetooth technology is found in many Palm PDAs, Pocket PCs, and about 10% of all cellphones. There is some concern that the security systems used in many such devices are susceptible to snooping by eavesdroppers.[33]

- **Ultra wideband (UWB):** Developed for use in military radar systems, **_ultra wideband (UWB)_ is a promising technology operating in the**

Bluetooth symbol on a notebook computer

● PANEL 6.15
Bluetooth

Bluetooth 1.1 allows two devices to interact wirelessly in sometimes novel ways. A Bluetooth-equipped notebook, for example, can connect through a similarly enabled cellphone or internet access point to send and receive email. In addition to the links shown, Bluetooth can be used to network similar devices—for example, to send data from PC to PC, as long as they are up to 33–330 feet apart.

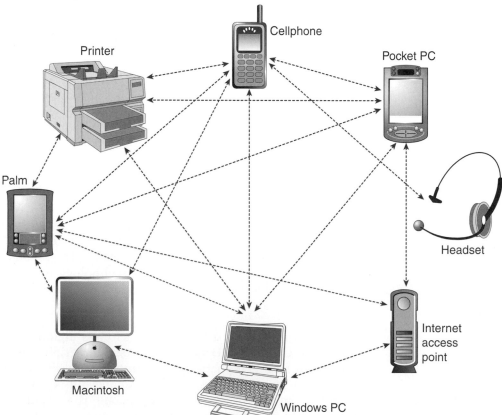

range of 480 megabits per second up to 30 feet that uses a low power source to send out millions of bursts of radio energy every second over many different frequencies, which are then reassembled by a UWB receiver.** It operates over low frequencies not used by other technologies, making it a candidate to replace many of the cables that now currently connect household and office devices. While UWB is 100 times as fast as Bluetooth and should be a natural successor to Wi-Fi, unfortunately the industry committee that is supposed to be setting specifications for it has been deadlocked between two technical approaches.[34] Nevertheless, it is expected that by 2007 some cell-phones and MP3 players will be UWB-equipped, allowing customers to beam music and video from their computers without adapters.[35]

- **Wireless USB:** USB is already the most used interface among PC users (with 2 billion units installed), and a wireless version could combine the speed and security of wired technology with the ease of use of wireless. Based on ultra wideband, just described, _**wireless USB would have a typical range of 32 feet and a maximum data rate of 480 megabits per second.**_ It could be a natural replacement for USB cable links for printers, scanners, MP3 players, hard disks, and the like.

more info!

Meshing

The following websites provide more information about these mesh technologies: _www.insteon.net_, _www.zigbee.org_, and _www.z-wavealliance.com_.

SHORT-RANGE WIRELESS FOR HOME AUTOMATION NETWORKS: INSTEON, ZIGBEE, & Z-WAVE Home automation networks—those that link switches and sensors around the house and yard—use low power, narrowband wireless technology, which operate in a range of 100–250 feet but at relatively slow data rates of 13.1–128 kilobits per second. The current standards are Insteon, ZigBee, and Z-Wave, which, as one writer observes, "sound more like comic book characters than wireless home-automation standards."[36] All three are so-called _mesh technologies_—networked devices equipped with two-way radios that can communicate with each other rather than just with the controller, the device that serves as central command for the network.

- **Insteon:** _Insteon_ combines electric power line and wireless technologies and is capable of sending data at 13.1 kilobits a second at a typical range of 150 feet. With this kind of technology, says the CEO of California-based Smarthome, which invented Insteon, you might "drive up to your house, the garage door device recognizes your car, and opens to let you in. The lights come on and your favorite radio station starts playing."[37] Insteon replaces an older home automation technology known as X10, which has been around for decades.

more info!

Curious about Networked Homes?

For starters, try _www. dummies.com/WileyCDA/ DummiesArticle/id-1630.html_.

- **ZigBee:** _ZigBee_ is an entirely wireless, very power-efficient technology that can send data at 128 kilobytes per second at a range of about 250 feet. It is primarily touted as sensor network technology, and we expect to see it in everything from automatic meter readers and medical sensing and monitoring devices to wireless smoke detectors and TV remote controls.[38] One of the best features is that it can run for years on inexpensive batteries, eliminating the need to be plugged into an electric power line.

- **Z-Wave:** _Z-Wave_ is also an entirely wireless, power-efficient technology, which can send data at 127 kilobits per second to a range of 100 feet. "In a Z-Wave home," says one account, "you can program the lights to go on when the garage door opens; the blinds draw and the lights dim when the TV comes on. You can program devices remotely, turning the thermostat up on the drive home from work."[39]

The various short-range radio technologies are compared in the following table. (● _See Panel 6.16._)

Communications, Networks, & Safeguards

Technology	Typical Range	Maximum Data Rate	Frequency
Local area networks			
Wi-Fi a	150 feet	54 Mbps	5 GHz
Wi-Fi b	150 feet	11 Mbps	2.4 GHz
Wi-Fi g	150 feet	54 Mbps	2.4 GHz
Wi-Fi n	150 feet	200 Mbps	2.4/5 GHz
Personal area networks			
Bluetooth	30 feet	720 Kbps	2.4 GHz
Bluetooth 2.0	32 feet	3 Mbps	2.4 GHz
Ultra wideband (UWB)	30 feet	480 Mbps	3.1–5 GHz
Wireless USB	32 feet	480 Mbps	7.5 GHz
Home automation networks			
Insteon	150 feet	13.1 Kbps	902–925 MHz
ZigBee	250 feet	128 Kbps	868, 910 MHz; 2.4 GHz
Z-Wave	100 feet	127 Kbps	908.4 MHz

● **PANEL 6.16**
Short-range wireless technologies

6.5

Cyberthreats, Hackers, & Safeguards

What are areas I should be concerned about for keeping my computer system secure?

SECURITY

San Francisco internet executive Lew Tucker found his desktop computer so compromised by internet-borne infections that he was spending time every week dealing with them. Rather than get rid of the offending programs, however, he took a radical step: He threw out the whole computer. "I was losing the battle," he says. "It was cheaper and faster to go to the store and buy a low-end PC."[40]

Is throwing your virus-overrun computer in the dumpster a rational response? Many would disagree. However, there's no question that the ongoing dilemma of the Digital Age is balancing convenience against security. *Security* is a system of safeguards for protecting information technology against unauthorized access and systems failures that can result in damage or loss. We consider the following aspects of security as it relates to telecommunications. (Wider matters of computer security are discussed in Chapter 9.)

- **Cyber threats:** denial-of-service attacks, worms, viruses, and Trojan horses
- **Perpetrators of cyber mischief:** hackers and crackers
- **Computer safety:** antivirus software, firewalls, passwords, biometric authentication, and encryption

Cyber Threats: Denial-of-Service Attacks, Worms, Viruses, & Trojan Horses

What are denial-of-service attacks, worms, viruses, and Trojan horses?

SECURITY

Internet users, especially home users, are not nearly as safe as they believe, according to a study by America Online and the nonprofit National Cyber Security Alliance.[41] Consumers suffer from complacency and a lack of expert advice on keeping their computer systems secure, said the group's chief, Ken Watson. "Just like you don't expect to get hit by a car, you don't believe a computer attack can happen to you," he said.[42] The study found that 77% of 326 adults in 12 states felt they were safe from online threats. Yet two-thirds did not have current antivirus software and another two-thirds did not have any firewalls. As we will explain, these two kinds of protections are musts in guarding against important cyber threats: denial-of-service attacks, worms, viruses, and Trojan horses.

DENIAL-OF-SERVICE ATTACKS   A ***denial-of-service (DoS) attack*** **consists of making repeated requests of a computer system or network, thereby overloading it and denying legitimate users access to it.** Because computers are limited in the number of user requests they can handle at any given time, a DoS onslaught will tie them up with fraudulent requests that cause them to shut down. The assault may come from a single computer or from hundreds or thousands of computers that have been taken over by those intending harm.

WORMS   Worms, viruses, and Trojan horses are three forms of *malware,* or malicious software, that attack computer systems. There are about 57,000 known worms and viruses, not to mention others that could afflict your computer in the future. Researchers at Symantec say they typically discover about 100 new worms and viruses every week.[43]

    A ***worm*** **is a program that copies itself repeatedly into a computer's memory or onto a disk drive.** Sometimes it will copy itself so often it will cause a computer to crash. Among some famous worms are Code Red, Nimda, Klez, SQL Slammer, MyDoom, and Sasser. Code Red primarily affected Microsoft Windows NT and 2000 systems that functioned as web servers; it even tried to attack the White House website and many public Pentagon websites. Like Code Red, the Nimda worm attacked Windows NT and 2000 machines; it reportedly crawled through more than a million computers in the United States, Europe, and Asia, clogging internet traffic and resulting in some computer shutdowns. The 2002 worm Klez, dubbed the most common worm ever, spread its damage through Microsoft products by being inside email attachments or part of email messages themselves, so that merely opening an infected message could infect a computer running Outlook or Outlook Express. (● *See Panel 6.17.*) The 2003 worm called *SQL Slammer*

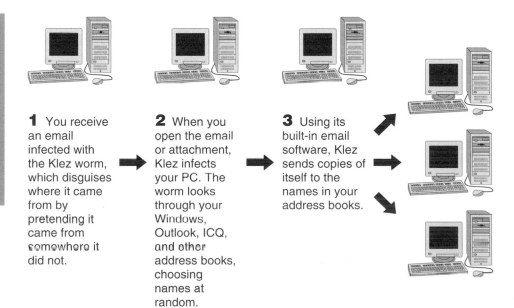

● PANEL 6.17
How one worm works
This is the Klez worm, but others work in similar ways. If you don't recognize the sender of an email message and you don't open the message, but instead just immediately delete it, your PC will not be infected.

1 You receive an email infected with the Klez worm, which disguises where it came from by pretending it came from somewhere it did not.

2 When you open the email or attachment, Klez infects your PC. The worm looks through your Windows, Outlook, ICQ, and other address books, choosing names at random.

3 Using its built-in email software, Klez sends copies of itself to the names in your address books.

(or *Slammer, SQ hell,* or *Sapphire*), which resembled Red Code, did not affect most home users. However, it did infect computers running Microsoft's database software SQL Server 2000, preventing customers from making withdrawals from ATM machines and interfering with email delivery and web browsing. MyDoom was a 2004 email worm that infected up to a million computers worldwide, mostly those belonging to relatively inexperienced users. The Sasser worm was estimated to account for 26% of all virus infections in the first half of 2004. Incredibly, one person, Sven Jaschan, 18, who admitted programming the Sasser and other worms and who was arrested in Germany in May 2004, was responsible for 70% of the virus infections in the early part of that year.[44]

VIRUSES   A _virus_ is a "deviant" program, stored on a computer floppy disk, hard drive, or CD, that can cause unexpected and often undesirable effects, such as destroying or corrupting data. (● *See Panel 6.18.*) The famous email Love Bug (its subject line was I LOVE YOU), which originated in the Philippines in May 2000 and did perhaps as much as $10 billion in damage worldwide, was both a worm and a virus, spreading faster and causing more damage than any other bug before it. The Love Bug was followed almost immediately by a variant virus. This new Love Bug didn't reveal itself with an I LOVE YOU line but changed to a random word or words each time a new computer was infected.

The virus usually attaches itself to your hard disk. It might then display annoying messages ("Your PC is stoned—legalize marijuana") or cause Ping-Pong balls to bounce around your screen and knock away text. More seriously, it might add garbage to your files and then erase or destroy your system software. It may evade your detection and spread its havoc elsewhere, infecting every floppy disk, CD/DVD, and even digital-camera flash memory used by the system.

TROJAN HORSES   If, as a citizen of Troy around 1200–1500 B.C.E., you looked outside your fortified city and saw that the besieging army of Greeks was gone but a large wooden horse was left standing on the battlefield, what would you think? Maybe you would decide it was a gift of the gods, as the Trojans did, and haul it inside the city gates—and be unpleasantly surprised when late at night several Greek soldiers climbed out of the horse and opened the gates for the invading army. This is the meaning behind the illegal program known as a Trojan horse, which, though not technically a virus, can act like one.

A _Trojan horse_ is a program that pretends to be a useful program, usually free, such as a game or screen saver, but carries viruses, or destructive instructions, that perpetrate mischief without your knowledge. One particularly malicious feature is that a Trojan horse may allow so-called backdoor programs to be installed. A *backdoor program* is an illegal program that allows illegitimate users to take control of your computer without your knowledge. An example is the kind of program that records what people type, logging individual keystrokes, paying particular attention to user names and passwords, which can be used to access and even open bank accounts online.

HOW MALWARE IS SPREAD   Worms, viruses, and Trojan horses are passed in the following ways:

- **By infected floppies or CDs:** The first way is via an infected floppy disk or CD, perhaps from a friend or a repair person.

- **Boot-sector virus:** The boot sector is that part of the system software containing most of the instructions for booting, or powering up, the system. The boot sector virus replaces these boot instructions with some of its own. Once the system is turned on, the virus is loaded into main memory before the operating system. From there it is in a position to infect other files. Any diskette that is used in the drive of the computer then becomes infected. When that diskette is moved to another computer, the contagion continues. Examples of boot-sector viruses: AntCMOS, AntiEXE, Form.A, NYB (New York Boot), Ripper, Stoned.Empire.Monkey.

- **File virus:** File viruses attach themselves to executable files—those that actually begin a program. (In DOS these files have the extensions .com and .exe.) When the program is run, the virus starts working, trying to get into main memory and infecting other files.

- **Multipartite virus:** A hybrid of the file and boot-sector types, the multipartite virus infects both files and boot sectors, which makes it better at spreading and more difficult to detect. Examples of multipartite viruses are Junkie and Parity Boot.

  A type of multipartite virus is the *polymorphic virus,* which can mutate and change form just as human viruses can. Such viruses are especially troublesome because they can change their profile, making existing antiviral technology ineffective.

  A particularly sneaky multipartite virus is the *stealth virus,* which can temporarily remove itself from memory to elude capture. An example of a multipartite, polymorphic stealth virus is One Half.

- **Macro virus:** Macro viruses take advantage of a procedure in which miniature programs, known as macros, are embedded inside common data files, such as those created by e-mail or spreadsheets, which are sent over computer networks. Until recently, such documents have typically been ignored by antivirus software. Examples of macro viruses are Concept, which attaches to Word documents and email attachments, and Laroux, which attaches to Excel spreadsheet files. Fortunately, the latest versions of Word and Excel come with built-in macro virus protection.

- **Logic bomb:** Logic bombs, or simply bombs, differ from other viruses in that they are set to go off at a certain date and time. A disgruntled programmer for a defense contractor created a bomb in a program that was supposed to go off two months after he left. Designed to erase an inventory tracking system, the bomb was discovered only by chance.

- **Trojan horse:** The Trojan horse covertly places illegal, destructive instructions in the middle of a legitimate program, such as a computer game. Once you run the program, the Trojan horse goes to work, doing its damage while you are blissfully unaware. An example of a Trojan horse is FormatC.

- **Email hoax:** These are really not virus programs; they are email messages sent by usually well-meaning people to warn others about a new virus they have read or heard of. These false warning messages often say "be sure to send this to everyone you know." Hoax virus warnings can cause huge amounts of internet traffic and unnecessary worry. Users should check with knowledgeable sources before forwarding such messages.

● PANEL 6.18
Types of worms and viruses

- **By opening unknown email attachments:** The second way is from an email attachment. This is why a basic rule of using the internet is: *Never* click on an email attachment that comes from someone you don't know. This advice also applies to unknown downloaded files, as for free video games or screen savers.

- **By clicking on infiltrated websites:** Some crackers "seed" web pages with contagious malware that enable them to steal personal data, so that by simply clicking on a website you can unwittingly compromise your PC. The risk can be minimized if you have a firewall and keep antivirus software on your computer up to date, as we describe below. (You might also consider switching to Mozilla Firefox from Internet Explorer as your browser.)

- **Through infiltrated Wi-Fi hot spots:** As mentioned earlier, if you're a user of Wi-Fi wireless access points, or hot spots, you have to be aware that your laptop or PDA could be exposed to wireless transmitted diseases from illegal users. "Many hot spots do not require passwords," says one account. "That lets anyone with a wireless connection and hacking know-how to hop aboard the network."[45] Here, too, having wireless firewalls can reduce risks.

A recent concern is that worms and viruses have been found to attack cellphones. One person was arrested in 2004 after breaking into a wireless carrier's network, where for 7 months he read email and personal computer files of hundreds of customers—including those of the Secret Service agent investigating him.[46]

We explore other threats to computer systems—fraud, theft, data manipulation, and the like—in Chapter 9. Now let us consider the people who perpetrate cyber threats—hackers and crackers.

Some Cyber Villains: Hackers & Crackers

What are the various kinds of hackers and crackers?

The popular press uses the word *hacker* to refer to people who break into computer systems and steal or corrupt data, but this is not quite the exact definition. Perhaps it helps to distinguish between hackers and crackers, although the term *cracker* has never caught on with the general public.

>>

PRACTICAL ACTION
Ways to Minimize Virus Attacks

Some tips for minimizing the chances of infecting your computer are as follows:

- Don't open, download, or execute any files, email messages, or email attachments if the source is unknown or if the subject line of an email is questionable or unexpected.

- Delete all spam and email messages from strangers. Don't open, forward, or reply to such messages.

- Use webmail (HTML email) sparingly, since viruses can hide in the HTML coding of the email. Even the simple act of previewing the message in your email program can activate the virus and infect your computer.

- Don't start your computer with a floppy disk in the floppy-disk drive (drive A), except for the recovery disk that came with your PC.

- Back up your data files regularly, and keep the backup floppy disks, CDs, or whatever in a location separate from your computer. Then if a virus (or a fire) destroys your work files, your data won't be totally devastated.

- Make sure you have virus protection software, such as McAfee VirusScan *(www.mcafee.com)* or Norton

AntiVirus *(www.symantec.com/nav)* activated on your machine. You can download it from these companies; then follow the installation instructions.

- Buying a new computer with antivirus software on it does not mean you are automatically protected. The software could be 6 months old and not cover new viruses. You have to register it with the designer company, and you need to receive antivirus updates.

- Scan your entire system with antivirus software the first time it's installed; then scan it regularly after that. Often the software can be set to scan each time the computer is rebooted or on a periodic schedule. Also scan any new floppy disks, CDs, Zip disks, and the like before using them.

- Update your antivirus software regularly. There are virus and security alerts almost every day. Any antivirus software is automatically linked to the internet and will add new virus diction code to your system whenever the software vendor discovers a new threat.

- If you discover you have a virus, you can ask McAfee or Norton to scan your computer online. Then follow the company's directions for cleaning or deleting it.

HACKERS _Hackers_ **are defined (1) as computer enthusiasts, people who enjoy learning programming languages and computer systems, but also (2) as people who gain unauthorized access to computers or networks, often just for the challenge of it.**

Considering the second kind of hacker, those who break into computers for relatively benign reasons, there are probably two types:

- **Thrill-seeker hackers:** _Thrill-seeker hackers_ are hackers who illegally access computer systems simply for the challenge of it. Although they penetrate computers and networks illegally, they don't do any damage or steal anything; their reward is the achievement of breaking in.

- **White-hat hackers:** _White-hat hackers_ are usually computer professionals who break into computer systems and networks with the knowledge of their owners to expose security flaws that can then be fixed. (The term "white hat" refers to the hero in old Western movies, who often wore a white hat, as opposed to the villain, who usually wore a black hat—see below.) Kevin Mitnik, for instance, became a tech security consultant after serving 5 years in prison for breaking into corporate computer systems in the mid-1990s.

CRACKERS As opposed to hackers who do break-ins for more or less positive reasons, _crackers_ **are malicious hackers, people who break into computers for malicious purposes**—to obtain information for financial gain, shut down hardware, pirate software, steal people's credit information, or alter or destroy data.

There seem to be four classes of crackers:

- **Script kiddies:** On the low end are _script kiddies,_ mostly teenagers without much technical expertise who use downloadable software or source code to perform malicious break-ins. Script kiddies (or "script bunnies") use published source code to construct viruses. Although releasing the viruses themselves is illegal, publishing the code for viruses is not; it is allowed under the First Amendment right to free speech.[47]

- **Hacktivists:** _Hacktivists_ are "hacker activists," people who break into a computer system for a politically or socially motivated purpose. For example, they might leave a highly visible message on the home page of a website that expresses a point of view that they oppose.

- **Black-hat hackers:** _Black-hat hackers_ are those who break into computer systems to steal or destroy information or to use it for illegal profit. They are usually not bored, crafty teenagers but often professional criminals, the people behind the increase in cyberattacks on corporate networks, which Symantec, the world's largest supplier of antivirus software, reported rose 332% between 2003 and 2004.[48]

- **Cyberterrorists:** Cyberterrorism, according to the FBI, is any "premeditated, politically motivated attack against information, computer systems, computer programs, and data which results in violence against noncombatant targets by sub-national groups or clandestine agents."[49] Thus, _cyberterrorists_ are politically motivated persons who attack computer systems so as to bring physical or financial harm to a lot of people or destroy a lot of information. Particular targets are power plants, water systems, traffic control centers, banks, and military installations.

The most flagrant cases of cracking are met with federal prosecution. For instance, Jeffrey Lee Parson of Hopkins, Minnesota, was an 18-year-old high school senior when he unleashed a variant of the Blaster internet worm in

2003 that crippled an estimated 48,000 computers—a crime for which he could have been sentenced to 10 years in prison and fined $250,000. Because Parson was a minor at the time of the crime, however, a federal district court judge gave him 18 months in prison.[50]

Online Safety: Antivirus Software, Firewalls, Passwords, Biometric Authentication, & Encryption

What should I be doing to keep my computer safe against cyber threats?

Now let us consider a few things you can do to try to keep your computer system safe: antivirus software, firewalls, passwords, biometric authentication, and encryption.

ANTIVIRUS SOFTWARE   A variety of virus-fighting programs are available. **_Antivirus software_ scans a computer's hard disk, floppy disks and CDs, and main memory to detect viruses and, sometimes, to destroy them.** Such virus watchdogs operate in two ways. First, they scan disk drives for "signatures," characteristic strings of 1s and 0s in the virus that uniquely identify it. Second, they look for suspicious viruslike behavior, such as attempts to erase or change areas on your disks.

Examples of antivirus programs are McAfee VirusScan, Norton AntiVirus, Pc-cillin Internet Security, and ZoneAlarm with Antivirus. Others worth considering are eTrust EZ Antivirus for Windows, Panda Antivirus Platinum, and Virex for Macs.

Other ways of protecting your computer against viruses are given in the Practical Action box on page 348.

FIREWALLS   **A _firewall_ is a system of hardware and/or software that protects a computer or a network from intruders.** The firewall software monitors all internet and other network activity, looking for suspicious data and preventing unauthorized access. Always-on internet connections such as cable modem and DSL, as well as some wireless devices, are particularly susceptible to unauthorized intrusion.

- **If you have one computer—software firewall:** If you have just one computer, a software firewall is probably enough to protect you while you're connected to the internet. (Windows XP has a built-in software firewall that can be activated quickly.)
- **If you have more than one computer—hardware firewall:** If you have more than one computer and you are linked to the internet by a cable modem or DSL, you probably need a hardware firewall, such as a hub/router, which is available for as little as $50.

PASSWORDS   When New York's World Trade Center was destroyed during the September 11, 2001, terrorist attack, debt-trading firm Cantor Fitzgerald lost 700 of its 1,000 employees—and no one knew the deceased workers' passwords, the special words, codes, or symbols required to access a computer system. Although records had been backed up and existed in another location, to maintain its customers' confidence, the company realized it had to be back up and running within 2 days. That meant discovering the passwords needed to get into essential files. What was it to do? According to one account, what surviving employees did was this: "They sat around in a group and recalled everything they knew about their colleagues, everything they had done, everywhere they had been, and everything that had ever happened to them. _And they managed to guess the passwords._"[51]

As this story shows, protecting your internet access accounts and files with a password isn't enough. Passwords (and PINs, too) can be guessed, forgotten, or stolen. To foil a stranger's guesses, experts say, you should never

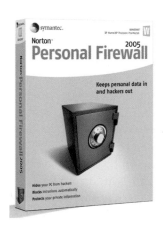

Windows XP comes with a fire-wall. Click on *Start*. Open *My Network Places*. Click on *View network connections*. Right-click the connection you use and choose *Properties*. Click the *Advanced tab*. Under *Internet Connections Firewall*, check *Protect my computer and network by limiting or preventing access to this computer from the internet*.

If you buy separate firewall software, turn off XP's firewall.

choose a real word or variations of your name, your birth date, or those of your friends or family. Instead you should mix letters, numbers, and punctuation marks in an oddball sequence of no fewer than eight characters. Examples of some good passwords: *2b/orNOT2b%*. *Alfred!E!Newman7*. Or you can also choose an obvious and memorable password but shift the position of your hands on the keyboard, creating a meaningless string of characters—the best kind of password. (Thus, *ELVIS* becomes *R;BOD* when you move your fingers one position right on the keyboard.)

Some other practical advice about passwords is given in the Practical Action box on the next page.

BIOMETRIC AUTHENTICATION A hacker or cracker can easily breach a computer system with a guessed or stolen password. But some forms of identification can't be easily faked—such as your physical traits. **Biometrics, the science of measuring individual body characteristics,** tries to use these in security devices. *Biometric authentication devices* authenticate a person's identity by comparing his or her physical or behavioral characteristics with digital code stored in a computer system. For example, before University of Georgia students can use the all-you-can-eat plan at the campus cafeteria, they must have their palms read. As one writer described the system, "a camera automatically compares the shape of a student's hand with an image of the same hand pulled from the magnetic strip of an ID card. If the patterns match, the cafeteria turnstile automatically clicks open. If not, the would-be moocher eats elsewhere."[52]

There are several kinds of devices for verifying physical or behavioral characteristics that can be used to authenticate a person's identity. (● *See Panel 6.19.*)

- **Hand-geometry systems:** Also known as *full-hand palm scanners*, these are like the University of Georgia cafeteria system—devices to verify a person's identity by scanning the entire hand, which, for each person, is as unique as a fingerprint and changes little over time.

- **Fingerprint scanners:** These range from optical readers, in which you place a finger over a window, to swipe readers, such as those built into laptops and some handhelds, which allow you to run your finger across a barlike sensor. Microsoft offers optical fingerprint readers to go with Windows XP.

- **Iris-recognition systems:** Because no two people's eyes are alike, iris scans are very reliable identifiers. In Europe, some airports are using iris-scanning systems as a way of speeding up immigration controls. The Nine Zero, an upscale hotel in Boston, has experimented with letting guests enter one of its more expensive suites by staring into a camera that analyzes iris patterns.

● PANEL 6.19
Biometric device
Iris recognition.

Wi-Fi connection in a Seattle park

more info!

Encryption Software

Some kinds of encryption software worth investigating:
PGP Personal Desktop
(www.pgp.com)
WinMagic's SecureDoc
(www.winmagic.com)

Examples of the two types are shown below. (● *See Panel 6.20.*)

Besides using encryption to secure email messages or transactions on e-commerce websites, you may also want to protect files on your computer—especially laptops and PDAs, in case you leave them behind on a train or bus. You can use special software to encrypt individual files, create "vaults" with many files or entire hard disks, or integrate smart cards to add additional security.[62] To shield your laptop against people snooping on your Wi-Fi connections, you can encrypt your data using such systems as Wired Equivalent Privacy (WEP) or the newer Wi-Fi Protected Access (WPA).[63] The Wi-Fi Alliance's website *(www.wi-fi.org)* offers information about WPA.

We return to the subject of security in Chapter 9.

● **PANEL 6.20**
Examples of two types of encryption

Private-key encryption

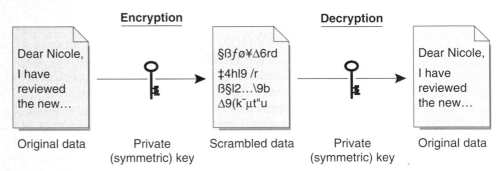

Original data   Private (symmetric) key   Scrambled data   Private (symmetric) key   Original data

Public-key encryption

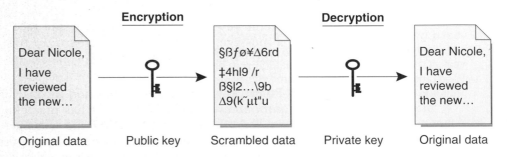

Original data   Public key   Scrambled data   Private key   Original data

The Future of Communications

What are the characteristics of the next generation of wired and wireless communications?

Clearly, we are very near the time when voices, images, and information can be transmitted to any place on earth. Indeed, these advances already are revolutionizing how we talk to each other and how we relate to each other. Now let's consider the outlook for the future. New kinds of wired and wireless services are rapidly being put in place, promising to offer us lots of choices. The following are a few such developments.

Satellite-Based Systems

What are possible new developments in satellites?

The first communications satellite, AT&T's Telstar, went up in 1962. Now all of a sudden it looks like we will have a traffic jam in space. In the next few years, four kinds of satellite systems will dot the skies to provide a variety of consumer services. The first is the TV direct-broadcast system, and the second is the GPS system, both of which we described earlier. The third type is designed to handle cellphone and paging services, using satellite transmissions in place of tower-to-tower microwave transmissions.

Probably most interesting is the fourth, which consists of global high-speed, low-orbital satellite networks that will let users exchange a much broader range of data, including videophone calls and satellite-delivered internet services. As we mentioned earlier, today there are 20 million consumers, primarily in rural areas, who can't get high-speed internet service from a telephone or cable-TV company. Clearly, they would benefit from a nationally promoted satellite-based internet service.[64]

Beyond 3G to 4G

How might I benefit from 4G wireless technology?

Third-generation phones, you'll recall, are supposed to send and receive voice and data at speeds of up to 2 megabits per second eventually, which is similar to rates offered by broadband internet connections. Hewlett-Packard is already looking beyond 3G to 4G wireless technology. Its partner in this endeavor is NTT DoCoMo Inc., a unit of the Japanese company Nippon Telephone & Telegraph Corp. DoCoMo has become famous for helping to raise interest in 3G with its commercially popular "i-mode" platform. This is a wildly successful service in Japan that features an always-on connection and delivery of all kinds of internet data from news to games to music.

Among the 4G issues H-P scientists are tackling, according to one account, are scalable coding, error-resisting coding, and transcoding. *Scalable coding* is a way to code video streams to work on many different devices. *Error-resisting coding* prevents transmissions from breaking up. *Transcoding* is a process of converting compressed video to any playback format.[65]

Photonics: Optical Technologies at Warp Speed

How would photonics speed up fiber-optic lines?

Photonics is the science of sending data bits by means of light pulses carried on hair-thin glass fibers. For 15 years, the glass fibers of fiber-optic lines have been used to carry light pulses representing voice and data in long-distance telephone lines. Photonics has achieved breakthroughs that enable glass fibers to carry more light signals than ever before.

Older fiber-optic technologies were limited to only a few dozen miles. Then, the light beams had to be converted to electrical signals, amplified, and converted back into light signals. This made the technique slow, unreliable, and expensive.

Then, in 1988, researcher David Payne at the University of Southampton in England developed an optical amplifier. This device boosts light signals without converting them first to electrical signals. The amplifier gives a huge push to the incoming signal, letting it carry on for dozens of more miles to the next amplifier. Engineers also devised another technology (called *wave-division multiplexing* or *dense wavelength-division multiplexing*) that allows laser pulses of different hues to be sent down the same tiny fiber. This allows carriers to transmit at least 16 channels per fiber, which may grow to 100 channels per fiber in a few years. The upshot is that in the laboratory researchers have been able to produce glass-fiber systems that carry 2 trillion bits (2 terabits) per second. That is six times the volume of all phone calls in the United States on an average day.

Software-Defined Radio

What would software-defined radio do?

Every new wireless communications device that comes along—cellphone, cordless home phone, radio, baby monitor, or whatever—has processing designed for a particular frequency band and technical standard, limiting it to communicating with similar devices. A field called *software-defined radio* aims to use software rather than hardware to control how such wireless gadgets recognize and manage signals, an approach providing many advantages.

The idea is to create software that can recognize various waveforms in any frequency and choose the appropriate applications to process them. With this, you could communicate with any wireless device and upgrade to advanced systems, like new generations of cellular technology, without having to replace your hardware.[66]

A New Way to Compute: The Grid

How would "the grid" operate?

At present, most business servers are used only 30% of the time, and the typical PC uses only about 5% of its capacity.[67] It's only a matter of time, many computer professionals believe, before the internet gives way to a new computing model known as *the grid*. This refers to the linking of many servers into a single system for the purpose of breaking down complex computing tasks and doing work previously possible only with expensive supercomputers. Even now geologists are using grid networks in a limited way to simulate the effects of earthquakes, and biochemists use it to simulate virus attacks on the human body. But the grid will become truly revolutionary when it links many institutions together—consumers, companies, and governments, for example—and makes possible broadband videoconferencing, telecommuting, e-banking, and other innovations.

Visitors to the Hong Kong Fashion Week expo surf the internet at computer terminals provided by a local business portal and trade provider.

Experience Box
Virtual Meetings: Linking Up Electronically

"I don't usually put on lipstick before making a phone call," wrote Anita Hamilton, "but this time was different. I was placing my first video phone call over the internet."[68]

Although video online chatting has been available for a long time, it has been hampered by the slowness of low-bandwidth dial-up modems, so pictures have not translated well. Now that more people have broadband (DSL, cable-modem, or even T1 connections) the online video quality is much better.

The videophone is more than a luxury for personal use, however. In business, organizational ideas about distance are changing, influencing when people meet face-to-face and when they link up electronically. If you're of a generation that's already used to instant-message chats (and multitasking by simultaneously watching TV, talking on the phone, and doing your homework), you may find yourself in a comfortable relationship with virtual meetings at work.

Types of Virtual Meetings: Remote Conferencing

Since the 9/11 terrorist attacks and the long economic downturn of the early 2000s, both of which led to the cancellation of a lot of business travel, videoconferencing has become a more accepted way of meeting over long distances. There are several ways in which people can conduct "virtual meetings"—meetings that don't entail physical travel, as follows. (For more information, see *www.coworking.com/resources.*)

Audioconferencing. *Audioconferencing* (or *teleconferencing*) is simply telephone conferencing. A telephone conference-call operator can arrange this setup among any three or more users. Users don't need any special equipment beyond a standard telephone. Audioconferencing is an inexpensive way to hold a long-distance meeting and is often used in business.

Computer-based audio, involving microphones, headsets or speakers, and internet-based telephony, is also used for audioconferencing.

Videoconferencing Using Closed-Circuit TV. "I was a little nervous about going in front of the camera," said job applicant Mark Dillard, "but I calmed down pretty quickly after we got going, and it went well."[69] Interviewing for a job can be uncomfortable for many people. However, Dillard had just talked to a job recruiter in New York while sitting in front of a video camera in a booth at a local Kinko's store in Atlanta.

This kind of closed-circuit television videoconferencing requires dedicated hardware and can be expensive. Companies will sometimes outfit conference rooms with this kind of equipment, and conference participants will then gather in these special rooms in different locations.

Videoconferencing Using a Webcam. A *webcam* is a tiny, often eyeball-shaped camera that sits atop a computer monitor and displays its output on a web page. The Logitech QuickCam Orbit Camera is able to automatically follow your face as you move. Headsets with built-in speakers and microphones are required. Yahoo and MSN offer free video chatting through their instant-message programs.

Videoconferencing Using PC Video Cameras. This kind of videoconferencing involves people making a video telephone call over the internet, with both parties being able to see each other as they talk. PC video cameras (which are available for less than $100) are positioned atop computer monitors and are connected to the computer through the USB or FireWire port.

Videoconferencing Using Videophones. Videophones don't involve use of a PC, although the calls take place on the internet. The internet company 8×8 Inc., for example, makes a VoIP (voice over internet protocol) videophone that has a built-in 4- by 3-inch display screen with camera mounted above, sells for $99, connects to a cable modem or DSL modem, and comes with a service package costing $19.95 a month for unlimited local and long-distance calls (which means you can use the phone both for video calls and regular voice calls).[70]

Workgroup Computing or Web Conferencing. The use of groupware allows two or more people on a network to share information, collaborating on graphics, slides, and spreadsheets while linked by computer and telephone. WebEx and PlaceWare are two of several web-based services that allow participants to conduct meetings using tools such as "whiteboards," which can display drawings or text, along with group presentations (using PowerPoint, which displays slides on everyone's screen) and online chatting. Usually people do these things while talking to one another on the telephone. Users can collaborate by writing or editing word processing documents, spreadsheets, or presentations. A relatively new kind of software, UGS collaboration software, allows thousands of people to cooperate on a single project.[71]

Ethernet (p. 324) LAN technology that can be used with almost any kind of computer and that describes how data can be sent in packets in between computers and other networked devices usually in close proximity. Why it's important: *Ethernet has become the most popular and most widely deployed network technology in the world.*

extranet (p. 320) Private intranet that connects not only internal personnel but also selected suppliers and other strategic parties. Why it's important: *Extranets have become popular for standard transactions such as purchasing.*

fiber-optic cable (p. 325) Cable that consists of dozens or hundreds of thin strands of glass or plastic that transmit pulsating beams of light rather than electricity. Why it's important: *These strands, each as thin as a human hair, can transmit up to 2 billion pulses per second (2 Gbps); each "on" pulse represents 1 bit. When bundled together, fiber-optic strands in a cable 0.12 inch thick can support a quarter-million to a half-million voice conversations at the same time. Moreover, unlike electrical signals, light pulses are not affected by random electromagnetic interference in the environment. Thus, they have much lower error rates than normal telephone wire and cable. In addition, fiber-optic cable is lighter and more durable than twisted-pair wire and co-ax cable. A final advantage is that it cannot easily be wiretapped, so transmissions are more secure.*

firewall (p. 350) System of hardware and/or software that protects a computer or a network from intruders. Always-on internet connections such as cable modem and DSL, as well as some wireless devices, are particularly susceptible to unauthorized intrusion and so need a firewall. Why it's important: *The firewall monitors all internet and other network activity, looking for suspicious data and preventing unauthorized access.*

gateway (p. 322) Interface permitting communication between dissimilar networks. Why it's important: *Gateways permit communication between a LAN and a WAN or between two LANs based on different network operating systems or different layouts.*

Global Positioning System (GPS) (p. 334) A series of earth-orbiting satellites continuously transmitting timed radio signals that can be used to identify earth locations. Why it's important: *A GPS receiver—handheld or mounted in a vehicle, plane, or boat—can pick up transmissions from any four satellites, interpret the information from each, and calculate to within a few hundred feet or less the receiver's longitude, latitude, and altitude. Some GPS receivers include map software for finding one's way around, as with the Guidestar system available with some rental cars.*

hacker (p. 349) A hacker can be either (1) a computer enthusiast, a person who enjoys learning programming languages and computer systems, or (2) a person who gains unauthorized access to computers or networks, often just for the challenge of it. Why it's important: *Unlike crackers who have malevolent purposes, a hacker may break in to computers for more or less positive reasons.*

home area network (HAN) (p. 317) Network using wired, cable, or wireless connections to link a household's digital devices. Why it's important: *A HAN can connect not only multiple computers, printers, and storage devices but also VCRs, DVDs, televisions, fax machines, video game machines, and home security systems.*

home automation network (p. 318) Network that relies on very inexpensive, very short-range, low-power (AA batteries) wireless technology in the under-200-Kbps range, such as Insteon, ZibBee, and Z-Wave. Why it's important: *This network can be used to link switches and sensors around the house and can control lights and switches, thermostats and furnaces, smoke alarms and outdoor floodlights.*

HomePlug (p. 326) Technological standard that allows users to send data over a home's existing electrical (AC) power lines, which can be transmitted at 14 megabits per second. Why it's important: *HomePlug can use a home's existing electric-power wiring to create a computer network.*

HomePNA (HPNA) (p. 326) Technological standard that allows users to use a home's existing telephone wiring for a home network, transmitting data at speeds of 10–240 megabits per second. No hubs are needed. Why it's important: *This standard allows people to use a home's existing telephone wiring to develop a home network.*

host computer (p. 320) A computer in a client/server network that controls a network and the devices on it, called nodes. Why it's important: *The host computer controls access to the hardware, software, and other resources on the network.*

hub (p. 321) A common connection point for devices in a network—a place of convergence where data arrives from one or more directions and is forwarded out in one or more other directions. Why it's important: *Hubs are commonly used to connect segments of a LAN.*

infrared wireless transmission (p. 330) Transmission of data signals using infrared-light waves. Why it's important: *Infrared ports can be found on some laptop computers and printers, as well as wireless mice. The advantage is that no physical connection is required among devices. The drawbacks are that line-of-sight communication is required—there must be an unobstructed view between transmitter and receiver—and transmission is confined to short range.*

intranet (p. 319) An organization's internal private network that uses the infrastructure and standards of the internet and the web. Why it's important: *When an organization creates an intranet, it enables employees to have quicker access to internal information and to share knowledge so that they can do their jobs better. Information exchanged on intranets may include employee email addresses and telephone numbers, product information, sales data, employee benefit information, and lists of jobs available within the organization.*

local area network (LAN) (p. 317) Communications network that connects computers and devices in a limited geographic area, such as one office, one building, or a group of buildings close together (for instance, a college campus). Why it's important: *LANs have replaced large computers for many functions and are considerably less expensive.*

medium band (p. 329) Bandwith used for transmitting data over long distances. Transmission rates are 100 kilobits to 1 megabit per second. Why it's important: *Medium band is used to transmit pictures, video, and high-fidelity sound. Compare with* **broadband** *and* **narrowband.**

metropolitan area network (MAN) (p. 317) Communications network covering a city or a suburb. Why it's important: *The purpose of a MAN is often to bypass local telephone companies when accessing long-distance services. Many cellphone systems are MANs.*

microwave radio (p. 331) Transmission of voice and data through the atmosphere as super-high-frequency radio waves called *microwaves.* These frequencies are used to transmit messages between ground-based stations and satellite communications systems. Why it's important: *Microwaves are line-of-sight; they cannot bend around corners or around the earth's curvature, so there must be an unobstructed view between transmitter and receiver. Thus, microwave stations need to be placed within 25–30 miles of each other, with no obstructions in between. In a string of microwave relay stations, each station receives incoming messages, boosts the signal strength, and relays the signal to the next station. Nowadays dish- or horn-shaped microwave reflective dishes, which contain transceivers and antennas, are nearly everywhere.*

modem (p. 314) Short for "modulate/demodulate"; device that converts digital signals into a representation of analog form (modulation) to send over phone lines. A receiving modem then converts the analog signal back to a digital signal (demodulation). Why it's important: *The modem provides a means for computers to communicate with one another using the standard copper-wire telephone network, an analog system that was built to transmit the human voice but not computer signals.*

narrowband (p. 329) Also known as *voiceband.* Bandwidth used for short distances. Transmission rates are usually 100 kilobits per second or less. Why it's important: *Narrowband is used in telephone modems and for regular telephone communications—for speech, faxes, and data. Compare with* **broadband** *and* **medium band.**

network (p. 315) Also called *communications network;* system of interconnected computers, telephones, or other communications devices that can communicate with one another and share applications and data. Why it's important: *The tying together of so many communications devices in so many ways is changing the world we live in.*

node (p. 320) Any device that is attached to a network. Why it's important: *A node may be a microcomputer, terminal, storage device, or peripheral device, any of which enhance the usefulness of the network.*

packet (p. 320) Fixed-length block of data for transmission. The packet also contains instructions about the destination of the packet. *Why it's important: By creating blocks, in the form of packets, a transmission system can deliver the data more efficiently and economically.*

pager (p. 336) Commonly known as *beeper;* simple radio receiver that receives data sent from a special radio transmitter. The pager number is dialed from a phone and travels via the transmitter to the pager. *Why it's important: Pagers have become a common way of receiving notification of phone calls so that the user can return the calls immediately; some pagers can also display messages of up to 80 characters and send preprogrammed messages.*

peer-to-peer (P2P) network (p. 319) Type of network in which all computers on the network communicate directly with one another rather than relying on a server, as client/server networks do. *Why it's important: Every computer can share files and peripherals with all other computers on the network, given that all are granted access privileges. Peer-to-peer networks are less expensive than client/server networks and work effectively for up to 25 computers, making them appropriate for home networks. Compare with* **client/server network.**

personal area network (PAN) (p. 317) Type of network that uses short-range wireless technology such as Bluetooth, ultra wideband, and wireless USB, to connect a person's personal electronics. *Why it's important: PANs are used to connect such items on a personal electronics as cellphone, PDA, MP3 player, notebook PC, and printer.*

protocol (p. 320) Also called *communications protocol;* set of conventions governing the exchange of data between hardware and/or software components in a communications network. *Why it's important: Protocols are built into hardware and software. Every device connected to a network must have an internet protocol (IP) address so that other computers on the network can route data to that address.*

radio-frequency (RF) spectrum (p. 329) The part of the electromagnetic spectrum that carries most communications signals. *Why it's important: The radio spectrum ranges from low-frequency waves, such as those used for aeronautical and marine navigation equipment, through the medium frequencies for CB radios, cordless phones, and baby monitors, to ultra-high-frequency bands for cellphones and also microwave bands for communications satellites.*

ring network (p. 323) Type of network in which all communications devices are connected in a continuous loop and messages are passed around the ring until they reach the right destination. There is no central server. *Why it's important: The advantage of a ring network is that messages flow in only one direction and so there is no danger of collisions. The disadvantage is that if a connection is broken, the entire network stops working.*

router (p. 322) Special computer that directs communicating messages when several networks are connected together. *Why it's important: High-speed routers can serve as part of the internet backbone, or transmission path, handling the major data traffic.*

star network (p. 324) Type of network in which all microcomputers and other communications devices are connected to a central hub, such as a file server. Electronic messages are routed through the central hub to their destinations. The central hub monitors the flow of traffic. *Why it's important: The advantage of a star network is that the hub prevents collisions between messages. Moreover, if a connection is broken between any communications device and the hub, the rest of the devices on the network will continue operating.*

switch (p. 322) A device that connects computers to a network. Unlike a hub, it sends messages only to a computer that is the intended recipient. *Why it's important: A switch is a full-duplex device, meaning data is transmitted back and forth at the same time, which improves the performance of the network.*

topology (p. 322) The logical layout, or shape, of a network. The five basic topologies are star, ring, and bus. *Why it's important: Different topologies can be used to suit different office and equipment configurations.*

Trojan horse (p. 346) A program that pretends to be a useful program, such as a game or screen saver, but that carries viruses, or destructive instructions. *Why it's important: A Trojan horse can perpetrate mischief without your knowledge, such as allow installation of backdoor programs, illegal programs that allow illegitimate users to take control of your computer without your knowledge.*

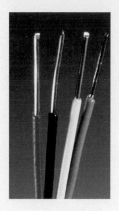

twisted-pair wire (p. 325) Two strands of insulated copper wire, twisted around each other. Why it's important: *Twisted-pair wire has been the most common channel or medium used for telephone systems. However, it is relatively slow and does not protect well against electrical interference.*

ultra wideband (UWB) (p. 342) A promising technology operating in the range of 480 megabits per second up to 30 feet that uses a low power source to send out millions of bursts of radio energy every second over many different frequencies, which are then reassembled by a UWB receiver. Why it's important: *UWB operates over low frequencies not used by other technologies, making it a candidate to replace many of the cables that now currently connect household and office devices.*

virtual private network (VPN) (p. 320) Private network that uses a public network, usually the internet, to connect remote sites. Why it's important: *Because wide area networks use leased lines, maintaining them can be expensive, especially as distances between offices increase. To decrease communications costs, some companies have established their own VPNs. Company intranets, extranets, and LANS can all be parts of a VPN.*

virus (p. 346) Deviant program that can cause unexpected and often undesirable effects, such as destroying or corrupting data. Why it's important: *Viruses can cause users to lose data and/or files or can shut down entire computer systems.*

wide area network (WAN) (p. 316) Communications network that covers a wide geographic area, such as a country or the world. Why it's important: *Most long-distance and regional telephone companies are WANs. A WAN may use a combination of satellites, fiber-optic cable, microwave, and copper-wire connections and link a variety of computers, from mainframes to terminals.*

Wireless Application Protocol (WAP) (p. 329) Communications protocol designed to link nearly all mobile devices to your telecommunications carrier's wireless network and content providers. Why it's important: *Wireless devices such as cellphones use the Wireless Application Protocol for connecting wireless users to the World Wide Web.*

wireless USB (p. 343) Wireless standard with a typical range of 32 feet and a maximum data rate of 480 megabits per second. Why it's important: *Because USB is already the most used interface among PC users, a wireless version could combine the speed and security of wired technology with the ease of use of wireless. It could be a natural replacement for USB cable links for printers, scanners, MP3 players, hard disks, and the like.*

worm (p. 345) Program that copies itself repeated into a computer's memory or onto a disk drive until no space is left. Why it's important: *Worms can shut down computers.*

Chapter Review

stage **LEARNING** **MEMORIZATION**

"I can recognize and recall information."

Self-Test Questions

1. A(n) _____ converts digital signals into analog signals for transmission over phone lines.

2. A(n) _____ network covers a wide geographic area, such as a state or a country.

3. _____ cable transmits data as pulses of light rather than as electricity.

4. _____ refers to waves continuously varying in strength and/or quality; _____ refers to communications signals or information in a binary form.

5. A _____ is a private network that uses a public network (usually the internet) to connect remote sites.

6. A(n) _____ is a computer that acts as a disk drive, storing programs and data files shared by users on a LAN.

7. The _____ is the system software that manages the activities of a network.

8. Modem is short for _____.

9. _____ is a short-range wireless digital standard aimed at linking cellphones, PDAs, computers, and peripherals up to distances of 30 feet; _____ is a short-range wireless digital standard aimed at helping portable computers and handheld wireless devices to communicate at high speeds and share internet connections at distances of 50–150 feet.

10. _____ programs can screen out objectionable material on the internet.

11. Any device that is attached to a network is called a _____.

12. A set of conventions governing the exchange of data between hardware and software components in a communications network is called a _____.

Multiple-Choice Questions

1. Which of the following best describes the telephone line that is used in most homes today?
   a. coaxial cable
   b. modem cable
   c. twisted-pair wire
   d. fiber-optic cable
   e. LAN

2. Which of the following do local area networks enable?
   a. sharing of peripheral devices
   b. sharing of programs and data
   c. better communications
   d. access to databases
   e. all of these

3. Which of the following is *not* a type of server?
   a. file server
   b. print server
   c. mail server
   d. disk server
   e. database server

4. Which of the following is *not* a short-distance wireless standard?
   a. Bluetooth
   b. PDA
   c. Wi-Fi
   d. WAP

5. Which type of local area network connects all devices through a central hub, such as a file server?
   a. bus network
   b. star network
   c. mesh network
   d. ring network
   e. router

6. Which is the type of local area network in which all microcomputers on the network communicate directly with one another without relying on a server?
   a. client/server
   b. domain
   c. peer to peer
   d. MAN
   e. WAN

7. How do fiber-optic cables transmit data?
   a. via copper wire
   b. via infrared
   c. via AC electric current
   d. via radio waves
   e. via pulsating beams of light

True/False Questions

T F   1. In a LAN, a bridge is used to connect the same type of networks, whereas a gateway is used to enable dissimilar networks to communicate.

T F 2. Frequency and amplitude are two characteristics of analog carrier waves.

T F 3. A range of frequencies is called a *spectrum*.

T F 4. Twisted-pair wire commonly connects residences to external telephone systems.

T F 5. Photonics is the science of sending data bits by means of light pulses carried on hair-thin glass fibers.

T F 6. Wi-Fi signals can travel up to 300 feet.

T F 7. Microwave transmissions are a line-of-sight medium.

T F 8. 2G is the newest cellphone standard.

T F 9. Wi-Fi is formally known as a 802.11 network.

T F 10. A Trojan horse copies itself repeatedly into a computer's memory or onto a hard drive.

 stage **LEARNING** COMPREHENSION

"I can recall information in my own terms and explain it to a friend."

Short-Answer Questions

1. What is the difference between an intranet and an extranet?

2. What is the difference between a LAN and a WAN?

3. Why is bandwidth a factor in data transmission?

4. What is a firewall?

5. What do *2G* and *3G* mean?

6. Explain the differences between ring, bus, and star networks.

7. What is the electromagnetic spectrum?

8. What does a protocol do?

9. What are the three basic rules for creating passwords?

10. What do we need encryption for?

 stage **LEARNING** APPLYING, ANALYZING, SYNTHESIZING, EVALUATING

"I can apply what I've learned, relate these ideas to other concepts, build on other knowledge, and use all these thinking skills to form a judgment."

Knowledge in Action

1. Are the computers at your school connected to a network? If so, what kind of network(s)? What types of computers are connected? What hardware and software allows the network to function? What department(s) did you contact to find the information you needed to answer these questions?

2. Using current articles, publications, and/or the web, research cable modems. Where are they being used? What does a residential user need to hook up to a cable modem system? Do you think you will use a cable modem in the near future?

3. Research the role of the Federal Communications Commission in regulating the communications industry. How are new frequencies opened up for new communications services? How are the frequencies determined? Who gets to use new frequencies?

4. Are you a fan of analog or of digital? Explain.

5. Research the Telecommunications Act of 1996. Do you think it has had a positive or a negative effect? Why?

6. Would you like to have a job for which you telecommute instead of "going in to work"? Why or why not?

7. From your experience with cellphones, do you think it is wise to continue also paying for a "land-line" POTS phone line, or is cellphone service reliable enough to use as your sole means of telephony? As cell service advances, how do you think the POTS infrastructure will be used?

Web Exercises

1. Compare digital cable and satellite TV in your area. Which offers more channels? Which offers more features? How much do the services cost? Does either allow internet connectivity? Does either use a telephone line to download the programming listings?

2. Calculate the amount of airborne data transmission that travels through your body. To do this, research the amount of radio and television station broadcast signals in your area, as well as the estimated number of mobile-phone users. Imagine what the world would look like if you could see all the radio-wave signals the way you can see the waves in the ocean.

3. On the web, go to *www.trimble.com* and work through some of the tutorial on "About GPS Technology." Then write a short report on the applications of a GPS system.

4. As mentioned in Chapter 4, microchips are implanted in animals to track them with GPS technology. Humans can now be tracked when they use their cellphones. Visit the following websites for more details:

   www.wired.com/news/business/0,1367,21781,00.html
   http://europe.cnn.com/2001/TECH/ptech/04/20/location.services.idg
   www.pcworld.com/news/article/0,aid,55986,00.asp

5. What is wardriving? Visit these sites to see what is being done to fix this security hole:

    www.wardriving.com/
    http://online.securityfocus.com/news/192/
    http://en.wikipedia.org/wiki/Wardriving
    www.wardrive.net/

6. Test your ports and shields. Is your internet connection secure, or is it inviting intruders to come in? Use the buttons at the bottom of the page at *https://grc.com/x/ne.dll?bh0bkyd2* to test your port security and your shields and receive a full report on how your computer is communicating with the internet.

7. Learn about the method of triangulation. This method helps when you use your cellphone to locate, for example, the nearest movie theater or restaurant. Visit the sites below to learn more. Run your own search on *triangulation.*

    http://searchnetworking.techtarget.com/sDefinition/0,,sid7_gci753924,00.html
    www.al911.org/wireless/triangulation_location.htm
    www.usatoday.com/tech/bonus/qa/2001/10/23/cell-phone-tracking.htm
    http://rfdesign.com/ar/radio_managing_wireless_internet/

8. TDMA, GSM, CDMA, and iDEN. What do these terms mean? The following websites do a good job of explaining these technologies and other cellphone-related issues.

    http://w3.iarc.org/~ronen/cellular.html
    www.arcx.com/sites/CDMAvsTDMA.htm

9. Why AM radio during the day is different from AM radio at night: Read the information at the following websites to learn more about the propagation of AM radio signals and how AM is governed by the FCC.

    www.wmox.net/power.htm
    www.grc.nasa.gov/WWW/MAEL/ag/agprop4.htm
    http://musicradio.computer.net/transm.html

10. Security issue—Wi-Fi security issues: As Wi-Fi becomes more widespread, how will users protect their networks? Wi-Fi is hacked almost as a harmless hobby to detect its vulnerabilities. Visit the following websites to learn about some issues of the Wi-Fi future.

    www.acsac.org/2002/case/wed-c-330-Miller.pdf
    www.wi-fi.org/OpenSection/secure.asp?TID=2
    http://www.osnews.com/story.php?news_id=11468

Personal Technology

The Future Is You

Chapter Topics & Key Questions

7.1   **Convergence, Portability, & Personalization** What are the pros and cons of the major trends in personal technology?

7.2   **MP3 Players** What should I know about digital audio players?

7.3   **High-Tech Radio: Satellite, High Definition, & Internet** How do the three forms of high-tech radio differ?

7.4   **Digital Cameras: Changing Photography** What are some of the things I can do with digital cameras?

7.5   **Personal Digital Assistants & Tablet PCs** How could I use a PDA and tablet PC to help me in college?

7.6   **The New Television** What's new about the new television?

7.7   **Smartphones: More Than Talk** How are smartphones different from basic cellphones?

7.8   **Videogame Systems: The Ultimate Convergence Machine?** How do the three principal videogame consoles compare?

The march of digital technologies through cameras, videos, and a flurry of gadgets is reshaping old standbys and broadening their use," says Gary McWilliams.

"Clock radios are turning into überstereos," continues *The Wall Street Journal* writer, "and cameras into slim wonders that beam images to a TV or printer. Television, which had evolved only gradually since the advent of color in the 1950s, is now a digital jungle of new screen types, shapes, and technologies."[1]

It's all leading to what Jerry Yang and David Filo, the founders of Yahoo, call the internet's "second act," when broadband extends out to every electronic gadget. The internet's first period, they say, involved shifting real-world activities such as shopping and dating into a virtual world, one dominated by the PC. In the net's second act, they believe, "creative power will shift into the hands of individuals, who will be just as likely to generate and share their own content as to consume someone else's."[2]

Indeed, this was readily demonstrated with the cellphone and blog-based outpouring of breaking news and shared information about the December 2004 tsunami in Asia and the July 2005 terrorist subway bombings in London.[3] It's also evident in the phenomenon known as *mash-ups*, in which a person takes a set of data from one website and attaches it to a map file on Google Maps, thus, for example, showing the day's cheapest gas stations anywhere in the United States, the location of potholes in New York, inches of snowfall for skiers, and where trucks that serve tacos are parked in Seattle.[4] (See *www.googlemapsmania.blogspot.com.*) "What we're seeing," says one technology observer, "is nothing less than the future of the World Wide Web. Suddenly, hordes of volunteer programmers are taking it upon themselves to combine and remix the data and services of unrelated, even competing sites. The result: entirely new offerings."[5]

7.1

Convergence, Portability, & Personalization

What are the pros and cons of the major trends in personal technology?

Here's a concept: In South Korea, Samsung Electronics introduced a cellphone (the SCH-S310) with "3-dimensional movement recognition." You wave the phone in the air, and it interprets your wandlike gestures. Draw a "3," and the phone types the digit. Make an "X," and it generates the voice response "no." Shake the phone twice, and it ends the call. Jerk the device to the right while listening to music on the built-in MP3 player, and it will skip to the next track.[6]

Is a phone with motion sensing really useful? What happens if you're calling while on a bumpy road? Regardless, the device seems to embody three principal results of the fusion of computers and communications that we mentioned in Chapter 1—namely, *convergence, portability,* and *personalization.* Let's see what these developments, which began to build in strength during the 1990s, are like today.

Convergence

How can I evaluate whether a hybrid convergence device is good or bad?

Ming Ma, 28, a Los Angeles–based systems administrator, bought an Xbox to play videogames. But soon after he brought it home, Ma realized it could be used for multiple purposes. "I initially purchased it to play games," he said, "but now I use it to watch DVD movies, play music CDs, and to chat with friends while we play multiplayer games. I like how I can do everything on one device—I have a small apartment, so this is ideal, for me—plus, I don't have to buy individual devices, so it also saves me money along with space."[7]

As we said in Chapter 1, **_convergence_, or _digital convergence_, describes the combining of several industries—computers, communications, consumer electronics, entertainment, and mass media—through various devices that exchange data in digital form.** Long predicted but not fulfilled, convergence began showing signs of happening in 2003, as more households in the United States—and throughout the world—began going to broadband, with the faster internet connections changing computer, cable, telephone, music, and movie businesses.[8] Convergence, as we've pointed out, has led to electronic products that perform multiple functions, such as TVs with internet access, cellphones that are also digital cameras, and a refrigerator that allows you to send email.

Hybrid convergence devices have pros and cons.

CONVERGENCE: THE UPSIDE Ma's multiple uses of the Xbox are an instance in which convergence makes sense. So, perhaps, is Sony's PlayStation Portable, which you can use not only to play videogames but also to surf the web, watch video, and read e-books.[9] So are cellphones with address books or digital cameras that also shoot video. The new Holy Grail of convergence seems to be a "digital Swiss Army knife," a universal device that will perform several functions.[10] The Nokia 9500 Communicator, for instance, is a flip phone that opens like a book to provide a wide-screen, high-resolution color display and standard keyboard plus the following multiple features: conference call support, Wi-Fi access, text messaging, email, web browser, personal digital assistant, game machine, and high-end digital camera. As a "mobile office" device, it's comparable in size to many personal digital assistants, though it won't fit easily in a shirt or trouser pocket (unless you wear cargo pants).

Convergence. Home entertainment centers bring together several electronic functions to access music, video, TV, and the internet, as well as support some computer functions.

CONVERGENCE: THE DOWNSIDE   Not all hybrid gadgets are necessarily practical. "Just because companies can converge technologies into one device doesn't [always] mean they should do it," says the research director of Jupiter Research. "Any convergent device where, as a result, a primary feature is compromised is just not a good idea."[11] Is an internet refrigerator, for instance, really the best way to get your email? Is a camera cellphone that takes tiny, grainy pictures worth the purchase price? One reviewer said the Nokia 9500 Communicator was "chock full of promising concepts" but it "almost invariably fails in execution."[12]

Portability

What are the positives and negatives of portability?

Smart cellphones are, of course, an example of portability. Not too many years ago, having a phone in your car was a badge of affluence, affordable mainly by Hollywood movie producers and Manhattan real estate developers. Now, thanks to increasing miniaturization, faster speeds, and declining costs, more and more mobilephone and other electronic components can be crammed into smaller and smaller gadgets. But, as today's students are well aware, a host of other devices are also portable: MP3 players, digital cameras, BlackBerries, notebook computers, PDAs, tablet PCs, and the like.

Portability also has its upside and downside.

PORTABILITY: THE UPSIDE   The advantages of portability seem obvious: being able to do phone calls and emails from anywhere that you can make a connection, keeping up with your social networks, listening to hundreds of songs on a digital music player such as an iPod, taking photos or video anywhere on a whim, watching TV anywhere anytime. One reason that manufacturers have targeted turning cellphones into hybrids, incidentally, is that people carry them wherever they go, unlike laptop computers, MP3 players, or digital cameras.[13]

iPod portability. Apple's iPod, available in 30- and 60-gigabyte models, is a handheld device that allows users to watch TV shows, music videos, and home movies, as well as listen to music and audiobooks.

PORTABILITY: THE DOWNSIDE   The same portable technology that enables you to access information and entertainment anytime anywhere often means that others can find *you* equally conveniently (for them). Thus, you have to become disciplined about preventing others from wasting your time, as by screening your cellphone calls with **_Caller ID_, a feature that shows the name and/or number of the calling party on the phone's display when you receive an incoming call.** You may also find yourself bombarded throughout the day by unnecessary incoming emails and those web services known as RSS aggregators that automatically update you on new information from various sites.

Nonstop connectivity may rain digital information on you, but it can also have the paradoxical result of removing you from real human contact. Consequently, the professor whose recommendation you may need someday for an employment or graduate school reference may never get to know you, because your interaction will never be face-to-face, as during his or her normal office hours. Businesses, incidentally, have recently had to deal with recurring complaints from employees about their inability to reach coworkers to get critical information they need—because their coworkers don't respond in timely fashion via email—and so some companies have taken measures to increase person-to-person contact, as by banning email use on Fridays.[14]

Personalization

What are the pluses and minuses of personalization?

Telecommunications can be organized through two kinds of arrangements: the *tree-and-branch model* and the *switched-network model*. In the ***tree-and-branch telecommunications model*, a centralized information provider sends out messages through many channels to thousands of consumers.** This is the model of most mass media, such as AM radio and network television broadcasting.

In the ***switched-network telecommunications model*, a common carrier provides circuit switching among public users; that is, a temporary connection is established by closing a circuit.** This is the model of the telephone system, of course, and also of the internet. In a telephone network, a connection for voice transmission is made by dialing; in a packet switching network (Chapter 6), a temporary connection is established between points for transmitting data in the form of packets. People on the system are not only consumers of information ("content") but also possible providers of it. During the past several years, mass-media radio and television have been losing listeners and viewers to the more personalized media based on the switched-network model. This has opened the door to more personalized uses of information technology.

PERSONALIZATION: THE UPSIDE As a consumer, you may have downloaded hundreds or thousands of songs, so that you have your own personalized library of music on your computer and MP3 player. You may also have created your own list of "favorites" or "bookmarks" on your PC so that you can readily access your own favorite websites. And you may have accessed or contributed to certain blogs, or personalized online diaries. In addition, of course, PC software can be used to create all kinds of personal projects, ranging from artwork to finances to genealogy.

PERSONALIZATION: THE DOWNSIDE The downside of personalization is that people may feel overburdened with *too much choice.* Indeed, Swarthmore College psychology professor Barry Schwartz, author of *The Paradox of Choice: Why More Is Less,* has identified several factors by which choice overload hurts us. As available options increase, he says, the following may happen:[15]

- **Regret:** People are more likely to regret their decisions.
- **Inaction:** People are more likely to anticipate regretting decisions, and the anticipated regret prevents people from actually deciding.
- **Excessive expectations:** Expectations about how good the decision will be go up. Thus, reality has a hard time living up to the expectations.
- **Self-blame:** When decisions have disappointing results, people tend to blame themselves, because they feel the unsatisfying results must be their fault.

One result of having several choices is that many people do ***multitasking*—performing several tasks at once,** such as studying while eating, listening to music, talking on the phone, and handling email. You may think you're one of those people who has no trouble juggling all this, but medical and learning experts say the brain has limits and can do only so much at one time. For instance, it has been found that people who do two demanding tasks simultaneously—drive in heavy traffic and talk on a cellphone, for example—do neither task as well as they do each alone.[16] Indeed, the result of constantly shifting attention is a sacrifice in quality for any of the tasks with which one is engaged. The phenomenon of half-heartedly talking to someone on the phone while simultaneously surfing the web, reading emails, or

Personal Technology

371

trading instant messages has been called "surfer's voice." Experts who study these things call it "absent presence."[17] This would seem to have some implications for how students handle their studying.

Popular Personal Technologies

What personal technologies am I familiar with?

Where are you, technologically speaking, with the personal electronic devices you use every day? Are you still carrying a "basic" cellphone—no camera, music player, or TV content? You haven't tuned in satellite radio? You're not up on HDTV? It's time to move on. Let's explore what's out there (knowing full well that this book won't be able to keep up with all the new devices).

In the coming sections, we consider the following kinds of personal technology:

- MP3 audio players
- Satellite, high-definition, and internet radios
- Digital cameras
- Personal digital assistants and tablet PCs
- The new television
- Smartphones
- Videogame systems

MP3 Players

What should I know about digital audio players?

MP3 digital audio players **are portable devices that enable you to play MP3 digital audio files.** **_MP3_** **is a format that allows audio files to be compressed so they are small enough to be sent over the internet or stored as digital files.** (MP3 is short for MPEG Audio Layer 3, an audio compression technology.) MP3 files are about one-tenth the size of uncompressed audio files. For example, a 4-minute song on a CD takes about 40 megabytes of space, but an MP3 version of that song takes only about 4 megabytes.

How MP3 Players Work

What's useful to know about digital audio players?

The most famous of digital music players is Apple Computer's iPod. Although in the past the Apple iPod, iPod mini, and iPod nano seemed to attract the most attention, there are many other players as well—with more coming along all the time, making it impossible to do an up-to-date comparison here. Competitors include Archos, Creative, Dell, iRiver, Panasonic, RCA, Samsung, Sandisk, Sony, and Virgin Electronics. (Rio fell by the wayside.[18]) These companies have been challenging the iPods on price, looks, and extras such as built-in FM tuners and color LCD screens that can display photo clips and arcade-style videogames, as well as record voice and music.

Basically, however, digital music players can be divided into those that have hard-disk drives for storage (which store more songs but are more expensive) and those that have flash memory (fewer songs but less expensive). Some, such as the Archos Gmini 400, have both hard-drive and flash memory.

DATA STORAGE—HARD DRIVE Apple dominates the digital music player market with 75%–80% of hard-drive-based players.[19] Apple's full-size iPod models, which all have hard drives, come with 15, 20, 30, 40, or 60 gigabytes of

MP3 player

Software for soft wear. Instead of wearing your heart on your sleeve, you can wear MP3 and cellphone technology, incorporated into this German-made jacket. The MP3 player is controlled through cloth buttons on the left sleeve, and the headphones are built into the collar.

storage (holding 3,700–15,000 songs). The iPod mini stores either 4 gigabytes (good for 1,000 songs) or 6 gigabytes (1,500 songs). Some hard-drive MP3 competitors, all with at least 20 gigabytes of storage, are Creative's Zen Touch, the iRiver H320, and the Archos Gmini 400. Sony's new Walkman also has a 20-gigabyte hard drive.[20] Mini hard-drive devices (5 gigabytes) are available from Creative, Dell, and Virgin.

DATA STORAGE—FLASH MEMORY If you jog or do other strenuous activity that might make an MP3 player's hard drive crash, you might want a player with flash memory, which usually takes the form of a card that can be slipped into a player's memory slot and is nonvolatile, holding its contents even after the power is turned off. For example, the iPod nano—described as "about the size of a playing card and thin enough to slip under a door"—has 4 gigabytes of flash memory and holds as much music (as well as album art and photographs) as some hard-drive players.[21] The Apple iPod shuffle is a finger-size, low-priced "mini-mini" music player that holds about 20 songs on a 512-megabyte, solid-state flash memory chip, or twice that many songs on a 1-gigabyte flash chip.[22] Rival iRiver offers a 256-megabyte flash player weighing 2.2 ounces. The capacity of flash memory cards varies between 64 megabytes and 1 gigabyte for most MP3 players.

SAMPLING RATE How many songs your audio player holds is affected not only by the storage capacity but also by the sound quality selected, which you can determine yourself when you're downloading songs from your computer to your player. Downloading and converting a digital audio track from a music CD to the MP3 format is called *ripping*. (Ripping software is available as a stand-alone program or a function in a software-based media player such as Windows Media Player 10.) If you want high-quality sound when you're ripping (converting), the files will be relatively large, which means your MP3 player will hold fewer songs. If you're willing to have lower-quality sound, perhaps for the device you use while jogging, you can carry more songs on your player. The quality is determined by the ___sampling rate___, **the number of times, expressed in kilobits per second, that a song is measured (sampled) and converted to a digital value.** For instance, a song sampled at 192 kilobits per second is three times the size of a song sampled at 64 kilobits per second and will be of better sound quality.

TRANSFERRING FILES When you buy an audio player, it comes with software that allows you to transfer MP3 files to it from your personal computer, using a high-speed port such as a FireWire or universal serial bus port (USB 1.0 or 2.0).

BATTERY LIFE Battery life varies. The iPod shuffle will run up to 12 hours before recharging, the iPod nano 14 hours, and the iPod mini 15 hours. The 5-gigabyte Rio Carbon is advertised as having a 20-hour battery life.

COLOR SCREENS & PHOTO VIEWING The top-of-the-line 40- or 60-gigabyte iPod comes with a crisp 2-square-inch color screen and allows you (with a $29 attachment) to download photos directly from your digital camera. If you want, you can even hook up the iPod to a TV and view a slide show. Some flash-based audio players have color screens that can display JPEG and BMP digital images.

OTHER FEATURES Many MP3 players, particularly flash-based devices, offer FM radio reception. Many players also are able to make high quality music recordings, using an extra microphone. Some also have a small internal

MP3 sunglasses. These sunglasses, called Thump, plug into your computer's USB jack so that you can download MP3s from the computer to the glasses. Then, just put on the glasses, put the earbuds in your ears, and enjoy the music.

microphone for voice recording, appropriate for capturing conversation or a lecture but usually unsuitable for music. Some devices come with a pair of earphone jacks so that two people can listen together.

MP3 IN YOUR CAR It's dangerous, and in most places illegal, to use earphones while driving. However, there are gadgets that allow you to listen to your MP3 collection while you're behind the wheel. There are two ways to do this—car-stereo adapters and FM transmitters. One way, although an expensive one, is to get a car-stereo adapter, which uses a dock connector on the bottom of dockable iPods and iPod minis. A less expensive way is to get a device that turns your audio player into a miniature radio station that broadcasts an FM signal to your car's radio antenna and thus through your car radio.[23]

The Societal Effects of MP3 Players

How have audio players changed listening habits?

"I've called the iPod the first cultural icon of the 21st century," says Michael Bull, an instructor in media and cultural studies at the University of Sussex in England. Bull has become known as "Professor iPod" because he has spent more than a decade researching the societal effects of portable audio devices, starting with Sony's Walkman portable cassette player and extending to Apple's iPod, which was introduced in 2001. The iPod, he says, "permits you to join the rhythm of your mind with the rhythm of the world," causing a cultural shift away from large communal areas—such as a cathedral, "a space we could all inhabit"—to the world of the iPod, "which exists in our heads."[24]

One in ten U.S. adults owns an MP3 player, and for Americans under age 30 about one in five have such devices.[25] Although clearly young people have taken to MP3 players with gusto, so have many older people. Bob Levens, 48, and his wife own three iPods between them, and the device, he says, has helped him to enjoy his music collection more. "Whereas before, I would stick a CD into the hi-fi and sit and listen or maybe read," he reports, "the iPod has allowed me to carry my music around with me and also to listen to some CDs that had been relegated to the back of the shelf. I rarely listen to mainstream commercial radio."[26]

Note: With about 12% of children and teens in the U.S. suffering from noise-induced hearing loss, hearing experts are concerned about the effect of hours of listening to MP3 players. The threshold for safe, extended listening is 85 decibels for 8 hours, according to the National Institute of Occupational Safety and Health, but the volume of most portable compact-disk players has been found to be 91–121 decibels, with in-the-ear earphones (earbuds) adding another 7–9 decibels.[27]

Using MP3 Players in College

How could I use an audio player to help me in my college work?

College instructors and students have found ways to expand the uses of the iPod beyond just the enjoyment of music. For instance, some MP3 players can also be used to store schedules, phone-number lists, and other personal information management software. At Duke University, which experimented with passing out iPods to its entire freshman class in fall 2004, instructors have helped their students to employ audio players as storage drives or as sound devices with the help of microphone attachments. "A journalism

- Professors record lectures and post them online for students to download for later review, as when students have dead time (such as riding the bus).
- Interviews with guest speakers and messages from administrators may also be sent straight to students' MP3 players.
- Some classes, such as those in music, foreign languages, or radio broadcasting, can especially benefit from use of audio players.
- Students record their own study group sessions and interviews, using their music players with microphones.
- Students record "audio web logs" during off-campus jobs to connect with people on campus.
- Students subscribe to feeds of frequently updated audio content from podcasts; students produce their own audio shows and podcast them. A "PodPage" website can be created that is devoted to such podcasts.
- Los Angeles software maker Pod2Mob is offering software for Windows and Macintosh computers that promises to enable iPod and cellphone users to stream recorded podcasts directly from their home computer.

Source: Adapted from Brock Read, "Seriously, iPods Are Educational," *The Chronicle of Higher Education,* March 18, 2005, pp. A30–A32; and Associated Press, "Software Makes Podcasts Mobile," *Reno Gazette-Journal,* August 25, 2005, p. 3D.

professor encourages students to use them to record interviews instead of simply taking notes," says one report. "An economics professor records lectures for her 300-person survey course on her iPod and posts them online as a pre-exam study. An engineering professor has students use their iPods to transfer research data from dormitory computers to laboratory machines."[28] (● See Panel 7.1.)

7.3

High-Tech Radio: Satellite, High Definition, & Internet

How do the three forms of high-tech radio differ?

"Portable satellite radio receivers . . . offer a real alternative to the iPod in the fight for your pocket space," says one reviewer.[29] He described taking a bike ride around New York City's Central Park listening to comedy legends Eddie Murphy, Billy Crystal, and Richard Pryor—who are not usually available on standard "terrestrial" radio—on a satellite radio clipped to his jersey. Another writer reported driving 50,000 miles around the United States in a car equipped with a Sirius satellite radio, only rarely hearing "feeble, quivering bursts" of signals from local AM and FM stations, which were largely overridden by Sirius's "clear, unvarying signal" from outer space—and its slightly surreal programming in which all music and news were national, not local, and seemed to emanate as an "invisible digital bubble of information located somewhere in the fifth dimension." "Having passed through the canyonlands of Utah while listening to Caribbean pop," he wrote, "and having crossed the Black Hills of South Dakota immersed in a disco channel called the Strobe, I feel after a year of nonstop driving . . . that I haven't, in fact, gone anywhere except deeper and deeper inside my radio."[30]

The $21 billion radio industry, with its 277 million regular listeners, is being transformed by a number of trends, of which satellite radio is only one.[31] Others are *high-definition radio, internet radio,* and *podcasting.* Let's distinguish among them.

Satellite Radio

How would satellite radio be different for me?

Satellite radio, also called *digital radio,* is a radio service in which digital signals are sent from satellites in orbit around the Earth to subscribers owning

The Sirius Starmate Replay, equipped to receive satellite radio

special radios that can decode the encrypted signals. The CD-quality sound is much better than that of regular radio, and there are many more channels available than there are on regular radio. Unlike standard broadcasters, satellite radio broadcasters are for the most part not regulated by the Federal Communications Commission, although the FCC established the playing field back in the early 1990s when it began selling parts of the radio-wave spectrum for this new class of service.

TWO PROVIDERS   In the United States, there are two satellite digital audio radio service (SDARS) providers, market leader XM Satellite Radio, which started delivering content in 2001, and runner-up Sirius Satellite Radio, which followed in 2002. Most of the content is commercial-free, and the two providers, which each offer more than 120 channels, support themselves by monthly subscription fees (XM's is currently $9.99, Sirius charges $12.99; some people subscribe to both). There are currently about 4 million subscribers, and the number is expected to increase to 35 million by 2010.[32]

ADVANTAGES OF SATELLITE RADIO   A particular attraction with satellite radio is that listeners avoid the endless commercials and limited number of formats of regular AM/FM stations, and the broadcasters serve a diverse group of niche markets, such as reggae, salsa, African folk songs, NASCAR racing, and other formats not available on existing radio stations. To lure more subscribers, both XM and Sirius are trying to improve technology and acquire exclusive content. Both have made deals with automakers to equip new cars with their respective radios. Sirius made headlines when it lured shock-jock radio host Howard Stern with a package of $500 million over 5 years, and it has also made splashy deals with the National Football League and other sports organizations. The ultimate in "ultra-niche" programming, Sirius also presents all-Elvis radio, offering 24/7 play from 2,700 Elvis Presley songs.[33] For its part, XM offers Bob Edwards, Opie and Anthony, and major-league baseball.[34] Besides car receivers, portable satellite radio receivers are available for both services. (● *See Panel 7.2.*)

● PANEL 7.2
Two satellite radio services

	XM Satellite Radio (www.xmradio.com)	Sirius Satellite Radio (www.sirius.com)
Monthly subscription fee	$9.99	$12.99
Number of subscribers	3.2 million	1.24 million
Music channels	67	65
News/talk/sports stations	85	55
Exclusive content	Bob Edwards, Opie and Anthony, major-league baseball	Howard Stern, National Football League, National Basketball Association, all-Elvis radio
Portable radios and prices	Delphi XM MyFi XM2Go, $192–$299; Pioneer XM2Go, $299; Tao XM2Go, $249.	XACT Stream Jockey, $190 (including battery pack); slimmed-down portable due by 2006

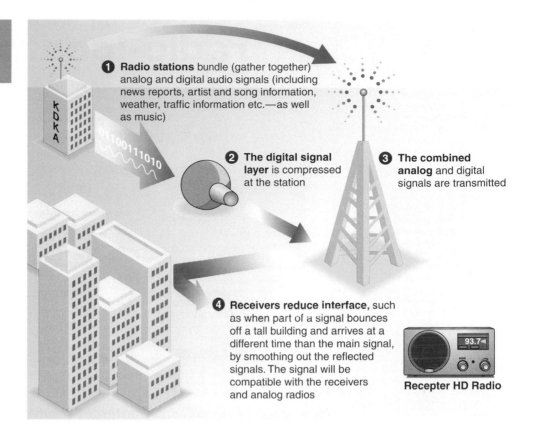

1. **Radio stations** bundle (gather together) analog and digital audio signals (including news reports, artist and song information, weather, traffic information etc.—as well as music)

2. **The digital signal layer** is compressed at the station

3. **The combined analog** and digital signals are transmitted

4. **Receivers reduce interface,** such as when part of a signal bounces off a tall building and arrives at a different time than the main signal, by smoothing out the reflected signals. The signal will be compatible with the receivers and analog radios

Recepter HD Radio

High-Definition Radio

How does high-definition radio differ from traditional radio?

Traditional radio broadcasters "were quite nervous back in 1995 when the FCC was considering licensing two satellite digital audio radio service . . . providers," says one account. "This was the potential death knell (or so they thought) of traditional radio. Something had to be done."[35] Enter **_high-definition (HD) radio_, which provides CD-quality sound and allows broadcasters to squeeze two digital and one analog station on the same frequency.** (● *See Panel 7.3.*) That is, HD combines digital and analog broadcast signals, enabling stations to offer an analog main channel and digital "sidebands," so that multiple types of content can be broadcast from the same position on the dial. An additional advantage over satellite radio, at least so far, is that broadcasts are free; there are no subscription charges.

Traditional broadcasters are hoping HD radio can introduce more local or innovative programming and blunt the threat from satellite radio. They also say that AM digital will have FM-like audio quality, and both AM and FM digital will be static-free and crystal-clear. Wireless data services could include on-demand audio, featuring services such as on-demand news, weather, and traffic reports.

Currently about 300 of the roughly 13,500 stations in the United States are broadcasting digitally, with industry executives saying in early 2005 that there would be about 2,000 stations during the next 3 years.[36] The switch is well under way in the largest cities, with Los Angeles and Chicago, for instance, featuring 10 HD stations each. To receive HD radio on your car, you don't need a new antenna, but an HD-compatible radio is required, such as the KTC-HR100 HD radio tuner or the Panasonic CQ-CB9900U integrated HD radio and CD player. Boston Acoustic's Recepter HD clock radio is available at retailers such as Best Buy.

Internet Radio

How could I benefit from internet radio?

"With all the talk about satellite radio services Sirius and XM," says one writer, "it's easy to forget a much bigger competitor to regular radio: the internet."[37] At the end of 2004, about 3.4 million Americans subscribed to satellite radio, but about 19 million were listening to internet radio each week—and the number of listeners is growing at an average of 43% a year as more people get broadband connections at home.[38] With ad sales rising, now traditional radio giants such as Infinity and Clear Channel Communications are focusing more and more on online broadcasting. Portable services from Cingular and Motorola are coming soon.

The most popular internet-only radio sites are owned by online companies such as Yahoo, AOL, and MSN. Yahoo's Musicmath service, for instance, allows listeners to pick a category, such as alternative rock, and then play a number of songs in that genre, with fewer ads than are found on traditional radio. However, there are also numerous mom-and-pop web broadcasters, such as Radioparadise.com, described as "a format-busting station that spins a tasteful mix of music ranging from the Beatles to Norah Jones to the Strokes," which is run by a couple from their home in Paradise, California, as a commercial-free operation. Although the station has fewer than 5,000 listeners at peak times, they sent the couple $120,000 in contributions in a recent year, enough to cover the cost of bandwidth, song royalties, and other expenses, leaving them enough for "a comfortable lifestyle."[39]

At present, most internet radio listeners tune in at the office. People who listen at home can obtain networking products that enable them to stream music from their computers to their stereo systems. Eventually, as the wireless internet expands, consumers may be able to go online from moving vehicles, with internet radio receivers on their dashboards.

Podcasting

How could I get involved in podcasting?

As we stated in Chapter 2, **_podcasting_ involves the recording of internet radio or similar internet audio programs.** Unlike traditional radio, podcasting requires no studio or broadcast tower, and there's no Federal Communications Commission regulation (so hosts can say whatever they want).[40]

Podcasting software now allows amateur deejays and hobbyists to create their own radio shows and offer them free over the internet. Listeners can then download shows onto their MP3 players like the iPod. As of early 2005, according to *BusinessWeek*, there were over 3,500 podcasts in existence.[41]

Some people describe podcasts as "TiVo for radio." Just as TiVo technology allows you to capture a television program while you're away from the set and then watch it at a time of your convenience, podcasting enables people to download their favorite radio shows to an MP3 device and then listen to them later whenever and wherever they please.[42] To subscribe to a podcast, you need not only a computer and MP3 player but also podcast-receiving software called an *aggregator*.

7.4

Digital Cameras: Changing Photography

What are some of the things I can do with digital cameras?

About four out of every five cameras sold in 2005 were expected to be digital rather than film cameras, and by now both professional and amateur photographers alike have pretty much abandoned film. "The evolution is having profound and unforeseen effects on society," says one analysis, "from changing the way that people record their daily lives to making it harder to trust the images we see."[43]

How Digital Cameras Work

What would motivate me to buy a digital camera?

We described general principles of digital cameras in Chapter 5. Here we consider the subject in more detail.

An Apple Powerbook G4 with a connected digital camera

The most obvious statement to make about digital cameras (or "digicams") is this: They do not use film. Instead, a digital camera uses a light-sensitive processor chip to capture photographic images in digital form and store them on a flash memory card. You can review your just-shot picture on the camera's LCD monitor, the little screen that displays what the camera lens sees, and decide whether to keep it or to try another angle in your next shot.

In general, digital cameras seem to be getting smaller, thinner (the size but not the thickness of a credit card), and less expensive, but they are also capable of performing more tricks. As with MP3 players, product releases occur so frequently that it is impossible to present an up-to-date picture of the various models. However, we can outline certain general guidelines.

POINT-AND-SHOOT VERSUS SINGLE-LENS REFLEX (SLR) "Point-and-shoots are fun to tote around," says David Ritz, who runs the 1,200-store Ritz Camera Centers, "but with an SLR, the picture is brighter and crisper."[44]

- Point-and-shoot: A **_point-and-shoot camera_ is a camera, either film or digital, that automatically adjusts settings such as exposure and focus.** Generally such cameras cost under $500. Manufacturers of digital point-and-shoots include Canon, Casio, Fuji, Hewlett-Packard, Kodak, Konika, Nikon, Olympus, Panasonic, Pentax, Samsung, Sony, and Vivitar, whose prices range from $197 up to about $600. While a point-and-shoot autofocus will do most of the work for you automatically, you may find it useful to get a camera that also has manual controls, so that you can take over if you want. Of course, there are

AMA **The Photography People**

= 800-223-2500 =

Shopping Cart | Account Info | Log In

ecials | Links | FAQ | Site Map | Wish List | Closeouts | Overstock

Home | Digital | Cameras & Accessories | Digital Cameras | Nikon Coolpix 5200 Point and Shoot Digital Camera ..

Nikon

Nikon Coolpix 5200 Point and Shoot Digital Camera - Refurbished By Nikon U.S.A.

SKU #	INKCP5200R
Mfr. Part #	25516B
Our Price:	$224.95

Add To Cart 🛒

Email this item
Add to Wish List

Click for large image

Item Includes

Strap, USB cable, Audio Video cable, EN-EL5 Rechargeable Battery, MH-61 Battery Charger, PictureProject CD-ROM, Quick start Guide, and Instruction manual
90 Day Warranty by Nikon U.S.A

Description

The Nikon COOLPIX 5200 is an easy-to-operate compact digital camera that boasts high-level comprehensive quality, 5.1 effective megapixel performance, and a selection of u

also disposable point-and-shoot cameras, such as the Kodak Zoom and the Fugifilm QuickSnap True Definition, selling for $12–$16, which are useful if you're afraid you might leave an expensive camera behind on the beach (and if you don't need high-quality images).

• **Single-lens reflex: A _single-lens reflex (SLR) camera_ is a camera, either film or digital, that has a reflecting mirror that reflects the incoming light in such a way that the scene viewed by the viewer through the viewfinder is the same as what's framed by the lens.** A digital SLR, which may cost anywhere from $450 on up, is the choice of "prosumers" (professional consumers, such as professional photographers and serious amateurs) because it provides more manual options and better image quality and allows the use of interchangeable lenses, from wide angle to telephoto. Some manufacturers in this category include Canon, Nikon, Pentax, and Olympus, whose prices range from $789 to $1,148.

Digicam. HP Photosmart R707 digital camera

RESOLUTION & MEGAPIXELS *Resolution* refers to image sharpness. A digital camera's resolution is expressed in *__megapixels__*, **or millions of picture elements, the electronic dots making up an image.** The more megapixels a digital camera has, the better the resolution and the higher the quality of the image, and so the higher the price of the camera. The millions of pixels are tightly packed together on the camera's image sensor, a half-inch-wide silicon chip. When light strikes a pixel, it generates an electric current that is converted into the digital data that becomes your photograph.[45]

- **How many megapixels are best?** Megapixels measure the maximum resolution of an image taken by the camera at its *top settings*. Thus, what the quantity of megapixels does is determine how big you should make your prints. If you are mainly interested in producing 4 × 6 (4-inch by 6-inch) prints (or in emailing images to friends), a 2-megapixel camera will do just fine. For 5 × 7 or 8 × 10 prints, 3 or 4 megapixels are better. For 11 × 14 or poster-size prints, or if you want to crop and zoom in your pictures on your computer screen before printing them, you should move up to 4–8 megapixels. (Discounter Concord Camera markets a 5-megapixel camera for just $99, using plastic casing instead of metal and no zoom lens or frills.)

- **Not all megapixels are alike:** Some cameras use larger pixels than others, as measured in microns. (A micron is 1/24,500 of an inch.) For example, says one analysis, "The pixels on the HP Photosmart R707 measure just 2.8 microns wide, whereas those on the Nikon D70 are 7.8 microns wide. . . . The advantage of a larger pixel is that it is able to pick up more information about the image it is sensing."[46] Thus, especially when buying an SLR camera, you should inquire about the size of the pixels in microns.

LENSES People talk as though megapixels are the main factor affecting photographic quality, but that's not all. You need to have a lens that "ensures that your picture is properly focused and [that] pull[s] in enough light to get good exposure," says one writer.[47]

As for zoom lenses, you should be aware of the difference between digital zoom and optical zoom:

- **Digital zoom:** Manufacturers like to tout *digital zoom,* but, says one description, it's just another way of saying "we'll crop the image for you in the camera."[48] Indeed, it actually lowers the resolution and often can produce a grainier photo.

- **Optical zoom:** Only an *optical zoom* will bring you closer to your subject without your having to move. That is, the lens actually extends to make distant objects seem larger and closer. Optical zooms may be of the telescoping type or internal (untelescoping) type; the latter allows a camera to start up fast.

Most cameras in the $250–$350 range come with a 3X optical zoom, which is good for wide-angle, normal, and telephoto shots. If you're an experienced photographer, you should look for a 4X, 5X, or even 10X optical zoom.

STORAGE Instead of being stored on film, the camera's digital images are stored on flash memory cards inside the camera. *(See ● Panel 7.4 on the next page.)* Cards come in a variety of formats, including Secure Digital (SD), Multimedia, Compact Flash, Memory Stick, and Smart Media. Most cameras come with "starter" cards, usually 16 or 32 megabytes, which hold only a

S u r v i v a l   T i p
Reformat Your Memory Card
to Avoid Losing Your Photos

Some picture takers delete
unwanted photos as they go
along. But they can lose
("crash") their saved images if
they aren't careful. The
reason: Using the camera's
Delete function to remove
unwanted images simply
spreads the pictures around
the card, causing fragmentation that leads to the same
errors that can plague a
computer's hard drive. To
avoid this problem, you
should download all your
images to a computer each
night and then reformat the
memory card. Or if you don't
have access to a PC, you
should bring extra memory
cards.

handful of photos. You'll need a card (which is reusable) with at least 128 mega-bytes of storage space, which will hold about 80 images from a 3-megapixel camera (depending on the resolution you set it at). A 1-gigabyte card will hold 600 still images at top resolution or more than 40 minutes of video from a 5-megapixel camera. Technology writer Walter Mossberg points out that many portable devices, such as music players, PDAs, and some smartphones, are standardized around SD cards. However, you may want to choose a memory card that's the same as the one on your other portable devices, so that you can share the cards among the gadgets.[49]

OPTICAL VIEWFINDERS & LCD SCREENS   The original digital cameras had tiny screens, which made it difficult to review photos before deciding whether to save or delete them. Now more cameras come with both optical viewfinders and LCD screens.

- **Optical viewfinders:** The *optical viewfinder* is the eye-level optical glass device on the camera that, when you look through it, shows the image to be photographed. Some digital cameras omit the viewfinder, forcing you to use only the LCD, which forces you to hold the camera with arms extended, making you more apt to shake the camera.

- **LCD screens:** The *LCD (liquid crystal display) screens* usually measure 2 inches or more diagonally and allow you to review the photos you take. As Mossberg points out, it's best to get a camera with an LCD that can be seen well in daylight if you do a lot of outdoor shooting.

START-UP TIME, SHUTTER LAG, & CONTINUOUS SHOOTING   A digital camera, like a PC, needs time to start up. You should look for one that takes no longer than a second or two, so that you won't miss that one-of-a-kind shot that suddenly appears out of nowhere. Another recommendation: You should look for a digicam with the least shutter lag, that annoying delay between the time you press the shutter-release button and the time the exposure is complete.[50] Many digital cameras have a special setting called "burst" or "continuous" mode, which allows you to squeeze off a limited number of shots without pausing—helpful if you're taking pictures at sporting events.[51]

BATTERY LIFE   "For digital photographers," says one writer, "film is no longer a worry. But batteries are more important than ever."[52] Some cameras

more info!

Reviews of Digital Cameras

For independent reviews of digital cameras, try the following sites:
www.cnet.com
www.dpreview.com
www.pcmag.com

come with rechargeable batteries, although they aren't always replaceable at the nearest drugstore (and replacements can be expensive, perhaps $40–$60). Some require you to recharge the battery inside the camera; others, in a separate charger; others, in both ways. Some cameras also require proprietary batteries, which also aren't as readily available. Regardless, you should be sure you have lots of batteries before setting out on a trip.

"A good rule of thumb," says Mossberg, "is that a camera should be able to get a day's worth of shooting out of a single battery charge. Depending on how you use your camera, this could mean 20, 50, 100, or 200 shots. Battery life varies depending on the size and quality of the images, how much of the time you use the flash, and whether you keep the LCD screen on all the time or use it sparingly."[53]

SHOOTING VIDEO CLIPS Although at one time digital cameras could shoot only about 30 seconds of video, today some digicams are available that, with a 1-gigabyte memory card, can shoot as much as 44 minutes at 30 frames

>>

PRACTICAL ACTION
Online Viewing & Sharing of Digital Photos

You have a great photo of the sunset at Lake Tahoe. How do you share it with others? You could always print it out and mail it via U.S. mail. Or you could try the following alternatives.[54]

- **Send as an email attachment:** You can download your photos to a PC and then send them as attached computer files to an email message. The drawback, however, is that if you have a lot of photos, it can quickly fill up the recipient's email inbox or take a lot of time if you are sending multiple images to multiple people.

- **Use an online photo-sharing service:** Web-based photo-sharing services provide online storage space for your photos and make it easier to share, especially if they are high-resolution files. Examples are Shutterfly, Kodak EasyShare Gallery (formerly Ofoto), Yahoo-Photos, Snapfish, Flickr, You've Got Pictures (AOL), Funtigo, Smugmug, and PhotoSite. (● See Panel 7.5.) No special software is required for you to upload from your computer, but a broadband connection is strongly recommended. Some charge fees, some do not.

Be sure to find out whether you have to renew membership or make a purchase in order to keep your photos from being deleted after 6 or 12 months.

- **Use an online service that allows direct transfer between computers:** This third approach uses peer-to-peer technology, which avoids your having to upload your photos on the company's server. An example is OurPictures.

- **If you have a camera with Wi-Fi transmitter, transmit your photos wirelessly:** Some kinds of cameras (such as the Nikon P1 and P2 or Kodak EasyShare-One) are built with Wi-Fi transmitters for wireless networking. Thus, if you're within range of a Wi-Fi hotspot in a coffee shop or airport lounge, you can post pictures to your website or to a photo-sharing service such as Flickr.

● PANEL 7.5
Photo-sharing services

AOL Pictures, *pictures.aol.com*	PhotoSite, *www.photosite.com*
Flickr, *www.flickr.com*	Shutterfly, *www.shutterfly.com*
Funtigo, *www.funtigo.com*	Smugmug, *www.smugmug.com*
Kodak EasyShare Gallery, *www.kodakgallery.com*	Snapfish, *www.snapfish.com*
OurPictures, *www.ourpictures.com*	YahooPhotos, *photosyahoo.com*

Personal Technology

per second—the same frame rate as camcorders. However, other cameras shoot at only 15 frames per second, which may store as little as 20 seconds on a clip and present a slightly jerky image. Digicams have begun to become popular for their video features because they are smaller in size than camcorders, which won't fit in a pocket.

Yet despite the advances in video quality, digital cameras still have limitations compared with video camcorders:[55]

- **Zoom:** Most camcorders allow you to zoom in and out while shooting. Most digital cameras don't.
- **Sound:** Camcorders shoot in stereo sound. Digicams record in mono sound from tiny microphones of lower quality.
- **Storage:** A camcorder can store images on a 60-minute MiniDV tape cassette costing about $5. A digital camera storing 44 minutes requires a 1-gigabyte memory card costing $200–$300.

TRANSFERRING IMAGES Let's assume you've spent a day taking pictures. How do you get them out of your camera? Here are the principal methods:[56]

- **Use a direct connection between your camera and your computer:** You'll need to have used the installation CD included with your digital camera to install the drivers and software on your PC. Then you can connect your camera to the computer using the USB or FireWire cable that came with it. (One end attaches to a slot on the camera, the other end to an open USB or FireWire port on the computer.) Most newer cameras support USB 2.0. After connecting the camera (which must be turned on during the process), you open its software and use it to transfer the photos into the PC, typically placing the pictures into a default folder, such as My Pictures in the My Documents folder.
- **Insert the memory card into your computer or a card reader:** Assuming your PC has a built-in slot for your memory card (USB card) or has a card reader attached to a USB port, you can remove the memory card from the camera and insert it into the slot. External card readers cost $20 or more.
- **Put your camera in a cradle attached to your PC:** Many camera manufacturers now include cradles into which you can set your camera and use it to transfer photos to your PC. Some cradles also are able to charge your camera, if it has rechargeable batteries.
- **Use a photo printer with a built-in card slot:** If you've bought one of the newer photo printers, you can skip using a PC entirely and insert the memory card from your camera into a slot in the printer.
- **Use a portable hard drive:** If you're traveling and don't have access to a PC (not even a laptop), what do you do? Austin-based photographer Andy Biggs, who leads photographic safaris to Africa, packs a portable hard drive called a Tripper. The device, which comes in 20- and 80-gigabyte versions and which operates on batteries or AC current, has a slot for a memory card; once he gets home, Biggs transfers the images from the Tripper to his PC. "We [travelers] spend a lot of money on our trips to go places," says Biggs. "If we can afford the time and money, we should take care to pack a device that makes sure our images are safe, too."[57]
- **Use a portable CD burner:** There are portable CD burners available that can record images directly from your camera memory card.
- **Use an MP3 player:** If you have an MP3 player, such as an Apple iPod or Dell DJ, you can use its several megabytes or even gigabytes of hard drive to store your photos. What's needed is a device (such as

Delkin Devices' USB Bridge) that will directly transfer data from the memory card to the MP3 player.

- **Use a photo-printing kiosk:** Various photo-printing kiosks are available at fast-food restaurants, big retailers, amusement parks, scrapbook-making stores, cruise ships, hospitals, and other high-traffic areas and can produce a standard 4-by-6-inch print in 4 seconds.[58]

- **Use a photo lab:** Photo stores and labs, as well as Costco, Kinko's, Target, Walgreens, and Wal-Mart, often sell inexpensive services (perhaps $3–$7) in which they will transfer the images on a memory card onto a CD and then clear the card for reuse. They will also print photos for you.

- **Bring along your own card reader and CDs and use others' computers:** If you bring along your own card reader and blank CDs, you can use others' computers to store your images. These PCs could belong to friends, of course, but also to hotel business centers and internet cafés.

The Societal Effects of Digital Cameras

How have digital cameras changed the way people take pictures?

One result of the evolution in photography is that people are taking their cameras everywhere, either in backpacks or purses or as mobile phones with built-in camera. (We describe cellphones and smartphones in another few pages.) Another is that people take far more pictures than they used to—perhaps 20 or more of a single scene, instead of three or four—since they don't have to worry about the cost of film and processing, and with digital it's easy to simply delete bad shots. (About 23% of all digital images captured by cameras are deleted.[59]) A third consequence may be that photography is becoming more casual, with more off-the-cuff snapshots being taken instead of subjects being asked to pose stiffly for formal portraits. The digital camera "leads to more openness," says sociologist John Grady of Wheaton College in Massachusetts, whose research specialty involves photography. "People are getting more documentary in style."[60] Yet Grady also thinks that people are taking more care than ever with the photos they keep, shooting a flurry of images but then spending time selecting the best one and touching it up to eliminate small blemishes.

All these developments may be bringing about different sensibilities about the value of photographs. On the one hand, images are now being viewed as less important. People take a quick look at photos emailed to them, for instance, and then immediately trash them. Or, since only a fraction of all digital photos are ever printed out, they park the images on their computer hard drives and then forget about them. (Only about 13% of digital images captured ever end up on paper—compared to 98% of film images.[61]) On the other hand, those photos that do end up being saved and treasured may have been so doctored to improve reality, with the help of photo-altering programs such as Apple's iPhoto or Adobe's Photoshop Elements, that tomorrow's generations may question their authenticity. Muses one writer, will future anthropologists wonder "Why do all those people look so good?"[62]

We consider questions of images and trustworthiness again in Chapter 9.

7.5 Personal Digital Assistants & Tablet PCs

How could I use a PDA and tablet PC to help me in college?

There you are in an unfamiliar city, looking for a restaurant. Fortunately, you have your Palm or Pocket PC with software, Zagat to Go, that not only has a review of the restaurant but also a map of how to get there and even a formula for calculating tips. Maybe you also have an application called World-Mate, software offering other tools for the sophisticated traveler, such as "weather forecasts, a clock that can display times around the world, a currency converter, a tax and tip calculator, clothing size converters, and international dialing codes," as one report describes it.[63]

Palm and Pocket PC are two of the best-known *handhelds* or *palmtops*, more usually called *personal digital assistants*. As we mentioned in Chapter 1, a ***personal digital assistant (PDA)*** **is a portable device that stores personal organization tools, such as schedule planner, address book, and to-do list, along with other, more specialized software,** if you want it. In 2004, the top vendors of PDAs were PalmOne (39.6% market share) and Hewlett-Packard (27.1%), followed by Dell, Sony, and Medion.[64] Because many of the functions of the PDA are also now incorporated in smartphones, sales of general-purpose PDAs have declined, and technology pundits have therefore predicted that palmtop computing will die out. However, handheld makers have continually been trying to reinvent their industry.

more
info!

Downloads for Handhelds

Websites offering software downloads for PDAs and smartphones:
www.download.com
www.handango.com
www.mobileplanet.com
www.pocketgear.com

How a PDA Works

What things should I know about PDA operation?

As we've mentioned elsewhere, a PDA is basically a small computer, with specially designed processors and operating system software, the principal ones being Palm OS (found on Palm and Sony PDAs) and Microsoft Pocket PC (found on Hewlett-Packard, Casio, and Toshiba PDAs). All feature touch-sensitive screens that allow you to enter data with a stylus, by tapping or writing on the screen.

DATA STORAGE Data is stored in RAM (16–64 megabytes), ordinarily a volatile form of storage that loses all contents when shut off (as in a desktop PC) but which in PDAs is kept functioning with a small amount of power running off the battery even after you've turned the machine off. The built-in RAM can be augmented by storage on flash memory cards (256 megabytes up to 1 gigabyte), which slide into a slot in the PDA. The latest versions, from Sharp and PalmOne, contain an actual miniature hard drive (4 gigabytes), which enables users to view hours of video and play hundreds of songs. The PMA 430 from Archos has even more storage, allowing users to record and play 120 hours of television shows or 60 full-length movies.[65]

PDA. LifeDrive mobile managers feature gigabytes of storage and both Wi-Fi and Bluetooth. They allow you to carry files from your desktop computer and access them wherever you go. You can also access the internet and email, record voice messages, and listen to MP3 files.

POWER SOURCES Batteries in PDAs are usually lithium ion, which can be recharged. Adapters also enable you to plug your PDA into a standard AC wall electrical outlet.

TRANSFERRING FILES To transfer files from your PDA to your desktop PC or laptop (and the reverse), you can do three things: (1) You can pull out your PDA's flash card and insert it into your computer's card reader, either built in or connected through a USB port. (2) You can put your PDA in a special cradle that is plugged into a USB port. Finally, (3) you can transfer data wirelessly, using (in the latest models) the PDA's built-in Wi-Fi or Bluetooth capability.

The Future of PDAs

In what ways could PDAs evolve?

As smart cellphones continue to usurp many of the personal information management kinds of features that characterized the original PDAs, it seems likely that new PDAs will take on specialized functions. Examples:

DISPLAYING TELEVISION & PHOTOGRAPHS Instead of being personal information organizers, the original purpose of PDAs, handhelds may migrate toward single-purpose uses such as playing TV shows or displaying photographs. Microsoft announced software that will allow handhelds to display PowerPoint presentations.

HANDHELD WEATHER METERS A handheld weather meter can range from a simple and inexpensive wind meter, useful for golfers and windsurfers, to a high-tech device that tracks almost all other aspects of weather, including figuring out the wind chill, useful for skiers and snowboarders.[66]

GPS LOCATORS Global Positioning System locators have already become popular with hiking buffs. "With a GPS, rediscovering a swell secret swimming hole year after year can be about as stressful as making your way to your own mailbox," says one writer.[67] Wristwatch models are also available to help runners pace themselves and track performance. However, GPS is yet another feature that has been made available with some smart cellphones.

Tablet PCs

How useful would a tablet PC be to me?

The PDA seems to be evolving into a different kind of appliance, perhaps a smartphone. Could the same be true of the tablet PC?

As we stated in Chapter 2, a *tablet PC* is a special notebook computer outfitted with a digitizer tablet and a stylus that allows a user to handwrite text on the unit's screen. The stylus can take the place of a keyboard when users use an on-screen input panel or tap letters and numbers directly on an on-screen keyboard. (● *See Panel 7.6.*)

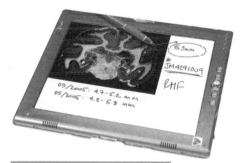

● PANEL 7.6
Tablet PC

7.6

Tablet PCs have not had the success that was anticipated for them, with only about 1% of laptops sold being tablets, despite being backed by Microsoft.[68] One problem is that traditionally tablets have cost $200–$400 more than notebook PCs.[69] Even so, these portable devices have found their uses in a handful of schools for student use in classes on such subjects as English, foreign languages, math, science, and social studies.[70]

More recently, Nokia, the company known for making cellphones, announced a mini-tablet internet appliance that looks much like a PDA and is designed to connect to the web using Wi-Fi and Bluetooth. For writing purposes, you can use the stylus either to tap the on-screen keyboard or to write on the screen and make use of handwriting recognition capability, after having trained the device to understand your handwriting style.[71]

The New Television

What's new about the new television?

"It's a transformation as significant as when we went from black-and-white to color," says *Newsweek* technology writer Steven Levy, "and it's already underway." The transformation is television, and we are nearing the day when we can watch almost any kind of programming anywhere, on a huge high-definition screen or on a smart cellphone.[72]

The Alienware DHS5 media center comes with a wireless keyboard and mouse and a remote control unit. It uses Microsoft Media Center Edition OS to integrate audio and visual functions in a single unit. It can store, play back, and burn CDs/DVDs, organize digital photos, work as TiVo does, play and record live FM radio broadcasts, play computer games, and download movies and music.

Interactive, Personalized, Internet, & Smart TVs & Entertainment PCs

How do experts distinguish between different technological ways TV can be used?

Today experts differentiate between interactive TV, personalized TV, internet TV, and smart TVs as well as entertainment PCs that allow you to do several things using your TV and sound system.

INTERACTIVE TV   *Interactive TV* **lets you interact with the show you're watching,** so that you can request information about a product or play along with a game show. We see this type of TV used with shows involving audience voting, such as that for *American Idol.*

PERSONALIZED TV   *Personalized TV* **consists of hard-drive-equipped** *personal video recorders (PVRs)*, **also known as** *digital video recorders (DVRs)*, such as TiVo and ReplayTV, which let you not only record shows but also pause, rewind, and replay live TV programs.

INTERNET TV   The MSN TV Service (formerly WebTV) is an example of *internet TV*, **which lets you read email, internet text, and web pages on your television set, using a set-top box.**

SMART TVs   Hewlett-Packard has introduced so-called *smart TVs*, **television sets equipped with hard drives and Wi-Fi capability that allows users to connect to wireless networks.** These sets won't be able to surf the web but are supposed to be able to connect to certain internet sites so that viewers can download movies, videos, and TV schedules. This is not building a PC into a TV but rather simply enhancing the functionality of the TV.[73]

ENTERTAINMENT PCs   There are also **multifunctional computers based on Microsoft's Media Center Edition 2005 that are generically called** *entertainment PCs*. You connect this computer to your TV and sound system and then watch any TV program you want, when you want, or play DVDs and CDs, pick songs from a digital library, or watch a slide show of your vacation photos, using a remote for the PC or a wireless keyboard.[74] (Examples are Hewlett-Packard's Digital Entertainment Center, Alienware's DHS5, and Gateway's 7200S.)

In the future, all such kinds of technology will probably come together in a single box that goes under the umbrella name of "digital television."

Three Kinds of Television: DTV, HDTV, SDTV

Why is it important to differentiate among the three types of TV?

When most of us tune in our TV sets, we get analog television, a system of varying signal amplitude and frequency that represents picture and sound elements. Since 1996, however, things have become more complicated.

DIGITAL TELEVISION (DTV)   In 1996, broadcasters and their government regulator, the Federal Communications Commission (FCC), adopted a standard called *digital television (DTV)*, **which uses a digital signal, or series of 0s and 1s.** DTV is much clearer and less prone to interference than analog TV. You see digital satellite TV systems and digital cable TV systems widely advertised, but they are not digital TV. These systems take normal analog broadcast signals, convert them to digital signals for transmission purposes, and then convert them back (using a set-top box) to analog signals for your TV viewing. Real digital TV, by contrast, is completely digital: It uses digital cameras, digital transmission, and digital receivers.

The FCC has mandated that all TV stations be capable of broadcasting DTV by 2006. This could have several benefits. Not only will you be able to get movie-quality pictures and CD-quality sound, but you'll also be able to receive various kinds of information services, such as announcements from public safety and fire departments.

HIGH-DEFINITION TELEVISION (HDTV)   HDTV is a subset of digital TV. "All HDTVs are digital TVs," points out technology writer Edward Baig. "But not all digital TVs are HDTVs."[75]

A form of real digital TV, ***high-definition television (HDTV)*** **works with digital broadcasting signals and has a wider screen and higher resolution than standard television.** Whereas standard analog TV has a width-to-height ratio, or *aspect ratio,* of 4 to 3, HDTV has an aspect ratio of 16 to 9, which is similar to the wide-screen approach used in movies. In addition, compared to analog display screens, an HDTV display has 10 times more pixels on a screen—1,920 × 1,080 pixels or more. Thus, HDTV could have 1,080 lines on a screen, compared with 525-line resolution for analog TV.

Analog TV

HDTV

Extra image area

Why don't more people have HDTV sets? In 2005, only about 15% of U.S. households had them, but the number was expected to rise to 40% by 2007.[76] In the past, the biggest reason for the lag in sales was expense, although some models sell for as low as $700. Another reason was that broadcasters and content producers didn't fully back the HDTV standards. Indeed, many seemed to favor another DTV standard known as SDTV, as we describe next. Finally, it costs 10%–20% more to create high-definition TV commercials, and advertisers have been unwilling to pay the extra tariff when the audience numbers were still low. Most advertisers are satisfied with making standard-definition ads look passable for HDTV.[77]

Even so, there are a number of "high-def" channels currently available to HDTV fans. For instance, Cablevision's satellite service known as Voom offers commercial-free channels such as Rush, an extreme-sports channel; Gallery, which has continually changing close-ups of paintings; MOOV, a 24-hour screen saver of video art; and several round-the-clock movie channels.[78] HDNet offers productions like "Bikini Destinations" and "What's Kewl" (about consumer electronics products).

STANDARD-DEFINITION TELEVISION (SDTV)   HDTV takes a lot of bandwidth that broadcasters could use instead for ***standard-definition television (SDTV),*** **which has a lower resolution, a minimum of 480 vertical lines, and a picture quality similar to that required to watch DVD movies.** What's important about the SDTV standard is that it enables broadcasters to transmit more information within the HDTV bandwidth. That is, broadcasters can *multicast* their products, transmitting up to five SDTV programs simultaneously—and getting perhaps five times the revenue—instead of just one HDTV program. (Analog broadcasts only one program at a time.) Thus, instead of beaming high-definition pictures, some broadcasters are splitting their digital streams into several SDTV channels. That may change by 2009, depending on recent legislation.[79]

info!

Choosing a TV

For HDTV FAQs, go to http://hometheater.about.com/od/televisionbasics/.

Personal Technology

389

PRACTICAL ACTION
Buying the Right HDTV

The most popular high-definition television sets are of three basic types, as follows:[80]

- **LCD—most expensive:** Liquid-crystal display (LCD) screens are thin (only about 2 inches thick) and lightweight, so they can be hung on a wall. LCDs are also more energy-efficient than the other two types. They have a longer life than plasmas, about 50,000 hours, and a wide viewing angle, so viewers sitting to the side don't have a distorted picture. At one point, LCDs had a reputation for being too slow for fast-paced sporting events, so a demonstration of a fast-moving event is advised before you buy. A 50-inch system costs roughly $9,000.

- **Plasma TV—next most expensive:** This big-screen display is so clear that it can make outdoor scenes appear as if you were looking at the real thing out a window. Life span is about 30,000 hours. A 50-inch system costs about $5,500.

- **DLP—least expensive:** Digital light processing (DLP) technology from Texas Instruments is used in rear-projection TVs. They aren't thin enough to hang on a wall but at 15 inches and 100 pounds they won't take up most of the living room either. Such sets aren't as sharp and bright as the other two types. About 10% of viewers see a "rainbow" smudge of color in certain parts of the screen (something you will notice when you first look at it, if you are one of these people). Over time, users may have to replace the lamp at the back of the unit. A 50-inch model costs about $3,000, about one-third the cost of an LCD.

You should be sure to ask the salesperson to show any set you're considering in both HDTV and SDTV formats.

The Societal Effects of the New TV

How has technology changed the way people experience television?

"The ethos of New TV can be captured in a single sweeping mantra," says Steven Levy, "*anything you want to see, any time, on any device*" (his emphasis).[81] The devices can range from Samsung's $100,000 102-inch HDTV screen that hangs on your living room wall to a $150 multifunction cellphone with a 2-inch display area, as we'll describe in the next section.

Some highlights of the "new TV" are as follows.[82]

TIME SHIFTING: CHANGING WHEN YOU WATCH TV Personal video recorders like those marketed by TiVo allow viewers to watch favorite shows at their own convenience rather than following a broadcast schedule. They also enable viewers to freeze-frame action sequences and to skip commercials. (What this could eventually do to advertising as the economic underpinning for free-to-viewers television can only be guessed at.)

Another technology affecting the "when" is ***video on demand* (*VOD* or *VoD*), which consists of a wide set of technologies that enable viewers to select videos or TV programs from a central server to watch when they want,** rather than when TV programmers offer them. The technology is already here in its rudimentary form. A company called Akimbo markets a TiVo-like set-top box with a hard drive that can hold 200 hours of video and offers a library of programs that, for a fee, can be downloaded. Although initially the fee structure seems high and the program offerings limited, that may change.

info!

Fast-Forward Tags

Go to www.newstarget.com/006595.html to find out what these are.

(Here's the entire list of sports categories, for instance: Billiards, Extreme Sports, Golf, Martial Arts, Documentaries, and Yachting.)[83]

SPACE SHIFTING: CHANGING WHERE YOU WATCH TV   New technology is allowing you to download or receive TV programs, either stored or real-time, and watch them on some sort of handheld device. A device from Sling Media enables you to watch a program playing on your living room TV on your laptop computer—anywhere in the world. MobiTV will send TV programs such as those on CNN and the Discovery Channel to the display screen on your smartphone. Verizon Communications also has a mobile TV service, called V CAST, that will let you watch (in certain metropolitan areas) TV shows, as well as news and sports highlights, stock market updates, city-specific weather reports, and movie trailers.[84] In South Korea, TU Media has launched a satellite-based service that beams seven video channels to cellphones, using digital multimedia broadcasting; Nokia and Qualcomm are backing similar technologies.[85]

CONTENT SHIFTING: CHANGING THE NATURE OF TV PROGRAMS   Perhaps the most important development is the movement of television to the internet. This is made possible by _**IPTV**_, **short for _Internet Protocol Television_, in which television and video signals are sent to viewers using internet protocols.**[86] Cable and satellite channels have limited capacity, but the internet "has room for everything," Levy points out. As a consequence, you may be able to cram even *more* programs on your screen simultaneously—and what does that do to the human attention span? (Multiple-channel TV sets are now available with what are known as "mosaic" screens that allow sports fans, for instance, to watch eight separate games on one screen.[87])

Already TV is being redesigned for cellphone viewers. For instance, the live-concert show "Pepsi Smash" failed on regular broadcast television because it drew only a modest audience. However, it was picked up by Yahoo and edited into a cluster of short segments to be provided on demand to online TV watchers, with many segments running under 4 minutes.[88] Another matter that future content providers may have to think about: Sony has offered technology on some of its TV sets that lets viewers manipulate the images they see on their screens, allowing them "to zoom into and pan around the picture as well as sharpen the resolution," in one description.[89]

7.7

Smartphones: More Than Talk

How are smartphones different from basic cellphones?

"There's a digital land rush going on," says *New York Times* technology writer Steve Lohr, "driven by rapid advances in technology that make it possible to put more and more tools of higher and higher quality into phones."[90]

Sanyo. LG. Sony Ericsson. Samsung. Nokia. Motorola. These Asian companies have been well known for making TVs, cameras, and stereos. Now they have applied their consumer-electronics experience to producing _**smartphones**_—**cellular telephones with microprocessor, memory, display screen, and built-in modem.** These multimedia phones, which combine some of the capabilities of a PC with a handset, offer a wealth of gadgetry: text messaging, cameras, music players, videogames, email access, digital-TV viewing, search tools, personal information management, GPS locators, internet phone service, and even phones doubling as credit cards (which, among other things, can be used to pay for parking time on a new generation of wireless parking meters).[91] As one newspaper headline put it, "It's Not Just a Phone, It's an Adventure."[92]

Let's see what some of these features are.

How a Mobile Phone Works

What are the basic elements of a mobile phone?

In Chapter 6, we described how wireless services use their networks of cell towers to send voice and data over the airwaves in digital form to your cellphone, handing off your call from one cell to another as you move through a series of geographically overlapping cells. The cellphone or smartphone contains many of the same attributes as a personal computer: processor, memory, input/output devices, and operating system (such as the Symbian OS or Microsoft Windows Mobile OS). The OS, which is stored in read-only memory (ROM) and is run by the processor, provides you with the interface that allows you to store data, change settings, and so on.

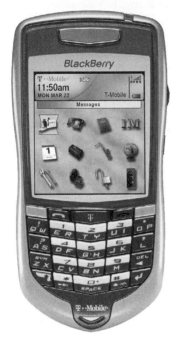

STORAGE The data you store in your phone, such as telephone numbers, is stored in ROM, which means that when you turn off the phone, the data does not disappear. Some smartphones, such as the Samsung SGH-P777, have as much as 100 megabytes of memory. Others, such as the Sanyo MM-5600, which has a high-quality MP3 player and camcorder, come with a 16-megabyte memory card on which to store your files.

INPUT At minimum, cellphones have a keypad for entering numbers and text (and doing text messaging) and a microphone for picking up your voice. Some are also mobile speakerphones, giving you a hands-free option and offering voice-activated dialing.[93] (The sound quality isn't as good as that with a normal call, but at least it gets the job done, especially if you make calls while driving.) Some phones also offer a touch-sensitive screen or a screen with a stylus. Xerox has developed software to turn a phone into a portable document scanner. Of course, there are phones with built-in digital cameras and camcorders, as we'll describe.

OUTPUT A cellphone includes a receiver or speaker, of course, for picking up voice calls. Mobile phones also have displays ranging from LCD to full-color, high-resolution plasma, suitable for watching TV and playing videogames, as we'll describe. Some phones act as MP3 players and offer FM radio and stereo sound. Others are also able to tap into Wi-Fi and Bluetooth networks.

Smartphone Services

What smartphone services would be most useful to me?

Proof that the phone is morphing into "the everything device," in the words of one analyst, is everywhere.[94] The percentages of people who use nonvoice applications on their cellphones, according to one survey, are as follows: text messages, 27%; downloaded ringtones, 14%; email, 11%; internet, 9%; and photography, 6%.[95] (For the 33% of people in the United States ages 8–18 who own cellphones, the percentage who downloaded extras were as follows: ringtones, 91%; games, 53%; screensavers, 44%; MP3s, 20%; and videos, 2%.[96]) Let's consider these and other services offered for mobile phones.

TEXT MESSAGING **_Text messaging_, or _texting_, is the sending of short messages, generally no more than a couple of hundred characters in length, to a pager, PDA, smartphone, or other handheld device,** including notebook computer. They can also be sent to desktop computers and fixed-line phones. Originally text messaging appeared during the days of mainframe computers, when workers sitting at terminals would send short text messages to each other. In the internet world, these evolved into instant messaging and live text chat sessions.[97] Today, says one writer, "text messaging combines the portability of cellphones with the convenience of email and instant messaging."[98]

To translate text lingo, go to www.transl8it.com.

info!

Voice-to-Text Dictation

The Samsung P207 features built-in "speech-to-text" technology that turns what you dictate into text on the screen. For more information:
www.cingularwireless.com
www.voicesignal.com

info!

Downloading Ringtones

Some are free. Check if your phone is compatible with the site.
www.3gforfree.com
www.3gupload.com
www.mbuzzy.com
www.myphonefiles.com
www.matrix.com

Text messaging is particularly appropriate for situations in which making a cellphone call is intrusive, as when you want to tell someone "I'm 15 minutes late."

Sometimes text messages ("texts") are called *SMSes*. **SMS stands for _Short Message Service_, a text message service originally designed for GSM mobile phones but now available on a range of networks,** including 3G networks. Text messaging has long been wildly popular in Europe and Asia but in recent years has become more mainstream in North America. One consequence is that as teenagers have become accustomed to it, there has developed a streamlined language, so-called *text message lingo*, with terms such as *XLnt* ("excellent"), *fbi* ("I'll look into it"), and *PCM* ("please call me"). (● See *Panel 7.7*.) As one writer put it, "It's like trying to read a sentence made of vanity license plates."[99] Some people program into their phone such standard reply phrases as "Yes," "No," "Call me," and "Will call you later."

Incidentally, texters need to be aware that it's unwise to assume that when they hit the Delete button, the messages are gone forever. Like email and instant messages, text messages are often saved on servers, at least for a while.[100]

DOWNLOADED RINGTONES When Shannon Dunham's phone rings, it goes "Knick, knack, patty whack, give your dog a bone"—the Reno, Nevada, woman's favorite childhood jingle—and it indicates to her that the person trying to call her is her mother. Dunham also has 17 other distinct ringtone tunes, including "Maple Leaf Rag" to signal it's her boyfriend calling.[101] If your phone's **_ringtone_, the audible sound a phone makes to announce that a call is coming in,** is still an old-fashioned "ring, ring," perhaps you have some catching up to do. Now, as one writer puts it, your phone could be "vibrating, flashing a photo of the caller across the phone's screen, and, above all, playing music—preferably an actual clip from a song like 'Drop It Like It's Hot' by Snoop Dogg."[102]

Some ringtones may be had for free; others, which you may download from your wireless carrier (such as Verizon Wireless, Cingular, or Sprint), may cost anywhere from about $1.25 to $4 per tune (not per ring). Ringtones may be *monophonic*, sounding more like beeps or the basic sounds that come on your cellphone when you buy it; *polyphonic*, sounding digitized, sort of like elevator music; or *music tones* (also called *master tones* or *true tones*), real audio clips from songs. Ringtones have become so big that in October 2004 *Billboard* magazine, which follows the music industry, even created a chart of popular hits for them.

● **PANEL 7.7**
Text message lingo

Words		Phrases	
bcum	become	ruok	are you OK?
b4	before	brb	be right back
bf/gf	boyfriend/girlfriend	bcnu	be seeing you
cn	can	cm	call me
gr8	great	eod	end of discussion
l8r	later	g2g	gotta go
msg	message	h2cus	hope to see you soon
2moro	tomorrow	idk	I don't know
ppl	people	j4f	just for fun
snd	send	lol	laughing out loud
2dA	today	cul8r	see you later
wot	what	wan2	want to
u	you	werv	where have

Source: www.transl8it.com.

EMAIL The most well-known hand-held device for sending wireless email is the BlackBerry, produced by Canadian company Research in Motion (RIM) Ltd., which practically created the industry. However, as the market for corporate wireless email is estimated in the United States alone to total 35 million users, RIM is being joined by a host of rivals—most of which also offer telephone and other smartphone capabilities.[103] Examples are the palmOne Treo 650, T-Mobile's Sidekick II (a favorite of Paris Hilton), and the Audiovox PPC-6600.

As with text messaging, something that most users need to pay attention to before buying is the layout and size of the keys. The BlackBerry 7100 series smartphones, for example, cram all letters and numbers onto just 14 keys. (As you type, built-in software anticipates the correct word, something that takes some getting used to.) The Sidekick II, by contrast, has a roomy keyboard, although it's hidden under a flip-up screen, which can make dialing a little awkward.[104]

Incidentally, as we mentioned, many of the personal organizing functions formerly found in personal digital assistants—such as address book, schedule planner, and to-do list—have now migrated to smartphones. What happens if you break or lose your cellphone (as people do all the time)? How do you replace your built-in telephone numbers and contacts? Some carriers and technology companies, such as Verizon Wireless and Nextel Communications, offer services that, for a small monthly fee, back up such data wirelessly.[105]

INTERNET ACCESS Many smartphone owners use their devices to perform web searches of one sort or another, accessing not only Yahoo and Google but also such things as maps and directions, using the Global Positioning System, if their phones are so equipped. For instance, Zipdash uses the growing number of cellphones with GPS receivers to gather the speeds of thousands of drivers and display highway conditions as maps on cellphone screens.[106] MapQuest enables you to pull up a map on your mobile phone while you're in transit and then navigate as you go along. Yahoo, which also provides mapping services, has a local search system for cellphones. If you type in the word *pizza* and an address in Chicago, for example, your phone will display all pizza restaurants in that neighborhood, as well as maps and driving directions.[107]

An important capability for wireless internet access is the ability to make connections to Wi-Fi networks. For instance, the HP iPaq Pocket PC h6315, sold through T-Mobile, looks like a PDA but also acts as a cellphone and Wi-Fi connection. Thus, you can surf the web using the phone network, but that can be slow compared with Wi-Fi. However, if you're near a Wi-Fi network on campus or in a coffee shop, you can switch over to the faster network; the device will automatically switch you back to the cellular network if you walk out while still online.[108]

Other smartphones, such as the Motorola CN620, allow you to use the VoIP (voice-over IP) technology, which we described in Chapter 2, to make phone calls on the internet. Thus, the device will work where cell service is offered as well as in Wi-Fi-accessible areas.

PHOTOGRAPHY   Camera phones are making the spontaneous snapshot an everyday part of life, and, says one observer, "turning ordinary citizens into documentarians, fine-art photographers, and, in cases such as hit-and-run accidents, community watchdogs."[109] Indeed, the devices have even inspired a trend called *mobile blogging*, or "moblogging," in which phones are used to visually record all kinds of events, both mundane and highly personal, which are then displayed and transmitted to websites such as Mobog, Textamerica, and Yafro.[110] In one experiment, 52 patients with leg ulcers were examined in person by doctors on-site and remotely by other doctors who had only pictures of the wounds taken by camera phone, and both sets of physicians were in remarkable agreement in their diagnoses and suggested treatment.[111]

Camera phone manufacturers are learning that consumers will take more pictures if the quality is improved and the printing made easier.[112] Thus, there are now available 5-megapixel phones (as from Samsung) and phones (as from Sanyo and Nokia) that can connect directly to many home printers. Camera phones are also being made with memory card slots, so that images stored on a memory card can then be taken directly to a retailer to have prints made. And the wireless carriers themselves are getting in on the act. Sprint, for instance, charges for using its network to transfer photos from camera phones. Many wireless carriers also make kiosks available for downloading and printing out camera images.

Phone games. WildTangent's online video game "24."

GAMES   In the past, cellphones used as game devices left something to be desired, with "flat, cartoonish graphics and simple scenes," in one description.[113] More recent smartphones, however, have powerful graphics chips and better screen quality that allow three-dimensional images, faster computing capabilities, realistic drawings, and fast-moving action.

RADIO & MUSIC   Some smartphones allow users to listen to FM radio. Sprint and RealNetworks have introduced an internet radio service for Sprint wireless customers.[114]

In addition, cellphone makers and wireless carriers are now crowding into the market for mobile music. For instance, Apple's iTunes phone, the music-playing cellphone called the Rokr that's made by Motorola, can store up to 100 songs on a 512-megabyte flash memory.[115] The Nokia N91 cellphone can store more than 1,000 MP3 digital songs (or hundreds of digital photos or video clips) on its 4-gigabyte hard drive[116]

Some phones are not multifunction phones so much as dual purpose: Motorola's E725, for instance, is designed specifically for music, in addition to making phone calls. Thus, these devices have stereo-quality sound, as well as large memories for storing songs.[117] Some of the biggest music companies and phone companies have been pairing up to make available song downloads—or songs and music videos to watch on mobile display screens—that customers may purchase through their phones.[118]

Short-range wireless Bluetooth technology enables cellphone users to put on headphones so they can connect to their cellphones without cords.[119]

TV & VIDEO   One mother confesses she uses her cellphone's TV feature to keep her 3-year-old son occupied while she waits in line at the post office. After tuning in the Cartoon Network, she says, "I just give him the phone and he's quiet."[120] The Walt Disney Company and Warner Brothers are also developing "mobi-toons" that can play as mini-television shows on smartphone displays.[121] Major-league baseball has arranged for TV clips to be posted to phones, offering video coverage of games.[122] Popular video services are VCAST from Verizon Wireless, MoviTV for Cingular customers, and Spring PCS Vision.

TV programs made for a handheld mobile device with a 2-inch or 2¼-inch display screen pose particular challenges. For one thing, although cellphone screen quality has gotten better, there are still limitations in battery life, processing power, and storage capacity.[123] In addition, most viewers don't have the patience to watch a 90-minute feature film on a screen this size. "Cell phone cinema has to hold your attention before your mind jets off to do something else with your phone, such as surf the web, check your email, or, gasp, make a phone call," says one analysis. "No time to develop character. No room for special effects. Not enough screen resolution for moody shadows or shades of meaning."[124] Whether it's a sitcom, news show, or sports highlights, TV programs designed for cellphones have to rely heavily on close-ups, more static shots, and little movement within the frame—the opposite of MTV video. In the end, then, the producers of such fare have to rely more on writing and storytelling.

OTHER FEATURES   We have mentioned GPS locators and video cameras, but there seems to be no end of other features existing and forthcoming for smartphones. In the United States, students at the University of California, Santa Barbara, pay for parking stalls by charging the fee to an account through their cellphones.[125] In Japan, customers can pay for purchases not with cash or credit cards but by making electronic payments through a smartphone.[126] Also in Japan, mobile phones have been turned into controllers for model racing cars, for use as television remotes, and as devices for fingerprint recognition.[127] In South Korea, cellphones have been modified to allow diabetics to check their blood-sugar levels, and the data can be sent to their physicians.[128] Finally—and unbelievably—researchers at the Massachusetts Institute of Technology have developed cellphone technology, called the "Jerk-O-Meter," that analyzes speech patterns and voice tones and claims to be able to tell a cellphone caller if the person on the other end really cares what you are talking about.[129] Is there anything that hasn't been thought of?

People are using electronic devices everywhere: digital cameras, computers, cellphones.

more info!

Big Boss Is Watching?

Read the article at
http://news.com.com/
Big+boss+is+watching/
2100-1036_3-5379953.html.

The Societal Effects of Cellphones

Have mobile phones been a positive or negative force?

The effects of the cellphone—and the smartphone now becoming more widespread—are mixed. The positive attributes are many: Parents can more easily monitor the safety of their children, police dispatchers can help people who are lost, information and amusements of all kinds are readily available, you can phone your apologies when you'll be late for an appointment, and so on.

However, people's personal behavior in using phones has not necessarily been improved. One survey, the 2004 Sprint Wireless Courtesy Report, found that 78% of those interviewed said that people are less polite, courteous, and respectful in cellphone manners than they were 5 years earlier. (Yet 95% thought their own cellphone manners were just fine.)[130] Cellphones regularly ring—and are answered—in theaters, despite movie screen advisories. Bus and train travelers become enraged by the loud conversations of fellow passengers.[131] The serenity of nature's wonders in national parks is disrupted by people yakking on their phones.[132] Bosses feel obliged to rebuke their employees whose cellphones ring in meetings.[133]

Phone use by car drivers makes even young people drive erratically, moving and reacting more slowly and increasing their risk of accidents.[134] People inadvertently dial numbers in their phone's address book when they've had too much to drink (a phenomenon known as "drunk dialing").[135] Camera-equipped phones have been used to take pictures of people in bathroom stalls, to cheat on school tests, and to allow thieves to capture credit card numbers.[136] Pornography companies see cellphones as a new frontier.[137] Employers can hire companies that can keep track of their employees' out-of-office locations through GPS tracking of their cellphones.[138]

But just as technology can create unforeseen problems, perhaps technology can also provide solutions. Some countries—but not the United States yet—are permitting the use of radio-jamming equipment that keeps nearby cellphones from working whether the phone user likes it or not.[139] This would certainly allow restaurants, theaters, and the like to impose a "cone of silence" that would provide relief to long-suffering members of the public who have been forced to unwillingly share in others' cell conversations.

7.8

Videogame Systems: The Ultimate Convergence Machine?

How do the three principal videogame consoles compare?

The "convergence of computing, communications, and entertainment has been promised before," says a 2005 *Wall Street Journal* article, "only to evaporate because of consumer indifference and technology that wasn't ready for prime time. But now the pieces are finally coming together. And corporations are scrambling to make sure they aren't left behind."[140]

The principal strategy under which this is happening is the placement of entertainment devices in your living room. Microsoft Corp., for example, is attempting to put its new Xbox 360 into people's homes under the guise of providing an online videogame system, but what it is really aiming at is installing a "miniature electronic ecosystem" there, with Microsoft at the center. Someday you may wake up and find that Xbox "entered your house under the humble pretense of being a game machine, a toy for the kids, but it just ate your CD player and your DVD player, and it's looking hungrily at your telephone," suggests a *Time* article. "It's talking to your iPod, your digital camera, your TV, your stereo, your PC, your credit card, and the Internet."[141]

Several industries—videogame console makers, consumer-electronics companies, PC companies, cable companies, and telephone companies—are

Multiplier. Could videogame machines such as the Microsoft Xbox 360, Sony PlayStation 3, and the Nintendo Revolution become the "miniature electronic ecosystem" that finally represents the convergence of computing, communications, and entertainment that has long been anticipated?

interested in this area. Here let us consider just the first: the makers of game-boxes. Already a company called Tiger Telematics has produced a handheld gaming device called the Gizmondo that can play videogames, movies, and music files; take photos; wirelessly send and receive text and email messages; display web pages; and function as a GPS navigation and tracking unit.[142] However, the big videogame hardware makers are Microsoft (until now mainly a software rather than hardware business), Sony, and Nintendo. Let's see how their consoles compare.[143] (● *See Panel 7.8.*)

Microsoft's Xbox 360

Why might I be inclined to buy an Xbox?

Scheduled for release in fall 2005, the Xbox 360 is double the power of the previous Xbox. As we said, the ostensible principal intended use of the new Xbox is for online videogames; Microsoft has invested heavily in its subscription service, Xbox Live, which allows Xbox players to meet online—and also to talk to people with whom they are playing via voice chat (a possible forerunner to a general internet telephone system). However, Microsoft really is trying to make the device a living-room hub for all kinds of digital media—not only video games but also movies, music, and online content. The Xbox enables users to connect cameras, digital music players, high-definition

● PANEL 7.8
Principal videogame consoles compared.
All are supposed to be backward-compatible, so that users can play games developed for earlier gameboxes.

	Microsoft Xbox 360	Sony PlayStation 3	Nintendo Revolution
DVD format	Dual-layer DVD disks	Blu-ray high-capacity DVD disks	New DVD-size optical disk
CPU	IBM PowerPC-based three-in-one processing chip	IBM and Toshiba Cell processor	"Broadway" processor from IBM
Storage	20-gigabyte hard disk; two memory card slots	Memory card slots for SD, MemoryStick, Compact Flash; optional hard disk	SD memory cards
Types of online connections	Wi-Fi; built-in Ethernet connection; Xbox Live online service included; USB 2.0 ports; ability to stream from Windows PCs, digital cameras, music portables	Wi-Fi; built-in Ethernet connection; Bluetooth wireless connections; USB 2.0 ports	Wi-Fi; Nintendo Wi-Fi Connection gaming service; USB 2.0

televisions, surround-sound stereo systems, and the like, and a 20-gigabyte hard drive is available on which users can copy music CDs, photos, and other digital content. With this gadget, Microsoft is also positioning itself to be a supplier of video on demand in competition with cable operators.

Sony's PlayStation 3

Is the PS3 a better game machine?

PlayStation 3

Scheduled for release in spring 2006, the PlayStation 3, or PS3, is intended to exploit the dominance of the TV-console market that Sony established with the PlayStation 2. PS3 is also said to be more powerful than Microsoft's new Xbox. Sony's new machine uses an ultra-powerful multiprocessor, the Cell chip from International Business Machines and Toshiba Corp, which is 35 times more powerful than the one in PS2. It's called a "cell" because software instructions can be divided up in cells and distributed among different processors and devices (TVs, music players, and the like). This means a network can be organized without human intervention, such as, in a recent demonstration, 48 separate high-definition video feeds to a single display screen. PS3 will enable gamers not only to videoconference over the internet but also to have movie-quality graphics, with lifelike textures, colors, and motion. Users will also be able to stream and download music and movies, handling several streams at once (so that you can play a videogame and watch a movie at the same time).

Nintendo's Revolution

What distinguishes the Nintendo console from the others?

Scheduled for release at some point in 2006, Revolution by Nintendo doesn't emphasize power and graphics so much as innovative, networked game play and simpler, cheaper game development. Thus, the company is positioning itself with new versions of two decades' worth of tried-and-true videogames, such as Mario, Zelda, and Metroid. Its one-handed remote controller, which is fitted with motion sensors, allows players to wave it around frenetically, swish it gently through the air, or kick, punch, jump, or steer their way through on-screen action.[144] Nintendo is also introducing the Nintendo DS, an advanced portable game console that will enable users to wirelessly connect via Wi-Fi to other users with Wi-Fi capabilities. Unlike Microsoft and Sony, however, Nintendo's gamebox won't play movies on DVD unless users buy a separate device.

info!

Prepared for the Workplace?

Some people believe that NetGeners are not prepared for today's corporate workplace. Read the article at www.zazamedia.de/pages/Commentary21.html.

The Results of Personal Technology: The "Always On" Generation

What technological "generation" am I part of?

With all these personal digital devices, it's no wonder that anyone born since 1982 is called a member of the "Always On" generation. They are the generation that has for their entire lives been surrounded by and using the toys and tools of the Digital Age. If you belong to this group, does it make you different from previous generations? We consider this question in the following Experience Box.

Experience Box
The "Always On" Generation

Videogames, computers, cellphones, portable music players, email, the internet, instant messaging, the web, video cameras, and similar personal technology. Is this what you grew up with? Then you're a member of the Always On generation, also known as *NetGeners* (for "Net Generation") and *Millennials* (for "Millennial Generation").

"Seventy-four percent of teens use instant messaging," says high-tech magazine editor Tony Perkins. "No matter what numbers you look at, the Always On generation is in full swing. They communicate through the computer."[145] A survey by Sprint finds that 76% of teens ages 15–19 and 90% of people in their early 20s regularly use their cellphones for text messaging, ringtones, and games.[146] Even poor teens may regularly owe well over $100 a month in phone charges—and skip lunch so they can pay the bills.[147] Members of this cohort "have never known life without computers and the internet," says one report. "To them the computer is not a technology—it is an assumed part of life." [148] As a result, say observers such as Marc Prensky, "today's students *think and process information fundamentally differently* from their predecessors" (his emphasis).[149]

Not every person in college fits this profile. Indeed, a high proportion of students in higher education, 39%, are over age 25, many going to school part-time, quite often having dependents or being single parents. Those most likely to have the mindset and characteristics of Always On were born on or after the year 1982. Unlike older students (members of the "Baby Boomers" and the generation that followed them, "Generation X"), Millennials tend to exhibit distinct learning styles. "Their learning preferences tend toward teamwork, experiential activities, structure, and the use of technology," says Diana Oblinger, executive director of Higher Education for Microsoft Corporation.[150] With such students, listening to music, sending instant messages, chatting on the phone, and doing homework is second nature.

Let's consider some of the characteristics of these students.

Staying Connected Is Essential

After a college class lets out, students step into the hallway and immediately activate their cellphones, PDAs, or notebook computers, talking to or IMing friends or checking their email. At rock concerts, when the lights dim, one may see the glow of tiny blue screens, as concert goers text-message friends or have them listen to music via their cellphones—or observe the action through camera phones.

"Connecting is what the current generation of students is all about," says one report. "College students use the internet as much for social reasons as for academic reasons."[151] At the University of North Carolina, for example, sophomore Dax Varkey carries his laptop computer everywhere, using it not only to take notes in class but also to instant-message or

email friends. In his dorm room, he even IMs his roommate a few feet a way. He uses his cellphone to call his buddy who lives one floor above him. "I always feel like I have a means of communications—in class, out of class," says Varkey, 19.[152]

Besides maintaining social contacts, students get homework assignments and lecture outlines online and participate in class discussions. They instant-message friends to brainstorm class projects and other assignments. They email professors with questions at any time—a boon for students too shy (or lazy) to visit during an instructor's regular office hours.[153] All in all, the lines between work, social life, and studying are blurring.

Multitasking Is a Way of Life

Multitasking is second nature. "It's perfectly normal for a group of college students to watch TV with their laptops in front of them," said one recent college graduate. "They'll check their campus mailbox once a day but their email every five minutes."[154] Says Prensky, such students "are used to receiving information really fast. They like to parallel process and multi-task. They prefer their graphics *before* their text rather than the opposite. They prefer random access (like hypertext)."[155] They are also, having grown up with videogames, accustomed to learning how to take in many sources of information at once and how to incorporate peripheral information. Finally, games have taught them how to use and manage a large database of information.[156]

Students Are Impatient & Results-Oriented

Today's students have grown up on the "twitch speed" of videogames, instant messaging, MTV, and a customer-service kind of culture. Thus, they "have a strong demand for immediacy and little tolerance for delays," says one analysis. "They expect that services will be available 24×7 in a variety of modes (web, phone, in person) and that responses will be quick."[157] One result of this is that many students prefer typing to handwriting: It's faster. Another result: Students expect to be engaged, which is why online forums, blogs, and use of RSS aggregators to update subjects are popular.[158] They don't want to be preached to, ignored, or bored. They are also experiential learners—they prefer to learn by doing rather than learn by listening.[159]

They Respect Differences & Gravitate toward Group Activity

NetGeners are racially and ethnically diverse and consequently "accept differences that span culture, ability—or disability—and style," says one report. Indeed, they "are more comfortable with their learning differences than any other generation has been."[160] They also gravitate toward group and team activities, owing to their group-gaming experiences and their constant communication with friends through cellphones, IM, and email.

Summary

Caller ID (p. 370) Feature that shows the name and/or number of the calling party on the phone's display when you receive an incoming call, enabling you to screen telephone calls. Why it's important: *Caller ID helps you become disciplined about preventing others from wasting your time.*

convergence (p. 369) Also known as *digital convergence;* the combining of several industries—computers, communications, consumer electronics, entertainment, and mass media—through various devices that exchange data in digital form. Why it's important: *Convergence has led to electronic products that perform multiple functions, such as TVs with internet access, cellphones that are also digital cameras, and a refrigerator that allows you to send email.*

digital television (DTV) (p. 388) Television standard adopted in 1996 that uses a digital signal, or series of 0s and 1s. It uses digital cameras, digital transmission, and digital receivers and is capable of delivering movie-quality pictures and CD-quality sound. The Federal Communications Commission has mandated that all TV stations be capable of broadcasting DTV by 2006. Why it's important: *DTV is much clearer and less prone to interference than analog TV. It may allow viewers to receive various kinds of information services, such as announcements from public safety and fire departments.*

entertainment PC (p. 388) Multifunctional computer based on Microsoft's Media Center Edition 2005. Why it's important: *When viewers connect this computer to their TVs and sound systems, they can then watch any TV program they want, when they want, or play DVDs and CDs, pick songs from a digital library, or watch a slide show of vacation photos, using a remote for the PC or a wireless keyboard.*

high-definition (HD) radio (p. 377) Form of radio that provides CD-quality sound and allows broadcasters to squeeze two digital and one analog station on the same frequency. Why it's important: *HD radio combines digital and analog broadcast signals, enabling stations to offer an analog main channel and digital "sidebands," so that multiple types of content can be broadcast from the same position on the dial.*

high-definition television (HDTV) (p. 389) A form of television that works with digital broadcasting signals. Why it's important: *HDTV has a wider screen and higher resolution than standard television. Whereas standard analog TV has a width-to-height ratio, or aspect ratio, of 4 to 3, HDTV has an aspect ratio of 16 to 9, which is similar to the wide-screen approach used in movies. In addition, compared to analog display screens, an HDTV display has 10 times more pixels on a screen—1,920 × 1,080 pixels or more. Thus, HDTV could have 1,080 lines on a screen, compared with 525-line resolution for analog TV.*

interactive TV (p. 388) Type of television that lets viewers interact with the show they're watching, allowing audience voting, such as that for *American Idol.* Why it's important: *Interactive TV allows viewers to request information about a product or play along with a game show.*

internet TV (p. 388) Type of television service, such as the MSN TV Service (formerly WebTV), that enables users to receive the internet through their television sets, using a set-top box. Why it's important: *Internet TV lets viewers read email, internet text, and web pages on their television sets.*

IPTV (Internet Protocol Television) (p. 391) Technology in which television and video signals are sent to viewers using internet protocols. Cable and satellite channels have limited capacity, but the internet has much more room. Why it's important: *Enables the movement of television to the internet.*

megapixels (p. 381) In a digital camera, the millions of picture elements, the electronic dots making up an image; the number of megapixels expresses the camera's resolution, or image sharpness. The millions of pixels are tightly packed together on the camera's image sensor, a half-inch-wide silicon chip. When light strikes a pixel, it generates an electric current that is converted into the digital data that becomes your photograph. Why it's important: *The more megapixels a digital camera has, the better the resolution and the higher the quality of the image.*

MP3 (p. 372) Format that allows audio files to be compressed. MP3 files are about one-tenth the size of uncompressed audio files. For example, a 4-minute song on a CD takes about 40 megabytes of space, but an MP3 version of that song takes only about 4 megabytes. Why it's important: *MP3 allows audio files to be made small enough to be sent over the internet or stored as digital files.*

MP3 digital audio player (p. 372) Portable device that enables users to play MP3 digital audio files. Why it's important: *MP3 players allow users to have access to hundreds of songs while they are on the go.*

multitasking (p. 371) Performing several tasks at once, such as studying while eating, listening to music, talking on the phone, and handling email. Why it's important: *Medical and learning experts say the brain has limits and can do only so much at one time. For instance, it has been found that people who do two demanding tasks simultaneously, such as driving in heavy traffic and talking on a cellphone, do neither task as well as they do each alone. Indeed, the result of constantly shifting attention is a sacrifice in quality for any of the tasks with which one is engaged.*

personal digital assistant (PDA) (p. 386) Also known as *handheld* or *palmtop;* a portable device that stores personal organization tools, such as schedule planner, address book, and to-do list, along with other, more specialized software. Examples are Palm and Pocket PC. Why it's important: *Because many of the functions of the PDA are also now incorporated in smartphones, sales of general-purpose PDAs have declined, and it's been predicted that the device will die out.*

personalized TV (p. 388) Also known as *digital video recorder (DVR).* Hard-drive-equipped device, such as TiVo and ReplayTV. Why it's important: *This device not only lets you record shows but also pause, rewind, and replay live TV programs.*

podcasting (p. 378) The recording of internet radio or similar internet audio programs. Why it's important: *Podcasting software allows amateur deejays and hobbyists to create their own radio shows and offer them free over the internet. Listeners can then download shows onto their MP3 players like the iPod.*

point-and-shoot camera (p. 379) A camera, either film or digital, that automatically adjusts settings such as exposure and focus. Generally such cameras cost under $500. Why it's important: *A point-and-shoot autofocus will do most of the work for you automatically, although you may find it useful to get a camera that also has manual controls, so that you can take over if you want.*

ringtone (p. 393) The audible sound a phone makes to announce that a call is coming in. A ringtone may be an old-fashioned "ring, ring" or a clip from a popular song. Why it's important: *Users can customize the ringtones on their cellphones.*

sampling rate (p. 373) The number of times, expressed in kilobits per second, that a song is measured (sampled) and converted to a digital value when it is being recorded as a digital file. Why it's important: *The sampling rate affects the audio quality. For instance, a song sampled at 192 kilobits per second is three times the size of a song sampled at 64 kilobits per second and will be of better sound quality.*

satellite radio (p. 375) A radio service in which signals are sent from satellites in orbit around the Earth to subscribers owning special radios that can decode the encrypted signals. Why it's important: *The CD-quality sound is much better than that of regular radio, and because the signals are digital, there are many more channels available than those on traditional radio. Also, unlike standard broadcasters, satellite radio broadcasters are for the most part not regulated by the Federal Communications Commission.*

single-lens reflex (SLR) camera (p. 380) A camera, either film or digital, that has a mirror that reflects the incoming light in such a way that the scene viewed by the viewer through the viewfinder is the same as what's framed by the lens. Why it's important: *A digital SLR, which may cost anywhere from $450 on up, is used by professional photographers and serious amateurs because it provides more manual options and better image quality and allows the use of interchangeable lenses, from wide angle to telephoto.*

smart TV (p. 388) Television set equipped with hard drives and Wi-Fi capability that allows users to connect to wireless networks. Why it's important: *It enables users to connect to certain internet sites so that they can download movies, videos, and TV schedules.*

smartphone (p. 391) Cellular telephone with microprocessor, memory, display screen, and built-in modem. Why it's important: *Some types of this multimedia phone, which combine some of the capabilities of a PC with a handset, offer a wealth of gadgetry: text messaging, cameras, music players, videogames, email access, digital TV viewing, search tools, personal information management, GPS locators, internet phone service, and even phones doubling as credit cards.*

SMS (Short Message Service) (p. 393) A text message service originally designed for GSM mobile phones but now available on a range of networks, including 3G networks. Why it's important: *Text messaging has long been wildly popular in Europe and Asia but in recent years has become more mainstream in North America.*

standard-definition television (SDTV) (p. 389) A TV standard that has a lower resolution than HDTV, a minimum of 480 vertical lines, and a picture quality similar to that required to watch DVD movies. Why it's important: *The SDTV standard enables broadcasters to transmit more information within the HDTV bandwidth, allowing them to multicast their products, transmitting up to five SDTV programs simultaneously—and getting perhaps five times the revenue—instead of just one HDTV program. (Analog broadcasts only one program at a time.) Thus, instead of beaming high-definition pictures, some broadcasters are splitting their digital streams into several SDTV channels.*

switched-network telecommunications model (p. 371) Model of telecommunications whereby a common carrier provides circuit switching among public users; that is, a temporary connection is established by closing a circuit. Why it's important. *This is the model of the telephone system and also of the internet. In a telephone network, a connection for voice transmission is made by dialing; in a packet switching network , a temporary connection is established between points for transmitting data in the form of packets. People on the system are not only consumers of information ("content") but also possible providers of it. (Compare* **tree-and-branch telecommunications model.***)*

text messaging (p. 392) Also known as *texting;* the sending of short messages, generally no more than a couple of hundred characters in length, to a pager, PDA, smartphone, or other handheld device, including notebook computer, as well as to desktop computers and fixed-line phones. Why it's important: *Text messaging combines the portability of cellphones with the convenience of email and instant messaging. Text messaging is particularly appropriate for situations in which making a cellphone call is intrusive.*

tree-and-branch telecommunications model (p. 371) A model of telecommunications in which a centralized information provider sends out messages through many channels to thousands of consumers. This is the model of most mass media, such as AM radio and network television broadcasting. (*Compare* **switched-network telecommunications model.***)* Why it's important: *Mass-media radio and television have been losing listeners and viewers to the more personalized media based on the switched-network model.*

video on demand (VOD or **VoD)** (p. 390) Set of technologies that enable viewers to select videos or TV programs from a central server to watch when they want. An example is the device offered by Akimbo, a TiVo-like set-top box with a hard drive that can hold 200 hours of video and offers a library of programs that, for a fee, can be downloaded. Why it's important: *VOD allows viewers to watch programs when they want rather than when TV programmers offer them.*

Chapter Review

"I can recognize and recall information."

Self-Test Questions

1. The combining of several industries through various devices that exchange data in digital form is called _____.

2. _____ allows amateur deejays and hobbyists to create their own radio shows and offer them free over the internet.

3. The type of communications model used for the telephone system and the internet is the _____ model.

4. The electronic dots that make up a digital-camera image are called _____.

5. Performing several tasks at once is called _____.

6. A portable, handheld computer that manages personal tasks is called a _____.

7. _____ is a format that allows files to be compressed so they are small enough to be sent over the internet as digital files.

8. Converting a CD audio track to play on an MP3 player is called _____.

9. _____ works with digital broadcasting signals and has a wider screen and higher resolution than standard TV.

10. A _____ camera is one that automatically adjusts settings such as exposure and focus.

Multiple-Choice Questions

1. The number of times that a song is measured and converted to a digital value is called
   a. SLR.
   b. ripping.
   c. sampling rate.
   d. mash-up.

2. The two U.S. satellite radio signal suppliers are
   a. Sirius; Worldscope.
   b. XM; Sirius.
   c. Casio; Konika.
   d. XM; Treo.
   e. Worldscope; Tivo.

3. A digital camera's resolution is expressed in
   a. dpi.
   b. rpm.
   c. pda.
   d. megapixels.
   e. betapixels.

4. Smartphones have
   a. a microprocessor.
   b. a display screen.
   c. memory.
   d. a modem.
   e. all of these

5. Which of the following concerns does NOT apply to smartphone use?
   a. often causes erratic driving
   b. people can track users' movements without their knowledge
   c. users can take photos in inappropriate situations
   d. signals can open locked car doors
   e. loud conversations and ringtones can aggravate people in the user's vicinity

True/False Questions

T F   1. The most well-known handheld device for sending wireless email is the Blackberry.

T F   2. The increased availability of broadband connections slowed down the process of digital convergence.

T F   3. MP3 increases the size of digital audio files in order to improve the sound quality.

T F   4. For 8 × 10 photo prints, 2-megapixel digital cameras are best.

T F   5. Memory Stick is a type of flash memory card.

T F   6. Sending photos as email attachments to many people is more efficient than using an online photo-sharing service.

T F   7. One cannot transfer files from a PDA to a desktop computer.

T F   8. HDTV uses analog signals.

"I can recall information in my own terms and explain it to a friend."

Short-Answer Questions

1. Briefly explain how satellite radio works. What are two advantages of satellite radio?

2. What could you use a mash-up for?

3. Why are MP3 files smaller than regular audio files, such as a purchased CD?

4. Why would you use an online photo-sharing service?

5. How does sampling rate relate to the size of MP3 files?

6. What is the difference between satellite radio and high-definition radio?

7. Which digital camera would you choose: point-and-shoot or single-lens reflex? Why?

stage **LEARNING** APPLYING, ANALYZING, SYNTHESIZING, EVALUATING

"I can apply what I've learned, relate these ideas to other concepts, build on other knowledge, and use all these thinking skills to form a judgment."

Knowledge in Action

1. Does almost everyone you know download and listen to MP3 files? Describe a few ways in which the iPod-types whom you have observed are being distracted from certain activities and some responsibilities. Do you perceive a problem?

2. List five situations in which you often find yourself multitasking. How any things have you done at once? How do you think your manner of multitasking affects the quality of what you achieve?

3. What questions would you ask a friend in order to determine which digital camera, with which characteristics, you would advise her or him to buy?

4. Describe how one can transfer images from a digital camera to a computer.

5. What does an entertainment PC do?

Web Exercises

1. Current rules of both the FCC and the U.S. Federal Aviation Administration ban in-flight cellular calling. The primary FCC concern has been possible disruption of cellphone communication on the ground. The FAA's worry is how cell phones might interfere with a plane's navigation and electrical systems. However, plans are in the works to assign new bandwidths to allow cellphone use in the air.

    Do several keyword searches and find three reasons to allow cellphone use during airplane flights, and three reasons NOT to allow it. Provide information to support each reason. What is your opinion on this issue?

2. In October 2005, Connecticut joined the list of 22 U.S. states that have restricted cellphone use by automobile drivers since New York passed the first such law in 2001. It is now illegal in Connecticut, as well as New York, New Jersey, and major cities such as Washington and Chicago for drivers to use handheld cellphones. Connecticut drivers can be ticketed only if they are pulled over for another moving violation, although in many states, including New York, police can stop drivers using hand-held cellphones. However, Connecticut's broad law bans "any activity not related to the actual operation of a motor vehicle in a manner that interferes with the safe operation of such vehicle." That can include eating, fiddling with the CD player, personal grooming, attending to children, reading, and the like.

    In addition, according to the National Conference of State Legislatures, 10 states have enacted legislation restricting use of electronics by teenage drivers.

    Do an internet search and find out what the law says about driving and cellphone use in the state where you are going to school and in the state where you come from.

    The rules on driving and cellphone use in many other countries are much stricter than they are in the United States; for more information, go to *www.cellular-news.com/car_bans/*.

    According to the Automobile Association of America (AAA), hands-free phones are not risk free. The hands-free feature is simply a convenience: it does not increase safety. Studies show that hands-free cellphones distract drivers the same as handheld phones do. Why? Because it's the conversation that distracts the driver, not the device. Search the internet and find some reports that support this view.

    Why do you think people don't just turn off their cellphones while they are driving and let voicemail collect messages to be retrieved later?

3. The Convergence Center (*http://dcc.syr.edu/index.htm*) supports research on and experimentation with media convergence. The Center is a joint effort of the Syracuse University School of Information Studies and the Newhouse School of Public Communications. Its mission is to understand the future of digital media and to engage students and faculty in the process of defining and shaping that future. Go to their site and click on What's New, Articles; choose one article to read; then write a couple of paragraphs that summarize the article.

4. Now that cellphones have cameras, camera voyeurism is becoming a problem. The U.S. Video Voyeurism Prevention Act of 2004 has made this behavior a federal offense. This act prohibits photographing or videocapturing (including with cellphones) a naked person without his or

her consent in any place where there can be "a reasonable expectation of privacy." Punishment includes fines of up to $100,000 or up to a year in prison, or both.

Cell phone vendors say this law may be hard to enforce and may even be a deterrent to promising technology and create a false sense of security. Some cellphone manufacturers deny that voyeurs are any more likely to snoop using a cellphone camera than using other technologies such as digital cameras. But other people say the opportunity differs, that most people don't carry digital cameras around with them. With a cellphone camera there is more opportunity to take snapshots of interesting images, and unfortunately this can include images than can threaten privacy. Voyeurs using cellphone cameras could easily pretend to be doing something else, such as dialing or talking.

California Assemblywoman Sarah Reyes (D-Fresno) favors a technological solution to the privacy problems presented by cellphone cameras; she urges state legislation requiring camera phones sold in California after 2008 to emit an audible noise or flash a light when users press the shutter. But such noise and light would disturb happy occasions such as weddings, where people use cellphone cameras to take pictures and send them instantly to loved ones who couldn't attend.

Research the cameraphone voyeurism problem on the internet. How could you protect your privacy in public situations?

5. Schools are starting to offer online digital photography courses. Check out *www.worldwidelearn.com/online-courses/digital-photography-course.htm*. Are such courses offered at your school? If not, should they be?

6. The Cellular Telecommunications & Internet Association (CTIA) reminds us that text messaging can be a fast, efficient, and reliable way to communicate in the event of an emergency. And, if more wireless users rely on text messaging in crisis situations, the people who need to make voice calls the most—emergency responders and 911 callers—can get through more easily.

"Everyone should have a plan for communicating in times of emergencies, and text messaging can be an efficient way to reach your friends, family or loved ones. In the time it takes one person to make a 1-minute voice call, hundreds of thousands of text messages can be exchanged," said Tom Wheeler, President and CEO of CTIA. "In these days of increased terrorist threats and heightened awareness, learning all the options on your wireless phone is an important piece of being prepared."

Text messages have also been used around the world to alert citizens in times of danger. In Kuwait, during a recent conflict, Kuwaitis were warned about imminent attacks from Iraq via text messaging. And, in Hong Kong, a few years ago, a wireless carrier set up a system to provide SARS updates via mobile phone. Users punched in a three-digit number and received a text message indicating if they were near any buildings where SARS victims lived or worked. The information was obtained from a daily list released by the Health Department.

In the frantic days leading up to the landfall of Hurricane Rita in September 2005, Houston radio station KRBE offered to deliver hurricane alerts via text messaging across cellphones to listeners, enabling information to be delivered anytime, anywhere, regardless of whether a person was near a radio or a computer. At a time when traditional media networks experienced coverage issues from storm damage, power lines were down, and cars were running out of gas, KRBE was able to provide a continuous two-way stream of information via the cellphone. Many listeners expressed gratitude for information on road closures, fuel availability, evacuation orders, and storm damage during a frightening experience.

Do an internet search on how text messaging has been used to help in various types of emergencies around the world. What types of suppliers have provided the messages? Using what types of systems? How do cellphone users know where/how to retrieve the messages?

Databases & Information Systems

Digital Engines for Today's Economy

Chapter Topics & Key Questions

8.1 **Managing Files: Basic Concepts** What are the data storage hierarchy, the key field, types of files, and some methods of data access and storage?

8.2 **Database Management Systems** What are the benefits of database management systems, and what are the main types of database access?

8.3 **Database Models** What are five types of database models?

8.4 **Data Mining** How does data mining work, and how could it be useful to me?

8.5 **Databases & the Digital Economy: E-Business & E-Commerce** What are e-business and e-commerce, and what are three types of e-commerce systems?

8.6 **Information Systems in Organizations: Using Databases to Help Make Decisions** How does information flow within an organization, and what are different types of information systems?

8.7 **Artificial Intelligence** What are the main areas of artificial intelligence?

8.8 **The Ethics of Using Databases: Concerns about Privacy & Identity Theft** What are two ethical concerns about the uses of databases?

The Library of Congress, the largest library in the world, has 130 million items on about 530 miles of bookshelves. How does the information on the internet compare?

Marketing research firm IDC predicts that by 2007 internet users worldwide will access, download, and share information equivalent to the entire Library of Congress more than 64,000 times over—*every day*. This represents a doubling every year from the internet traffic of 2002, rising from 180 petabits to 5,175 petabits per day.[1] (A petabit is 1,000,000,000,000,000 bits—a quadrillion bits.) Researchers at the University of California at Berkeley's School of Information Management estimated that the amount of new information stored in 2002 by the human species was five exabytes (an exabyte is a billion gigabytes), equivalent to "all words ever spoken by humans since the dawn of time," according to one report. And that didn't include telephone conversations, which added up to about 17 exabytes, more than three times all the words ever spoken by humans.[2]

How will all such information be organized and made accessible? The answer has to do with databases. The arrival of databases has changed the nature of many of our business and social institutions. Today, for instance, because of the huge increase in online internet sales, credit card fraud has boomed, and so slight changes in your spending patterns—buying much more clothing than usual or filling your car's gas tank twice in one day—will trigger an alert that may freeze your card.[3] But databases are everywhere now and promise to bring even more changes in the future. Wikipedia *(www.wikipedia.org)*, for instance, is a web-based, free-content, multilingual encyclopedia that anyone can log on to and add to or edit as he or she sees fit; it has 1.5 million entries in 76 languages and receives more than 60 million hits a day.[4] The National Geographic Society and IBM's Watson Research Labs launched a massive database called the Geographic Project that is cataloging genetic markers and is capable of tracing the geographic origins of your and other people's ancestors back 10,000 years.[5] Google has been trying to create a database consisting of millions of digitally scanned books from several university libraries.[6]

Traditional electronic databases include those that handle airline reservation systems, many library catalogs, magazine subscriptions for large publishing companies, patient tracking in large hospitals, and inventories for supermarkets and big-box stores like Wal-Mart. These databases typically use text- and numeric-based data. Newer types of databases include multimedia data and formulas for data analysis.

Before we discuss traditional types of databases and recent innovations in database management, we need to discuss some basic concepts of file management.

Instant ID. The IBIS Mobile Identification System can verify a person's identity in minutes from a remote location. It uses fingerprint information captured by the handheld device; the information is transmitted to the Automated Fingerprint Identification System (AFIS), a huge database used by law enforcement officers, among others.

8.1

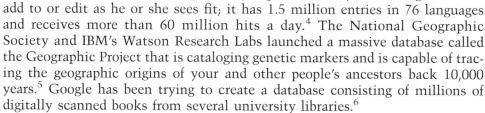

Managing Files: Basic Concepts

What are the data storage hierarchy, the key field, types of files, and some methods of data access and storage?

An electronic database is not just the computer-based version of what used to go into manila folders and filing cabinets. A ___database___ is a logically organized collection of related data designed and built for a specific purpose, a

technology for pulling together facts that allows the slicing and dicing and mixing and matching of data in all kinds of ways. The data in a database has some inherent meaning. In other words, a random assortment of data cannot correctly be called a *database*.[7] A database can be of any size and of any degree of complexity, and it can be maintained manually or by software on computers. This chapter, of course, covers computer databases.

How Data Is Organized: The Database Storage Hierarchy

What are the units of data, from smallest to largest?

Data in a database can be grouped into a hierarchy of categories, each increasingly more complex. **The _data storage hierarchy_ consists of the levels of data stored in a computer database: bits, characters (bytes), fields, records, and files.** (● *See Panel 8.1 on the next page.*)

BITS   Computers, as we have said, are based on the principle that electricity may be on or off. Thus, the *bit* is the smallest unit of data the computer can store in a database—represented by 0 for off or 1 for on.

CHARACTERS   A *character (byte)* **is a letter, number, or special character.** *A, B, C, 1, 2, 3, #, $, %* are all examples of single characters. A combination of bits represents a character.

Bits and bytes are the building blocks for representing data, whether it is being processed, telecommunicated, or stored in a database. The computer deals with the bits and bytes; you, however, will need to deal mostly with fields, records, and files.

FIELD   A *field* **is a unit of data consisting of one or more characters (bytes).** It is the smallest unit of meaningful information in the database. Each field has a *field name* that describes the kind of data that should be entered into the field. An example of a field is your first name, your street address, *or* your Social Security number.

Fields can be designed to be a certain maximum length or a variable length, and they can also be designed to hold different types of data, such as text only, numbers only, dates only, time, a "yes" or "no" answer only, web links only, or pictures, sound, or video.

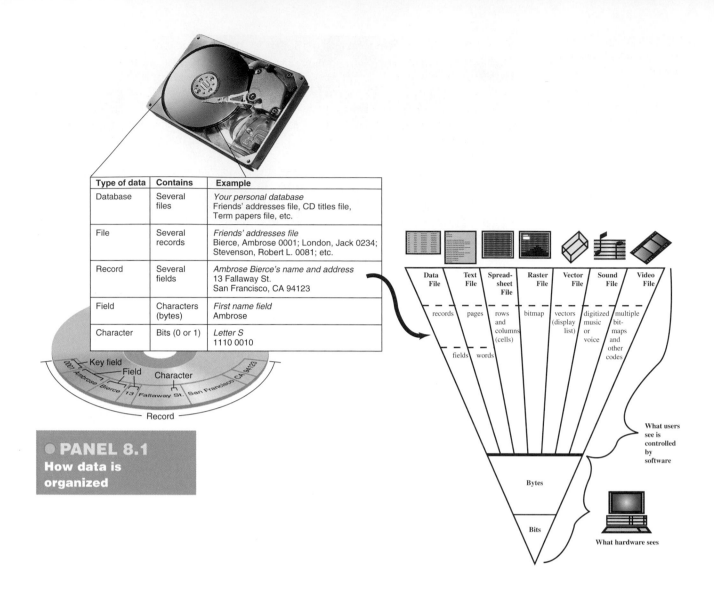

Type of data	Contains	Example
Database	Several files	*Your personal database* Friends' addresses file, CD titles file, Term papers file, etc.
File	Several records	*Friends' addresses file* Bierce, Ambrose 0001; London, Jack 0234; Stevenson, Robert L. 0081; etc.
Record	Several fields	*Ambrose Bierce's name and address* 13 Fallaway St. San Francisco, CA 94123
Field	Characters (bytes)	*First name field* Ambrose
Character	Bits (0 or 1)	*Letter S* 1110 0010

● PANEL 8.1
How data is organized

RECORD A _record_ **is a collection of related fields.** Each record stores data about only one entity, which can be a person, a place, a thing, an occurrence, or a phenomenon. An example of a record would be your name *and* address *and* Social Security number.

FILE A _file_ **is a collection of related records.** An example of such a file is data collected on everyone employed in the same department of a company, including all names, addresses, and Social Security numbers. You use files a lot because the file is the collection of data or information that is treated as a unit by the computer.

The file is at the top of the data hierarchy. A collection of related files forms the database. A company database might include files on all past and current employees in all departments. There would be various files for each employee: payroll, retirement benefits, sales quotas and achievements (if in sales), and so on.

The Key Field

What is a key field, and what is probably my most important key field?

An important concept in data organization is that of the key field. (● *See Panel 8.1.*) As we said in Chapter 3, a *key field* is a field that is chosen to uniquely identify a record so that it can be easily retrieved and processed.

The key field is often an identification number, Social Security number, customer account number, or the like or a combination of letters and numbers set up as a meaningful code. The primary characteristic of the key field is that it is *unique.* Thus, numbers are clearly preferable to names as key fields because there are many people with common names like James Johnson, Susan Williams, Ann Wong, or Roberto Sanchez, whose records might be confused. Student records are often identified by student ID numbers used as key fields.

As we mentioned in Chapter 3, most database management systems allow you to have more than one key field, so that you can sort records in different ways. One of the keys is designated the *primary key* and must hold a unique value for each record. A key field that identifies records in different tables is called a *foreign key.* Foreign keys are used to cross-reference data.

Types of Files: Program Files & Data Files

What's the difference between a program file and a data file?

As we said, the *file* is the collection of data or information that is treated as a unit by the computer. **Files are given names—_filenames_.** If you're using a word processing program to write a psychology term paper, you might name it "Psychreport." In a database, a filename might be "AccountingPersonnel."

Filenames also have *extensions,* or *extension names,* usually three letters added after a period following the filename. For example, the *.doc* in *Psychreport.doc* is recognized by Microsoft Word as a "document." Extensions are usually inserted automatically by the application software.

When you look up the filenames listed on your hard drive, you will notice a number of extensions, such as *.doc, .exe,* and *.com.* There are many kinds of files, but two principal ones are *program files* and *data files.*

PROGRAM FILES: FOR SOFTWARE INSTRUCTIONS **_Program files_ are files containing software instructions.** Examples are word processing or spreadsheet programs, which are made up of several different program files. The two most important are source program files and executable files.

Source program files contain high-level computer instructions in the original form written by the programmer. Some source program files have the extension of the language in which they are written, such as *.bas* for BASIC, *.pas* for Pascal, or *.jav* for Java. (Programming languages are described in Chapter 10.)

For the processor to use source program instructions, they must be translated into an *executable file,* which contains the instructions that tell the computer how to perform a particular task. You can identify an executable file by its extension, *.exe* or *.com.* You use an executable file by running it—as when you select Microsoft Excel from your on-screen menu and run it. (There are some executable files that you cannot run—other computer programs called *runtime libraries* cause them to execute. These are identified by such extensions as *.dll, .drv, ocx, .sys,* and *.vbx.*)

DATA FILES: FOR HOLDING DATA **_Data files_ are files that contain data—** words, numbers, pictures, sounds, and so on. These are the files used in databases. Unlike program files, data files don't instruct the computer to do anything. Rather, data files are there to be acted on by program files. Examples of common extensions in data files are *.txt* (text) and *.xls* (Excel worksheets). Certain proprietary computer programs apply their own extensions, such as *.ppt* for PowerPoint and *.mdb* for Access.

Three types of files worth particular attention are graphics, audio, and video files.

- **Graphics files:** Some important ones are *.bmp, .tiff, .gif, .jpeg,* and *.png.*

Survival Tip
Some Records Have to Be Hardcopy

You could scan your birth certificate, will, or car ownership title into your computer to make a digital record. But such records printed off a hard drive aren't always legally acceptable. Original documents are often required by government agencies and the court system.

- **Audio files:** The ones you're most apt to encounter are *.mp3, .wav,* and *.mid.*
- **Animation/video files:** Common files are *.qt, .mpg, .wmv, .avi.,* and *.rm.*

The box at right describes these and other common types of files. (● *See Panel 8.2.*)

Compression & Decompression: Putting More Data in Less Space

How are large files compressed and decompressed?

The vast streams of text, audio, and visual information threaten to overwhelm us. To fit large multimedia files into less space and increase the speed of data transmission, a technique called compression/decompression, or *codec,* is used. <u>**Compression**</u> **is a method of removing repetitive elements from a file so that the file requires less storage space and therefore less time to transmit.** Later the data is decompressed—the repeated patterns are restored.

There are two principal methods of compressing data—lossless and lossy. In any situation, which of these two techniques is more appropriate will depend on whether data quality or storage space is more critical.

Lossless compression uses mathematical techniques to replace repetitive patterns of bits with a kind of coded summary. During decompression, the coded summaries are replaced with the original patterns of bits. In this method, the data that comes out is exactly the same as what went in; it has merely been repackaged for purposes of storage or transmission. Lossless techniques are used when it's important that nothing be lost—for instance, for computer data, database records, spreadsheets, and word processing files.

Lossy compression techniques permanently discard some data during compression. Lossy data compression involves a certain loss of accuracy in exchange for a high degree of compression (to as little as 5% of the original file size). This method of compression is often used for graphics files and sound files. Thus, a lossy codec might discard subtle shades of color or sounds outside the range of human hearing. Most users wouldn't notice the absence of these details. Examples of two lossy compression file formats are *.jpeg* and *.mpeg.*

8.2

Database Management Systems

What are the benefits of database management systems, and what are the main types of database access?

As we've said, a database is an organized collection of related (integrated) files. A database may be small, contained entirely within your own personal computer, or it may be massive, available through online connections. Such massive databases are of particular interest because they offer phenomenal resources that until recently were unavailable to most ordinary computer users.

In the 1950s, when commercial use of computers was just beginning, a large organization would have different files for different purposes. For example, a university might have one file for course grades, another for student records, another for tuition billing, and so on. In a corporation, people in the accounting, order-entry, and customer-service departments all had their own separate files. Thus, if an address had to be changed, for example, each file would have to be updated separately. The database files were stored on magnetic tape and had to be accessed in sequence in what was called a *file-processing system.*

Graphics files

- **.bmp (BitMaP):** Bitmap graphic format native to Microsoft Windows. Some Macintosh programs can also read .bmp files.
- **.gif (Graphic Interchange Format):** Pronounced "jiff." Format used on web pages and downloadable images.
- **.jpeg or .jpg (Joint Photographic Experts Group):** Pronounced "jay-peg." Used for web images and for digital photography, especially for high-resolution images.
- **.pcx:** Format introduced for PC Paintbrush. Used for other graphics packages as well.
- **.pict (PICTure):** Format used by Apple for use on Macintosh computers.
- **.png (Portable Network Graphic):** Pronounced "ping." Patent-free alternative to .gif.
- **.tiff or .tif (Tagged Image File Format):** High-resolution bitmap graphics file widely used on both Macintosh and PC computers. Used in exchanging bitmap files that will be printed.

Audio files

- **.au:** Low-fidelity monaural format now often used to distribute sample sounds online.
- **.mid (MIDI, Musical Instrument Digital Interface):** Format meant to drive music synthesizers.
- **.mp3 (MPEG-3):** File format used to compress CD-quality music while preserving much of the original sound quality.
- **.wav (WAVe):** Waveform file format that contains all the digital information needed to play speaker-quality music.

Video files

- **.avi (Audio Video Interleaved):** Video file format recognized by Windows Media Player. Not good for broadcast-quality video.
- **.mov or .qt (QuickTime):** Video file formats developed by Apple for QuickTime video player. Can play broadcast-quality video.
- **.mpg or .mpeg (Motion Picture Experts Group):** Video formats for full-motion video. MPEG-2 format used by DVD-ROM disks. MPEG-4 recognized by most video player software.
- **.rm (RealMedia):** Popular file format for streaming video.
- **.wmv (Windows Media Video):** Video format recognized by Windows Media Player.

Other files

- **ASCII files:** Text-only files containing no graphics and no formatting such as boldface or italic. ASCII format is used to transfer documents between computers, such as PC and Macintosh. Such files may use the *.txt* (for text) extension.
- **Web files:** Files carried over World Wide Web. Extensions include *.html, .htm, .xml,* and *.asp* (active server page).
- **Desktop publishing files:** Include PostScript commands, which instruct a PostScript printer how to print a file and use *.eps* (encapsulated PostScript).
- **Drivers:** Software drivers often have the extension *.drv.*
- **Windows operating system files**: Files such as *Autoexec.bat* and *Config.sys* relate to OS setup.
- **PDF (Portable Document Format) files:** Files that use Adobe Acrobat's format for all types of document exchange as well as for publishing documents on the web that are downloaded and read independently of the HTML pages. These files use the extension *pdf.* Adobe Reader is Adobe's free download for displaying and printing PDF files, and hundreds of millions of users have downloaded this software from *www.adobe.com.* Adobe Reader lets you view and print PDF files, but not create or edit them. Editable PDF files are created with Adobe's Acrobat software. Acrobat can convert a wide variety of document types on Windows, Mac, and Unix to the PDF format.

Solutions   Support   Purchase   Company

is Adobe PDF?

cure, reliable electronic document distribution and excha

by Adobe, Portable Document Format (PDF) is the specification used by standards bodies around the more secure, reliable electronic document distribution nge. Adobe® PDF has been adopted by enterprises, and governments around the world to streamline exchange, increase productivity, and reduce reliance it is the standard format for the electronic submission provals to the U.S. Food and Drug Administration (FDA) ctronic case filings in U.S. federal courts.

he look and integrity of your original documents — files look exactly like original documents and he fonts, images, graphics, and layout of any source file ss of the application and platform used to create it.

ments with anyone — Adobe PDF documents can be ewed, and printed by anyone, on any system, using free der® software — regardless of the operating system, plication, or fonts.

e — Adobe PDF files are compact and easy to Creation can be as simple as clicking a button from cations including Microsoft Word, Excel, and

Find ou

- Try Acrob days
- Buy Acro
- Learn mo 7.0
- Adobe PD (PDF, 1.5

Get Adobe
Reade

Later, magnetic-disk technology came along, allowing any file to be accessed randomly. This permitted the development of new technology and new software: the database management system. **A _database management system (DBMS)_, or _database manager_, is software written specifically to control the structure of a database and access to the data.** In a DBMS, an address change need be entered only once, and the updated information is then available in any relevant file. (Strictly speaking, the _database_ is the collection of the data, and the _database management system_ is the software—but many professionals use "database" to cover both meanings.)

The Benefits of Database Management Systems

What are four advantages of a DBMS?

The advantages of database management systems are as follows:

REDUCED DATA REDUNDANCY   _Data redundancy_, or _repetition_, means that the same data fields (a person's address, say) appear over and over again in different files and often in different formats. In the old file-processing system, separate files would repeat the same data, wasting storage space. In a database management system, the information appears just once, freeing up more storage capacity. In the old data storage systems, if one field needed to be updated, someone had to make sure that it was updated in _all_ the places it appeared—an invitation to error.

IMPROVED DATA INTEGRITY   _Data integrity_ means that data is accurate, consistent, and up to date. In the old system, when a change was made in one file, it might not have been made in other necessary files. The result was that some reports were produced with erroneous information. In a DBMS, reduced redundancy increases the chances of data integrity—the chances that the data is accurate, consistent, and up to date—because each updating change is made in only one place.

Also, many DBMSs provide built-in check systems that help ensure the accuracy of the data that is input. The expression "garbage in, garbage out" (abbreviated _GIGO_) refers to the fact that a database with incorrect data cannot generate correct information.

INCREASED SECURITY   Although various departments may share data, access to specific information can be limited to selected users. Thus, through the use of passwords, a student's financial, medical, and grade information in a university database is made available only to those who have a legitimate need to know.

Image database. Corbis is a database of more than 3 million images. Many of them are available for free to be used in many types of documents and publications.

EASE OF DATA MAINTENANCE   Database management systems offer standard procedures for adding, editing, and deleting records, as well as validation checks to ensure that the appropriate type of data is being entered properly and completely into each field type. Data backup utilities ensure availability of data in case of primary system failure.

Three Database Components

What do the three principal components of a database do?

A database management system may have three components integrated into the software—a _data dictionary_, _DBMS utilities_, and a _report generator_.

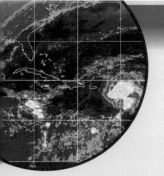

PRACTICAL ACTION

Storing Your Stuff:
How Long Will Digitized Data Last?

How long are those floppy disks or CD-ROMs on which you're storing your important documents going to last? Will you, or anyone else, be able to make use of them 15 or 25 or 50 years from now?

In 1982, software pioneer Jaron Lanier created a video game called *Moondust* for the then popular Commodore 64 personal computer. Fifteen years later, when asked by a museum to display the game, he couldn't find a way to do it—until he tracked down an old microcomputer of exactly that brand, type, and age, along with a joystick and video interface that would work with it.[8]

Would this have been a problem if Lanier had originally published a game in a *book*? Probably not. Books have been around since about 1453, when Johannes Gutenberg developed the printing press and used it to print 150 copies of the Bible in Latin. Some of these Gutenberg Bibles still exist—and are still readable (if you can read Latin).[9]

Digital storage has a serious problem: It isn't as long-lived as older forms of data storage. Today's books printed on "permanent" (low-acid, buffered) paper may last up to 500 years. Even books printed on cheap paper that crumbles will still be readable.

By contrast, data stored on diskettes, magnetic tape, and optical disks is subject to two hazards:[10]

- **Short life span of storage media:** The storage media themselves have a short life expectancy, and often the degradation is not apparent until it's too late. The maximum time seems to be 50 years, the longevity of a high-quality CD. Some average-quality CDs won't last 5 years, according to tests run at the National Media Laboratory.

  The magnetic tapes holding government records, which are stored in the National Archives in Washington, D.C., need to be "refreshed"—copied onto more advanced tapes—every 10 years.

- **Hardware and software obsolescence:** As Jaron Lanier found out, even when tapes and disks remain intact, the hardware and software needed to read them may no longer be available. Without the programs and computers used to encode data, digital information may no longer be readable.

  Eight-inch floppy disks and drives, popular 25 years ago, are now extinct, and their $5^1/_4$-inch successors are now a rarity as well. Even $3^1/_2$-inch floppy disks are rapidly disappearing in favor of CD/DVDs. "Optical and magnetic disks recorded under nonstandard storage schemes will be increasingly useless," one writer points out, "because of the lack of working equipment to read them."[11]

What about the personal records you store on your own PC, such as financial records, inventories, genealogies, and photographs? *New York Times* technology writer Stephen Manes has a number of suggestions:[12]

1. Choose your storage media carefully. CD-R disks are probably best for archiving—especially if you also keep a paper record.
2. Keep it simple. Store files in a standard format, such as text files and uncompressed bitmapped files.
3. Store data along with the software that created it.
4. Keep two copies, stored in separate places, preferably cool, dry environments.
5. Use high-quality media, not off-brands.
6. When you upgrade to a new hardware or software product, have a strategy for migrating the old data.

DATA DICTIONARY: FOR DEFINING DATA DEFINITIONS & STRUCTURE A _data dictionary_, also called a *repository,* is a document or file that stores the data definitions and descriptions of the structure of data used in the database. Data dictionaries contain no actual data from the database, only information for managing it. Without a data dictionary, however, a DBMS cannot access data from the database. The data dictionary defines the basic organization of the database and contains a list of all files in the database, the number of records in each file, and the names and types of each field. The data dictionary may also help protect the security of the database by indicating who has the right to access it. Most database management systems keep the data dictionary hidden from users to prevent them from accidentally destroying its contents.

UTILITIES: FOR MAINTAINING THE DATABASE *DBMS utilities* **are programs that allow you to maintain the database by creating, editing, and deleting data, records, and files.** The utilities enable you to monitor the types of data being input and to sort your database by key fields, making searching and organizing information much easier.

REPORT GENERATOR: FOR PRODUCING DOCUMENTS A *report generator* **is a program for producing an on-screen or printed document from all or part of a database.** You can specify the format of the report in advance—row headings, column headings, page headers, and so on. With a report generator, even nonexperts can create attractive, readable reports on short notice.

The Database Administrator

What does a DBA do?

Large databases are managed by a specialist called a database administrator. The *database administrator (DBA)* **coordinates all related activities and needs for an organization's database,** ensuring the database's recoverability, integrity, security, availability, reliability, and performance. Database administrators determine user access privileges; set standards, guidelines, and control procedures; assist in establishing priorities for requests; prioritize conflicting user needs; and develop user documentation and input procedures. They are also concerned with security—establishing and monitoring ways to prevent unauthorized access and making sure data is backed up and recoverable should a failure occur—and to establish and enforce policies about user privacy.

8.3

Database Models

What are five types of database models?

Just as files can be organized in different ways, so can database management systems. Organizations may use one kind of DBMS for daily processing of transactions (such as sales figures) and then move the processed data into another DBMS that's better suited for random inquiries and analysis. Older DBMS models, introduced in the 1960s, are *hierarchical* and *network*. New models are *relational, object-oriented, and multidimensional.* (● See Panel 8.3.)

● PANEL 8.3
Timeline:
Developments in database technology

4000–1200 BCE	3000 BCE–1400 CE	1086	1621	1642	1666
Inhabitants of the first known civilization in Sumer keep records of commercial transactions on clay tablets	Inca civilization creates the khipu coding system to store the results of mathematical calculations	Domesday Book: William I orders a survey of England to assess value for taxing purposes	Slide rule invented (Edmund Gunther)	First mechanical adding machine (Blaise Pascal)	First mechanical calculator that can add and subtract (Samuel Morland)

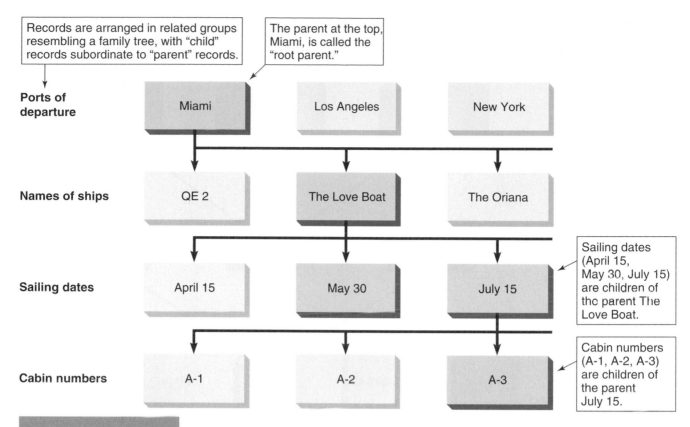

Records are arranged in related groups resembling a family tree, with "child" records subordinate to "parent" records.

The parent at the top, Miami, is called the "root parent."

Ports of departure

Miami | Los Angeles | New York

Names of ships

QE 2 | The Love Boat | The Oriana

Sailing dates

April 15 | May 30 | July 15

Sailing dates (April 15, May 30, July 15) are children of the parent The Love Boat.

Cabin numbers

A-1 | A-2 | A-3

Cabin numbers (A-1, A-2, A-3) are children of the parent July 15.

● **PANEL 8.4**
Hierarchical database
Example of a cruise ship reservation system

Hierarchical Database

How does a hierarchical database resemble a family tree?

In a *hierarchical database*, **fields or records are arranged in related groups resembling a family tree, with child (lower-level) records subordinate to parent (higher-level) records.** The parent record at the top of the database is called the *root record* or *root parent*. (● *See Panel 8.4.*)

The hierarchical database is the oldest and simplest of the five models. It lent itself well to the tape storage systems used by mainframes in the 1970s. It is still used in some types of passenger reservation systems. In hierarchical databases, accessing or updating data is very fast, because the relationships have been predefined. However, because the structure must be defined in advance, it is quite rigid. There can be only one parent per child, and no relationships among the child records are possible. Moreover, adding new fields to database records requires that the entire database be redefined. A new database model was needed to address the problems of data redundancy and complex data relationships.

1820	1843	1854	1890
The first mass-produced calculator, the Thomas Arithnometer	World's first computer programmer, Ada Lovelace, publishes her notes	George Boole publishes "An Investigation on the Laws of Thought," a system for symbolic and logical reasoning that will become the basis for computer design	Hollerith's automatic census-tabulating machine (used punched cards) tabulates U.S. Census in 2–3 months (compared to 7 years needed previously)

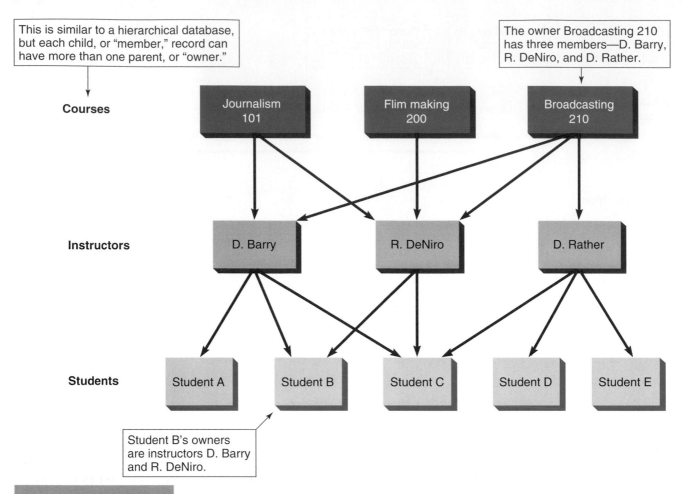

This is similar to a hierarchical database, but each child, or "member," record can have more than one parent, or "owner."

The owner Broadcasting 210 has three members—D. Barry, R. DeNiro, and D. Rather.

Courses

Journalism 101

Flim making 200

Broadcasting 210

Instructors

D. Barry

R. DeNiro

D. Rather

Students

Student A

Student B

Student C

Student D

Student E

Student B's owners are instructors D. Barry and R. DeNiro.

Network Database

How does a network database differ from a hierarchical database?

The network database was in part developed to solve some of the problems of the hierarchical database model. **A _network database_ is similar to a hierarchical database, but each child record can have more than one parent record.** (● *See Panel 8.5.*) Thus, a child record, which in network database terminology is called a *member,* may be reached through more than one parent, which is called an *owner.*

Also used principally with mainframes, the network database is more flexible than the hierarchical arrangement, because different relationships may be established between different branches of data. However, it still requires that the structure be defined in advance, and, as with the hierarchical model,

1961	1962	1964	1967	1967–1968	1969	1970
Prototype of first DBMS	Stanford and Purdue Universities establish the first departments of computer science	IBM introduces 360 line of computers	Hand-held calculator	National Crime Information Center (NCIC) goes online with 95,000 pieces of information in five databases, handling 2 million transactions in its first year	ARPANet established, led to internet	Micro-processor chips come into use; floppy disk introduced for storing data; E.F. Codd develops the relational database, which evolved into IBM's System R project, which in turn evolved into SQL

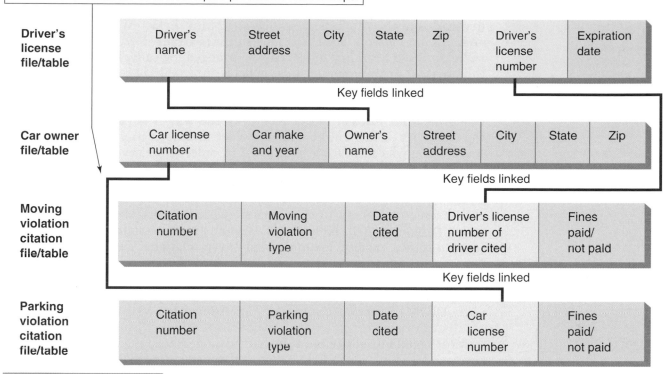

This kind of database relates, or connects, data in different files through the use of a key, or common data element. The relational database does not require predefined relationships.

Driver's license file/table

| Driver's name | Street address | City | State | Zip | Driver's license number | Expiration date |

Key fields linked

Car owner file/table

| Car license number | Car make and year | Owner's name | Street address | City | State | Zip |

Key fields linked

Moving violation citation file/table

| Citation number | Moving violation type | Date cited | Driver's license number of driver cited | Fines paid/ not paid |

Key fields linked

Parking violation citation file/table

| Citation number | Parking violation type | Date cited | Car license number | Fines paid/ not paid |

● **PANEL 8.6**
Relational database
Example of a state department of motor vehicles database

the user must be familiar with the structure of the database. Moreover, there are limits to the number of possible links among records, and to examine a field one must retrieve the entire record.

Although the network database was an improvement over the hierarchical database, some people in the database community believed there must be a better way to manage large amounts of data.[13]

Relational Database

How does a relational database work?

More flexible than hierarchical and network database models, the **_relational database_ relates, or connects, data in different files through the use of a key, or common data element.** (● *See Panel 8.6.*) Examples of microcomputer DBMS programs, all of which are relational, are Paradox and Access. Examples of relational models used on larger computer systems are Oracle, Informix, and Sybase.

1971	1975	1976	1976	1978	1980	1981
First pocket calculator; all U.S. states and Washington, D.C. can access NCIC criminal data	First micro-computer (MITS Altair 8800)	Queen Elizabeth sends the first royal email	Apple I computer (first personal computer sold in assembled form)	Introduction of SQL (Structured Query Language); publication of OSI model by the International Standards Organization	Wayne Ratliff develops dBASE II, the first database program for personal computers; during the 1980s, the hierarchical and network database models fade into the background	IBM introduces personal computer

HOW A RELATIONAL DATABASE WORKS In the relational database, there are no access paths down through a hierarchy. Instead, data elements are stored in different tables made up of rows and columns. In database terminology, the tables are called *relations* (files), the rows are called *tuples* (records), and the columns are called *attributes* (fields). Whereas in the hierarchical and network database models data is arranged according to physical address, in the relational model data is arranged logically, by content. Hence, the physical order of the records or fields in a table is completely immaterial. Each record in the table is identified by a field—the primary key—that contains a unique value. These two characteristics allow the data in a relational database to exist independently of the way it is physically stored on the computer. Thus, unlike with the older models, a user is not required to know the physical location of a record in order to retrieve its data.[14]

USING STRUCTURED QUERY LANGUAGE To retrieve data in a relational database, you specify the appropriate fields and the tables to which they belong in a query to the database, using a query language. **Structured** **_query language (SQL_, pronounced "sequel") is the standard query language used to create, modify, maintain, and query relational databases.** The three components of a basic SQL query are the SELECT . . . FROM statement, the WHERE clause, and the ORDER BY clause. The fields used in the query are specified with SELECT, and the tables to which they belong are specified with FROM. Selection criteria are determined by WHERE, and the query results can be sorted in any sequence with ORDER.[15]

Most popular database programs provide a graphical query-building tool, so the user does not need to have a thorough knowledge of SQL. (● *See Panel 8.7.*)

An example of an SQL query is as follows:

SELECT PRODUCT-NUMBER, PRODUCT-NAME
 FROM PRODUCT
 WHERE PRICE < 100.00
 ORDERBY PRODUCT-NAME;

This query selects all records in the product file for products that cost less than $100.00 and displays the selected records alphabetically according to product name and including the product number—for example:

C-50   Chair
A-34   Mirror
D-168   Table

QUERY BY EXAMPLE One feature of most query languages is query by example. Often a user will seek information in a database by describing a procedure for finding it. However, in **_query by example (QBE)_, the user asks for information in a database by using a sample record form, or table, to define the qualifications he or she wants for selected records;** in other words, the user fills in a form. (● *See Panel 8.8.*)

For example, a university's database of its student-loan records might have the column headings (field names) NAME, ADDRESS, CITY, STATE, ZIP, AMOUNT OWED. When you use the QBE method, the database would

1982	1984	Early 1990s	1993	1994	2003
Portable computers	Apple Macintosh; first personal laser printer; release of Ashton-Tate dBase III	The first object-oriented DBMSs are released	Multimedia desktop computers	Apple and IBM introduce PCs with full-motion video built in; wireless data transmission for small portable computers; web browser first invented	Human Genome Project completes mapping all 90,000 or so genes for massive database

Query Wizard

Access' Query Wizard will easily assist you to begin creating a select query.

- Click the **Create query by using wizard** icon in the database window to have Access step you through the process of creating a query.

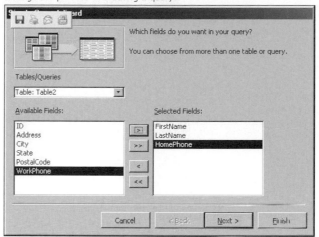

- From the first window, select fields that will be included in the query by first selecting the table from the drop-down **Tables/Queries** menu. Select the fields by clicking the > button to move the field from the Available Fields list to Selected Fields. Click the double arrow button >> to move all of the fields to Selected Fields. Select another table or query to choose from more fields and repeat the process of moving them to the Selected Fields box. Click **Next >** when all of the fields have been selected.

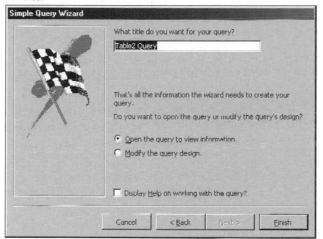

- On the next window, enter the name for the query and click **Finish**.

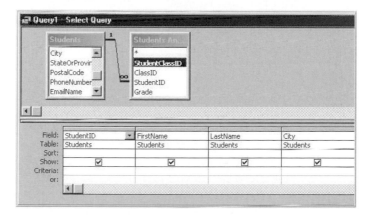

display an empty record with these column headings. You would then type in the search conditions that you want in the appropriate columns.

Thus, if you wanted to find all Beverly Hills, California, students with a loan balance due of $3,000 or more, you would type *BEVERLY HILLS* in the CITY column, *CA* in the STATE column, and *>=3000* ("greater than or equal to $3,000") in the AMOUNT OWED column.

Database technology is crucial to managing inventory in large warehouses.

Some DBMSs, such as Symantec's Q&A, use natural language interfaces, which allow users to make queries in any spoken language, such as English. With this software, you could ask your questions—either typing or speaking (if the system has voice recognition)—in a natural way, such as "How many sales reps sold more than one million dollars' worth of books in the Western Region in January?"

Object-Oriented Database

How would I describe the principle behind an object-oriented database?

Traditional database models, including the relational model, have worked well in traditional business situations. However, they fall short in areas such as engineering design and manufacturing, scientific experiments, telecommunications, geographic information systems, and multimedia.[16] The object-oriented database model was developed to meet the needs of these applications. **An _object-oriented database_ uses "objects," software written in small, reusable chunks, as elements within database files.** An *object* consists of (1) data in any form, including graphics, audio, and video, and (2) instructions on the action to be taken on the data. Examples of object-oriented databases are FastObjects, GemStone, Objectivity DB, Jasmine Object Database, and KE Texpress. Many high-tech companies exist that can create custom databases.

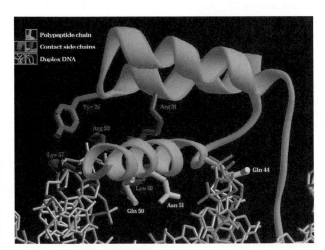

Computer-generated model of DNA; it was constructed from data in a multimedia database.

A MULTIMEDIA DATABASE An object-oriented database is a multimedia database; it can store more types of data than a relational database can. For example, an object-oriented student database might contain each student's photograph, a "sound bite" of his or her voice, and even a short piece of video, in addition to grades and personal data. Moreover, the object would store operations, called *methods*, the programs that objects use to process themselves. For example, these programs might indicate how to calculate the student's grade-point average or how to display or print the student's record.

TYPES OF OBJECT-ORIENTED DATABASES One type of object-oriented database is a *hypertext database*, or *web database*, which contains text links to other documents. Another type is a *hypermedia database*, which contains these links as well as graphics, sound, and video. These two types of object-oriented databases, which are created by software such as Cold Fusion, are accessible via the web.

In addition, companies such as Microsoft, IBM, Informix, Sybase, and Oracle have developed *object-relational*, or *enhanced-relational*, database models. Examples are DB2, Cloudscape, and Oracle9i. These handle both hierarchical and network data (structured data) and relational and object-oriented data.

Multidimensional Database

How is a multidimensional database different from the other four types of databases?

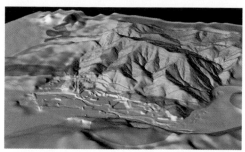 

(*Left*) Geological multimedia databases are used in the mining industry. (*Right*) A weather map generated from a multimedia meteorological database.

The multidimensional database was developed during the past decade for use when the purpose is to analyze data rather than perform online transactions. **A _multidimensional database (MDA)_ models data as facts, dimensions, or numerical measures for use in the interactive analysis of large amounts of data for decision-making purposes.** Examples are InterSystems Caché, ContourCube, and Cognoa PowerPlay. A multidimensional database uses the idea of a cube to represent the dimensions of data available to a user, using up to four dimensions. (● *See Panel 8.9.*) For example, "sales" could be viewed in the dimensions of (1) product model, (2) geography, (3) time, or (4) some additional dimension. In this case, "sales" is known as the main attribute (or measure) of the data cube and the other dimensions are seen as "feature" attributes.

Unlike relational databases, which often require SELECT . . . FROM and other types of SQL queries to provide information, multidimensional databases allow users to ask questions in more colloquial English, such as "How

● **PANEL 8.9**
Multidimensional database cube
Sample cube capturing sales data. Data cubes support viewing of up to four dimensions simultaneously.

Sample sales spreadsheet

Product	Number of purchases by city			
	Aalborg	Copenhagen	Los Angeles	New York City
Milk	123	555	145	5,001
Bread	102	250	54	2,010
Jeans	20	89	32	345
Light bulbs	22	213	32	9,450

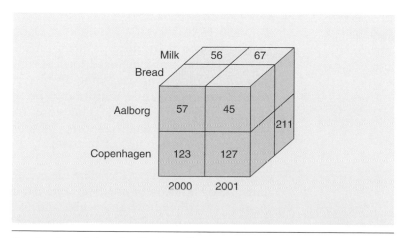

Sample cube capturing sales data. Data cubes support viewing of up to four dimensions simultaneously. In this case, the cube generalizes the spreadsheet from the table to three dimensions—product, city, year.

many type Z dog leashes have been sold in New Jersey so far this year?" The means for doing so is *online analytical processing (OLAP) software,* which can quickly provide answers to complex database queries. OLAP software is used in business reporting for sales, marketing, management reporting, trend analysis, and similar areas. An OLAP application that accesses data from a multidimensional database is known as a *MOLAP (multidimensional OLAP)* application.

Data Mining

How does data mining work, and how could it be useful to me?

A personal database, such as the address list of friends you have on your microcomputer, is generally small. But some databases are almost unimaginably vast, involving records for millions of households and trillions of bytes of data. Some record-keeping activities require the use of so-called massively parallel database computers that cost $1 million or more. "These machines gang together scores or even hundreds of the fastest microprocessors around," says one description, "giving them the oomph to respond in minutes to complex database queries."[17]

These large-scale efforts go under the name "data mining." **_Data mining (DM)_ is the computer-assisted process of sifting through and analyzing vast amounts of data in order to extract hidden patterns and meaning and to discover new knowledge.** The purpose of DM is to describe past trends and predict future trends. Thus, data-mining tools might sift through a company's immense collections of customer, marketing, production, and financial data and identify what's worth noting and what's not.

The Process of Data Mining

What are the steps in the data-mining process?

In data mining, data is acquired and prepared for what is known as a "data warehouse" through the following steps.[18] (● *See Panel 8.10.*)

1. DATA SOURCES   Data may come from a number of sources: (a) point-of-sale transactions in files (flat files) managed by file management systems on mainframes, (b) databases of all kinds, and (c) other—for example, news articles transmitted over newswires or online sources such as the internet. To the mix may also be added (d) data from data warehouses, as we describe below.

2. DATA FUSION & CLEANSING   Data from diverse sources, whether from inside the company (internal data) or purchased from outside the company (external data), must be fused together and then put through a process known as *data cleansing,* or *scrubbing.* Even if the data comes from just one source, such as one company's mainframe, the data may be of poor quality, full of errors and inconsistencies. Therefore, for data mining to produce accurate results, the source data has to be "scrubbed"—that is, cleaned of errors and checked for consistency of formats.

3. DATA & META-DATA   The cleansing process yields both the cleaned-up data and a variation of it called meta-data. *Meta-data* is essentially data about data; it describes how and when and by whom a particular set of data was collected and how the data is formatted. (The data in a data dictionary, for example, is meta-data.)

Meta-data is essential for understanding information stored in data warehouses. Meta-data shows the origins of the data, the transformations it has

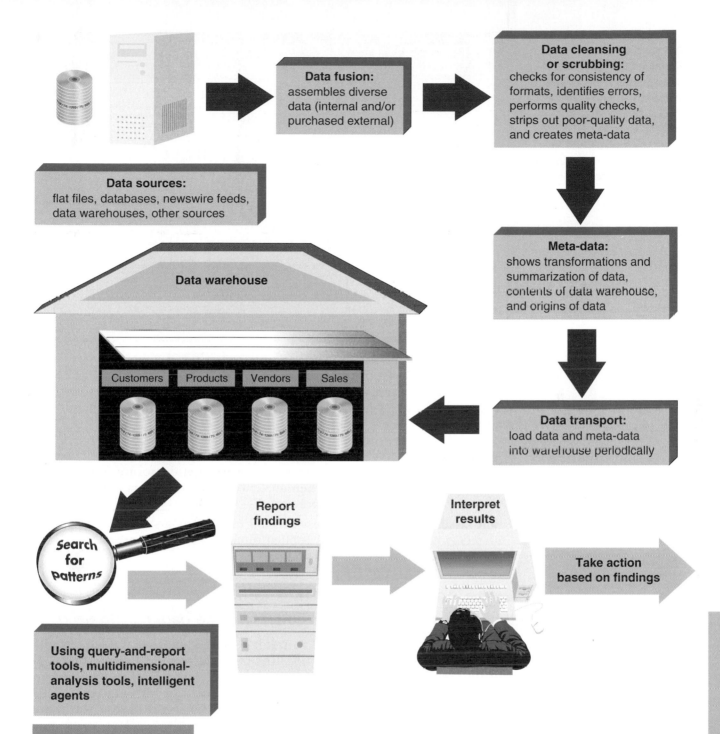

Data fusion:
assembles diverse
data (internal and/or
purchased external)

Data cleansing
or scrubbing:
checks for consistency of
formats, identifies errors,
performs quality checks,
strips out poor-quality data,
and creates meta-data

Data sources:
flat files, databases, newswire feeds,
data warehouses, other sources

Data warehouse

Customers | Products | Vendors | Sales

Meta-data:
shows transformations and
summarization of data,
contents of data warehouse,
and origins of data

Data transport:
load data and meta-data
into warehouse periodically

Search for patterns

Report findings

Interpret results

Take action
based on findings

Using query-and-report
tools, multidimensional-
analysis tools, intelligent
agents

● PANEL 8.10
The data-mining
process

undergone, and summary information about it, which makes it more useful than the cleansed but unintegrated, unsummarized data. The meta-data also describes the contents of the data warehouse.

4. DATA TRANSPORT TO THE DATA WAREHOUSE Both the data and the meta-data are sent to the data warehouse. **A _data warehouse_ is a special database of cleaned up data and meta-data**. It is a replica, or close reproduction, of a mainframe's data.

The data warehouse is stored on disk using storage technology such as RAID (redundant arrays of independent disks). Small data warehouses may hold 100 gigabytes of data or less. Once 500 gigabytes are reached, massively parallel processing computers are needed. Projections call for large data warehouses holding hundreds of terabytes within the next few years.

Data mining. This type of credit card analysis determines the most influential factors common to non-profitable customers. In this case, BusinessMiner from Business Objects determined that the credit limit had the greatest effect on profitability and prioritized the results in graphical form.

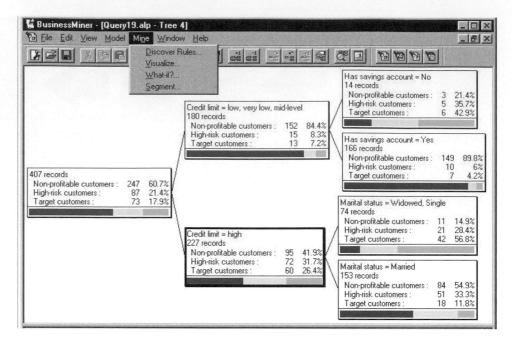

MINING THE DATA: SEARCHING FOR PATTERNS & INTERPRETING THE RESULTS Data in the warehouse is usually analyzed—or mined—using one of two popular *algorithms*, or step-by-step problem-solving procedures:

- **Regression analysis:** Basically, *regression analysis* takes a particular set of numerical data and develops a mathematical formula that fits the data. This formula is then applied to new sets of data of the same type to predict future situations.

- **Classification analysis:** *Classification analysis* is a statistics pattern-recognition process that is applied to data sets with more than just numerical data.

Some Applications of Data Mining

What are some ways in which I might see data mining used?

Some short-term payoffs from data mining can be dramatic. One telephone company, for instance, mined its existing billing data to identify 10,000 supposedly "residential" customers who spent more than $1,000 a month on their phone bills. When it looked more closely, the company found that these customers were really small businesses trying to avoid paying the more expensive business rates for their telephone service.[19]

However, the payoffs in the long term could be truly astonishing. Sifting medical-research data or subatomic-particle information may reveal new treatments for diseases or new insights into the nature of the universe.[20] Text-mining programs, which build on data-mining principles, may scour 250,000 pages an hour of scholarly articles, automatically categorizing information, making links between unconnected documents, and providing visual maps.[21]

Some other applications of data mining are as follows.[22]

SPORTS Brian James, coach of the Toronto Raptors, has used data mining to "rack and stack" his team against the rest of the National Basketball Association (NBA). A coach in the U.S. Gymnastics Federation used a DM tool called *IDIS* (Iowa Drug Information Service) to discover what long-term factors contributed to athletes' performance, so as to know what problems to treat early on.

MARKETING Marketers use DM tools to mine point-of-sale databases of retail stores, which contain facts (such as prices, quantities sold, and dates of sale) for thousands of products in hundreds of geographic areas. By understanding customer preferences and buying patterns, marketers hope to target consumers' individual needs.

One way that grocery stores create profiles is by the use of data mining. Data mining is a process that is used to predict future behaviors and trends by discerning patterns and relationships from large amounts of warehoused raw (from disparate sources) data through the use of statistics and other mathematical techniques.*

HEALTH A Los Angeles hospital used IDIS to see what subtle factors affect success and failure in back surgery. Another system helps health care organizations pinpoint groups whose costs are likely to increase in the near future, so that medical interventions can be made.

SCIENCE DM techniques are being employed to find new patterns in genetic data, molecular structures, global climate changes, and more. For instance, one DM tool (called *SKICAT*, for "SKy Image CATaloging") has been used to catalog more than 50 million galaxies.

There are positive aspects to data mining such as fraud detection and identification of items that are commonly bought together. But there is a darker side to data mining too, as we discuss in Section 8.8.

more info!

IDIS & IDIN

The University of Iowa's Division of Drug Information Service, within the College of Pharmacy, offers IDIS, a bibliographic database, and IDIN (Iowa Drug Information Network), which provides drug information. Go to *www.uiowa.edu/~idis*.

8.5

Databases & the Digital Economy: E-Business & E-Commerce

What are e-business and e-commerce, and what are three types of e-commerce systems?

At one time there was a difference between the Old Economy and the New Economy. The first consisted of traditional companies—car makers, pharmaceuticals, retailers, publishers. The second consisted of computer, telecommunications, and internet companies (AOL, Amazon, eBay, and a raft of "dot-com" firms). Now, however, most Old Economy companies have absorbed the new internet-driven technologies, and the differences between the two sectors have dwindled. In other words, most companies are now engaged in *e-business,* using the internet to facilitate every aspect of running a business. As one article puts it, "At bottom the internet is a tool that dramatically lowers the cost of communication. That means it can radically alter any industry or activity that depends on the flow of information."[23]

One sign of growth is that the number of internet host computers has been almost doubling every year. But the mushrooming of computer networks and the booming popularity of the World Wide Web are only the most obvious signs of the digital economy. Behind them lies something equally important: the growth of vast stores of information in databases.

How are databases underpinning the digital economy? Let us consider two aspects: *e-commerce* and *types of e-commerce systems—B2B, B2C,* and *C2C.*

Harrah's casino empire uses data mining to identify gambling patron preferences for marketing and special promotions.

E-Commerce: Online Buying & Selling

What is e-commerce, and what technologies have made it possible?

The internet might have remained the text-based province of academicians and researchers had it not been for the creative contributions of Tim Berners-Lee, whom we introduced in Chapter 2. He was the computer scientist who came up with the coding system (hypertext markup language, or HTML), linkages, and addressing scheme (URLs) that debuted in 1991 as the graphics-laden and multimedia World Wide Web. "It's hard to overstate the impact

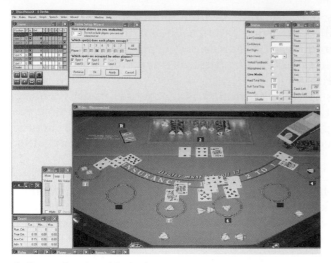

Online poker is a kind of e-commerce

of the global system he created," states technology writer Joshua Quittner. "He took a powerful communications system [the internet] that only the elite could use, and turned it into a mass medium."[24]

The arrival of the web, aided by the growth and increasing sophistication of enormous databases, has led to **_e-commerce_, or _electronic commerce_, the buying and selling of products and services through computer networks.** E-commerce is reshaping entire industries and revamping the very notion of what a company is. Indeed, online shopping is growing even faster than the increase in computer use, which has been fueled by the falling price of personal computers.

Two well-known e-firms are Amazon.com, which sells books and other goods, and Priceline.com, which lets you name the price you're willing to pay for airline tickets and hotel rooms. (eBay is considered an online shopping bazaar rather than an online retailer.) Probably the foremost example of e-commerce is Amazon.com.[25]

AN EXAMPLE OF E-COMMERCE: AMAZON.COM   In 1994, seeing the potential for electronic retailing on the World Wide Web, Jeffrey Bezos left a successful career on Wall Street to launch an online bookstore called Amazon.com. Why the name "Amazon"?

"Earth's biggest river, Earth's biggest bookstore," said Bezos in a 1996 interview. "The Amazon River is ten times as large as the next largest river, which is the Mississippi, in terms of volume of water. Twenty percent of the world's fresh water is in the Amazon River Basin, and we have six times as many titles as the world's largest physical bookstore."[26] A more hardheaded reason is that, according to consumer tests, words starting with "A" show up on search-engine lists first.[27]

Still, Bezos realized that no bookstore with four walls could possibly stock the more than 2.5 million titles that are now active and in print. Moreover, he saw that an online bookstore wouldn't have to make the same investment in retail clerks, store real estate, or warehouse space (in the beginning, Amazon.com ordered books from the publisher *after* it took the book buyer's order), so it could pass savings along to customers in the form of discounts. In addition, he appreciated that there would be opportunities to obtain demographic information about customers in order to offer personalized services. For example, Amazon could let customers know of books that might be of interest to them. Such personalized attention is difficult for traditional large bookstores. Finally, Bezos saw that there could be a good deal of online interaction: Customers could post reviews of books they read and could reach authors by email to provide feedback. All this was made possible on the web by the recording of information on giant databases.

Amazon.com sold its first book in July 1995, and by the end of 2004 had sales of $6.92 billion, ranking it at the top of online businesses, well ahead of Dell Inc., which posted $3.25 billion in online business-to-consumer sales.[28] What began as Earth's biggest bookstore also has rapidly become Earth's biggest anything store, offering CDs, DVDs, videos, electronics, toys, tools, home furnishings, clothing, prescription drugs, and film-processing services. For a long time the company put market share ahead of profits, but now it's focusing on profits.

ONLINE RETAILERS: E-TAILERS   Some e-commerce businesses are also known as *e-tailers*, or electronic retailers. The effectiveness of e-tailing was

more info!

Comparing Prices at the Local Mall

Some online companies allow you to compare prices in local stores:
www.cairo.com
www.shoplocal.com
www.stepup.com

evident as early as 1997, when Dell Computer reported taking multimillion-dollar orders at its website. E-tailing has resulted in the development of *e-tailware,* software tools for creating online catalogs and managing the business connected with doing online retailing. Some e-commerce groups (such as shopping.com and shopzilla) offer price-comparison websites that can quickly compare prices from a number of different e-tailers and link you to them.

Some particularly interesting technologies that have helped online retailers survive are the following:

- **The "one-click" option:** Initially, online shopping promised the idea that, as one writer described it, "You see something you want, click and it's yours. Unfortunately, the reality was more like this: You see something you want, you click and click and click, you type in a bunch of information, click again, type in some more information, click a few more times."[29] That changed when in September 1997 Amazon introduced its "one click" option, in which shoppers agree to skip several of the review and confirmation steps in the buying process.

- **360-degree images:** People might well buy a book or a computer off the web on the basis of a blurry photograph, but a sweater or a sofa? Today zoom technologies allow for close inspection of nearly anything, from the detail on jewelry to the rooms in a house. With technology from iPix, 360-degree images can be created using just two standard photographs.

- **Order tracking:** The average Federal Express package is scanned between 15 and 20 times between pickup and delivery, with the location of each package stored in a database. For a long time, FedEx used such order tracking strictly for internal use, but in 1994 the company turned it into a customer-service application—a brilliant marketing stroke.

- **Shop bots:** The buying of online wares has also been facilitated by the arrival of shopping robots—*shop bots,* or *trading agents,* concierge programs that help users search the internet for a particular product or service, then bring up price comparisons, locations, and other information. Examples of shop bots are my-Simon, Deal-Time, Price-Grabber.com, Shop-Bot.com, and Book-Finder.com.

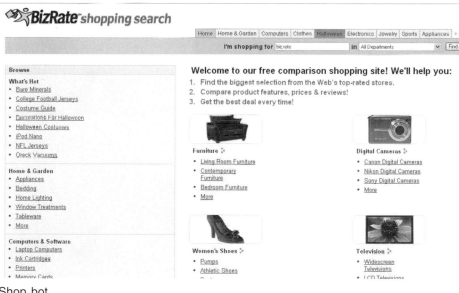

Shop bot

Types of E-Commerce Systems: B2B, B2C, & C2C

How would I distinguish the three types of e-commerce systems?

Three main types of e-commerce systems are business-to-business (B2B), business-to-consumer (B2C), and consumer-to-consumer (C2C).

BUSINESS-TO-BUSINESS (B2B) SYSTEMS In a ***business-to-business (B2B) system***, **a business sells to other businesses, using the internet or a private network to cut transaction costs and increase efficiencies.** Besides selling products, a business might sell advertising, employee training, market research, technical support, banking services, and other business support services.

One of the most famous examples of B2B is the auto industry online exchange developed by the big three U.S. automakers—General Motors, Ford, and DaimlerChrysler. The companies have put their entire system of purchasing, involving more than $250 billion in parts and materials and 60,000 suppliers, on the internet. The result is that so-called reverse auctions, in which suppliers bid to provide the lowest price, have driven down the cost of parts such as tires and window sealers. This system replaces the old-fashioned bureaucratic procurement process built on phone calls and fax machines and provides substantial cost savings.[30]

Online B2B exchanges have been developed to serve a variety of businesses, from manufacturers of steel and airplanes to convenience stores to olive oil producers. B2B exchanges help business by moving beyond pricing mechanisms and encompassing product quality, customer support, credit terms, and shipping reliability, which often count for more than price. The name given to this system is the *business web,* or *b-web,* in which suppliers, distributors, customers, and e-commerce service providers use the internet for communications and transactions. In addition, b-webs are expected to provide extra revenue from ancillary services, such as financing and logistics. (See Panel 8.11.)

None of these innovations is possible without databases and the communications lines connecting them. B2B used to be conducted mostly

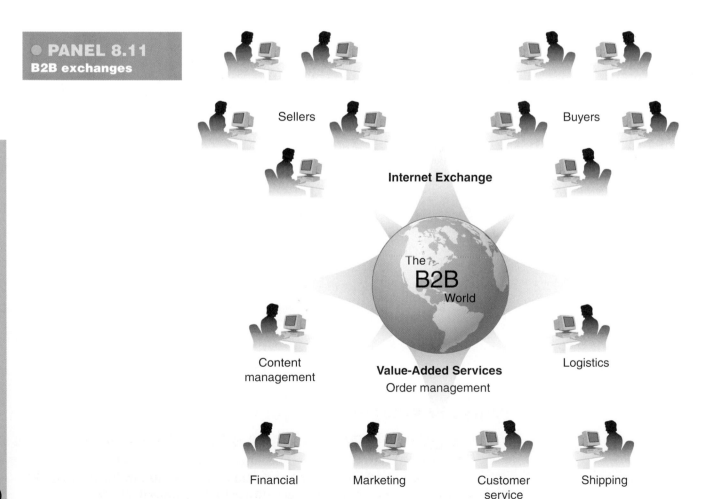

PANEL 8.11
B2B exchanges

Sellers

Buyers

Internet Exchange

The
B2B
World

Content management

Value-Added Services
Order management

Logistics

Financial

Marketing

Customer service

Shipping

WHO WE ARE

Mission and Guiding Values

Mission Statement
Heritage Foods USA exists to promote genetic diversity, small family farms, and a fully traceable food supply. We are committed to making wholesome, delicious and sustainably produced heritage foods available to all Americans. In doing so, we will foster the link between sustainable land use, small-scale food production and preservation of the foods of past generations for future generations.

Guiding Values

American

- We uphold the American ideals of equity and dignity for all producers and their foods.
- We are committed to heritage foods of all of the Americas, North, South and Central.
- We believe that our inalienable rights of life, liberty and the pursuit of happiness include access to high quality, sustainably raised and traceable foods

> A Brief Overview
> Mission and Values
> Meet the Staff

A B2C business

over leased communications lines; now, however, many B2B businesses use the internet, an extranet, and/or a virtual private network.

BUSINESS-TO-CONSUMER (B2C) SYSTEMS

In a **business-to-consumer (B2C) system, a business sells goods or services to consumers, or members of the general public.** This kind of e-commerce system essentially removes the middleman and often even the need for a physical ("bricks-and-mortar") store. Examples of B2C systems are those of Amazon.com and BarnesandNoble.com, as well as many financial institutions and the U.S. government.

CONSUMER-TO-CONSUMER (C2C) SYSTEMS In a **consumer-to-consumer (C2C) system, consumers sell goods or services directly to other consumers,** often with the help of a third party, such as eBay, the online auction company. Such intermediaries mediate between consumers who want to buy and sell, taking a small percentage of the seller's profit as a fee. Examples of C2C exchanges are classified ads (such as TradingPost.com), music and file sharing (AllCoolMusic.com), career and job websites (monster.com), social or dating websites (myspace.com, friendster.com), and online communities (craigslist.org). The advantage of C2C e-commerce is most often the reduced costs and a smaller but profitable customer base. It also gives many small business owners a way to sell their goods without running a costly bricks-and-mortar store.

8.6

Information Systems in Organizations: Using Databases to Help Make Decisions

How does information flow within an organization, and what are different types of information systems?

The data in databases is used to build information, and information—and how it is used—lies at the heart of every organization. Of course, how useful information is depends on the quality of it, as well as the information systems used to distribute it. Let us consider these subjects.

The Qualities of Good Information

What are the qualities of good information?

In general, all information to support intelligent decision making within an organization must be as follows:

- **Correct and verifiable:** This means information must be accurate and checkable.
- **Complete yet concise:** *Complete* means information must include *all* relevant data. *Concise* means it includes *only* relevant data.

- **Cost effective:** This means the information is efficiently obtained and understandable.
- **Current:** *Current* means timely yet also time sensitive, based on historical, present, or future information needs.
- **Accessible:** This means the information is quickly and easily obtainable.

Information Flows within an Organization: Horizontally between Departments & Vertically between Management Levels

What are the six departments and three management levels to which information must flow?

Consider any sizable organization with which you are familiar. Its purpose is to perform a service or deliver a product. If it's nonprofit, for example, it may deliver the service of educating students or the product of food for famine victims. If it's profit-oriented, it may, for example, sell the service of fixing computers or the product of computers themselves. Information—whether computer-based or not—has to flow within an organization in a way that will help managers, and the organization, achieve their goals. To this end, organizations are often structured horizontally and vertically—horizontally to reflect functions and vertically to reflect management levels.

THE HORIZONTAL FLOW OF INFORMATION BETWEEN SIX DEPARTMENTS Depending on the services or products they provide, most organizations have departments that perform six functions: *research and development (R&D), production (or operations), marketing and sales, accounting and finance, human resources (personnel),* and *information systems (IS).* (● *See Panel 8.12.*)

● **PANEL 8.12**
Organization chart
The six functional responsibilities are shown on the opposite page at the bottom of the pyramid. The three management levels are shown along the sides.

- **Research and development:** The research and development (R&D) department does two things: (1) It conducts basic research, relating discoveries to the organization's current or new products. (2) It does product development and tests and modifies new products or services created by researchers. Special software is available to aid in these functions.

- **Production (operations):** The production department makes the product or provides the service. In a manufacturing company, it takes the raw materials and has people or machinery turn them into finished goods. In many cases, this department uses CAD/CAM software and workstations, as well as robots (described on p. 443). In another type of company, this department might manage the purchasing, handle the inventories, and control the flow of goods and services.

- **Marketing and sales:** The marketing department oversees advertising, promotion, and sales. The people in this department plan, price, advertise, promote, package, and distribute the services or goods to customers or clients. The sales reps may use laptop computers, cell-phones, wireless email, and faxes in their work while on the road.

- **Accounting and finance:** The accounting and finance department handles all financial matters. It handles cash management, pays bills and taxes, issues paychecks, records payments, makes investments, and compiles financial statements and reports. It also produces financial budgets and forecasts financial performance after receiving information from other departments.

- **Human resources:** The human resources, or personnel, department finds and hires people and administers sick leave and retirement matters. It is also concerned with compensation levels, professional development, employee relations, and government regulations.

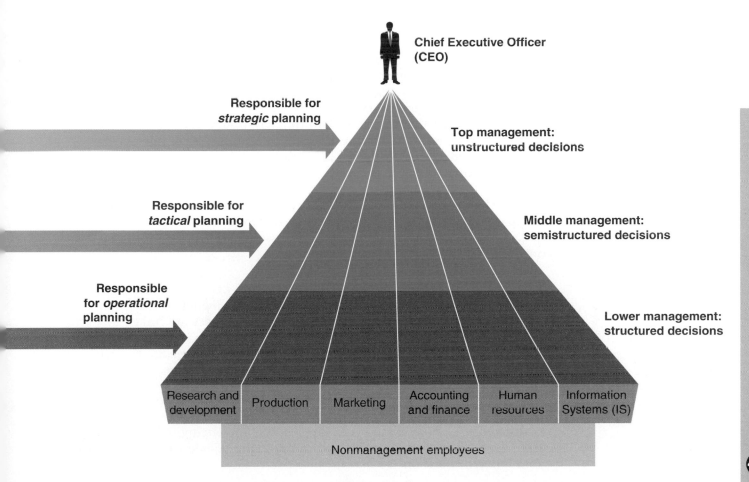

Databases & Information Systems

- **Information systems (IS):** The IS department manages the organization's computer-based systems and plans for and purchases new ones.

THE VERTICAL FLOW OF INFORMATION BETWEEN MANAGEMENT LEVELS Large organizations traditionally have three levels of management— *strategic management, tactical management,* and *operational management.* These levels can be shown on an *organization chart,* a schematic drawing showing the hierarchy of formal relationships among an organization's employees. Managers on each of the three levels have different levels of responsibility and are therefore required to make different kinds of decisions. (● *See Panel 8.12 again.*)

- **Strategic-level management:** Top managers are concerned with long-range, or strategic, planning and decisions. This top level is headed by the chief executive officer (CEO) along with several vice presidents or managers with such titles as chief financial officer (CFO), chief operating officer (COO), and chief information officer (CIO). *Strategic* decisions are complex decisions rarely based on predetermined routine procedures; they involve the subjective judgment of the decision maker. For instance, strategic decisions relate to how growth should be financed and what new markets should be tackled first. Determining the company's 5-year goals, evaluating future financial resources, and formulating a response to competitors' actions are also strategic decisions.

- **Tactical-level management:** Tactical, or middle-level managers, make tactical decisions to implement the strategic goals of the organization. A *tactical* decision is made without a base of clearly defined informational procedures; it may require detailed analysis and computations. Examples of tactical-level managers are plant manager, division manager, sales manager, branch manager, and director of personnel.

- **Operational-level management:** Operational, or low-level (supervisory level), managers make *operational* decisions—predictable decisions that can be made by following well-defined sets of routine procedures. These managers focus principally on supervising non-management employees, monitoring day-to-day events, and taking corrective action where necessary. An example of an operational-level manager is a warehouse manager in charge of inventory restocking.

Note that, because there are fewer people at the level of top management and many people at the bottom, this management structure resembles a pyramid. It's also a hierarchical structure because most of the power is concentrated at the top.

NEW INFORMATION FLOW: THE DECENTRALIZED ORGANIZATION The hierarchical, pyramid-oriented structure is changing in the computer-network era to a decentralized form in which employees are linked to a centralized database. Although the different responsibilities of the various types of managers remain, the pyramid is flattened somewhat owing to increased participation of all employees via computer-enabled systems.

Nowadays, for instance, organizations that have computer networks often use groupware to enable cooperative work by groups of people—what are known as *computer-supported cooperative work (CSCW) systems.* Through the shared use of databases, software, videoconferencing, email, intranets, organization forms and reports, and so on, many people can work together from different locations to manage information.

Computer-Based Information Systems

What are six computer-based information systems?

The purpose of a computer-based information system is to provide managers (and various categories of employees) with the appropriate kind of information to help them make decisions. It is used to collect and analyze data from all departments and is designed to provide an organization's management with up-to-date information at any time. There are several types of computer-based information systems, which serve different levels of management:

- Office information systems
- Transaction processing systems
- Management information systems
- Decision support systems
- Executive support systems
- Expert systems

Let us consider these.

Office Information Systems

What is an office information system?

Office information systems (OISs), **also called** *office automation systems (OASs)*, **combine various technologies to reduce the manual labor required in operating an efficient office environment and to increase productivity.** Used throughout all levels of an organization, OIS technologies include fax, voice mail, email, scheduling software, word processing, and desktop publishing, among others. (● *See Panel 8.13.*)

The backbone of an OIS is a network—LAN, intranet, extranet—that connects everything. All office functions—dictation, typing, filing, copying, fax, microfilm and records management, telephone calls and switchboard operations—are candidates for integration into the network.

Transaction Processing Systems

What is a TPS, and what are its principal features?

In most organizations, particularly business organizations, most of what goes on consists largely of structured information known as transactions. A *transaction* is a recorded event having to do with routine business activities.

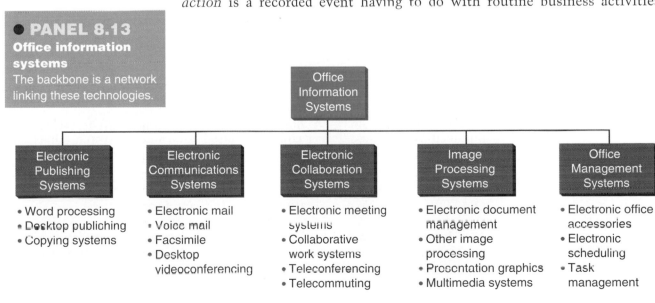

● PANEL 8.13
Office information systems
The backbone is a network linking these technologies.

A transaction may be recorded manually or via a computer system and includes everything concerning the product or service in which the organization is engaged: production, distribution, sales, orders. It also includes materials purchased, employees hired, taxes paid, and so on. Today in most organizations, the bulk of such transactions is recorded in a computer-based information system. These systems tend to have clearly defined inputs and outputs, and there is an emphasis on efficiency and accuracy. Transaction processing systems record data but do little in the way of converting data into information.

TPS: THE DEFINITION A _transaction processing system (TPS)_ **is a computer-based information system that keeps track of the transactions needed to conduct business.** The transactions can be handled via _batch processing,_ also known as _offline processing_—that is, the data is gathered and processed in batches at periodic intervals, such as at the end of the day or once a week. Or they may be handled via _real-time processing,_ also known as _online transaction processing (OLTP)_—that is, each transaction is processed immediately as it is entered. The data collected by a TPS is typically stored in databases.

FEATURES OF A TPS Some features of a TPS are as follows:

- **Input and output:** The inputs to the system are transaction data: bills, orders, inventory levels, and the like. The output consists of processed transactions: bills, paychecks, and so on.
- **For operational managers:** Because the TPS deals with day-to-day matters, it is principally of use to operational-level or supervisory managers, although it can also be helpful to tactical-level managers.
- **Produces detail reports:** A manager at the operational level typically receives information in the form of detail reports. A _detail report_ contains specific information about routine activities. An example might be the information needed to decide whether to restock inventory.
- **One TPS for each department:** Each department or functional area of an organization usually has its own TPS. For example, the accounting and finance TPS handles order processing, accounts receivable, inventory and purchasing, accounts payable, and payroll.
- **Basis for MIS and DSS:** The database of transactions stored in a TPS provides the basis for management information systems and decision support systems, as we describe next.

Management Information Systems

What are the characteristics of an MIS?

The next level of information system is the management information system.

DEFINITION OF MIS A _management information system (MIS)_ (pronounced "em-eye-ess") **is a computer-based information system that uses data recorded by a TPS as input into programs that produce routine reports as output.**

FEATURES OF AN MIS Features of an MIS are as follows:

- **Input and output:** Inputs consist of processed transaction data, such as bills, orders, and paychecks, plus other internal data. Outputs consist of summarized, structured reports: budget summaries, production schedules, and the like.

- **For tactical managers:** An MIS is intended principally to assist tactical-level managers. It enables them to spot trends and get an overview of current business activities.

- **Draws from all departments:** The MIS draws from all six departments or functional areas, not just one.

- **Produces several kinds of reports:** Managers at this level usually receive information in the form of several kinds of reports: *summary, exception, periodic, demand.*

   Summary reports show totals and trends. An example is a report showing total sales by office, by product, and by salesperson, as well as total overall sales.

   Exception reports show out-of-the-ordinary data. An example is an inventory report listing only those items of which fewer than 10 are in stock.

   Periodic reports are produced on a regular schedule. Such daily, weekly, monthly, quarterly, or annual reports may contain sales figures, income statements, or balance sheets. They are usually produced on paper, such as computer printouts.

   Demand reports produce information in response to an unscheduled demand. A director of finance might order a demand credit-background report on an unknown customer who wants to place a large order. Demand reports are often produced on a terminal or microcomputer screen, rather than on paper.

Decision Support Systems

What are the characteristics of a DSS?

A more sophisticated information system is the decision support system.

DEFINITION OF DSS   A *decision support system (DSS)* is a computer-based information system that provides a flexible tool for analysis and helps managers focus on the future. It gathers and presents data from a wide range of sources in a way that can be interpreted by humans. Some decision support systems come very close to acting as artificial intelligence agents (covered in the next section). DSS applications are not single information resources, such as a database or a program that graphically represents sales figures, but a combination of integrated resources working together. Whereas a TPS records data and an MIS summarizes data, a DSS analyzes data. To reach the DSS level of sophistication in information technology, an organization must have established TPS and MIS systems first.

FEATURES OF A DSS   Some features of a DSS are as follows:

- **Inputs and outputs:** Inputs include internal data—such as summarized reports and processed transaction data—and also data that is external to the organization. External data may be produced by trade associations, marketing research firms, the U.S. Bureau of the Census, and other government agencies. The outputs are demand reports on which a top manager can make decisions about unstructured problems.

- **Mainly for tactical managers:** A DSS is intended principally to assist tactical-level managers in making tactical decisions. Questions addressed by the DSS might be, for example, whether interest rates will rise or whether there will be a strike in an important materials supplying industry.

- **Produces analytic models:** The key attribute of a DSS is that it uses models. A *model* is a mathematical representation of a real system. The models use a DSS database, which draws on the TPS and MIS

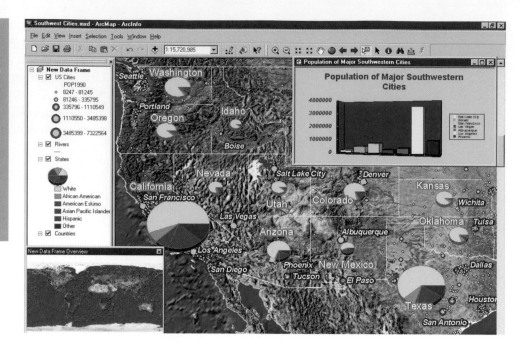

files, as well as external data such as stock reports, government reports, and national and international news. The system is accessed through DSS software. The model allows the manager to do a simulation—play a "what-if" game—to reach decisions. Thus, the manager can simulate an aspect of the organization's environment in order to decide how to react to a change in conditions affecting it. By changing the hypothetical inputs to the model, the manager can see how the model's outputs are affected.

SOME USES OF DSSs Many DSSs are developed to support the types of decisions faced by managers in specific industries, such as airlines or real estate. Curious how airlines decide how many seats to sell on a flight when so many passengers are no-shows? American Airlines developed a DSS, the yield management system, that helps managers decide how much to overbook and how to set prices for each seat so that a plane is filled and profits are maximized. Wonder how owners of those big apartment complexes set rents and lease terms? Investors in commercial real estate use a DSS called RealPlan to forecast property values up to 40 years into the future, based on income, expense, and cash-flow projections. Ever speculate about how insurance carriers set different rates or how Arby's and McDonald's decide where to locate a store? Many companies use DSSs called *geographic information systems (GISs)*, such as MapInfo and Atlas GIS, which integrate geographic databases with other business data and display maps. (● *See Panel 8.14.*)

Executive Support Systems

How does an ESS differ from information systems used by lower-level managers?

Also called an *executive information system (EIS)*, an **_executive support system (ESS)_ is an easy-to-use DSS made especially for strategic managers; it specifically supports strategic decision making.** It draws on data not only from systems internal to the organization but also from those outside, such as news services or market-research databases. (● *See Panel 8.15.*)

An ESS might allow senior executives to call up predefined reports from their personal computers, whether desktops or laptops. They might, for instance, call up sales figures in many forms—by region, by week, by anticipated year, by projected increases. The ESS includes capabilities for analyzing

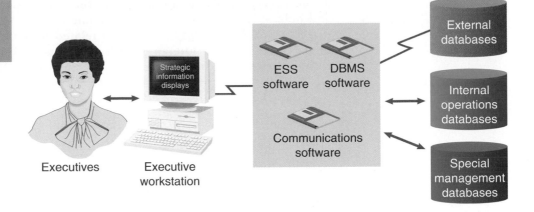

Executives   Executive
workstation

data and doing "what-if" scenarios. ESSs also have the capability to browse through summarized information on all aspects of the organization and then zero in on ("drill down" to) detailed areas the manager believes require attention.

ESSs are relatively user-friendly and require little training to use.

Expert Systems

Would I be apt to use an expert system?

An *expert system*, or *knowledge-based system*, **is a set of interactive computer programs that helps users solve problems that would otherwise require the assistance of a human expert.** Expert systems are created on the basis of knowledge collected on specific topics from human specialists, and they imitate the reasoning process of a human being. As we describe in the next section, expert systems have emerged from the field of artificial intelligence, the branch of computer science that is devoted to the creation of computer systems that simulate human reasoning and sensation.

Expert systems are used by both management and nonmanagement personnel to solve specific problems, such as how to reduce production costs, improve workers' productivity, or reduce environmental impact. Because of their giant appetite for memory, expert systems are usually run on large computers, although some microcomputer expert systems also exist. For example, Negotiator Pro for IBM and Macintosh computers helps executives plan effective negotiations by examining the personality types of the other parties and recommending negotiating strategies.

8.7

Artificial Intelligence

What are the main areas of artificial intelligence?

You're having trouble with your new computer program. You call the customer "help desk" at the software maker. Do you get a busy signal or get put on hold to listen to music (or, worse, advertising) for several minutes? Technical support lines are often swamped, and waiting is commonplace. Or, to deal with your software difficulty, do you find yourself dealing with . . . other software?

The odds are good that you will. For instance, a software technology called *automated virtual representatives*, or *vReps*, offers computer-generated images (animation or photos of real models) that answer customer questions in real time, using natural language.[31] vReps, which are becoming more and more common on self-service websites, are examples of expert systems, which,

Hank, the Coca-Cola vRep

as we suggested above, are programs that can walk you through a problem and help solve it. Expert systems are one of the most useful applications of **_artificial intelligence (AI)_**, **a group of related technologies used for developing machines to emulate human qualities, such as learning, reasoning, communicating, seeing, and hearing.** As will become clear, AI would not be possible without developments in database technology.

Today the main areas of AI are as follows:

- Expert systems
- Natural language processing
- Intelligent agents
- Pattern recognition
- Fuzzy logic
- Virtual reality and simulation devices
- Robotics

We will consider these areas, and then distinguish between *weak AI* and *strong AI*. We will also consider an area known as *artificial life*.

Expert Systems

How does an expert system work?

As we said in the last section, an *expert system* is an interactive computer program used to solve problems that would otherwise require the assistance of a human specialist. As the name suggests, it is a system imbued with knowledge by human experts.

EXAMPLES OF EXPERT SYSTEMS   Expert systems have been designed both for microcomputers and for larger computer systems with huge databases. One of the earliest expert systems, MYCIN helped diagnose infectious diseases. PROSPECTOR assesses geological data to locate mineral deposits. MailJail uses more than 600 rules to screen out junk email, or spam. Business Insight helps businesses find the best strategies for marketing a product. CARES (Computer Assisted Risk Evaluation System) helps nonprofit organizations evaluate risks and protect clients and staff. CLUES (Countrywide Loan

Expert system. This software company offers strategic-planning software that uses an expert knowledge base in the area of business strategy.

Business Strategy Software

Business Insight®

"Any business can spot problems and opportunities by submitting their business situation to the disciplined appraisal provided by Business Insight."

Philip Kotler,
Northwestern University

$795   Buy now!   More info...

Intelligent software has been used to:

- diagnose diseases
- anticipate the unknowns of space exploration
- predict weather patterns
- plan battle strategies
- **analyze any competitive market**

Intelligent software can be taught and can apply that knowledge to solve real problems.

more information:

Overview
1. Audit Your Business Model
2. Structure Your Planning
3. Think Outside-the-Box
4. Identify Risks
5. Strategy based on Fact

Demo
Charts
Testimonials

...MORE ⊙

Expert Knowledge
Business Insight is intelligent software with a knowledge base that includes the primary curriculum in strategic planning published in the last 20 years. The software has been taught the principles of business strategy based upon the published works of 100+ experts. Business Insight applies that expert system technology to analyze any competitive market.

Smarts and Experience
The knowledge base continues to grow as it is used by over half of the Fortune 500 companies in America.

Dialog with an Expert
Your dialog with Business Insight is equivalent to an in-depth engagement with a consulting MBA trained in strategic planning. You describe your business and the experts apply their specialized

Same task, different costs
You can hire an MBA to analyze your business, or you can use Business Insight to do the same things:

- ask every question relevant to the success of your venture
- question your assumptions and validate or challenge your thinking
- point out inconsistencies in your strategy
- be exhaustive; warning of factors that could blind-side you
- identify competitive strengths and strategies to leverage them
- identify competitive weaknesses and ways to compensate
- discover competitive opportunities in your market
- anticipate competitive threats

Underwriting Expert System) evaluates home-mortgage loan applications. CLIPS (C Language Integrated Production System), widely used in government, industry, and academia, is an expert system used to build other expert systems. Whale Watch is an expert system used to identify whales. STREAMES assists water managers to evaluate the effect of large stream nutrient loads on stream nutrient retention.

HOW EXPERT SYSTEMS DRAW ON KNOWLEDGE All these programs simulate the reasoning process of experts in certain well-defined areas. That is, professionals called *knowledge engineers* interview the expert or experts and determine the rules and knowledge that must go into the system. For example, to develop Muckraker, an expert system to assist newspaper reporters with investigative reporting, the knowledge engineers interviewed journalists.

Programs incorporate not only the experts' surface knowledge ("textbook knowledge") but also their deep knowledge ("tricks of the trade"). What, exactly, is *deep knowledge*? "An expert in some activity has by definition reduced the world's complexity by his or her specialization," say some authorities. One result is that "much of the knowledge lies outside direct conscious awareness."[32]

THE THREE COMPONENTS OF AN EXPERT SYSTEM An expert system consists of three components. (● *See Panel 8.16.*)

- **Knowledge base:** A *knowledge base* is an expert system's database of knowledge about a particular subject, including relevant facts, information, beliefs, assumptions, and procedures for solving problems. The basic unit of knowledge is expressed as an IF-THEN-ELSE rule. ("IF this happens, THEN do this, ELSE do that.") Programs can have many thousands of rules. A system called ExperTAX, for example, which helps accountants figure out a client's tax options, consists of more than 3,000 rules. Other systems have 80,000 or 100,000 rules, which means that very large computer systems are required.

- **Inference engine:** The *inference engine* is the software that controls the search of the expert system's knowledge base and produces conclusions. It takes the problem posed by the user and fits it into the rules in the knowledge base. It then derives a conclusion from the facts and rules contained in the knowledge base.

- **User interface:** The *user interface* is the display screen. It gives the user the ability to ask questions and get answers. It also explains the reasoning behind the answer.

People fearful about machines taking over our lives need to understand that expert systems are designed to be users' assistants, not replacements. Also, the success of these systems depends on the quality of the data and rules obtained from the human experts.

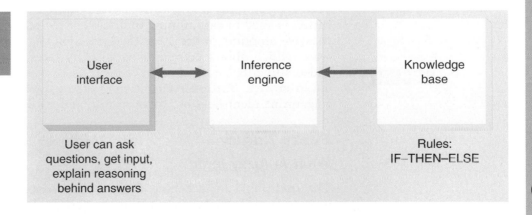

● **PANEL 8.16**
Components of an expert system

User interface → Inference engine ← Knowledge base

User can ask questions, get input, explain reasoning behind answers

Rules: IF–THEN–ELSE

Natural Language Processing

When have I encountered the use of natural language processing?

Natural languages are ordinary human languages, such as English. (A second definition, discussed in Chapter 10, is that natural languages are fifth-generation programming languages.) **_Natural language processing_ is the study of ways for computers to recognize and understand human language,** whether in spoken or written form. Major advances in natural language processing have occurred in *speech recognition,* in which computers translate spoken speech into text.

Think how challenging it is to make a computer translate English into another language. In one instance, the English sentence "The spirit is willing, but the flesh is weak" came out in Russian as "The wine is agreeable, but the meat is spoiled." The problem with human language is that it is often ambiguous; different listeners may arrive at different interpretations.

Most existing language systems run on large computers, although scaled-down versions are now available for microcomputers. A product called Intellect uses a limited English vocabulary to help users orally query databases on both mainframes and microcomputers. LUNAR, developed to help analyze moon rocks, answers questions about geology on the basis of an extensive database. Verbex, used by the U.S. Postal Service, lets mail sorters read aloud an incomplete address and will reply with the correct ZIP code.

Intelligent Agents

Have I ever used an intelligent agent?

How do you find information in the vast sea of the internet? As one solution, computer scientists have been developing so-called intelligent agents to find information on computer networks and filter it. **An _intelligent agent_ is a form of software with built-in intelligence that monitors work patterns, asks questions, and performs work tasks on your behalf,** such as roaming networks and compiling data. One type of intelligent agent is a kind of electronic assistant that will filter messages, scan news services, and perform similar secretarial chores. Microsoft Office has an agent called Office Assistant that can answer your questions, offer tips, and provide help for a variety of features specific to the program you're using. We mentioned another kind in the shop bots, or shopping robots, used in online shopping. Also known as *bots* or *network agents,* these intelligent agents search the internet and online databases for information and bring the results back to you.

A Bit about Bots

More information about bots is available at BotKnowledge:
http://psychology.about.com/ gi/dynamic/offsite.htm ?site=http%3A%2F%2F www .botknowledge.com%2F.

Pattern Recognition

Should I be concerned about pattern recognition?

Pattern recognition involves a camera and software that identify recurring patterns in what they are seeing and recognize the connections between the perceived patterns and similar patterns stored in a database. Pattern recognition is used in data mining to discover previously unnoticed patterns from massive amounts of data. Another principal use is in facial-recognition software, which allows computers to identify faces, using a digital "faceprint." Video surveillance cameras have been used to pick out suspicious individuals in crowds. Pattern recognition is also used for handwriting recognition, fingerprint identification, robot vision, and automatic voice recognition.

SECURITY

Fuzzy Logic

What is fuzzy logic?

The traditional logic behind computers is based on either-or, yes-no, true-false reasoning. Such computers make "crisp" distinctions, leading to precise decision making. _**Fuzzy logic**_ **is a method of dealing with imprecise data and uncertainty, with problems that have many answers rather than one.** Unlike classical logic, fuzzy logic is more like human reasoning: It deals with probability and credibility. That is, instead of being simply true or false, a proposition is mostly true or mostly false, or more true or more false.

For example, fuzzy logic has been applied in running elevators. How long will most people wait for an elevator before getting antsy? About a minute and a half, say researchers at the Otis Elevator Company. The Otis artificial intelligence division has thus done considerable research into how elevators may be programmed to reduce waiting time.[33] Ordinarily, when someone on a floor in the middle of the building pushes the call button, the system will send whichever elevator is closest. However, that car might be filled with passengers, who will be delayed by the new stop (perhaps making them antsy), whereas another car that is farther away might be empty. In a fuzzy-logic system, the computer assesses not only which car is nearest but also how full the cars are before deciding which one to send.

Fuzzy-logic circuitry also enables handheld autofocus video cameras to focus properly. If your hand is unsteady, the circuitry in the camera determines which parts in the visual field should stand still and which should move and makes the necessary adjustments in the image.

Virtual Reality & Simulation Devices

How would I define VR and simulation?

**Virtual reality (VR)**, **a computer-generated artificial reality, projects a person into a sensation of three-dimensional space.** (● _See Panel 8.17 on the next page._) To put yourself into virtual reality, you need software and special headgear; then you can add gloves, and later perhaps a special suit. The headgear—which is called a _head-mounted display_ (marketed as a VR Headset)—has two small video display screens, for each eye, to create the sense of three-dimensionality. Headphones pipe in stereophonic sound or even 3-D sound so that you think you are hearing sounds not only near each ear but also in various places all around you. The glove has sensors for collecting data about your hand movements. Once you are wearing this equipment, software gives you interactive sensory feelings similar to real-world experiences.

You may have seen virtual reality used in arcade-type games, such as Atlantis, a computer simulation of _The Lost Continent._ However, there are far more important uses—for example, in simulators for training. _**Simulators**_ **are devices that represent the behavior of physical or abstract systems.** Virtual-reality simulation technologies are applied a great deal in training. For instance, to train bus drivers, they create lifelike bus control panels and various scenarios such as icy road conditions. They are used to train pilots on various aircraft and to prepare air-traffic controllers for equipment failures. Surgeons in training can develop their skills through simulation on "digital patients." Architects create virtual walkthroughs of the structures they are designing. Virtual-reality therapy has been used for autistic children and in the treatment of phobias, such as extreme fear of public speaking or of being in public places or high places.

Robotics

How would I define robotics?

Nearly a half-century ago, in the film _Forbidden Planet_, Robby the Robot could sew, distill bourbon, and speak 187 languages. We haven't caught up with science-fiction movies, but maybe we'll get there yet.

**Robotics** **is the development and study of machines that can perform work that is normally done by people.** The machines themselves are called _robots_.

more info!

Be a Train Engineer

How does it feel to drive a train? You can find out with Microsoft's Train Simulator, pretending to direct trains over routes in the United States, United Kingdom, Austria, and Japan: www.microsoft.com/games/trainsimulator/info.asp.

more info!

Virtual Reality Products

For more about VR products: www.vrealities.com/main.html.

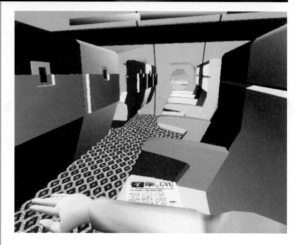

● PANEL 8.17
Virtual reality

(*Top left*) A research assistant works on a virtual reality program at the Virtual Reality in Medicine laboratory at the University of Illinois at Chicago. Surgeons use such programs to practice surgery on patients' "virtual organs" before making the first real cut. (*Top right*) Virtual rendering of a factory floor. (*Middle left*) Professor and Cultural Virtual Reality Lab Director at UCLA Dr. Bernard Fischer shows a virtual reality rendering of the ancient Roman forum. The rendering allows the user to enter a computerized recreation of what the archeological site appeared like at its peak in 400. C.E. (*Middle right*) Simulation training session dealing with terrorist activity.

(*Bottom right*) Passenger airline flight simluation that allows people who are afraid to fly to experience an airplane trip without leaving the ground.

(● *See Panel 8.18.*) As we said in Chapter 1, a *robot* is an automatic device that performs functions ordinarily executed by human beings or that operates with what appears to be almost human intelligence. (The word *robot*, derived from the Czech word for "compulsory labor," was first used by Karel Câpek in an early 1920s play.)

Shakey, developed in 1970, was the first robot to use artificial intelligence to navigate. Today Rosie and Roscoe are R2D2-like robots that perform a variety of courier duties for hospitals, saving nurses from having to make trips to supply areas, pharmacies, and cafeterias. ScrubMate—a robot equipped with computerized controls, ultrasonic "eyes," sensors, batteries, three different cleaning and scrubbing tools, and a self-squeezing mop—can clean bathrooms. Robodoc is used in surgery to bore the thighbone so that a hip implant can be attached. A driverless harvester, guided by satellite signals and an artificial vision system, is used to harvest alfalfa and other crops. A robot dog named AIBO is able to learn how to sit, roll over, fetch, and do other activities. You can buy your own robot vacuum cleaner, Roomba, or the more advanced Scooba, which will not just pick up dirt but also wash, scrub, and dry the floor.[34]

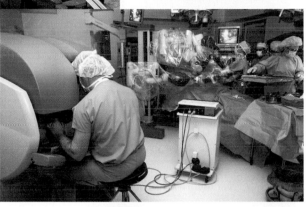

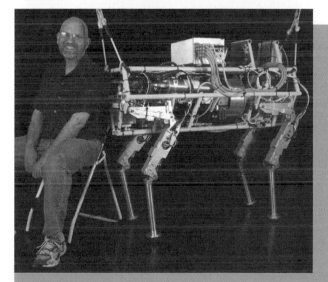

Robots are also used for more exotic purposes such as fighting oil-well fires, doing nuclear inspections and cleanups, and checking for land mines and booby traps. An eight-legged, satellite-linked robot called Dante II was used to explore the inside of Mount Spurr, an active Alaskan volcano, sometimes without human guidance. A six-wheeled robot vehicle called Sojourner was used in NASA's 1997 Pathfinder exploration of Mars to sample the planet's atmosphere and soil and to radio data and photos back to Earth. A similar robot has been designed for use on the next Mars expedition.

Two Approaches to Artificial Intelligence: Weak versus Strong AI

What is the difference between weak and strong AI?

Two principal approaches to artificial intelligence are *weak AI* and *strong AI.*

WEAK AI *Weak AI* makes the claim that computers can be programmed to *simulate* human cognition—that some "thinking-like" features can be added to computers to make them more useful tools. We have already seen this kind of AI—call it "AI lite"—in expert systems, speech-recognition software, computer games, and the like.

A useful concept for considering weak AI is that of brute force. In programming and in AI, *brute force* is a technique for solving a complex problem by using a computer's fast processing capability to repeat a simple procedure many times. For example, a spelling checker in a word processing program doesn't really check the spelling of words; rather, it compares all the words you type into your document to a dictionary of correctly spelled words. Similarly, a chess-playing program will calculate all the possible moves that can apply to a given situation and then choose the best one; it will not analyze and strategize the way a human chess player would. Even IBM's Deep Blue program, which defeated Russian chess master Garry Kasparov in May 1997, basically took advantage of a supercomputer's fast processing abilities to examine 200 million possible plays per second.

STRONG AI *Strong AI* makes the claim that computers can be made to think on a level that is at least equal to humans and possibly even be conscious of themselves. So far, most AI advances have been piecemeal and single-purpose, such as factory robots, but proponents of strong AI believe that it's possible for computers to have the kind of wide-ranging, problem-solving ability that people have. Two different approaches are *Cyc* and *Cog.*

- **Cyc—stuffing a database with facts and rules:** *Cyc* (pronounced "psych") is a project based in Austin, Texas, that was launched by former Stanford University professor Douglas Lenat. Since 1984, teams of programmers, linguists, theologians, mathematicians, and philosophers have been feeding a database named Cyc 1.4 million basic truths and generalities about daily life. The idea, in one description, is to enable Cyc to "automatically make assumptions humans make: Creatures that die stay dead. Dogs have spines. Scaling a cliff requires intense physical effort."[35] The challenge, then, is to create a computer database that knows as much as a 12-year-old. Besides stuffing the database with straightforward facts, researchers have also instructed Cyc to ask questions if it needs clarity and to generalize as much as possible, based on what it already knows, until further generalization is false. Eventually, Cyc may learn on its own.

- **Cog—using sensory systems to search for patterns:** *Cog* takes a different approach. A project backed by MIT professor Rodney Brooks in Cambridge, Massachusetts, Cog doesn't follow predetermined sets of rules but rather tries to identify and search for patterns. Controlled by a network of many different microcontrollers, Cog is a humanoid robot with sensory systems that mimic the visual, hearing, and tactile systems of humans. The philosophy of the researchers is that the form of our bodies is important in the development of internal thought and language as we interact with our environments.

Some other forms of strong AI are described in the box at right. (● *See Panel 8.19.*)

AI Conversation

Want to talk to Cyc? Go to *www.itellibuddy.com.*

COGnition

For more about Cog, go to *www.ai.mit.edu/projects/ humanoid-robotics-group/ cog/cog.html.*

Neural Networks. *Neural networks* use physical electronic devices or software to mimic the neurological structure of the human brain. Because they are structured to mimic the rudimentary circuitry of the cells in the human brain, they learn from example and don't require detailed instructions.

To understand how neural networks operate, we may compare them to the operation of the human brain.

- *The human neural network:* The human body has a neural network consisting of neurons, or nerve cells. The neurons are connected by a three-dimensional lattice called *axons*. Electrical connections between neurons are activated by *synapses*. The human brain is made up of about 100 billion neurons. However, these cells do not act as "computer memory" sites. No cell holds a picture of your dog or the idea of happiness. You could eliminate any cell—or even a few million—in your brain and not alter your "mind." Where do memory and learning lie? In the electrical connections between cells, the synapses. Using electrical pulses, the neurons send on/off messages along the synapses.

- *The computer neural network:* In a hardware neural network, the nerve cell is replaced by a transistor, which acts as a switch. Wires connect the cells (transistors) with one another. The synapse is replaced by an electronic component called a *resistor*, which determines whether a cell should activate the electricity to other cells. A software neural network emulates a hardware neural network, although it doesn't work as fast. Computer-based neural networks use special AI software and complicated fuzzy-logic processor chips to take inputs and convert them to outputs with a kind of logic similar to human logic.

Neural networks are already being used in a variety of situations. At a San Diego hospital emergency room in which patients complained of chest pains, the same information was given to a neural-network program and to doctors. The network correctly diagnosed patients with heart attacks 97% of the time, compared to 78% for the human physicians. Banks that use neural-network software to spot irregularities in purchasing patterns within individual accounts often notice when a credit card is stolen before its owner does.

Genetic Algorithms. A *genetic algorithm* is a program that uses Darwinian principles of random mutation to improve itself. The algorithms are lines of computer code that act like living organisms. Different sections of code haphazardly come together, producing programs. As in Darwin's rules of evolution, many chunks of code compete to see which can best fulfill the goal of the program. Some chunks will even become extinct. Those that survive will combine with other survivors to produce offspring programs.

Expert systems can capture and preserve the knowledge of expert specialists, but they may be slow to adapt to change. Neural networks can sift through mountains of data and discover obscure causal relationships, but if there is too much data, or too little, they may be ineffective. Genetic algorithms, by contrast, use endless trial and error to learn from experience—to discard unworkable approaches and grind away at promising approaches with the kind of tireless energy of which humans are incapable.

In 2000, scientists at Brandeis University reached a major milestone when they created a computerized robot that designs and builds other robots, automatically evolving without any significant human intervention. The "robotic life forms" are only a few inches long and are composed of a few plastic parts with rudimentary nervous systems made of wire. They do only one thing: inch themselves, worm-like, along a horizontal surface, using miniature motors. With no idea what a successful design might look like, the Brandeis computer was given the goal of moving on a horizontal surface; a list of possible parts to work with; a group of 200 randomly constructed, nonworking designs; and the physical laws of gravity and friction. Mimicking evolution through classic survival-of-the-fittest selection, the computer changed pieces in the designs, mutated the programming instructions for controlling movements, and ran simulations to test the designs. After 300–600 generations of evolution, the computer sent the design to a machine to build the robot. The robots currently have the brainpower of bacteria, say researchers, who say they hope to get up to insect level in a couple of years.

Cyborgs. *Cyborgs* are hybrids of machine and organisms. One example is a hockey-puck-sized robot on wheels controlled by living tissue, an immature lamprey eel brain. The brain had been removed from the eel, kept alive in a special solution, and attached to the robot by wires. The brain is thus able to receive signals from the robot's electronic eyes and in turn can send commands to move the machine's wheels.

A number of cyborgs are in existence today—for instance, the estimated 10% of the American population with electronic heart pacemakers, artificial joints, implanted corneal lenses, and drug implant systems. Future cyborgs, however, might consist of bacteria attached to computer chips to map pollutants, insects used as parts of sensors to detect land mines and chemical weapons, and rodent brains to help identify new medicines. In his book *I, Cyborg*, British cybernetics professor Kevin Warwick described having an electrode in his arm pick up neural signals and send them to a computer, which converted them into instructions for a three-fingered robot hand elsewhere.

Artificial Life, the Turing Test, & AI Ethics

How could I tell if a computer actually achieved human intelligence?

What is life, and how can we replicate it from silicon chips, networks, and software? We are dealing now not with artificial intelligence but with artificial life. *Artificial life,* or *A-life,* is the field of study concerned with "creatures"—computer instructions, or pure information—that are created, replicate, evolve, and die as if they were living organisms. Thus, A-life software (such as LIFE) tries to simulate the responses of a human being.

Of course, "silicon life" does not have two principal attributes associated with true living things—it is not water- and carbon-based. Yet in other respects such creatures mimic life: If they cannot learn or adapt, then they perish.

Cyber human. With the help of new scanning and computing technologies, medical imaging technology can now show real-time activities inside the human body. One such system is 4-D Imaging System's "Cyber Human," discussed here by Dr. Suzuki of the National Museum of Emerging Science and Innovation in Tokyo.

HOW CAN WE KNOW A MACHINE IS TRULY INTELLIGENT?   How can we know when we have reached the point at which computers have achieved human intelligence? How will you know, say, whether you're talking to a human being on the phone or to a computer? Clearly, with the strides made in the fields of artificial intelligence and artificial life, this question is no longer just academic.

Interestingly, Alan Turing, an English mathematician and computer pioneer, addressed this very question in 1950. Turing predicted that by the end of the century computers would be able to mimic human thinking and to conduct conversations indistinguishable from a person's. Out of these observations came the *Turing test,* which is intended to determine whether a computer possesses "intelligence" or "self-awareness."

In the Turing test, a human judge converses by means of a computer terminal with two entities hidden in another location—one a person typing on a keyboard, the other a computer program. Following the conversation, the judge must decide which entity is human. In this test, intelligence—the ability to think—is demonstrated by the computer's success in fooling the judge. (● *See Panel 8.20.*)

● PANEL 8.20
The Turing test

"**Computer:** As a psychoanalyst, I am familiar with discussing such marital troubles with you as the patient on the couch. If you don't have a bad marriage just pretend that you do. Please start with 'My wife . . .' or 'My husband . . .'.
Judge: OK, I'll pretend I have a bad marriage. Let's say the reason is because he never picks up his dirty underwear. Is this marriage irreparable?
Computer: Well, if you think I'm hopeless, you can always just swear at me."

—Judith Anne Gunther, "An Encounter with A.I.," *Popular Science*

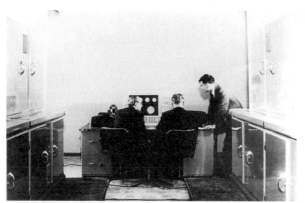

Alan Turing (*right*) in 1951, working on the Mark I computer.

info!

Take a Few Captcha Tests

To get a feel for captchas, go to *www.captcha.net.*

Judith Anne Gunther participated as one of eight judges in the third annual Loebner Prize Competition, which is based on Turing's ideas.[36] (There have been other competitions since.) The "conversations"—each limited to 15 minutes—are restricted to predetermined topics, such as baseball, because even today's best programs have neither the databases nor the syntactical ability to handle an unlimited number of subjects.

Gunther found that she wasn't fooled by any of the computer programs. The winning program, for example, relied as much on deflection and wit as it did on responding logically and conversationally. (For example, to a judge trying to discuss a federally funded program, the computer said: "You want logic? I'll give you logic: shut up, shut up, shut up, shut up, shut up, now go away! How's that for logic?") However, Gunther *was* fooled by one of the five humans, a real person discussing abortion. "He was so uncommunicative," wrote Gunther, "that I pegged him for a computer."

A new sort of Turing test, one that's simple for humans but that will baffle sophisticated computer programs, has been devised in the form of cognitive puzzles called *Captchas,* an acronym for "Completely Automated Public Turing Test to Tell Computers and Humans Apart."[37] Captchas are most commonly used to prevent rogue bots from spamming comments or signing up for web services. Yahoo, for example, uses a Captcha for screening when you sign up for an account.

ETHICS IN AI Behind everything to do with artificial intelligence and artificial life—just as it underlies everything we do—is the whole matter of ethics. In his book *Ethics in Modeling,* William A. Wallace, professor of decision sciences at Rensselaer Polytechnic Institute, points out that computer software, including expert systems, is often subtly shaped by the ethical judgments and assumptions of the people who create it.[38] In one instance, he notes, a bank had to modify its loan-evaluation software on discovering that the software rejected certain applications because it unduly emphasized old age as a negative factor. Another expert system, used by health maintenance organizations (HMOs), tells doctors when they should opt for expensive medical procedures, such as magnetic resonance imaging tests. HMOs like such systems because they help control expenses, but critics are concerned that doctors will have to base decisions not on the best medicine but simply on "satisfactory" medicine combined with cost constraints.[39]

Clearly, there is no such thing as completely "value-free" technology. Human beings build it, use it, and have to live with the results.

8.8

The Ethics of Using Databases: Concerns about Privacy & Identity Theft

What are two ethical concerns about the uses of databases?

If you're not long out of high school, you may have found that you were made part of a database without your consent, the U.S. Defense Department's database on 16- to 25-year-olds, designed to help recruiters find young people to join the military. The detailed information in that database includes Social Security numbers, grade-point averages, fields of study, ethnicity, and email addresses. Were you surprised to find this out? Fortunately, you can request, if you're 18 or older (as can your parents, if you're 16 or 17), that personal information not be used for recruitment purposes, although the information will remain in the database.[40]

The Threat to Privacy

What are some ways my privacy is being threatened?

Privacy is the right of people not to reveal information about themselves. Who you vote for in a voting booth and what you say in a letter sent through the U.S. mail are private matters. However, the ease of pulling together information from databases and disseminating it over the internet has put privacy under extreme pressure. Many people are worried about the loss of their right to privacy, fearing they will lose all control of the personal information being collected and tracked by computers. There are good reasons for this concern, as the following discussion shows.

NAME MIGRATION   As you've no doubt discovered, it's no trick at all to get your name on all kinds of mailing lists. Theo Theoklitas, for instance, received applications for credit cards, invitations to join video clubs, and notification of his finalist status in Ed McMahon's $10 million sweepstakes. Theo is a black cat who's been getting mail ever since his owner sent in an application for a rebate on cat food.

Once you're in one database, clearly, your name seems to migrate to others. In a recent year, retired language-school owner Paul Kameny, 64, of San Francisco, along with his wife and two grown children, received 217 credit card solicitations, or about four a week, worth somewhere between $2.7 million and $8.5 million in credit. "This is criminal," said Kameny. "Banks are doing everything they can to saddle people with debt."[41]

RÉSUMÉ RUSTLING & ONLINE SNOOPING   When you post your résumé on an internet job board, you assume that it's private, and indeed it is supposed to be restricted to recruiters and other employers. But job boards that are just starting up may poach résumés from other job websites as a way of building up their own databases.[42] Another problem is that much of your life may already be online and available to anyone. For instance, Camberley Crick, 24, a part-time computer tutor, went to the apartment of a prospective client and discovered he had pulled all kinds of facts about her simply by typing her name into the search engine Google. Among the things he found: "her family Web site, a computer game she had designed for a freshman college class, a program from a concert she had performed in, and a short story she wrote in elementary school called 'Timmy the Turtle.'"[43]

GOVERNMENT PRYING & SPYING   As part of its war on terrorism, the government has been taking steps that make defenders of privacy worried. One, for instance, was a controversial data-mining project for the U.S. Defense Department's Total Information Awareness program that would have compiled electronic dossiers on Americans. The U.S. Senate forced restrictions on the TIA program, saying that it would amount to a domestic spying apparatus.[44]

Another concern to some is the Computer Assisted Passenger Prescreening Program (CAPPS II) designed to screen out supposedly dangerous commercial airline passengers. This program involves looking through commercial and government data on all air travelers without their knowledge or permission and using the information to give every flyer a security-risk ranking.[45]

Librarians and booksellers worry that the USA PATRIOT Act, passed by Congress six weeks after the September 11, 2001, attacks, allows FBI agents (through Section 215 of the act) to obtain a warrant from a secret federal court for library or bookstore records of anyone connected to an investigation of international terrorism or spying. The law prohibits librarians and booksellers from telling patrons that the FBI has requested their records. Unlike conventional search warrants, agents need not show that the target is suspected of a crime or possesses evidence of a crime. Nearly 60% of 906 librarians surveyed in a University of Illinois study felt that the law went too far.[46]

Freedom of Information Act (1970)
Fair Credit Reporting Act (1970)
Privacy Act (1974)
Family Educational Rights and Privacy Act (1974)
Right to Financial Privacy Act (1978)
Privacy Protection Act (1980)
Cable Communications Policy Act (1984)
Computer Fraud and Abuse Act (1984)
Electronic Communications Privacy Act (ECPA) (1986)
Computer Security Act (1987)
Computer Matching and Privacy Protection Act (1988)
Video Privacy Protection Act (1988)
Telephone Consumer Protection Act (1988)
Cable Act (1992)
Computer Abuse Amendments Act (1994)
Communications Assistance for Law Enforcement Act (1994)
National Information Infrastructure Protection Act (1996)
No Electronic Theft (NET) Act (1997)
Children's Online Privacy Protection Act (1998)
Identity Theft and Assumption Deterrence Act (1998)
Digital Millennium Copyright Act (1998)
Financial Services Modernization Act (1999)
Children's Online Privacy Protection Act (COPPA) (2000)
USA PATRIOT Act (2001)
Health Insurance Portability and Accountability Act (2003)

■ more info!

Checking Your Personal Records

Browse the full text of FBI documents and files released under the Freedom of Information Act at *http//foia.fbi.gov/room.htm.*

For instructions on how to check and repair your personal credit reports, go to *www.fair-credit-reporting.com.*

For more information on the privacy of personal health records, go to the U.S. Department of Health and Human Services at *www.hhs.gov/ocr/hipaa.*

PRIVACY LAWS   Over the years, concerns about privacy have led to the enactment of a number of laws to protect individuals from invasions of privacy. (● *See Panel 8.21.*) Still, the tension between security and privacy worries many observers. Recently, an idea has been to use *cost-benefit analysis* to weigh the benefits of tighter domestic security against the costs of lost privacy and freedom. The idea has attracted an unusual array of supporters, ranging from conservatives who fret about "big government" to civil-liberties lawyers to consumer advocate Ralph Nader.[47]

A NATIONAL ID CARD?   Demand for photo identification has transformed driver's licenses into a primary identification document, with a number of states now using digital fingerprinting and facial-recognition security systems for their licenses. Although there is no present nationwide identity card, Congress has required (under the 2002 Enhanced Border Security and Visa Entry Reform Act) the definition of a standard under which the State Department and immigration service can issue machine-readable documents that can be used at most porous border entry points.[48] Foreign governments are also required to use improved identity-system (biometric) technology in their passports.[49]

Does this mean that we're on our way to some national ID card, a sort of internal passport like those carried by citizens in some European countries? Critics from Republican conservatives to civil-liberties defenders so dislike the idea that even efforts to get state governments to standardize driver's licenses have been met with extreme resistance. One answer, perhaps, is to have a voluntary nationally accepted ID card: Private companies would issue cards that adhere to strict government standards, and anyone who signs up for one would be checked against government criminal and watch-list databases that are constantly updated.[50] Still, Congress has weighed legislation in which applicants for driver's licenses would be required to show four types of identification, such as a photo ID, a birth certificate or passport, a Social Security card, or some other document that would prove name, address, and residency. The purpose would be to turn state issued driver's licenses into a kind of national identity card.[51]

Identity Theft

What's the definition of identity theft?

Identity theft, or _theft of identity (TOI)_, is a crime in which thieves hijack your name and identity and use your good credit rating to get cash or buy things. Anywhere from 3.2 million to 10 million people are victimized by this crime every year, and losses run into the hundreds of millions of dollars annually.[52] However, as one victim noted, these are just statistics. "It's a whole other thing when you're suddenly faced with the reality of a stranger living your life."[53] Often this begins with someone getting hold of your Social Security number. We discuss identity theft further in the Experience Box.

Bank of America Higher Standards   Online Banking

Here's How SiteKey Works

By passing back and forth secret information that only you and Bank of America know, you can feel even more secure with your Online Banking experience. We recognize you and you recognize us.

Online Banking Sign In
View demo | Learn more | Enroll

Enter Online ID

☐ Save this online ID

Sign In

1 Enter your online ID.

2 Click **Sign In.**

Your SiteKey Image and Message:

cute dog

3 If we recognize your computer:
We will show you your secret SiteKey.

What was your high school mascot?

* Answer:

4 If we don't recognize your computer:
We will ask you one of your secret SiteKey Confirmation Questions.
After you answer your question correctly, we will show you your SiteKey.

Passcode: ********

5 Once you view your valid SiteKey, you can then safely enter your Passcode and continue onto your Online Banking account.

To take advantage of this new security feature now, sign in to **Online Banking** and select the **Learn more and sign up for SiteKey** link from either the Accounts Overview page or Customer Service tab. Then follow the simple on-screen instructions to set up your SiteKey and Confirmation Questions.

Return to sign in page

Bank of America, N.A. Member FDIC. Equal Housing Lender
© 2005 Bank of America Corporation. All rights reserved.

Experience Box
Preventing Your Identity from Getting Stolen

One day, Kathryn Rambo, 28, of Los Gatos, California, learned that she had a new $35,000 sports utility vehicle listed in her name, along with five credit cards, a $3,000 loan, and even an apartment—none of which she'd asked for. "I cannot imagine what would be weirder, or would make you angrier, than having someone pretend to be you, steal all this money, and then leave you to clean up all their mess later," said Rambo, a special-events planner.[54] Added to this was the eerie matter of constantly having to prove that she was, in fact, herself: "I was going around saying, 'I am who I am!'"[55]

Identity Theft: Stealing Your Good Name— and More

Theft of identity (TOI) is a crime in which thieves hijack your very name and identity and use your good credit rating to get cash or to buy things. To begin, all they need is your full name and Social Security number. Using these, they tap into internet databases and come up with other information—your address, phone number, employer, driver's license number, mother's maiden name, and so on. Then they're off to the races, applying for credit everywhere.

In Rambo's case, someone had used information lifted from her employee-benefits form. The spending spree went on for months, unbeknownst to her. The reason it took so long to discover the theft was that Rambo never saw any bills. They went to the address listed by the impersonator, a woman, who made a few payments to keep creditors at bay while she ran up even more bills. For Rambo, straightening out the mess required months of frustrating phone calls, time off from work, court appearances, and legal expenses. One survey found that a third of victims of identity theft had still been unable to repair their tainted identities even a year after the information was stolen, and most victims on average spent 81 hours trying to resolve their cases.[56]

How Does Identity Theft Start?

Identity theft typically starts in one of several ways:[57]

- **Wallet or purse theft:** There was a time when a thief would steal a wallet or purse, take the cash, and toss everything else. No more. Everything from keys to credit cards can be parlayed into further thefts.

- **Mail theft:** Thieves also consider mailboxes fair game. The mail will yield them bank statements, credit card statements, new checks, tax forms, and other personal information. (So it's best to receive your mail in a locked box—and mail your outgoing letters at a postbox.)

- **Mining the trash:** You might think nothing of throwing away credit card offers, portions of utility bills, or old canceled checks. But "dumpster diving" can produce gold for thieves. Credit card offers, for instance, may have limits of $5,000 or so.

- **Telephone solicitation:** Prospective thieves may call you up and pretend to represent a bank, credit card company, government agency, or the like in an attempt to pry loose essential data about you.

- **Insider access to databases:** You never know who has, or could have, access to databases containing your personnel records, credit records, car-loan applications, bank documents, and so on. This is one of the harder TOI methods to guard against.

- **Outsider access to databases:** In recent years, data burglars and con artists have tapped into big databases at such companies as ChoicePoint, Reed Elsevier, DSW Shoe Warehouse, Bank of America/ Wachovia. Other institutions, such as CitiFinancial, the University of California at Berkeley, Time Warner, and MCI have exposed files of data through carelessness. These losses have put millions of people at risk.[58]

What to Do Once Theft Happens

If you're the victim of a physical theft (or even loss), as when your wallet is snatched, you should immediately contact—first by phone, and then in writing—all your credit card companies, other financial institutions, the Department of Motor Vehicles, and any other organizations whose cards you use that are now compromised. Be sure to call utility companies—telephone, electricity, and gas; identity thieves can run up enormous phone bills. Also call the local police and your insurance company to report the loss. File a complaint with the Federal Trade Commission (1-877-ID-THEFT, or *www.ftc.gov*), which maintains an ID theft database.

It's important to notify financial institutions within 2 days of learning of your loss because then you are legally responsible for only the first $50 of any theft. If you become aware of fraudulent transactions, immediately contact the fraud units of the three major credit bureaus: Equifax, Experian, and TransUnion. (See Panel 8.22.)

● PANEL 8.22
The three major credit bureaus

Equifax	Experian	TransUnion
To check your credit report: 800-685-1111	To check your credit report: 800-397-3742	To check your credit report: 800-888-4213
www.equifax.com	*www.experian.com*	*www.tuc.com*

If your Social Security number has been fraudulently used, alert the Social Security Administration (800-772-1213).

If you have a check guarantee card that was stolen, if your checks have been lost, or if a new checking account has been opened in your name, there are two organizations to notify so that payment on any fraudulent checks will be denied. They are Telecheck (800-366-5010, or *www.telecheck.com*) and National Processing Company (800-526-5380, or *www.npc.net*).

If your mail has been used for fraudulent purposes or if an identity thief filed a change of address form, look in the phone directory under U.S. Government Postal Service for the local postal inspector's office.

In some states, you can freeze your credit reports, preventing lenders and others from reviewing your credit history, which will prevent identity thieves from opening fraudulent accounts using your name.

How to Prevent Identity Theft

One of the best ways to keep your finger on the pulse of your financial life is, on a regular basis—once a year, say—to get a copy of your credit report from one or all three of the main credit bureaus. This will show you whether there is any unauthorized activity. Reports are free from the major credit bureaus (Equifax, Experian, and Transunion). To learn more, go to *www.annualcreditreport.com* or call toll-free 877-322-8228.

In addition, there are some specific measures you can take to guard against personal information getting into the public realm.

- ***Check your credit card billing statements:*** If you see some fraudulent charges, report them immediately. If you don't receive your statement, call the creditor first. Then call the post office to see if a change of address has been filed under your name.

- ***Treat credit cards and other important papers with respect:*** Make a list of your credit cards and other important documents and a list of numbers to call if you need to report them lost. (You can photocopy the cards front and back, but make sure the numbers are legible.)

Carry only one or two credit cards at a time. Carry your Social Security card, passport, or birth certificate only when needed.

Don't dispose of credit card receipts in a public place.

Don't give out your credit card numbers or Social Security number over the phone, unless you have some sort of trusted relationship with the party on the other end.

Tear up credit card offers before you throw them away. Even better, buy a shredder and shred these and other such documents.

Keep a separate credit card for online transactions, so that if anything unusual happens you're more likely to notice it and you can more easily close the account.

Keep tax records and other financial documents in a safe place.

- ***Treat passwords with respect:*** Memorize passwords and PINs. Don't use your birth date, mother's maiden name, or similar common identifiers, which thieves may be able to guess.

- ***Treat checks with respect:*** Pick up new checks at the bank. Shred canceled checks before throwing them away. Don't let merchants write your credit card number on the check.

- ***Watch out for "shoulder surfers" when using phones and ATMs:*** When using PINs and passwords at public telephones and automated teller machines, shield your hand so that anyone watching through binoculars or using a video camera—"shoulder surfers"—can't read them.

Summary

artificial intelligence (p. 440) Group of related technologies used for developing machines to emulate human qualities, such as learning, reasoning, communicating, seeing, and hearing. Why it's important: *Today the main areas of AI are expert systems, natural language processing, intelligent agents, pattern recognition, fuzzy logic, virtual reality and simulation devices, and robotics.*

business-to-business (B2B) system (p. 430) Direct sales between businesses that involve using the internet or a private network to cut transaction costs and increase efficiencies. Why it's important: *Business-to-business activity helps business by moving beyond pricing mechanisms and encompassing product quality, customer support, credit terms, and shipping reliability, which often count for more than price.*

business-to-consumer (B2C) system (p. 431) System in which a business sells goods or services to consumers, or members of the general public. An example is Amazon.com. Why it's important: *This kind of e-commerce system essentially removes the middleman and often even the need for a physical ("bricks-and-mortar") store.*

character (p. 409) Also called *byte;* a single letter, number, or special character. Why it's important. *Characters—such as A, B, C, 1, 2, 3, #, $, %—are part of the data storage hierarchy.*

compression (p. 412) Method of removing repetitive elements from a file so that the file requires less storage space, then later decompressing the removed data, or restoring the repeated patterns. Why it's important: *Compression/decompression makes storage and transmission of large files, such as multimedia files, more feasible.*

consumer-to-consumer (C2C) system (p. 431) System in which consumers sell goods or services directly to other consumers, often with the help of a third party, such as eBay. Why it's important: *The advantage of C2C e-commerce is most often the reduced costs and a smaller but profitable customer base. It also gives many small business owners a way to sell their goods without running a costly bricks-and-mortar store.*

data dictionary (p. 415) Also called *repository;* a procedures document or disk file that stores data definitions and descriptions of database structure. It may also monitor new entries to the database as well as user access to the database. Why it's important: *The data dictionary monitors the data being entered to make sure it conforms to the rules defined during data definition. The data dictionary may also help protect the security of the database by indicating who has the right to gain access to it.*

data files (p. 411) Files that contain data—words, numbers, pictures, sounds, and so on. Why it's important: *Unlike program files, data files don't instruct the computer to do anything. Rather, data files are there to be acted on by program files. Examples of common extensions in data files are .txt (text) and .xls (spreadsheets). Certain proprietary software programs have their own extensions, such as .ppt for PowerPoint and .mdb for Access.*

Meta-data: shows transformations and summarization of data, contents of data warehouse, and origins of data

Data transport. load data and meta-data into warehouse periodically

data mining (DM) (p. 424) Computer-assisted process of sifting through and analyzing vast amounts of data in order to extract hidden patterns and meaning and to discover new knowledge. Why it's important: *The purpose of DM is to describe past trends and predict future trends. Thus, data-mining tools might sift through a company's immense collections of customer, marketing, production, and financial data and identify what's worth noting and what's not.*

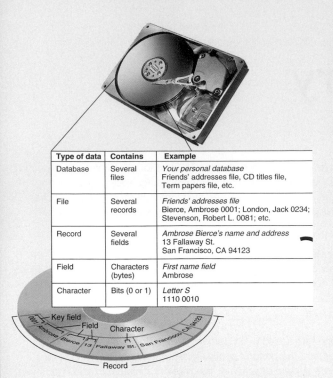

Type of data	Contains	Example
Database	Several files	*Your personal database* Friends' addresses file, CD titles file, Term papers file, etc.
File	Several records	*Friends' addresses file* Bierce, Ambrose 0001; London, Jack 0234; Stevenson, Robert L. 0081; etc.
Record	Several fields	*Ambrose Bierce's name and address* 13 Fallaway St. San Francisco, CA 94123
Field	Characters (bytes)	*First name field* Ambrose
Character	Bits (0 or 1)	*Letter S* 1110 0010

data storage hierarchy (p. 409) The levels of data stored in a computer database: bits, bytes (characters), fields, records, and files. Why it's important: *Understanding the data storage hierarchy is necessary to understand how to use a database.*

data warehouse (p. 425) A database containing cleaned-up data and meta-data (information about the data) stored using high-capacity-disk storage technology. Why it's important: *Data warehouses combine vast amounts of data from many sources in a database form that can be searched, for example, for patterns not recognizable with smaller amounts of data.*

database (p. 408) Logically organized collection of related data designed and built for a specific purpose, a technology for pulling together facts that allows the slicing and dicing and mixing and matching of data. Why it's important: *Businesses and organizations build databases to help them keep track of and manage their affairs. In addition, online database services put enormous research resources at the user's disposal.*

database administrator (DBA) (p. 416) Person who coordinates all related activities and needs for an organization's database. Why it's important: *The DBA determines user access privileges; sets standards, guidelines, and control procedures; assists in establishing priorities for requests; prioritizes conflicting user needs; develops user documentation and input procedures; and oversees the system's security.*

database management system (DBMS) (p. 414) Also called *database manager;* software that controls the structure of a database and access to the data. It allows users to manipulate more than one file at a time. Why it's important: *This software enables sharing of data (same information is available to different users); economy of files (several departments can use one file instead of each individually maintaining its own file, thus reducing data redundancy, which in turn reduces the expense of storage media and hardware); data integrity (changes made in the files in one department are automatically made in the files in other departments); and security (access to specific information can be limited to selected users).*

DBMS utilities (p. 416) Programs that allow users to maintain databases by creating, editing, and deleting data, records, and files. Why it's important: *DBMS utilities allow people to establish what is acceptable input data, to monitor the types of data being input, and to adjust display screens for data input.*

decision support system (DSS) (p. 437) Computer-based information system that helps managers with nonroutine decision-making tasks. Inputs consist of some summarized reports, some processed transaction data, and other internal data plus data from sources outside the organization. The outputs are flexible, on-demand reports. Why it's important: *A DSS is installed to help top managers and middle managers make strategic decisions about unstructured problems.*

e-commerce (p. 428) Electronic commerce; the buying and selling of products and services through computer networks. Why it's important: *U.S. e-commerce and online shopping are growing even faster than the increase in computer use.*

executive support system (ESS) (p. 438) Also called an *executive information system (ESS);* DSS made especially for top managers. It draws on data from both inside and outside the organization. Why it's important: *The ESS includes capabilities for analyzing data and doing "what if" scenarios to help with strategic decision making.*

expert system (p. 439) Also called *knowledge-based system;* set of interactive computer programs that helps users solve problems that would otherwise require the assistance of a human expert. Expert systems are created on the basis of knowledge collected on specific topics from human specialists, and they imitate the reasoning process of a human being. Why it's important: *Expert systems are used by both management and nonmanagement personnel to solve specific problems, such as how to reduce production costs, improve workers' productivity, or reduce environmental impact.*

field (p. 409) Unit of data consisting of one or more characters (bytes). Examples of fields are your first name, your street address, or your Social Security number. Why it's important: *A collection of fields makes up a record.*

file (p. 410) Collection of related records. An example of a file is a stored listing of everyone employed in the same department of a company, including all names, addresses, and Social Security numbers. Why it's important: *A file is the collection of data or information that is treated as a unit by the computer; a collection of related files makes up a database.*

filename (p. 411) The name given to a file. Why it's important: *Files are given names so that they can be differentiated. Filenames also have extension names of up to three or four letters added after a period following the filename.*

fuzzy logic (p. 443) Method of dealing with imprecise data and uncertainty, with problems that have many answers rather than one. Why it's important: *Unlike "crisp," yes/no digital logic, fuzzy logic deals with probability and credibility.*

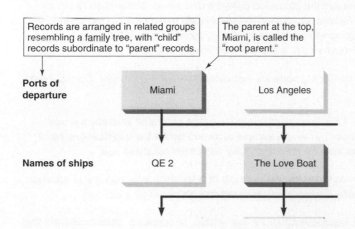

Records are arranged in related groups resembling a family tree, with "child" records subordinate to "parent" records.

The parent at the top, Miami, is called the "root parent."

Ports of departure — Miami — Los Angeles

Names of ships — QE 2 — The Love Boat

hierarchical database (p. 417) Database in which fields or records are arranged in related groups resembling a family tree, with child (lower-level) records subordinate to parent (higher-level) records. The parent record at the top of the database is called the *root record*. Why it's important: *The hierarchical database is one of the common database structures.*

identity theft (p. 452) Also called *theft of identity (TOI)*; crime in which thieves hijack a person's name and identity and use his or her good credit rating to get cash or buy things. Why it's important: *Identity theft is on the rise in developed countries; it can ruin a person's life for many years.*

intelligent agent (p. 442) Software with built-in intelligence that monitors work tasks, asks questions, and performs work tasks, such as roaming networks, on the user's behalf. Why it's important: *An intelligent agent can filter messages, scan news services, travel over communications line to databases, and collect files to add to a personal database.*

management information system (MIS) (p. 436) Computer-based information system that uses data recorded by TPS as input into programs that produce summary, exception, periodic, and on-demand reports of the organization's performance. Why it's important: *An MIS principally assists middle managers, helping them make tactical decisions—spotting trends and getting an overview of current business activities.*

multidimensional database (MDA) (p. 423) Type of database that models data as facts, dimensions, or numerical measures for use in the interactive analysis of large amounts of data for decision-making purposes. A multidimensional database uses the idea of a cube to represent the dimensions of data available to a user, using up to four dimensions. Why it's important: *A multidimensional database allows users to ask questions in colloquial English.*

natural language processing (p. 442) Study of ways for computers to recognize and understand human language, whether in spoken or written form. Why it's important: *Natural languages make it easier to work with computers.*

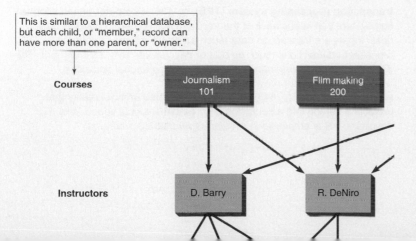

This is similar to a hierarchical database, but each child, or "member," record can have more than one parent, or "owner."

Courses — Journalism 101 — Film making 200

Instructors — D. Barry — R. DeNiro

network database (p. 418) Database similar in structure to a hierarchical database; however, each child record can have more than one parent record. Thus, a child record, which in network database terminology is called a *member,* may be reached through more than one parent, which is called an *owner*. Why it's important: *The network database is one of the common database structures.*

object-oriented database (p. 422) Database that uses "objects," software written in small, reusable chunks, as elements within database files. An object consists of (1) data in any form, including graphics, audio, and video, and (2) instructions for the action to be taken on the data. Why it's important: *A hierarchical or network database might contain only numeric and text data. By contrast, an object-oriented database might also contain photographs, sound bites, and video clips. Moreover, the object would store operations, called methods, the programs that objects use to process themselves.*

office automation system (OAS) (p. 435) Also called *office information system (OIS);* computer information system that combines various technologies to reduce the manual labor needed to operate an office efficiently and increase productivity; used at all levels of an organization. Why it's important: *An OAS uses a network to integrate such technologies as fax, voice mail, email, scheduling software, word processing, and desktop publishing and make them available throughout the organization.*

pattern recognition (p. 442) Use of camera and software to identify recurring patterns and to recognize the connections between the perceived patterns and similar patterns stored in a database. Why it's important: *Pattern recognition is used in data mining to discover previously unnoticed patterns; in facial recognition software to identify faces; and in handwriting recognition, fingerprint identification, robot vision, and automatic voice recognition.*

program files (p. 411) Files containing software instructions. Why it's important: *Contrast **data files.***

query by example (QBE) (p. 420) Feature of query-language programs whereby the user asks for information in a database by using a sample record to define the qualifications he or she wants for selected records. Why it's important: *QBE simplifies database use.*

record (p. 410) Collection of related fields. An example of a record is your name and address and Social Security number. Why it's important: *Related records make up a file.*

relational database (p. 419) Database structure that relates, or connects, data in different files through the use of a key field, or common data element. In this arrangement there are no access paths down through a hierarchy. Instead, data elements are stored in different tables made up of rows and columns. In database terminology, the tables are called *relations* (files), the rows are called *tuples* (records), and the columns are called *attributes* (fields). All related tables must have a key field that uniquely identifies each row; that is, the key field must be in all tables. Why it's important: *The relational database is one of the common database structures; it is more flexible than hierarchical and network database models.*

report generator (p. 416) In a database management system, a program users can employ to produce on-screen or printed-out documents from all or part of a database. Why it's important: *Report generators allow users to produce finished-looking reports without much fuss.*

robotics (p. 443) Development and study of machines that can perform work normally done by people. Why it's important: *See **robot.***

simulator (p. 443) Device that represents the behavior of physical or abstract systems. Why it's important: *Virtual-reality simulation technologies are widely applied for training purposes.*

structured query language (SQL) (p. 420) Standard language used to create, modify, maintain, and query relational databases. Why it's important: *SQL simplifies database use.*

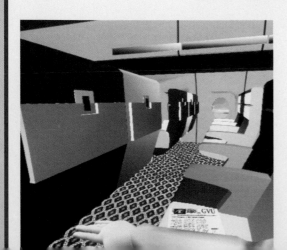

transaction processing system (TPS) (p. 436) Computer-based information system that keeps track of the transactions needed to conduct business. Inputs are transaction data (such as bill or orders). Outputs are processed transactions (such as bills or paychecks). Why it's important: *The TPS helps supervisory managers in making operational decisions.*

virtual reality (VR) (p. 443) Computer-generated artificial reality that projects a person into a sensation of three-dimensional space. Why it's important: *VR is employed in simulators for training programs.*

Chapter Review

"I can recognize and recall information."

Self-Test Questions

1. According to the data storage hierarchy, databases are composed of _____, _____, _____, _____, and _____.

2. An individual piece of data within a record is called a _____.

3. A(n) _____ coordinates all activities related to an organization's database.

4. _____ is the right of people not to reveal information about themselves.

5. The five types of database models are _____, _____, _____, _____, and _____.

6. _____ files contain software instructions; _____ files contain data.

7. _____ is done so that files require less storage space and take less time to transmit.

8. A single letter or number is considered a _____.

9. A _____ is a logically organized collection of related data designed and built for a specific purpose.

10. A _____ is a collection of related fields.

11. A(n) _____ is an automatic device that performs functions ordinarily performed by human beings.

12. The goal of _____ is to enable the computer to communicate with the user in the user's native language.

13. _____ is a group of related technologies used for developing machines to emulate human qualities such as learning, reasoning, communicating, seeing, and hearing.

14. Devices that represent the behavior of physical or abstract systems are called _____.

15. When unauthorized persons use your name to get cash and buy things, it is called _____.

16. Most organizations have six departments: _____, _____, _____, _____, _____, and _____.

17. _____ is used to create, modify, maintain, and query relational databases.

Multiple-Choice Questions

1. Which of the following is *not* an advantage of a DBMS?
   a. file sharing
   b. reduced data redundancy
   c. increased data redundancy
   d. improved data integrity
   e. increased security

2. Which of the following database models stores data in any form, including graphics, audio, and video?
   a. hierarchical
   b. network
   c. object-oriented
   d. relational
   e. offline

3. What are *.exe, .bas, .pas, .jav, .dll, .sys, pdf.,* and *.vbx* examples of?
   a. programming languages
   b. file extensions
   c. databases
   d. data files
   e. master files

4. Which of the following qualities are necessary for information to be "good"?
   a. concise
   b. complete
   c. verifiable
   d. current
   e. all of these

5. Which of the following database models relates, or connects, data in different files through the use of a key, or common data element?
   a. hierarchical
   b. network
   c. object-oriented
   d. relational
   e. offline

6. Which of the following areas of study is not included in AI?
   a. pattern recognition
   b. fuzzy logic
   c. transaction processing systems
   d. natural language processing
   e. robotics

7. Which of the following levels is not a management level?

    a. strategic

    b. tactical

    c. hierarchical

    d. operational

    e. all of these are management levels

True/False Questions

T F 1. The use of key fields makes it easier to locate a record in a database.

T F 2. A network database models data as facts, dimension, or numerical measures for use in interactive analysis of large amounts of data for decision-making purposes.

T F 3. A database is an organized collection of integrated files.

T F 4. Data mining is used only for small databases.

T F 5. A character is smaller than a field.

T F 6. A report generator is a machine that produces electricity by calculating complex algorithms.

T F 7. A data dictionary defines the basic organization of the database and contains a list of all files in the database.

T F 8. Most types of storage media last indefinitely.

T F 9. Expert systems are not interactive.

T F 10. You need only software to create virtual reality.

T F 11. A knowledge base is part of a natural language system.

T F 12. Program files contain all the documents you have created.

 stage **LEARNING** COMPREHENSION

"I can recall information in my own terms and explain it to a friend."

Short-Answer Questions

1. Name three responsibilities of a database administrator.

2. What is data mining?

3. Briefly explain what a data warehouse is.

4. List four basic advantages provided by database management systems.

5. What is the difference between a data file and a program file?

6. What are expert systems used for?

7. What is SQL? QBE? Oracle? Access? DB2?

8. Explain e-commerce.

9. What are some things you can do if identity theft happens to you?

10. What are two problems regarding long-term digital storage?

11. What are intelligent agents used for?

12. What do you need to experience virtual reality?

13. What is the main difference between weak AI and strong AI?

14. What are the four main areas of artificial intelligence?

15. What is a DSS used for? An ESS?

 stage **LEARNING** APPLYING, ANALYZING, SYNTHESIZING, EVALUATING

"I can apply what I've learned, relate these ideas to other concepts, build on other knowledge, and use all these thinking skills to form a judgment."

Knowledge in Action

1. Interview someone who works with or manages an organization's database. What types of records make up the database? Which departments use it? What database structure is used? What are the types and sizes of storage devices? Are servers used? Was the database software custom-written?

2. Are you comfortable with giving away some of your privacy for increased security? Why or why not? How far would you let the government go in examining people's private lives?

3. Can identity theft be prevented? Propose some new solutions to this problem.

4. If you could design a robot, what kind would you create? What would it do?

5. If you could build an expert system, what would it do? What kinds of questions would you ask experts in order to elicit the appropriate information?

6. Should internet-purchased products or services have a sales tax (like items purchased in regular stores)? Why or why not?

7. Do you know someone who refuses to use the internet in any manner that requires entering any personal information whatsoever? Give some reasons for disagreeing and for agreeing with this person.

Web Exercises

1. Have you ever thought of starting your own B2C business? Visit *www.cio.com/ec/edit/b2cabc.html* and read what the three major obstacles are for a successful B2C business. Search for other articles on this topic. If you decide to go ahead with your venture, will you use Pay-Pal? Why or why not? Search for answers.

2. Visit these sites about data mining and the search for meaningfulness in large quantities of data:

   www.dmbenchmarking.com
   www.dmg.org
   www.spss.com/datamine
   www.dwreview.com/Data_mining/index.html

   What do you think the next technology will be for handling and analyzing massive amounts of data?

3. Grocery store loyalty cards ask consumers to trade away extremely detailed personal information in return for the promise of savings—which may in fact not exist. Some experts say the reason that stores offer cards is so that they can profile and target their customers more accurately—not to give you savings but to increase their bottom line. Also, your personal information can be sold or traded to third parties.

   One way that grocery stores create customer profiles is by the use of data mining. There are positive aspects to data mining, such as fraud detection, but there is a darker side to data mining, too.

   Stores may use the address information from your loyalty card application to match up your shopping history with data from other databases or public records (income, how much you paid for your house) so that the store knows what kinds of specials to offer you. Information about your shopping habits can be accessed with a subpoena or warrant and used against you in court proceedings. In a "trip-and-fall" case in California, a man shopping at a Southern California grocery store sued after falling in one of the aisles. It was reported (although the store denied it) that the store threatened to use his shopping history—which included large amounts of alcohol—against him in the proceedings.

   Some states limit the types of information that a grocery store can collect from you when you register for a loyalty card. For example, California state law prohibits a grocery store from requiring that you turn over your social security card or your driver's license number. However, data matching techniques mean that this provides very little protection to your privacy rights.

   Find out more about loyalty cards. Start with the *Seattle Press* article at *www.seattlepress.com/article-9715.html*. Then read the *BusinessWeek* article at *www.businessweek.com/bwdaily/dnflash/jun2002/nf20020620_7007.htm*. Find a few more articles and then write a short report on the ethical aspects and the practical aspects of using loyalty cards.

4. How much information about you is out there? Run various search strings about yourself to see just how private your life is.

5. An extensive database featuring photographs and detailed information of each inmate of the Florida Department of Corrections can be found at the following website:

   www.dc.state.fl.us/inmateinfo/inmateinfomenu.asp

   Click on Search All Corrections Offenders Databases. Then, in the text boxes, type in a common last name and a common first name, and anything else you care to enter, and click on Submit Request.

   Click on any of the following links to see the offender records that matched your search criteria:

   Inmate Population Search Results
   Inmate Release Search Results
   Supervised Population Search Results
   Absconder/Fugitive Search Results

   What kind of databases do you think these are? Do you think such databases are useful? Ethical? If you think they are useful, give your reasons.

6. DNA databases: What is a DNA database? Visit the following websites to learn more:

   www.ebi.ac.uk/embl/
   http://ndbserver.rutgers.edu/
   www.msnbc.com/news/710648.asp?cp1=1

7. Homeland Security Database: Read the following articles regarding the Homeland Security Database:

   www.foxnews.com/story/0,2933,70992,00.html
   http://ppri.tamu.edu/homeland_security/prepare.htm
   www.twotigersonline.com/resources.html
   www.securityinfowatch.com/online/Standards-and-Legislation/4516SIW320

   What is your opinion of the efficacy of such a database? Have you any concerns about the ethics or constitutionality about aspects of this database?

8. Security issue—identity theft: Listen to the online audio report (parts 1 and 2) about what to do if you become a victim of identity theft.

   http://radio.ucanr.org/cut.cfm?storynum=243&releasenum=59
   http://radio.ucanr.org/cut.cfm?storynum=244&releasenum=59

   Then visit these websites for more information on identity theft:

   www.fool.com/ccc/check/check06.htm
   http://illinoisissues.uis.edu/features/2002mar/name.html
   http://info.everydaywealth.com/seg15/?PARTNERID=IDT-IdentityTheft-5
   www.consumer.gov/idtheft/

   Are you putting yourself at risk? What can you do to improve your situation?

9. Virtual reality is being used in some surprising areas. Visit:

   http://coe.sdsu.edu/eet/articles/VRApps/start.htm
   www-vrl.umich.edu/
   www.doc.ic.ac.uk/~nd/surprise_96/journal/vol4/kcgw/report.html
   www.answers.com/virtual%20reality?gwp=11

The Challenges of the Digital Age

Society & Information Technology Today

Chapter Topics & Key Questions

9.1   **Truth Issues: Manipulating Digital Data** What are some ways digitized output can be manipulated to fool people?

9.2   **Security Issues: Threats to Computers & Communications Systems** What are some key threats to computers?

9.3   **Security: Safeguarding Computers & Communications** What are the characteristics of the five components of security?

9.4   **Quality-of-Life Issues: The Environment, Mental Health, Child Protection, & the Workplace** How does information technology create environmental, mental-health, pornography, and workplace problems?

9.5   **Economic & Political Issues: Employment & the Haves/Have-Nots** How may technology affect employment and the gap between rich and poor?

C omputers have invaluable uses for specialized work," says San Francisco historian and commentator Harold Gilliam, "but we need to question the assumption that whatever ails modern society can be cured by more information."[1]

Indeed, he goes on, many users "are hypnotized by the computer's power to summon endless arrays of facts—information without context, data without values, knowledge without perspective."[1]

Such matters take on more urgency since the beginning of the war on terrorism that started on September 11, 2001, when terrorist-hijacked planes destroyed the World Trade Center Twin Towers in New York City and part of the Pentagon in Washington, D.C. The sense of urgency was reinforced by the July 7, 2005, terrorist bombings in London, considered the worst attack on that city since World War II. Will the resulting tougher security rules imposed everywhere (including on use of information technology) be beneficial? How can we evaluate the effect of more names in more databases, for instance, on lost privacy and even lost liberty? What kind of context or perspective will help us? As we mentioned in Chapter 8, one possible tool used by advocates at both ends of the political spectrum is cost-benefit analysis—to analyze the trade-offs of heightened security on privacy, convenience, and ease of movement. Even if we can't always assign precise dollar amounts, it's important to weigh issues in a way that will prevent security goals from overtaking common sense.[2]

London, 2005, after terrorist bombings.

College students continually face issues of information technology about which benefits need to be balanced against costs. For instance, should the Department of Defense be allowed to collect information about students for a database designed for military recruiting purposes, and does that invade your privacy rights?[3] Should your Social Security number be used as a key identifier in campus records if there's a chance low-paid clerks might steal it to apply for bogus credit cards?[4] Should laptops and wireless technology be allowed in classrooms if students will mainly use them to web surf instead of attending to the lecture?[5] Should online instructor rating systems be allowed if prospective employers can also access them to see whether you mainly took easy courses?[6] Should there be tighter government scrutiny of foreign students, even though this may result in their going to other countries instead of to the United States?[7]

Such are the examples of the many infotech challenges that confront us. Elsewhere in the book we have considered ergonomics (Chapter 5) and privacy (Chapter 8). In this chapter, we consider some other major issues:

- Truth issues—manipulation of sound and images in digital data
- Security issues—accidents, natural hazards, terrorist hazards, and crime
- Quality-of-life issues—environment, mental health, child protection, the workplace
- Economic and political issues—employment and the haves/have-nots

9.1

Truth Issues: Manipulating Digital Data

What are some ways digitized output can be manipulated to fool people?

The enormous capacities of today's storage devices have given photographers, graphics professionals, and others a new tool—the ability to manipulate images at the pixel level. For example, photographers can easily do *morphing*—transforming one image into another, using image-altering software such as Adobe Photoshop. In morphing, a film or video image is displayed on a computer screen and altered pixel by pixel, or dot by dot. As a result, the image metamorphoses into something else—a pair of lips morphs into the front of a Toyota, for example, or an owl into a baby.

The ability to manipulate digitized output—images and sounds—has brought a wonderful new tool to art. However, it has created some big new problems in the area of credibility, especially for journalism. How can we know that what we're seeing or hearing is the truth? Consider the following.

Manipulation of Sound

Could I tell whether or not sound has been manipulated?

In 2004, country music artist Anita Cochran released some new vocals, including a duet, "(I Wanna Hear) a Cheatin' Song," with Conway Twitty—who had died a decade before the song was written. The producers pulled snippets of Twitty's voice from his recording sessions, put them on a computer hard drive in digital form, and used software known as Pro Tools to patch the pieces together.[8] Ten years earlier Frank Sinatra's 1994 album *Duets* paired him through technological tricks with singers like Barbra Streisand, Liza Minnelli, and Bono of U2. Sinatra recorded solos live in a recording studio. His singing partners, while listening to his taped performance on earphones, dubbed in their own voices. These second voices were recorded not only at different times but often, through distortion-free phone lines, from different places. The illusion in the final recording is that the two singers are standing shoulder to shoulder.

Newspaper columnist William Safire called *Duets* "a series of artistic frauds." Said Safire, "The question raised is this: When a performer's voice and image can not only be edited, echoed, refined, spliced, corrected, and enhanced—but can be transported and combined with others not physically present—what is performance? . . . Enough of additives, plasticity, virtual venality; give me organic entertainment."[9] Some listeners feel that the technology changes the character of a performance for the better. Others, however, think the practice of assembling bits and pieces in a studio drains the music of its essential flow and unity.

Whatever the problems of misrepresentation in art, however, they pale beside those in journalism. What if, for example, a radio station were to edit a stream of digitized sound so as to misrepresent what actually happened?

Manipulation of Photos

How can digitally distorted photographs be detected?

When in 2005 editors at *Newsweek* ran a cover photo of style guru Martha Stewart, who had been sent to prison for lying to federal investigators, they put a photo of her face on someone else's body, making the 63-year-old look terrific after five months behind bars.[10] When O. J. Simpson was arrested in 1994 on suspicion

The Challenges of the Digital Age

of murder, a *Time* artist working with a computer modified the police department mug shot and darkened the image so that O.J.'s face had a sinister cast to it. (*Newsweek* ran the mug shot unmodified.)[11] Should a magazine that reports the news be taking such artistic license? At the least, shouldn't magazines be running credit lines (as the *New York Times Magazine* often does) that say something such as "Photographic illustration by X" or "Photomontage by Y, with digital manipulation by Z"?[12]

Creator of the image-editing program known as Photoshop in 1989, John Knoll said later, "Mostly we saw the possibilities, the cool things, not how it would be abused."[13] But Photoshop and similar programs designed to edit ("morph") digital images can also be used to distort and falsify them. In one case, to show what can be done, a photographer digitally and seamlessly manipulated a famous 1945 photo showing the meeting of the leaders of the World War II Allied powers at Yalta; joining Stalin, Churchill, and Roosevelt are some digitally added newcomers: movie star Sylvester Stallone and the late comedian Groucho Marx. In another instance, a Danish computer company employee created an image purporting to show a photo from a 1954 issue of *Popular Mechanics* of what a future home computer was expected to look like. (● *See Panel 9.1.*) Before the 2004 presidential election, opposition groups, in an attempt to harm Democratic candidate John Kerry's election bid, circulated a composite photo, made of two photos taken a year apart, that purported to show Kerry and actress Jane Fonda, known for her anti–Vietnam War views, on a speaker's platform at a 1971 antiwar rally. The *Boston Globe* published manipulated photos of alleged sexual abuse of Iraqi women by U.S. soldiers, images taken from a pornographic website.[14]

The potential for abuse is clear. "For 150 years, the photographic image has been viewed as more persuasive than written accounts as a form of 'evidence,'" says one writer. "Now this authenticity is breaking down under the assault of technology."[15] Asks a former photo editor of the *New York Times Magazine*, "What would happen if the photograph appeared to be a straightforward recording of physical reality, but could no longer be relied upon to depict actual people and events?"[16]

● PANEL 9.1
Photo manipulation
A manipulated photo of a mock submarine console was passed off as a 1950s projection of the 2004 home computer. A well-known CEO of a computer technology company actually used the photo, believing it was real, during a speech in 2004. The image was created by a Danish hardware and software distributor, who originally entered it in a photo-manipulation contest. He said he had not intended to create a believable fake.

Scientists from the RAND Corporation have created this model to illustrate how a "home computer" could look like in the year 2004. However the needed technology will not be economically feasible for the average home. Also the scientists readily admit that the computer will require not yet invented technology to actually work, but 50 years from now scientific progress is expected to solve these problems. With teletype interface and the Fortran language, the computer will be easy to use.

Survival Tip
Is It True?

Snopes.com, a website for debunking urban legends and hoaxes, includes a section on composite photos, some true, some fake. Go to *www.snopes.com*.

Fortunately, some steps are being taken to combat this kind of digital deception. Corbis, the online stock photo agency, which is owned by Microsoft's Bill Gates, places an imperceptible digital watermark on the 3.5 million images it makes available, which encodes information about the owner of the image within the pixels of the photograph and can be traced even if the image has been modified.[17] In another technique, Dartmouth College computer science professor Hany Farid has devised algorithms that can detect changes in photographic images, provided the pictures are uncompressed. (Compressed files are harder.) Electrical and computer engineering professor Jessica Fridrich, of the State University of New York at Binghamton, is doing research in which a digital camera would take a picture of a subject and at the same time a picture of the photographer's iris, a distinctive form of identification, that would be hidden inside the larger picture just taken.[18]

Manipulation of Video & Television

Can videos and films be manipulated to present a different reality?

The technique of morphing, used in still photos, takes a massive jump when used in movies, videos, and television commercials. Digital image manipulation has had a tremendous impact on filmmaking. Director and digital pioneer Robert Zemeckis *(Death Becomes Her)* has compared the technology to the advent of sound in Hollywood.[19] It can be used to erase jet contrails from the sky in a western and to make digital planes do impossible stunts. It can even be used to add and erase actors. For instance, actor Steve McQueen, who died in 1980, was revived in his character of Lt. Frank Bullitt, from the film *Bullitt,* to promote the 2005 redesigned Ford Mustang in a commercial called "Cornfield."[20]

Films and videotapes are widely thought to accurately represent real scenes. Thus, the possibility of digital alterations raises some real problems. Videotapes supposed to represent actual events could easily be doctored. Another concern is for film archives: Because digital videotapes suffer no loss in resolution when copied, there are no "generations." Thus, it will be impossible for historians and archivists to tell whether the videotape they're viewing is the real thing or not.[21]

Indeed, it is possible to create virtual images during live television events. These images—such as a Coca-Cola logo in the center of a soccer field—don't exist in reality but millions of viewers see them on their TV screens.[22]

Accuracy & Completeness

What are five limitations I should be aware of when using databases for research?

Databases—including public data banks such as LexisNexis—can provide you with more facts and faster facts but not always better facts. Penny Williams, professor of broadcast journalism at Buffalo State College in New York and formerly a television anchor and reporter, suggests five limitations to bear in mind when using databases for research:[23]

YOU CAN'T GET THE WHOLE STORY   For some purposes, databases are only a foot in the door. There may be many facts or facets of the topic that are not in a database. Reporters, for instance, find a database is a starting point. It may take intensive investigation to get the rest of the story.

IT'S NOT THE GOSPEL   Just because you see something on a computer screen doesn't mean it's accurate. Numbers, names, and facts must be verified in other ways.

KNOW THE BOUNDARIES One database service doesn't have it all. For example, you might find full text articles from the *New York Times* on one service, from the *Wall Street Journal* on another, and from the *San Jose Mercury News* on yet another, but no service carrying all three.

FIND THE RIGHT WORDS You have to know which keywords (search words) to use when searching a database for a topic. As Lynn Davis, a professional researcher with ABC News, points out, if you're searching for stories on guns, the keyword "can be guns, it can be firearms, it can be handguns, it can be pistols, it can be assault weapons. If you don't cover your bases, you might miss something."[24]

HISTORY IS LIMITED Most public databases, Davis says, have information going back to 1980, and a few into the 1970s, but this poses problems if you're trying to research something that happened or was written about earlier.

9.2

Security Issues: Threats to Computers & Communications Systems

What are some key threats to computers?

Internet users just don't have "street smarts" about online safety, a 2004 survey found, and that makes them vulnerable.[25] And a 2005 study suggested that while users think they are able to recognize when they are being manipulated, either legally or illegally, in fact they are quite naïve.[26] Because the result has been a deluge of online scams and breaches of financial databases, yet another recent study has found that 42% of online shoppers and 28% of people who bank online have begun cutting back on their online financial dealings—sobering news for e-commerce.[27] As one might expect, most Americans now believe that the government should do more to make the internet safe, even though they don't trust the federal institutions, such as Congress and the Federal Trade Commission, that are responsible for creating and enforcing the laws online.[28] But there is much that we as individuals can and must do to protect our own security.

SECURITY Security issues go right to the heart of the workability of computer and communications systems. Here we discuss the following threats to computers and communications systems. (● *See Panel 9.2.*)

- Errors and accidents
- Natural hazards
- Computer crimes
- Computer criminals

Errors & Accidents

What are the kinds of errors and accidents I need to be alert for in computer systems?

In general, errors and accidents in computer systems may be classified as *human errors, procedural errors, software errors, electromechanical problems,* and *"dirty data" problems.*

HUMAN ERRORS Which would you trust—human or computer? If you were a pilot and your plane's collision-avoidance computer tells you to ascend but a human air-traffic controller tells you to descend, which order would you follow? In 2001, a Russian pilot near the Swiss-German border

Threat Category	Types of Threats
Errors and accidents	• Human errors • Procedural errors • Software errors • Electromechanical problems • "Dirty data" problems
Natural hazards	
Computer crimes	• Theft of hardware • Theft of software • Theft of online music and movies • Theft of time and services • Theft of information • Internet-related fraud • Taking over your PC: zombies, botnets, and blackmail • Crimes of malice: crashing entire systems
Computer criminals	• Individuals or small groups • Employees • Outside partners and suppliers • Corporate spies • Foreign intelligence services • Organized crime • Terrorists

ignored his computer (against mandatory regulations) and complied with erroneous human orders, resulting in a collision with another plane.[29]

Human errors can be of several types. Quite often, when experts speak of the "unintended effects of technology," what they are referring to are the unexpected things people do with it. Among the ways in which people can complicate the workings of a system are the following:[30]

- **Humans often are not good at assessing their own information needs:** For example, many users will acquire a computer and communications system that either is not sophisticated enough or is far more complex than they need.

- **Human emotions affect performance:** For example, one frustrating experience with a computer is enough to make some people abandon the whole system. But smashing your keyboard isn't going to get you any closer to learning how to use it better.

- **Humans act on their perceptions, which may not be fast enough to keep up:** In modern information environments, human perceptions are often too slow to keep up with the equipment. Decisions influenced by information overload, for example, may be just as faulty as those based on too little information.

PROCEDURAL ERRORS   Some spectacular computer failures have occurred because someone didn't follow procedures. In 1999, the $125 million Mars Climate Orbiter was fed data expressed in pounds, the English unit of force, instead of newtons, the metric unit (about 22% of a pound). As a result, the spacecraft flew too close to the surface of Mars and broke apart.[31] A few years earlier, Nasdaq, the nation's second-largest stock market, was shut down for $2^{1}/_{2}$ hours by an effort, ironically, to make the computer system more user-friendly. Technicians were phasing in new software, adding technical improvements a day at a time. A few days into this process, technicians trying to add more features to the software flooded the data storage capability of the computer system. The result was to delay the opening of the stock market and shorten the trading day.[32] In 2001, a failed software upgrade halted trading on the New York Stock Exchange for an hour and a half.[33]

SOFTWARE ERRORS We are forever hearing about "software glitches" or "software bugs." A *software bug* is an error in a program that causes it not to work properly. In 2001, the nonprofit American Medical College Application Service launched a new web-based application service that was supposed to make medical-school applications for 115 medical schools easier and more efficient than the old paper version. Instead, it was plagued by seemingly endless software bugs. One applicant, Yale University student Amit Sachdeva, found himself spending night after night logging on again and again. His routine: Stay awake until 2 a.m. when internet traffic slowed down, plod through a few pages until an error message appeared, then watch as the system froze and crashed. Time for completion: an estimated 24–36 hours, instead of the 5–8 hours the process was supposed to take. "It was incredibly frustrating," said Sachdeva.[34]

ELECTROMECHANICAL PROBLEMS: ARE "NORMAL ACCIDENTS" INEVITABLE? Mechanical systems, such as printers, and electrical systems, such as circuit boards, don't always work. They may be faultily constructed, get dirty or overheated, wear out, or become damaged in some other way. Power failures (brownouts and blackouts) can shut a system down. Power surges can also burn out equipment.

Modern systems are made up of thousands of parts, all of which interrelate in ways that are impossible to anticipate. Because of that complexity, argues Yale University sociologist Charles Perrow, what he calls "normal accidents" are inevitable.[35] That is, it is almost certain that some combinations of minor failures will eventually amount to something catastrophic. Indeed, it is just such collections of small failures that led to catastrophes such as the blowing up of the *Challenger* space shuttle in 1986 and the near-meltdown of the Three Mile Island nuclear-power plant in 1979. On a smaller scale, we also see this phenomenon in the rise in car crashes caused by drivers talking on their wireless phones.[36] (Indeed, an Australian study found that drivers using cellphones were four times as likely to be involved in a serious crash, even if they used hands-free phones or speakerphones.[37]) In the Digital Age, "normal accidents" will not be anomalies but are to be expected. They will also, suggests urbanist and philosopher Paul Virilio, be more global in their impact.[38]

"DIRTY DATA" PROBLEMS When keyboarding a research paper, you undoubtedly make a few typing errors (which, hopefully, you clean up). So do all the data-entry people around the world who feed a continual stream of raw data into computer systems. A lot of problems are caused by this kind of "dirty data." *Dirty data* is incomplete, outdated, or otherwise inaccurate data.

A good reason for having a look at your records—credit, medical, school— is so that you can make any corrections to them before they cause you complications. As the president of a firm specializing in business intelligence writes, "Electronic databases, while a time-saving resource for the information seeker, can also act as catalysts, speeding up and magnifying bad data."[39]

Natural Hazards

What kind of natural hazards in my area might be a threat to my computer system?

Some disasters do not merely lead to temporary system downtime; they can wreck the entire system. Examples are natural hazards.

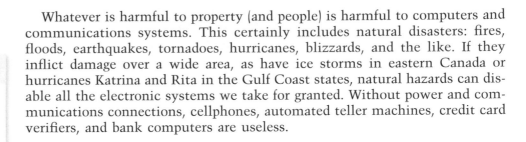

Whatever is harmful to property (and people) is harmful to computers and communications systems. This certainly includes natural disasters: fires, floods, earthquakes, tornadoes, hurricanes, blizzards, and the like. If they inflict damage over a wide area, as have ice storms in eastern Canada or hurricanes Katrina and Rita in the Gulf Coast states, natural hazards can disable all the electronic systems we take for granted. Without power and communications connections, cellphones, automated teller machines, credit card verifiers, and bank computers are useless.

Computer Crimes

What are examples of various crimes involving computers and computer systems?

A _computer crime_ can be of two types. (1) It can be an illegal act perpetrated against computers or telecommunications, or (2) it can be the use of computers or telecommunications to accomplish an illegal act. The following are crimes of both types.

THEFT OF HARDWARE   Hardware theft can range from shoplifting an accessory in a computer store to removing a laptop or cellular phone from someone's car. Professional criminals may steal shipments of microprocessor chips off a loading dock or even pry cash machines out of shopping-center walls.

Eric Avila, 26, a history student at the University of California at Berkeley, had his doctoral dissertation—involving 6 years of painstaking research—stored on the hard drive of his Macintosh laptop when a thief stole it from his apartment. Although he had copied an earlier version of his dissertation (70 pages entitled "Paradise Lost: Politics and Culture in Post-War Los Angeles") onto a floppy disk, the thief stole that, too. "I'm devastated," Avila said. "Now it's gone, and there is no way I can recover it other than what I have in my head." To make matters worse, he had no choice but to pay off the $2,000 loan for a computer he no longer possessed.[40] The moral, as we've emphasized in this book: Always make backup copies of your important data, and store them in a safe place—away from your computer.

Laptop on a car seat.
Computers left in cars are temptations for smash-and-grab thieves.

THEFT OF SOFTWARE   _Pirated software,_ as we stated in Chapter 3, is software obtained illegally, as when you make an illegal copy of a commercial videogame. This is so commonplace that software makers secretly prowl electronic bulletin boards in search of purloined products and then try to get a court order to shut down the bulletin boards. They also look for organizations that "softlift"—companies, colleges, or other institutions that buy one copy of a program and make copies for many computers. In addition, software pirates often operate in China, Taiwan, Mexico, Russia, and various parts of Asia and Latin America, where the copying or counterfeiting of well-known software programs is practiced on a large scale. In some countries, most of the U.S. microcomputer software in use is thought to be illegally copied.

Many software pirates are reported by coworkers or fellow students to such organizations as the Software and Information Industry Association, the Interactive Digital Software Association, and the Entertainment and Leisure Software Publishers Association.

THEFT OF ONLINE MUSIC & MOVIES   Many students may feel that illegally downloading music and movies is a victimless crime, but to the entertainment industry it is just plain piracy or theft.

- **Stealing music:** College students were quick to discover the internet music service Napster, which in its original form allowed millions of people to exchange songs for free. Then, as illegal file swapping shifted from client/server services like Napster to peer-to-peer services like Kazaa, Grokster, Limeware, and StreamCast and music-CD sales shrank, music companies decided to go after downloaders. They did so by getting their names and addresses from internet access providers and by deploying electronic "robots" to monitor traffic on file-swapping networks and seek out Internet Protocol (IP) addresses. In 2004, for example, the record industry filed copyright infringement lawsuits against students illegally downloading songs on file-sharing networks at 21 college campuses. Some settled, at fees ranging from $2,000 to $10,000 and fines of $12,000 to $17,500.[41] In June 2005, the U.S. Supreme Court also ruled against a pair of file-sharing networks, and 2 days later record companies sued 784 people for illegally distributing songs from the networks.[42] (People who download music illegally are liable for damages as high as $150,000 per song. Fortunately, student sharing of music files has declined drastically.[43])

- **Stealing movies:** The film industry has also taken aggressive aim at pirated movies. For instance, in 2005 the government announced an 11-nation crackdown on organizations called "warez" (pronounced "wares"), groups that are sort of underground internet co-ops set up to trade in copyrighted materials. Four men were charged with conspiring to violate copyright laws for operating an internet site that offered stolen movies such as *Batman Begins*.[44]

Piracy of intellectual property, whether software, music, or movies, ultimately harms everyone, directly or indirectly, because, says one writer, "thieves do not invest in research, design, production, development, or advertising." The result is "fewer advances in science, fewer new products, fewer new music CDs, fewer new movies, less new software, and higher prices for whatever is created."[45] While piracy will never entirely disappear, however, it seems likely that new business models, such as those in which songs are sold at 99 cents per download, will come to replace the older systems of distribution.[46]

THEFT OF TIME & SERVICES The theft of computer time is more common than you might think. Probably the biggest instance is people using their employer's computer time to play games, do online shopping, or dip into web pornography. Some people even operate sideline businesses.

For years "phone phreaks" have bedeviled the telephone companies. For example, they have found ways to get into company voice mail systems and then use an extension to make long-distance calls at the company's expense. They have also found ways to tap into cellular phone networks and dial for free. Satellite-TV piracy has also grown at an alarming rate, and firms such as DirecTV have reported soaring losses.[47]

THEFT OF INFORMATION Paris Hilton, the heiress and reality-TV personality, woke up one day to report that her Sidekick—a high-tech device that combines phone, organizer, and camera and also allows users to send and receive email—had been hacked, and all her contact information for her social network (and some racy photos) had been posted on the internet. During the same week, ChoicePoint Inc., one of the largest buyers and sellers of personal data, announced that criminals posing as legitimate buyers had purchased sensitive information on about 145,000 people.[48]

Clearly, information thieves are having a field day. They have infiltrated the files of the Social Security Administration, stolen confidential personal records, and sold the information. On college campuses, they have snooped on or stolen private information such as grades. They have broken into computers of the major credit bureaus and stolen credit information and have then used the information to charge purchases or have resold it to other people.[49]

INTERNET-RELATED FRAUD In Iraq in 2003, the day after the enormous statue of Saddam Hussein was toppled in downtown Baghdad, war souvenirs were already being offered on eBay, the online auction house. One ad posted was for a full-size statue of the deposed Iraqi leader, although any buyer was supposed to have to pay to have it shipped to the United States. The fraudulent listing was pulled from eBay in about an hour.[50]

Fraud on the internet is a runaway problem, accounting for 53% of all consumer-fraud complaints to the Federal Trade Commission in 2004. The most common complaints in that year, according to the Internet Fraud Complaint Center (IFCC), were auction fraud (71.2% of complaints), nondelivered merchandise and/or payment, credit card or debit card fraud, check fraud, investment fraud, confidence fraud, and identity theft.[51]

- **The Nigerian letter scam:** An example of a classic kind of confidence fraud is the persistent one known as the *Nigerian letter scam,* which

in a recent year was the cause of 16,000 complaints with the IFCC *(www1.ifccfbi .gov/index.asp).* Most Nigerian-letter perpetrators claim to have discovered inactive or delinquent accounts that hold vast amounts of money ready to be claimed. Victims are given a chance to receive nonexistent government money, often from the "Government of Nigeria," as long as they pay a fee to help transfer the money to an overseas account.[52] Similar kinds of cyberspace fraud involve nonexistent investment deals and phony solicitations.

- **Other scams, including Evil Twin attacks:** The majority of scams involve ruses such as those we described in Chapters 2 and 6. Among them are *phishing* (sending emails that appear to come from a trusted source, which direct you to a website where you're asked to reveal personal information), *pharming* (in which malicious software is implanted in your computer that redirects you to an impostor web page), and *Trojan horses* (a program such as a screen saver that carries viruses that perpetrate mischief without your knowledge).

A variant on conventional phishing is *Wi-Fi phishing*, known as the ___Evil Twin attack___, **in which a hacker or cracker sets up a Wi-Fi hot spot or access point that makes your computer think it's accessing a safe public network or your home network and then monitors your communications and steals data you enter into a website,** if it doesn't have the right security measures. Some internet access providers offer phishing-site blockers. "Wireless users should enter private information only into sites that protect data with encryption technology," advises one article, "which is signified by a little lock on the bottom of the page."[53]

TAKING OVER YOUR PC: ZOMBIES, BOTNETS, & BLACKMAIL   Betty Carty, 54, a New Jersey grandmother of three, was flabbergasted when her internet access provider curtailed her outbound email privileges. The reason: an intruder had taken over her PC without her knowledge and turned it into a device for disseminating as many as 70,000 emails a day.[54]

Carty's machine had become what is known as a ___zombie___, **or** *drone*, **a computer taken over covertly and programmed to respond to instructions sent remotely,** often by instant-messaging channels. The New Jersey woman's PC, however, was only one of several home and business personal computers in what is known as a ___botnet___, **short for "robot network," a network of computers compromised by means of a Trojan horse that plants instructions within each PC to wait for commands from the person controlling that network.** These remote-controlled networks are best detected by the internet access provider, which can block the illicit network connections and help users disinfect their PCs.

The zombie computers and botnet are used to launch phishing attacks or send spam messages. They can also be used to launch denial-of-service attacks, perhaps to extort money from the targeted sites in return for halting the attacks. For instance, one cyber-blackmailer threatened to paralyze the servers of a small online-payment processing company unless it sent a $10,000 bank wire—and when the company refused, its servers were bombarded with barrages of data for four days.[55] Blackmail has also been used in conjunction with the theft of credit card numbers or documents.[56] One thief broke into the systems of internet retailer CD Universe, stole 300,000 customers' credit card numbers, and, when the company's executives refused to pay a ransom demand sold them off piecemeal on the internet (for others to illegally charge on) until he was stopped. More recently, security researchers uncovered a plot in which someone locked up another's electronic documents, effectively holding them hostage, and demanded $200 (to be sent by email) as ransom to deliver the digital keys to unlock the files.[57]

CRIMES OF MALICE: CRASHING ENTIRE SYSTEMS   Sometimes criminals are more interested in abusing or vandalizing computers and telecommunications systems than in profiting from them. For example, a student at a Wisconsin campus deliberately and repeatedly shut down a university computer system, destroying final projects for dozens of students; a judge sentenced him to a year's probation, and he left the campus. In 2003, the website of WeaKnees, an online seller of digital video recorders, was overwhelmed by an electronic attack that knocked out its email system for weeks, the result of cyber-mercenaries allegedly hired by an entrepreneur who had been rebuffed by WeaKnees over a proposed business deal.[58]

Other kinds of malevolent attacks could be far more serious, as follows.

- **Attacks on power-control systems:** One possibility that concerns security specialists is cyberattacks on the nation's water, power, transportation, and communications systems, causing them to crash and disrupting services to thousands, even millions of people. None of this has happened yet, although we have experienced regionwide blackouts from power failures and natural disasters owing to other causes. Even if such systems are not the target of terrorists (who may be more interested in generating spectacular casualty counts), it must be assumed that they are vulnerable.

- **Attacks on the internet—could the entire net crash?** Also quite worrisome is the possibility of an attack on the internet that could actually crash the whole worldwide network. This would involve crackers' tampering with something called the *border gateway protocol*, which individual networks use to announce their routes so they can carry each other's messages. By falsifying the announcements, a cracker could direct messages to nonexistent routes, thereby overloading and perhaps crashing parts of the internet.

  On a few occasions, parts of a large internal networking company have been made to disappear, so that no contact could be made for several hours, which shows that a real threat may exist. The reason it hasn't happened yet, speculates one pair of writers, is that "hackers, thieves, and terrorists have come to depend on the internet just like everyone else, and don't want it wrecked."[59]

Computer Criminals

What kind of people are threats to computers?

Known as the most remote city in the lower 48 states, Ely is a desert town of 4,000 or so residents located on the eastern edge of Nevada. Yet the day after the 2003 war with Iraq began, computers at a 40-bed hospital in Ely came under electronic attack. Initially this was traced to an Arab news network, suggesting cyberterrorism, but later it was pinned to a source in the former Soviet Union. "Here's tiny Ely, a place where people leave doors unlocked, and we get hacked by the Russian Mafia, who are pretending to be Arab," said a hospital technology manager. "We may be remote in geography, the most distant city from any metropolitan area, but with the internet we might as well be in downtown New York or Los Angeles."[60]

As this report shows, there's almost no telling where the next threat to computer security will come from. Let us consider the various sources.

INDIVIDUALS OR SMALL GROUPS Earlier we discussed phishers, pharmers, and creators of spyware and viruses, along with various types of crackers and hackers. These include individuals or members of small groups who use fraudulent email and websites to obtain personal information that can be exploited, either for monetary gain or sometimes simply to show off their power and give them bragging rights with other members of the hacker/cracker community.

EMPLOYEES In a 2002 survey of 1,009 chief security officers and security experts, the opinion was that the people who posed the greatest threat to their company's technology infrastructure as a result of cyberattacks were current employees (53%), former employees (10%), and nonemployees (28%).[61] Earlier Michigan State University criminal justice professor David

Carter surveyed companies about computer crime and concluded that "75% to 80% of everything happens from inside"—that is, most crimes are committed by employees.[62] Unfortunately, although crime is soaring in cyberspace, many companies keep it quiet—in part because cybercrime is harder to detect than crime in the real world.[63]

Workers may use information technology for personal profit or to steal hardware or information to sell. They may also use it to seek revenge for real or imagined wrongs, such as being passed over for promotion; indeed, the disgruntled employee is a principal source of computer crime.[64] Most common frauds, Carter found, involved credit cards, telecommunications, employees' personal use of computers, unauthorized access to confidential files, and unlawful copying of copyrighted or licensed software.

OUTSIDE PARTNERS & SUPPLIERS Suppliers and clients may also gain access to a company's information technology and use it to commit crimes, especially since intranets and extranets have become more commonplace. Partners and vendors also may be the inadvertent source of hacker mischief because their systems may not be as well protected as the larger partner's networks and computers, and so a third party may penetrate their security.

CORPORATE SPIES Competing companies or individuals may break into a company's computer system to conduct industrial espionage—obtain trade secrets that they can use for competitive advantages.

FOREIGN INTELLIGENCE SERVICES Just as the various U.S. intelligence services try to ferret out the secrets of other governments, both hostile and friendly, so foreign intelligence services are trying to do the same with us. "In addition," says one account, "several countries are working to develop the capability to disrupt the supply chains, communications, and economic infrastructure that support the military power of an enemy."[65]

ORGANIZED CRIME Members of organized crime rings not only steal hardware, software, and data; they also use spam, phishing, and the like to commit identity theft and online fraud. Even street gangs now have their own websites, most of them perfectly legal, but some of them possibly used as chat rooms for drug distribution.[66] In addition, gangs use computers the way legal businesses do—as business tools—but they use them for illegal purposes, such as keeping track of gambling debts and stolen goods.

TERRORISTS Even before September 11, 2001, the United States was not immune to terrorism, as was seen in the first (1993) bombing of New York's World Trade Center and the 1995 bombing of the Murrah Federal Building in Oklahoma City. But what really focused Americans' attention on terrorism, of course, was the 2001 hijacked-plane crashes into the World Trade Center and the Pentagon.

The Pentagon alone has 650,000 terminals and workstations, 100 WANs, and 10,000 LANs, although the 9/11 attack damaged only a few of them. More serious was the destruction to companies occupying the top floors of the Twin Towers of the World Trade Center, such as Cantor Fitzgerald, which lost 658 of its approximately 1,000 employees there as well as its main data center. Although the bond brokerage firm had kept duplicate files at a secondary site, it lost all the passwords needed to access them, which had been in the heads of its dead employees. For the first 12 hours following the attack, therefore, as we mentioned in Chapter 6, the survivors spent hours trying to guess passwords by recalling the deceased employees' hobbies, endearments, names of pets, and other likely associations.[67]

A member of the Dutch bomb squad with a security robot at Schiphol Airport terminal 1 after examining a suspicious suitcase found in the luggage department.

PRACTICAL ACTION
Is the Boss Watching You? Trust in the Workplace

SECURITY

Peter Whitney, 27, launched an internet blog chronicling his life, friends, and job handling mail and the front desk at a division of Wells Fargo—and he probably would have been all right until he began criticizing a few people he worked with and he was fired.[68] That's when he learned that his employers had been monitoring his online activity.

More and more employers are using surveillance techniques to monitor their employees, from using cameras and phone taps to recording website visits and visiting the files that employees delete. Snooping on email is now standard workplace procedure, with 60% of companies in one survey saying they now use some type of software to watch incoming and outgoing employee electronic mail.[69] In most states, the companies don't have to inform employees of this fact (Connecticut and Delaware are exceptions). Now many companies are also instituting guidelines about what is and is not permissible in employee blogs, or web logs.

What are employers looking for that you should be aware of? Here are the principal areas.

Less Productivity

There's a word for all those hours employees spend each workday talking on the phone with friends, getting coffee, or gossiping—"undertime." Some such goofing off is tolerated in the name of workplace morale. The internet, however, has created whole new ways to slack off at work. One 2002 survey of 305 human resource managers and 250 employees found that workers spent an average of 8.3 hours a week (more than an entire workday) peeking at non-work-related websites. The sites that employees said they visited most were news (67%), shopping (37%), gambling (2%), and pornography (2%).[70]

Misleading Résumés

Employers are carrying out more rigorous background checks on prospective workers since the September 11, 2001, terrorist attacks in New York City and Washington, D.C. Among the problems not listed on résumés or application forms that turned up in 1.8 million background checks by one screening company: driving violations (35%—moving violations, driving while intoxicated, or suspended license), poor credit records (26%—judgment, lien, bankruptcy, or collection agency activity), lack of job and/or education (24%), and criminal record (6%).[71]

Fraud

Fraud is a major concern of employers. For example, a survey of 53 computer security officers and senior security executives found that they are most concerned about fraud by employees (55%), hackers (38%), and business partners or vendors (8%), in that order.[72] A poll of 617 workers found that one in five employees say they've witnessed fraud in the workplace, saying their coworkers are doing the following: taking office supplies and shoplifting (37%), stealing product or cash (25%), claiming extra hours worked (18%), inflating expense accounts (8%), accepting kickbacks from suppliers (6%), and doing phony bookkeeping (3%).[73] Of great concern to employers is web fraud, as when companies are ripped off by people with stolen credit card numbers.[74]

Risk of Lawsuits & Viruses

One reason corporate executives are becoming more aggressive about spying on employees is that, besides ferreting out job shirkers and office-supply thieves, "they have to worry about being held accountable for the misconduct of their subordinates," says one report. "Even one offensive email message circulated around the office by a single employee can pose a liability risk for a company" under federal anticorruption and corporate governance laws.[75]

In addition, employers are worried about workers using company computers to download copyrighted music and movies, because these acts not only are a drain on corporate resources but also expose employers to the risk of lawsuits from record and movie companies as well as open company networks to viruses.[76]

What Employers Must Do: Shred Personal Data

On the other hand, employees catch a break under part of the Fair and Accurate Credit Transactions Act passed in late 2003. Regardless of size, employers must destroy—by "shredding or burning" or "smashing or wiping"—any personal information about employees before they throw it out if they got the information from a credit report.[77]

Security: Safeguarding Computers & Communications

What are the characteristics of the five components of security?

The ongoing dilemma of the Digital Age is balancing convenience against security. **_Security_ is a system of safeguards for protecting information technology against disasters, system failures, and unauthorized access that can result in damage or loss.**

We consider five components of security.

- Deterrents to computer crime
- Identification and access
- Encryption
- Protection of software and data
- Disaster-recovery plans

Deterrents to Computer Crime

What are some ways to deter computer crime?

As information technology crime has become more sophisticated, so have the people charged with preventing it and disciplining its outlaws.

ENFORCING LAWS   Campus administrators are no longer being quite as easy on offenders and are turning them over to police. Industry organizations such as the Software Publishers Association are going after software pirates large and small. (Commercial software piracy is now a felony, punishable by up to 5 years in prison and fines of up to $250,000 for anyone convicted of stealing at least 10 copies of a program, or more than $2,500 worth of software.) Police departments as far apart as those in Medford, Massachusetts, and San Jose, California, now have officers patrolling a "cyber beat." They regularly cruise online bulletin boards and chat rooms looking for pirated software, stolen trade secrets, child molesters, and child pornography.

CERT: THE COMPUTER EMERGENCY RESPONSE TEAM   In 1988, after one widespread internet break-in, the U.S. Defense Department created the Computer Emergency Response Team (CERT). Although it has no power to arrest or prosecute, CERT provides round-the-clock international information and security-related support services to users of the internet. Whenever it gets a report of an electronic snooper, whether on the internet or on a corporate email system, CERT stands ready to lend assistance. It counsels the party under attack, helps thwart the intruder, and evaluates the system afterward to protect against future break-ins.

TOOLS FOR FIGHTING FRAUDULENT & UNAUTHORIZED ONLINE USES   Among the tools used to detect fraud are the following:

- **Rule-based-detection software:** In this technique, users such as merchants create a "negative file" that states the criteria each transaction must meet. These criteria include not only stolen credit card numbers but also price limits, matches of the cardholder's billing address and shipping address, and warnings if a large quantity of a single item is ordered.

- **Predictive-statistical-model software:** In this technique, tons of data from previous transactions are examined to create mathematical descriptions of what a typical fraudulent transaction is like. The software then rates incoming orders according to a scale of risk based on their resemblance to the fraud profile.[78] Thus, for example, if some thief overhears you giving out your phone company calling-card number and he or she makes 25 calls to a country that you never have occasion to call, AT&T's software may pick up the unusual activity and call you to see if it is you who is making the calls.

- **Employee internet management (EIM) software:** Programs made by Websense, SurfControl, and SmartFilter, are used to monitor how much time workers spend on the web and even block access to gambling and porn sites.

- **Internet filtering software:** Some employers use special filtering software to block access to pornography, bootleg-music download, and other unwanted internet sites that employees may want to access.

- **Electronic surveillance:** Employers use various kinds of electronic surveillance that includes visual and audio monitoring technologies, reading of email and blogs, and recording of keystrokes. Some companies even hire undercover agents to pretend to be coworkers.

Identification & Access

What are three ways a computer system can verify legitimate right of access?

Are you who you say you are? The computer wants to know.

There are three ways a computer system can verify that you have legitimate right of access. Some security systems use a mix of these techniques. The systems try to authenticate your identity by determining (1) what you have, (2) what you know, or (3) who you are.

WHAT YOU HAVE—CARDS, KEYS, SIGNATURES, & BADGES   Credit cards, debit cards, and cash-machine cards all have magnetic strips or built-in computer chips that identify you to the machine. Many require that you display your signature, which may be compared with any future signature you write. Computer rooms are always kept locked, requiring a key. Many people also keep a lock on their personal computers. A computer room may also be guarded by security officers, who may need to see an authorized signature or a badge with your photograph before letting you in.

Of course, credit cards, keys, and badges can be lost or stolen. Signatures can be forged. Badges can be counterfeited.

WHAT YOU KNOW—PINS & PASSWORDS   To gain access to your bank account through an automated teller machine (ATM), you key in your PIN. A *PIN (personal identification number)* is the security number known only to you that is required to access the system. Telephone credit cards also use a PIN. If you carry either an ATM or a phone card, never carry the PIN written down elsewhere in your wallet (even disguised).

As we stated earlier in the book, *passwords* are special words, codes, or symbols required to access a computer system. Passwords are one of the weakest security links, and most can be easily guessed or stolen. We gave some suggestions on passwords in Chapter 6.

Some computer security systems have a "callback" provision. In a callback system, the user calls the computer system, punches in the password, and hangs up. The computer then calls back a certain preauthorized number. This measure will block anyone who has somehow got hold of a password but is calling from an unauthorized telephone.

Sigmatek's facial identification technology for corporate security at the International Security & Defense Exhibition in Tel Aviv, Israel.

Unattractive passport photos may become mandatory. Why? Because computers do not like smiles. A United Nations agency that sets standards for passports wants all countries to switch to a document that includes a digital representation of the bearer's face recorded on an embedded computer chip. In airports and at border crossings, a machine will read the chip—but the machine can be fooled by smiles, which introduce teeth, wrinkles, lines, and other distortions.

WHO YOU ARE—PHYSICAL TRAITS Some forms of identification can't be easily faked—such as your physical traits. <u>**Biometrics**</u>, **the science of measuring individual body characteristics,** tries to use these in security devices, as we pointed out in Chapter 6. *Biometric authentication devices* authenticate a person's identity by verifying his or her physical or behavioral characteristics with a digital code stored in a computer system.

Encryption

Encryption sounds like a good idea, but how could it be harmful to us?

<u>*Encryption*</u>, as we said in Chapter 6, **is the process of altering readable data into unreadable form to prevent unauthorized access,** and it is what has given people confidence to do online shopping and banking. Encryption is clearly useful for some organizations, especially those concerned with trade secrets, military matters, and other sensitive data. Recently, many financial organizations, such as Bank of America, Time Warner, and Citigroup's CitiFinancial division, stung by misplaced data of nearly 6 million people, decided to encrypt the backup tapes of customer information that they store with third-party vendors.[79]

A very sophisticated form of encryption is used in most personal computers and is available with every late-model web browser to provide for secure communications over the internet. However, from the standpoint of society, encryption is a two-edged sword. For instance, the 2001 attacks on the World Trade Center and Pentagon raised the possibility that the terrorists might have communicated with each other using unbreakable encryption programs. (There is no evidence they did.) Should the government be allowed to read the coded email of overseas terrorists, drug dealers, and other enemies? What about the email of all American citizens?

The U.S. government has maintained that it needs access to scrambled data for national security and law enforcement. Indeed, during the 1990s, officials urged that encryption companies be required to include a "back door" in their products that would allow the government to peek at messages exchanged by criminals and terrorists. Companies and consumers said they would not use such a product and contended that criminals surely would not either. It was also argued that many people with the most basic

Chapter 9

education in mathematics could write their own encryption systems. Ultimately, the back-door idea was dropped.[80]

The 2001 terrorist incidents resurrected the debate. Some academics who, over the objections of the government, had freely published their research on how to make unbreakable codes were haunted by the idea that law enforcement might have figured out terrorist plans if the encryption techniques had been kept secret. Although publicly available encryption allows ordinary people to protect their privacy and businesses to protect their data, it is clear that a by-product is limitation on the ability to fight lawbreakers and terrorists.[81]

Protection of Software & Data

What are three ways that organizations can protect software and data?

Organizations go to tremendous lengths to protect their programs and data. As might be expected, this includes educating employees about making backup disks, protecting against viruses, and so on. Other security procedures include the following:

CONTROL OF ACCESS Access to online files is restricted to those who have a legitimate right to access—because they need them to do their jobs. Many organizations have a system of transaction logs for recording all accesses or attempted accesses to data.

AUDIT CONTROLS Many networks have audit controls for tracking which programs and servers were used, which files opened, and so on. This creates an audit trail, a record of how a transaction was handled from input through processing and output.

PEOPLE CONTROLS Because people are the greatest threat to a computer system, security precautions begin with the screening of job applicants. Résumés are checked to see if people did what they said they did. Another control is to separate employee functions, so that people are not allowed to wander freely into areas not essential to their jobs. Manual and automated controls—input controls, processing controls, and output controls—are used to check if data is handled accurately and completely during the processing cycle. Printouts, printer ribbons, and other waste that may contain passwords and trade secrets to outsiders are disposed of through shredders or locked trash barrels.

Disaster. Wreckage and debris line the streets on January 1, 2005, following a massive tsunami that struck the area in Aceh, Indonesia.

Disaster-Recovery Plans

What is a disaster-recovery plan?

A *disaster-recovery plan* is a method of restoring information-processing operations that have been halted by destruction or accident. "Among the countless lessons that computer users have absorbed in the hours, days, and weeks after the [1993 New York] World Trade Center bombing," wrote one reporter, "the most enduring may be the need to have a disaster-recovery plan. The second most enduring lesson may be this: Even a well-practiced plan will quickly reveal its flaws."[82] Although the second (2001) attack on the World Trade Center reinforced these lessons in a spectacular way, as did Hurricane Katrina in New Orleans 4 years later, interestingly many companies have not gotten the message. A 2005 survey of 1,200 businesses found that nearly one-third of them did not have emergency continuity plans in place—up from 25% a year earlier. The survey also found that two-thirds of companies that suffered through a disaster lost business.[83] Many small companies may not have any backup systems at all, since they may see installing a system as pricey and difficult.[84]

Mainframe computer systems are operated in separate departments by professionals, who tend to have disaster plans. Whereas mainframes are usually backed up, many personal computers, and even entire local area networks, are not, with potentially disastrous consequences. It has been reported that, on average, a company loses as much as 3% of its gross sales within 8 days of a sustained computer outage. In addition, the average company struck by a computer outage lasting more than 10 days never fully recovers.[85]

A disaster-recovery plan is more than a big fire drill. It includes a list of all business functions and the hardware, software, data, and people that support those functions, as well as arrangements for alternate locations. The disaster-recovery plan also includes ways for backing up and storing programs and data in another location, ways of alerting necessary personnel, and training for those personnel.

9.4

Quality-of-Life Issues: The Environment, Mental Health, Child Protection, & the Workplace

How does information technology create environmental, mental-health, pornography, and workplace problems?

The worrisome effects of technology on intellectual-property rights and truth in art and journalism, on censorship, on health matters and ergonomics, and on privacy were explained earlier in this book. Here are some other quality-of-life issues related to information technology.

Environmental Problems

What environmental concerns are attached to information technology?

In 1999, it was estimated that the share of all U.S. electricity consumed by computer-based microprocessors was 13% and that half of the electrical grid would be powering the digital-internet economy within the first decade of the 21st century.[86] More recently, a team at the University of California's Lawrence Berkeley National Laboratory estimated that computers in all forms—not just those tied to the internet—used only 3%, not 13%, of all electricity in 1999.[87]

If electricity demand is no longer thought to be quite the problem it once was, other environmental challenges remain—specifically manufacturing by-products, disposal by-products, environmental blight, and possible risks of nanotechnology.

MANUFACTURING BY-PRODUCTS Many communities are eager to have computer and chip manufacturers locate there because they perceive them to be "clean" industries. But there have been lawsuits charging that the semiconductor industry has knowingly exposed workers to a variety of hazardous toxins, some of which were linked to miscarriages, and there is speculation that others may be linked to cancer and birth defects.[88]

DISPOSAL BY-PRODUCTS What to do with the "e-waste"—2.2 million tons (in 2001, the last year with numbers available) or over 100 million obsolete or broken PCs, monitors, printers, cellphones, TVs, and other electronic gadgetry?[89] Ninety-one percent of it is discarded, according to the

info!

Composition of a Desktop Microcomputer

Plastics Lead Aluminum Germanium Gallium Iron Tin Copper Barium Nickel Zinc Tantalum Indium Vanadium Terbium Beryllium Gold Europium Titanium Ruthenium Cobalt Palladium Manganese Silver Antimony Bismuth Chromium Cadmium Selenium Niobium Yttrium Rhodium Platinum Mercury Arsenic Silica

For more information, go to www.galtglobalreview.com/business/toxic_pcs.html.

Chemical	Source
Lead	• Cathode-ray tubes (CRTs) in amounts of 4–8 pounds. Also found in solder in circuit boards. Discarded electronics account for 30–40% of lead in the waste stream.
Cadmium	• Circuit boards, semiconductors. More than 2 million pounds estimated to exist in discarded computers in 2005.
Mercury	• Switches, batteries, fluorescent bulbs in liquid-crystal displays. By 2005, 400,000 pounds discarded.
Chromium	• Circuit boards, corrosion protection in steel. By 2005, 1.2 million pounds discarded.
PVC (polyvinyl chloride) plastics	• Connectors, cables, housings, plastic covers; about 250 million pounds discarded each year.
Brominated flame retardants	• Printed circuit boards, connectors, plastic covers, cables.

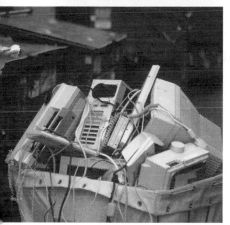

Environmental Protection Agency; only 9% is recycled. Only a scant 10% of dead PCs are recovered for recycling. Much of the electronic waste that winds up in the nation's 2,200 landfills contains large amounts of lead and other toxins that can leach into groundwater or produce dioxins and other cancer-causing agents when burned.[90] (● *See Panel 9.3.*) The problem is worsening as new computer models are introduced on faster cycles. Many people feel that the United States should follow the lead of Europe and Japan and pass laws forcing manufacturers to reduce the use of toxic materials in new products and take back old computers for recycling.

Don't always assume you can get a school or local charity to take your old personal computer; some will, but many are tired of being stuck with junk. Fortunately, the Electronic Industries Alliance Consumer Education Initiative offers a website *(www.eiae.org)* that helps consumers locate donation programs and recycling companies. Dell, Gateway, and Hewlett-Packard have all created programs whereby users can send unwanted computer gear for donation, refurbishing, or recycling. Cellphones may also be turned in for recycling or reuse.[91] Staples takes back old printer cartridges for recycling.

ENVIRONMENTAL BLIGHT Call it "techno-blight." This is the visual pollution represented by the forest of wireless towers, roof antennas, satellite dishes, and all the utility poles topped with transformers and strung with electric, phone, cable-TV, and other wires leading off in all directions. As the nation's electrical grid becomes more pervasive, so, people worry, will the obtrusive, ugly technology in our physical environment. Environmentalists worry about its impact on vegetation and wildlife, such as the millions of birds and bats that collide with cellular towers. Residents worry about the effect on views and property values, although antennas can be made into "stealth towers"—fake water towers, flagpoles, trees, and the like—to blend in better with the environment. Some people worry that there may be unknown health effects. Few of the thousands of miles of lines are buried underground, and most of those are in large cities. Ultimately the decision to approve or deny wireless towers resides with the Federal Communications Commission.[92]

POSSIBLE RISKS OF NANOTECHNOLOGY Although scientists call the fears speculation, some environmentalists worry that the spread of *nanotechnology*—manipulating materials such as carbon, zinc, and gold at the molecular level—could create contaminants whose tiny size makes them ultrahazardous. "If they get in the bloodstream or into the groundwater," says

info!
more

Donating Old PCs & Cellphones

Want to get rid of that old PC or Mac? For guidance, try:
www.nsc.org/ehc/epr2/recycler.htm
www.sharetechnology.org/donate.html
www.worldcomputerexchange.org
www.microweb.com/pepsite/Recycle/recycle_index.html.
For unwanted cellphones:
www.collectivegood.com
www.donateaphone.com
www.cellforcash.com
www.oldcellphone.com
www.phonefund.com

The Powermate Eco, a PC engineered specifically with the environment in mind.

Fighting techno-blight.
Environmentally friendly cell-phone towers disguised as plam trees and cacti.

a researcher for ETC Group, an environmental organization, "they could react with other things that are harmful."[93] However, experiments using carbon nanotube molecules, which might replace silicon in ever tinier transistors, have not proved risky in studies of mice and guinea pigs and so probably pose little risk to humans.

Mental-Health Problems

What problems with isolation, gambling, and stress can derive from information technology?

Some of the mental-health problems linked to information technology are the following:

ISOLATION?   Automation allows us to go for days without actually speaking with or touching another person, from buying gas to playing games. Even the friendships we make online in cyberspace, some believe, "are likely to be trivial, short lived, and disposable—junk friends."[94]

A few years ago, a Stanford University survey led by Norman H. Nie of the Stanford Institute for the Quantitative Study of Society found that, as people spent more time online, they had less time for real-life relationships with family and friends. "As internet use becomes more widespread," predicted a study coauthor, "it will have an increasingly isolating effect on society."[95] The inclination of heavy internet users to experience isolation, as well as loneliness and depression, was also supported by a 1998 study by Robert Kraut.

Three years later, however, Kraut reported the opposite—that the study's subjects experienced fewer feelings of loneliness and isolation. And he found that, for a new group, the more they used computers, the more integrated they were with others, the better their sense of well-being, and the more positive their moods.[96] Indeed, the internet has led to a great many web communities, from cancer survivors to Arab immigrants. Nevertheless, further research by the Stanford group found that the average internet user, who spends 3 hours a day online, watches less television, spends less time sleeping, and also spends less time socializing with friends. Indeed, according to the study, an hour of time spent using the internet reduces face-to-face contact with friends, coworkers, and family by 23.4 minutes.[97]

GAMBLING   Gambling is already widespread in North America, but information technology makes it almost unavoidable. Instead of driving 40 minutes to a Connecticut casino, for example, Rhode Island resident Beverly Richard discovered that slots, roulette, and blackjack were just a mouse click away, and she quickly found herself $13,000 in debt. "It was too convenient," she said. "I don't have to leave home."[98]

Although gambling by wire is illegal in the United States, host computers for internet casinos and sports books have been established in Caribbean tax havens. Satellites, decoders, and remote-control devices allow TV viewers to do racetrack wagering from home. An Australian firm is developing gambling games for mobile phones.[99] In these circumstances, law enforcement is extremely difficult. Despite U.S. efforts to curb gambling on the web, bettors in the United States generate an estimated 70% of global online gambling revenues, and the legal sites operating offshore are fighting back against U.S. Justice Department crackdowns as being unconstitutional.[100]

STRESS   In one survey of 2,802 American PC users, three-quarters of the respondents (ranging in age from children to retirees) said personal computers had increased their job satisfaction and were a key to success and learning. However, many found PCs stressful: Fifty-nine percent admitted getting

angry at their PCs within the previous year, and 41% said they thought computers had reduced job opportunities rather than increased them.[101] Another survey found that 83% of corporate network administrators reported "abusive and violent behavior" by employees toward computers—including smashing monitors, throwing mice, and kicking system units.[102] Yet another—but probably not very scientific—survey by the Institute of Psychiatry at the University of London for Hewlett-Packard found that excessive day-to-day use of technology, whether using cellphones or sending emails, produced what they called "info-mania," which resulted in a 10-point drop in IQ.[103]

Protecting Children: Pornography, Sexual Predators, & Online Bullies

What kind of steps can parents take to protect their children from inappropriate material?

Since computers are simply another way of communicating, it is no surprise that they are used to communicate about pornography, sexual solicitations, and threats and violence. Let's consider these.

PORNOGRAPHY   One of the biggest cultural changes in the United States of the past quarter century has been the widespread distribution of sexually explicit material, making adult entertainment or pornography a $12 billion yearly business. Indeed, some of America's most well known companies, such as General Motors, Time Warner, and Marriott, are now said to make millions selling pornography, as on closed-circuit TV channels in hotels.[104] With the recruitment of shock jock Howard Stern to XM Satellite Radio, the two nationwide satellite radio stations, Sirius and XM, seem poised to become the radio equivalent of cable TV, with racy unregulated content.[105] Audio files called "porncasts" are available for downloading for adult podcasts.[106] On the highways, motorists have complained of "dirty driving" as they found themselves looking at X-rated fare playing in other cars outfitted with DVD players.[107]

As for the internet, the word *sex*, according to Yahoo, is the most popular search word on the net.[108] The number of pornography websites rose from 88,000 in 2000 to 1.6 million 4 years later.[109] In April 2005, it was found that two of five internet users in the United States had visited a porn site.[110] Some of the most innovative and experimental of web entrepreneurs, in fact, have been online pornographers, who had been among the first to exploit video streaming, pop-up ads, and electronic billing.[111] Indeed, pornographers are now targeting cellphones as a technology for expanding their industry.[112]

Of course, all this is of great concern to parents, who fear their children will be exposed to inappropriate sexual and also violent material, both on the web and in videogames. However, in 2004 the U.S. Supreme Court ruled in a 5-4 decision that a law intended to keep children away from internet pornography was unconstitutional and suggested that parents should take the lead in screening children's web access.[113]

Following are some steps being taken to shield children from adult material.

- Online blocking software: Some software developers have discovered a golden opportunity in making programs like Cybersitter, Cyber Patrol, and Net Nanny. These blocking programs are designed to screen out objectionable material, typically by identifying certain unapproved keywords in a user's request or comparing the user's request for information against a list of prohibited sites. The leading online access providers also offer software filters.[114]

- **DVD filters:** A DVD player made by RCA includes filtering technology developed by ClearPlay that allows parents to edit out the inappropriate parts—such as disturbing images, violence, nudity, swear words, and ethnic and social slurs—of DVD movies they own or rent.[115] (The RCA player comes with built-in filters for 100 movies, and owners can get updated filters for a monthly fee.)

- **Videogame rating systems:** The Entertainment Software Ratings Board has a videogame ratings system that is linked to children's ages. (● *See Panel 9.4.*) To search for game ratings, parents may go to *www.esrb.org/esrbratings_search.asp* and type in the name of a game they are considering buying. For example, if you type in the popular "Grand Theft Auto," you will get ratings for 11 variations on the game, most of them T (Teen) or M (Mature) but one of them AO (Adults Only—this version is Grand Theft Auto San Andreas, which has strong sexual content and use of drugs). However, compliance among retailers selling videogames is voluntary and spotty, and one study found that 69% of children ages 13–16 were able to purchase M-rated games (supposedly reserved for those age 17 and up).[116]

- **The V-chip:** The 1996 Telecommunications Law officially launched the era of the *V-chip*, a device that is required equipment in all new television sets with screen size of 13 inches or larger sold after January 2000. The V-chip allows parents to automatically block out programs that have been labeled as high in violence, sex, or other objectionable material. Unfortunately, to turn on the controls, owners have to follow complicated instructions and menu settings, and then they have to remember their passwords when they want to watch *The Sopranos*.[117]

- **"xxx" web addresses:** In mid-2005, the internet's primary oversight body, the Internet Corporation for Assigned Names and Numbers, approved a plan for pornographic websites to use new addresses ending in *.xxx*. The purpose of the new domain would be to enable parents to more easily apply filtering software and so more effectively block access to porn sites.[118] However, some experts argue that pornographers are likely to keep their existing *.com* domain names, even as they set up shop with *.xxx* domain names, thereby diminishing the effectiveness of software filters set up to block the *.xxx* names.[119] As of this writing, the xxx domain system is on hold.

ONLINE SEXUAL PREDATORS Parents are also concerned about internet predators reaching their children. One in five children received a sexual

● **PANEL 9.4**
Videogame industry's rating system
These ratings are supposed to appear on videogame packages.

- **EC (Early Childhood):** Ages 3 and older. Contains no inappropriate material.
- **E (Everyone):** Ages 6 and older. May contain minimal or mild violence and/or infrequent mild language.
- **E10+ (Everyone 10 and older):** May contain more mild violence, mild language, and/or minimal suggestive themes.
- **T (Teen):** Ages 13 and older. May contain violence, suggestive themes, crude humor, minimal blood, and/or infrequent use of strong language.
- **M (Mature):** Ages 17 and older. May contain intense violence, blood and gore, sexual content, and/or strong language.
- **AO (Adults Only):** Ages 18 and over. May contain prolonged scenes of intense violence and/or graphic sexual content and nudity.
- **RP (Rating Pending):** Means game has been submitted to ERSB and is awaiting rating (used in advertising prior to game's release).

Source: Entertainment Software Ratings Board, ESRB Game Ratings,
www.esrb.org/esrbratings_guide.asp.

solicitation in 1998 and 1999, according to a survey of 1,501 regular internet users ages 10–17.[120] Girls were the primary target, receiving two-thirds of the solicitations. Moreover, it has been found that, despite the popular view, child molesters looking for victims online typically aren't pursuing young children. According to a study of 129 suspected molesters, most victims are girls 13–15 who willingly meet their predators and also usually willingly agree to have sex.[121] In addition, it was found that only 5% of the molesters pretended to be peers, that they messaged online future victims for more than a month, and only one in five mentioned wanting sex before the first meeting.

Some suggested prevention strategies are as follows:[122]

- **Monitor internet use:** Parents are advised to monitor their children's internet use and to install filters.
- **Be candid:** Parents should also make children aware that having sex with an adult is a crime and explain that molesters capitalize on teenagers' needs for acceptance.

CYBERBULLIES In yet another example of how information technology can negatively affect the social lives of children, there have been numerous reports of so-called *cyberbullying*, in which children in the 9–14 age range use the internet to unleash merciless taunting, nasty rumors, humiliating pictures, and other put-downs of fellow adolescents who have fallen into disfavor. The tactics include stealing each other's screen names, forwarding private material to people for whom it was not intended, and posting derogatory material on blogs.[123]

Some suggested tactics for dealing with cyberbullies are as follows:[124]

- **Save the evidence:** Children should print out offending messages and show them to their parents, who should then contact parents of the bully as well as inform school officials.
- **Block messages:** Victims should use the Block function to block further online communication from the bully. Parents should also contact email services and change the victim's screen name (and the victim should tell only his or her friends the new name).
- **Contact an attorney or police:** If there are threats of violence or sexual harassment, parents should contact a lawyer or the police.

Cyberbullying

Mobilizing educators, parents, students, and others to combat online social cruelty

Welcome to this web site! Cyberbullying is sending or posting harmful or cruel text or images using the Internet or other digital communication devices. The stories are heart breaking. Teens who are:

- Sending cruel, vicious, and sometimes threatening messages.
- Creating web sites that have stories, cartoons, pictures, and jokes ridiculing others.
- Posting pictures of classmates online and asking students to rate them, with questions such "Who is the biggest ___ (add a derogatory term)?"
- Breaking into an e-mail account and sending vicious or embarrassing material to others.
- Engaging someone in IM (instant messaging), tricking that person into revealing sensitive personal information, and forwarding that information to others.
- Taking a picture of a person in the locker room using a digital phone camera and sending that picture to others.

Cyberbullying is emerging as one of the more challenging issues facing educators and parents as young people embrace the Internet and other mobile communication technologies.

Cyberbully.org is provided by the Center for Safe and Responsible Internet Use. CSRIU provides resources for educators and others to promote the safe and responsible use of the Internet. The Cyberbully.org web site provides:

Professional Development
Information and Resources for Educators

- Educator's Guide to Cyberbullying
- New Book for School Leaders! *Cyberbully: Mobilizing Educators, Parents, Students and Others to Combat Online Social Cruelty* (coming soon!)
- Cyberbullying Needs Assessment Survey
- Cyberbullying Workshops and Online Course

Workplace Problems: Impediments to Productivity

What are some ways information technology does not make us productive?

First the mainframe computer, then the desktop stand-alone PC, and recently the networked computer were all brought into the workplace for one reason only: to improve productivity. How is it working out? Let's consider three aspects: misuse of technology, fussing with computers, and information overload.

MISUSE OF TECHNOLOGY "For all their power," said an economics writer, "computers may be costing U.S. companies tens of billions of dollars a year in downtime, maintenance and training costs, useless game playing, and information overload."[125]

Employees may look busy, as they stare into their computer screens with brows crinkled. But sometimes they're just hard at work playing videogames. Or browsing online malls (forcing corporate mail rooms to cope with a deluge of privately ordered parcels). Or looking at their investments or pornography sites. Indeed, one study found that recreational web surfing accounts for nearly one-third of office workers' time online.[126]

FUSSING WITH COMPUTERS Another reason for so much wasted time is all the fussing that employees do with hardware, software, and online connections. One study in the early 1990s estimated that microcomputer users wasted 5 billion hours a year waiting for programs to run, checking computer output for accuracy, helping coworkers use their applications, organizing cluttered disk storage, and calling for technical support.[127] And that was before most people had to get involved with making online connections work or experience the frustrations of untangling complications wrought by spam, phishing, viruses, and other internet deviltry. The Stanford study led by Nie, referred to above, found that junk email and computer maintenance take up a significant amount of time spent online each day. Indeed, people surveyed said they spent 14 minutes daily dealing with computer problems, which would add up to a total of 10 days a year.[128] Comments technology writer Dan Gillmor, "We would never buy a TV that forced us to reboot the set once a month, let alone once a week or every other day."[129]

INFORMATION OVERLOAD "It used to be considered a status symbol to carry a laptop computer on a plane," says futurist Paul Saffo. "Now anyone who has one is clearly a working dweeb who can't get the time to relax. Carrying one means you're on someone's electronic leash."[130]

The new technology is definitely a two-edged sword. Cellphones, pagers, and laptops may untether employees from the office, but these employees tend to work longer hours under more severe deadline pressure than do their tethered counterparts who stay at the office, according to one study.[131] Moreover, the devices that once promised to do away with irksome business travel by ushering in a new era of communications have done the opposite. They have created the office-in-a-bag that allows businesspeople to continue to work from airplane seats, hotel desks, and their own kitchen tables. The diminishing difference between work and leisure is what Motorola calls the "blurring of life segments."[132]

Some studies have found an increase in labor productivity in the early years of the 21st century, most of it attributed to more investment in information technology and efficiency improvements made possible by this technology.[133] However, says Stephen Roach, chief economist of Morgan Stanley Dean Witter, "the dirty little secret of the information age is that an increasingly large slice of work goes on outside the official work hours the government recognizes. . . . The '24/7' culture of nearly round-the-clock work is endemic to the wired economy. . . . But improving productivity is not about working longer; it's about adding more value per unit of work time."[134]

David M. Levy, a computer scientist at the University of Washington, is concerned that "our workload and speed [do] not leave room for thoughtful reflection."[135] Adds Bill McKibben, author of *Enough: Staying Human in an Engineered Age*, "There is a real danger that one is absorbing and responding to bursts of information, rather than having time to think."[136]

PRACTICAL ACTION
When the Internet Isn't Productive: Online Addiction & Other Time Wasters

There is a handful of activities that can drain hours of time, putting studying—and therefore college—in serious jeopardy. They include excessive television watching, partying, and working too many hours while going to school. They also include misuse of the computer.

The Great Campus Goof-Off Machine?

"I have friends who have spent whole weekends doing nothing but playing Quake or Warcraft or other interactive computer games," reported Swarthmore College sophomore Nate Stulman, in an article headed "The Great Campus Goof-Off Machine." He went on: "And many others I know have amassed overwhelming collections of music on their computers. It's the searching and finding they seem to enjoy: some of them have more music files on their computers than they could play in months."[137]

In Stulman's opinion, having a computer in the dorm is more of a distraction than a learning tool for students. "Other than computer science or mathematics majors, few students need more than a word processing program and access to email in their rooms."

Most educators wouldn't banish computers completely from student living quarters. Nevertheless, it's important to be aware that your PC can become a gigantic time sink, if you let it. Reports Rutgers communications professor Robert Kubey: "About 5% to 10% of students, typically males and more frequently first- and second-year students, report staying up late at night using chat lines and email and then feeling tired the next day in class or missing class altogether."[138]

Internet Addiction/Dependency

"A student emails friends, browses the World Wide Web, blows off homework, botches exams, flunks out of school."[139] This is a description of the downward spiral of the "net addict," often a college student—because schools give students no-cost/low-cost linkage to the internet—but it can be anyone. Some become addicted to chat groups, some to online pornography, some simply to the escape from real life.[140]

Stella Yu, 21, a college student from Carson, California, was rising at 5 a.m. to get a few hours online before school, logging on to the internet between classes and during her part-time job, and then going home to web surf until 1 a.m. Her grades dropped and her father was irate over her phone bills. "I always make promises I'm going to quit; that I'll just do it for research," she said. "But I don't. I use it for research for 10 minutes, then I spend two hours chatting."[141]

College students are unusually vulnerable to internet addiction, which is defined as "a psychological dependence on the internet, regardless of type of activity once 'logged on,'" according to psychologist Jonathan Kandell.[142] The American Psychological Association, which officially recognized "pathological internet use" as a disorder in 1997, defines the internet addict as anyone who spends an average of 38 hours a week online.[143] (The average internet user spends 5½ hours a week on the activity.[144]) More recently, psychologist Keith J. Anderson of Rensselaer Polytechnic Institute found that internet-dependent students, who make up at least 10% of college students, spent an average of 229 minutes a day online for nonacademic reasons, compared with 73 minutes a day for other students. As many as 6% spend an average of more than 400 minutes a day—almost 7 hours—using the internet.[145]

Internet addiction is often accompanied by another kind of dependency—data addiction—say Harvard University psychiatrists Edward Hallowell and John Ratey. The constant stimulation provided by frequent incoming data and the constant multitasking that goes with it have produced a condition they call "pseudo-attention deficit disorder." Its sufferers, says one description, "become frustrated with long-term projects, thrive on the stress of constant fixes of information, and physically crave the bursts of stimulation from checking e-mail or voice mail or answering the phone."[146]

What are the consequences of internet addiction? A study of the freshman dropout rate at Alfred University in New York found that nearly half the students who quit the preceding semester had been engaging in marathon, late-night sessions on the internet.[147] The University of California, Berkeley, found that some students linked to excessive computer use neglected their course work.[148] A survey by Viktor Brenner of State University of New York at Buffalo found that some internet addicts had "gotten into hot water" with their school for internet-related activities.[149] "Grades decline, mostly because attendance declines," says psychologist Anderson about internet-dependent students. "Sleep patterns go down. And they become socially isolated."

Online Gambling

A particularly risky kind of internet dependence is online gambling. David, a senior at the University of Florida, Gainesville, owed $1,500 on his credit cards as a result of his online gambling habit, made possible by easy access to offshore casinos in cyberspace. He is not alone. A survey of 400 students at Southern Methodist University found that 5% said they gambled frequently via the internet.[150] Some students have even gambled away their tuition money.

Economic & Political Issues: Employment & the Haves/Have-Nots

How may technology affect employment and the gap between rich and poor?

In recent times, a number of critics have provided a counterpoint to the hype and overselling of information technology to which we have long been exposed. Some critics find that the benefits of information technology are balanced by a real downside. Other critics make the alarming case that technological progress is actually no progress at all—indeed, it is a curse. The two biggest charges (which are related) are, first, that information technology is killing jobs and, second, that it is widening the gap between the rich and the poor.

Technology, the Job Killer?

Could technology put me out of a job?

Certainly, ATMs do replace bank tellers, E-Z pass electronic systems do replace turnpike-toll takers, and internet travel agents do lure customers away from small travel agencies. Hundreds of companies are replacing service representatives with voice software.[151] In new so-called *lights-out factories,* machines make things—for example, the tiny cutting devices you see mounted on dental-floss containers—even when no one is there; as much as possible is done with no labor.[152] The contribution of technological advances to economic progress is steady, but the contribution to social progress is not purely positive.

A lights-out factory. The ABA-PGT plastics engineering plant.

But is it true, as technology critic Jeremy Rifkin has said, that intelligent machines are replacing humans in countless tasks, "forcing millions of blue-collar and white-collar workers into temporary, contingent, and part-time employment and, worse, unemployment"?[153]

This is too large a question to be fully considered in this book. We can say for sure that the U.S. economy is undergoing powerful structural changes, brought on not only by the widespread diffusion of technology but also by greater competition, increased global trade, the sending of jobs offshore, the shift from manufacturing to service employment, the weakening of labor unions, more flexible labor markets, more rapid immigration, partial deregulation, and other factors.[154] One futurist argues that by 2100 perhaps 2% of the U.S. non-farm workforce will be needed to handle white-collar "know-how" functions, such as those of most of today's office workers, but that "hyper-human" service workers, who will be required for creativity, social skills, conscious perception, positive feelings, hypothesizing, and the like, will zoom to over 90%.[155]

A counterargument is that jobs don't disappear—they just change. According to some observers, the jobs that do disappear represent drudgery. "If your job has been replaced by a computer," says Stewart Brand, "that may have been a job that was not worthy of a human."[156] Of course, that means little to someone who has no job at all.

Gap between Rich & Poor

What is the "digital divide"?

"In the long run, improvements in technology are good for almost everyone," economist and columnist Paul Krugman has said. "Unfortunately, what is true in the long run need not be true over shorter periods."[157] We are now, he believes, living through one of those difficult periods in which technology doesn't produce widely shared economic gains but instead widens the gap between those who have the right skills and those who don't.

More disadvantaged people may have access to computers than ever before, particularly at work, where many computer skills are, in point of fact, not very high tech or demanding. However, it may be a different story at home. According to a Ford Foundation–financed report, "Bringing a Nation Online: The Importance of Federal Leadership," there still exists a so-called *digital divide* "between high and low income households, among different racial groups, between Northern and Southern states, and rural and urban households."[158] For instance, only 29% of school-age children from households with incomes of less than $15,000 a year use a home computer to complete school assignments. By contrast, 77% of children from households with annual incomes of $75,000 or more use a home computer for homework.[159] In college, says one report, "some minority students are still puttering along in the breakdown lane of the information superhighway."[160] Although computer hardware is getting cheaper, high-speed internet connections can be prohibitively expensive for poor and working-class families.

The digital divide also extends to the nations of the world. As *New York Times* columnist Thomas Friedman has written, in *The World Is Flat*, the world is "flattening," becoming more interconnected as the result of the internet, wireless technology, search engines, file sharing, digital photography, and other cutting-edge technologies.[161] This may be good for many advanced nations. But half the world—Africa, much of Latin America, and rural areas of India and China—isn't flattening at all. And the future may be risky for America as well, because the fiber optic cables that have been laid across the Pacific Ocean have made it easy to send offshore the jobs of all kinds of workers—accountants, radiologists, illustrators, and so on—whose work is knowledge-based.[162]

Whom Does the Internet Serve?

How could corporate and foreign government control affect internet access?

Because of its unruly nature, perhaps the internet truly serves no one—and that is both its blessing and its curse. Business executives, for instance, find the public internet so unreliable that they have moved their most critical applications to semiprivate networks, such as intranets and extranets. That means that more of the network is increasingly brought under commercial control, which for consumers might mean more fees for special services and the stifling of cultural empowerment and free speech. Despite early euphoria that the internet would unleash a democratic spirit, nonprofit uses, and progressive websites, some see increasing corporate consolidation, at least in the United States.

For many restrictive governments outside the United States—China, Saudi Arabia, Iran, Singapore, United Arab Emirates, Bahrain, for example—that try to control internet access by their citizens, the net poses threats to their authority. China, for instance, is second only to the United States in the number of people online; its nondemocratic government has cracked down on some kinds of internet political activism but tolerated others—largely because there is so much online chatter that it can't censor it all.[163] (Some of the more than 1,000 words and phrases filtered by the Chinese instant-messaging service include "democracy," "human rights," and "oppose corruption.") But if such governments want to join the global economy, an important goal for most, perhaps they will find that less control and regulation will translate into greater e-commerce.

In a World of Breakneck Change, Can You Still Thrive?

Where will you be in the future?

"The most surprising thing that could happen in the future," writes futurist Edward Cornish, "is if it offered no surprises, especially since the future has surprised us so often in the past."[164]

Clearly, information technology is driving the new world of jobs, services, and leisure, and nothing is going to stop it. People pursuing careers find the rules are changing very rapidly. Up-to-date skills are becoming ever more crucial. Job descriptions of all kinds are metamorphosing, and even familiar jobs are becoming more demanding.

Where will you be in all this? Today, experts advise, you must be willing to continually upgrade your skills, to specialize, and to market yourself. In a world of breakneck change, you can still thrive.

Experience Box
Student Use of Computers: Some Controversies

Information technology is very much a part of the college experience, of course. Elsewhere we discussed such matters as distance learning, web research, online plagiarism, and internet addiction. Here we describe some other issues regarding students' computer use.

Using Computers in the Classroom

Although they're more expensive than desktop computers, laptops are useful because you can take them not only to libraries, to help with reading or term-paper notes, but also to class to use in taking lecture notes.[165] You might even try using a palmtop computer, which costs half as much as a laptop and comes with word processing and spreadsheet software.[166] Be aware, however, that the small palmtop keyboard requires some dexterity. With either laptop or palmtop, battery life may also be a factor.

The use of computers in classrooms is still controversial in certain quarters. Some campuses allow students to bring laptops to class but are imposing rules about what they are allowed to do. "More students are sending instant messages to one another (chatting and note passing, 21st-century style)," says one article about business-school campuses, "day trading (as opposed to daydreaming), and even starting their own companies, all in class."[167] Computer-generated voices that accompany the downloading of email attachments are another classroom annoyance.

Notes Posted on the Web

Wouldn't it be nice, when you're out sick, to be able to go to a website and get the notes of lectures for the classes you missed?

This is possible on campuses served by some commercial firms. Many such services are free, since the firms try to generate revenue by selling online advertising, but some charge a fee. Some colleges even have their own operations, such as Black Lightning Notes at the University of California, Berkeley.

Such services can be a real help to students who learn best by reading rather than by hearing. They also provide additional reinforcement to students who feel they have not been able to grasp all of a professor's ideas during the lecture. However, they are no substitute for the classroom experience, with its spontaneous exchange of ideas. Moreover, as one writer points out, "the very act of taking notes—not reading somebody else's notes, no matter how stellar—is a way of engaging the material, wrestling with it, struggling to comprehend or take issue."[168] In other words, you'll be better able to remember the lecture if you've reinforced the ideas by writing them down yourself.

Some faculty members have no problem with note-taking operations. Indeed, for some time, some professors have been posting their lecture notes, practice exams, and reading lists on web pages linked to the World Lecture Hall *(www.texas.edu).* Others disapprove, however. Among the criticisms: (1) Note-taking services don't always ask permission. (2) Instructors are reluctant to share their unpublished research in class if they think their ideas might end up posted in a public place and ripped off. (3) They might not wish to share controversial opinions with students if their views might be criticized in a worldwide forum. (4) Students might not come to class, especially if they think the lectures are boring or if they are given to chronic oversleeping or hangovers. (5) The notes may be sloppy, inaccurate, or incomplete.

Sloppy notes are a serious concern. "The notes were very, very inaccurate and included gross errors," said one Yale University professor of environmental studies who checked out a website with material on her course. "For example, there was a review of an educational film that had obviously been shown in another course—or perhaps it was shown by a visiting professor last year."[169] A newspaper reporter investigating lecture-note websites took 118 pages of notes during her seven class visits, whereas the notes posted on the web of those same events "were, at most, a few paragraphs per lecture."[170]

Bottom line: These websites may be helpful, but they are certainly no substitute for going to class.

Online Student Evaluations

Student evaluations of courses and professors have moved to the internet. These can be useful, but since such evaluations are often expressed anonymously, they can also be inaccurate and unfair—even vicious, if students receiving poor grades take revenge by vilifying their instructors online. According to a spokesperson for the American Federation of Teachers, there has been a rise in false-accusation cases, which can have a severe impact on instructors' careers.[171]

Justine Heinze Giardello, 18, whose father is a professor, says students don't understand how hard instructors work. And she feels that students should not be able to bash teachers in a public forum. "I think it is horrible," she said, commenting on postings to a student-run teacher-evaluation website in which an openly gay community-college instructor was anonymously called "homomanic," "racist," and "mentally ill." "How would we feel," she asks, "if there was a student review judging us on a personal basis?"[172]

In reading teacher evaluations, it's useful to pay close attention to how civil and fair-minded the reports seem. And, of course, you should try to be as considerate as possible when writing them.

Summary

biometrics (p. 480) Science of measuring individual body characteristics. Why it's important: *Biometric technology is used in some computer security systems to restrict user access. Biometric devices, such as those that use fingerprints, eye scans, palm prints, and face recognition, authenticate a person's identity by verifying his or her physical or behavioral characteristics.*

botnet (p. 474) Short for *robot network*. A network of computers compromised by means of a Trojan horse that plants instructions within each PC to wait for commands from the person controlling that network. Why it's important: *If your PC becomes part of a botnet, it may be victimized by spam, phishing attacks, or denial-of-service attacks.*

computer crime (p. 471) Crime of two types: (1) an illegal act perpetrated against computers or telecommunications; (2) the use of computers or telecommunications to accomplish an illegal act. Why it's important: *Crimes against information technology include theft—of hardware, of software, of computer time, of cable or telephone services, or of information. Other illegal acts are crimes of malice and destruction.*

disaster-recovery plan (p. 481) Method of restoring information-processing operations that have been halted by destruction or accident. Why it's important: *Such a plan is important if an organization desires to resume computer operations quickly.*

encryption (p. 480) Process of altering data so that it is not usable unless the changes are undone. Why it's important: *Encryption is clearly useful for some organizations, especially those concerned with trade secrets, military matters, and other sensitive data. Some maintain that encryption will determine the future of e-commerce, because transactions cannot flourish over the internet unless they are secure.*

Evil Twin attack (p. 474) A variant on conventional phishing, in which a hacker or cracker sets up a Wi-Fi hot spot or access point that makes your computer think it's accessing a safe public network or your home network and then monitors your communications. Why it's important: *The hacker can steal data you enter into a website, if it doesn't have the right security measures.*

security (p. 478) System of safeguards for protecting information technology against disasters, system failures, and unauthorized access that can result in damage or loss. Five components of security are deterrents to computer crime, identification and access, encryption, protection of software and data, and disaster-recovery plans. Why it's important: *With proper security, organizations and individuals can minimize information technology losses from disasters, system failures, and unauthorized access.*

zombie (p. 474). Also known as a *drone*. A computer taken over covertly and programmed to respond to instructions sent remotely, often by instant-messaging channels. Why it's important: *People whose computers become zombies may be victimized by spam messages, phishing attacks, or denial-of-service attacks.*

Chapter Review

stage 1 LEARNING MEMORIZATION

"I can recognize and recall information."

Self-Test Questions

1. The five limitations to remember when using databases for research are _____, _____, _____, _____, and _____.

2. So that information-processing operations can be restored after destruction or accident, organizations should adopt a _____.

3. _____ is the altering of data so that it is not usable unless the changes are undone.

4. _____ is incomplete, outdated, or otherwise inaccurate data.

5. An error in a program that causes it not to work properly is called a _____.

6. CERT, created by the U.S. Department of Defense, stands for _____.

7. Software obtained illegally is called _____ software.

Multiple-Choice Questions

1. Which of the following are crimes against computers and communications?
   a. natural hazards
   b. software theft
   c. information theft
   d. software bugs
   e. procedural errors

2. Which of the following methods or means are used to safeguard computer systems?
   a. signatures
   b. keys
   c. physical traits of users
   d. worms
   e. Internet2

3. Which of the following are threats to computer systems?
   a. zombies
   b. evil twins
   c. crackers
   d. Trojan horses
   e. all of these

True/False Questions

T F   1. The category of computer crimes includes dirty-data problems.

T F   2. Software bugs include procedural errors.

T F   3. Two types of encryption are public-key and private-key.

T F   4. It is impossible to detect when a photo has been morphed.

T F   5. If they got it from an employee's credit report, employers must destroy any personal information about the employee before they throw it out.

stage 2 LEARNING COMPREHENSION

"I can recall information in my own terms and explain it to a friend."

Short-Answer Questions

1. What is the difference between a hacker and a cracker?
2. Give some examples of dirty data.
3. Briefly describe how encryption works.
4. What is a procedural error?
5. Name five threats to computers and communications systems.
6. The definition of computer crime distinguishes between two types. What are they?
7. What is phishing?
8. If your employer is checking on you, what might he or she be watching for?
9. What are three ways of verifying legitimate right of access to a computer system?
10. Name four environmental problems caused by computers.
11. What is the "digital divide"?

"I can apply what I've learned, relate these ideas to other concepts, build on other knowledge, and use all these thinking skills to form a judgment."

Knowledge in Action

1. What, in your opinion, are the most significant disadvantages of using computers? What do you think can be done about these problems?

2. What's your opinion about the issue of free speech on an electronic network? Research some recent legal decisions in various countries, as well as some articles on the topic. Should the contents of messages be censored? If so, under what conditions?

3. Research the problems of stress and isolation experienced by computer users in the United States, Japan, and one other country. Write a brief report on your findings.

4. How can you ensure that information is accurate and complete? Make a list of things to remember when you are doing research on the internet for a term paper.

Web Exercises

1. What is 128-bit encryption? Is your browser equipped? Visit the following websites to learn more about browser security:

   www.consumersenergy.com/welcome.htm
   products/index.asp?ASID=106
   http://iasweb.com/articles/webshopping.html
   www.us-cert.gov/cas/tips/ST05-001.html

   Go to *http://bcheck.scanit.be/bcheck/* and *www.jasons-toolbox.com/BrowserSecurity/* to run a browser security check on your computer. What are the results? What do you plan to do about any problems that were detected?

2. Here's a way to semi-encrypt an email message to a friend. Type out your message; then go to the Edit menu and choose *Select All.* Next go to the Format menu and select *Font;* then choose a font such as Wingdings that doesn't use letters. When your friend receives your message, he or she need only change the font back to an understandable one (such as Times New Roman) in order to read it. This certainly isn't high-level encryption, but it's a fun activity to try.

3. Visit the following website, which discusses general Internet addiction:

   www.netaddiction.com/
   http://library.albany.edu/briggs/addiction.html
   www.addictions.org/internet.htm
   www.addictionrecov.org/wrkguide_www.htm

   Do you have an internet problem? What do you plan to do about it?

4. If you're spending too much time indoors using a PC or watching TV, you might want to consider going outside. Visit the following websites to learn about indoor air versus outdoor air:

   www.epa.gov/iaq/ia-intro.html
   www.pp.okstate.edu/ehs/links/iaq.htm
   www.lbl.gov/Education/ELSI/pollution-main.html

   Also, to read more about internet addiction that may cause isolation from the real world, visit

   www.sciencenews.org/20000226/fob8.asp
   http://news-service.stanford.edu/news/2005/february23/internet-022305.html
   http://archives.cnn.com/2000/HEALTH/06/13/internet.addiction.wmd/
   www.eurekalert.org/pub_releases/2004-02/uoa-nhi020404.php
   www.psychadvisor.com/inthenews/viewart.cfm?ItemID=9

   What's your opinion on this issue?

   The following two websites discuss computer health and safety:

   www.ics.uci.edu/~chair/comphealth2.html
   www.osha.gov/SLTC/computerworkstation/
   www.usernomics.com/workplace-ergonomics.html

5. How much do you know about the U.S. Department of Homeland Security's rules and policies concerning computer users' privacy? Go to *www.dhs.gov* and find out how you are affected.

Systems Analysis & Programming

Software Development, Programming, & Languages

Chapter Topics & Key Questions

10.1   **Systems Development: The Six Phases of Systems Analysis & Design** What are the six phases of the systems development life cycle?

10.2   **Programming: A Five-Step Procedure** What is programming, and what are the five steps in accomplishing it?

10.3   **Five Generations of Programming Languages** What are the five generations of programming languages?

10.4   **Programming Languages Used Today** What are some third-generation languages, and what are they used for?

10.5   **Object-Oriented & Visual Programming** How do OOP and visual programming work?

10.6   **Markup & Scripting Languages** What do markup and scripting languages do?

Organizations can make mistakes, of course, and big organizations can make really big mistakes.

California's state Department of Motor Vehicles' databases needed to be modernized, and Tandem Computers said it could do it. "The fact that the DMV's database system, designed around an old IBM-based platform, and Tandem's new system were as different as night and day seemed insignificant at the time to the experts involved," said one writer who investigated the project.[1] The massive driver's license database, containing the driving records of more than 30 million people, first had to be "scrubbed" of all information that couldn't be translated into the language used by Tandem computers. One such scrub yielded 600,000 errors. Then the DMV had to translate all its IBM programs into the Tandem language. "Worse, DMV really didn't know how its current IBM applications worked anymore," said the writer, "because they'd been custom-made decades before by long departed programmers and rewritten many times since." Eventually the project became a staggering $44 million loss to California's taxpayers.

10.1

Systems Development: The Six Phases of Systems Analysis & Design

What are the six phases of the systems development life cycle?

Needless to say, not all mistakes are so huge. Computer foul-ups can range from minor to catastrophic. But this example shows how important planning is, especially when an organization is trying to launch a new kind of system. The best way to avoid such mistakes is to employ systems analysis and design.

But, you may say, you're not going to have to wrestle with problems on the scale of motor-vehicle departments. That's a job for computer professionals. You're mainly interested in using computers and communications to increase your own productivity. Why, then, do you need to know anything about systems analysis and design?

In many types of jobs, you may find that your department or your job is the focus of a study by a systems analyst. Knowing how the procedure works will help you better explain how your job works or what goals your department is supposed to achieve. In progressive companies, management is always interested in suggestions for improving productivity. Systems analysis provides a method for developing such ideas.

The Purpose of a System

How would I define a system, and what is its purpose?

A _system_ is defined as a collection of related components that interact to perform a task in order to accomplish a goal. A system may not work very well, but it is nevertheless a system. The point of systems analysis and design is to ascertain how a system works and then take steps to make it better.

An organization's computer-based information system consists of hardware, software, people, procedures, and data, as well as communications setups. These components work together to provide people with information for running the organization.

Getting the Project Going: How It Starts, Who's Involved

Who are the three kinds of users of a project?

A single individual who believes that something badly needs changing is all it takes to get the project rolling. An employee may influence a supervisor. A customer or supplier may get the attention of someone in higher management. Top management may decide independently to take a look at a system that seems inefficient. A steering committee may be formed to decide which of many possible projects should be worked on.

Participants in the project are of three types:

- **Users:** The system under discussion should *always* be developed in consultation with users, whether floor sweepers, research scientists, or customers. Indeed, if user involvement in analysis and design is inadequate, the system may fail for lack of acceptance.
- **Management:** Managers within the organization should also be consulted about the system.
- **Technical staff:** Members of the company's information systems (IS) department, consisting of systems analysts and programmers, need to be involved. For one thing, they may have to execute the project. Even if they don't, they will have to work with outside IS people contracted to do the job.

Complex projects will require one or several systems analysts. A **_systems analyst_ is an information specialist who performs systems analysis, design, and implementation.** The analyst's job is to study the information and communications needs of an organization and determine what changes are required to deliver better information to the people who need it. "Better" information means information that is summarized in the acronym *CART*—complete, accurate, relevant, and timely. The systems analyst achieves this goal through the problem-solving method of systems analysis and design.

The Six Phases of Systems Analysis & Design

What are the six phases of the systems development life cycle?

Systems analysis and design is a six-phase problem-solving procedure for examining an information system and improving it. The six phases make up what is called the *systems development life cycle*. The **_systems development life cycle (SDLC)_ is the step-by-step process that many organizations follow during systems analysis and design.**

Whether applied to a Fortune 500 company or a three-person engineering business, the six phases in systems analysis and design are as shown in the illustration. (● *See Panel 10.1.*) Phases often overlap, and a new one may start before the old one is finished. After the first four phases, management must decide whether to proceed to the next phase. User input and review is a critical part of each phase.

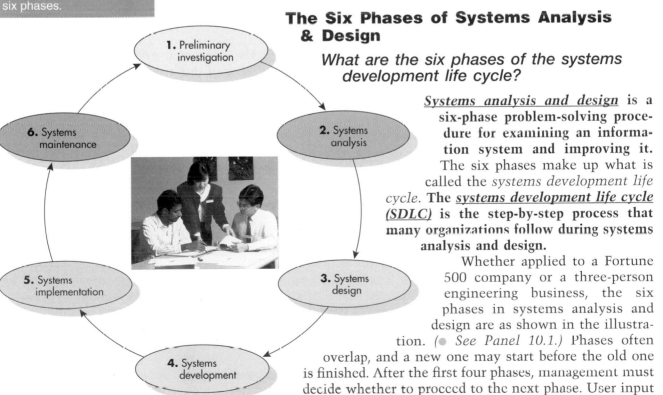

1. Preliminary investigation
2. Systems analysis
3. Systems design
4. Systems development
5. Systems implementation
6. Systems maintenance

Systems Analysis & Programming

499

● **PANEL 10.2**

First phase:
Preliminary
investigation

1. Conduct preliminary analysis. This includes stating the objectives, defining nature and scope of the problem.
2. Propose alternative solutions: leave system alone, make it more efficient, or build a new system.
3. Describe costs and benefits of each solution.
4. Submit a preliminary plan with recommendations.

The First Phase: Conduct a Preliminary Investigation

What four steps are involved in the first phase?

The objective of **Phase 1**, _preliminary investigation_, is to conduct a preliminary analysis, propose alternative solutions, describe costs and benefits, and submit a preliminary plan with recommendations. (● *See Panel 10.2.)*

1. CONDUCT THE PRELIMINARY ANALYSIS In this step, you need to find out what the organization's objectives are and the nature and scope of the problem under study. Even if a problem pertains only to a small segment of the organization, you cannot study it in isolation. You need to find out what the objectives of the organization itself are. Then you need to see how the problem being studied fits in with them.

2. PROPOSE ALTERNATIVE SOLUTIONS In delving into the organization's objectives and the specific problem, you may have already discovered some solutions. Other possible solutions can come from interviewing people inside the organization, clients or customers affected by it, suppliers, and consultants. You can also study what competitors are doing. With this data, you then have three choices. You can leave the system as is, improve it, or develop a new system.

3. DESCRIBE THE COSTS & BENEFITS Whichever of the three alternatives is chosen, it will have costs and benefits. In this step, you need to indicate what these are. Costs may depend on benefits, which may offer savings. A broad spectrum of benefits may be derived. A process may be speeded up, streamlined through elimination of unnecessary steps, or combined with other processes. Input errors or redundant output may be reduced. Systems and subsystems may be better integrated. Users may be happier with the system. Customers' or suppliers' interactions with the system may be more satisfactory. Security may be improved. Costs may be cut.

4. SUBMIT A PRELIMINARY PLAN Now you need to wrap up all your findings in a written report. The readers of this report will be the executives who are in a position to decide in which direction to proceed—make no changes, change a little, or change a lot—and how much money to allow the project. You should describe the potential solutions, costs, and benefits and indicate your recommendations.

The Second Phase: Do an Analysis of the System

What three steps are involved in analyzing a system?

The objective of **Phase 2**, _systems analysis_, is to gather data, analyze the data, and write a report. (● *See Panel 10.3.)* In this second phase of the SDLC, you will follow the course that management has indicated after having read your Phase 1 feasibility report. We are assuming that management has ordered you to perform Phase 2—to do a careful analysis or study of the existing system in order to understand how the new system you proposed would differ. This analysis will also consider how people's positions and tasks will have to change if the new system is put into effect.

1. GATHER DATA In gathering data, you will review written documents, interview employees and managers, develop questionnaires, and observe people and processes at work.

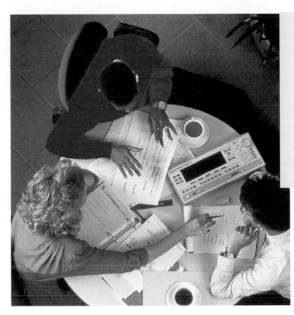

1. Gather data, using tools of written documents, interviews, questionnaires, and observations.
2. Analyze the data, using modeling tools: grid charts, decision tables, data flow diagrams, systems flow-charts, connectivity diagrams.
3. Write a report.

2. ANALYZE THE DATA Once the data has been gathered, you need to come to grips with it and analyze it. Many analytical tools, or modeling tools, are available. **_Modeling tools_ enable a systems analyst to present graphic, or pictorial, representations of a system. An example of a modeling tool is a _data flow diagram (DFD)_, which graphically shows the flow of data through a system—that is, the essential processes of a system, along with inputs, outputs, and files.** (● *See Panel 10.4.*)

CASE tools may also be used during the analysis phase, as well as in most other phases. **_CASE (computer-aided software engineering) tools_ are programs that automate various activities of the SDLC.** This technology is intended to speed up the process of developing systems and to improve the quality of the resulting systems. Such tools can generate and store diagrams, produce documentation, analyze data relationships, generate computer code, produce graphics, and provide project management functions. (Project management software, discussed in Chapter 3, consists of programs used to plan, schedule, and control the people, costs, and resources required to complete a project on time.) Examples of such programs are Analyst Pro, Visible Analyst, and System Architect.

● **PANEL 10.4**
Data flow diagram
Example of data flow diagram and explanation of symbols.

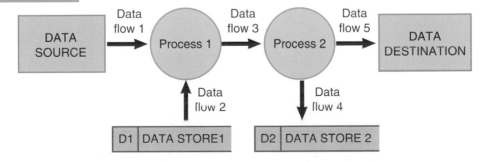

Explanation of standard data flow diagram symbols

Name or Name
Terminator Symbols (entity name)
(person or organization outside the system boundaries)

File Name or File Name
Data Store Symbol

Name or Name
Process Symbol

Name
Data Flow Symbol
(inputs and outputs)

3. WRITE A REPORT Once you have completed the analysis, you need to document this phase. This report to management should have three parts. First, it should explain how the existing system works. Second, it should explain the problems with the existing system. Finally, it should describe the requirements for the new system and make recommendations on what to do next.

At this point, not a lot of money will have been spent on the systems analysis and design project. If the costs of going forward seem prohibitive, this is a good time for the managers reading the report to call a halt. Otherwise, you will be asked to move to Phase 3.

The Third Phase: Design the System

What three steps are involved in designing a system?

The objective of **Phase 3, _systems design_, is to do a preliminary design and then a detail design and to write a report.** (● *See Panel 10.5.*) In this third phase of the SDLC, you will essentially create a "rough draft" and then a "detail draft" of the proposed information system.

1. DO A PRELIMINARY DESIGN A *preliminary design* describes the general functional capabilities of a proposed information system. It reviews the system requirements and then considers major components of the system. Usually several alternative systems (called *candidates*) are considered, and the costs and the benefits of each are evaluated.

Tools used in the design phase may include CASE tools and project management software.

Prototyping is often done at this stage. **_Prototyping_ refers to using workstations, CASE tools, and other software applications to build working models of system components so that they can be quickly tested and evaluated.** Thus, **a _prototype_ is a limited working system , or part of one, developed to test design concepts.** A prototype, which may be constructed in just a few days, allows users to find out immediately how a change in the system might benefit them. For example, a systems analyst might develop a menu as a possible screen display, which users could try out. The menu can then be redesigned or fine-tuned, if necessary.

2. DO A DETAIL DESIGN A *detail design* describes how a proposed information system will deliver the general capabilities described in the preliminary design. The detail design usually considers the following parts of the system in this order: output requirements, input requirements, storage requirements, processing requirements, and system controls and backup.

● **PANEL 10.5**
Third phase: Systems design

1. Do a preliminary design, using CASE tools, prototyping tools, and project management software, among others.
2. Do a detail design, defining requirements for output, input, storage, and processing and system controls and backup.
3. Write a report.

3. WRITE A REPORT All the work of the preliminary and detail designs will end up in a large, detailed report. When you hand over this report to senior management, you will probably also make some sort of presentation or speech.

The Fourth Phase: Develop the System

What are the three steps required in developing a system?

In **Phase 4, _systems development_, the systems analyst or others in the organization develop or acquire the software, acquire the hardware, and then test the system.** (● *See Panel 10.6.)* Depending on the size of the project, this phase will probably involve the organization in spending substantial sums of money. It could also involve spending a lot of time. However, at the end you should have a workable system.

1. DEVELOP OR ACQUIRE THE SOFTWARE During the design stage, the systems analyst may have had to address what is called the "make-or-buy" decision, but that decision certainly cannot be avoided now. In the *make-or-buy decision,* you decide whether you have to create a program—have it custom-written—or buy it, meaning simply purchase an existing software package. Sometimes programmers decide they can buy an existing program and modify it rather than write it from scratch.

If you decide to create a new program, then the question is whether to use the organization's own staff programmers or to hire outside contract programmers (outsource it). Whichever way you go, the task could take many months.

Programming is an entire subject unto itself, which we discuss later in this chapter, along with programming languages.

2. ACQUIRE HARDWARE Once the software has been chosen, the hardware to run it must be acquired or upgraded. It's possible your new system will not require any new hardware. It's also possible that the new hardware will cost millions of dollars and involve many items: microcomputers, mainframes, monitors, modems, and many other devices. The organization may find it's better to lease than to buy some equipment, especially since, as we mentioned (Moore's law), chip capability has traditionally doubled every 18 months.

3. TEST THE SYSTEM With the software and hardware acquired, you can now start testing the system. Testing is usually done in two stages: unit testing, then system testing.

- **Unit testing:** In *unit testing,* the performance of individual parts is examined, using test (made-up or sample) data. If the program is written as a collaborative effort by multiple programmers, each part of the program is tested separately.

● **PANEL 10.6**
Fourth phase: Systems development

1. Develop or acquire the software.
2. Acquire the hardware.
3. Test the system.

- **System testing:** In *system testing,* the parts are linked together, and test data is used to see if the parts work together. At this point, actual organization data may be used to test the system. The system is also tested with erroneous data and massive amounts of data to see if the system can be made to fail ("crash").

At the end of this long process, the organization will have a workable information system, one ready for the implementation phase.

The Fifth Phase: Implement the System

What's required to implement a new system?

Whether the new information system involves a few handheld computers, an elaborate telecommunications network, or expensive mainframes, the fifth phase will involve some close coordination in order to make the system not just workable but successful. **Phase 5, *systems implementation*, consists of converting the hardware, software, and files to the new system and training the users.** (● *See Panel 10.7.*)

1. CONVERT TO THE NEW SYSTEM   Conversion, the process of transition from an old information system to a new one, involves converting hardware, software, and files. There are four strategies for handling conversion: direct, parallel, phased, and pilot.

- **Direct implementation:** This means that the user simply stops using the old system and starts using the new one. The risk of this method should be evident: What if the new system doesn't work? If the old system has truly been discontinued, there is nothing to fall back on.

- **Parallel implementation:** This means that the old and new systems are operated side by side until the new system has shown it is reliable, at which time the old system is discontinued. Obviously there are benefits in taking this cautious approach. If the new system fails, the organization can switch back to the old one. The difficulty with this method is the expense of paying for the equipment and people to keep two systems going at the same time.

- **Phased implementation:** This means that parts of the new system are phased in separately—either at different times (parallel) or all at once in groups (direct).

- **Pilot implementation:** This means that the entire system is tried out, but only by some users. Once the reliability has been proved, the system is implemented with the rest of the intended users. The pilot approach still has its risks, since all of the users of a particular group

1. Convert hardware, software, and files through one of four types of conversions: direct, parallel, phased, or pilot.
2. Train the users.

are taken off the old system. However, the risks are confined to a small part of the organization.

2. TRAIN THE USERS Various tools are available to familiarize users with a new system—from documentation (instruction manuals) to videotapes to live classes to one-on-one, side-by-side teacher-student training. Sometimes training is done by the organization's own staffers; at other times it is contracted out.

The Sixth Phase: Maintain the System

Is maintaining a system an elaborate process?

Phase 6, _systems maintenance_, adjusts and improves the system by having system audits and periodic evaluations and by making changes based on new conditions. (● *See Panel 10.8.*) Even with the conversion accomplished and the users trained, the system won't just run itself. There is a sixth—and continuous—phase in which the information system must be monitored to ensure that it is successful. Maintenance includes not only keeping the machinery running but also updating and upgrading the system to keep pace with new products, services, customers, government regulations, and other requirements.

After some time, maintenance costs will accelerate as attempts continue to keep the system responsive to user needs. At some point, these maintenance costs become excessive, indicating that it may be time to start the entire SDLC again.

10.2

Programming: A Five-Step Procedure

What is programming, and what are the five steps in accomplishing it?

To see how programming works, we must understand what constitutes a program. **A _program_ is a list of instructions that the computer must follow to process data into information.** The instructions consist of statements used in a programming language, such as BASIC. Examples are programs that do word processing, desktop publishing, or payroll processing.

The decision whether to buy or create a program forms part of Phase 4 in the systems development life cycle. Once the decision is made to develop a new system, the programmer goes to work.

A program, we said, is a list of instructions for the computer. **_Programming_ is a five-step process for creating that list of instructions.** Programming is sometimes called *software engineering*; however, the latter term includes in its definition the use of best-practice processes to create and/or maintain

The sixth phase is to keep the system running through system audits and periodic evaluations.

Who wrote this, and when? "Systems are seductive. They promise to do a hard job faster, better, and more easily than you could do it by yourself. But if you set up a system, you are likely to find your time and effort now being consumed in the care and feeding of the system itself. New problems are created by its very presence. Once set up, it won't go away, it grows and encroaches. It begins to do strange and wonderful things. Breaks down in ways you never thought possible. It kicks back, gets in the way, and opposes its own proper function. Your own perspective becomes distorted by being in the system. You become anxious and push on it to make it work. Eventually you come to believe that the misbegotten product it so grudgingly delivers is what you really wanted all the time. At that point encroachment has become complete . . . you have become absorbed . . . you are now a systems person!"

software, whether for groups or individuals, in an attempt to get rid of the usual haphazard methods that have plagued the software industry. Software engineering involves the establishment and use of recognized engineering principles to obtain software that is reliable and works efficiently. It requires the application of a systematic, disciplined, quantifiable approach to the development, operation, and maintenance of software.

The five steps in the programming process are as follows:

1. Clarify/define the problem—include needed output, input, processing requirements.
2. Design a solution—use modeling tools to chart the program.
3. Code the program—use a programming language's syntax, or rules, to write the program.
4. Test the program—get rid of any logic errors, or "bugs," in the program ("debug" it).
5. Document and maintain the program—include written instructions for users, explanation of the program, and operating instructions.

Coding—sitting at the keyboard and typing words into a computer—is how many people view programming. As we see, however, it is only one of the five steps.

The First Step: Clarify the Programming Needs

How are programming needs clarified?

The *problem clarification* (definition) step consists of six mini-steps— clarifying program objectives and users, outputs, inputs, and processing tasks; studying the feasibility of the program; and documenting the analysis. (● *See Panel 10.9.*) Let us consider these six mini-steps.

1. CLARIFY OBJECTIVES & USERS You solve problems all the time. A problem might be deciding whether to take a required science course this term or next or selecting classes that allow you also to fit a job into your schedule. In such cases, you are specifying your objectives. Programming works the same way. You need to write a statement of the objectives you are trying to accomplish—the problem you are trying to solve. If the problem is that your company's systems analysts have designed a new computer-based payroll-processing proposal and brought it to you as the programmer, you need to clarify the programming needs.

You also need to make sure you know who the users of the program will be. Will they be people inside the company, outside, or both? What kind of skills will they bring?

2. CLARIFY DESIRED OUTPUTS Make sure you understand the outputs— what the system designers want to get out of the system—before you specify

● **PANEL 10.9**
First step: Clarify programming needs

1. Clarify objectives and users.
2. Clarify desired outputs.
3. Clarify desired inputs.
4. Clarify desired processing.
5. Double–check feasibility of implementing the program.
6. Document the analysis.

the inputs. For example, what kind of hardcopy is wanted? What information should the outputs include? This step may require several meetings with systems designers and users to make sure you're creating what they want.

3. CLARIFY DESIRED INPUTS Once you know the kind of outputs required, you can then think about input. What kind of input data is needed? In what form should it appear? What is its source?

4. CLARIFY THE DESIRED PROCESSING Here you make sure you understand the processing tasks that must occur in order for input data to be processed into output data.

5. DOUBLE-CHECK THE FEASIBILITY OF IMPLEMENTING THE PROGRAM Is the kind of program you're supposed to create feasible within the present budget? Will it require hiring a lot more staff? Will it take too long to accomplish?

Occasionally programmers suggest to managers that they buy an existing program and modify it rather than having one written from scratch.

6. DOCUMENT THE ANALYSIS Throughout program clarification, programmers must document everything they do. This includes writing objective specifications of the entire process being described.

The Second Step: Design the Program

How is a program designed?

Assuming the decision is to make, or custom-write, the program, you then move on to design the solution specified by the systems analysts. To design the solution, one first needs to create an algorithm. **An _algorithm_ is a formula or set of steps for solving a particular problem.** To be an algorithm, a set of rules must be unambiguous and have a clear stopping point. We use algorithms every day. For example, a recipe for baking bread is an algorithm. Most programs, with the exception of some artificial intelligence applications, consist of algorithms. In computer programming, there are often different algorithms to accomplish any given task, and each algorithm has specific advantages and disadvantages in different situations. Inventing elegant algorithms—algorithms that are simple and require the fewest steps possible—is one of the principal challenges in programming.

Algorithms can be expressed in various ways. **In the _program design_ step, the software is designed in two mini-steps. First, the program logic is determined through a top-down approach and modularization, using a _hierarchy chart_. Then it is designed in detail, either in narrative form, using _pseudocode_, or graphically, using _flowcharts_. (●** _See Panel 10.10._)

info!

Algorithm: What's the Origin of This Word?

The word _algorithm_ originates in the name of an Arab mathematician, Al-Khowarizmi, of the court of Mamun in Baghdad in the 800s. His treatises on Hindu arithmetic and on algebra made him famous. He is also said to have given algebra its name. Much of the mathematical knowledge of medieval Europe was derived from Latin translations of his works.

● **PANEL 10.10**
Second step:
Program design

1. Determine program logic through top-down approach and modularization, using a hierarchy chart.
2. Design details using pseudocode and/or flowcharts, preferably involving control structures.

It used to be that programmers took a kind of seat-of-the-pants approach to programming. Programming was considered an art, not a science. Today, however, most programmers use a design approach called *structured programming*. **Structured programming takes a top-down approach that breaks programs into modular forms.** It also uses standard logic tools called *control structures (sequencel, selection, case, and iteration).*

The point of structured programming is to make programs more efficient (with fewer lines of code) and better organized (more readable) and to have better notations so that they have clear and correct descriptions.

The two mini-steps of program design are as follows.

1. DETERMINE THE PROGRAM LOGIC, USING A TOP-DOWN APPROACH Determining the program logic is like outlining a long term paper before you proceed to write it. **Top-down program design proceeds by identifying the top element, or module, of a program and then breaking it down in hierarchical fashion to the lowest level of detail.** The top-down program design is used to identify the program's processing steps, or modules. After the program is designed, the actual coding proceeds from the bottom up, using the modular approach.

- **Modularization:** The concept of modularization is important. Modularization dramatically simplifies program development, because each part can be developed and tested separately. **A module is a processing step of a program. Each module is made up of logically related program statements.** (Sometimes a module is called a *subprogram* or *subroutine*.) An example of a module might be a programming instruction that simply says "Open a file, find a record, and show it on the display screen." It is best if each module has only a single function, just as an English paragraph should have a single, complete thought. This rule limits the module's size and complexity.

- **Top-down program design:** Top-down program design can be represented graphically in a hierarchy chart. **A hierarchy chart, or structure chart, illustrates the overall purpose of the program, by identifying all the modules needed to achieve that purpose and the relationships among them.** (● *See Panel 10.11.*) It works from the general down to the specific, starting with the top-level (high-level) view of what the program is to do. Then each layer refines and expands the previous one until the bottom layer can be made into specific programming modules. The program must move in sequence from one module to the next until all have been processed. There must be three principal modules corresponding to the three principal computing operations—input, processing, and output. (In Panel 10.11 they are "Read input," "Calculate pay," and "Generate output.")

2. DESIGN DETAILS, USING PSEUDOCODE AND/OR FLOWCHARTS Once the essential logic of the program has been determined, through the use of top-down programming and hierarchy charts, you can go to work on the details.

There are two ways to show details—write them or draw them; that is, use *pseudocode* or use *flowcharts*. Most projects use both methods.

- **Pseudocode: Pseudocode is a method of designing a program using normal human-language statements to describe the logic and the processing flow.** (● *See Panel 10.12.*) Pseudocode is like an outline or summary form of the program you will write.

Sometimes pseudocode is used simply to express the purpose of a particular programming module in somewhat general terms. With the use of such terms as *IF, THEN*, or *ELSE*, however, the pseudocode follows the rules of

● PANEL 10.11
A hierarchy chart

This represents a top-down design for a payroll program. Here the modules, or processing steps, are represented from the highest level of the program down to details. The three principal processing operations—input, processing, and output—are represented by the modules in the second layer: "Read input," "Calculate pay," and "Generate output." Before tasks at the top of the chart can be performed, all the ones below must be performed. Each module represents a logical processing step.

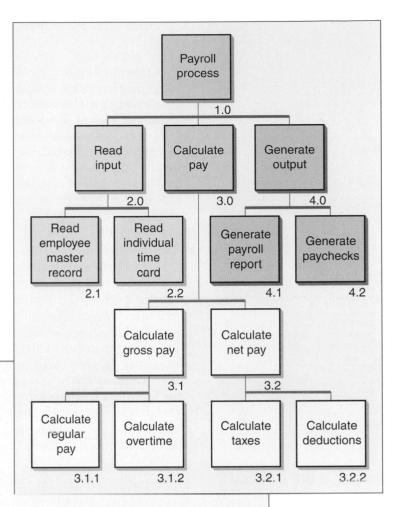

1. Each module must be of manageable size.

2. Each module should be independent and have a single function.

3. The functions of input and output are clearly defined in separate modules.

4. Each module has a single entry point (execution of the program module always starts at the same place) and a single exit point (control always leaves the module at the same place).

5. If one module refers to or transfers control to another module, the latter module returns control to the point from which it was "called" by the first module.

● PANEL 10.12
Pseudocode

```
START
DO WHILE (so long as) there are records
        Read a customer billing account record
        IF today's date is greater than 30 days from
        date of last customer payment
            Calculate total amount due
            Calculate 5% interest on amount due
            Add interest to total amount due to calculate
            grand total
            Print on invoice overdue amount
        ELSE
            Calculate total amount due
        ENDIF
        Print out invoice
END DO
END
```

control structures, an important aspect of structured programming, as we shall explain.

- **Program flowcharts:** We described system flowcharts in the previous chapter. Here we consider program flowcharts. **A _program flowchart_ is a chart that graphically presents the detailed series of steps (algorithm, or logical flow) needed to solve a programming problem.** The flowchart uses standard symbols—called *ANSI symbols,* after the American National Standards Institute, which developed them. (● *See Panel 10.13.)*

  The symbols at the left of the drawing might seem clear enough. But how do you figure out the logic of a program? How do you reason the program out so that it will really work? The answer is to use control structures, as explained next.

- **Control structures:** When you're trying to determine the logic behind something, you use words like *if* and *then* and *else.* (For example, without using these exact words, you might reason along these lines: "If she comes over, then we'll go out to a movie, else I'll just stay in and watch TV.") Control structures make use of the same words. **A _control structure_, or _logic structure_, is a structure that controls the logical sequence in which computer program instructions are executed. In structured program design, three control structures are used to form the logic of a program: sequence, selection, and iteration (or loop).** (● *See Panel 10.14 on page 512.)* These are the tools with which you can write structured programs and take a lot of the guesswork out of programming. (Additional variations of these three basic structures are also used.)

COMPARING THE THREE CONTROL STRUCTURES One thing that all three control structures have in common is *one entry* and *one exit.* The control structure is entered at a single point and exited at another single point. This helps simplify the logic so that it is easier for others following in a programmer's footsteps to make sense of the program. (In the days before this requirement was instituted, programmers could have all kinds of variations, leading to the kind of incomprehensible program known as *spaghetti code.*)

Let us consider the three control structures:

- **Sequence control structure:** In the *sequence control structure,* one program statement follows another in logical order. For instance, in the example shown in Panel 10.14, there are two boxes ("Statement" and "Statement"). One box could say "Open file," the other "Read a record." There are no decisions to make, no choices between "yes" or "no." The boxes logically follow one another in sequential order.

- **Selection control structure:** The *selection control structure*—also known as an IF-THEN-ELSE *structure*—represents a choice. It offers two paths to follow when a decision must be made by a program. An example of a selection structure is as follows:

  IF a worker's hours in a week exceed 40
  THEN overtime hours equal the number of hours exceeding 40
  ELSE the worker has no overtime hours.

  A variation on the usual selection control structure is the *case control structure.* This offers more than a single yes-or-no decision. The case structure allows several alternatives, or "cases," to be presented. ("IF Case 1 occurs, THEN do thus-and-so. IF Case 2 occurs, THEN follow an alternative course. . . ." And so on.) The case control structure saves the programmer the trouble of having to indicate a lot of separate IF-THEN-ELSE conditions.

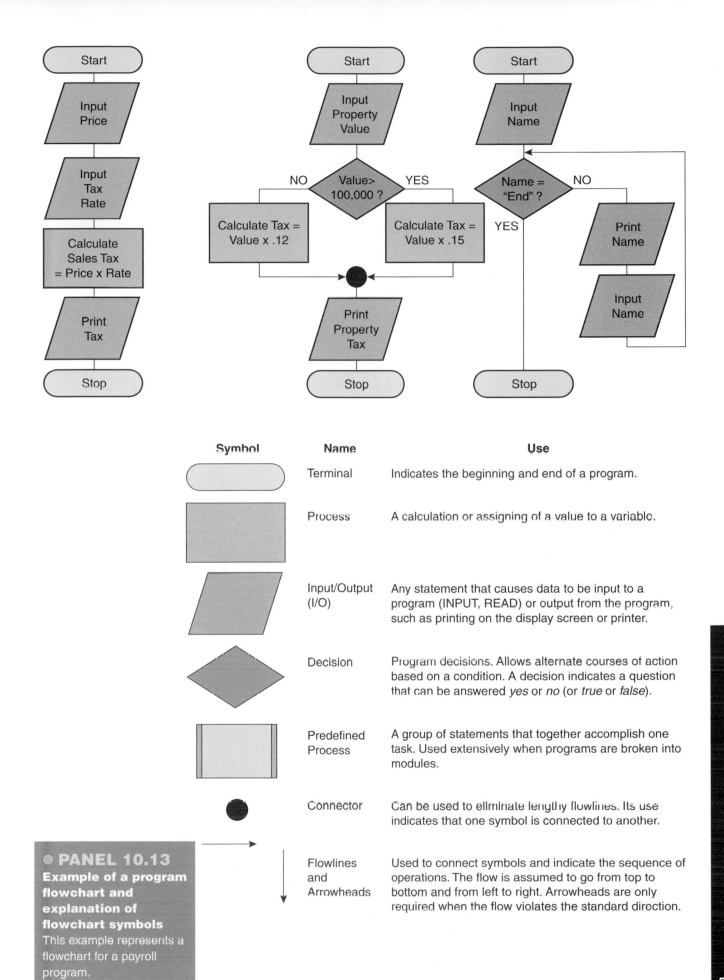

Symbol	Name	Use
	Terminal	Indicates the beginning and end of a program.
	Process	A calculation or assigning of a value to a variable.
	Input/Output (I/O)	Any statement that causes data to be input to a program (INPUT, READ) or output from the program, such as printing on the display screen or printer.
	Decision	Program decisions. Allows alternate courses of action based on a condition. A decision indicates a question that can be answered *yes* or *no* (or *true* or *false*).
	Predefined Process	A group of statements that together accomplish one task. Used extensively when programs are broken into modules.
	Connector	Can be used to eliminate lengthy flowlines. Its use indicates that one symbol is connected to another.
	Flowlines and Arrowheads	Used to connect symbols and indicate the sequence of operations. The flow is assumed to go from top to bottom and from left to right. Arrowheads are only required when the flow violates the standard direction.

● PANEL 10.13
Example of a program flowchart and explanation of flowchart symbols
This example represents a flowchart for a payroll program.

Sequence control structure
(one program statement follows another
in logical order)

Iteration control structures:
DO UNTIL and DO WHILE

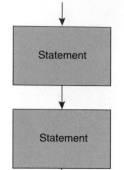

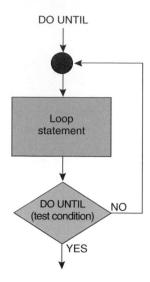

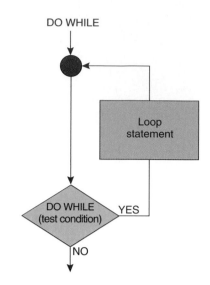

Selection control structure
(IF-THEN-ELSE)

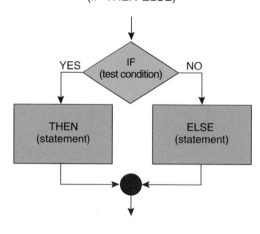

● **PANEL 10.14**
The three control structures
The three structures used in structured program design to form the logic of a program are sequence, selection, and iteration. Case is an important version of selection.

Variation on selection: the case control structure
(more than a single yes-or-no decision)

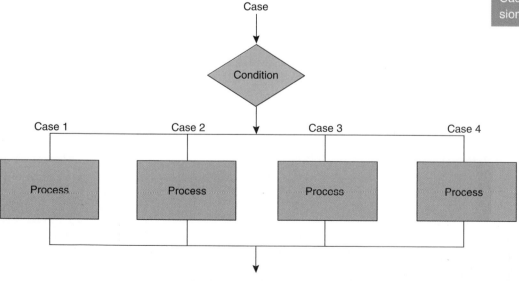

● **Iteration control structure:** In the *iteration,* or *loop, control structure,* a process may be repeated as long as a certain condition remains true. There are two types of iteration structures—DO UNTIL and DO WHILE. Of these, DO UNTIL is more often encountered.

An example of a DO UNTIL structure is as follows:
DO read in employee records UNTIL there are no more employee records.

An example of a DO WHILE structure is as follows:
DO read in employee records WHILE [that is, as long as] there continue to be employee records.

What is the difference between the two iteration structures? It is simply this: If several statements need to be repeated, you must decide when to stop repeating them. You can decide to stop them at the beginning of the loop, using the DO WHILE structure. Or you can decide to stop them at the end of the loop, using the DO UNTIL structure. The DO UNTIL iteration means that the loop statements will be executed at least once, because in this case the iteration statements are executed before the program checks whether to stop.

The Third Step: Code the Program

What is involved in coding a program?

Once the design has been developed, the actual writing of the program begins. **Writing the program is called _coding_.** (● *See Panel 10.15.*) Coding is what many people think of when they think of programming, although it is only one of the five steps. Coding consists of translating the logic requirements from pseudocode or flowcharts into a programming language—the letters, numbers, and symbols that make up the program.

1. SELECT THE APPROPRIATE HIGH-LEVEL PROGRAMMING LANGUAGE A _programming language_ **is a set of rules that tells the computer what operations to do.** Examples of well-known programming languages are C, C++, and Java. These are called *high-level languages*, as we explain in a few pages.

    Not all languages are appropriate for all uses. Some, for example, have strengths in mathematical and statistical processing. Others are more appropriate for database management. Thus, in choosing the language, you need to consider what purpose the program is designed to serve and what languages are already being used in your organization or in your field. We consider these matters in the second half of this chapter.

2. CODE THE PROGRAM IN THAT LANGUAGE, FOLLOWING THE SYNTAX For a program to work, you have to follow the _syntax_, **the rules of the programming language.** Programming languages have their own grammar just as human languages do. But computers are probably a lot less forgiving if you use these rules incorrectly.

● PANEL 10.15
Third step: Program coding
The third step in programming is to translate the logic of the program worked out from pseudocode or flowcharts into a high-level programming language, following its grammatical rules.

1. Select the appropriate high-level programming language.
2. Code the program in that language, following the syntax carefully.

The Fourth Step: Test the Program

How is a program tested?

Program testing involves running various tests and then running real-world data to make sure the program works. (● *See Panel 10.16.*) Two principal activities are desk-checking and debugging. These steps are called *alpha testing.*

1. PERFORM DESK-CHECKING **_Desk-checking_ is simply reading through, or checking, the program to make sure that it's free of errors and that the logic works.** In other words, desk-checking is like proofreading. This step should be taken before the program is actually run on a computer.

2. DEBUG THE PROGRAM Once the program has been desk-checked, further errors, or "bugs," will doubtless surface. **To _debug_ means to detect, locate, and remove all errors in a computer program.** Mistakes may be syntax errors or logical errors. **_Syntax errors_ are caused by typographical errors and incorrect use of the programming language. _Logic errors_ are caused by incorrect use of control structures.** Programs called *diagnostics* exist to check program syntax and display syntax-error messages. Diagnostic programs thus help identify and solve problems.

3. RUN REAL-WORLD DATA After desk-checking and debugging, the program may run fine—in the laboratory. However, it needs to be tested with real data; this is called *beta testing.* Indeed, it is even advisable to test the program with bad data—data that is faulty, incomplete, or in overwhelming quantities—to see if you can make the system crash. Many users, after all, may be far more heavy-handed, ignorant, and careless than programmers have anticipated.

Several trials using different test data may be required before the programming team is satisfied that the program can be released. Even then, some bugs may persist, because there comes a point where the pursuit of errors is uneconomical. This is one reason why many users are nervous about using the first version (version 1.0) of a commercial software package.

The Fifth Step: Document & Maintain the Program

What is involved in documenting and maintaining a program?

Writing the program documentation is the fifth step in programming. The resulting _documentation_ consists of written descriptions of what a program

● **PANEL 10.16**
Fourth step: Program testing
The fourth step is to test the program and "debug" it of errors so that it will work properly. The word "bug" dates from 1945, when a moth was discovered lodged in the relay of the Mark I computer. The moth disrupted the execution of the program.

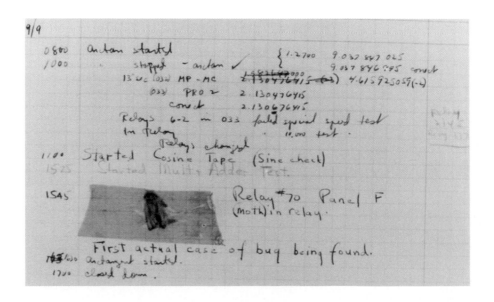

1. Write user documentation.
2. Write operator documentation.
3. Write programmer documentation.
4. Maintain the program.

is and how to use it. Documentation is not just an end-stage process of programming. It has been (or should have been) going on throughout all programming steps. Documentation is needed for people who will be using or be involved with the program in the future. (● *See Panel 10.17.*)

Documentation should be prepared for several different kinds of readers—users, operators, and programmers.

1. WRITE USER DOCUMENTATION When you buy a commercial software package, such as a spreadsheet, you normally get a manual with it. This is user documentation. Nowadays manuals are usually on the software CD.

2. WRITE OPERATOR DOCUMENTATION The people who run large computers are called *computer operators.* Because they are not always programmers, they need to be told what to do when the program malfunctions. The *operator documentation* gives them this information.

3. WRITE PROGRAMMER DOCUMENTATION Long after the original programming team has disbanded, the program may still be in use. If, as is often the case, a fifth of the programming staff leaves every year, after 5 years there could be a whole new bunch of programmers who know nothing about the software. *Program documentation* helps train these newcomers and enables them to maintain the existing system.

4. MAINTAIN THE PROGRAM A word about maintenance: *Maintenance* includes any activity designed to keep programs in working condition, error-free, and up to date—adjustments, replacements, repairs, measurements, tests, and so on. The rapid changes in modern organizations—in products, marketing strategies, accounting systems, and so on—are bound to be reflected in their computer systems. Thus, maintenance is an important matter, and documentation must be available to help programmers make adjustments in existing systems.

The five steps of the programming process are summarized in the table on the next page. (● *See Panel 10.18.*)

10.3

Five Generations of Programming Languages

What are the five generations of programming languages?

As we've said, a programming language is a set of rules that tells the computer what operations to do. Programmers, in fact, use these languages to create other kinds of software. Many programming languages have been written, some with colorful names (SNOBOL, HEARSAY, DOCTOR, ACTORS, EMERALD, JOVIAL). Each is suited to solving particular kinds of problems. What do all these languages have in common? Simply this: Ultimately they

Step	Activities
Step 1: Problem clarification	1. Clarify progam objectives and progam users 2. Clarify desired outputs. 3. Clarify desired inputs. 4. Clarify desired processing. 5. Double-check feasibility of implementing the program. 6. Document the analysis.
Step 2: Program design	1. Determine program logic through top-down approach and modularization, using a hierarchy chart. 2. Design details using pseudocode and/or using flowcharts, preferably on the basis of control structures. 3. Test design with structured walkthrough.
Step 3: Program coding	1. Select the appropriate high-level programming language. 2. Code the program in that language, following the syntax carefully.
Step 4: Program testing	1. Desk-check the program to discover errors. 2. Run the program and debug it (alpha testing). 3. Run real-world data (beta testing).
Step 5: Program documentation and maintenance	1. Write user documentation. 2. Write operator documentation. 3. Write programmer documentation. 4. Maintain the program.

must be reduced to digital form—a 1 or 0, electricity on or off—because that is all the computer can work with.

To see how this works, it's important to understand that there are five levels, or generations, of programming languages, ranging from low level to high level. **The five _generations of programming languages_ start at the lowest level with (1) machine language. They then range up through (2) assembly language, (3) high-level languages (procedural languages and object-oriented languages), and (4) very-high-level languages (problem-oriented languages). At the highest level are (5) natural languages.** Programming languages are said to be _lower-level_ when they are closer to the language that the computer itself uses—the 1s and 0s. They are called _higher level_ when they are closer to the language people use—more like English, for example.

Beginning in 1945, the five levels, or generations, have evolved over the years, as programmers gradually adopted the later generations. The births of the generations are as follows. (● _See Panel 10.19._)

- First generation, 1945—_machine language_
- Second generation, mid-1950s—_assembly language_
- Third generation, mid-1950s to early 1960s—_high-level languages (procedural languages and object-oriented)_; for example, FORTRAN, COBOL, BASIC, C, and C++
- Fourth generation, early 1970s—_very-high-level languages (problem-oriented languages)_; for example, SQL, Intellect, NOMAD, FOCUS
- Fifth generation, early 1980s—_natural languages_

Let's consider these five generations.

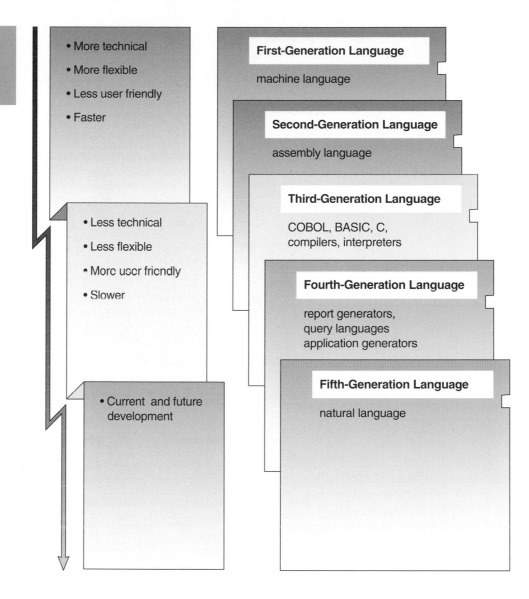

- More technical
- More flexible
- Less user friendly
- Faster

First-Generation Language

machine language

Second-Generation Language

assembly language

Third-Generation Language

COBOL, BASIC, C, compilers, interpreters

- Less technical
- Less flexible
- More user friendly
- Slower

Fourth-Generation Language

report generators, query languages application generators

- Current and future development

Fifth-Generation Language

natural language

First Generation: Machine Language

How would I characterize first-generation programming language?

Machine language is the basic language of the computer, representing data as 1s and 0s. (● *See Panel 10.20 on the next page.*) Each CPU model has its own machine language. Machine-language programs vary from computer to computer; that is, they are *machine-dependent*.

Machine-language binary digits, which correspond to the on and off electrical states of the computer, are clearly not convenient for people to read and use. Believe it or not, though, programmers *did* work with these mind-numbing digits. There must have been great sighs of relief when the next generation of programming languages—assembly language—came along.

Second Generation: Assembly Language

Could I write a computer program in assembly language?

Assembly language is a low-level programming language that allows a programmer to write a program using abbreviations or more easily remembered words instead of numbers, as in machine language; essentially, assembly language is a mnemonic version of machine language. *(Refer to Panel 10.20*

● **PANEL 10.20**

Three generations of programming languages

(*Top*) Machine language is all binary 0s and 1s—difficult for people to work with. (*Middle*) Assembly language uses abbreviations for major instructions (such as *MP* for "MULTIPLY"). This is easier for people to use, but still challenging. (*Bottom*) COBOL, a third-generation language, uses English words that people can understand.

First generation
Machine language

```
11110010 01110011 1101 001000010000 0111 00000101011
11110010 01110011 1101 001000011000 0111 00000101111
11111100 01010010 1101 001000010010 1101 001000011101
11110000 01000101 1101 001000010011 0000 00000111110
11110011 01000011 0111 000001010000 1101 001000010100
10010110 11110000 0111 000001010100
```

Second generation
Assembly language

```
PACK  210(8,13),02B(4,7)
PACK  218(8,13),02F(4,7)
MP    212(6,13),21D(3,13)
SRP   213(5,13),03E(0),5
UNPK  050(5,7),214(4,13)
OI    054(7),X'F0'
```

Third generation
COBOL

```
MULTIPLY HOURS-WORKED BY PAY-RATE GIVING GROSS-PAY ROUNDED.
```

again.) For example, the letters "MP" could be used to represent the instruction MULTIPLY and "STO" to represent STORE.

As you might expect, a programmer can write instructions in assembly language more quickly than in machine language. Nevertheless, it is still not an easy language to learn, and it is so tedious to use that mistakes are frequent. Moreover, assembly language has the same drawback as machine language in that it varies from computer to computer—it is machine-dependent.

We now need to introduce the concept of *language translator*. Because a computer can execute programs only in machine language, a translator or converter is needed if the program is written in any other language. **A _language translator_ is a type of system software that translates a program written in a second-, third-, or higher-generation language into machine language.**

Language translators are of three types:

- Assemblers
- Compilers
- Interpreters

An _assembler_, or _assembler program_, is a program that translates the assembly-language program into machine language. We describe compilers and interpreters in the next section.

Third Generation: High-Level or Procedural Languages

Could I learn to read third-generation languages?

A *high-level*, or *procedural/object-oriented*, *language* resembles some human language such as English; an example is COBOL, which is used for business applications. *(Refer again to Panel 10.20.)* A procedural language allows users to write in a familiar notation, rather than numbers or abbreviations. Also, unlike machine and assembly languages, most are not machine-dependent—that is, they can be used on more than one kind of computer. Familiar languages of this sort include FORTRAN, COBOL, BASIC, Pascal, and C. (We cover object-oriented languages shortly.)

For a procedural language to work on a computer, it needs a language translator to translate it into machine language. Depending on the procedural language, either of two types of translators may be used—a *compiler* or an *interpreter*.

COMPILER—EXECUTE LATER A *compiler* **is a language-translator program that converts the entire program of a high-level language into machine language *before* the computer executes the program.** The programming instructions of a procedural language are called the *source code*. The compiler translates it into machine language, which in this case is called the *object code*. The important point here is that the object code can be saved and thus can be executed later (as many times as desired), rather than run right away. (● *See Panel 10.21.)* These executable files—the output of compilers—have the *.exe* extension.

Examples of procedural languages using compilers are COBOL, FORTRAN, Pascal, and C.

INTERPRETER—EXECUTE IMMEDIATELY An *interpreter* **is a language-translator program that converts each procedural language statement into machine language and executes it *immediately*, statement by statement.** In contrast to the compiler, an interpreter does not save object code. Therefore, interpreted code generally runs more slowly than compiled code. However,

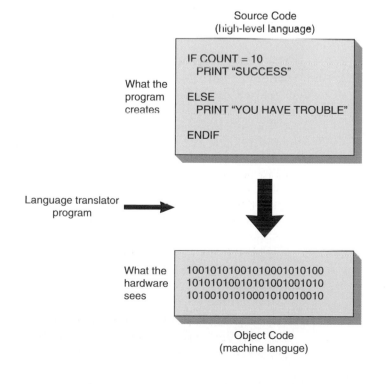

● **PANEL 10.21**
Compiler
This language translator converts the procedural language *(source code)* into machine language *(object code)* before the computer can execute the program.

Source Code
(high-level language)

What the program creates

```
IF COUNT = 10
    PRINT "SUCCESS"

ELSE
    PRINT "YOU HAVE TROUBLE"

ENDIF
```

Language translator program

What the hardware sees

```
10010101001010001010100
10101010010101001001010
10100101010001010010010
```

Object Code
(machine langue)

code can be tested line by line. BASIC is a procedural language using an interpreter.

WHY DOES IT MATTER? Who cares, you might say, whether you can run a program now or later? (After all, "later" could be only a matter of seconds or minutes.) Here's the significance: When a compiler is used, it requires two steps (the source code and the object code) before the program can be executed. The interpreter, on the other hand, requires only *one* step. The advantage of a compiler language is that, once you have obtained the object code, the program executes faster. The advantage of an interpreter language, on the other hand, is that programs are easier to develop. Some language translators—such as those with C++ and Java, covered shortly—can both compile and interpret.

Some of the most popular procedural languages are Visual BASIC, C, and C++. (• *See Panel 10.22.*)

Fourth Generation: Very-High-Level or Problem-Oriented Languages

What are three categories of fourth-generation languages?

Third-generation languages tell the computer *how* to do something. Fourth-generation languages, in contrast, tell the computer *what* to do. **_Very-high-level_, or *problem-oriented* or *nonprocedural*, _languages_, also called *fourth-generation languages (4GLs)*, are much more user-oriented and allow users to develop programs with fewer commands compared with procedural languages,** although they require more computing power. These languages are called *problem-oriented* because they are designed to solve specific problems, whereas procedural languages are more general-purpose languages.

Three types of problem-oriented languages are report generators, query languages, and application generators.

REPORT GENERATORS A *report generator*, also called a *report writer*, is a program for end-users that produces a report. The report may be a printout or a screen display. It may show all or part of a database file. You can specify the format in advance—columns, headings, and so on—and the report generator will then produce data in that format. Report generators (an example is RPGIII) were the precursor to today's query languages.

QUERY LANGUAGES A *query language* is an easy-to-use language for retrieving data from a database management system. The query may be expressed in the form of a sentence or near-English command. Or the query may be obtained from choices on a menu.

Examples of query languages are SQL (for "Structured Query Language") and Intellect. For example, with Intellect, which is used with IBM mainframes, you can make an English-language request such as "Tell me the number of employees in the sales department."

APPLICATION GENERATORS An *application generator* is a programmer's tool consisting of modules that have been preprogrammed to accomplish various tasks. The benefit is that the programmer can generate applications programs from descriptions of the problem rather than by traditional programming, in which he or she has to specify how the data should be processed.

Programmers use application generators to help them create parts of other programs. For example, the software is used to construct on-screen menus or types of input and output screen formats. NOMAD and FOCUS, two database management systems, include application generators.

● **PANEL 10.22**
Third-generation languages compared
This shows how five languages handle the same statement. The statement specifies that a customer gets a discount of 7% of the invoice amount if the invoice is greater than $500; if the invoice is lower, there is no discount.

FORTRAN

```
            IF (XINVO .GT. 500.00) THEN
                DISCNT = 0.07 * XINVO
            ELSE
                DISCNT = 0.0
            ENDIF
            XINVO = XINVO – DISCNT
```

COBOL

```
OPEN-INVOICE-FILE.
    OPEN I-O INVOICE FILE.

READ-INVOICE-PROCESS.
    PERFORM READ-NEXT-REC THROUGH READ-NEXT-REC-EXIT UNTIL END-OF-FILE
    STOP RUN.

READ-NEXT-REC.
    READ INVOICE-REC
        INVALID KEY
            DISPLAY 'ERROR READING INVOICE FILE'
            MOVE 'Y' TO EOF-FLAG
            GOTO READ-NEXT-REC-EXIT.
    IF INVOICE-AMT > 500
        COMPUTE INVOICE-AMT = INVOICE-AMT   (INVOICE-AMT * .07)
        REWRITE INVOICE-REC.

READ-NEXT-REC-EXIT.
    EXIT.
```

BASIC

```
10  REM       This Program Calculates a Discount Based on the Invoice Amount
20  REM            If Invoice Amount is Greater Than 500, Discount is 7%
30  REM            Otherwise Discount is 0
40  REM
50  INPUT "What is the Invoice Amount"; INV.AMT
60  IF INV.AMT > 500 THEN LET DISCOUNT = .07 ELSE LET DISCOUNT = 0
70  REM          Display results
80  PRINT "Original Amt", "Discount", "Amt after Discount"
90  PRINT INV.AMT, INV.AMT * DISCOUNT, INV.AMT – INV.AMT * DISCOUNT
100 END
```

Pascal

```
if INVOICEAMOUNT > 500.00 then
    DISCOUNT := 0.07 * INVOICEAMOUNT
else
    DISCOUNT := 0.0;
INVOICEAMOUNT := INVOICEAMOUNT – DISCOUNT
```

C

```
if (invoice_amount > 500.00)
    discount = 0.07 * invoice_amount;
else
    discount = 0.00;
invoice_amount = invoice_amount – discount;
```

Fifth Generation: Natural Languages

Have I ever seen or used a fifth-generation language?

<u>*Natural languages*</u> are of two types. The first comprises ordinary human languages: English, Spanish, and so on. The second type comprises programming languages that use human language to give people a more natural connection with computers.

With a problem-oriented language, you can type in some rather routine inquiries, such as (in the language known as FOCUS) the following:

SUM SHIPMENTS BY STATE BY DATE.

Natural languages, by contrast, allow questions or commands to be framed in a more conversational way or in alternative forms—for example:

I WANT THE SHIPMENTS OF PERSONAL DIGITAL ASSISTANTS FOR ALABAMA AND MISSISSIPPI BROKEN DOWN BY CITY FOR JANUARY AND FEBRUARY. ALSO, MAY I HAVE JANUARY AND FEBRUARY SHIPMENTS LISTED BY CITIES FOR PERSONAL COMMUNICATORS SHIPPED TO WISCONSIN AND MINNESOTA.

Natural languages are part of the field of study known as *artificial intelligence* (discussed in Chapter 8). Artificial intelligence (AI) is a group of related technologies that attempt to develop machines capable of emulating human qualities, such as learning, reasoning, communicating, seeing, and hearing.

The dates of the principal programming languages are shown in the time line running along the bottom of these facing pages. (● *See Panel 10.23.*)

10.4

Programming Languages Used Today

What are some third-generation languages, and what are they used for?

Let us now turn back and consider some of the third-generation, or high-level, languages in use today.

● **PANEL 10.23**
Timeline: Brief history of the development of programming languages and formatting tools

FORTRAN: The Language of Mathematics & the First High-Level Language

What is FORTRAN useful for?

Developed from 1954 to 1956 by John Backus and others at IBM, ___**FORTRAN**___ (for "FORmula TRANslator") was the first high-level language. *(Refer back*

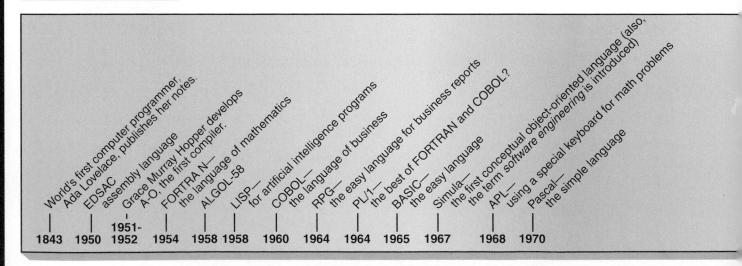

to Panel 10.22.) Originally designed to express mathematical formulas, it is still the most widely used language for mathematical, scientific, and engineering problems. It is also useful for complex business applications, such as forecasting and modeling. However, because it cannot handle a large volume of input/output operations or file processing, it is not used for more typical business problems.

FORTRAN has both advantages and disadvantages:

ADVANTAGES (1) FORTRAN can handle complex mathematical and logical expressions. (2) Its statements are relatively short and simple. (3) FORTRAN programs developed on one type of computer can often be easily modified to work on other types.

DISADVANTAGES (1) FORTRAN does not handle input and output operations to storage devices as efficiently as some other higher-level languages. (2) It has only a limited ability to express and process nonnumeric data. (3) It is not as easy to read and understand as some other high-level languages.

COBOL: The Language of Business

How would I characterize the purpose of COBOL?

Developed under the auspices of the U.S. Department of Defense, with Grace Murray Hopper as a major contributor, and formally adopted in 1960, **_COBOL_ (for COmmon Business-Oriented Language) is the most frequently used business programming language for large computers.** (Refer again to Panel 10.22.) Its most significant attribute is that it is extremely readable. For example, a COBOL line might read:

MULTIPLY HOURLY-RATE BY HOURS-WORKED GIVING GROSS-PAY

Writing a COBOL program resembles writing an outline for a research paper. The program is divided into four divisions—Identification, Environment, Data, and Procedure. The divisions in turn are divided into sections, which are divided into paragraphs, which are further divided into sections. The Identification Division identifies the name of the program and the author (programmer) and perhaps some other helpful comments. The Environment Division describes the computer on which the program will be compiled and executed. The Data Division describes what data will be processed. The Procedure Division describes the actual processing procedures.

Some people believe that COBOL is becoming obsolete. However, others disagree:

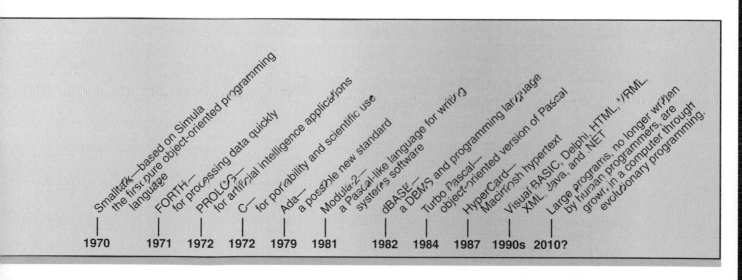

| 1970 | 1971 | 1972 | 1972 | 1979 | 1981 | 1982 | 1984 | 1987 | 1990s | 2010? |

Few things that were around in the IT industry in 1960 are still around today, except perhaps in museums and in basements. Yet the COBOL language, albeit highly evolved from its origins, remains relevant. The corporate assets represented by the billions (trillions?) of lines of COBOL code still running on commercial computers aren't about to be abandoned. Nor should they be. New tools that help integrate legacy (read: COBOL) systems with PC applications, web services, new data format, and protocols have been available since distributive systems emerged years ago. As today's application environment has evolved, .NET for instance, so has COBOL's capability to link to it, merge with it, and interact with it. Stretching the lifespan of an enterprise's legacy systems increases the value and productivity of its IT development staff and of the assets they produce.[2]

COBOL, too, has both advantages and disadvantages.

ADVANTAGES (1) It is machine-independent. (2) Its English-like statements are easy to understand, even for a nonprogrammer. (3) It can handle many files, records, and fields. (4) It easily handles input/output operations.

DISADVANTAGES (1) Because it is so readable, it is wordy. Thus, even simple programs are lengthy, and programmer productivity is slowed. (2) It cannot handle mathematical processing as well as FORTRAN.

BASIC: The Easy Language

Have I ever written a program in BASIC or could I?

BASIC was developed by John Kemeny and Thomas Kurtz in 1964 for use in training their students at Dartmouth College. By the late 1960s, it was widely used in academic settings on all kinds of computers, from mainframes to PCs.

BASIC (Beginner's All-purpose Symbolic Instruction Code) used to be the most popular microcomputer language and is considered the easiest programming language to learn. *(Refer to Panel 10.22.)* Although it is available in compiler form, the interpreter form is more popular with first-time and casual users. This is because it is interactive, meaning that user and computer can communicate with each other during the writing and running of the program. Today there is no one version of BASIC. One of the popular current evolutions is Visual BASIC, discussed shortly.

The advantage and disadvantages of BASIC are as follows:

ADVANTAGE The primary advantage of BASIC is its ease of use.

DISADVANTAGES (1) Its processing speed is relatively slow, although compiler versions are faster than interpreter versions. (2) There is no one version of BASIC, although in 1987 ANSI adopted a new standard that eliminated portability problems—that is, problems with running it on different computers.

Pascal: The Simple Language

When might I program in Pascal?

Named after the 17th-century French mathematician Blaise Pascal and developed in 1970 by Niklaus Wirth, **_Pascal_ is an alternative to BASIC as a language for teaching purposes and is relatively easy to learn.** *(Refer to Panel 10.22.)* A difference from BASIC is that Pascal uses structured programming.

A compiled language, Pascal offers the following advantages and disadvantage:

ADVANTAGES (1) Pascal is easy to learn. (2) It has extensive capabilities for graphics programming. (3) It is excellent for scientific use.

Pascal has limited input/output programming capabilities, which limits its business applications.

C: For Portability & Scientific Use

What is C mainly used for?

C is the successor of B, which was the successor of BCPL, which was the successor of CPL (Computer Programming Language), an early programming language that was not implemented. Developed by Dennis Ritchie at Bell Laboratories in the early 1970s, **C is a general-purpose, compiled language that was developed for midrange computers (minicomputers) but that works well for microcomputers and is portable among many computers.** *(Refer to Panel 10.22.)* It was originally developed for writing system software. The first major program written in C was the Unix operating system, and for many years C was considered to be inextricably linked with Unix. Now, however, C is an important language independent of Unix. Today it is widely used for writing applications, including word processing, spreadsheets, games, robotics, and graphics programs. It is now considered a necessary language for programmers to know.

Here are the advantages and disadvantages of C:

ADVANTAGES (1) C works well with microcomputers. (2) It has a high degree of portability—it can be run without change on a variety of computers. (3) It is fast and efficient. (4) It enables the programmer to manipulate individual bits in main memory. (5) It requires less memory than many other programming languages.

DISADVANTAGES (1) C is considered difficult to learn. (2) Because of its conciseness, the code can be difficult to follow. (3) It is not suited to applications that require a lot of report formatting and data file manipulation.

LISP: For Artificial Intelligence Programs

What is LISP used for?

LISP (LISt Processor) is a third-generation language used principally to construct artificial intelligence programs. Developed at the Massachusetts Institute of Technology in 1958 by mathematician John McCarthy, LISP is used to write expert systems and natural language programs. As we saw in Chapter 8, expert systems are programs that are imbued with knowledge by a human expert; the programs can walk you through a problem and help solve it.

10.5

Object-Oriented & Visual Programming

How do OOP and visual programming work?

Consider how it was for the computer pioneers, programming in machine language or assembly language. Novices putting together programs in BASIC or C can breathe a collective sigh of relief that they weren't around at the dawn of the Computer Age. Even some of the simpler third-generation languages represent a challenge, because they are procedure-oriented, forcing the programmer to follow a predetermined path.

Fortunately, two developments have made things easier—object-oriented programming and visual programming.

Conventional Programs

Object-Oriented Programs

Object-Oriented Programming: Block by Block

What is an object?

Imagine you're programming in a traditional third-generation language, such as BASIC, creating your coded instructions one line at a time. As you work on some segment of the program (such as how to compute overtime pay), you may think, "I'll bet some other programmer has already written something like this. Wish I had it. It would save a lot of time." Fortunately, a kind of recycling technique exists. This is object-oriented programming, an improved version of 3GLs.

HOW OOP WORKS: OBJECT, MESSAGE, & METHOD Object-oriented programming consists of the following components:

1. *What OOP is:* **In *object-oriented programming* (*OOP*, pronounced "oop"), data and the instructions for processing that data are combined into a self-sufficient "object" that can be used in other programs.** The important thing here is the object.

2. *What an "object" is:* **An *object* is a self-contained module consisting of preassembled programming code.** The module contains, or encapsulates, both (1) a chunk of data and (2) the processing instructions that may be performed on that data.

3. *When an object's data is to be processed—sending the "message":* Once the object becomes part of a program, the processing instructions may or may not be activated. A particular set of instructions is activated only when the corresponding "message" is sent. A *message* is an alert sent to the object when an operation involving that object needs to be performed.

4. *How the object's data is processed—the "methods":* The message need only identify the operation. How it is actually to be performed is embedded within the processing instructions that are part of the object. These processing instructions within the object are called the *methods.*

RECYCLING BLOCKS OF PROGRAM CODE Once you've written a block of program code (that computes overtime pay, for example), it can be reused in any number of programs. Thus, with OOP, unlike traditional programming, you don't have to start from scratch—that is, reinvent the wheel—each time.

Object-oriented programming takes longer to learn than traditional programming because it means training oneself to a new way of thinking. However, the beauty of OOP is that an object can be used repeatedly in different applications and by different programmers, speeding up development time and lowering costs.

Three Important Concepts of OOP

How would I define the three basic concepts of OOP?

Object-oriented programming has three important concepts, which go under the jaw-breaking names of *encapsulation, inheritance,* and *polymorphism.* Actually, these are not as fearsome as they look:

ENCAPSULATION *Encapsulation* means an object contains (encapsulates) both (1) data and (2) the relevant processing instructions, as we have seen. Once an object has been created, it can be reused in other programs. An object's uses can also be extended through concepts of *class* and *inheritance.*

INHERITANCE Once you have created an object, you can use it as the foundation for similar objects that have the same behavior and characteristics. All objects that are derived from or related to one another are said to form a *class*. Each class contains specific instructions (methods) that are unique to that group.

Classes can be arranged in hierarchies—classes and subclasses. *Inheritance* is the method of passing down traits of an object from classes to subclasses in the hierarchy. Thus, new objects can be created by inheriting traits from existing classes.

Writer Alan Freedman gives this example: "The object MACINTOSH could be one instance of the class PERSONAL COMPUTER, which could inherit properties from the class COMPUTER SYSTEMS."[3] If you were to add a new computer, such as COMPAQ, you would need to enter only what makes it different from other computers. The general characteristics of personal computers could be inherited.

POLYMORPHISM Polymorphism means the presence of "many shapes." In object-oriented programming, *polymorphism* means that a message (generalized request) produces different results based on the object that it is sent to.

Polymorphism has important uses. It allows a programmer to create procedures about objects whose exact type is not known in advance but will be at the time the program is actually run on the computer. Freedman gives this example: "A screen cursor may change its shape from an arrow to a line depending on the program mode." The processing instructions "to move the cursor on screen in response to mouse movement would be written for 'cursor,' and polymorphism would allow that cursor to be whatever shape is required at runtime." It would also allow a new cursor shape to be easily integrated into the program.

EXAMPLES OF OOP LANGUAGES: C++ & JAVA Two important examples of OOP languages are C++ and Java.

- C++: *C++*—the plus signs stand for "more than C"—**combines the traditional C programming language with object-oriented capability.** C++ was created by Bjarne Stroustrup. With C++, programmers can write standard code in C without the object-oriented features, use object-oriented features, or do a mixture of both.
- Java: A high-level programming language developed by Sun Microsystems in 1995, *Java*, **also an object-oriented language, is used to write compact programs that can be downloaded over the internet and immediately executed on many kinds of computers.** Java is similar to C++ but is simplified to eliminate language features that cause common programming errors.

  Java source code files (files with a *.java* extension) are compiled into a format called *bytecode* (files with a *.class* extension), which can then be executed by a Java interpreter. Compiled Java code can run on most computers because Java interpreters and runtime environments, known as *Java Virtual Machines (VMs)*, exist for most operating systems, including Unix, the Macintosh OS, and Windows.

  Small Java applications are called Java *applets* and can be downloaded from a web server and run on your computer by a Java-compatible web browser, such as Netscape Navigator or Microsoft Internet Explorer. Java applets make websites more interactive and attractive, adding features such as animation and calculators—but only if a browser is capable of supporting Java. Users also can download free Java applets from various sites on the internet.

Visual Programming: The Example of Visual BASIC

How would I characterize visual programming?

Essentially, visual programming takes OOP to the next level. The goal of visual programming is to make programming easier for programmers and more accessible to nonprogrammers by borrowing the object orientation of OOP languages but exercising it in a graphical or visual way. Visual programming enables users to think more about the problem solving than about handling the programming language. There is no learning of syntax or actual writing of code

Visual programming is a method of creating programs in which the programmer makes connections between objects by drawing, pointing, and clicking on diagrams and icons and by interacting with flowcharts. Thus, the programmer can create programs by clicking on icons that represent common programming routines.

An example of visual programming is *Visual BASIC*, a Windows-based, object-oriented programming language from Microsoft that lets users develop Windows and Office applications by (1) creating command buttons, textboxes, windows, and toolbars, which (2) then may be linked to small BASIC programs that perform certain actions. Visual BASIC is *event-driven*, which means that the program waits for the user to do something (an "event"), such as click on an icon, and then the program responds. At the beginning, for example, the user can use drag-and-drop tools to develop a graphical user interface, which is created automatically by the program. Because of its ease of use, Visual BASIC allows even novice programmers to create impressive Windows-based applications.

Since its launch in 1990, the Visual BASIC approach has become the norm for programming languages. Now there are visual environments for many programming languages, including C, C++, Pascal, and Java. Visual BASIC is sometimes called a *rapid application development (RAD)* system because it enables programmers to quickly build prototype applications.

10.6

Markup & Scripting Languages

What do markup and scripting languages do?

A **_markup language_ is a kind of coding, or "tags," inserted into text that embeds details about the structure and appearance of the text.** Markup languages have codes for indicating layout and styling (such as boldface, italics, paragraphs, insertion of graphics, and so on) within a text file—for example, HTML. The name "markup" is derived from the traditional publishing practice of "marking up" a manuscript, that is, adding printer's instructions in the margins of a paper manuscript.

Some early examples of markup languages available outside the publishing industry could be found in typesetting tools on Unix systems. In these systems, formatting commands were inserted into the document text so that typesetting software could format the text according to the editor's specifications. After a time it was seen that most markup languages had many features in common. This led to the creation of *SGML (Standard Generalized Markup Language)*, which specified a syntax for including the markup in documents, as well as another system (a so-called *metalanguage*) for separately describing what the markup meant. This allowed authors to create and use any markup they wished, selecting tags that made the most sense to them. SGML was developed and standardized by the International Organization for Standards (ISO) in 1986.

SGML is used widely to manage large documents that are subject to frequent revisions and need to be printed in different formats. Because it is a large and complex system, it is not yet widely used on personal computers.

This changed dramatically when Tim Berners-Lee used some of the SGML syntax, without the metalanguage, to create HTML (Hypertext Markup Language, Chapter 2). HTML may be the most used document format in the world today.

For some people, the term *script* may conjure up images of actors and actresses on a sound stage practicing lines from a book of text. In web terms, however, a **_script_ is a short list of self-executing commands embedded in a web page that perform a specific function or routine.** Scripts are similar to the macros used in Microsoft Word or the batch files used in the early days of DOS, both of which performed functions ranging from generating text to displaying the date and time. Because they are self-executing, scripts can perform their work without user involvement, although some are initiated by an action on the part of the user (such as a mouse click) and others require user input to complete a task.[4]

On web pages, scripting languages are used to perform many duties. For example, they create traffic counters and scrolling text, set cookies so that websites can remember user preferences, and switch out graphics and text when users click buttons or pass their mouse over items. Any time you see something interesting occurring on a website, there is a good chance that a script is involved. Scripting languages are often designed for interactive use.

Following are some popular markup and scripting languages.

HTML: For Creating 2-D Web Documents & Links

What would I use HTML for?

more info!

HTML Tutorial

For an easy-to-follow tutorial on using HTML, go to *http:// computer.howstuffworks.com/ web-page2.htm*. After you work through the tutorial, create a sample HTML-tagged document for what could become your personal website.

As we discussed in Chapter 2, **_HTML (Hypertext Markup Language)_ is a markup language that lets people create on-screen documents for the internet that can easily be linked by words and pictures to other documents.** HTML is a type of code that embeds simple commands within standard ASCII text documents to provide an integrated, two-dimensional display of text and graphics. In other words, a document created in any word processor and stored in ASCII format can become a web page with the addition of a few HTML commands.

One of the main features of HTML is the ability to insert hypertext links into a document. *Hypertext links* enable you to display another web document simply by clicking on a link area—usually underlined or highlighted—on your current screen (Chapter 2). One document may contain links to many other related documents. The related documents may be on the same server as the first document, or they may be on a computer halfway around the world. A link may be a word, a group of words, or a picture.

VRML: For Creating 3-D Web Pages

How is VRML used?

more info!

VRML Plug-In

Two places to download VRML add-ins are *www .parallelgraphics.com/products/ cortona/download/iexplore* and *http://cic.nist.gov/vrml/ cosmoplayer.html.*

Mark Pesce and Tony Parisi created VRML at Silicon Graphics in 1994. *VRML* rhymes with "thermal." **_VRML (Virtual Reality Modeling [or Markup] Language)_ is a type of programming language used to create three-dimensional web pages including interactive animation.** Even though VRML's designers wanted to let nonprogrammers create their own virtual spaces quickly and painlessly, it's not as simple to describe a three-dimensional scene as it is to describe a page in HTML. However, many existing modeling and CAD tools now offer VRML support, and new VRML-centered software tools are arriving. An example of a VRML scene might be a virtual room where the viewer can use controls to move around inside the room (or move the room itself) as though she or he were walking through it in real space.

To view VRML files, users need a special VRML browser (in addition to an internet connection and a web browser). The VRML browser is what interprets VRML commands and lets the user interact with the virtual world.

VRML browsers typically work as plug-ins for traditional web browsers, but newer browsers may already have the appropriate VRML browser plug-in installed. There are several VRML browsers available for Windows users, such as Cosmo and Cortona.

Web designers are starting to use new software such as WireFusion to cerate 3-D web page components that will no longer require a plug-in for users to view them with their browser. (The successor to VRML is called *X3D*.)

XML: For Making the Web Work Better

How does XML improve on HTML?

The chief characteristics of HTML are its simplicity and its ease in combining plain text and pictures. But, in the words of journalist Michael Krantz, "HTML simply lacks the software muscle to handle the business world's endless and complex transactions."[5] Another, newer markup language is XML, a standard maintained by the World Wide Web Consortium for creating special-purpose markup languages. **_XML (eXtensible Markup Language)_ is a metalanguage** (a language used to define another language) **written in SGML that allows one to facilitate the easy interchange of documents on the internet.**

Unlike HTML, which uses a set of "known" tags, XML allows you to create any tags you wish (thus it's extensible) and then describe those tags in a metalanguage known as the *DTD (Document Type Definition)*. XML is similar to the concept of SGML, and in fact XML is a subset of SGML in general terms. The main purpose of XML (as opposed to SGML) is to keep the system simpler by focusing on a particular problem—documents on the internet.

Whereas HTML makes it easy for humans to read websites, XML makes it easy for machines to read websites by enabling web developers to add more "tags" to a web page. At present, when you use your browser to locate a website, search engines can turn up too much, so it's difficult to find the specific site you want—say, one with a recipe for a low-calorie chicken dish for 12. According to Krantz, "XML makes websites smart enough to tell other machines whether they're looking at a recipe, an airline ticket, or a pair of easy-fit blue jeans with a 34-inch waist." XML lets website developers put "tags" on their web pages that describe information in, for example, a food recipe as "ingredients," "calories," "cooking time," and "number of portions." Thus, your browser no longer has to search the entire web for a low-calorie poultry recipe for 12.

Following are examples of XML and HTML tags. Note that the XML statements define data content, whereas the HTML lines deal with fonts and display (boldface). XML defines "what it is," and HTML defines "how it looks."

```
XML
<firstName>Maria</firstName>
<lastName>Roberts</lastName>
<dateBirth>10-29-52</dateBirth>
```

```
HTML
<font size="3">Maria Roberts</font>
<b>October 29, 1952<b>
```

JavaScript: For Dynamic Web Pages

How can JavaScript be helpful?

JavaScript is a popular object-oriented scripting language that is widely supported in web browsers. It adds interactive functions to HTML pages, which are otherwise static, since HTML is a display language, not a programming

info!

What Are Lateral Thinking Tools?

Lateral thinking tools also help people think clearly, logically, and creatively. Do a keyword search and find out what they are.

language. JavaScript is easier to use than Java but not as powerful and deals mainly with the elements on the web page. JavaScript was originally developed by Netscape Communications under the name "LiveScript" but was then renamed to "JavaScript" and given a syntax closer to that of Sun Microsystems' Java language. The change of name happened at about the same time Netscape was including support for Java technology in its Netscape Navigator browser. Consequently, the change proved a source of much confusion. There is no real relation between Java and JavaScript; their only similarities are some syntax and the fact that both languages are used extensively on the World Wide Web.

Many website designers use JavaScript technology to create powerful dynamic web applications. One major use of JavaScript is to write little functions that are embedded in HTML pages and interact with the browser to perform certain tasks not possible in static HTML alone, such as opening a new window and changing images as the mouse cursor moves over them.

ActiveX: For Creating Interactive Web Pages

Would I ever need to use ActiveX?

ActiveX was developed by Microsoft as an alternative to Java for creating interactivity on web pages. Indeed, Java and ActiveX are the two major contenders in the web war for transforming the World Wide Web into a complete interactive environment.

ActiveX is a set of controls, or reusable components, that enable programs or content of almost any type to be embedded within a web page. Whereas Java requires that you download an applet each time you visit a website, with ActiveX the component is downloaded only once (to a Windows-based computer) and then stored on your hard drive for later and repeated use. ActiveX is built into Microsoft's Internet Explorer and is available as a plug-in for Netscape Navigator.

The chief characteristic of ActiveX is that it features reusable components—small modules of software code that perform specific tasks (such as a spelling checker), which may be plugged seamlessly into other applications. With ActiveX, you can obtain from your hard disk any file that is suitable for the web—such as a Java applet, animation, or pop-up menu—and insert it directly into an HTML document.

Programmers can create ActiveX controls or components in a variety of programming languages, including C, C++, Visual BASIC, and Java. Thousands of ready-made ActiveX components are now commercially available from numerous software development companies.

Perl: For CGI Scripts

What is Perl used for?

Perl (Practical Extraction and Report Language) is a general-purpose programming language developed for text manipulation and now used for web development, network programming, system administration, GUI development, and other tasks. Perl is widely used to write web server programs for such tasks as automatically updating user accounts and newsgroup postings, processing removal requests, synchronizing databases, and generating reports. The major features of Perl is that it is easy to use and supports both procedural and object-oriented programming.

Perl was developed in 1987 by Larry Wall, and it combines syntax from several Unix utilities and languages. Perl has also been adapted to non-Unix platforms.

Systems Analysis & Programming

Experience Box
Critical Thinking Tools

Clear thinkers aren't born that way. They work at it.

The systems development life cycle is basically an exercise in clear thinking—critical thinking. Critical thinking is fundamental to systems analysis and design—particularly in the first phase, preliminary analysis. Reaching for the truth may not come easily; it is a stance toward the world, developed with practice. To achieve this, we have to wrestle with obstacles that are mostly of our own making: *mindsets*. By the time we are grown, our minds have become "set" in various patterns of thinking that affect the way we respond to new situations and new ideas. Such mindsets determine what ideas we think are important and, conversely, what ideas we ignore.

To break past mindsets, we need to learn to think critically. *Critical thinking* means sorting out conflicting claims, weighing the evidence for them, letting go of personal biases, and arriving at reasoned conclusions. Critical thinking means actively seeking to understand, analyze, and evaluate information in order to solve specific problems.

Learning to identify fallacious (incorrect) arguments will help you avoid patterns of faulty thinking in your own writing and thinking and to identify it in others'.

Jumping to Conclusions   In the fallacy called *jumping to conclusions,* also known as *hasty generalization,* a decision maker reaches a conclusion before all the facts are available. *Example:* A company instituted the strategy of total quality management (TQM) 12 months ago. As a new manager of the company, you see that TQM has not improved profitability over the past year, and you order TQM junked, in favor of more traditional business strategies. However, what you don't know is that the traditional business strategies employed prior to TQM had an even worse effect on profitability.

Irrelevant Reason or False Cause   In the faulty reasoning known as *non sequitur* (Latin for "it does not follow"), which might better be called *false cause* or *irrelevant reason,* the conclusion does not follow logically from the supposed reasons stated earlier. There is no causal relationship. *Example:* You receive an A on a test. However, because you felt you hadn't been well prepared, you attribute your success to your friendliness with the professor. Or to your horoscope. Or to wearing your "lucky shirt." None of these supposed reasons have anything to do with the result.

Irrelevant Attack on a Person or Opponent   Known as an *ad hominem* argument (Latin for "to the person"), the irrelevant attack on an opponent attacks a person's reputation or beliefs rather than his or her argument. *Example:* Your boss insists you may not hire a certain person as a programmer because he or she has been married and divorced nine times, although the person's marital history plainly has no bearing on his or her skills as a programmer.

Slippery Slope   *Slippery slope* is a failure to see that the first step in a possible series of steps does not lead inevitably to the rest. *Example:* The "domino theory," under which the United States waged wars against Communism for half a century, was a slippery-slope argument. It assumed that if Communism triumphed in one country (for example, Nicaragua), then it would inevitably triumph in other regions (the rest of Central America), finally threatening the borders of the United States itself.

Appeal to Authority   The *appeal-to-authority* argument (known in Latin as *argumentum ad verecundiam*) uses authority in one area in an effort to validate claims in another area where the person is not an expert. *Example:* You see the appeal-to-authority argument used all the time in advertising. But how qualified is a professional golfer to speak about headache remedies?

Circular Reasoning   In *circular reasoning,* a statement to be proved true is rephrased, and then the new formulation is offered as supposed proof that the original statement is in fact true. *Example:* You declare that you can drive safely at high speeds with only inches separating you from the car ahead because you have driven this way for years without an accident.

Straw Man Argument   In the *straw man* argument, you misrepresent your opponent's position to make it easier to attack, or you attack a weaker position while ignoring a stronger one. In other words, you sidetrack the argument from the main discussion. *Example:* Politicians use this argument all the time. If you attack a legislator for being "fiscally irresponsible" in supporting funds for a gun-control bill, when what you really object to is the fact of gun control, you're using a straw man argument.

Appeal to Pity   In the *appeal-to-pity* argument, the advocate appeals to mercy rather than making an argument on the merits of the case itself. *Example:* Begging the dean not to expel you for cheating because your parents are poor and made sacrifices to put you through college would be a blatant appeal to pity.

Questionable Statistics   Statistics can be misused in many ways as supporting evidence. The statistics may be unknowable, drawn from an unrepresentative sample, or otherwise suspect. *Example*: Stating how much money is lost to taxes because of illegal drug transactions is speculation because such transactions are hidden or unrecorded.

Summary

ActiveX (p. 531) A set of controls, or reusable components, that enable programs or content of almost any type to be embedded within a web page. *Why it's important: ActiveX features reusable components—small modules of software code that perform specific tasks, such as a spelling checker, which may be plugged seamlessly into other applications.*

algorithm (p. 507) Formula or set of steps for solving a particular problem. *Why it's important: All programs consist of algorithms.*

assembler (p. 518) Also called *assembler program;* language-translator program that translates assembly-language programs into machine language. *Why it's important: Language translators are needed to translate all upper-level languages into machine language, the language that actually runs the computer.*

First generation
Machine language

```
11110010 01110011 1101 001000010000 0111 00000101011
11110010 01110011 1101 001000011000 0111 00000101111
11111100 01010010 1101 001000010010 1101 001000011101
11110000 01000101 1101 001000010011 0000 00000111110
11110011 01000011 0111 000001010000 1101 001000010100
10010110 11110000 0111 000001010100
```

assembly language (p. 517) Second-generation programming language; it allows a programmer to write a program using abbreviations instead of the 0s and 1s of machine language. *Why it's important: A programmer can write instructions in assembly language faster than in machine language.*

Second generation
Assembly language

```
PACK 210(8,13),02B(4,7)
PACK 218(8,13),02F(4,7)
MP   212(6,13),21D(3,13)
SRP  213(5,13),03E(0),5
UNPK 050(5,7),214(4,13)
OT   054(7),X'F0'
```

BASIC (Beginner's All-purpose Symbolic Instruction Code) (p. 524) Developed in 1965, BASIC was once the most popular microcomputer language and is still the easiest programming language to learn. *Why it's important: The interpreter form of BASIC is popular with first-time and casual users because it is interactive—user and computer can communicate during writing and running of a program.*

C (p. 525) High-level general purpose programming language that works well for microcomputers and is portable among many computers. *Why it's important: C was originally developed for writing system software. Today it is widely used for writing applications, including word processing, spreadsheets, games, robotics, and graphics programs. It is now considered a necessary language for programmers to know.*

Third generation
COBOL

```
MULTIPLY HOURS-WORKED BY PAY-RATE GIVING GROSS-PAY ROUNDED.
```

C++ (p. 527) High-level programming language that combines the traditional C programming language with object-oriented capability (the plus signs mean "more than C"). *Why it's important: With C++, programmers can write standard code in C without the object-oriented features, use object-oriented features, or do a mixture of both.*

CASE (computer-aided software engineering) tools (p. 501) Software that provides computer-automated means of designing and changing systems. *Why it's important: CASE tools may be used in almost any phase of the SDLC.*

COBOL (Common Business-Oriented Language) (p. 523) High-level programming language of business. First standardized in 1968, the language has been revised three times. Why it's important: *COBOL is extremely readable. Traditionally it is the language used by the majority of mainframe users.*

coding (p. 513) In the programming process, the third step, consisting of translating logic requirements from pseudocode or flowcharts into a programming language. Why it's important: *Coding is the actual writing of a computer program, although it is only the third of five steps in programming.*

compiler (p. 519) Language translator that converts the entire program of a high-level language (called the *source code*) into machine language (called the *object code*) for execution later. Examples of compiler languages: COBOL, FORTRAN, Pascal, and C. Why it's important: *Unlike other language translators (assemblers and interpreters), a compiler program enables the object code to be saved and executed later rather than run right away. The advantage of a compiler is that, once the object code has been obtained, the program executes faster.*

control structure (p. 510) Also called *logic structure;* in structured program design, the programming structure that controls the logical sequence in which computer program instructions are executed. Three control structures are used to form the logic of a program: sequence, selection, and iteration (or loop). Why it's important: *One thing that all three control structures have in common is one entry and one exit. The control structure is entered at a single point and exited at another single point. This helps simplify the logic so that it is easier for others following in a programmer's footsteps to make sense of the program.*

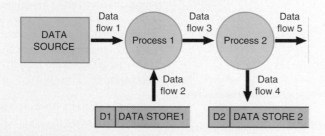

data flow diagram (DFD) (p. 501) Modeling tool that graphically shows the flow of data through a system—essential processes, including inputs, outputs, and files. Why it's important: *A DFD diagrams the processes that change data into information.*

Explanation of standard data flow diagram symbols

Terminator Symbols (entity name)
(person or organization outside the
system boundaries)

Data Store Symbol

Proces

debug (p. 514) Part of program testing; the detection, location, and removal of syntax and logic errors in a program. Why it's important: *Debugging may take several trials using different data before the programming team is satisfied the program can be released. Even then, some errors may remain, because trying to remove them all may be uneconomical.*

desk-checking (p. 514) Form of program testing; programmers read through a program to ensure it's error-free and logical. Why it's important: *Desk-checking should be done before the program is actually run on a computer.*

documentation (p. 514) Written descriptions of a program and how to use it; supposed to be done during all programming steps. Why it's important: *Documentation is needed for all people who will be using or be involved with the program—users, operators, programmers, and future systems analysts.*

FORTRAN (FORmula TRANslator) (p. 522) Developed in 1954, FORTRAN was the first high-level language and was designed to express mathematical formulas. Why it's important: *The most widely used language for mathematical, scientific, and engineering problems, FORTRAN is also useful for complex business applications, such as forecasting and modeling.*

generations of programming languages (p. 516) Five increasingly sophisticated levels (generations) of programming languages: (1) machine language, (2) assembly language, (3) high-level languages, (4) very-high-level languages, (5) natural languages. Why it's important: *Programming languages are said to be lower-level when they are closer to the language used by the computer (0s and 1s) and higher-level when closer to the language used by people. High-level languages are easier for most people to use than are lower-level languages.*

hierarchy chart (p. 508) Also called *structure chart;* a diagram used in programming to illustrate the overall purpose of a program, identifying all the modules needed to achieve that purpose and the relationships among them. Why it's important: *In a hierarchy chart, the program* *must move in sequence from one module to the next until all have been processed. There must be three principal modules corresponding to the three principal computing operations—input, processing, and output.*

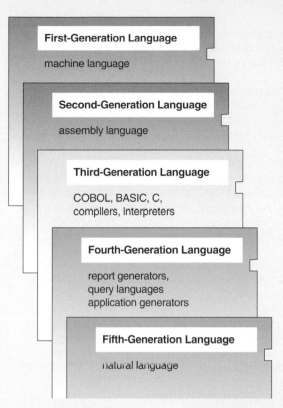

First-Generation Language

machine language

Second-Generation Language

assembly language

Third-Generation Language

COBOL, BASIC, C,
compilers, Interpreters

Fourth-Generation Language

report generators,
query languages
application generators

Fifth-Generation Language

natural language

high-level languages (p. 519) Also known as *procedural or object-oriented languages;* third-generation programming languages. They somewhat resemble human languages. Examples: FORTRAN, COBOL, BASIC, Pascal, and C. Why it's important: *High-level languages allow programmers to write in a familiar notation rather than numbers or abbreviations. Most can also be used on more than one kind of computer.*

HTML (Hypertext Markup Language) (p. 529) Type of formatting language that embeds commands within standard ASCII text documents to provide an integrated, two-dimensional display of text and graphics. Hypertext is used to link the displays. Why it's important: *HTML is used to create web pages.*

interpreter (p. 519) Language translator that converts each high-level language statement into machine language and executes it immediately, statement by statement. An example of a high-level language using an interpreter is BASIC. Why it's important: *Unlike a compiler translator, an interpreter does not save object code. The advantage of an interpreter is that programs are easier to develop.*

Java (p. 527) Object-oriented programming language used to write compact programs that can be downloaded over the internet and immediately executed on many kinds of computers. Java is similar to C++ but is simplified to eliminate language features that cause common programming errors. Why it's important: *Compiled Java code can run on most computers. Small Java applications, called applets, which can be downloaded from a web server and run on a personal computer, make websites more interactive and attractive, adding features such as animation.*

JavaScript (p. 530) Object-oriented scripting language used, for example, to write little functions that are embedded in HTML pages and interact with the browser to perform certain tasks not possible in static HTML alone. There is no real relationship between JavaScript and Java. Why it's important: *JavaScript is used to create powerful dynamic web applications.*

language translator (p. 518) Type of system software that translates a program written in a second-, third-, or higher-generation language into machine language. Language translators are of three types: (1) assemblers, (2) compilers, and (3) interpreters. Why it's important: *Because computers run only using machine language, all higher-level languages must be translated.*

LISP LISt Processor) (p. 525) Developed in 1958, LISP is a third-generation language used principally to construct artificial intelligence programs. Why it's important: *LISP is used to write expert systems and natural language programs. Expert systems are programs that are imbued with knowledge by a human expert; the programs can walk you through a problem and help solve it.*

logic errors (p. 514) Programming errors caused by incorrect use of control structures. Why it's important: *Logic errors prevent a program from running properly.*

machine language (p. 517) Lowest-level (first-generation) programming language; the language of the computer, representing data as 1s and 0s. Most machine-language programs vary from computer to computer—they are machine-dependent. Why it's important: *Machine language is the language that actually runs the computer.*

markup language (p. 528) A kind of coding, or "tags," inserted into text that embeds details about the structure and appearance of the text within a text file—for example, HTML. The word "markup" is derived from the traditional publishing practice of "marking up" a manuscript, that is, adding printer's instructions in the margins of a paper manuscript. An example of a markup language is SGML (Standard Generalized Markup Language). Why it's important: *Markup languages have codes for indicating layout and styling, such as boldface, italics, paragraphs, and insertion of graphics.*

modeling tools (p. 501) Analytical tools such as charts, tables, and diagrams that are used by systems analysts. Examples are data flow diagrams, decision tables, systems flowcharts, and object-oriented analysis. Why it's important: *Modeling tools enable a systems analyst to present graphic, or pictorial, representations of a system.*

module (p. 508) Sometimes called *subprogram* or *subroutine;* a processing step of a program. Each module is made up of logically related program statements. Why it's important: *Each module has only a single function, which limits the module's size and complexity.*

natural languages (p. 522) (1) Ordinary human languages (for instance, English, Spanish); (2) fifth-generation programming languages that use human language to give people a more natural connection with computers. Why it's important: *Natural languages are part of the field of artificial intelligence; these languages are approaching the level of human communication.*

object (p. 526) In object-oriented programming, block of preassembled programming code that is a self-contained module. The module contains (encapsulates) both (1) a chunk of data and (2) the processing instructions that may be called on to be performed on that data. Once the object becomes part of a program, the processing instructions may be activated only when a "message" is sent. Why it's important: *The object can be reused and interchanged among programs, thus making the programming process much easier, more flexible and efficient, and faster.*

object-oriented programming (OOP) (p. 526) Programming method in which data and the instructions for processing that data are combined into a self-sufficient *object*—a piece of software that can be used in other programs. Why it's important: *Objects can be reused and interchanged among programs, producing greater flexibility and efficiency than is possible with traditional programming methods.*

Pascal (p. 524) High-level programming language; an alternative to BASIC as a language for teaching purposes that is relatively easy to learn. Why it's important: *Pascal has extensive capabilities for graphics programming and is excellent for scientific use.*

Perl (Practical Extraction and Report Language) (p. 531) A general-purpose programming language developed for text manipulation and now used for web development, network programming, system administration, GUI development, and other tasks. Why it's important: *The major features of Perl is that it is easy to use and supports both procedural and object-oriented programming.*

preliminary investigation (p. 500) Phase 1 of the SDLC; the purpose is to conduct a preliminary analysis (determine the organization's objectives, determine the nature and scope of the problem), propose alternative solutions (leave the system as is, improve the efficiency of the system, or develop a new system), describe costs and benefits, and submit a preliminary plan with recommendations. Why it's important: *The preliminary investigation lays the groundwork for the other phases of the SDLC.*

problem clarification/definition (p. 506) Step 1 in the programming process. The problem-definition step requires performing six mini-steps: specifying objectives and users, outputs, inputs, and processing tasks and then studying the feasibility of the program and documenting the analysis. Why it's important: *Problem definition is the forerunner to step 2, program design, in the programming process.*

program (p. 505) List of instructions the computer follows to process data into information. The instructions consist of statements written in a programming language (for example, BASIC). Why it's important: *Without programs, computers could not process data into information.*

program design (p. 507) Step 2 in the programming process; programs are designed in two ministeps: (1) the program logic is determined through a top-down approach and modularization, using a hierarchy chart; (2) the program is designed in detail, using pseudocode or flowcharts with logical tools called *control structures.* Why it's important: *Program design is the forerunner to step 3, writing (coding), in the programming process.*

program flowchart (p. 510) Chart that graphically presents the detailed series of steps needed to solve a programming problem; it uses standard symbols called *ANSI symbols.* Why it's important: *The program flowchart is an important program design tool.*

program testing (p. 514) Step 4 in the programming process; involves running various tests and then running real-world data to make sure the program works. Why it's important: *The program must be tested before it is released to be sure that it works properly.*

programming (p. 505) Five-step process for creating software instructions: (1) Clarify/define the problem; (2) design a solution; (3) write (code) the program; (4) test the program; (5) document and maintain the program. Why it's important: *Programming is one step in the systems development life cycle.*

programming language (p. 513) Set of rules (words and symbols) that allow programmers to tell the computer what operations to follow. The five levels (generations) of programming languages are (1) machine language, (2) assembly language, (3) high-level (procedural) languages, (4) very-high-level (nonprocedural) languages, and (5) natural languages. Why it's important: *Not all programming languages are appropriate for all uses. Thus, languages must be chosen to suit the purpose of the program and to be compatible with other languages being used.*

prototype (p. 502) A limited working system, or part of one, developed to test design concepts. Why it's important: *A prototype, which may be constructed in just a few days, allows users to find out immediately how a change in the system might benefit them.*

prototyping (p. 502) Using workstations, CASE tools, and other software applications to build working models of system components so that they can be quickly tested and evaluated. Why it's important: *Prototyping is part of the preliminary design stage of Phase 3 of the SDLC.*

```
START
DO WHILE (so long as) there are records
        Read a customer billing account record
        IF today's date is greater than 30 days from
        date of last customer payment
                Calculate total amount due
                Calculate 5% interest on amount due
                Add interest to total amount due to calculate
                grand total
                Print on invoice overdue amount
        ELSE
                Calculate total amount due
        ENDIF
        Print out invoice
END DO
END
```

pseudocode (p. 508) Tool for designing a program in narrative form using normal human-language statements to describe the logic and processing flow. Using pseudocode is like doing an outline or summary form of the program to be written. Why it's important: *Pseudocode provides a type of outline or summary of the program.*

script (p. 529) A short list of self-executing commands embedded in a web page that perform a specific function or routine, often without user involvement. Why it's important: *Because they are self-executing, scripts can perform their work without user involvement.*

structured programming (p. 508) Method of programming that takes a top-down approach, breaking programs into modular forms and using standard logic tools called *control structures* (sequence, selection, case, iteration). Why it's important: *Structured programming techniques help programmers write better-organized programs, using standard notations with clear, correct descriptions.*

syntax (p. 513) "Grammar" rules of a programming language. Why it's important: *Each programming language has its own syntax, just as human languages do.*

syntax errors (p. 514) Programming errors caused by typographical errors and incorrect use of the programming language. Why it's important: *If a program has syntax errors, it will not run correctly or perhaps not run at all.*

system (p. 498) Collection of related components that interact to perform a task in order to accomplish a goal. Why it's important: *Understanding a set of activities as a system allows one to look for better ways to reach the goal.*

systems analysis (p. 500) Phase 2 of the SDLC; the purpose is to gather data (using written documents, interviews, questionnaires, and observation), analyze the data, and write a report. Why it's important: *The results of systems analysis determine whether the system should be redesigned.*

systems analysis and design (p. 499) Problem-solving procedure for examining an information system and improving it; consists of the six-phase systems development life cycle. Why it's important: *The point of systems analysis and design is to ascertain how a system works and then take steps to make it better.*

systems analyst (p. 499) Information specialist who performs systems analysis, design, and implementation. Why it's important: *The systems analyst studies the information and communications needs of an organization to determine how to deliver information that is complete,* *accurate, timely, and useful. The systems analyst achieves this goal through the problem-solving method of systems analysis and design.*

systems design (p. 502) Phase 3 of the SDLC; the purpose is to do a preliminary design and then a detail design and to write a report. Why it's important: *Systems design is one of the most crucial phases of the SDLC.*

systems development (p. 503) Phase 4 of the SDLC; consists of acquiring and testing hardware and software for the new system. This phase begins once management has accepted the report containing the design and has approved the way to development. Why it's important: *This phase may involve the organization in investing substantial time and money.*

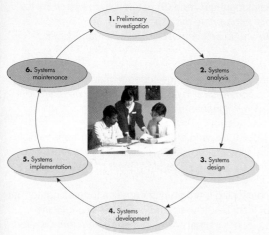

systems development life cycle (SDLC) (p. 499) Six-phase process that many organizations follow during systems analysis and design: (1) preliminary investigation; (2) systems analysis; (3) systems design; (4) systems development; (5) systems implementation; (6) systems maintenance. Phases often overlap, and a new one may start before the old one is finished. After the first four phases, management must decide whether to proceed to the next phase. User input and review is a critical part of each phase. Why it's important: *The SDLC is a comprehensive tool for solving organizational problems, particularly those relating to the flow of computer-based information.*

systems implementation (p. 504) Phase 5 of the SDLC; consists of converting the hardware, software, and files to the new system and training the users. Why it's important: *This phase involves putting design ideas into operation.*

systems maintenance (p. 505) Phase 6 of the SDLC; consists of keeping the system working by having system audits and periodic evaluations and by making changes based on new conditions. Why it's important: *This phase is important for keeping a new system operational and useful.*

top-down program design (p. 508) Method of program design; a programmer identifies the top or principal processing step, or module, of a program and then breaks it down in hierarchical fashion to the lowest level of detail.

very-high-level languages (p. 520) Also known as *problem-oriented* and *nonprocedural languages* and *fourth-generation languages (4GLs)*; more user-oriented than third-generation languages, 4GLs require fewer commands. 4GLs consist of report generators, query languages, and application generators. Why it's important: *Programmers can write programs that need to tell the computer only what they want done, not all the procedures for doing it, which saves them the time and the labor of having to write many lines of code.*

visual programming (p. 528) Method of creating programs in which the programmer makes connections between objects by drawing, pointing, and clicking on diagrams and icons and by interacting with flowcharts. Programming is made easier because the orientation of object-oriented programming is used in a graphical or visual way. Why it's important: *Visual programming enables users to think more about the problem solving than about handling the programming language.*

VRML (Virtual Reality Modeling [Markup] Language) (p. 529) Type of programming language used to create three-dimensional (3-D) web pages. Why it's important: *VRML expands the information-delivering capabilities of the web.*

writing the program documentation (p. 514) Step 5 in the programming process; programmers write procedures explaining how the program was constructed and how it is to be used. Why it's important: *Program documentation is the final stage in the five-step programming process, although documentation should also be an ongoing task accompanying all steps.*

XML (eXtensible Markup Language) (p. 530) Programming language used to make it easy for machines to read web sites by allowing web developers to add more "tags" to a web page. Why it's important: *XML is more powerful than HTML, allowing information on a web site to be described by general tags—for example, identifying one piece of information in a recipe as "cooking time" and another as "ingredients."*

Chapter Review

"I can recognize and recall information."

Self-Test Questions

1. The _____ comprises six phases of examining an information system and improving it.

2. The first major program written in C was the _____ operating system.

3. _____ is a method of creating programs in which the programmer makes connections between objects by drawing, pointing, and clicking on diagrams and icons and by interacting with flowcharts.

4. Software engineering, or _____, refers to creating instructions for computers.

5. A _____ is a formula or a set of steps for solving a particular problem.

6. A _____ is a collection of related components that interact to perform a task in order to accomplish a goal.

7. _____ is the basic language of the computer, representing data as 0s and 1s.

Multiple-Choice Questions

1. One of the following activities is *not* an objective of Phase 1 of the SDLC, preliminary investigation. Which one?
   a. conduct preliminary analysis
   b. describe costs and benefits
   c. acquire new software and hardware
   d. submit a preliminary plan
   e. propose alternative solutions

2. One of the following activities is *not* an objective of Phase 4 of the SDLC, systems development. Which one?
   a. convert files to the new system
   b. acquire software
   c. acquire hardware
   d. test the system
   e. address the make-or-buy decision

3. Third-generation programming languages include all the following languages except which one?
   a. FORTRAN
   b. BASIC
   c. COBOL
   d. XML
   e. Pascal

True/False Questions

T F 1. Programming errors caused by incorrect use of control structures are called *logic errors*.

T F 2. CASE tools—programs that automate various activities of the SDLC—are used only in Phase 3.

T F 3. Four methods of systems implementation are direct, parallel, phased, and pilot.

T F 4. User training takes place during Phase 1 of the SDLC.

T F 5. JavaScript is an object-oriented scripting language used in web browsers to add interactive functions to HTML pages.

T F 6. There are four generations of programming languages.

"I can recall information in my own terms and explain it to a friend."

Short-Answer Questions

1. What is the straw man argument? Appeal to pity? What are some of the other elements of critical thinking?

2. What does a systems analyst do?

3. What are the fours ways to implement s new system?

4. What are the five steps in the programming process?

5. What are the six phases of the SDLC?

6. What is a software bug?

7. What is a prototype, and what does it do?

"I can apply what I've learned, relate these ideas to other concepts, build on other knowledge, and use all these thinking skills to form a judgment."

Knowledge in Action

1. Using an internet search tool, identify a company that develops CASE tools. In a few paragraphs, describe what this company's CASE tools are used for.

2. Design a system that would handle the input, processing, and output of a simple form of your choice. Use a data flow diagram to illustrate the system.

3. Have you participated in a project that failed? Why did it fail? Based on what you know now, what might you have done to help the project succeed?

4. Which step of the SDLC do you find most interesting? Why?

5. Are you interested in learning how to program? Which languages would you choose to learn? Why?

Web Exercises

1. The waterfall model and the spiral model are variations of the SDLC. Do keyword searches to find out how these models differ from the basic approach.

2. Compare the SDLC process to the scientific method. How are they alike? How do they differ? If you need to be refreshed on the scientific method, you can visit

   http://koning.ecsu.ctstateu.edu/Plants_Human/ scimeth.html or run a search on *"scientific method."*

3. One of the most beneficial courses you can take in college is critical thinking. The Experience Box in this chapter has an introduction to fallacious arguments. A

background in logic and identification of fallacies is essential not only for making valid arguments but for life in general. After reading the Experience Box, visit the following websites:

www.datanation.com/fallacies/index.htm
www.don-lindsay-archive.org/skeptic/arguments.html
http://uwc.fac.utexas.edu/media/Handouts/ Rhetorical%20Fallacies.pdf

For a humorous account of how not to argue, read the Monty Python Argument sketch:

www.infidels.org/news/atheism/sn-python.html.

After familiarizing yourself with the subject matter, watch TV for an hour (especially commercials) and identify 10 fallacies. Write them down, and discuss them with your class the next day. Are our collective reality assumptions valid and cohesive? Logic and critical thinking will also help tremendously if you decide to get into computer programming.

4. Create a prototype invention (of anything) and have classmates test and evaluate it. Once the evaluations are received, write a detailed report, as you would at the end of Phase 3 in the SDLC process.

5. Curious about the educational requirements for a systems designer/analyst? Go to

   www.umuc.edu/prog/ugp/majors/ifsm.shtml
   www.bls.gov/oco/ocos042.htm
   http://jobguide.thegoodguides.com.au/text/ jobdetails.cfm?jobid=266

Extra exercise: Here are simple examples of algorithms. Choose another simple statement and develop an algorithm for it.

Your friend is arriving at the airport and needs to get to your house:

Taxi algorithm

1. Go to the taxi stand outside the airport.

2. Get in a taxi.

3. Give the taxi driver my address.

Call-me algorithm

1. After you retrieve your luggage, call my cellphone number.

2. I'll pick you up outside, in front of the baggage-claim area; meet me there.

Bus algorithm

1. Catch bus #70 at the bus stop outside the baggage-claim area.

2. Get off bus #70 at the Main Street stop and transfer to bus #14 in the direction of downtown.

3. Get off at Elm Street.

4. Walk two blocks north to my house at 230 Elm Street.

5. Ring the doorbell.

6. I'll let you in.

Notes

Chapter 1

1. Kevin Maney, "Net's Next Phase Will Weave through Your Life," *USA Today*, March 2, 2001, pp. 1B, 2B.
2. Peter Lyman and Hal R. Varian, *How Much Information 2003?* School of Information Management and Systems, University of California at Berkeley, *www.sims.berkeley .edu/research/projects/how-much-info* (accessed January 9, 2005.)
3. Dave Wilson, "Some Are Losing It, Bit by Bit," *Los Angeles Times*, July 17, 2001, pp. A1, A8.
4. Sherry Turkle, "How Computers Change the Way We Think," *The Chronicle of Higher Education*, January 30, 2004, pp. B26–B28.
5. Howard Rheingold, *Smart Mobs: The Next Social Revolution* (New York: Perseus, 2002).
6. Katie Hafner, "In Class, the Audience Weighs In," *New York Times*, April 29, 2004, pp. E1, E6.
7. Greg Toppo, "Schools Achieving a Dream: Near-Universal Net Access," *USA Today*, June 9, 2004, p. 6D.
8. Steve Jones, *Pew Internet & American Life Project: The Internet Goes to College*, September 15, 2002, *www.pewinternet.org/ pdfs/PIP_College_Report.pdf* (accessed January 4, 2005).
9. Catherine Yang, "Big Program on Campus," *BusinessWeek*, September 20, 2004, pp. 96, 98.
10. Anne McGrath, "Bricks & Clicks," *U.S. News & World Report*, October 18, 2004, pp. 64–68.
11. Nanette Asimov, "Home-Schoolers Plug into the Internet for Resources," *San Francisco Chronicle*, January 29, 1999, pp. A1, A15.
12. John Schwartz, "A Different Course," *New York Times*, April 25, 2004, sec. 4A, pp. 28–30, 42.
13. Dan Carnevale, "Many Online Courses Work Best at No Distance at All," *The Chronicle of Higher Education*, July 30, 2004, p. A22.
14. Brenda Wade Schmidt, "Learning by Remote, in a Remote Prairie Town," *USA Today*, November 29, 2004, p. 7D; Laura Vanderkam, "Online Learning: A Smart Way to Nurture Gifted Kids," *USA Today*, September 28, 2004, p. 13A; McGrath, 2004; Katie Hafner, "Software Tutors Offer Help and Customized Hints," *New York Times*, September 16, 2004, p. E3; Branda Dela Vega, "Business Simulation Games Used to Teach at Colleges," *Reno Gazette-Journal*, August 2, 2004, p. 3E; and Jeanette Borzo, "Almost Human," *The Wall Street Journal*, May 24, 2004, pp. R4, R10.
15. April Lynch, "Bleeding Sailor Performs Self-Surgery via E-Mail," *San Francisco Chronicle*, November 19, 1998, pp. A1, A10.
16. Bernadine Healy, "2004: A Medical Odyssey," *U.S. News & World Report*, August 2, 2004, p. 61.
17. Rob Turner, "Remote Intensive Care That's More Intensive," *U.S. News & World Report*, August 2, 2004, pp. 56, 60.
18. E. A. Kerr, E. A. McGlynn, J. Adams et al., "Profiling the Quality of Care in Twelve Communities: Results from the CQI Study," *Health Affairs*, May/June 2004, pp. 247–256.
19. Denise Grady, "Software to Compute Women's Cancer Risk," *New York Times*, January 26, 1999, p. D4.
20. Laura Landro, "Push Grows for Online Health Data," *The Wall Street Journal*, March 11, 2004, p. D6; "The Doctor Is Online: Secure Messaging Boosts the Use of Web Consultations," *The Wall Street Journal*, September 2, 2004, p. D1.
21. Bernard Wysocki Jr., "Robots in the OR," *The Wall Street Journal*, February 26, 2004, pp. B1, B6; Mike Crissey, "Hospital Robots Aid the Staff," *San Francisco Chronicle*, July 12, 2004, pp. F1, F5; and Robert Davis, "Robo Doc: Medicine by 'Extension,'" *USA Today*, August 4, 2004, p. 8D.
22. Nancy Ann Jeffrey, "Hydraulics and Computers Help Artificial Limbs Get 'Smarter,'" *The Wall Street Journal*, August 14, 1998, p. B1.
23. Associated Press, "Implant Transmits Brain Signals Directly to Computer," *New York Times*, October 22, 1998, p. A22.
24. Associated Press, "Dad Uses Internet to Find Wonder Drug for His Son," *San Francisco Chronicle*, May 29, 1998, p. G9.
25. Jim Hudak, quoted in Heather Green and Linda Himelstein, "A Cyber Revolt in Health Care," *BusinessWeek*, October 19, 1998, pp. 154–156.
26. Laura Landro, "Your Medical History on a Microchip: Having Key Data Ready in an Emergency," *The Wall Street Journal*, July 27, 2004, p. D1; Claudia Kalb, "Get Ready for E-Medicine," *Newsweek*, August 9, 2004, p. 53; and Banaby J. Feder and Tom Zeller Jr., "Identity Badge Worn Under Skin Approved for Use in Health Care," *New York Times*, October 14, 2004, pp. A1, C12.
27. Cynthia G. Wager, "Money's Digital Future," *The Futurist*, January–February 2003, pp. 14–15.
28. The Nilson Report, cited in "Electronic Transactions on the Rise," *USA Today*, October 6, 2003, p. 1B.
29. Jupiter Research, cited in Jason Strazluso, "After Flashy Failures, Internet Grocery Quietly Grows to $2.4 Billion Industry," *Reno Gazette-Journal*, May 17, 2004, pp. 1E, 3E.
30. Eve Tahmincioglu, "Banks Try to Pave the Way to Online Bill Paying," *New York Times*, January 18, 2004, sec. 3, p. 7.
31. Paul Saffo, quoted in Jared Sandverg, "CyberCash Lowers Barriers to Small Transactions at Internet Storefronts," *The Wall Street Journal*, September 30, 1996, p. B6.
32. Daniel Nasaw, "A Dollar Here, a Dollar There," *The Wall Street Journal*, May 24, 2004, p. R6; and Dan Fost, "Tiny Bills = Big Deal," *San Francisco Chronicle*, September 8, 2004, pp. C1, C4.
33. Heather Green, "Kissing Off Big Labels," *BusinessWeek*, September 6, 2004, pp. 90, 92.
34. Susan Wloszczyna, "When Free Is Profitable," *USA Today*, May 21, 2004, p. 1E; and Jon Parclcs, "No Fears: Laptop D.J.'s Have a Feast," *New York Times*, September 10, 2004, pp. B1, B8.
35. James Barron, "Best Musical Score (by a Laptop)," *New York Times*, June 26, 2004, p. A13.
36. Bill Werde, "We've Got Algorithm, but How about Soul?" *New York Times*, March 21, 2004, sec. 4, p. 12.
37. Peter Stack, "An Animated Future," *San Francisco Chronicle*, May 19, 1999, p. E1.
38. William Keck, "Their World of Tomorrow Revolves Around Playtime," *USA Today*, September 6, 2004, p. 10D.
39. Laura M. Holson, "Out of Hollywood, Rising Fascination with Video Games," *New York Times*, April 10, 2004, pp. A1, B2; and Robert A. Guth and Merissa Marr, "Videogames Go Hollywood," *The Wall Street Journal*, May 10, 2004, pp. B1, B4.
40. *The State of "Electronically Enhanced Democracy": A Survey of the Internet* (New Brunswick, NJ: Rutgers University, Douglass Campus, Walt Whitman Center, Department of Political Science, 1998).
41. John Horrigan, *Pew Internet & American Life Project: How Americans Get in Touch with Government*, May 24, 2004, *www .pewinternet.org/pdfs/PIP_E-Gov_Report_ 0504.pdf* (accessed January 7, 2005).
42. Jan Dempsey, "Government Contact via Internet Increases," *San Francisco Chronicle*, July 5, 2004, p. C4; and Valerie Alvord, "It's the Era of Big.government," *USA Today*, August 19, 2004, p. 3A.
43. Susan Stellin, "Getting Your Resume to the Top of the Electronic Pile Can Be a Matter of Paying an Extra Fee. But Does It Translate into Better Results?" *New York Times*, January 20, 2003, p. C3.
44. Graeme Weardon, "Nokia: 2 Billion Cell Phone Users by 2006," CNET News.com, December 9, 2004, *http://news.com.com/ Nokia+2+billion+cell+phone+users+ by+2006/2100-1039_3-5485543.html* (accessed December 29, 2004).
45. ForceNine Consulting/Wirthlin Worldwide, and Harris Interactive, in "Cellphones Add Features," *USA Today*, November 29, 2004, p. 1D.
46. Michael Specter, "Your Mail Has Vanished," *The New Yorker*, December 6, 1999, pp. 96–103.
47. IDC, Nua Internet Surveys, March 2003, *www.nua.com/surveys*.
48. "Like It or Not, You've Got Mail," *BusinessWeek*, October 4, 1999, pp. 178–184.
49. Robert Rossney, "E-Mail's Best Asset— Time to Think," *San Francisco Chronicle*, October 5, 1995, p. E7.
50. Adam Gopnik, "The Return of the Word," *The New Yorker*, December 6, 1999, pp. 49–50.
51. Gopnik, 1999.
52. David A. Whittler, quoted in "Living Online," *The Futurist*, July–August 1997, p. 54.
53. Lee Rainie, *Pew Internet & American Life Project: New Data on Internet Use & Demographics* December 22, 2004, *www .pewinternet.org/PPF/p/1026/pipcomments .asp* (accessed January 3, 2005).
54. Anick Jesdanun, "High-Speed Growth Changing Lifestyles on the Net—and Off," *Reno Gazette-Journal*, December 27, 2004, pp. 1E, 3E.
55. Kevin Murphy, "The Net Effect: Evolution or Revolution?" *USA Today*, August 9, 1999, pp. 1B, 2B.
56. *The UCLA Internet Report—Surveying the Digital Future, Year Three*, UCLA Center

for Communication Policy, University of California, Los Angeles, January 2003, *http://ccp.ucla.edu/pages/internet-report .asp* (accessed January 3, 2005).

57. *Computer Industry Almanac*, cited in "Population Explosion!," ClickZ Network, September 10, 2004, www.clickz.com/ stats/sectors/geographics/article.php/ 5911_151151 (accessed January 2, 2005).

58. Rainie, 2004.

59. Stephen Baker, "Channeling the Future," *BusinessWeek*, July 12, 2004, pp. 70–72.

60. Deborah Fallows, *Pew Internet & American Life: Email at Work*, December 8, 2002, *www.pewinternet .org/pdfs/PIP_Work_ Email_Report .pdf* (accessed January 4, 2005).

61. Donald Tapscott, *Growing Up Digital: The Rise of the Net Generation* (New York: McGraw-Hill, 1998).

62. David M. Ewalt, "The Next (Not So) Big Thing," *InformationWeek*, May 13, 2002, *www.informationweek.com/story/ IWK20020510S0005* (accessed January 4, 2005).

63. David Einstein, "Custom Computers," *San Francisco Chronicle*, April 15, 1999, pp. B1, B3.

64. Einstein, 1999.

65. Laurence Hooper, "No Compromises," *The Wall Street Journal*, November 16, 1992, p. R8.

66. John Markoff, "By and for the Masses," *New York Times*, June 29, 2005, pp. C1, C5.

67. Robert D. Hof, "The Power of Us," *BusinessWeek*, June 20, 2005, pp. 74–82.

68. Verne Kopytoff, "Citizen Journalism Takes Root Online," *San Francisco Chronicle*, June 6, 2005, pp. E1, E5.

69. Tom Forester and Perry Morrison, *Computer Ethics: Cautionary Tales and Ethical Dilemmas in Computing* (Cambridge, MA: MIT Press, 1990), pp. 1–2.

70. F. P. Robinson, *Effective Study*, 4th ed. (New York: Harper & Row, 1970).

71. B. K. Broumage and R. E. Mayer, "Quantitative and Qualitative Effects of Repetition on Learning from Technical Text," *Journal of Educational Psychology*, 1982, 78, 271–278.

72. R. J. Palkovitz and R. K. Lore, "Note Taking and Note Review: Why Students Fail Questions Based on Lecture Material," *Teaching of Psychology*, 1980, 7, 159–161.

Chapter 2

1. Graham T. T. Molitor, "Five Forces Transforming Communications," *The Futurist*, September–October 2001, pp. 32–37.

2. Molitor, 2001, p. 33.

3. *Computer Industry Almanac*, reported in "Population Explosion!" *ClickZ Network*, September 10, 2004, www.clickz.com/ stats/sectors/geographics/article.php/ 5911_151151 (accessed January 11, 2005).

4. John Horrigan, "Pew Internet & American Life Project: Broadband Penetration on the Upswing," April 19, 2004, *www .pewinternet.org/PPF/r/121/report_ display.asp* (accessed January 11, 2005).

5. Nielsen/NetRatings, reported in Vauhini Vara, "High-Speed Surpasses Dial-Up at Top Home Web Access in U.S.," *The Wall Street Journal*, August 18, 2004, p. D4.

6. Vincent Kiernan, "2 High-Speed Network Groups Discuss Merger," *The Chronicle of Higher Education*, July 15, 2005, p. A38.

7. Scarlett Pruitt, "ICANN Works on Going Global," *InfoWorld*, March 26, 2003, *www.infoworld.com*.

8. Victoria Shannon, "U.S. Seeks to Keep Role on Internet," *New York Times*, July 4, 2005, p. C6, reprinted from *International Herald Tribune*.

9. Kitty Burns Florey, "Around the (Virtual) World in Half a Second" [letter], *New York Times*, March 2, 2004, p. A26.

10. Rick Mathieson, "The Wireless Web Goes Yahoo! An Interview with Christopher Wu, Head of Product Strategy for Yahoo

Everywhere," *Impulse Magazine*, April 2002, *http://cooltown.hp.com/cooltown/ mpulse/0402-christopherwu.asp*; and Cameron Crouch, "Yahoo Boosts Wireless Portal," *PC World*, March 21, 2000, *www.pcworld.com/news/article/ 0,aid,15857,00.asp* (all accessed January 24, 2005).

11. Ellen Chamberlain, "Bare Bones 101: A Basic Tutorial on Searching the Web," University of South Carolina, Beaufort Library, September 27, 2004, *www.sc.edu/ beaufort/library/pages/bones/bones.shtml* (accessed January 21, 2005).

12. Fred Langa, "Track It Down," *Windows*, July 1998, *www.winmag.com/library/1998/ 0701/fea0077.htm* (accessed January 23, 2005).

13. Walter S. Mossberg, "Unlike Search Engines, Answers.com Responds with Data, Not Links," *The Wall Street Journal*, January 27, 2005, *http://ptech.wsj.com/ archive/ptech-20050127.html* (accessed January 29, 2005).

14. Clusty, Next-Generation Search . . . Today, http//clusty.com (accessed August 3, 2005). See also Heather Green, "Building a Smarter Search Engine," BusinessWeek online, January 11, 2005, *www .businessweek.com/print/technology/ content/jan2005/tc2005014_2937 .htm!cha* (accessed July 18, 2005).

15. John Markoff, "Your Internet Search Results, in the Round," *New York Times*, May 9, 2005, p. C3.

16. Chamberlain, 2004.

17. Chris Taylor, "It's a Wiki, Wiki World," *Time*, June 6, 2005, pp. 40–42. See also Stacy Schiff, "The Interactive Truth," *New York Times*, June 15, 2005, p. A29.

18. Chamberlain, 2004.

19. Adam Gregerman, "Online Research Is So Easy, So Unreliable" [letter], *New York Times*, June 23, 2004, p. A26.

20. Janet Hogan, "The ABCDs of Evaluating Internet Resources," Binghamton University Libraries, November 2, 2004, *http:// library.lib.binghamton.edu/search/ evaluation.html*; Hope N. Tillman, "Evaluating Quality on the Net," Babson College. March 28, 2003, *www.hopetillman .com/findqual.html*; and Esther Grassian, "Thinking Critically about World Wide Web Resources," UCLA College Library, September 6, 2000, *www.library.ucla.edu/ libraries/college/help/critical/index.htm* (all accessed January 24, 2005).

21. Dan Fost, "Google Tunes to Video Search," *San Francisco Chronicle*, January 25, 2005, pp. E1, E4; Gary Price, "Yahoo Video Search Expands, on Home Page," *SearchEngineWatch*, January 24, 2005, *http://blog.searchenginewatch.com/blog/ 050124-180435* (accessed January 26, 2005).

22. Peter Suciu, "Amazon's Search Engine: Book It!" *Newsweek*, October 4, 2004, p. 15.

23. E. Schwartz, "Google, Yahoo Video Search Is the Tip of the Iceberg," *InfoWorld*, January 26, 2005, *http://weblog.infoworld .com/techwatch/archives/001024.html* (accessed January 26, 2005).

24. Daniel Terdiman, "A Tool for Scholars Who Like to Dig Deep," *New York Times*, November 25, 2004, p. B6. See also Jeffrey R. Young, "Google Unveils a Search Engine Focused on Scholarly Materials," *The Chronicle of Higher Education*, December 3, 2004, p. A34.

25. Jefferson Graham, "Google's Library Plan 'a Huge Help,'" *USA Today*, December 15, 2004, p. 3B. See also Kevin DeLaney and Jeffrey A. Trachtenberg, "Google Goes to College," *The Wall Street Journal*, December 14, 2004, pp. D1, D4.

26. Stephen H. Wildstrom, "Search-Boosters for Your PC," *BusinessWeek*, April 26, 2004, p. 26.

27. Verne Kopytoff, "Seeking a Fuller Search Engine," *San Francisco Chronicle*, August 16, 2004, pp. C1, C2. See also Chris Sherman, "Yahoo Launches Desktop Search," *SearchEngineWatch*, January 11, 2005, *http://searchenginewatch.com/ searchday/article.php/3457011* (accessed January 26, 2005).

28. Matt Lake, "Desperately Seeking Susan OR Suzie NOT Sushi," *New York Times*, September 3, 1998, p. D1.

29. Heather Green and Robert D. Hof, "Picking Up Where Search Leaves Off," *BusinessWeek*, April 11, 2005, pp. 88–89; and Steven Levy, "In the New Game of Tag, All of Us Are It," *Newsweek*, April 18, 2005, p. 14.

30. Tony Kontzer, "IM Users Will Double This Year," *Internet Week*, April 7, 2003, *www.internetweek.com/story/showArticle .jhtml!articleID=8600178* (accessed January 21, 2005).

31. Edward C. Baig, "Yahoo Jazzes Up Instant Messaging," *USA Today*, May 18, 2005, p. 3B.

32. Micheolle Slatalla, "The Office Meeting That Never Ends," *New York Times*, September 23, 1999, pp. D1, D8.

33. Jeanne Hinds, quoted in Slatalla, 1999.

34. John Schwartz, "Blogs Provide Raw Details from Scene of the Disaster," *New York Times*, December 28, 2004, p. A14. See also John Schwartz, "A Catastrophe Strikes, and the Cyberworld Responds," *New York Times*, January 2, 2005, sec. 3, p. 4; John Schwartz, "Myths Run Wild in Blog Tsunami Debate," *New York Times*, January 3, 2005, p. A9; Scott Shane and Nicholas Confessore, "In Seeking Help and Giving It, Computers Become a Lifeline," *New York Times*, January 5, 2005, p. A8; Kevin Maney, "Cellphones, Net Could Have Saved Thousands from Waves," *USA Today*, January 5, 2005, p. 5B; and Rebecca Buckman, "Relief, High-Tech Style," *The Wall Street Journal*, January 5, 2005, pp. B1, B6.

35. Lee Gomes, "How the Next Big Thing in Technology Morphed into a Really Big Thing," *The Wall Street Journal*, October 4, 2004, p. B1.

36. Antonio Regalado and Jessica Mintz, "Video Blogs Break Out with Tsunami Scenes," *The Wall Street Journal*, January 3, 2005, pp. B1, B5.

37. Study by Keynote Systems Inc., reported in Bruce Meyerson, "Study: Web-Based Phone Service Inferior," Yahoo! News, July 13, 2005, *http://news.yahoo.com/s/ap/ 20050714/ap_on_hi_te/techbits_internet_ phones&printer* (accessed July 14, 2005).

38. Stephen H. Wildstrom, Google's Magic Carpet Ride," *BusinessWeek*, July 18, 2005, p. 22.

39. Steven Levy, "The Earth Is Ready for Its Close-up," *Newsweek*, June 6, 2005, p. 13. See also James Fallows, "An Update on Stuff That's Cool (Like Google's Photo Maps," *New York Times*, April 17, 2005, sec. 3, p. 5; Verne Kopytoff, "Google's Free 3-D Service Brings Views of Earth Down to the PC," *San Francisco Chronicle*, June 29, 2005, pp. A1, A16; John Markoff, "Marrying Maps to Data for a New Web Service," *New York Times*, July 18, 2005, pp. C1, C8; and Verne Kopytoff, "Microsoft, Google in Sky Fight," *San Francisco Chronicle*, July 26, 2005, pp. D1, D2.

40. Anick Jesdanun, "Streaming Video Comes of Age in AOL's Webcasts of Live 8," *San Francisco Chronicle*, July 6, 2005, p. C3.

41. Jefferson Graham, "Websites Act More Like TV to Keep Users 'Tuned In,'" *USA Today*, June 16, 2005, pp. 1B, 2B.

42. Kevin J. Delaney, "Yahoo's Big Play in Online Music," May 11, 2005, pp. D1, D10; and Laurie J. Flynn, "Yahoo's New Music Service Will Cost About a Third of What Similar

Services Charge. Can It Earn a Profit?" *New York Times*, May 12, 2005, p. C7.

43. Heather Green, "All the News You Choose—on One Page," *BusinessWeek*, October 25, 2004, p. 46; and Kim Peterson, "Software Programs Called RSS Readers Creating a Blog Jam," *Seattle Times*, September 20, 2004, http://seattletimes .nwsource.com/html/businesstechnology/ 2002040564_rssclog20.html (accessed January 31, 2005).

44. Dan Carnevale, "A New Technology Lets Colleges Spread Information to People Who Want It," *The Chronicle of Higher Education*, February 13, 2004, pp. A31–A32.

45. Chris Taylor, "Let RSS Go Fetch," *Time*, May 30, 2005, p. 82.

46. Janet Kornblum, "Welcome to the Blogo-sphere," *USA Today*, July 8, 2003, p. 7D.

47. Pew Internet & American Life Project study, reported in Kevin Maney, "Once Blogs 'Change Everything,' Fascination with Them Will Chill," *USA Today*, May 25, 2005, p. 3B.

48. Megan Doscher, "The Best Way to Find Love," *The Wall Street Journal*, December 6, 1999, p. R34.

49. Gary Rivin, "Skeptics Take Another Look at Social Sites," *New York Times*, May 9, 2005, pp. C1, C6.

50. Timothy L. O'Brien and Saul Hansell, "Barbarians at the Digital Gate," *New York Times*, September 19, 2004, sec. 3, pp. 1, 4.

51. Jared Sandberg, "Monitoring of Workers Is Boss's Right but Why Not Include Top Brass," *The Wall Street Journal*, May 18, 2005, p. B1.

52. "Egregious Email," *Smart Computing*, October 2002, pp. 95–97.

53. Rob McGann, "The Deadly Duo—Spam and Viruses—2004 Year-End Recap," *ClickZ Network*, January 11, 2005, www .clickz.com/stats/sectors/email/article .php/3456921 (accessed January 27, 2005).

54. Tom Spring, "Spam Fighting Tips for the New Year," *PC World*, January 24, 2005, www.pcworld.com/news/article/ 0,aid,119358,00.asp (accessed January 27, 2005); Sean Carroll, "Slam the Spam," *PC Magazine*, February 25, 2003, pp. 74–97.

55. Julian Haight, quoted in David Lazarus, "Fax Spam Is Hard to Shake," *San Francisco Chronicle*, February 7, 2000, pp. C1, C2.

56. Riva Richmond, "Companies Target E-Mail 'Spoofing,'" *The Wall Street Journal*, June 9, 2004, p. D9; and Amey Stone, "How to Avoid the 'Phish' Hook," *BusinessWeek online*, May 24, 2004, www.businessweek .com/technology/content/may2004/ tc20040524_8133_tc024.htm (accessed January 29, 2005).

57. "Email Spoofing," SearchSecurity.com Definitions, November 20, 2003, http:// searchsecurity.techtarget.com/sDefinition/ 0,,sid14_gci840262,00.html (accessed January 28, 2005).

58. Brian Grow, "Spear-Phishers Are Sneaking In," *BusinessWeek*, July 11, 2005, p. 13.

59. David F. Gallagher, "Users Find Too Many Phish in the Internet Sea," *New York Times*, September 20, 2004, p. C4; ; Ann Grimes, "No Phishing Allowed," *The Wall Street Journal*, September 16, 2004, p. B4; and Stone, 2004.

60. Kevin J. Delaney, "'Evil Twins' and 'Pharming,'" *The Wall Street Journal*, May 17, 2005, pp. B1, B2; and Jon Swartz, "Thieves Hit Internet with Sneakier Software," *USA Today*, May 18, 2005, p. 1B.

61. Kim Komando, "5 Tips for Spurning Spyware and Browser Hijackers," *Microsoft Small Business Center*, 2005, www.microsoft.com/smallbusiness/ issues/marketing/privacy_spam/ 5_tips_for_spurning_spyware_and_ browser_hijackers.mspx (accessed January 28, 2005).

62. David Kesmodel, "Marketers Seek to Make Cookies More Palatable," *The Wall Street Journal*, June 17, 2005, pp. B1, B2.

63. America Online and the National Cyber Security Alliance, *AOL/NCSA Online Safety Study*, October 2004, www .staysafeonline.info/news/safety_study_ v04.pdf (accessed January 27, 2005).

64. Joseph Telafici, quoted in Vincent Kiernan, "The Next Plague," *The Chronicle of Higher Education*, January 28, 2005, pp. A36–A38.

65. See Rachel Dodes, "Terminating Spyware with Extreme Prejudice," *New York Times*, December 30, 2004, pp. E1, E6; Walter S. Mossberg, "A Primer on Fighting Spyware," *The Wall Street Journal*, December 29, 2004, pp. D1, D3; Byron Acohido and Jon Swartz, "Market to Protect Consumer PCs Seems Poised for Takeoff," *USA Today*, December 27, 2004, pp. 1B, 3B; Anick Jesdanun, Associated Press, "Spyware Assault Gets More Aggressive," *Reno Gazette-Journal*, October 31, 2004, pp. 1A, 14A; J. D. Biersdorfer, "Leave No Footprints in Online Transactions," *New York Times*, October 24, 2004, p. E6; Toddi Gutner, "What's Lurking in Your PC?" *Business-Week*, October 4, 2004, pp. 108 – 110; James R. Hagerty and Dennis K. Berman, "New Battleground in Web Privacy War: Ads That Snoop," *The Wall Street Journal*, August 27, 2003, pp. A1, A8.

66. Lee Gomes, "Spyware Is Easy to Get, Difficult to Remove, Increasingly Mali-cious," *The Wall Street Journal*, July 12, 2004, p. B1.

67. Brock Read, "As Security Concerns Rise, New Web Browsers Gain Favor at Colleges," *The Chronicle of Higher Education*, July 15, 2005, p. A37.

68. Ellen Laird, "Internet Plagiarism: We All Pay the Price," *The Chronicle of Higher Education*, June 13, 2001, p. B5.

69. Bruce Leland, quoted in Peter Applebome, "On the Internet, Term Papers Are Hot Items," *New York Times*, June 8, 1997, sec. 1, pp. 1, 20.

70. Eugene Dwyer, "Virtual Term Papers" [letter], *New York Times*, June 10, 1997, p. A20.

71. John Yankey, "Cheating, Digital Plagiarism Rise as Tech Hits Classroom," *Reno Gazette-Journal*, July 30, 2001, p. 23E.

72. Scott Carlson, "Journal Publishers Turn to Software to Root Out Plagiarism by Schol-ars," *The Chronicle of Higher Education*, June 10, 2005, p. A27.

73. William L. Rukeyser, "How to Track Down Cyber-Cheaters" [letter], *New York Times*, June 14, 1997, sec. 6, p. 14.

74. David Rothenberg, "How the Web Destroys the Quality of Students' Research Papers," *The Chronicle of Higher Education*, August 15, 1997.

Chapter 3

1. Alan Robbins, "Why There's Egg on Your Interface," *New York Times*, December 1, 1996, sec. 3, p. 12.

2. Stephen H. Wildstrom, "Tiger Makes Mac's Edge Even Sharper," *BusinessWeek*, May 9, 2005, p. 28.

3. Laurie J. Flynn, "New Version of Windows Coming in '06," *New York Times*, July 23, 2005, p. B3; Robert A. Guth, "Microsoft Opens New Windows with a Less-Technical Name," *The Wall Street Journal*, July 25, 2005, p. B3; and Yardena Arar, "Microsoft Unveils Windows Vista Beta," *Digit*, July 28, 2005, www.digitmag.co.uk/news/index .cfm?NewsID=5017 (accessed July 28, 2005).

4. Walter S. Mossberg, "Free Security Upgrade to Windows XP Has Value but Falls Short," *The Wall Street Journal*, August 19, 2004, p. B1.

5. Byron Acohido, "Microsoft Tells Users to Patch 21 New Flaws in Windows," *USA Today*, October 13, 2004, p. 5B.

6. Karen Dearne, "Microsoft Gets Tough on Malware," *Australian IT*, February 22, 2005, http://australianit.news.com.au/ articles/0,7204,12326425% 5E15343% 5E%5Enbv%5E15306-15317,00.html (accessed February 22, 2005).

7. Greg Sullivan, quoted in Edward C. Baig, "Windows Upgrade Makes Strides to Outrun the Bad Guys," *USA Today*, August 12, 2004, p. 3B.

8. Kevin Maney, "Mac or PC? Windows' Security Issues Help Some Users Choose," *USA Today*, September 22, 2004, p. 4B.

9. Al Fasoldt, "Mac Covers Most Software Needs," *San Francisco Chronicle*, August 30, 2004, p. C4.

10. Chris Reiter, "New Software Poses Threat to Windows," *The Wall Street Journal*, July 6, 2005, p. B2B.

11. Irving Wladawsky-Berger, quoted in Deborah Solomon, "Could Linux Outdo Windows?" *USA Today*, March 9, 2000, pp. 1B, 2B.

12. Glenn Rifkin, "Designing Tools for De-signers," *New York Times*, June 18, 1992, p. C6.

13. Study by Kent Norman, Laboratory of Automation Psychology and Decision Processes, University of Maryland, cited in Katherine Seligman, "Computer Crashes Booming Business," *San Francisco Chroni-cle*, April 17, 2005, pp. A1, A21.

14. James Aley, "Software That Doesn't Work. Customers Are in Revolt. Here's the Plan," *Fortune*, November 25, 2002, pp. 147–158.

15. Jane Black, "Usability Is Next to Profitabil-ity," *BusinessWeek online*, December 4, 2002, www.businessweek.com/ technologycontent/dec2002.

16. Harris Interactive Poll, May 2002, reported in Andrew Park, "Not Much Help from the Help Line," *BusinessWeek* , July 15, 2002, p. 10.

17. Edward C. Baig, "Have You Tried to Get Tech Support Lately? Arrgh!#*!!" *USA Today*, August 27, 2004.

18. Edward C. Baig, "Tech Support: Not Quite at Our Beck and Call," *USA Today*, July 17, 2002, p. 3D.

19. Adapted from Jane Spencer, "Computer Glitches? Rent Your Own Tech," *The Wall Street Journal*, August 22, 2002, table, p. D3.

20. Stan Sams, quoted in Spencer, 2002, p. D3.

Chapter 4

1. Michael S. Malone, "The Tiniest Trans-former," *San Jose Mercury News*, Septem-ber 10, 1995, pp. 1D, 2D; excerpted from *The Microprocessor: A Biography* (New York: Telos/Springer Verlag, 1995).

2. Malone, 1995.

3. Laurence Hooper, "No Compromises," *The Wall Street Journal*, November 16, 1992, p. R8.

4. "Exabyte," The Sharpened Glossary: Defini-tions of Computer Terms, www.sharpened .net/glossary/definition.php!exabyte (accessed August 1, 2005).

5. Philip Robinson, "When the Power Fails," *San Jose Mercury News*, December 17, 1995, pp. 1F, 6F.

6. Matthew Yi, "Faster and Faster, Smaller and Smaller—It's the Law," *San Francisco Chronicle*, April 18, 2005, p. E1.

7. Scott Wasson, "Intel's Pentium Extreme Edition 840 Processor," *The Tech Report*, May 5, 2005, http://techreport.com/ reviews/2005q2/pentium-xe-840/ index.x?pg=1

8. Nick Wingfield and Don Clark, "With Intel Inside Apple, Macs May Be Faster, Smaller," *The Wall Street Journal*, June 7, 2005, pp. B1, B7; Mattew Yi and Benny Evangelista, "Apple Gains Room to Grow," *San Francisco Chronicle*, June 7, 2005, pp. A1, A8; Walter S. Mossberg, "What the Apple Plan to Switch to Intel Chips Means for Consumers," *The Wall Street Journal*,

June 9, 2005, p. B1; John Markoff, "Think Similar," *New York Times*, June 11, 2005, pp. B1, B13; and Peter Burrows, "Tougher Days, Bolder Apple," *BusinessWeek*, June 20, 2005, pp. 38, 40.

9. Mossberg, 2005.

10. Stephen H. Wildstrom, "Chips with Two Brains," *BusinessWeek*, August 1, 2005, p. 20.

11. Adam Aston, "More Life for Moore's Law," *BusinessWeek*, June 20, 2005, pp. 108–109.

12. Matthew Yi, "AMD Chip Faster, Costlier," *San Francisco Chronicle*, June 27, 2005, pp. E1, E6; and Timothy Prickett Morgan, "Intel Previews Dual-Core Montecito Itanium Performance," *The Linux Beacon*, July 12, 2005, *www.itjungle.com/tlb/ tlb071205-story03.html* (accessed August 2, 2005).

13. Don Clark and Robert A. Guth, "Sony, IBM, Toshiba to Offer First Peek of 'Cell' Chip Design," *The Wall Street Journal*, February 7, 2005, pp. B1, B4; John Markoff, "Smaller than a Pushpin, More Powerful than a PC," *New York Times*, February 7, 2005, p. C3; Michelle Kessler, "Chipmakers Create Processors That Handle Many Tasks," *USA Today*, February 8, 2005, p. 3B; and Otis Port, "Mighty Morphing Power Processors," *BusinessWeek*, June 6, 2005, pp. 60–61.

14. James Gorbold, "Product Reviews: Intel CPUs," *Custom PC*, June 5, 2005, *http://www.custompc.co.uk/custompc/ processors/reviews/72870/intel-pentium-extreme-edition-840.html*

15. Kevin Maney, "Concept of $100 Laptop for World's Poor Is a Winner," *USA Today*, February 9, 2005, p. 3B.

16. Cisco Cheng and Erik Rhey, "Ports," *PC Magazine*, April 26, 2005, p. 66.

17. Iomega 2003 study, cited in Henry Norr, "Data Files Need to Be Backed Up," *San Francisco Chronicle*, March 13, 2003, p. E1.

18. Edward Baig, "Be Happy, Film Freaks," *BusinessWeek*, May 26, 1997, pp. 172–173.

19. Michio Kaku, "What Will Replace Silicon?" *Time*, June 19, 2000, p. 99.

20. October 2004 study by America Online and the National Cyber Security Alliance, cited in Rachel Dodes, "Terminating Spyware with Extreme Prejudice," *New York Times*, December 30, 2004, pp. E1, E6.

21. Alan Luber, "Hard Drive Backup & Restore Basics, Part 1," *Smart Computing*, August 2004, p. 96; Dodes, 2004; Rachel Dodes, "Tools to Make Your Hard Drive Forget Its Past," *New York Times*, December 30, 2004, p. E6; Jeff Dodd, "How to Install Operating Systems," *Smart Computing*, February 2005, pp. 60–63; and Nigel Powell, "Archive Your Drive," *Popular Science*, March 2005, pp. 73–74.

22. Gordon Moore, quoted in "Gordon Moore Q&A," *Time*, June 19, 2000, p. 99.

23. Gordon Moore, quoted in Yi, 2005.

24. John Markoff, "IBM Plans to Announce Tiny Transistor," *New York Times*, December 9, 2002, p. C4; and Don Clark, "IBM, Xilinx Pass a Milestone in Race to Shrink Circuitry," *The Wall Street Journal*, December 16, 2002, p. 5.

25. David Legard, "AMD Fabricates First 10-Nanometer Transistor," *IDG News Service/Australia Bureau*, September 10, 2002, *www.idg.net*.

26. Michael Kanellos, "HP Nanotech Takes Chips beyond Transistors," *CNET News.com*, February 10, 2005.

27. Aston, 2005; John Markoff, "Chip Maker Develops Denser Storage Method," *New York Times*, May 9, 2005, p. C3.

28. Byron Acohido, "IBM to Sell Units of Computer Power," *USA Today*, July 1, 2002, p. 5B.

29. Stephen Baker and Adam Aston, "Universe in a Grain of Sand," *BusinessWeek*, October 11, 2004, pp. 138–140. See also Stephen Baker and Adam Aston, "The Business of Nanotech," *BusinessWeek*, February 14, 2005, pp. 64–71; and Lee Bruno, "Smallville," *Stanford*, May/June 2005, pp. 44–49.

30. David Voss, "Silicon Lasters," *Technology Review*, June 2001, p. 35; Peter Pietromanaco, "Optical Computing: The Wave of the Future," *Poptronics*, October 7, 2002, p. 12; Antonio Regalado, "Can Optical Chips Replace Telecom Gear?" *The Wall Street Journal*, January 9, 2003, pp. B5, B6; and Benjamin Pimentel, "HP Makes Futuristic Discovery," *San Francisco Chronicle*, July 1, 2005, pp. C1, C6.

31. Roland Piquepaille, "DNA Computing," *Roland Piquepaille's Technology Trends*, August 15, 2002, *http://radio.weblogs .com/0105910/2002/08/15.html*; Roland Piquepaille, "DNA Computer Plays Tic-Tac-Toe," *Roland Piquepaille's Technology Trends*, August 18, 2003, *http://www .primidi.com/2003/08/18.html*; and Stefan Lovgren, "Computer Made from DNA and Enzymes," *National Geographic News*, February 24, 2003, *http://news .nationalgeographic.com/news/2003/02/ 0224_030224_DNAcomputer.html*.

32. Josh Wolf, "Nanotech on the Front Lines," *Forbes.com*, March 19, 2003, *www.forbes .com*; Goerge Johnson, *A Shortcut Through Time: The Path to the Quantum Computer* (New York: Knopf, 2003); and Lee Gomes, "Quantum Computing May Seem Too Far Out, but Don't Count on It," *The Wall Street Journal*, April 25, 2005, p. B1.

33. James Canton, quoted in Barnaby J. Feder, "Scientists of Very Small Draw Disciplines Together," *New York Times*, February 10, 2003, p. C4.

34. Barnaby J. Feder, "Nanotechnology Has Arrived; a Serious Opposition Is Forming," *New York Times*, August 19, 2002, p. C3; Barnaby J. Feder, "From Nanotechnology's Sidelines, One More Warning," *New York Times*, February 1, 2003, pp. C1, C3; James M. Pethokoukis, "Devil in the Details?" *U.S. News & World Report*, January 27/February 3, 2003, p. 44; Mike Treder, "Molecular Nanotech: Benefits and Risks," *The Futurist*, January–February 2004, pp. 42–46; Dan Vergano, "Creating a Monster?" *USA Today*, September 28, 2004, p. 6D; Fred Krupp and Chad Holliday, "Let's Get Nanotech Right," *The Wall Street Journal*, June 14, 2005, p. B2; and Kevin Maney, "Scared of New Nano-Pants? Hey, You May Be Onto Something," *USA Today*, June 22, 2005, p. 3B.

35. Don Clark, "Seagate Introduces Disk Drive with More Storage Capacity," *The Wall Street Journal*, June 9, 2005, p. B7.

36. John Markoff, "Hitachi Achieves Storage Record for Disk Drives," *New York Times*, April 4, 2005, p. C4.

37. Walter S. Mossberg, "Shopping for a Laptop? Expect Lots of Choices, and a Range of Prices," *The Wall Street Journal*, April 14, 2005, p. B1.

38. Stephen H. Wildstrom, "How to Shop for a Laptop," *BusinessWeek*, April 3, 2000, p. 25.

39. "Notebooks," *PC Magazine*, November 30, 2004, pp. 121–134; Mossberg, "Shopping for a Laptop?" 2005; "Inside Notebooks," *PC Magazine*, April 26, 2005, pp. 62–68; Bill Howard, "Multimedia Notebooks," *PC Magazine*, July 2005, pp. 89–102; and Wilson Rothman, "Laptops and Desktops: Basics, Bells and Whistles," *New York Times*, August 3, 2005, p. E6.

40. John Schwartz, "Back to School," *New York Times*, August 3, 2005, pp. E1, E6.

41. Mossberg, "Shopping for a Laptop?" 2005.

Chapter 5

1. David F. Gallagher, "2 Rooms, River View, ATM in Lobby," *New York Times*, June 6, 2002, p. E1.

2. Mary Kathleen Flynn, "Banking with a Big Gulp," *U.S. News & World Report*, April 25, 2005, pp. EE10, EE12.

3. Jeffrey Selingo, "Want to Rent a Movie? Help Yourself," *New York Times*, September 2, 2004, p. E3; Linda Stern, "Here Come the Kiosks," *Newsweek*, March 22, 2004, p. E4; and Christopher Elliott, "Invasion of the Kiosks," *March 5, 2004*, p. 82.

4. John Tierney, "Shop Till Eggs, Diapers, Toothpaste Drop," *New York Times*, August 28, 2002, pp. A1, A14.

5. Ben Bederson, quoted in Tom Zeller Jr., "A Great Idea That's All in the Wrist," *New York Times*, June 5, 2005, sec. 4, p. 5.

6. Kevin J. Delaney, "Dumber PCs That Use the Net for Processing, Storage Get Hot—Again," *The Wall Street Journal*, December 27, 2004, p. B1.

7. Matthew Yi, "Apple's New Mouse Works on Macs and Windows PCs," *San Francisco Chronicle*, August 3, 2005, pp. C1, C3; and Walter S. Mossberg, "Apple's New Mouse Is Not as Mighty as Rival's Magnifier," *The Wall Street Journal*, August 4, 2005, p. B1.

8. Steven Kent, "Nintendo DS: Doubly Good," *USA Today*, November 22, 2004, p. 4D.

9. Thomas J. Fitzgerald, "The Tablet PC Takes Its Place in the Classroom," *New York Times*, September 9, 2004, p. E5; and Christopher Elliott, "The Tablet Turns for Computing," *U.S. News & World Report*, May 3, 2004, p. 74. See also Edward C. Baig, "Pen Could Be Mightier than Keyboard with ThinkPad X41," *USA Today*, June 9, 2005, p. 3B.

10. Edward C. Baig, "Will Pen Be Mightier than Other Toys?" *USA Today*, January 18, 2005, p. 3B.

11. Stephanie Kang, "New Computer Pen Reads Handwriting and Can Talk Back," *The Wall Street Journal*, January 12, 2005, pp. B1, B4.

12. Peter Svensson, "Portable Scanner Stores Text on the Fly," *San Francisco Chronicle*, March 15, 2004, p. E2.

13. Marlon Manuel, "Scan-It-Yourself Part, Parcel of Life," *San Francisco Chronicle*, January 30, 2005, p. E3.

14. Jeffrey Selingo, "You're the Shopper and the Cashier," *New York Times*, May 4, 2005, p. E3.

15. Robert Johnson, "The Fax Machine: Technology That Refuses to Die," *New York Times*, March 27, 2005, sec. 3, p. 8.

16. Katie Hafner, "At Your Service (or Wits' End)," *New York Times*, September 9, 2004, pp. E1, E7.

17. William J. Broad, "A Web of Sensors Taking Earth's Pulse," *New York Times*, May 10, 2005, pp. D1, D4.

18. John Schwartz, "Graduate Cryptographers Unlock Code of 'Thiefproof' Car Key," *New York Times*, January 29, 2005, p. A10.

19. Peter Sanders, "Casinos Bet on Radio-ID Gambling Chips," *The Wall Street Journal*, May 13, 2005, pp. B1, B7.

20. Jenny Strasburg, "Chase Introduces No-Swipe Plastic," *San Francisco Chronicle*, May 20, 2005, pp. C1, C6; and Mindy Fetterman, "Wave-and-Pay Credit Cards May Make Buying Too Easy," *USA Today*, May 24, 2005, p. 1B.

21. James M. Pethokoukis, "Big Box Meets Big Brother," *U.S. News & World Report*, January 24, 2005, pp. 46–47.

22. Barnaby J. Feder, "Is There Anyone Out There Who Can Read My Tag?" *New York Times*, July 23, 2005, p. B5.

23. Patricia Yollin, "Ted's Excellent Adventure," *San Francisco Chronicle*, September 25, 2003, pp. A1, A11.

24. Geri Smith, "These ID Tags Get under Your Skin," *BusinessWeek*, August 2, 2004, p. 77.

25. Diedtra Henderson, "Implantable Chip Sparks Privacy Concerns," *Reno Gazette-Journal*, October 14, 2004, p. 4A.

26. Benjamin Pimentel, "Biometrics Puts Security at a User's Fingertips," *San Francisco Chronicle*, July 11, 2005, pp. E1, E2; Andrew Morse, "New Biometric Identifier Is at Hand," *The Wall Street Journal*, July 21, 2005, p. B4; and Thomas Frank, "Biometric IDs Could See Massive Growth," *USA Today*, August 15, 2005, pp. 1B, 2B.

27. Ian Austen, "Crisp, Bright Images on Flat-Panel Screens," *New York Times*, January 13, 2005, p. E7.

28. Andrew Park, "Monitors," *BusinessWeek*, August 8, 2005, p. 80.

29. J. D. Biersdorfer, "An Inexpensive Laser Printer Gives the Inkjets a Run for Their Money," *New York Times*, June 30, 2005, p. C8; and Walter S. Mossberg, "The $99 Laser Printer: Home Options Get Close to Office Quality," *The Wall Street Journal*, July 13, 2005, pp. D1, D4.

30. Michell Kessler, "New HP Printers Pop Out Photos in 14 Seconds Flat," *USA Today*, July 12, 2005, p. 3B; and David Pogue, "An Inkjet Built for Speed," *New York Times*, July 21, 2005, pp. C1, C10.

31. Larry Armstrong, "Printers," *BusinessWeek*, August 8, 2005, p. 77.

32. Danielle Weatherbee, quoted in Steve Friess, "Laptop Design Can Be a Pain in the Posture," *USA Today*, April 13, 2005, p. 8D.

33. Tom Albin, quoted in Friess, 2005.

34. Tamara James, quoted in Friess, 2005.

35. Rachel Konrad, "Logitech Spawns New Breed of Mice That Is Twice as Nice," *San Francisco Chronicle*, September 27, 2004, p. F4.

36. Albert R. Karr, "An Ergo-Unfriendly Home Office Can Hurt You," *The Wall Street Journal*, September 30, 2003, p. D6.

37. J. H. Andersen, J. F. Thomsen, E. Overgaard, C. F. Lassen, L. P. A. Brandt, I. Vilstrup, A. I. Kryger, and S. Mikkelsen, "Computer Use and Carpal Tunnel Syndome: A 1-Year Follow-up Study," *Journal of the American Medical Association*, 283(2003): 2963–2969; and J. C. Stevens, J. C. Witt, B. E. Smith, and A. L. Weaver, "The Frequency of Carpal Tunnel Syndrome in Computer Users at a Medical Facility," *Neurology*, 56(2001):1568–1570.

38. Cornell University Ergonomics Web, "Carpal Tunnel Syndrome and Computer Use—Is There a Link?" http://ergo .human.cornell.edu/JAMAMayoCTS.html (accessed March 31, 2005).

39. "Cellular Phone Health," *www .cellularphonehealth.com* (accessed March 31, 2005).

40. Minouk Schoemaker, Anthony Swerdlow, et al., "Mobile Phone Use and Risk of Acoustic Neuroma: Results of the Interphone Case-Control Study in Five North European Countries," *British Journal of Cancer*, August 31, 2005, reported in Matt Moore, "No Link Between Cellphones, Tumors: Study," *GlobeandMail.com*, August 31, 2005, http://www .globetechnology.com/servlet/story/ RTGAM.20050831.gttumouraug31/ BNStory/Technology.

41. Benjamin Fulford, "Sensors Gone Wild," *Forbes*, October 28, 2002, pp. 306–308.

42. Don Clark, "Low-Cost Device for Monitoring Is Set for Market," *The Wall Street Journal*, September 20, 2004, p. B5; and Karen F. Schmidt, "'Smart Dust' Is Way Cool," *U.S. News & World Report*, February 16, 2004, pp. 56–57.

43. Clark Nguyen, professor of electrical engineering, quoted in Fulford, 2002.

44. Mimi Hall, "Sensors to Provide Early Warning for Bioterror Attacks," *USA Today*, January 23, 2003, p. 3A.

45. David Perlman, "H-Bomb Sensors Yield New Benefits," *San Francisco Chronicle*, December 9, 2002, p. A9.

46. "Biosensors for Underwear," *The Futurist*, January–February 2004, p. 13.

47. Ken Brown, "New Technologies Bring the Sense of Touch to Computers," *The Wall Street Journal*, November 26, 2004, pp. B1, B3.

48. Faith Keenan, "PCs and Speech: A Rocky Marriage," *BusinessWeek*, September 9, 2002, pp. 64–66; and Anne Eisenberg, "Teaching Machines to Hear Your Prose and Your Pain," *New York Times*, August 1, 2002, p. E9.

49. Douglas Heingartner, "Attention, Cows: Please Speak into the Microphone," *New York Times*, August 1, 2002, p. E5.

50. David Pogue, "Camcorders (Size Small) Offer a Fit," *New York Times*, March 10, 2005, pp. E1, E6; Mike Langberg, "Sony Offers Respectable Digital Camera/ Camcorder Combo," *San Jose Mercury News*, February 14, 2005; and "Happy New Gear!" Time, December 27, 2004.

51. Benny Evangelista, "Microsoft Team Works on Getting Ahead of Its Time," *San Francisco Chronicle*, June 14, 2004, p. D4.

52. David Pogue, "Sizing Up a New Species: Camera-Binoculars," *New York Times*, March 24, 2005, pp. E1, E4.

53. Anita Hamilton, "Who Needs a Mouse?" *Time*, July 21, 2003, p. 76.

54. Anne Eisenberg, "Fleeting Experience, Mirrored in Your Eyes," *New York Times*, July 29, 2004, p. E8.

55. Kevin Maney, "Scientists Gingerly Tap into Brain's Power," *USA Today*, October 11, 2004, pp. 1B, 2B. See also Andrew Pollack, "With Tiny Brain Implants, Just Thinking May Make It So," *New York Times*, April 13, 2004, p. D5.

56. Sandra Blakeslee, "Imagining Thought Controlled Movement for Humans," *New York Times*, October 14, 2003, p. D3.

57. Don Clark, "Mind Games: Soon You'll Be Zapping Bad Guys without Lifting a Finger," *The Wall Street Journal*, June 16, 1995, p. B12; Malcolm W. Browne, "How Brain Waves Can Fly a Plane," *New York Times*, March 7, 1995, pp. B1, B10; and Dan Vergano, "Paralyzed Use Brain Waves to Move," *USA Today*, December 7, 2004, p. 8D.

58. Brandon Michener, "Controlling a Computer by the Power of Thought," *New York Times*, March 14, 2001, pp. B1, B4.

59. Phillip Robinson, "Home Theater No Longer Just for the Rich and Famous," *San Jose Mercury News*, November 30, 1997, p. 9S.

60. Don Clark, "Videogames Get Real," *The Wall Street Journal*, April 14, 2004, pp. B1, B2.

61. Harry Somerfield, "3-D Is Headed for Home Television," *San Francisco Chronicle*, March 2, 1994, p. Z5.

62. Peter Svensson, "A Flat Panel That Also Goes Deep," *San Francisco Chronicle*, November 1, 2004, p. D2; and Suzanne Kantra Kirschner, "Into the 3rd Dimension," *Popular Science*, February 2004, p. 22.

63. Adam Ashton, "If You Can Draw It, They Can Make It," *BusinessWeek*, May 23, 2005, pp. MTL7–MTL8.

64. William Van Winkle, "Under Development: A Peek at What's Brewing in the Laboratory," *SmartComputing*, April 2003, pp. 106–107.

65. Michelle Rama, "Burn Researchers Use Old Inkjet Printers to Make Human Skin," *The Wall Street Journal*, August 11, 2004, pp. B1, B10.

66. "Boot Up or Die," *PC Computing*, April 1998, pp. 172–186.

67. Peter Svensson, "Tech Goes Rugged at Trade Show," *San Francisco Chronicle*, May 31, 2004, p. D2.

Chapter 6

1. "What Does 'Digital' Mean in Regard to Electronics?" *Popular Science*, August 1997, pp. 91–94.

2. Ken Belson, "Dial-Up Internet Starts to Go the Way of Rotary Phones," *New York Times*, June 21, 2005, pp. C1, C5.

3. Lisa Guernsey, "Wire Wires in the Walls, the Cyberhome Hums," *New York Times*, March 27, 2003, pp. D1, D7.

4. Shawn Young, "Why the Glut in Fiber Lines Remains Huge," *The Wall Street Journal*, May 12, 2005, pp. B1, B10.

5. Siobhan McAndrew, "As Living Room Fades, Tech Takes Center Stage," *Reno Gazette-Journal*, January 14, 2005, p. 1D.

6. Christie Thomas, quoted in Stephanie Armour, "Job Opening? Work-at-Home Moms Fill Bill," *USA Today*, July 20, 2005, p. 3B.

7. Kris Maher, "Corner Office Shift: Telecommuting Rises in Executive Ranks," *The Wall Street Journal*, September 21, 2004, pp. B1, B10.

8. David Kline, quoted in W. James Au, "The Lonely Long-Distance Worker," *PC Computing*, February 2000, pp. 42–43.

9. IDC, cited in Carolyn Said, "Work Is Where You Hang Your Coat," *San Francisco Chronicle*, July 18, 2005, pp. E1, E5.

10. Hewitt Associates survey, cited in Said, 2005.

11. Jim Hopkins, "How Solo Workers Keep from Getting Depressed," *USA Today*, May 9, 2001; Stephanie Armour, "Telecommuting Gets Stuck in the Slow Lane," *USA Today*, June 25, 2001, pp. 1A, 2A; Stephanie Armour, "More Bosses Keep Tabs on Telecommuters," *USA Today*, July 24, 2001, p. 1B; Katherine Reynolds Lewis, "Working from Home Not Always the Right Fit," *San Francisco Chronicle*, July 14, 2003, p. E2; and Sue Shellenbarger, "'Shed Boy Is on Line One': Some Tales from the Growing World of Home Offices," *The Wall Street Journal*, June 17, 2004, p. D1.

12. Said, 2005, p. E5.

13. Robert Smith, quoted in Said, 2005, p. E5.

14. Evan Ramstad, "Works in Progress," *The Wall Street Journal*, September 11, 1997, pp. 1E, 4E.

15. Leslie Cauley, "iTown Is Gearing Up to Deliver Broadband to Your Town," *USA Today*, July 25, 2005, p. 4B.

16. Andy Pasztor, "Web by Satellite in Round Two: Will It Fly Now?" *The Wall Street Journal*, June 3, 2005, pp. B1, B2.

17. Jathon Sapsford, "A New Direction," *The Wall Street Journal*, July 25, 2005, p. R4.

18. Aaron Weiss, "Digital Breadcrumbs for Your Photos," *New York Times*, December 16, 2004, p. E4.

19. Richard Moreno, "Cache Me If You Can," *Nevada Magazine*, July/August 2005, pp. 26–27.

20. Christopher Elliott, "Online Maps That Steer You Wrong," *New York Times*, June 28, 2005, p. C8.

21. Louise Rafkin, "We Are Here. Or Are We There? Just Where Are Global Positioning Systems Leading Us?" *San Francisco Chronicle Magazine*, May 23, 2004, pp. 10–13.

22. Anne Marie Squeo, "Cellphone Hangup: When You Dial 911, Can Help Find You?" *The Wall Street Journal*, May 12, 2005, pp. A1, A10; Shawn Young, "Some Internet Phone Firms Fail to Link Callers to 911," *The Wall Street Journal*, May 12, 2005, pp. D1, D4; and Anne Marie Squeo, "Tests Show Many Cellphone Calls to 911 Go Unlocated," *The Wall Street Journal*, May 19, 2005, pp. B1, B6.

23. Steve Rosenbush and Heather Green, "Looking to Pick Off BlackBerry," *BusinessWeek*, May 23, 2005, p. 104; Sarmad Ali, "Wireless Email's New Fans," *The Wall Street Journal*, pp. B1, B4; and Roger O. Crockett, Cliff Edwards, and Spencer E. Ante, "How Motorola Got Its Groove Back," *BusinessWeek*, August 8, 2005, pp. 68–70.

24. Kevin Maney, "A Very Different Future Is Calling—on Billions of Cellphones," *USA Today*, July 27, 2005, p. 3B.
25. Gartner Group, cited in Maney, 2005.
26. Jesse Drucker, "New Standard for Wireless to Be Studied," *The Wall Street Journal*, January 5, 2005, p. B6.
27. David Pogue, "Beyond Wi-Fi: Laptop Heaven but at a Price," *New York Times*, June 23, 2005, pp. C1, C10.
28. Craig Ellison, "The Power and Promise of WiMax," *PC Magazine*, March 22, 2005, p. 102.
29. Davis D. Janowski, "New Ways to Go Wireless," *PC Magazine*, March 22, 2005, pp. 98–102.
30. Oliver Kaven, "Wi-Fi's Promising New Wave: 802.11n," *PC Magazine*, March 22, 2005, p. 100; Edward C. Baig, "New Wireless Technology Kills 'Dead Spots' at Home," *USA Today*, March 31, 2005, p. 8B; and Stephen H. Wildstrom, "Wi-Fi: Pumping Up the Volume," *BusinessWeek*, May 16, 2005, p. 18.
31. Marshall Loeb, "Securing Your Home Wi-Fi Network Saves Money," *Reno Gazette-Journal*, June 30, 2005, p. 3D.
32. Oliver Kaven, "More than Just Flashes in the PAN," *PC Magazine*, March 22, 2005, p. 101.
33. Keith J. Winstein, "Bluetooth Gear May Be Open to Snooping," *The Wall Street Journal*, June 16, 2005, pp. B1, B6.
34. Lee Gomes, "Making Connections: Bluetooth, Copper Wire Are Showing New Life," *The Wall Street Journal*, January 10, 2005, p. B1.
35. Judy Lam, "Ultrawideband Promises Boost to Wireless World," *The Wall Street Journal*, June 16, 2005, pp. B1, B6.
36. Davis D. Janowski, "The Battle for Your Automated Home," *PC Magazine*, March 22, 2005, pp. 101–102.
37. Jessica Davis, "ZigBee Gets Some Company," *Electronic News*, June 10, 2004, *www.reed-electronics.com/electronicnews/article/CA424448* (accessed April 11, 2005).
38. William A. Bulkeley, "Wireless's New Hookup," *The Wall Street Journal*, February 24, 2005, pp. B1, B2; and John K. Waters, "ZigBee Alliance Opens Membership to Adopter Class," *ADTmag.com*, March 17, 2005, *www.adtmag.com/article.asp?id=10771* (accessed April 9, 2005).
39. Tony Kistner, "HomePlug Networking Charges Up," *PCWorld.com*, March 28, 2005, *www.pcworld.com/news/article/0,aid,120175,00.asp* (accessed April 9, 2005).
40. Lew Tucker, quoted in Matt Richtel and John Markoff, "Corrupted PC's Discover a Home: The Dumpster," *New York Times*, July 17, 2005, sec. 1, p. 13.
41. America Online and the National Cyber Security Alliance, *AOL/NCSA Online Safety Study*, October 2004, *www.staysafeonline.info/news/safety_study_v04.pdf* (accessed April 14, 2005).
42. Ken Watson, quoted in Ted Bridis, "Internet Users Virtually Clueless about Security," *San Francisco Chronicle*, October 25, 2004, p. A2.
43. David Bank, "Keeping Information Safe," *The Wall Street Journal*, November 11, 2004, pp. B1, B4.
44. Munir Kotadia, "One Man Created Most PC Viruses," *San Francisco Chronicle*, July 29, 2004, p. C3.
45. Jon Swartz, "Hackers Lurk through Holes in Hot Spots," *USA Today*, April 13, 2004, p. 1B.
46. Daniel Nasaw, "Viruses Lurk as a Threat to 'Smart' Cellphones," *The Wall Street Journal*, March 18, 2004, pp. B4, B5.; J. D. Biersdorfer, "Worms Like Wireless Phones, Too," *New York Times*, June 24, 2004, p. E1; and "Hacker Cracks T-Mobile, Reads Secret Service E-Mail," *San Francisco Chronicle*, January 13, 2005, p. C2.
47. Cassell Bryan-Low and Gary Fields, "www.infect.com," *The Wall Street Journal*, March 31, 2005, pp. B1, B5.
48. Byron Acohido, "Cyberattacks on Corporate Networks Rising, Surveys Show," *USA Today*, March 21, 2005, p. 6B; see also Riva Richmond, "Money Increasingly Is Motive for Computer-Virus Attacks," *The Wall Street Journal*, September 20, 2004, p. B5.
49. Federal Bureau of Investigation, quoted in TechTarget Security Media, "Glossary," *http://searchsecurity.techtarget.com/gDefinition/0,294236,sid14_gci771061,00.html* (accessed April 14, 2005).
50. "Teen Is Sentenced to 18 Months in Jail for Computer Worm," *The Wall Street Journal*, January 31, 2005, p. B3.
51. Duncan J. Watts, "Unraveling the Mysteries of the Connected Age," *The Chronicle of Higher Education*, February 14, 2003, pp. B7–B9.
52. Eugene Carlson, "Some Forms of Identification Can't Be Handily Faked," *The Wall Street Journal*, September 14, 1993, p. B2.
53. Mary Dalrymple, "Employees Prove Vulnerable to Hackers Posing as IT Employees," *Information Week*, March 17, 2005; *www.informationweek.com/story/showArticle.jhtml?articleID=159901703* (accessed April 15, 2005).
54. Scott Thurm and Mylene Mangalindan, "Trying to Remember New Passwords Isn't as Easy as ABC123," *The Wall Street Journal*, December 9, 2004, pp. A1, A9.
55. Anick Jesdanun, "Simple Passwords Don't Suffice Online," *San Francisco Chronicle*, June 1, 2004, pp. C1, C5.
56. Carl Herberger, cited in Thurm and Mangalindan, 2004.
57. Symantec study, cited in Thurm and Mangalindan, 2004.
58. Jesdanun, 2004.
59. David Einstein, "Safekeeping for All of Those Passwords," *San Francisco Chronicle*, August 16, 2004, p. C2.
60. Vincent Kiernan, "Stanford U. Researchers Develop a Tool to Help Prevent Password Theft," *The Chronicle of Higher Education*, August 12, 2005, p. A33.
61. Stephen H. Wildstrom, "Security at the Touch of a Finger," *BusinessWeek*, December 13, 2004, p. 28; and Larry Armstrong, "Let Your Fingers Do the Log-In," *BusinessWeek*, September 1, 2003, p. 20.
62. Thomas J. Fitzgerald, "Protecting Your Files When a Password Isn't Enough," *New York Times*, September 2, 2004, p. E7.
63. Sean Captain, "Empowering the Wi-Fi User to Foil the Snoop," *New York Times*, March 11, 2004, p. E5.
64. Andy Pasztor, "Web by Satellite in Round Two: Will It Fly Now?" *The Wall Street Journal*, June 3, 2005, pp. B1, B2.
65. "Braving 4G," *The Wall Street Journal*, June 21, 2001, p. B8; Steven Komarrow," Oh, the Places Your Cellphone Can Go," *USA Today*, June 26, 2001, p. 14E.
66. Barnaby I. Feder, "For the Gadget Universe, a Common Tongue," *New York Times*, January 2, 2003, pp. E1, E6; and David Marsh, "Software-Defined Radio Tunes In," *EDN*, March 3, 2005, *http://www.edn.com/article/CA505082.html* (accessed September 6, 2005).
67. Rana Foroohar, "A New Way to Compute," *Newsweek*, September 16, 2002, p. 34J; and Ian Foster, "The Grid," *ClusterWorld*, January 2004, *http://www.globus.org/alliance/publications/clusterworld/0104Head.pdf*. (accessed September 6, 2005).
68. Anita Hamilton, "Can You See Me Now?" *Time*, March 15, 2004, p. 96.
69. Mark Dillard, quoted in Marcia Vickers, "Don't Touch That Dial: Why Should I Hire You?" *New York Times*, April 13, 1997, sec. 3, p. 11.
70. Marlon A. Walker and Almar Latour, "The Videophone Goes Mass Market," *The Wall Street Journal*, August 24, 2004, pp. D1, D4.
71. Spencer E. Ante, "The World Wide Work Space," *BusinessWeek*, June 6, 2005, pp. 106–108.

Chapter 7

1. Gary McWilliams, "It All Connects—and Converges," *The Wall Street Journal*, January 31, 2005, p. R3.
2. Jerry Yang and David Filo, cited in Cliff Edwards, "The Web's Future Is You," *BusinessWeek*, April 25, 2005, p. 18.
3. Joyce Cohen, "Armed with Right Cellphone, Anyone Can Be a Journalist," *New York Times*, July 18, 2005, p. C3; and Mark Memmott, "Disaster Photos: Newsworthy or Irresponsible?" *USA Today*, August 5, 2005, p. 4A.
4. Kevin Maney, "Multiple Mash-ups Ensue as Google Maps Mate with Net Info," *USA Today*, August 16, 2005, *www.usatoday.com/tech/columnist/kevinmaney/2005-08-16-maney-google-mashups_x.htm#* (accessed August 29, 2005).
5. Robert D. Hof, "Mix, Match, and Mutate," *BusinessWeek*, July 25, 2005, pp. 72, 75.
6. Wilson Rothman, "Phone Sense," *Time*, April 2005, p. A4; and "Samsung Introduces World's First '3-Dimensional Movement Recognition' Phone," press release, Samsung Press Center, January 12, 2005, *www.samsung.com/PressCenter/PressRelease/PressRelease.asp?seq=20050112_0000094230* (accessed May 10, 2005).
7. Ming Ma, quoted in Marc Saltzman, "Convergence Is King with Today's Devices," *USA Today*, May 17, 2004, p. 4E.
8. "Media and Technology in 2004," *New York Times*, December 29, 2003, p. C1.
9. Peter Svensson, "It's Called PlayStation, but It Can Do More than Play," *San Francisco Chronicle*, April 11, 2005, p. E5.
10. Christopher Rhoads, "Cellphones Become 'Swiss Army Knives' as Technology Blurs," *The Wall Street Journal*, January 4, 2005, pp. B1, B8.
11. Michael Gartenberg, quoted in Saltzman, 2004.
12. Jørgen Sundgot, "Nokia 9500 Communicator," infoSync World: Review Centre, April 18, 2005, *www.infosyncworld.com/reviews/n/5930.html* (accessed May 10, 2005).
13. Rhoads, 2005.
14. Stephanie Armour, "Some Firms Trade Email for Face Time," *USA Today*, December 7, 2004, p. 1B.
15. Barry Schwartz, *The Paradox of Choice: Why More Is Less* (New York: HarperCollins, 2005). See also Barry Schwartz, "Choice Overload Burdens Daily Modern Life," *USA Today*, January 5, 2004, p. 13A.
16. Robin Marantz Henig, "Driving? Maybe You Shouldn't Be Reading This," *New York Times*, July 13, 2004, p. D5; and Jeremy Peters, "Hi, I'm Your Car. Don't Let Me Distract You," *New York Times*, November 26, 2004, pp. C1, C3.
17. Dennis K. Berman, "Technology Has Us So Plugged into Data, We Have Turned Off," *The Wall Street Journal*, November 10, 2003, p. B1.
18. "MP3 Player Pioneer Rio Goes Silent," *San Francisco Chronicle*, August 27, 2005, p. C2; and Evan Ramstad, "An iPod Casualty: The Rio Digital-Music Player," *The Wall Street Journal*, September 1, 2005, p. B3.
19. Creative Strategies, cited in Jefferson Graham, "New iPod Mini Holds More Tunes," *USA Today*, February 24, 2005, p. 3B.
20. Phred Dvorak, "Sony Introduces New Walkman in Next Bid to Outgun the iPod," *The Wall Street Journal*, September 9, 2005, p. B3; and Martin Fackler, "Sony Says It's an iPod Killer, Not Just Another Walkman," *New York Times*, September 15, 2005, p. C6.

21. David Pogue, "Defying Odds, One Sleek iPod at a Time," *New York Times*, September 15, 2005, pp. C1, C12. See also Benny Evangelista, "Playing on the 'Ooh' Factor," *San Francisco Chronicle*, September 19, 2005, pp. F1, F5; and Lev Grossman, "Stevie's Little Wonder," *Time*, September 19, 2005, pp. 63–64.

22. Peter Lewis, "Play That Funky Music, White Toy," *Fortune*, February 7, 2005, pp. 38–40; and Walter S. Mossberg, "iPod's Latest Siblings," *The Wall Street Journal*, September 8, 2005, pp. B1, B5.

23. Jay Greene, "iPod Gizmos for the Car," *BusinessWeek*, May 16, 2005, pp. 112–114.

24. Michael Bull, quoted in Benny Evangelista, "The iPod Generation," *San Francisco Chronicle*, December 27, 2004, pp. E1, E6.

25. Marie Madden and Lee Rainie, *Music and Video Downloading Moves Beyond P2P*, Pew/Internet & American Life Project, March 23, 2005, www.pewinternet.org/pdfs/PIP_Filesharing_March05.pdf (accessed May 25, 2005).

26. Bob Levins, quoted in Evangelista, 2004.

27. Shirley Wang, "Experts: Loud iPods Can Damage Ears," *Reno Gazette-Journal*, August 9, 2005, pp. 1E, 5E.

28. Brock Read, "Seriously, iPods Are Educational," *The Chronicle of Higher Education*, March 18, 2005, pp. A30–A32.

29. Adam Aston, "Satellite Radio," *Business-Week*, May 16, 2005, p. 116.

30. Walter Kirn, "Stuck in the Orbit of Satellite Radio," *Time*, May 23, 2005, p. 82.

31. Glenn Fleishman, "Revolution on the Radio," *New York Times*, July 28, 2005, p. C11; and Paul Davidson, "Radio Ready to Join Digital Revolution," *USA Today*, August 24, 2005, pp. 1B, 2B.

32. Betsy Streisand, "Radio Shock Waves," *U.S. News & World Report*, February 14, 2005, pp. 50–51; Kate Murphy, "To Win Subscribers, Satellite Radio Sells Sur prises," *New York Times*, February 20, 2005, sec. 3, p. 6; Heather Green, Tom Lowry, and Catherine Yang, "The New Radio Revolution," *BusinessWeek*, March 14, 2005, pp. 32–35; and Sarah McBride, "Two Upstarts Vie for Dominance in Satellite Radio," *The Wall Street Journal*, March 30, 2005, pp. A1, A9.

33. Martin Miller, "All Elvis Radio? It's Now or Never," *Los Angeles Times*, August 29, 2005, www.latimes.com/technology/la-et-elvis29aug29,1,6001749.story?coll=la-headlines-technology (accessed August 31, 2005).

34. Sabrina Tavernise, "The Broad Reach of Satellite Radio," *New York Times*, October 4, 2004, p. C8; Nat Ives, "Howard Stern's Switch Could Help Satellite Radio Grow, Especially If Men Aged 18 to 34 Go Along, Too," *New York Times*, October 7, 2004, p. C4; Ellen Simon, "Satellite Radio Uses Shock Jocks to Lure Male Buyers, Listeners," *San Francisco Chronicle*, October 11, 2004, p. H3; and McBride, 2005.

35. Anthony Armstrong, "Satellite Radio vs. High-Definition Radio for the Layperson: The Battle for America's Ears Has Begun," http://stereos.about.com/od/homestereotechnologies/a/radio.htm (accessed May 28, 2005).

36. Sarah McBride, "Hit by iPod and Satellite, Radio Tries New Tune: Play More Songs," *The Wall Street Journal*, March 18, 2005, pp. A1, A10.

37. Sarah McBride, "Where the Listeners Are," *The Wall Street Journal*, December 13, 2004, p. B4.

38. Daren Fonda, "The Revolution in Radio," *Time*, April 19, 2004, pp. 55–56.

39. Fonda, 2004.

40. Laura A. Locke, "The Podfather: Part One," *Time Inside Business*, October 2005, pp. A20–A24.

41. Green, Lowry, and Yang, 2005, p. 34.

42. Byron Acohido, "Radio to the MP3 Degree: Podcasting," *USA Today*, February 9, 2005, pp. 1B, 2B.

43. Todd Wallack, "Torrent of Images Is Leaving Film in the Dust," *San Francisco Chronicle*, May 23, 2005, pp. A1, A5.

44. David Ritz, quoted in Jefferson Graham, "Digital Cameras Get Smaller but Do More Tricks," *USA Today*, February 22, 2005, p. 6B.

45. Anita Hamilton, "The Myth of Megapixels," *Time*, July 5, 2004, p. 95.

46. Hamilton, 2004.

47. Hamilton, 2004.

48. Jefferson Graham, "Tips for Buyers," *USA Today*, May 17, 2004, p. 9E.

49. Walter S. Mossberg, "Missed Moments: Pitfalls of Buying a Digital Camera," *The Wall Street Journal*, March 3, 2004, pp. D1, D12.

50. "Are You Ready to Go Digital?" *Via*, September–October 2004, p. 24.

51. Mossberg, 2004, p. D12.

52. Jefferson Graham, "Get Lots of Batteries before You Go," *USA Today*, July 12, 2004, p. 2B.

53. Mossberg, 2004, p. D12.

54. Walter S. Mossberg, "Logging On to the Family Album," *The Wall Street Journal*, September 22, 2004, p. D4; Tracy Baker, "Print & Share Your Photos," *Smart-Computing*, July 2005, p. 72; and David Pogue, "A Baby Step Toward Wi-Fi Photos," *New York Times*, September 1, 2005, pp. C1, C10.

55. Jefferson Graham, "Newer Digital Cameras Raise Bar on Video," *USA Today*, May 26, 2004, p. 4B.

56. Jefferson Graham, "Don't Get Stuck with a Full Digital Memory Card on Your Trip," *USA Today*, July 12, 2004, p. 2B; Mossberg, 2004, p. D12; and Christian Perry, "Picture Passage," *SmartComputing*, July 2005, p. 65.

57. Andy Biggs, quoted in Graham, July 12, 2004.

58. Pui-Wing Tam, "Digital Snaps in a Snap," *The Wall Street Journal*, August 4, 2005, pp. B1, B5.

59. International Data Corporation, cited in Maria Puente, "Memories Gone in a Snap," *USA Today*, January 21, 2005, pp. 1D, 2D.

60. John Grady, quoted in Wallack, 2005, p. A5.

61. Certified Digital Photo Processors, cited in Puente, 2005, p. 2D.

62. Puente, 2005, p. 1D.

63. Susan Stellin, "Fitting the World's Biggest Travel Guide in a Pocket," *New York Times*, March 8, 2005, p. C8.

64. Johan Bostrom, "Handheld Market Still in Decline," *Techworld*, February 3, 2005, www.techworld.com/mobility/news/index.cfm?NewsID=3068 (accessed May 18, 2005).

65. Pui-Wing Tam, "The Hand-Helds Strike Back," *The Wall Street Journal*, May 18, 2005, pp. D1, D6; and David Pogue, "A New Spin on a Palmtop (or Inside It)," *New York Times*, May 19, 2005, pp. C1, C16.

66. Larry Armstrong, "Baby, Is It Cold Outside?" *BusinessWeek*, December 20, 2004, p. 102.

67. Louise Rafkin, "We Are Here," *San Francisco Chronicle Magazine*, May 23, 2004, pp. 10–13.

68. Michelle Kessler, "Nokia Brings Web Tablets Back to Life," *USA Today*, May 26, 2005, p. 2B.

69. Bruce Einhorn and Jay Greene, "Headaches for Tablet PCs," *BusinessWeek*, March 1, 2004, p. 14.

70. Thomas J. Fitzgerald, "The Tablet PC Takes Its Place in the Classroom," *New York Times*, September 9, 2004, p. E5.

71. Kessler, 2005, p. 1B.

72. Steven Levy, "Television Reloaded," *Newsweek*, May 30, 2005, pp. 49–55.

73. Terril Yue Jones, "HP Ready to Introduce 'Smart' Television Sets," *San Francisco Chronicle*, September 9, 2005, p. C3; reprinted from *Los Angeles Times*.

74. John R. Quain, "PCs Made for TV," *U.S. News & World Report*, November 22, 2004, p. D10; and Saul Hansell, "Logging On to Tune in TV," *New York Times*, August 1, 2005, pp. C1, C6.

75. Edward C. Baig, "Great TV Picture Can Often Lead to a Confused Viewer," *USA Today*, February 24, 2005, p. 5B.

76. Consumer Electronics Association, cited in Sarah McBride, "Now a Word from Our Low-Definition Sponsor," *The Wall Street Journal*, April 25, 2005, p. R6.

77. Sarah McBride, Phred Dvorak, and Don Clark, "Why HDTV Hasn't Arrived in Many Homes," *The Wall Street Journal*, January 5, 2005, pp. B1, B3; and McBride, April 25, 2005.

78. David Pogue, "For High-Definition Sets, Channels to Match," *New York Times*, June 3, 2004, pp. E1, E7.

79. Paul Davidson, "TV Changes Are Signal for Debate," *USA Today*, May 26, 2005, p. 3B.

80. Marc Saltzman, "HDTV: Choices from Plasma to LCD to DLP," *USA Today*, December 10, 2004, p. 6B; and Lee Gomes, "A Buyer's Guide to HDTV," *The Wall Street Journal*, January 17, 2005, p. R6.

81. Levy, 2005, p. 50.

82. Levy, 2005, pp. 49–55.

83. David Pogue, "TV's Future Is Here. It Needs Work," *New York Times*, June 2, 2005, pp. C1, C12.

84. Burt Helm, "Cellular Television," *Business-Week*, May 16, 2005, p. 106.

85. Donald Macintyre, "TV Anywhere, Anytime," *Time Bonus Section*, June 2005, p. A4.

86. Leslie Cauley, "Telecoms' Quest for Customers Follows Path to Internet TV," *USA Today*, August 17, 2005, pp. 1B, 2B.

87. Peter Grant, "New on TV: The Multiple-Channel Screen," *The Wall Street Journal*, August 30, 2005, pp. B1, B5.

88. Jeff Leeds, "A Failed TV Show Attempts New Life as a Yahoo Webcast," *New York Times*, June 2, 2005, p. C3.

89. Phred Dvorak, "New Sony TV Chips Give Viewers Control; 'Can 'Pan and Scan,'" *The Wall Street Journal*, August 12, 2004, pp. B1, B2.

90. Steve Lohr, "How Much Is Too Much?" *New York Times*, May 4, 2005, pp. E1, E9.

91. Roger O. Crockett, "Multimedia Phones," *BusinessWeek*, May 16, 2005, pp. 100–104; Christopher Conkey, "Parking Meters Get Smarter," *The Wall Street Journal*, June 30, 2005, pp. B1, B5; and Kevin Maney, "Melding of Cellphones and Wi-Fis Will Be Cosmic, Man," *USA Today*, August 24, 2005, p. 3B.

92. Michael Marriott and Kathie Hafner, "It's Not Just a Phone, It's an Adventure," *New York Times*, March 17, 2005, p. C9.

93. "Look Ma, No Hands—Mobile Speakerphones," *San Francisco Chronicle*, May 2, 2005, p. 3!.

94. Scott Ellison, quoted in Jefferson Graham, "Cellphones Add TV, Radio to Repertoire," *USA Today*, March 15, 2005, p. 4B.

95. ForceNine Consulting/Wirthlin Worldwide and Harris Interactive, reported in "Cellphones Add Features," *USA Today*, November 29, 2004, p. 1D.

96. Harris Interactive for Verisign, reported in "Downloading on the Go," *USA Today*, January 20, 2005, p. 1D.

97. John R. Quain, "Instant Messaging Moves Beyond Chat to Multimedia," *New York Times*, May 15, 2003, p. E6; Kevin Maney, "Surge in Text Messaging Makes Cell Operators :-)" *USA Today*, July 28, 2005, pp. 1B, 2B; Li Yuan, "Text Messages Sent by Cellphone Finally Catch On in U.S.," *The Wall Street Journal*, August 11, 2005, pp. B1, B3.

98. Janet Kornblum, "Cellphones Do a New Number with 'Texting,'" *USA Today,* June 3, 2003, p. 7D.

99. Stephanie Dunnewind, "Text Lingo: What ppl Are Using 2da," *Reno Gazette-Journal,* May 5, 2003, pp. 1D, 2D; reprinted from *Seattle Times.*

100. Jon Sarche, "Think Before You Text Message," *San Francisco Chronicle,* June 14, 2004, p. D2.

101. "Who's Calling? Check the Ring," *Reno Gazette-Journal,* October 18, 2004, pp. 1B, 3B.

102. Walter S. Mossberg, "Sorting Out the Three-Ring Circus of Ringtones," *The Wall Street Journal,* January 5, 2005, p. D8.

103. Heather Green, Cliff Edwards, and Roger O. Crockett, "The Squeeze on Black-Berry," *BusinessWeek,* December 6, 2004, pp. 80, 85.

104. Kelly DiNardo, "Smartphones Do It All to Keep Savvy Users in the Know," *USA Today,* November 22, 2004, p. 4E.

105. Carl Bialik, "Losing Cellphone No Longer Means Loss of Contacts," *The Wall Street Journal,* October 6, 2004, p. D11.

106. John Markoff, "That's the Weather, and Now, Let's Go to the Cellphone for Traffic," *New York Times,* March 1, 2004, p.C3.

107. Avery Johnson, "Cellphone Directions," *The Wall Street Journal,* December 28, 2004, p. D5.

108. John R. Quain, "Phones That Love Wi-Fi," *U.S. News & World Report,* September 20, 2004, p. 75.

109. James Sullivan, "Time Waits for Everyone, Now That We've All Got Camera Phones," *San Francisco Chronicle,* May 20, 2004, p. E2.

110. Ann Grimes, "Moblog for the Masses," *The Wall Street Journal,* April 29, 2004, p. B4.

111. "Picture It: A Diagnosis via Cellphone," *USA Today,* February 22, 2005, p. 6D.

112. Jefferson Graham, "Camera Phones Get More User Friendly," *USA Today,* February 21, 2005, p. 5B.

113. Anne Eisenberg, "Cellphone Games Take a Big Leap Forward into 3-D," *New York Times,* June 17, 2004, p. E8.

114. Nick Wingfield, "RealNetworks, Sprint Will Offer Radio via Phones," *The Wall Street Journal,* September 19, 2005, p. B4.

115. Edward C. Baig, "New iTunes Phone a Snazzy Device," *USA Today,* September 8, 2005, p. 4B; Benny Evangelista, "Apple Unveils iTunes Cell Phone, New iPod," *San Francisco Chronicle,* September 8, 2005, pp. C1, C8; Michelle Kessler, "iPods Shrink, Phones Now Sing," *USA Today,* September 8, 2005, p. 1B; John Markoff, "Apple Unveils a New iPod and a Phone Music Player," *New York Times,* September 8, 2005, pp. C1, C11; Nick Wingfield, "New Apple-Motorola Cellphone May Be Just the Overture," *The Wall Street Journal,* September 8, 2005, pp. B1, B5; and Peter Burrows, Roger O. Crockett, and Heather Green, "Apple's Phone Isn't Ringing Any Chimes," *BusinessWeek,* September 19, 2005, pp. 58, 61.

116. Jefferson Graham, "Nokia Cellphone the First of Its Kind," *USA Today,* April 28, 2005, p. 3B; and David Pringle, "Nokia Unveils Digital-Music Player in a Cellphone," *The Wall Street Journal,* April 28, 2005, p. D3.

117. Christopher Rhoads and Nick Wingfield, "Apple's iPod Faces Challenge from Cellphones," *The Wall Street Journal,* April 11, 2005, pp. B1, B4.

118. David Pringle and Charles Goldsmith, "Cellphone Companies Chime in with Music Downloading," *The Wall Street Journal,* November 1, 2004, pp. B1, B4; and "Musiwave Expected to Offer Downloads for Cellphone Users," *The Wall Street Journal,* December 21, 2004, p. B3.

119. Edward C. Baig, "Headphones Do Double Duty," *USA Today,* September 15, 2005, p. 3B.

120. Crockett, 2005, p. 102.

121. Doreen Carvajal, "A Way to Calm Fussy Baby: 'Sesame Street' by Cellphone," *New York Times,* April 18, 2005, p. C10.

122. Graham, March 15, 2005.

123. Bruce Myerson, "Phone Businesses Start Gearing Up to Deliver TV," *Reno Gazette-Journal,* December 27, 2004, p. 3E.

124. "Cell Phones: The Next Movie Screens?" *Reno Gazette-Journal,* September 28, 2004, p. 3F; reprinted from *Chicago Tribune.*

125. Edward C. Baig, "Running Late? Feed the Meter by Phone," *USA Today,* February 14, 2005, p. 3B.

126. Yuri Kageyama, "Make Purchases without Plastic or Cash with Wallet Cell Phones," *San Francisco Chronicle,* July 26, 2004, p. F2; and Ian Rowley, "$5,000? Sure, Put It on My Cell Phone," *BusinessWeek,* June 6, 2005, p. 56.

127. Phred Dvorak, "Testing the TV Tuners and Fingerprint Checks in Cellphones in Japan," *The Wall Street Journal,* June 3, 2004, p. B1.

128. Rhoads, 2005, p. B8.

129. Michael Kunzelman, "Psst! Pay Attention," *San Francisco Chronicle,* August 12, 2005, p. A2.

130. 2004 Sprint Wireless Courtesy Report, cited in Steven Winn, "The Cell Phone Warning Has a Familiar Ring. Who's Listening?" *San Francisco Chronicle,* December 16, 2004, pp. E1, E4.

131. Elizabeth Olson, "Sound, Fury, and Cellphone Users and Abusers," *New York Times,* July 15, 2003, p. C6.

132. Becky Rohrer, "Cell Phones Bug National Parks," *San Francisco Chronicle,* April 19, 2004, p. D3.

133. Maggie Jackson, "Turn Off That Cellphone. It's Meeting Time," *New York Times,* March 2, 2003, sec. 3, p. 12.

134. "Phone Use Slows Driver Reaction," *The Wall Street Journal,* February 3, 2005, p. D5.

135. Carol S. Lede, "The New Social Etiquette: Friends Don't Let Friends Dial Drunk," *New York Times,* January 30, 2005, sec. 4, p. 16.

136. "Camera-Equipped Cell Phones Spreading New Types of Mischief," *Reno Gazette-Journal,* July 10, 2003, pp. 1A, 4A; Dennis K. Berman, David Pringle, and Phred Dvorak, "You're on Candid Cellphone!" *The Wall Street Journal,* September 30, 2003, pp. B1, B6; Carolyn Said, "Are Camera Phones Too Revealing?" *San Francisco Chronicle,* May 16, 2004, pp. A1, A10; and Ryan Kim, "School District Bans Taking Photos with Cell Phones," *San Francisco Chronicle,* August 23, 2004, pp. B1, B5.

137. Bob Keefe, "Pornography Industry Sees Cell Phones as New Frontier," *San Francisco Chronicle,* January 3, 2005, p. F3.

138. Steven Levy, "A Future with Nowhere to Hide?" *Newsweek,* June 7, 2004, p. 76.

139. Ken Belson, "When Etiquette Isn't Enough, a Cellphone Cone of Silence," *New York Times,* November 7, 2004, sec. 4, p. 2.

140. Michael Totty, "Who's Going to Win the Living-Room Wars?" *The Wall Street Journal,* April 25, 2005, pp. R1, R4.

141. Lev Grossman, "Out of the Xbox," *Time,* May 23, 2005, p. 51.

142. Eric A. Taub, "A Hand-Held That Doesn't Just Play Games," *New York Times,* August 29, 2005, p. C4.

143. Steven Kent, "Revolution Joins the Gaming Wars," *USA Today,* May 17, 2005, p. 1D; "A Buyer's Guide to the New Gameboxes," *The Wall Street Journal,* May 18, 2005, pp. D1, D6; Holman W. Jenkins Jr., "What the Fuss Is All About," *The Wall Street Journal,* May 18, 2005, p. A15; Mike Snider, "New Xbox, PlayStation, and Nintendo Jockey for Position," *USA Today,* May 18, 2005, p. 6D; and Grossman, 2005, pp. 44–53.

144. Karyn Poupee, "Nintendo Steals Xbox Thunder," News24.com, September 16, 2005, www.news24.com/News24/Technology/News/0,,2-13-1443_1771912,00.html (accessed September 16, 2005); AND Ginny Parker Woods, "This Isn't Your Basic Joystick," *The Wall Street Journal,* September 16, 2005, p. B3.

145. Tony Perkins, quoted in Bill O'Driscoll, "Expert:Business Needs to Tap the 'Always On,'" *Reno Gazette-Journal,* February 12, 2003, pp. 1D, 6D.

146. Jyoti Thottam, "How Kids Set the (Ring) Tone," *Time,* April 4, 2005, pp. 40–45.

147. Heather Knight, "Cell Phones a High Priority, Even among Poor Teens," *San Francisco Chronicle,* February 27, 2005, pp. A17, A21.

148. Diana Oblinger, "Boomers, Gen-Xers, and Millennials: Understanding the New Students," *EDUCAUSE Review,* July/August 2003, pp. 37–47.

149. Marc Prensky, "Digital Natives, Digital Immigrants," from *On the Horizon,* October 2001, © 2001 Marc Prensky, *www.marcprensky.com/writing/Prensky%20-%20Digital%20Natives,%20Digital%20Immigrants%20-%20Part1.pdf* (accessed May 15, 2005).

150. Oblinger, 2003, p. 38.

151. Wendy Ricard and Diana Oblinger, "The Next-Generation Student," Higher Education Leaders Symposium, Redmond, WA, June 17–18, 2003, p. 2, *http://download.microsoft.com/download/d/c/7/dc70bbbc-c5a3-48f3-855b-f01d5de42fb1/TheNextGenerationStudent.pdf* (accessed May 15, 2005).

152. Paul Davidson, "Gadgets Rule on College Campuses," *USA Today,* March 29, 2005, pp. 1B, 2B.

153. Davidson, 2005.

154. Andrew Payne, quoted in Ricard and Oblinger, 2003, p. 5.

155. Prensky, 2001, p. 2.

156. Marc Prensky, "What Kids Learn That's POSITIVE from Playing Videogames," © 2002 Marc Presnsky, *www.marcprensky.com/writing/Prensky%20-%20What%20Kids%20Learn%20Thats%20POSITIVE%20From%20Playing%20Video%20Games.pdf* (accessed June 12, 2005).

157. Oblinger, 2003, p. 40, citing Jason Frand, "The Information Age Mindset: Changes in Students and Implications for Higher Education," *EDUCAUSE Review,* September/October 2000, pp. 15–24.

158. Kevin J. Delaney, "Teaching Tools," *The Wall Street Journal,* January 17, 2005, pp. R4, R5.

159. Diana G. Oblinger, "The Next Generation of Educational Engagement," *Journal of Interactive Media in Education,* May 21, 2004, pp. 1–18.

160. Ricard and Oblinger, 2003, p. 4.

Chapter 8

1. "IDC Research: Worldwide Net Traffic to Rise," March 3, 2003, NUA Internet Surveys, *www.nua.com/surveys/index.cgi?f=vs&art_id=905358733&rel=true; accessed October 14, 2005.*

2. Verlyn Klinkenborg, "Trying to Measure the Amount of Information That Humans Create," *New York Times,* November 12, 2003, p. A22.

3. Ron Lieber, "When Visa Thinks You're a Thief," *The Wall Street Journal,* April 24, 2003, pp. D1, D2.

4. Chris Taylor, "It's a Wiki, Wiki World," *Time*, June 6, 2005, pp. 40–41.

5. Charles Forelle, "Project Hopes to Trace Your Ancestors Back 10,000 Years," *The Wall Street Journal*, April 13, 2005, pp. B1, B11; and Benjamin Pimentel, "DNA Study of Human Migration," *San Francisco Chronicle*, April 13, 2005, pp. A1, A6.

6. Mark Y. Herring, "Don't Get Goggle-Eyed over Google's Plan to Digitize," *The Chronicle of Higher Education*, March 11, 2005, p. B20; and Verne Kopytoff, "Google, 5 Big Libraries Team to Offer Books," *San Francisco Chronicle*, December 14, 2004, pp. D1, D4.

7. Ramez Elmasri and Shamkant Navathe, *Fundamentals of Database Systems*, 3rd ed. (Reading, MA: Addison-Wesley, 2000), p. 4.

8. Stephen Manes, "Time and Technology Threaten Digital Archives . . . ," *New York Times*, April 7, 1998, p. B15.

9. Mike Snider, "Obsolescence: The No. 1 Built-In Feature," *USA Today*, October 29, 1997, p. 6D.

10. Laura Tangley, "Whoops, There Goes Another CD-ROM," *U.S. News & World Report*, February 16, 1999, pp. 67–68.

11. Marcia Stepanek, "From Digits to Dust," *BusinessWeek*, April 20, 1998, pp. 128–130.

12. Manes, 1998.

13. Michael J. Hernandez, *Database Design for Mere Mortals* (Reading, MA: Addison-Wesley, 1997), p. 11. Copyright 1997 by Michael J. Hernandez.

14. Hernandez, 1997, p. 12.

15. Hernandez, 1997, p. 15.

16. Elmasri and Navathe, 2000, p. 4.

17. Jonathan Berry, John Verity, Kathleen Kerwin, and Gail DeGeorge, "Database Marketing," *BusinessWeek*, September 5, 1994, pp. 56–62.

18. Cheryl D. Krivda, "Data Mining Dynamite," *Byte*, October 1995, pp. 97–103.

19. Krivda, 1995.

20. Edmund X. DeJesus, "Data Mining," *Byte*, October 1995, p. 81.

21. Lisa Guernsey, "Digging for Nuggets of Wisdom," *New York Times*, October 16, 2003, pp. E1, E9.

22. Sara Reese Hedberg, "The Data Gold Rush," *Byte*, October 1995, pp. 83–88.

23. Michael J. Mandel and Robert D. Hof, "Rethinking the Internet," *BusinessWeek*, March 26, 2001, p. 118.

24. Joshua Quittner, "Tim Berners-Lee," *Time*, March 29, 1999, pp. 193–194.

25. Sarah E. Hutchinson and Stacey C. Sawyer, *Computers, Communications, and Information: A User's Introduction*, rev. ed. (Burr Ridge, IL: Irwin/McGraw-Hill, 1998), pp. E1.1–E1.3.

26. Jeff Bezos, quoted in K. Southwick, interview, October 1996, www.upside.com.

27. D. Levy, "On-line Gamble Pays Off with Rocketing Success," *USA Today*, December 24, 1998, pp. 1B, 2B.

28. Elizabeth M. Gillespie, "Amazon.com Sitting Pretty after 10 Years Online," *Reno Gazette-Journal*, July 5, 2005, pp. 1D, 2D.

29. Margaret Kane, "E-Commerce Tech—Here's What Clicks," *ZDNet*, September 12, 2002, http://zdnet.com.

30. Don Tapscot, "Virtual Webs Will Revolutionize Business," *The Wall Street Journal*, April 24, 2000, p. A38.

31. Sabra Chartrand, "Software to Provide 'Personal' Attention to Online Customers with Service Untouched by a Human," *New York Times*, August 20, 2001, p. C8.

32. Robert Benfer Jr., Louanna Furbee, and Edward Brent Jr., quoted in Steve Weinberg, "Steve's Brain," *Columbia Journal Review*, February 1991, pp. 50–52.

33. Jeanne B. Pinder, "Fuzzy Thinking Has Merits When It Comes to Elevators," *New York Times*, September 22, 1993, pp. C1, C7.

34. Robert Barker, "A Robot That Could Hit a Wall," *BusinessWeek*, September 5, 2005, p. 28.

35. Brian Bergstein, "Teaching Computer a Thing or Million," *San Francisco Chronicle*, June 10, 2002, pp. E1, E3.

36. Judith Anne Gunther, "An Encounter with AI," *Popular Science*, June 1994, pp. 90–93.

37. Sara Robinson, "Human or Computer? Take This Test," *New York Times*, December 10, 2002, pp. D1, D4.

38. Williams A. Wallace, *Ethics in Modeling* (New York: Elsevier Science, 1994).

39. Laura Johannes, "Meet the Doctor: A Computer That Knows a Few Things," *The Wall Street Journal*, December 18, 1995, p. B1.

40. Lenita Powers, "Mom's Group Fights Recruiting Database," *Reno Gazette-Journal*, July 5, 2005, pp. A1, A4; and Sara Lipka, "Pentagon System to Gather Student Data Raises Privacy Fears," *The Chronicle of Higher Education*, July 8, 2005, p. A30.

41. David Lazarus, "A Deluge of Credit," *San Francisco Chronicle*, March 7, 2003, pp. B1, B3.

42. Kris Maher, "Résumé Rustling Threatens Online Job Sites," *The Wall Street Journal*, February 25, 2003, pp. B1, B10; Adam Geller, "Identity Theft a Monster of a Problem for Job Site," *Reno Gazette-Journal*, February 28, 2003, pp. 1A, 4A; and Carrie Kirby, "Online Résumés Turn Risky," *San Francisco Chronicle*, July 4, 2005, pp. E1, E2.

43. Jennifer 8. Lee, "Trying to Elude the Google Grasp," *New York Times*, July 25, 2001, pp. E1, E6.

44. Declan McCullagh, "Senate Curbs Pentagon Data-Mining Plans," *CNET News.com*, January 24, 2003, http://zdnet.com. See also Leslie Walker, "Data-Mining Software Digs for Business Leads," *San Francisco Chronicle*, March 8, 2004, p. E6, reprinted from *Washington Post*.

45. "Plan to Snoop on Fliers Takes Intrusion to New Heights," editorial, and James M. Loy, "Privacy Will Be Protected," *USA Today*, March 12, 2003, p. 12A.

46. Bob Egelko and Maria Alicia Gaura, "Librarians Try to Alter Patriot Act," *San Francisco Chronicle*, March 10, 2003, pp. A1, A4.

47. Edmund L. Andrews, "New Scale for Toting Up Lost Freedom vs. Security Would Measure in Dollars," *New York Times*, March 11, 2003, p. A11.

48. Nick Anderson, "U.S. Extends Digital ID Technology to Land Points of Entry," *San Francisco Chronicle*, January 4, 2005, p. A4; reprinted from *Los Angeles Times*.

49. Jennifer 8. Leek, "Progress Seen in Border Tests of ID System," *New York Times*, February 7, 2003, p. A11.

50. Steven Brill, "The Biggest Hole in the Net," *Newsweek*, December 30, 2002/January 6, 2003, pp. 48–51.

51. "An Unrealistic 'Real ID,'" editorial, *New York Times*, May 4, 2005, p. A24; Donna Leinwand, "Congress Weighs 4 IDs for Licenses," *USA Today*, May 5, 2005, p. 1A; and Amanda Ripley, "Revamping Your Driver's License," *Time*, May 16, 2005, pp. 40–41.

52. Gary Rivlin, "Purloined Lives," *New York Times*, March 17, 2005, pp. C1, C8.

53. David Lazarus, "I Found an ID Thief on My Credit Report," *San Francisco Chronicle*, October 27, 2002, pp. G1, G4.

54. Kathryn Rambo, quoted in Ramon G. McLeod, "New Thieves Prey on Your Very Name," *San Francisco Chronicle*, April 7, 1997, pp. A1, A6.

55. Rambo, quoted in T. Trent Gegax, "Stick 'Em Up? Not Anymore. Now It's Crime by Keyboard," *Newsweek*, July 21, 1997, p. 14.

56. June 2005 survey commissioned by Nationwide Mutual Insurance, reported in Jon Swartz, "Survey: ID Thief Takes Time to Wipe Clean," *USA Today*, July 29, 2005, p. 1B.

57. McLeod, 1997; John Waggoner, "Dodging Some ID Theft Not So Easy," *USA Today*, February 23, 2005, p. 3B; Carrie Kirby, "New, Smarter Generation of Internet Crooks," *San Francisco Chronicle*, April 11, 2005, pp. A1, A8; Sandra Block, "Thieves Love Debit Cards, So Keep Them Safe—Here's How," *USA Today*, May 10, 2005, p. 3B; Alina Tugend, "Oh, No! My Identity's Gone! Call the Insurer," *New York Times*, May 28, 2005, p. B6; Sandra Block, "Is Freezing Your Credit the Way to Safeguard Your Identity?" *USA Today*, June 20, 2005, pp. 1A, 2A; and Stephen Levy and Brad Stone, "Grand Theft Identity," *Newsweek*, July 4, 2005, pp 38–47.

58. Paul J. Lim, "Gimme Your Name and SSN," *U.S. News & World Report*, March 7, 2005, pp. 46–47; Byron Cohido and Jon Swartz, "ID Thieves Search Ultimate Pot of Gold—Databases," *USA Today*, June 22, 2005, p. 3B; and Levy and Stone, 2005.

Chapter 9

1. Harold Gilliam, "Mind over Matter," *San Francisco Chronicle*, February 9, 2003, pp. D1, D6.

2. Edmund L. Andrews, "New Scale for Toting Up Lost Freedom vs. Security Would Measure in Dollars," *New York Times*, March 11, 2003, p. A11.

3. Sara Lipka, "Pentagon System to Gather Student Data Raises Privacy Fears," *The Chronicle of Higher Education*, July 8, 2005, p. A30.

4. Andrea L. Foster, "ID Theft Turns Students into Privacy Activists," *The Chronicle of Higher Education*, August 2, 2002, pp. A27–A28.

5. Ian Ayres, "Lectures vs. Laptops," *New York Times*, March 20, 2001, p. A29; John Schwartz, "Professors Vie with Web for Class's Attention," *New York Times*, January 2, 2003, pp. A1, A14; and Lisa Guernsey, "When Gadgets Get in the Way," *New York Times*, August 19, 2004, pp. E1, E7.

6. Tamar Lewin, "New Online Guides Rate Professors," *New York Times*, March 24, 2003, p. A9.

7. Ann Davis, "Some Colleges Balk at FBI Request for Data on Foreigners," *The Wall Street Journal*, November 25, 2002, pp. B1, B5; and Andrews, 2003.

8. Kevin Maney, "Late Performers Get Chance to Record Live," *USA Today*, July 7, 2004, p. 4B.

9. William Safire, "Art vs. Artifice," *New York Times*, January 3, 1994, p. A11.

10. "Newsweek Gives Stewart a New Body," *San Francisco Chronicle*, March 4, 2005, p. A14.

11. Cover, *Newsweek*, June 27, 1994, and cover, *Time*, June 27, 1994.

12. Byron Calame, "Pictures, Labels, Perception and Reality," *New York Times*, July 3, 2005, sec. 4, p. 10.

13. John Knoll, quoted in Katie Hafner, "The Camera Never Lies, but the Software Can," *New York Times*, March 11, 2004, pp. E1, E7.

14. Daryl Plummer, "Give Counterfeit Reality a Closer Look," *USA Today*, July 20, 2004, p. 13A.

15. Jonathan Alter, "When Photographs Lie," *Newsweek*, June 30, 1990, pp. 44–45.

16. Fred Ritchin, quoted in Alter, 1990.

17. Hafner, 2004, p. E7.

18. Noah Shachtman, "For Doctored Photos, a New Flavor of Digital Truth Serum," *New York Times*, July 22, 2004, p. E5.

19. Robert Zemeckis, cited in Laurence Hooper, "Digital Hollywood: How Computers Are Remaking Movie Making," *Rolling Stone*, August 11, 1994, pp. 55–58, 75.

20. Jeremy Peters and Danny Hakim, "Is That Steve McQueen in the Cornfield? Yes, Brought Back by Ford," *New York Times*, October 15, 2004, p. C3.

21. Woody Hochswender, "When Seeing Cannot Be Believing," *New York Times*, June 1992, pp. B1, B3.

22. Bruce Horowitz, "Believe Your Eyes? Ads Bend Reality," *USA Today*, April 24, 2000, pp. 1B, 2B.

23. Peggy Williams, "Database Dangers," *Quill*, July/August 1994, pp. 37–38.

24. Davis, quoted in Williams, 1994.

25. May 2004 survey commissioned by Wells Fargo & Co, reported in Julie Dunn, "Poor 'Street Smarts' Make Web Users Vulnerable," *San Francisco Chronicle*, August 17, 2004, p. C8; reprinted from *Denver Post*.

26. May 2005 study by Annenberg Public Policy Center, University of Pennsylvania, reported in Tom Zeller Jr., "You've Been Scammed Again? Maybe the Problem Isn't Your Computer," *New York Times*, June 6, 2005, p. C4.

27. June 2005 study from Garner Inc., reported in Riva Richmond, "Internet Scams, Breaches Drive Buyers Off-Line, Survey Finds," *The Wall Street Journal*, June 23, 2005, p. B3.

28. June 2005 survey funded by Cyber Security Industry Alliance of Washington, reported in Ted Bridis, "Users Want Net to Be Made Safer," *San Francisco Chronicle*, June 15, 2005, p. C3.

29. Scott McCartney, "Pilots Go to 'the Box' to Avoid Midair Collisions," *The Wall Street Journal*, July 18, 2002, p. D3; and George Johnson, "To Err Is Human," *New York Times*, July 14, 2002, sec. 4, pp. D1, D7.

30. We are grateful to Professor John Durham for contributing these ideas.

31. John Allen Paulos, "Smart Machines, Foolish People," *The Wall Street Journal*, October 5, 1999, p. A26.

32. Arthur M. Louis, "Nasdaq's Computer Crashes," *San Francisco Chronicle*, July 16, 1994, pp. D1, D3.

33. Alex Berenson, "Software Failure Halts Big Board Trading for Over an Hour," *New York Times*, June 9, 2001, pp. B1, B3.

34. Sally McGrane, "Glitches Stymie Medical School Applicants," *New York Times*, July 5, 2001, p. D3; and Katherine S. Mangan, "Online Medical-School Application Becomes a Nightmare for Students," *The Chronicle of Higher Education*, July 13, 2001, p. A33.

35. Charles Perrow, *Normal Accidents: Living with High-Risk Technologies* (New York: Basic Books, 1984).

36. Ricardo Alonso-Zaldivar, "More Car Crashes Tied to Cell Phones," *San Francisco Chronicle*, December 2, 2002, p. A15; Kathleen Kerwin, "Driving to Distraction," *BusinessWeek*, September 22, 2003, pp. IM5–IM6; Debbie Howlett, "Americans Driving to Distraction," *USA Today*, March 5, 2004, p. 3A; Robin Marantz Henig, "Driving? Maybe You Shouldn't Be Reading This," *New York Times*, July 13, 2004, p. D5; Jeremy Peters, "Hi, I'm Your Car. Don't Let Me Distract You," *New York Times*, November 26, 2004, pp. C1, C3; "Cellphone Use in Cars Is Subject of Study," *The Wall Street Journal*, June 22, 2005, p. A6; and Andres R. Martinez, "Careless Motorists Prevalent, Survey Finds," *Reno Gazette-Journal*, July 8, 2005, pp. 1C, 5C.

37. Suzanne P. McEvoy, Mark R. Stevenson, Anne T. McCartt, Mark Woodward, Claire Haworth, Peter Palamara, and Rina Cercarelli, "Role of Mobile Phones in Motor Vehicle Crashes Resulting in Hospital Attendance: A Case-Crossover Study," *The British Medical Journal*, July 12, 2005, *http://press.psprings.co.uk/bmj/july/mobilephones.pdf* (accessed October 5, 2005). See also Jeremy W. Peters, "Hands-Free Cellphone Devices Don't Aid Road Safety, Study Concludes," *New York Times*, July 12, 2005, p. C3.

38. Paul Virilio, reported in Alan Riding, "Expounding a New View of Accidents," *New York Times*, December 26, 2002, pp. B1, B2.

39. Leonard M. Fuld, "Bad Data You Can't Blame on Intel," *The Wall Street Journal*, January 9, 1995, p. A12.

40. Henry K. Lee, "UC Student's Dissertation Stolen with Computer," *San Francisco Chronicle*, January 27, 1994, p. A15.

41. Jefferson Graham, "College Students Sued Over Music Downloads," *USA Today*, March 24, 2005, p. 5B.

42. Larem Gullo, "Record Industry Sues 784 Users for Illegally Sharing Music Online," *San Francisco Chronicle*, June 30, 2005, pp. C1, C2.

43. Report by Pew Internet & American Life Project, reported in "Sharing of Music Files by Students Has Declined Drastically, New Pew Survey Finds," *The Chronicle of Higher Education*, January 16, 2004, p. A34.

44. "Raids Aim at Web Piracy," *New York Times*, July 1, 2005, p. C6; "Web Piracy Suspects Arraigned on Charges of Copyright Violation," *San Francisco Chronicle*, July 15, 2005, p. B1.

45. Pat Choate, *Hot Property: The Stealing of Ideas in an Age of Globalization* (New York: Alfred A. Knopf, 2005), quoted in Michael Lind, "Freebooters of Industry," *New York Times Book Review*, July 10, 2005, p. 34.

46. Tom Zeller Jr., "The Imps of File Sharing May Lose in Court, but They Are Winning in the Marketplace," *New York Times*, July 4, 2005, p. C3; Saul Hansell, "Forget the Bootleg, Just Download the Movie Legally," *New York Times*, July 4, 2005, pp. C1, C4; and "Dell, Napster to Promote Legal Downloads to Students," *San Francisco Chronicle*, July 7, 2005, p. C3.

47. David Lieberman, "Millions of Pirates Are Plundering Satellite TV," *USA Today*, December 2, 2002, pp. 1D, 2D.

48. John Schwartz, "Some Sympathy for Paris Hilton," *New York Times*, February 27, 2005, sec. 4, pp. 1, 14; and Jon Swartz and Sandra Block, "Underground Market for Stolen IDs Thrives," *USA Today*, March 3, 2005, pp. 1B, 2B.

49. Jon Swartz, "40 Million Credit Card Holders May Be at Risk," *USA Today*, June 20, 2005, p. 1B; and Eric Dash, "Take a Number," *New York Times*, June 30, 2005, pp. C1, C9.

50. "Iraq War Souvenirs Selling on eBay," ABC7News, April 11, 2003, *http://abclocal.go.com/kgo/news* (accessed April 11, 2003).

51. National White Collar Crime Center and the Federal Bureau of Investigation, *IC3 2004 Internet Fraud—Crime Report*, January 1, 2004–December 31, 2004, *http://www1.ifccfbi.gov/strategy/2004_IC3Report.pdf*. (accessed October 5, 2005).

52. Henry Norr, "Fast-Growing Fraud from Nigeria Uses Internet to Search for Suckers," *San Francisco Chronicle*, September 8, 2002, p. A6; Curt Anderson, "Internet Fraud Reports Triple in 2002," *San Francisco Chronicle*, April 10, 2003, p. B3; and Dulue Mbach, "Nigerian Scams Don't Stop," *San Francisco Chronicle*, August 7, 2005, pp. E1, E5.

53. Bank and Richmond, 2005.

54. Byron Acohido and Jon Swartz, "Are Hackers Using Your PC to Spew Spam and Steal," *USA Today*, September 8, 2004, pp. 1B, 4B.

55. Cassell Bryan-Low, "Tech-Savvy Blackmailers Hone a New Form of Extortion," *The Wall Street Journal*, May 5, 2005, pp. B1, B3.

56. Brian Grow, "Hacker Hunters," *BusinessWeek*, May 30, 2005, pp. 74–82; and Tom Zeller Jr., "Black Market in Credit Cards Thrives on Web," *New York Times*, June 21, 2005, pp. A1, C4.

57. Ted Bridis, "Hackers Demand $200 to Unlock Data Files," *San Francisco Chronicle*, May 25, 2005, p. A2.

58. Cassell Bryan-Low, "Seeking New Payoff, Hackers Now Strike Web Sites for Cash," *The Wall Street Journal*, November 30, 2004, pp. A1, A8.

59. David Bank and Riva Richmond, "Where the Dangers Are," *The Wall Street Journal*, July 18, 2005, pp. R1, R3.

60. Frank X. Mullen Jr., "Hackers into Ely Hospital's Computers Traced to Russia," *Reno Gazette-Journal*, April 7, 2003, pp. 1A, 6A.

61. CSO magazine survey, reported in "Companies and Cyber Attacks," *USA Today*, October 7, 2002, p. 1B.

62. David Carter, quoted in "Computer Crime Usually Inside Job," *USA Today*, October 25, 1995, p. 1B.

63. Bob Tedeschi, "Crime Is Soaring in Cyberspace, but Many Companies Keep It Quiet," *New York Times*, January 27, 2003, p. C4.

64. "Rogues' Gallery," *The Wall Street Journal*, July 18, 2003, p. R3.

65. "Rogues' Gallery," 2003.

66. Karen Kaplan, "Gangs Finding New Turf," *Los Angeles Times*, May 31, 2001, pp. A1, A8.

67. Tom Barbash, *On Top of the World: Cantor Fitzgerald, Howard Lutnick and 9/11: A Story of Loss and Renewal* (New York: HarperCollins, 2003); and Jimmy Gurulé, "Civil Society and Terrorism," November 14, 2002, *www.embusa.es/bilateral/cantor.html* (accessed October 5, 2005).

68. Stephanie Armour, "Beware the Blog: You May Get Fired," *Reno Gazette-Journal*, June 25, 2005, pp. 1E, 4E.

69. American Management Association survey of 840 U.S. companies, reported in Pai-Wing Tam, Erin White, Nick Wingfield, and Kris Maher, "Snooping E-Mail by Software Is Now a Workplace Norm," *The Wall Street Journal*, March 9, 2005, pp. B1, B3.

70. Websense and Harris Interactive, August 2002 survey, reported in Brad Stone, "Is the Boss Watching?" *Newsweek*, September 30, 2002, pp. 38J, 38L.

71. ADP Screening & Selection Services, reported in Joann S. Lublin, "Check, Please," *The Wall Street Journal*, March 11, 2002, p. R11; and Stephanie Armour, "Worker Background Checks Raise Privacy Concerns," *USA Today*, May 21, 2002, p. 1A.

72. CSO magazine survey, reported in "Cyber Security," *USA Today*, February 4, 2003, p. 1B.

73. Poll by Ernst & Young, June 3–6, 2002, reported in "Crime Spree," *BusinessWeek*, September 9, 2002, p. 8.

74. Erik Sherman, "Fighting Web Fraud," *Newsweek*, June 10, 2002, pp. 32B, 32D.

75. Marci Alboher Nusbaum, "New Kind of Snooping Arrives at the Office," *New York Times*, July 13, 2003, sec. 3, p. 12.

76. Stephanie Armour, "Workers' Downloading Puts Employers at Risk," *USA Today*, July 30, 2002, p. 1B.

77. Mindy Fetterman, "Employers Must Shred Personal Data," *USA Today*, June 1, 2005, p. 3B.

78. Erik Sherman, "Fighting Web Fraud," *Newsweek*, June 10, 2002, pp. 32B, 32D.

79. Jon Swartz, "Data Losses Push Businesses to Encrypt Backup Tapes," *USA Today*, June 13, 2005, p. 1B.

80. John Schwartz, "Disputes on Electronic Message Encryption Take on New Urgency," *New York Times*, September 25, 2001, pp. C1, C16.

81. Gina Kolata, "Scientists Debate What to Do When Findings Aid an Enemy," *New York Times*, September 25, 2001, pp. D1, D2.

82. John Holusha, "The Painful Lessons of Disruption," *New York Times*, March 17, 1993, pp. C1, C5.

83. Amanda Cantrell, "Business After Disaster," CNNMoney, October 4, 2005, *http://cnnmoney.com/2005/10/04/technology/disaster_recovery* (accessed October 5, 2005).

84. Michelle Kessler, "Backing Up Data Keeps Companies Running Even When Disaster Strikes," *USA Today*, September 6, 2005, p. 2B.

85. Enterprise Technology Center, 1992.

86. Peter W. Huber, "Dig More Coal — the PCs Are Coming," *Forbes*, May 31, 1999, *www.forbes.com*.

87. Report by University of California, Lawrence Berkeley National Laboratory, cited in David Wessel, "Bold Estimate of Web's Thirst for Electricity Seems All Wet," *The Wall Street Journal*, December 5, 2002, pp. B1, B3.

88. David Lazarus, "Toxic Technology," *San Francisco Chronicle*, December 3, 2000, pp. B1, B4, B8; Benjamin Pimentel, "The Valley's Toxic History," *San Francisco Chronicle*, January 30, 2004, pp. B1, B4; Benjamin Pimentel, "Tech to Study Cancer Data," *San Francisco Chronicle*, March 19, 2004, pp. C1, C3; and Benjamin Pimentel, "Dust on Gadgets Is Toxic," *San Francisco*, June 4, 2004, pp. C1, C3.

89. Alina Tugend, "When Just Putting It Out on the Curb Isn't Enough," *New York Times*, July 23, 2005, p. B6; Elizabeth Royte, "E-gad!" *Smithsonian*, August 2005, pp. 82–87; and Christine Nuzum, "Before You Throw It Out . . . ," *The Wall Street Journal*, September 12, 2005, p. R9.

90. Tina Kelley, "Socks? With Holes? I'll Take It," *New York Times*, March 16, 2004, p. A21; Laurie J. Flynn, "2 PC Makers Favor Bigger Recycling Roles," *New York Times*, May 19, 2004, p. C3; Cris Prystay, "Recycling 'E-Waste,'" *The Wall Street Journal*, September 23, 2004, pp. B1, B6; and Jesse Drucker, "Old Cellphones Pile Up by the Millions," *The Wall Street Journal*, September 23, 2004, pp. B1, B6; Jon Swartz, "Got an Old PC? Don't Trash It: Recycle It," *USA Today*, January 7, 2005, p. 7B; Andy Reinhardt and Rachel Tiplady, "Europe Says: Let's Get the Lead Out," *BusinessWeek*, February 7, 2005, p. 12; John R. Quinn, "How Do I Dump My PC?" *U.S. News & World Report*, April 11, 2005, p. 83; Rachel Konrad, "Activists Push Recycling to Fight 'E-Waste,'" *Reno Gazette-Journal*, April 24, 2005, p. 4C; Edward Epstein, "Congress Examines Disposal Options for Electronics," *San Francisco Chronicle*, June 6, 2005, pp. A1, A7; and Michael Cabanatuan, "Bay Area's Easy Endings for Electronics," *San Francisco Chronicle*, June 6, 2005, p. A7.

91. Rachel Metz, "Out with the Old Phone, in with the Cash," *New York Times*, July 7, 2005, p. C10.

92. Julie Rawe, "Cellular's New Camouflage," *Time Inside Business*, December 9, 2002, pp. A9–A10; Elaine Goodman, "Cell Towers Go Incognito," *Reno Gazette-Journal*, February 2, 2004, pp. 1B, 2B; Paul Davidson, "Cellphone Tower Rules May Loosen Up," *USA Today*, September 9, 2004, p. 3B; and Katie Hafner, "First Come Cellphone

Towers, Then the Babel," *New York Times*, May 1, 2005, sec. 1, pp. 1, 24.

93. Kathy Jo Wetter, ETC Group, quoted in Jim Krane, "Risks of Nanotechnology Debated," *San Francisco Chronicle*, September 9, 2002, p. E6. See also Bernadette Tansey, "An Insider's View of Nanotech," *San Francisco Chronicle*, February 1, 2004, p. I1; Dan Vergano, "Creating a Monster?" *USA Today*, September 28, 2004, p. 6D; Kevin Maney, "Nanotechnology's in Your Pants, on Your Face, Even Your Bling," *USA Today*, June 1, 2005, p. 4B; Fred Krupp and Chad Holliday, "Let's Get Nanotech Right," *The Wall Street Journal*, June 14, 2005, p. B2; and Kevin Maney, "Scared of New Nano-Pants? Hey, You May Be onto Something," *USA Today*, June 22, 2005, p. 3B.

94. Andrew Kupfer, "Alone Together," *Fortune*, March 20, 1995, pp. 94–104.

95. Lutz Erbring, coauthor of Stanford University survey of 4,113 people about internet impact on daily activities, quoted in Joellen Perry, "Only the Cyberlonely," *U.S. News & World Report*, February 28, 2000, p. 62.

96. Study by Robert Kraut, *Journal of Social Issues*, reported in Deborah Mendenhall, "Web Doesn't Promote Isolation, Study Says," *San Francisco Chronicle*, August 22, 2001, p. C3; reprinted from *Pittsburgh Post-Gazette*; and Lisa Guernsey, "Cyberspace Isn't So Lonely After All," *New York Times*, July 26, 2001, pp. D1, D5. Also see study by Jeffrey Cole, UCLA Center for Communication Policy, reported in Greg Miller and Ashley Dunn, "Net Does Not Exact a Toll on Social Life, New Study Finds," *Los Angeles Times*, October 26, 2000, pp. C1, C8.

97. Norman H. Nie, Alberto Simpser, Irena Stepanikova, and Lu Zheng, *Ten Years after the Birth of the Internet, How Do Americans Use the Internet in Their Daily Lives?* (Stanford, CA: Stanford Institute for the Quantitative Study of Society, 2004), *www.stanford.edu/group/siqss/SIQSS_Time_Study_04.pdf* (accessed July 20, 2005).

98. Tom Verdin, "Rapid Growth of Cyber-Gambling Prompts Calls for Regulation," *San Francisco Chronicle*, March 9, 2000, p. A9.

99. Stephen Wright, "Australian Firm Bets on Gambling by Mobile Phone," *The Wall Street Journal*, June 1, 2005, p. BS8.

100. Ian Urbina, "Online Poker: Hold 'Em and Hide 'Em," *New York Times*, March 14, 2004, p. C11; Julia Angwin, "Could U.S. Bid to Curb Gambling on the Web Go Way of Prohibition?" *The Wall Street Journal*, August 2, 2004, p. B1; Lorraine Woellert, "Can Online Betting Change Its Luck?" *BusinessWeek*, December 20, 2004, pp. 66–67; and Jon Swartz, "Online Gambling Sites Expect Big Payoffs," *USA Today*, February 8, 2005, p. 1B.

101. Survey by Microsoft Corp., reported in Don Clark and Kyle Pope, "Poll Finds Americans Like Using PCs, but May Find Them to Be Stressful," *The Wall Street Journal*, April 10, 1995, p. B3.

102. Survey by Concord Communications, reported in Matt Richtel, "Rage against the Machine: PCs Take Brunt of Office Anger," *New York Times*, March 11, 1999, p. D3.

103. Study by Glenn Wilson and scientists from the Institute of Psychiatry at the University of London for Hewlett-Packard, reported in Benjamin Pimentel, "E-mail Addles the Mind," *San Francisco Chronicle*, May 4, 2005, pp. C1, C6.

104. "Porn in the U.S.A.," *60 Minutes*, CBSNews.com, September 5, 2004, *www.cbsnews.com/stories/2003/11/21/*

60minutes/main585049.shtml (accessed July 20, 2005).

105. Ellen Simon, "Satellite Radio Could Benefit from Censorship," *Reno Gazette-Journal*, April 17, 2004, pp. 1D, 3D.

106. Vauhini Vara, "Now Playing on Apple's iTunes: Adult-Oriented Podcasts," *The Wall Street Journal*, July 22, 2005, pp. B1, B5.

107. Charisse Jones, "X-Rated DVDs in Vehicles Spark Outcry," *USA Today*, March 29, 2004, p. 3A.

108. Yahoo, cited in Del Jones, "Poses Workplace Threat," *USA Today*, November 27, 1995, p. B1.

109. Websense Inc., 2004, and Nielsen/Net Ratings, 2004, reported in "Internet Porn Grows," *USA Today*, June 20, 2004, p. 12A.

110. Tracking by comScore Media Metrix, reported in Anick Jesdanun, "Will a Virtual Red-Light District on the Web Help Parents Curb Online Porn?" *San Francisco Chronicle*, June 13, 2005, p. E2.

111. Jon Swartz, "Online Porn Often Leads High-Tech Way," *USA Today*, March 9, 2004, pp. 1B, 2B.

112. Matt Richtel, "Cellphone Entertainment, Yes, but Carriers Shy from X-Rated," *New York Times*, December 20, 2004, p. C8; Bob Keefe, "Pornography Industry Sees Cell Phones as New Frontier," *San Francisco Chronicle*, January 3, 2005, p. F3; and Matt Richtel and Michel Marriott, "Ring Tones, Cameras, Now This: Sex Is Latest Cellphone Feature," *New York Times*, September 17, 2005, pp. A1, B4.

113. Joan Biskupic, "Justices Block Limits on Net Porn," *USA Today*, June 30, 2004, p. 1A; and Bob Egelko, "Top Court Upholds Ban on Internet Porn Limits," *San Francisco Chronicle*, June 30, 2004, pp. A1, A13.

114. Anita Hamilton, "The Web-Porn Patrol," *Time*, July 12, 2004, p. 87.

115. Benny Evangelista, "New DVD Player Lets Parents Clean Up Movies," *San Francisco Chronicle*, April 12, 2004, pp. D1, D4; and John R. Quain, "Keeping Your Screen Clean," *U.S. News & World Report*, May 31, 2004, p. 71.

116. Federal Trade Commission 2003 study, cited in Anita Hamilton, "Video Vigilantes," *Time*, January 10, 2005, pp. 60, 63.

117. Quain, 2004.

118. Ted Bridis, "World Wide Web Gets Its Own Red-Light District," *Reno Gazette-Journal*, June 2, 2005, p. 1C; "Fence Off Internet Porn" (editorial), *USA Today*, September 15, 2005, p. 12A; and Patrick Trueman, "XXX Would Legitimize Porn," *USA Today*, September 15, 2005, p. 12A.

119. Jesdanun, 2005.

120. David Finkelhor, Kimberly J. Mitchell, and James Wolak, *Online Victimization: A Report on the Nation's Youth*, June 2000, Crimes Against Children Research Center, University of New Hampshire, Durham, NH. See also Jane L. Levere, "Blunt Ads for Teenagers Warn of Net Predators," *New York Times*, June 8, 2005, p. C10.

121. Janis Wolak, Kimberly Mitchell, and David Finkelhor, *Internet Sex Crimes against Minors: The Response of Law Enforcement*, November 2003, Crimes Against Children Research Center, University of New Hampshire, Durham, NH, *www.missingkids.com/en_US/publications/NC132.pdf* (accessed July 25, 2005). See also Marilyn Elias, "Survey Paints Different Portrait of Online Abuser," *USA Today*, August 2, 2004, p. 8D.

122. Elias, 2004; and Levere, 2005.

123. Amy Harmon, "Internet Gives Teenage Bullies Weapons to Wound from Afar," *New York Times*, August 26, 2004, pp. A1, A21; and Jon Swartz, "Schoolyard Bullies Get Nastier Online," *USA Today*, March 7, 2005, pp. 1A, 2A.

124. Swartz, "Schoolyard Bullies Get Nastier Online," 2005; "Cyberbullies Cause Real Pain" [letters], *New York Times*, August 30, 2004, p. A20; and Jeff Chu, "You Wanna Take This Online?" *Time*, August 8, 2005, pp. 52, 55.

125. Jonathan Marshall, "Some Say High-Tech Boom Is Actually a Bust," *San Francisco Chronicle*, July 10, 1995, pp. A1, A4.

126. Surfwatch Checknet, cited in Keith Naughton, Joan Raymond, Ken Shulman, and Diane Struzzi, "Cyberslacking," *Newsweek*, November 29, 1999, pp. 62–65.

127. STB Accounting Systems 1992 survey, reported in Del Jones, "On-line Surfing Costs Firms Time and Money," *USA Today*, December 8, 1995, pp. 1A, 2A.

128. Nie et al., 2004.

129. Dan Gillmor, "Online Reliability Will Carry a Price," *San Jose Mercury News*, July 18, 1999, pp. 1E, 7E.

130. Paul Saffo, quoted in Laura Evenson, "Pulling the Plug," *San Francisco Chronicle*, December 18, 1994, Sunday sec., p. 53.

131. Daniel Yankelovich Group report, cited in Barbara Presley Noble, "Electronic Liberation or Entrapment?" *New York Times*, June 15, 1994, p. C4.

132. Kevin Maney, "No Time Off? It's Tech Giants' Fault," *USA Today*, July 21, 2004, p. 4B.

133. U.S. Labor Department, reported in Sue Kirchoff, "Worker Productivity Rises at Fastest Rate since 1950," *USA Today*, February 7, 2003, p. 5B.

134. Stephen S. Roach, "Working Better or Just Harder?" *New York Times*, February 14, 2000, p. A27.

135. David Levy, quoted in Joseph Hart, "Technoskeptic Techie," *Utne*, January– February 2005, pp. 28–29.

136. Bill McKibben, quoted in Jeffrey R. Young, "Knowing When to Log Off," *The Chronicle of Higher Education*, April 22, 2005, pp. A34–A35.

137. Nate Stulman, "The Great Campus Goof-Off Machine," *New York Times*, March 15, 1999, p. A25.

138. Robert Kubey, "Internet Generation Isn't Just Wasting Time," *New York Times*, March 21, 1999, sec. 4, p. 14.

139. Marco R. della Cava, "Are Heavy Users Hooked or Just Online Fanatics?" *USA Today*, January 16, 1996, pp. 1A, 2A.

140. Kenneth Hamilton and Claudia Kalb, "They Log On, but They Can't Log Off," *Newsweek*, December 18, 1995, pp. 60–61; and Kenneth Howe, "Diary of an AOL Addict," *San Francisco Chronicle*, April 5, 1995, pp. D1, D3.

141. Stella Yu, quoted in Hamilton and Kalb, 1995.

142. Jonathan Kandell, quoted in J. R. Young, "Students Are Unusually Vulnerable to Internet Addiction, Article Says," *The Chronicle of Higher Education*, February 6, 1998, p. A25.

143. American Psychological Association, reported in R. Leibrock, "AOLaholic: Tales of an Online Addict," *Reno News & Review*, October 22, 1997, pp. 21, 24.

144. Hamilton and Kalb, 1995.

145. Keith J. Anderson, reported in Leo Reisberg, "10% of Students May Spend Too Much Time Online," *The Chronicle of Higher Education*, June 16, 2000, p. A43.

146. Matt Richtel, "The Lure of Data: Is It Addictive?" *New York Times*, July 6, 2003, sec. 3, pp. 1, 8, 9.

147. R. Sanchez, "Colleges Seek Ways to Reach Internet-Addicted Students," *San Francisco Chronicle*, May 23, 1996, p. A16; reprinted from *Washington Post*.

148. Sanchez, 1996.

149. Sanchez, 1996.

150. Ben Gose, "A Dangerous Bet on Campus," *The Chronicle of Higher Education*, April 7, 2000, pp. A49–A51.

151. Jon Swartz, "More Firms Replace Operators with Software," *USA Today*, May 20, 2002, p. 1B.

152. Timothy Aeppel, "Workers Not Included," *The Wall Street Journal*, November 10, 2002, pp. B1, B11.

153. Jeremy Rifkin, "Technology's Curse: Fewer Jobs, Fewer Buyers," *San Francisco Examiner*, December 3, 1995, p. C19.

154. Robert Kuttner, "The Myth of a Natural Jobless Rate," *BusinessWeek*, October 20, 1997, p. 26; and Clyde Prestowitz, "Globalization's Next Victim: Us," *San Francisco Chronicle*, July 10, 2005, pp. F1, F6.

155. Richard W. Samson, "How to Succeed in the Hyper-Human Economy," *The Futurist*, September– October 2004, pp. 38–43.

156. Stewart Brand, in "Boon or Bane for Jobs?" *The Futurist*, January–February 1997, pp. 13–14.

157. Paul Krugman, "Long-Term Riches, Short-Term Pain," *New York Times*, September 25, 1994, sec. 3, p. 9.

158. Leslie Harris & Associates, *Bringing a Nation Online: The Importance of Federal Leadership*, a report by the Leadership Conference on Civil Rights Education Fund and the Benton Foundation, with support from the Ford Foundation, July 2002, *www.civilrights.org/publications/ reports/nation_online/bringing_a_nation .pdf*; see also John Schwartz, "Report Disputes Bush Approach to Bridging 'Digital Divide,'" *New York Times*, July 11, 2002, p. A16.

159. Paul Lamb, "Spanning the New Digital Divide," *San Francisco Chronicle*, June 14, 2005, p. B7.

160. Elizabeth F. Farrell, "Among Freshmen, a Growing Digital Divide," *The Chronicle of Higher Education*, February 4, 2005, p. A32.

161. Thomas L. Friedman, *The World Is Flat: A Brief History of the Twenty-First Century* (New York: Farrar, Straus & Giroux, 2005).

162. Paul Magnusson, "Globalization Is Great—Sort Of," *BusinessWeek*, April 25, 2005, p. 25. See also Fareed Zakaria, "The Wealth of Yet More Nations," *New York Times Book Review*, May 1, 2005, pp. 10–11; Russ L. Juskalian, "Prospering in Brave New World Takes Adaptation," *USA Today*, May 2, 2005, p. 4B; and Roberto J. Gonzalez, "Falling Flat," *San Francisco Chronicle*, May 15, 2005, pp. B1, B4.

163. Charles Hutzler, "Yuppies in China Protest via the Web—and Get Away with It," *The Wall Street Journal*, March 10, 2004, pp. A1, A8; Howard W. French, "Despite an Act of Leniency, China Has Its Eye on the Web," *New York Times*, June 27, 2004, sec. 1, p. 6; Charles Hutzler, "China Finds New Ways to Restrict Access to the Internet," *The Wall Street Journal*, September 1, 2004, pp. B1, B2; Tom Zeller Jr., "Beijing Loves the Web until the Web Talks Back," *New York Times*, December 6, 2004, p. C15; Jim Yardley, "A Hundred Cellphones Bloom, and Chinese Take to the Streets," *New York Times*, April 25, 2005, pp. A1, A6; and Bruce Einhorn and Heather Green, "Blogs Under Its Thumb," *BusinessWeek*, August 8, 2005, pp. 42–43.

164. Edward Cornish, "The Wild Cards in Our Future," *The Futurist*, July–August 2003, pp. 18–22.

165. Stephen H. Wildstrom, "What Does a Freshman Need?" *BusinessWeek*, August 7, 2000, p. 22.

166. Richard A. Seigel, "Palmtop Computers Prove Useful in Class" [letter], *The Chronicle of Higher Education*, February 4, 2000, p. B10.

167. "Business Schools Struggle to Impose Laptop Etiquette," *San Francisco Chronicle*, April 20, 2000, p. B3; reprinted from *New York Times*.

168. Todd Gitlin, quoted in Dora Straus, "Lazy Teachers, Lazy Students" [letter], *New York Times*, September 12, 1999, sec. 4, p. 18.

169. Jean Richardson, quoted in Wendy R. Leibowitz, "At Yale's Demand, a Web Site Drops Lecture Notes from the University's Classes," *The Chronicle of Higher Education*, March 17, 2000, pp. A49–A50.

170. Tanya Schevitz, "Web Sites Snag Students," *San Francisco Chronicle*, October 18, 1999, pp. A1, A13.

171. Jamie Horwitz, American Federation of Teachers, reported in Marco R. della Cava, "Blackboard Jungle Turns Ugly Online," *USA Today*, May 8, 2000, p. 3D.

172. Tanya Schevitz, "Prof Fights Web Trash Talk," *San Francisco Chronicle*, April 6, 2000, pp. A1, A14.

Chapter 10

1. Gary Webb, "Potholes, Not 'Smooth Transition,' Mark Project," *San Jose Mercury News*, July 3, 1994, p. 18A.

2. Jerome Garfunkel, "COBOL: Still Relevant after All These Years," December 17, 2003, *www.cobolreport.com/columnists/ jermoe/02172003.asp*

3. Alan Freedman, *The Computer Glossary*, 6th ed. (New York: AMACOM, 1993), p. 370.

4. "How to Use Scripts on Your Site: The Commands That Perform Specific Functions," *Smart Computing*, August 2001, *www.smartcomputing.com/editorial/ article.asp?article=articles*

5. Michael Krantz, "Keeping Tabs Online," *Time*, November 10, 1997, pp. 81–82.

Credits

CHAPTER 1
p. 1, Coco Marlet/Photo Alto/Getty Images; **p. 2**, Ed Bock/Corbis; **p. 3**, CD is Absolute Couples Image 100/Punchstock; **p. 6** top, Courtesy of Oddcast.com; **p. 6** bottom left, Tim Ockenden/ EMPICS; **p. 6** bottom right, Helen King/ Corbis; **p. 7** right, Fujifotos/ The Image Works; **p. 7** left, Fujifotos/The Image Works; **p. 9** left, (c) Warner Brothers/Courtesy Everett Collection; **p. 9** right, Torin Boyd; **p. 10**, Jeff Greenberg/The Image Works; **p. 11**, Rob Crandall/The Image Works; **p. 12**, Mitch Wojnarowicz/Amsterdam Recorder/The Image Works; **p. 14** left, Courtesy of Unisys Archives; **p. 14** right, Mark Richards/PhotoEdit; **p. 15** left, Courtesy of Motorola; **p. 15** middle, Courtesy of Motorola; **p. 15** right, Katsumi Kasahara/AP Photo/Wide World Photos; **p. 17**, Coco Marlet/Photo Alto/Getty Images; **p. 21**, Courtesy of IBM; **p. 22** top, Courtesy of Unisys; **p. 22** middle, Courtesy of Sun Microsystems; **p. 22** bottom left, Courtesy of Hewlett-Packard; **p. 22** bottom right, Courtesy of Apple Computer; **p. 23** top left, Courtesy of Hewlett-Packard; **p. 23** top right, Courtesy of Hewlett-Packard; **p. 23** middle left, Courtesy of Hewlett-Packard; **p. 23** bottom left, Courtesy of Motorola Corp.; **p. 23** bottom right, Courtesy of Intel; **p. 24**, Courtesy of Hewlett-Packard; **p. 28** left middle, Courtesy of Motorola; **p. 28** right, Don Mason/Corbis; **p. 30** bottom left, Courtesy of Iomega; **p. 32** bottom, Courtesy of Microsoft Corporation; **p. 33** top left, Courtesy of Adobe Systems Inc.; **p. 33** bottom left, Courtesy of Adobe Systems Inc.; **p. 34**, Shizuo Kambayashi/AP Photo/Wide World Photo; **p. 36**, Courtesy of PR NewsFoto; **p. 40**, Courtesy of Hewlett-Packard; **p. 43** top, Courtesy of Hewlett-Packard; **p. 43** bottom, Courtesy of Hewlett-Packard; **p. 44** top, Helen King/Corbis; **p. 44** bottom, Courtesy of Sun Microsystems.

CHAPTER 2
p. 49, Royalty-Free/Corbis; **p. 56**, Judy Mason; **p. 57**, Courtesy of DirectTV; **p. 65**, Elise Amendola/AP Photo/Wide World; **p. 67**, Richard Drew/AP Photo/Wide World; **p. 82**, **Panel 2.26** Adapted from How Computers Work, www.smartcomputing.com; **p. 83**, Courtesy of Research in Motion Limited; **p. 105**, Judy Mason; **p. 110**, Courtesy of Research in Motion Limited.

CHAPTER 3
p. 117, Courtesy of Quark Inc.; **p. 127**, Courtesy of Symantec; **p. 135** bottom, Courtesy of Joe Auricchio; **p. 136**, Courtesy of Apple Computer; **p. 137**, Chris Farina/Corbis; **p. 138** top right, Courtesy of Microsoft; **p. 138** top left, Courtesy of Microsoft; **p. 138** bottom left, Courtesy of Hewlett-Packard; **p. 138** middle, Courtesy of Creative Technology Ltd.; **p. 141**, Courtesy of Joe Auricchio; **p. 142**, Courtesy of Joe Auricchio; **p. 144** top, Courtesy of Hewlett-Packard; **p. 144** bottom, Courtesy of Symbian Ltd.; **p. 158**, Peter Beck/Corbis; **p. 169**, Courtesy of Quark Inc.; **p. 172** top, Tom Wagner/Corbis; **p. 172** bottom, Courtesy of Alchemy Mindworks; **p. 173**, Courtesy of Macromedia; **p. 174** bottom, Courtesy of Mindjet Corporation; **p. 175**, Courtesy of Okino; **p. 178**, Courtesy of Okino; **p. 179**, Courtesy of Joe Auricchio; **p. 180**, Courtesy of Apple Computer; **p. 181**, Courtesy of Microsoft; **p. 182**, Courtesy of Mindjet Corporation; **p. 184**, Courtesy of MobileRobots. com, Amherst, NH (PRNewsFoto).

CHAPTER 4
p. 189, Courtesy of Apple Computer; **p. 191**, Courtesy of IBM Archives; **p. 192**, Courtesy of Intel; **p.193** all, Courtesy of Intel; **p. 202** top left, Courtesy of www.apc.com; **p. 202** top right, Courtesy of www.apc.com; **p. 202** bottom, Courtesy of Apple Computer; **p. 203**, Courtesy of Intel; **p. 204** top, Courtesy of Intel; **p. 204** bottom, Courtesy of AMD; **p. 205** top, Courtesy of Motorola; **p. 210**, Courtesy of Intel; **p. 211**, Courtesy of Intel; **p. 213**, Courtesy of The Jhai Foundation; **p. 214** top, Courtesy of Acer Inc.; **p. 214** bottom, Andy Resek/MHDIL; **p. 215** bottom, Courtesy of Adaptec; **p. 215** top, Courtesy of Hewlett-Packard; **p. 216** bottom, Courtesy of Hewlett-Packard; **p. 216** top, Courtesy of Hewlett-Packard; **p. 218**, Brian Williams; **p. 220**, John S. Reid; **p. 221**, Royalty-Free/ Corbis; **p. 222** left, Courtesy of Iomega; **p. 222** right, Courtesy of Acer Inc.; **p. 223**, Courtesy of IBM; **p. 224**, Courtesy of IBM; **p. 225** top, Courtesy of Apple Computer; **p. 225** bottom, Koji Sasahara/AP Photo/Wide World; **p. 227**, Brian Williams; **p. 228**, Toru Yamanaka/Getty Images; **p. 230**, Courtesy of Hewlett Packard; **p. 231**, Katsumi Kasahara/AP Photo/Wide World Photos; **p. 232** top, Tony Cenicola/New York Times; **p. 232** bottom, Fabian Bimmer/AP Photo/Wide World Photos; **p. 235** bottom, Kurt Stier/Corbis; **p. 235** top, Frank Franklin II/AP Photo/Wide World Photos; **p. 237** Adapted from pcworld.com; **p. 241**, Courtesy of Intel; **p. 242**, Tony Cenicola/New York Times; **p. 243**, Courtesy of IBM; **p. 243**, Courtesy of Hewlett-Packard; **p. 244**, Brian Williams; **p. 244** bottom, John S. Reid; **p. 245**, Courtesy of Hewlett-Packard; **p. 246**, Brian Williams; **p. 247** bottom, Courtesy of Iomega; **p. 247** top, Courtesy of Hewlett-Packard.

CHAPTER 5
p. 251, Charles Gupton/Stock Boston; **p. 252**, Charles Gupton/Stock Boston; **p. 252** bottom right, Jon Freilich/AP Photo/Wide World; **p. 252** left, Bell Atlantic/AP Photo/Wide World; **p. 255** top right, Judy Mason; **p. 255** bottom, Jochen Luebke/AFP/Getty Images; **p. 255** middle, Courtesy of Microsoft; **p. 255** top left, Courtesy of Hewlett-Packard; **p. 256** bottom left, Ryan McVay/Getty Images; **p. 256** bottom right, John Slater/Taxi/Getty Images; **p. 256** top right, Peter Cade/Getty Images; **p. 257** bottom left, Courtesy of Think Outside; **p. 257** bottom right, Courtesy of Ed Loera /FrogPad.com; **p. 257** top, Courtesy of QSI Corporation; **p. 259** left, Judy Mason; **p. 259** right, Courtesy of Logitech; **p. 260**, Courtesy of Logitech; **p. 261** top, Courtesy of IBM; **p. 261** middle, Courtesy of Hewlett Packard; **p. 261** bottom left, Courtesy of Motion Computing; **p. 261** bottom right, Joe Gill/The Express Times/AP Photo/Wide World; **p. 262** left, David Kohl/AP Photo/Wide World; **p. 262** right, Rich Pedroncelli/AP Photo/Wide World; **p. 263** top, Courtesy of Acer; **p. 263** bottom, Courtesy of FastPoint Technologies; **p. 264** top, Courtesy of Wacom Technology; **p. 264** all, Courtesy of Seiko Instruments USA Inc.; **p. 265** bottom left, Christopher Fitzgerald/The Image Works; **p. 265** bottom right, Wenatchee World, Mike Bonnicksen/AP Photo/Wide World; **p. 265** top, Courtesy of LeapFrog Enterprises; **p. 267**, Charles Gupton/Stock Boston; **p. 268** bottom, Courtesy of NCR Corporation; **p. 268** top, Courtesy of 3D Scanners Ltd./www.3dscanners.com; **p. 268** top, Courtesy of 3D Scanners Ltd./www.3dscanners.com; **p. 269**, Courtesy of Symbol Technology; **p. 270** top, Courtesy of Hewlett-Packard; **p. 270** bottom, Courtesy of IBM; **p. 270** middle, Courtesy of Hewlett-Packard; **p. 271** left, Courtesy of IBM; **p. 271** right, Left Lane Productions/Corbis; **p. 272**, Courtesy of Hewlett-Packard; **p. 273** right, Courtesy of Samsung; **p. 273** left, Courtesy of Samsung; **p. 273** top, Courtesy of Hewlett-Packard; **p. 275** top, William Thomas Cain/Getty Images; **p. 275**, CP/Aaron Harris/AP Photo/Wide World Photos; **p. 276** right, Courtesy of Texas Instruments; **p. 276** middle, Gilles Mingasson/Getty Images; **p. 276** left, Gilles Mingasson/Getty Images; **p. 277** top right, Courtesy of Identix; **p. 277** top left, Michael Probst/AP Photo/Wide World Photos; **p. 277** bottom, Tony Cenicola/New York Times Photo Service; **p. 278**, "Happy" from www.bergen.org/AAST/ComputerAnimation/

Graph-Pixel.html; **p. 279**, Courtesy of Hewlett-Packard; **p. 280**, Courtesy of Viewsonic; **p. 283**, Courtesy of Hewlett-Packard; **p. 284**, Courtesy of Hewlett-Packard; **p. 285** top, Peter Thompson/New York Times; **p. 285** bottom, Patrick Durand/Sygma/Corbis; **p. 286** top left, Courtesy of JCM American Corporation/PRNewsFoto; **p. 286** top right, Courtesy of Sony Corporation; **p. 287** top, Courtesy of Hewlett-Packard; **p. 287** middle, Courtesy of Hewlett-Packard; **p. 287** bottom, Steven Senne/AP Photo/Wide World Photos; **p. 292** top, Owen Franken/Corbis; **p. 292** bottom, Richard T.Nowitz/Corbis; **p. 293** top, Courtesy of Micron; **p. 293** bottom, Dr. Irfan Essa/Georgia Tech; **p. 294** top, Courtesy of IBM Research, Almaden Research Center. Unauthorized use prohibited; **p. 294** bottom, HO/AP Photo/Wide World; **p. 295**, Rick Friedman/Corbis; **p. 297**, **Panel 5.32**, from www.rsi.deas.harvard.edu/preventing .html; **p. 298**, Courtesy of Hewlett-Packard; **p. 299** top, Courtesy of Wacom Technology; **p. 299**, Courtesy of Hewlett-Packard; **p. 300** top, Courtesy of Hewlett-Packard; **p. 300** bottom, Courtesy of Microsoft; **p. 302**, Courtesy of IBM; **p. 302** bottom, Christopher Fitzgerald/The Image Works; **p. 303**, David Kohl/AP Photo/Wide World; **p. 304**, Courtesy of Hewlett-Packard.

CHAPTER 6
p. 309 right, Pallava Bagla/Corbis; **p. 309** left, HO/AP Photo/Wide World; **p. 316**, Fievet Laurent/AFP/Getty Images; **p. 325** top, Corbis; **p. 325** middle, Ed Degginger/Getty Images; **p. 325** bottom, Eric Myer/Getty Images; **p. 325** middle, Phil Degginger/Getty Images; **p. 326**, Steve Allen/Brand X Pictures/Getty Images; **p. 331** right, Brian Williams; **p. 331** left, Bob Rowan;Progressive Image/Corbis; **p. 332** top, Liu Liqun/Corbis; **p. 332** middle left, Olivier Prevosto/Corbis; **p. 332** bottom left, Pallava Bagla/Corbis; **p. 332** right, HO/AP Photo/Wide World; **p. 334** top right, Najtah Feanny/Corbis; **p. 334** bottom left, LM Otero/AP Photo/Wide World Photos; **p. 334** bottom right, James Leynse/Corbis; **p. 334** top left, Courtesy of Pulse Data; **p. 335**, Adapted from San Francisco Chronicle, Benny Evangelista, "GPS Essential for Success of U.S.'s Effort in Iraq," March 20, 2003, p. W6; **p. 336**, Corbis; **p. 337**, Courtesy of Blackberry; **p. 338**, Judy Mason; **p. 338**, Judy Mason; **p. 339**, Courtesy of Blackberry; **p. 340**, Courtesy of Kodak; **p. 342**, Courtesy of IBM; **p. 342**, Courtesy of IBM; **p. 342**, Bluetooth adapted from David Pogue, "Bluetooth, the Wireless Matchmaker, Tries Again," New York Times, May 16, 2002, p. E1; **p. 350**, Courtesy of Symantec; **p. 351**, Daniel Hulshizer/AP Photo/Wide World; **p. 354**, Chris Hondros/Getty Images; **p. 356**, Richard A. Brooks/AFP/Getty Images; **p. 358**, Daniel Hulshizer/AP Photo/Wide World; **p. 359**, Eric Myer/Getty Images; **p. 360**, Courtesy of Symantec; **p. 362**, Corbis; **p. 363**, Ed Degginger/Getty Images.

CHAPTER 7
p. 367, Peter Kramer/Getty Images; **p. 368**, Lee Jin-man/AP Photo/Wide World Photos; **p. 369**, Shizuo Kambayashi/AP Photo/Wide World Photos; **p. 370**, Courtesy of Apple Computer; **p. 372**, Courtesy of Creative Technology Ltd.; **p. 373**, Martin Meissner/AP Photo/Wide World Photos; **p. 374**, Reed Saxon/AP Photos/Wide World Photos; **p. 376**, Courtesy of Sirius/PRNewsFoto/NewsCom; **p. 379**, Courtesy of Apple Computer; **p. 381**, Courtesy of Hewlett-Packard; **p. 382**, Eric Risberg/AP Photo/Wide World Photos; **p. 386**, Courtesy of Palm Inc.; **p. 387**, Courtesy of Motion Computing; **p. 388**, Courtesy of Alienware; **p. 392**, Courtesy of Research in Motion Limited; **p. 394**, Courtesy of BlackBerry; **p. 395**, Courtesy of PRNewsFoto; **p. 396**, StockByte/ PunchStock; **p. 398**, Peter Kramer/Getty Images; **p. 399**, Katsumi Kasahara/AP Photo/Wide World Photos; **p. 401**, Shizuo Kambayashi/AP Photo/ Wide World Photos; **p. 402**, Courtesy of Apple Computer.

CHAPTER 8
p. 407, NOAA/Reuters/Corbis; **p. 408**, Courtesy of Identix; **p. 409**, Noboru Hashimoto/Corbis; **p. 410**, partially adapted from Computer Desktop Encyclopedia; bottom **p. 423** Christian S. Jensen, http://infolab.usc.edu/csci599/Fall2002/paper/I1_pederson_p40.pdf; **p. 414**, Courtesy of Corbis; **p. 422** bottom, Corbis; **p. 422** top, Charlie Westerman/Getty Images; **p. 423** left, Courtesy of TECH-BASE International; **p. 423** right, NOAA/Reuters/Corbis; **p. 426** http://computing-dictionary-thefreedictionary.com?BusinessMiner; **p. 427**, Brian Williams; **p. 428**, Courtesy of Shuffle Master Inc.; **p. 438**, Courtesy of ESRI Software; **p. 444** bottom right, Bob Mahoney/The Image Works; **p. 444** middle right, Ted Kawalerski Photography Inc./Getty Images; **p. 444** top right, Charlie Riedel/AP Photo/Wide World; **p. 444** top left, M.Spencer Green/AP Photo/Wide World; **p. 444** middle left, Nick Ut/AP Photo/Wide World; **p. 445** bottom left, Jodi Hilton/New York Times; **p. 445** top right, Kim Jae-Hwan/AFP/Getty Images; **p. 445** middle left, Lisa Poole/AP Photo/Wide World Photos; **p. 445** top left, Bertil Ericson/AP Photo/Wide World; **p. 445** top middle, Brett Coomer/AP Photo/Wide World; **p. 445** bottom right, Shizuo Kambayashi/AP Photo/Wide World; **p. 445** middle right, Mike Derer/AP Photo/Wide World; **p. 448** top left, AFP/Corbis; **p. 449**, Science Museum/SSPL/The Image Works; **p. 452**, PRNewsFoto/Photo via NewsCom; **p. 458**, Bob Mahoney/The Image Works.

CHAPTER 9
p. 463 bottom right, J.W.Burkey/Getty Images; **p. 463** top right, Bob Elsdale/Getty Images; **p. 463** left, LWA-Sharie Kennedy/Corbis; **p. 464** bottom, Dylan Martinez/Reuters/Corbis; **p. 465**, Bob Elsdale/Getty Images; **p. 466** top, J.W.Burkey/Getty Images; **p. 466** bottom, Courtesy of Troels Eklund Andersen; **p. 471**, Brian Williams; **p. 476**, Michael Kooren/Reuters/Corbis; **p. 480** left, Ricki Rosen/Corbis; **p. 480** right, Tony Cenicola/New York Times; **p. 481**, Michael L. Bak/Department of Defense/Getty Images; **p. 483**, table top, galtglobalreview.com; **p. 483** top, Brian Williams; **p. 483** bottom, Gabe Palmer/Corbis; **p. 484** middle, bottom, Courtesy of Larson, a DMB Company; **p. 484** top, Courtesy of NEC; **p. 490**, David M.Russell; **p. 494**, Michael L. Bak/Department of Defense/Getty Images.

CHAPTER 10
All photos in Chapter 10, PhotoDisc/Getty Images.

Index

Boldface page numbers indicate pages on which key terms are defined.

Abandonware, 147
Accelerator board, 218
Access points, **58**, **105**, 341
Access security
    control of access and, 481
    identification systems and, 480
    techniques used for, 479
Accounting and finance department, 433
Acronyms, 39
Acrostics, 39
Active-matrix display, 239, **281**, **298**
Active pixel sensor, 293
ActiveX controls, **531**, **533**
Addicts, internet, 489
Address-book feature, 83, 85
Addresses
    email, 83
    website, 66–68
Adware, **102**, **105**
Aerial maps, 94
Aggregator software, 379
AGP (accelerated graphics port) bus,
    218–219, **240**
Algorithms, 426, **507**, **533**
Always On generation, 399, 400
Amazon.com, 428
American Psychological Association
    (APA), 489
American Registry for Internet Numbers
    (ARIN), 64
America Online (AOL), 174
Amplitude modulation, 315
Analog cellphones, **337**, **358**
Analog signals, **312**, **358**
    converting into digital signals, 314–315
    modem conversion of, 313–314
Analog-to-digital converter, 314
Analytical graphics, **158–159**, **178**
Analyzing systems. *See* Systems analysis
Anderson, Keith J., 489
Andreessen, Marc, 67
Animation, 10, **94**, **105**, **172**, **178**
Animation files, 412
Animation software, 172
Anonymous FTP sites, 88
ANSI symbols, 510
Antivirus software, 126, **350**, **358**
Appeal-to-authority argument, 532
Appeal-to-pity argument, 532
Applets, **93–94**, **105**, 527
Application generator, 520
Application service providers (ASPs), 147
Application software, **33**, **40**, **118**,
    145–175, **178**
    animation software, 172
    audio editing software, 171
    computer-aided design programs, **175**
    database software, **159–163**
    desktop publishing programs, **168**–170
    documentation for, 147
    drawing programs, 171
    file types and, 148
    financial software, **166**–168
    IT timeline and, 162–169
    licenses for, 145–146
    methods for obtaining, 145–147
    multimedia authoring software, **173**

painting programs, 171
presentation graphics software, **164**–166
project management software, **174**–175
review questions/exercises on, 185–188
specialty software, 163–175
spreadsheet programs, **156–159**
tech support for, 176–177
tutorials for, 147
types of, 148–149
versions and releases of, 146
video editing software, 171
web page design software, **173**
word processing software, 149–155
Archival copy, 145
Arendt, Hannah, 310
Arithmetic/logic unit (ALU), **208**, **240**
Arithmetic operations, 208
ARPANET, 50
Artifacts, 171
Artificial intelligence (AI), 439, **440–449**, 455
    cyborgs, 447
    ethics of, 449
    expert systems, 440–441
    fuzzy logic, **442**–443
    genetic algorithms, 447
    intelligent agents, **442**
    natural language processing, **442**
    neural networks, 447
    pattern recognition, **442**
    review questions/exercises on, 459–461
    robotics, **443–446**
    simulators, **443**, 444
    Turing test and, 448–449
    virtual reality, **443**, 444
    weak vs. strong, 446–447
Artificial life (A-life), 448–449
Art programs, 170–171
ASCII coding scheme, **197**, 198, **240**
ASCII files, 413
Aspect ratio, 389
Assemblers, **518**, **533**
Assembly language, **517–518**, **533**
ATMs (automated teller machines), 252,
    256, 479
Attachments, email, 85–86, 87, 346
Attentive systems, 294
Auctions, online, 97–98
Audio
    digital sampling of, 314
    editing software for, 171
    input, 270
    MP3 format, **372**
    output, 288
    search tools for, 79
    streaming, **95**
    web-based, 79, 95
Audio CDs, 314
Audioconferencing, 357
Audio files, 412, 413
Audio-input devices, **270**, **298**
Audit controls, 481
Authentication process, 62, 351
Auto Correct function, 152
Automated teller machines (ATMs), 252,
    256, 479
Automated virtual representatives
    (vReps), 439
Automator feature, 137
Avatars, **6**, **40**
Avila, Eric, 471

B2B (business-to-business) commerce, **97**, **105**
Backbone, **358**
    internet, 61, **62**, 322
    network, **322**
Backdoor programs, 346
Backups
    application software, 145
    importance of, 203, 227
Backup utility, 125
Backus, John, 522
Baig, Edward C., 176
Bandwidth, **52**, **105**, **329**, **358**
Banking, online, 97
Bar charts, 159
Bar-code readers, **267–268**, **298**
Bar codes, 266–268, **298**
Baseband transmission, 52
BASIC programming language, 521, **524**, **533**
Batch processing, 436
Batteries
    digital camera, 382–383
    laptop, 238
    MP3 player, 373
    PDA, 386
Bays, **201**, **240**
Bederson, Ben, 254
Berners-Lee, Tim, 64, 65, 427, 529
Beta testing, 514
Bezos, Jeffrey, 428
Biggs, Andy, 384
Binary coding schemes, 197, 198
Binary system, **195–198**, **240**, 311
Biological nanocomputers, 22
Biometrics, **277**, 293, **298**, **351**, 353, **358**,
    480, **494**
BIOS programs, 120
Bit depth, 266, 279
Bit-mapped images, 171
Bitmaps, 266
Bits (binary digits), **196**, **240**
Bits per second (bps), **52**, **105**
Blackberry devices, 337, 394
Black-hat hackers, 349
Blackmail, 474
Blocking software, 485
Blogosphere, **96**, **105**
Blogs, 92, **96**, **105**
Bloom, Benjamin, 45
Blue Gene/L supercomputer, 21
Bluetooth ports, 217
Bluetooth technology, 331, **341–342**, **358**
Blu-ray Disc (BD), 227–228, 237, **240**
Bookmarks, 72
Boolean operators, 80
Boot disk, 121
Booting, **120–121**, **178**
Boot-sector virus, 347
Border gateway protocol, 475
Botnet, 474, **494**
Bots, 442
Brainwave-input technology, 294
Brand, Stewart, 490
Breazeal, Cynthia, 445
Brenner, Viktor, 489
Bricklin, Daniel, 156
Bridges, **322**, **358**
Broadband connections, **52**, 56, **105**, **329**, **358**
Broadband wireless services, 339
Broadcast radio, **330**–331, **358–359**
Brooks, Rodney, 446

Browser hijackers, **102**, **105**
Browser software. *See* Web browsers
Brute force technique, 446
BSD operating system, 141
Buffer, 121
Bugs, software, 470
Bull, Michael, 374
Bursting, 212
Buses, 209, 218, **240**
Business, online, 97–98, 427–431
    See also E-commerce
Business graphics, 158–159
Business-to-business (B2B) systems, **430**, **455**
Business-to-consumer (B2C) systems, **431**, **455**
Business webs (b-webs), 430
Bus network, **323**, **359**
Buying considerations
    for high-definition TVs, 390
    for laptop computers, 238–239
    for personal computers, 194, 195
    for printers, 286
Bytecode, 527
Bytes, 30, **196**, **240**

C programming language, 521, **525**, **533**
C++ programming language, **527**, **533**
Cable modems, 56–57, 59, **105**
Cache, **211–212**, **240**
Cache card, 218
CAD programs, **175**, **178**
CAD/CAM programs, 175
Callback systems, 479
Caller ID, **370**, **401**
Call waiting, 54
Camera phones, 273, 395
Cameras. *See* Digital cameras
Campus-area networks, 317
Canton, James, 236
Cantu, Homaro, 285
Capek, Karel, 444
Career information, 11–13
    See also Job searches
Carlin, Daniel, 6
Carpal tunnel syndrome (CTS), **290**, **298**
Carrier waves, 312
Carter, David, 475–476
Carty, Betty, 474
Cascading menus, 132
Case (system cabinet), **28**, **40**, 200, 201
Case control structure, 510, 512
CASE tools, **501**, **533**
CD drive, **31**, **40**
CDMA wireless standard, 338
CD-R disks, 226, **240**
CD-ROM disks, **225–226**, **240–241**
CD-ROM drive, 226
CD-RW disks, **227**, **241**
Cell pointer, 158
Cells, spreadsheet, **156**, **178**
Cellular telephones, 14–15, 328, 337–340
    analog, **337**
    broadband, 339
    digital, 338–340
    E-911 capability, 335, 336
    future of, 355
    health issues and, 290
    how they work, 392
    keeping track of, 471
    motion-sensing, 368
    recycling, 483
    services and features, 392–396
    smartphones as, 14–15, **391–397**
    societal effects of, 397
Central processing unit (CPU), 25, 28, **208**, **241**
    data processing by, 208–209
    internal management of, 121
Channel capacity, 52
Character-recognition devices, 268–269
Characters, **409**, **455**
Charge-coupled devices (CCDs), 273
Charts, 158–159

Cheating, 104
Check bit, 198
Children, protecting, 485–487
Chips, **192**, **241**
    CMOS, **211**
    flash memory, **211**
    making of, 193
    memory, 209–211
    microprocessor, **192**, 204–205, 206
    RAM, **210–211**
    ROM, 211
Chipsets, **204**, **241**
Circuit, 190
Circular reasoning, 532
Classes, 527
Classification analysis, 426
Classrooms, computer use in, 493
Clients, **24**, **40**, **60**, **105**
Client/server networks, **318–319**, **359**
Clip art, 154
Clipboard, 152
Clock speed, 206
Closed architecture, 217
Closed-circuit television, 357
Closing windows, 134
Clusters, 222
CMOS (complementary metal-oxide semi-conductor) chips, **211**, **241**
Coaxial cable, **325**, **359**
COBOL programming language, 518, 519, 521, **523–524**, **534**
Cochran, Anita, 465
Codd, E. F., 420
Codecs, 412
Coding, **513**, **534**
Cog project, 446
Cold boot, 121
Collaboration, 36–37
Color depth, 266, **279**, **298**
Column charts, 159
Columns
    spreadsheet, 156
    text, 154
Command-driven interface, 130
Commands, 127
Comments, inserting, 155
Commerce, electronic. *See* E-commerce
Commercial online services, 58
Commercial software, 145–146
Communications, 26
Communications hardware, 32
Communications media, **325**, **359**
    wired, 325–326
    wireless, 328–344
Communications satellites, 57, 59, **106**, **332–333**, 355, **359**
Communications technology, **5**, **40**, 309–366
    bandwidth and, **329**
    cellular telephones, 328, 337–340
    contemporary examples of, 5–13
    electromagnetic spectrum and, **329**
    future developments in, 355–356
    Global Positioning System, 333, **334–336**
    networks, **315–325**
    overview of developments in, 35
    pagers, 336–337
    radio-frequency spectrum and, **329**, 330–331
    review questions/exercises on, 364–366
    satellites and, 332–333, 355
    security issues and, 344–354
    telecommuting and, 327–328
    timeline of progress in, 14–21, 310–313
    videoconferencing and, 357
    virtual office and, 327–328
    wired, 325–326
    wireless, 328–344
    workgroup computing and, 357
    See also Information technology
Compilers, **519**, **534**
Comprehension, 45
Compression, **412**, **455**
    lossless vs. lossy, 412

utility programs for, 126
    video and audio, 171
Computer-aided design (CAD) programs, **175**, **178**
Computer-aided design/computer-aided manufacturing (CAD/CAM) programs, 175
Computer-aided software engineering (CASE) tools, **501**, **533**
Computer animation, 172
Computer Assisted Passenger Prescreening Program (CAPPS II), 450
Computer-based information systems, 435–439
    decision support systems, **437–438**
    executive support systems, **438–439**
    expert systems, **439**
    management information systems, 436–437
    office information systems, **435**
    review questions/exercises on, 459–460
    transaction processing systems, 435–**436**
Computer crime, **471–476**, **494**
    deterrents to, 478–479
    perpetrators of, 475–476
    types of, 471–475
Computer Emergency Response Team (CERT), 478
Computer Ethics (Forester and Morrison), 37
Computer operators, 515
Computers, **4**, **40**
    basic operations of, 25–26
    crimes related to, 471–476, 478–479
    custom-built, 25, 27, 33–34
    digital basis of, 311
    environmental issues related to, 482–484
    guarding against theft, 296
    hardware of, **25**, 27–32
    health issues related to, 289–291, 296–297
    recycling, 37, 483
    software of, **25**, 32–33
    student use of, 493
    types of, 20–24
    See also Personal computers
Computer savvy, **3–4**, **40**
Computer-supported cooperative work (CSCW) systems, 434
Computer technology, 4
    contemporary examples of, 5–13
    future developments in, 232, 234–237
    overview of developments in, 34–35
    publications about, 194
    timeline of progress in, 14–21, 310–313
    See also Information technology
Computer vision syndrome (CVS), **290**, **298**
Concurrent-use license, 146
Connectivity, **35**, **40**
Consumer-to-consumer (C2C) systems, **431**, **455**
Contactless smart cards, 231
Contact smart cards, 230
Content templates, 164
Context-sensitive help, 135
Control structures, 508, **510**, 512, **534**
Control unit, **208**, **241**
Convergence, 36, **369–370**, **401**
Conversion
    audio file, 373
    signal, 313–314
    system, 504–505
Cookies, **101–102**, **106**
Coprocessing, 123
Coprocessor board, 218
Copy command, 152
Copyright, 145
Cornish, Edward, 492
Corporate spies, 476
Cost-benefit analysis, 464
Course-management software, **5**, **40**
CPU. *See* Central processing unit
Crackers, 348, **349–350**, **359**
Cramming, 38
Creating documents, 150

Credit bureaus, 453, 454
Crichton, Michael, 236
Critical thinking, 45, 532
Crosstalk, 325
CRT (cathode-ray tube), **279**, 280, **298**
Cunningham, Ward, 78
Cursor, **150**, **178**
Custom-built PCs, 25, 27, 33–34
Custom software, 147
Cut command, 152
Cyberbullying, 487
Cyberspace, 16–**17**, **41**
Cyberterrorists, 349
Cyborgs, 447
Cycles, 206
Cyc project, 446

Daisy chains, 213–214, 215
Dashboard feature, 137
Data, **25**, **41**
    accuracy of, 467–468
    analyzing, 162, 501
    backing up, 125, 233
    completeness of, 467–468
    compressing, 126
    digital vs. analog, 311–312
    gathering, 500
    hierarchical organization of, 409–410
    importing and exporting, 148
    longevity of, 415
    manipulation of, 465–467
    permanent storage of, 26, 30–31
    privacy issues and, 450–451
    protecting, 296, 481
    recovering, 224
Data access area, 222
Database administrator (DBA), **416**, **456**
Database files, 148
Database management system (DBMS), 412, **414–416**, **456**
    advantages of, 414
    components of, 414–416
Databases, **159**, **178**, **408–439**, **456**
    accuracy of information on, 467–468
    components of, 414–416
    data mining and, **424–427**
    data storage hierarchy and, **409–410**
    decision-making process and, 431–439
    e-commerce and, 427, **428–431**
    ethical issues related to, 449–452
    file types and, 411–412, 413
    hierarchical, **417**
    identity theft and, 453
    IT timeline and, 416–420
    key field in, 410–411
    management systems for, 412, 414–416
    multidimensional, 422, **423–424**
    network, **418–419**
    object-oriented, **422**
    privacy issues and, 450–451
    querying, 420–422
    relational, **160**, **419–422**
    review questions/exercises on, 459–461
    security of, 414
Database servers, 319
Database software, **159–163**, **178**
    benefits of, 159–160
    features of, 160–162
    illustrated overview of, 161
    personal information managers, **162**–163
Data cleansing, 424
Data compression utilities, 126
Data dictionary, **415–416**, **455**
Data files, **455**
Data flow diagram (DFD), **501**, **534**
Data integrity, 414
Data mining (DM), **424–427**, **455**
    applications of, 426–427
    process of, 424–426
Data-recovery utility, 125
Data redundancy, 414
Data scrubbing, 424

Data storage hierarchy, **409–410**, **456**
Data transmission speeds, 52–53
Data warehouse, **425**, **456**
Davis, Lynn, 468
DBMS utilities, **416**, **456**
    See also Database management system
DDR-SDRAM chips, 210
Debugging programs, **514**, **534**
Decentralized organizations, 434
Decision making
    computer-based information systems for, 437–438
    make-or-buy decisions, 503
    management levels and, 433, 434
Decision support system (DSS), **437–438**, **456**
Dedicated fax machines, 270
Dedicated ports, 213
Deep knowledge, 441
Default settings, **154–155**, **179**
Defragmenter, 126
Deleting text, 150
Demand reports, 437
Democracy, online, 10–11
Denial-of-service (DoS) attacks, **345**, **359**
Departments, organizational, 432–433
Design templates, 164
Desk-checking programs, **514**, **534**
Desktop, **130**, **179**
DesktopLX, 142
Desktop PCs, **22**, **41**
Desktop publishing (DTP), **168**–170, **179**
Desktop scanners, 266
Desktop search engines, 79–81
Detail design, 502
Detail reports, 437
Developing information systems. *See* Systems development
Device drivers, 119, **124**, **179**
Diagnostic programs, 514
Diagnostic routines, 120
Dial-up connections, 53, 59, **106**
Digital audio players. *See* MP3 digital audio players
Digital cameras, 271, **272–273**, 293, **298**, 379–385
    features of, 381–383
    how they work, 273, 379
    improvement of, 293
    point-and-shoot, 379–380
    resolution of, 381
    sharing photos from, 383
    single-lens reflex, **380**
    societal effects of, 385
    transferring images from, 384–385
    video clips and, 383–384
    warnings about, 272–273, 382
Digital convergence, 310, 369–370
Digital divide, 491
Digital light processing (DLP) technology, 390
Digital radio, 375–376
Digital sampling, 314–315
Digital signals, **311**, **359**
    converting analog signals into, 314–315
    modem conversion of, 313–314
Digital subscriptions, 285
Digital television (DTV), **388–389**, **401**
Digital-to-analog converter, 314
Digital wireless services, **338–339**, **359**
Digital zoom, 381
Digitizers, **264**, **299**
Digitizing tablets, **264**, **299**
Dillard, Mark, 357
DIMMs, 211
Direct implementation, 504
Dirty data problems, 470
Disabled computer users, 292, 295
Disaster-recovery plans, **481–482**, **494**
Disk cleanup utility, 126
Disk controller card, 218
Diskettes, 221–222
Disk scanner utility, 126

Display screens, **278–282**, **299**
    future developments in, 294–295
    health issues related to, 290–291, 297
    laptop computer, 239
    touch screens, 261–262
Distance learning, 5–6, **41**
Distractions, 38
DNA computing, 236
Documentation, 147, **514–515**, **534**
Document files, 148
    creating, 150
    editing, 150, 152–153
    formatting, 153–155
    illustrated overview of, 151
    inserting comments into, 155
    printing, 155
    saving, 155
    toggling between, 155
    tracking changes in, 155
    web, 155
Document Type Definition (DTD), 530
Domains, **67**, 68, 83, **106**
DOS (Disk Operating System), **135**, **179**
Dot-matrix printers, 282
Dot pitch (dp), **278**, 279, **299**
Dotted quad, 63
Downlinking, 332
Downloading, 9, **41**, 53, **106**
Dpi (dots per inch), 266, 279, 282, **299**
Draft quality output, 282
DRAM chips, 210
Drawing programs, 171
Drive bays, 201
Drive gate, 222
Drivers. *See* Device drivers
Drone computers, 474
Drop-down menu, 130
Drum scanners, 266
DSL (digital subscriber line), 55–56, 57, 59, **106**
Dumb mobs, 3
Dumb terminal, **256**, **299**
Dunham, Shannon, 393
DuraPoint Mouse, 259
DVD drives, **31**
DVD filters, 486
DVD-R disks, **227**, **241**
DVD-ROM disks, **227**–228, **241**
DVD zones, 229
Dvorak, John, 486
Dwyer, Eugene, 104
Dynamic IP address, 63

E-911 (Enhanced 911), 335, 336
EBCDIC coding scheme, **197**, 198, **241**
E-business, 427
E-commerce (electronic commerce), 97–98, **106**, 427, **428–431**, **456**
    examples of, 428–429
    systems for, 429–431
Economic issues, 490–492
EDGE wireless standard, 339
Editing documents, 150, 152–153
Education
    distance learning and, **5–6**
    information technology and, 5–6
    plagiarism issues and, 104
    software for, 149
    See also Learning
EIDE controllers, 225
Einstein, David, 25, 33
E-learning, 5–6
Electrical power issues, 202, 203
Electromagnetic fields (EMFs), **290**, **299**
Electromagnetic spectrum, **329**, **359**
Electromechanical problems, 470
Electronic imaging, 265
Electronic Monitoring & Surveillance Survey, 476
Electronic "paper," 294
Electronic spreadsheets. *See* Spreadsheets
Electronic surveillance, 479
Electrostatic plotters, 287

Email (electronic mail), 5, 15–16, **41**, 81–88
  addresses, 83
  attachments, 85–86, 87, 346
  fake, 101
  filters for, 85
  hoax virus warnings, 347
  how to use, 83–85
  instant messaging and, **86**, 88
  junk, 99–100
  mailing lists, 90
  mobile devices for, 82, 83
  netiquette, **91**–92
  privacy issues, 99
  remote access to, 86
  replying to, 83, 84
  sending and receiving, 81–82, 84
  software and services, 81–82
  sorting, 85
  tips on managing, 19
  viruses passed via, 86
  web-based, 82
  wireless, 337, 394
Email program, 81–82, **106**
Embedded systems, 143
Emoticons, 91
Emotion-recognition devices, 293
Employee crime, 475–476
Employee internet management (EIM)
  software, 479
Employee monitoring, 477
Employment
  computer training and, 11–12
  contemporary changes in, 488, 490
  online resources for, 12–13
  technology issues and, 477
  See also Job searches
Emulator board, 218
Encapsulation, 526
Encryption, **353**–354, **359**, 480–481, **494**
Enhanced-relational databases, 422
ENIAC computer, 14, 190–191, 207
Entertainment PCs, **388, 401**
Entertainment software, 149
Environmental problems, 482–484
Environmental Protection Agency (EPA),
  291, 423, 483
Erasable optical disk, 227
Ergonomics, **291**, 296–297, **299**
Error-resisting coding, 355
Errors
  human, 468–469
  procedural, 469–470
  software, 470
E-tailing, 428–429
Ethernet, 217, **324**, 326, **360**
Ethics, 37, **41**
  artificial intelligence and, 449
  information accuracy and, 467–468
  media manipulation and, 465–467
  nanotechnology and, 236
  online file sharing and, 319
  plagiarism and, 104
  privacy and, 99, 450–451
  RFID tags and, 276
Ethics in Modeling (Wallace), 449
EV-DO wireless standard, 339
Even parity, 198
Event-driven program, 528
Evil Twin attack, **474, 494**
Exabyte (EB), **197, 241**
Exception reports, 437
Executable files, 411
Executive information system (EIS), 438–439
Executive support system (ESS), **438**–439, **456**
Expansion, **202**, 217, **241**
Expansion buses, 218
Expansion cards, 217–220, **241**
Expansion slots, **29, 41**, 217, 218, **242**
Expert systems, **439**, 440–441, **456**
Exporting data, **148, 179**
Extensible markup language (XML), **96, 530**
Extension names, 411
External cache, 212

External hard disks, 224
Extranets, **320, 360**
Eye-gaze technology, 294
Eyestrain, 290, 297

Face-recognition systems, 353
Fair and Accurate Credit Transactions Act
  (2003), 477
Fake email, 101
False cause argument, 532
FAQs (frequently asked questions), **91, 106**
Farid, Hany, 467
Fast Ethernet, 324
Fault-tolerant systems, 123
Favorites, 72
Fax machines, **269**–270, **299**
Fax modems, 218, 270
Feathering, 284
Federal Communications Commission (FCC),
  290, 329, 376, 388
Fiber-optic cable, **325**–326, **360**
Fibre Channel controllers, 225
Fields, 160, **409, 457**
Filenames, **411, 457**
File-processing system, 412
Files, **121, 179, 410, 457**
  backing up, 125
  compressing, **412**
  converting, 504–505
  data, **411**–412
  exporting, 148
  graphics, 171
  importing, 148
  managing, 121–122
  program, **411**
  transferring, 373, 384–385, 386
  types of, 148, 411–412, 413
File servers, 318
File sharing, 319
File virus, 347
Film making, 9
Filo, David, 368
Filters
  email, 85
  spam, 100
Finance, online, 8, 97
Financial software, **166**–168, **179**
Find command, 152
Fingerprint scanners, 351
Firewalls, **350**, 351, **360**
FireWire, **216, 242**
First-generation (1G) technology, 337
Fischer, Bernard, 444
Fixed-disk drives, 224
Flaming, **91, 106**
Flash memory, 231–232, 373
Flash memory cards, **232, 242**, 381–382
Flash memory chips, **211, 242**
Flash memory drive, **232, 242**
Flash memory sticks, **232, 242**
Flatbed scanners, **266, 299**
Flat-panel displays, **279**–280, **299**
Floppy-disk cartridges, **222, 242**
Floppy-disk drive, **30, 41**
Floppy disks, 221–222, **242**
Flops, 21, **207, 242**
Flowchart, 507
Foldable PDA keyboards, 257
Folders, 122, 130
Fonts, 154
Food and Drug Administration (FDA), 276
Footers, 154
Foreign keys, 160
Forester, Tom, 37
Formatting
  documents, 153–155
  hard drives, 233
Formulas, **158**, 162, **179**
FORTRAN programming language, 521,
  522–523, **534**
Fourth-generation (4G) technology, 355
Fourth-generation languages (4GLs), 520

Fragmentation, 126
Frame-grabber video card, 271
Frames, **72, 106**
Fraud
  internet, 473–474
  workplace, 477
Freedman, Alan, 527
Freedom of Information Act (1970), 451
Freeware, **146, 179**
Frequency
  modem technology and, 313–314
  radio signals and, 329
Frequency modulation, 315
Fridrich, Jessica, 467
Friedman, Thomas, 491
Frontside bus, 218, 219
FTP (File Transfer Protocol), 88, 89, **106**
Full-duplex transmission, 322
Full-hand palm scanners, 351
Full-motion video card, 271
Function keys, **128, 179**
Functions, **158, 179**
Fuzzy logic, **442**–443, **457**

Gambling, online, 484, 489
Games
  online game players, 257
  smartphone, 395
  videogame systems, 397–399
Garden area networks, 317
Gates, Bill, 137, 467
Gateways, **322, 360**
Generations of programming languages,
  516, 534
Genetic algorithms, 447, 531
Geocaching, 335–336
Geographic information system (GIS), 438
Geographic Project, 408
Geostationary earth orbit (GEO), 332
Geschke, Charles, 168
Gesture recognition, 293
Giardello, Justine Heinze, 493
Gibson, William, 16
GIF animation, 172
Gigabit Ethernet, 324
Gigabits per second (Gbps), **53, 106**
Gigabyte (G, GB), 30, **197, 242**
Gigahertz (GHz), 28, 206, **242**
Gilliam, Harold, 464
Gillmor, Dan, 488
Global Positioning System (GPS), 332, 333,
  334–336, **360**, 387
Gomes, Lee, 92
Google, 79, 94, 408
Government
  domestic spying by, 450
  information technology and, 10–11
  privacy laws and, 451
GPRS wireless standard, 339
GPS. *See* Global Positioning System
Grady, John, 385
Grammar checker, 152, 153
Graphical user interface (GUI), **130**–134, **180**
Graphics
  analytical, 158–159
  presentation, 163–166
Graphics cards, 218, **219, 243**, 281
Graphics files, 171, 411, 413
Greene, Alan, 8
Grid networks, 356
Groupware, 149
GSM wireless digital standard, 338–339
Gunther, Judith Anne, 449
Gutenberg, Johannes, 415

Hackers, 348, **349, 360**
Hacktivists, 349
Half-duplex transmission, 322
Hallowell, Edward, 489
Hamilton, Anita, 357
Hand-geometry systems, 351

Handheld computers, 23
    operating systems for, 143–144
    special keyboards for, 257
    See also Personal digital assistants
Handheld scanners, 266
Handshaking process, 62, 320
Handwriting recognition, 262–263, **299**
Hardcopy output, **277**, **299**
Hard-disk cartridges, 224
Hard-disk controller, 224–225
Hard-disk drive, 30, **41**
Hard disks, **222–225**, **243**
    crashing of, 223–224
    defragmenting, 126
    external, 224
    future developments in, 237
    large computer system, 225
    nonremovable, **224**
    removable, **224**
    starting over with, 233
Hard goods, 97
Hardware, **25**, 27–32, **41**, 189–250
    buying, 194, 195
    cache, **211–212**
    communications, 32
    computer case, 199–201
    converting, 504–505
    CPU, **208–209**
    ergonomics and, **291**, 296–297
    expansion cards, 217–220
    future developments in, 232, 234–237,
        291–295
    health issues related to, 289–291,
        296–297
    input, 27–28, **253**, 254–277
    IT timeline and, 192–201, 258–264
    memory, 28–29, 209–211
    microchips, 192, 193
    miniaturization of, 192
    mobility of, 192, 194
    motherboard, 200, 202, 203
    obtaining for systems, 503
    output, 31–32, **253**, 277–288
    ports, **213–217**
    power supply, **201**
    processing, 28–29, 204–205
    protecting, 296
    review questions/exercises on, 248–250,
        305–307
    secondary storage, 30–31, **220–232**
    system unit, 199, 200
    theft of, 471
    transistors, **191**
Hasty generalization, 532
Hawkins, Jeff, 143
Headaches, 290, 297
Head crash, 223–224
Headers, 154
Head-mounted display, 443
Health issues
    computer use and, 289–291, 296–297
    data mining and, 427
    medical technology and, 6–8
    online information about, 7–8
    storing data about, 416
Help command, **134–135**, **180**
Help programs, 177
Hepatic systems, 292
Hertz (Hz), 279, 329
Hertz, Heinrich Rudolf, 279
Hierarchical databases, **417**, **457**
Hierarchy chart, 507, **508**, 509, **534–535**
High-definition (HD) radio, **377**, **401**
High-definition television (HDTV), **389**,
    390, **401**
High-density disks, 236–237
High-level programming languages, 513,
    519–520, **521**, **535**
High-speed phone lines, 55–56
Hill, Orion E., 99
Hilton, Paris, 472
History list, 71
Hits, **75**, **106**

Hoax virus warnings, 347
Hof, Robert, 36
Home area networks (HANs), **317**, **360**
Home automation networks, **318**, 340,
    343, **360**
Home entertainment centers, 369
Home page, **66**, 70, **106**
    personalizing, 70
    web portal, 74
HomePlug technology, **326**, **360**
HomePNA (HPNA) technology, **326**, **360**
Home theater systems, 294
Hopper, Grace Murray, 523
Host computer, 60, **320**, **361**
Hotspots, **58**, **107**
HTML (Hypertext Markup Language), **68**,
    69, **107**, 116, 155, **529**, **535**
HTTP (HyperText Transfer Protocol), **67**,
    68, **107**
Hubs, **321–322**, **361**
Human-biology input devices, 277
Human-centric computing, 293
Human-computer interaction (HCI), 118
Human errors, 468–469
Human resources department, 433
Hyperlinks, 69, 72, 529
Hypermedia database, 422
Hypertext database, 422
Hypertext index, 74
Hypertext links, **69**, **107**, 529
Hypertext Markup Language (HTML), **68**,
    69, **107**, 116, 155, **529**, **535**
HyperText Transfer Protocol (HTTP), **67**,
    68, **107**
Hyperthreading, 205, 213

IBIS Mobile Identification System, 408
ICANN (Internet Corporation for Assigned
    Names & Numbers), 63–64
Icons, **130**, **180**
Identification systems, 351
Identity theft, **452**, 453–454, **457**
ID Mouse, 259
IF-THEN-ELSE structure, 510, 512
Illustration software, 171
Image-editing software, 171
Image files
    search tools for, 78
    transferring, 384–385
    web-based, 78, 94
Imagery, 39
Impact printers, **282**, **300**
Implementing systems. See Systems
    implementation
Importing data, **148**, **180**
Income gap, 491
Inference engine, 441
Info-mania, 485
Information, **25**, **41**
    accuracy of, 467–468
    completeness of, 467–468
    evaluation of, 78, 79
    flow within organizations, 432–434
    manipulation of, 465–467
    privacy issues and, 450–451
    qualities of good, 431–432
    theft of, 472–473
    See also Data
Information overload, 2, 488
Information systems. See Computer-based
    information systems
Information technology (IT), **4**, **41**
    business and, 8
    contemporary trends in, 34–37
    crime related to, 471–476, 478–479
    economic issues and, 490
    education and, 5–6
    employment resources and, 11–12
    entertainment industry and, 9–10
    ethics and, 37
    finances and, 8
    government and, 10–11

    health/medicine and, 6–8
    job/careers and, 11–13
    leisure and, 9–10
    modern examples of, 5–13
    quality-of-life issues and, 482–489
    security issues and, 468–482
    timeline of progress in, 14–21
    See also Communications technology;
        Computer technology
Infrared wireless transmission, **330**, **361**
Inheritance, 527
Ink-jet printers, **284–285**, **300**
Input, **25**, 26, **41**, 253
Input hardware, 27–28, **253**, 254–277, **300**
    audio-input devices, **270**
    digital cameras, 271, **272–273**
    digitizers, **264**
    disabled users and, 292, 295
    ergonomics and, **291**, 296–297
    future of, 291–294
    health issues related to, 289–291, 296–297
    human-biology input devices, 277
    IT timeline and, 258–264
    keyboards, **27**, **254–257**
    light pens, **263**
    mouse, **28**, **258–261**
    pen-based computer systems, **262–263**
    pointing devices, **258–264**
    radio-frequency identification tags, **276**
    remote locations and, 292
    review questions/exercises on, 305–307
    scanning/reading devices, 265–270
    sensors, **275**
    source data-entry devices, 265–270
    speech-recognition systems, **274–275**
    touch screens, **261–262**
    types of, **254**
    video-input cards, 271
    webcams, **271**
Inserting text, 150
Installation process, 33
Instant messaging (IM), **86**, 88, **107**
Insteon technology, 343
Instruction manuals, 177
Integrated circuits, 34, **191**, **243**
Intelligence, artificial. See Artificial
    intelligence
Intelligence services, 476
Intelligent agents, **442**, **457**
Intelligent sensors, 291
Intelligent terminal, **256**, **300**
Intel-type chips, **204**, **243**
Interactive TV, **388**, **401**
Interactivity, **35**, **41**, 72
Interfaces, 118
Interlacing, 279
Interleaving, 212
Internal cache, 211–212
International Telecommunications Union
    (ITU), 329
Internet, **17**, **41**, 49–116
    addiction to, 489
    attacks on, 475
    bandwidth and, **52**
    blogs on, **96**
    business conducted on, 97–98
    cellphone access to, 394–395
    citing sources from, 78
    discussion groups, 90
    distance learning via, 5–6
    domain abbreviations on, 68
    email and, 81–92
    evaluating information on, 75, 78, 79
    examples of uses for, 51
    fraud on, 473–474
    FTP sites on, 88, 89
    government resources on, 10–11
    historical emergence of, 50–55
    how it works, 60–64
    influence of, 18
    intrusiveness of, 98–103
    ISPs and, 58, 60
    job searching on, 12–13

Internet—*Cont.*
 netiquette, **91**–92
 newsgroups on, 88–89
 physical connections to, 52–58, 59
 plagiarism issues, 104
 programming for, 529–531
 protocol used on, 62
 real-time chat on, **90**
 regulation of, 63–64, 491–492
 review questions/exercises on, 112–116
 telephone calls via, 92–93
 tips for searching, 80
 workplace issues and, 477
 See also World Wide Web
Internet2, **62**, **107**
Internet access providers, 53, **107**
Internet Assistant, 174
Internet backbone, 61, **62**, **107**
Internet Corporation for Assigned Names &
 Numbers (ICANN), 63–64, **107**
Internet Explorer, 64, 65
Internet filtering software, 479
Internet Fraud Complaint Center
 (IFCC), 473
Internet Message Access Protocol (IMAP), 81
Internet Protocol (IP) address, **63**, **107**
Internet Protocol Television (IPTV), **391**,
 401–402
Internet radio, 378
Internet Service Providers (ISPs), **58**, 60, **107**
 comparison shopping for, 58
 questions to ask of, 60
Internet Society (ISOC), 63
Internet telephony, 92–93, **107**
Internet terminal, **256**–257, **300**
Internet Traffic Report, 63
Internet TV, **388**, **401**
Interpreters, **519**–520, **535**
Intranets, **319**, **361**
iPods, 370, 372–375
 how they work, 372–374
 societal effects of, 374
 using in college, 374–375
IPTV (Internet Protocol Television), **391**,
 401–402
IrDA ports, **216**, **243**
Iris-recognition systems, 351
Irrelevant attack, 532
Irrelevant reason, 532
ISDN (Integrated Services Digital Network),
 55, 59, **107**
Isolation, 484
ISPs. *See* Internet Service Providers
Iteration control structure, 512–513

Jacobs, Richard, 176
James, Brian, 426
Java programming language, **94**, **107**, **527**, **535**
JavaScript, **530**–531, **535**
Java Virtual Machines (VMs), 527
Jhai computer, 213
Job searches, 12–13
 posting your résumé online, 13
 websites for, 13, 98
 See also Employment
Joules, 202
Joystick, 254
Jumping to conclusions, 532
Junk email, 99–100
Justification, 154

Kahrs, Hank, 328
Kaku, Michio, 232
Kameny, Paul, 450
Kandell, Jonathan, 489
Kasparov, Garry, 446
Kemeny, John, 524
Kernel, 121
Keyboards, **27**, **42**, **254**–257, **300**
 laptop computer, 239
 layout and features, 127–129

repetitive stress injuries from,
 289–290, 296
 specialty, 256–257
 traditional, 255
Keyboard shortcuts, 128
Key field, **160**, **180**, 410–411
Key loggers, **103**, **108**
Keyword indexes, 75, 76, **108**
Keywords, **74**, **108**
Kilobits per second (Kbps), **52**, **108**
Kilobyte (K, KB), 30, **196**, **243**
Kiosks, 252
Knoll, John, 466
Knowledge base, 441
Knowledge engineers, 441
Krantz, Michael, 530
Kraut, Robert, 484
Krugman, Paul, 491
Kubey, Robert, 489
Kurtz, Thomas, 524

Labels
 mailing, 162
 spreadsheet, 156
Languages, programming. *See* Programming
 languages
Language translators, 119, 199, **518**, **535**
Lanier, Jaron, 415
LANs. *See* Local area networks
Laptop computers, 23
 classroom use of, 493
 durability of, 296
 guarding against theft, 296
 tips on buying, 238–239
 See also Handheld computers
Large-format plotters, 287
LaserCard technology, 231
Laser printers, **283**–284, **300**
Last-mile problem, 56, 325
Latency problem, 280
Lateral thinking tools, 530
Laws
 personal privacy, 451
 software piracy, 478
Lawsuits, 477
Learning
 critical-thinking skills for, 45
 distance, **5**–6
 information technology and, 5–6
 lectures and, 39
 memorization and, 38, 39
 note taking and, 39
 prime study time for, 38
 reading method for, 38–39
 See also Education
Lectures, 39
Legacy systems, 135
Leisure activities, 9–10
Lenat, Douglas, 446
Levens, Bob, 374
Levy, David M., 488
Levy, Steven, 387, 390
Library of Congress website, 409
Life
 analog basis of, 312
 artificial, 448–449
Light pens, **263**, **300**
Lights-out factories, 490
Line graphs, 159
Line-of-sight systems, 255, 330
Links. *See* Hyperlinks
Linspire operating system, 142
Linux, **142**–143, **180**
Liquid crystal display (LCD), **279**, **300**,
 382, 390
LISP programming language, **525**, **535**
Listserv, **90**, **108**
Local area networks (LANs), **22**, **42**, **317**, **361**
 client/server, **318**–319
 components of, 320–322
 operating systems for, 139
 peer-to-peer, 318, **319**

topology of, **322**–324
 wireless, 340–341
Local bus, 218
Location memory, 39
Logical operations, 208
Logic bomb, 347
Logic errors, **514**, **535**
Logic structures, 510
Log-on procedures, **60**, **108**
Lohr, Steve, 391
Long-distance wireless communications,
 333–340
Loop control structure, 512–513
Lord, Robert, 7
Lossless compression, 412
Lossy compression, 412
Lost clusters, 223
Low-earth orbit (LEO), 332
Lucas, George, 9
Lycos, 174

Ma, Ming, 369
Machine cycle, **208**, 209, **243**
Machine interfaces, 118
Machine language, **199**, **243**, **517**, 518, **535**
Macintosh operating system (Mac OS),
 136–137, 143, **180**
Macros, **128**, **180**
Macro virus, 347
Magnetic ink character recognition (MICR),
 268–269, **300**
Magnetic-strip card, 230
Magnetic tape, **229**–230, **243**
Mail, electronic. *See* Email
Mailing labels, 162
Mailing lists, 90
Mail servers, 81, 319
Mainframes, **22**, **42**
Main memory, 210
Maintenance
 data, 414
 program, 515
 system, 505
Make-or-buy decision, 503
Malone, Michael, 190
Management
 CPU, 121
 database, 412, 414–416
 file, 121–122
 memory, 121
 security, 124
 task, 122–123
Management information system (MIS),
 436–437, **457**
Managers
 computer-based information systems for,
 435–439
 levels and responsibilities of, 432,
 433, 434
 participation in systems development
 by, 499
Manes, Stephen, 415
Maney, Kevin, 18, 337
Margins, 154
Marketing
 data mining used in, 427
 organizational department for, 433
Markoff, John, 36
Mark-recognition devices, 268–269
Markup languages, **528**–529, **535**
Mash-ups, 368
Matthies, Brad, 78
Maximizing windows, 134
McCarthy, John, 525
McDermott, Trish, 98
McKibben, Bill, 488
McWilliams, Gary, 368
Media manipulation, 465–467
Medical technology, 6–8
 See also Health issues
Medium band communications, **329**, **361**
Medium-earth orbit (MEO), 332

Meetings, virtual, 357
Megabits per second (Mbps), 53, **108**
Megabyte (M, MB), 30, **196**, **243**
Megahertz (MHz), 28, 206, **243**
Megapixels, **381**, **402**
Memorization, 38, 39, 45
Memory, 209–211
    flash, 211, 373
    main, 210
    managing, 121
    read-only, 211
    virtual, 121, 212
Memory bus, 218
Memory chips, **28**, **42**, 209–211
Memory expansion board, 218
Memory hardware, 28–29
Memory modules, 210–211
Mental-health problems, 484–485
Menu, **130**, **132**, **180**
Menu bar, **132**, **180**
Menu-driven interface, 130
Mesh technologies, 343
Message, 526
Meta-data, 424–425
Metalanguage, 528
Metasearch engines, **76**, **108**
Metcalfe, Bob, 324
Methods, 526
Metropolitan area networks (MANs), **317**, **361**
Microchips. *See* Chips
Microcomputers, **22–23**, **42**
    networking, 22
    types of, 22–23
    See also Personal computers
Microcontrollers, **23**, **42**
Micropayments, **8**, **42**
Microphones, 270
Microprocessors, 20, 190, **192**, 204–205, 206, **243**
Microreplication, 294
Microsoft Internet Assistant, 174
Microsoft Internet Explorer, 64, 65
Microsoft Pocket PC, **144**, **180**
Microsoft Tablet PC, 263
Microsoft Windows, **137**–139, **180**
Microsoft Windows CE, **144**
Microsoft Windows NT, **139**, **180**
Microsoft Windows XP, **137**–138, 139, 143, **181**
Microsoft Xbox 360, 397, 398–399
Microwave radio, **331**–332, **361**
Middle-level managers, 434
MIDI board, **270**, **300**
MIDI ports, **216**, **244**
Midsize computers, 22
Millennials, 400
Mindsets, 532
Miniaturization, 34
Minicomputers, 22
Minimizing windows, 134
MIPS, **207**, **244**
Mobile blogs, 96, 395
Mobile data terminal (MDT), 256
Mobile phones. *See* Cellular telephones
Mobile-telephone switching office (MTSO), 337
Mobility, 190, 192, 194
Modeling tools, **501**, **536**
Models, 437–438
Modem cards, 220
Modems, **32**, **42**, **53**, **108**, **314**, **361**
    cable, **56**–57
    dial-up, 53–54
    fax, 270
    signal conversion by, 313–314
Modules, **508**, **536**
Molecular electronics, 237
Molitor, Graham, 50
Money
    information technology and, 8
    software for managing, 166–168
Monitors, 31, 42, 278–282
    See also Display screens

Monophonic ringtones, 393
Moore, Gordon, 204, 234
Moore's law, 204, 234
More Info! icons, 4
Morphing, 465, 467
Morrison, Perry, 37
Mossberg, Walter, 205, 238, 382, 383
Motherboard, **29**, **42**, 200, 202, 203
Motorola-type chips, **205**, **244**
Mouse, **28**, **42**, 128, **258**–261, **301**
    functions performed by, 129
    pros and cons of, 260
    setting properties for, 259
    specialty versions of, 259–260
    variant forms of, 260–261
    See also Pointing devices
Mouse pointer, 128, 258–259
Movie industry, 9–10, 472
MP3 digital audio players, **372**–375, **402**
    how they work, 372–374
    societal effects of, 374
    using in college, 374–375
MP3 format, **372**, **402**
M-RAM, 234
Multicasting, 389
Multicore processors, **205**, **244**
Multidimensional databases (MDAs), 422, **423**–424, **457**
Multidimensional spreadsheets, 158
Multifunction printers, **287**, **301**
Multimedia, 18, **35**, **42**
    databases for storing, 422
    search engines for, 77–79
    software for creating, 173
    World Wide Web and, 64, 77–79, 93–95
Multimedia authoring software, **173**, **181**
Multimedia computers, 194
Multimedia ports, 217
Multipartite virus, 347
Multiple-user license, 146
Multiprocessing, 123
Multiprogramming, 123
Multitasking, 2, **122**–123, **181**, **371**–372, 400, **402**
Music
    file sharing, 319
    MP3 players and, 372–375
    online sources of, 9, 95
    smartphones and, 395
    theft of, 471–472
Music tones, 393
My Computer icon, 132, 133
My Documents icon, 132, 133

Nagle, Matt, 295
Name migration, 450
Nanotechnology, **22**, **42**, 235, 236, 483–484
Napster, 319
Narrowband connections, **53**, **108**, **329**, **361**
National ID cards, 451
National Science Foundation Network (NSFNET), 317
National Telecommunications and Information Administration (NTIA), 329
Natural hazards, 470–471
Natural language processing, **442**, **457**
Natural languages, **522**, **536**
NBIC technology, 236
Near-letter-quality (NLQ) output, 282
Negroponte, Nicholas, 207
Neighborhood Link, 11
NetGeners, 399, 400
Netiquette, **91**–92, **108**
Netscape Navigator, 64, 65
Netware, **139**, **181**
Network access point (NAP), **61**, **108**
Network agents, 442
Network computers, 257
Network databases, **418**–419, **457**
Network interface cards (NICs), **220**, **244**, 322
Network operating system (NOS), 322

Networks, **5**, **43**, **315**–325, **361**
    benefits of, 316
    client/server, 24, **318**–319
    components of, 320–322
    Ethernet, **324**
    extranets, **320**
    firewalls for, **350**
    home area, **317**
    home automation, **318**, 340, 343
    intranets, **319**
    local area, **317**, 318–319, 340–341
    metropolitan area, **317**
    operating systems for, 139–143
    peer-to-peer, 318, **319**
    personal area, **317**, 340, 341–343
    topologies of, **322**–324
    types of, 316–318
    virtual private, **320**
    wide area, **316**–317
    wireless, **58**
Network service provider (NSP), 61
Neural networks, 447
Neuromancer (Gibson), 16
Newsgroups, **88**–89, **108**
Newsreader program, **89**, **108**
Nie, Norman H., 484
Nigerian letter scam, 473
Nintendo Revolution, 398, 399
Node, **320**, **361**
Noise-cancellation technology, 274
Noise-induced hearing loss, 374
Noise pollution, 291
Nonimpact printers, **283**, **301**
Nonremovable hard disks, **224**, **244**
Normal accidents, 470
Norton, Peter, 125
Notebook computers, **23**, **43**
    See also Laptop computers
Note taking
    tips for students, 39
    web-based services, 493

Object, **526**, **536**
Object code, 519
Object-oriented databases, **422**, **458**
Object-oriented programming (OOP), **526**–527, **536**
Object-relational databases, 422
Oblinger, Diana, 400
Odd parity, 198
OEM operating systems, 139
Office information systems (OISs), **435**, **458**
Office suite, 148–149
Offline processing, 436
OLAP software, 424
One-hand PDA keyboards, 257
One-way communications, 333–337
Online, **5**, **43**
Online analytical processing (OLAP) software, 424
Online colleges, 5
Online communities, 98
Online connections, 59
Online game player, 257
Online matchmaking, 98
Online retailers, 428–429
Online services. *See* Internet Service Providers
Online snooping, 450
Online storage, 232
Online transaction processing (OLTP), 436
Open architecture, 217
Open Source Development Labs (OSDL), 142
Open-source software, **142**, 143, **181**
Operating systems (OS), 32, **119**–124, 135–144, **181**
    comparison of, 143
    DOS, **135**
    functions of, 119–124
    handheld computer, 143–144
    Linux, **142**–143
    Macintosh, **136**–137, 143

Operating systems (OS)—*Cont.*
  Netware, **139**
  network, 139–143
  Palm OS, **143**–144
  Pocket PC, **144**
  researching, 144
  Symbian, 144
  Unix, **141**
  Windows, **137**–141, 143, 144
  See also System software
Operational-level management, 434
Operator documentation, 515
Operators, Boolean, 80
Optical cards, **231**, **244**
Optical character recognition (OCR), **269**, **301**
Optical computing, 235–236
Optical disks, **225**–229, **244**
Optical mark recognition (OMR), **269**, **301**
Optical mice, 258
Optical scanners, 265–266
Optical viewfinders, 382
Optical zoom, 381
Organic TFT, 281
Organization chart, 432, 434
Organizations
  decentralized, 434
  departments in, 432–433
  flow of information within, 432–434
  management levels in, 432, 433, 434
  review questions/exercises on, 459–461
Organized crime, 476
OUM (ovonic unified memory), 234
Outline View feature, 150
Output, **26**, **43**, 253, 277
Output hardware, 31–32, **253**, 277–288, **301**
  display screens, 278–282
  ergonomics and, **291**, 296–297
  future of, 294–295
  health issues related to, 289–291,
    296–297
  IT timeline and, 258–264
  printers, **282**–287
  remote locations and, 294
  review questions/exercises on, 305–307
  sound-output devices, **288**
  types of, 277–278
  video, **288**
  voice-output devices, **288**
Overlearning, 38

Packaged software, 145
Packets, **62**–63, **108**, **320**, **362**
Page description language (PDL), **283**, **301**
Page printers, 283
Pagers, **336**–337, **362**
Painting programs, 171
Palm OS, **143**–144, **181**
Palmtops, 23
Paradox of Choice: Why More Is Less, The
  (Schwartz), 371
Parallel implementation, 504
Parallel ports, **213**, **244**
Parallel processing, 123
Parent directory, 122
Parisi, Tony, 529
Parity bit, **198**, 199, **244**
Parson, Jeffrey Lee, 349
Pascal, Blaise, 524
Pascal programming language, 521,
  524–525, **536**
Passive-matrix display, 239, **281**, **301**
Passwords, **60**, **108**
  access security and, 124, 479
  online safety and, 350–351
  tips for managing, 352–353
Paste command, 152
Pathnames, 122
Pattern recognition, 293, **442**, **458**
Payne, David, 356
PC cards, **220**, **244**
PCI (peripheral component interconnect)
  bus, **218**, **244**

PCs. *See* Personal computers
PC/TV terminal, 257
PC video cameras, 357
PDAs. *See* Personal digital assistants
PDF files, 413
Peer-to-peer (P2P) networks, 318, **319**, **362**
Pen-based computer systems, 262–263, **301**
Pen plotters, 287
Periodic reports, 437
Peripheral devices, **31**, **43**
Perkins, Tony, 400
Perl (Practical Extraction and Report
  Language), **531**, **536**
Perpendicular recording technology, 237
Perrow, Charles, 470
Personal area networks (PANs), **317**, 340,
  341–343, **362**
Personal computers (PCs)
  buying, 194, 195
  custom-built, 25, 27, 33–34
  environmental issues related to, 482–484
  guarding against theft, 296
  health issues related to, 289–291,
    296–297
  problem chemicals in, 483
  recycling, 37, 483
  tune-ups for, 136
  types of, 22–23
  See also Computers; Laptop computers
Personal digital assistants (PDAs), **23**, **43**,
  386–387, **402**
  future of, 387
  how they work, 386
  special keyboards for, 257
  web resources, 386
Personal-finance managers, **167**, **181**
Personal information managers (PIMs),
  162–163, **181**
Personalization, 36, 371–372
Personalized TV, **388**, **402**
Personal software, 149
Personal technology, 367–406
  convergence and, **369**–370
  digital cameras, 379–385
  high-tech radio, 375–379
  MP3 digital audio players, **372**–375
  multitasking and, **371**, 400
  Netgeners and, **399**, 400
  new television, 387–391
  personal digital assistants, **386**–387
  personalization and, 371–372
  portability and, 370
  review questions/exercises on, 404–406
  smartphones, 391–397
  tablet PCs, 387
  videogame systems, 397–399
Personal video recorders, 390
Pervasive computing, 2, 293
Pesce, Mark, 529
Petabyte (P, PB), 30, **197**
Pharming, **101**, **109**, **245**, 473
Phased implementation, 504
Philips, Michelle, 313
Phishing, **101**, **109**, 473
Phonemes, 274
Photography
  camera phones and, 273, 395
  digital cameras and, 271, **272**–273, 293,
    298, 379–385
Photolithography, 193
Photo manipulation, 465–467
Photonics, 355–356
Photo printers, 286–287
Photo-sharing services, 383
Physical connections, 52–58, 59
  broadband, 52
  cable modems, 56–57
  dial-up modems, 53–54
  high-speed phone lines, 55–56
  wireless systems, 57–58
Pie charts, 159
Pilot implementation, 504–505
PIN (personal identification number), 479

Pipelining, 212
Pirated software, **147**, **181**, 471
Pixels, **278**, **301**
Plagiarism, 104
PlanetFeedback, 176
Plasma TVs, 390
Platform, **135**, **181**
PlayStation3, 398, 399
Plotters, **287**, **301**
Plug-ins, **93**, **109**
Plug and play, **215**, **245**
Podcasting, **96**–97, **109**, **378**–379, **402**
Point-and-shoot cameras, 379–380, **402**
Pointer, **128**, **182**
Pointing devices, 258–264, **301**
  disabled users and, 292
  laptop computer, 239
  mouse and its variants, **258**–261
  pen input devices, 262–264
  pros and cons of, 260
  touch screens, **261**–262
Pointing stick, **261**, **302**
Point-of-sale (POS) terminal, 256
Point of presence (POP), **61**, **109**
Polygons, 295
Polymer memory, 237
Polymorphic virus, 347
Polymorphism, 527
Polyphonic ringtones, 393
Pop-up ads, **102**, **109**
Pop-up menu, 132
Pornography, 485
Portability, 36, 190, 370
Portable computers. *See* Handheld computers;
  Laptop computers
Ports, 29, **213**–217, **245**
Post Office Protocol version 3 (POP3), 81
POTS connection, **55**, **109**
PowerPC chips, **205**, **245**
Power supply, **201**, **245**
Predictive-statistical-model software,
  478–479
Preliminary design, 502
Preliminary investigation, **500**, **536**
Prensky, Marc, 400
Presentation graphics software, 5, 163,
  164–166, **182**
  illustrated overview of, 165
  templates, 164, 166
  views, 165, 166
Previewing documents, 155
Primary key, 160
Primary storage, **25**–26, 28, **43**
Prime study time, 38
Printers, **32**, **43**, 282–287, **302**
  buying considerations, 286
  dot-matrix, 282
  ink-jet, **284**–285
  laser, **283**–284
  multifunction, **287**
  photo, 286–287
  plotter, **287**
  specialty, 287
  thermal, **285**–286
  three-dimensional, 295
Printing documents, 155
Print Scrn key, 256
Print servers, 319
Privacy, 450
  databases and, 450–451
  email messages and, 99
  federal laws on, 451
  government spying and, 450
  information on guarding, 450
Private-key encryption, 353, 354
Private/Peer NAPs (PNAPs), 61–62
Probe storage, 237
Problem clarification, **506**–507, **536**
Problem-oriented programming
  languages, **520**
Procedural errors, 469–470
Procedural programming languages,
  519–520, 521

Processing, **25**, 26, **43**
    future developments in, 235–236
    hardware components for, 28–29
    IT timeline and, 192–201
    speed of, 35, 206–207, 212–213
Processor chips, **28**, **43**, 192, 204–205, 206
Production department, 433
Productivity issues, 477
Productivity software, **148–149**, **182**
Program design, **507–513**, **536**
Program documentation, 515
Program files, **411**, **458**
Program flowchart, **510**, 511, **536**
Programming, **505–515**, **537**
    internet, 529–531
    languages used for, **513**, 515–525
    object-oriented, **526–527**
    review questions/exercises on, 539–540
    steps in process of, 506–515
    structured, **508–513**
    visual, **528**
Programming languages, **513**, **537**
    assembly language, **517–518**
    contemporary examples of, 522–525
    generations of, **516**
    high-level or procedural, **519–520**
    history and evolution of, 522–523
    machine language, 199, **517**
    markup, **528–529**
    natural, **522**
    scripting, **529**, 530–531
    very-high-level or problem-oriented, **520**
    See also names of specific languages
Programming process, 506–515
    clarifying programming needs, 506–507
    coding the program, 513
    designing the program, 507–513
    documenting the program, 514–515
    maintaining the program, 515
    summary of steps in, 516
    testing the program, 514
Programs, 25, 118, **505**, **536**
    creating, 505–515
    utility, 119, **124–127**
    See also Software
Program testing, **514**, **537**
Project management software, **174–175**, **182**
PROM (programmable read-only memory), 211
Proprietary software, 136, 145
Protocols, **62**, **109**, 320–321, **362**
Prototypes, **502**, **537**
Prototyping, **502**, **537**
Pseudocode, 507, **508**, 509, **537**
Public-domain software, **146**, **182**
Public-key encryption, 353–354
Pull-down menu, 130
Pull technology, 95
Pull-up menu, 132
Push technology, **95**, **109**

Quality-of-life issues, 482–489
    environmental problems, 482–484
    internet addiction/dependence, 489
    mental-health problems, 484–485
    protection of children, 485–487
    workplace problems, 487–488
Quantum computing, 236
Queries, database, 160–161, 420–422
Query by example (QBE), **420–421**, **458**
Query language, 520
Questionable statistics, 532
Queues, 121
Quittner, Joshua, 428
QXGA (quantum extended graphics array), **282**, **302**

Radio
    broadcast, **330–331**
    high-definition, **377**
    Internet, 378
    microwave, **331–332**
    podcasts, 378–379

    satellite, **375–376**
    smartphone, 395
    software-defined, 356
Radio buttons, **72**, **109**
Radio-frequency (RF) spectrum, **329**, 330–331, **362**
Radio-frequency identification (RFID) tags, 268, **276**, **302**
Raibert, Marc, 445
RAID storage system, **225**, **245**
RAM (random access memory), 28, 210–211
    See also Memory
Rambo, Kathryn, 453
RAM chips, **210**–211, **245**
Range, spreadsheet, **156**, **182**
Rapid application development (RAD) system, 528
Raster images, 171
Ratey, John, 489
Reading data, **211**, 221, **245**
Reading skills, 38–39
Read-only memory (ROM), 120, **211**
    CD-ROM disks, 226
Read/write head, **222**, **245**
Real-time chat (RTC), **90**, **109**
Real-time processing, 436
Recalculation, **158**, **182**
Records, 160, **410**, **458**
Recycling computers, 37, 483
Refresh rate, **279**, **302**
Regional Internet Registries (RIRs), 64
Registers, **209**, **245**
Regression analysis, 426
Relational databases, **160**, **182**, 419–422, **458**
Relationships, online, 98
Releases, software, 146
Remote locations
    input from, 292
    output in, 294
Removable hard disks, **224**, **245**
Rentalware, **147**, **182**
Repetitive stress injuries (RSIs), **289–290**, 296, **302**
Replace command, 152
Report generator, **416**, **458**, 520
Research, web-based, 104
Research and development (R&D) department, 433
Resolution, **266**, **278–279**, **302**, 381
Résumés
    misleading information on, 477
    posting on the internet, 13
    privacy issues and, 450
Review questions/exercises, 45
    on application software, 185–188
    on artificial intelligence, 459–461
    on communications technology, 364–366
    on computer-based information systems, 459–460
    on databases, 459–461
    on hardware, 248–250, 305–307
    on information technology, 45–48
    on the internet, 112–116
    on organizations, 459–461
    on personal technology, 404–406
    on programming, 539–540
    on security, 495–496
    on systems analysis and design, 539–540
    on system software, 185–188
Revolutions per minute (rpm), 224
Ribbon cable, 30
Richard, Beverly, 484
Richardson, Doug, 336
Rifkin, Jeremy, 490
Ring network, **323**, **362**
Ringtones, **393**, **402**
Ripping process, 373
Ritchie, Dennis, 525
Ritz, David, 379
Roach, Stephen, 488
Robbins, Alan, 118
Robinson, Phillip, 203, 294
Robotics, **443–446**, **458**

Robots, 7, **43**, 443–446
Rollover feature, **130**, **182**
ROM (read-only memory), 120, **211**, **245**
ROM chips, 211
Root directory, 122
Root record, 417
Rothenberg, David, 104
Routers, **322**, **362**
RSS newsreaders, **96**, **109**
Ruggiero, Kimberly, 252
Rukeyser, William, 104
Rule-based-detection software, 478
Runtime libraries, 411

Saffo, Paul, 8, 488
Safire, William, 465
Salford, Lief, 290
Sampling process, 314–315
Sampling rate, **373**, **402**
Sanger, Larry, 78
Satellite radio, **375–376**, **402**
Satellites, communications, **57**, 59, **332–333**, 355
Saving files, **155**, **182–183**
Scalable coding, 355
Scanners, **265–266**, **302**
Scanning and reading devices, 265–270
    bar-code readers, **267–268**
    character-recognition devices, 268–269
    fax machines, 269–270
    mark-recognition devices, 268–269
    optical scanners, 265–266
Scatter charts, 159
Scholarly works, 79
Schwartz, Barry, 371
Science, data mining used in, 427
Screens. *See* Display screens
Scripting languages, 529, 530–531
Script kiddies, 349
Scripts, **529**, **537**
Scroll arrows, **72**, **109**
Scrolling, **72**, **109**, 150, **183**
SCSI controllers, 225
SCSI ports, **213–214**, **245**
SDLC. *See* Systems development life cycle
SDRAM chips, 210
Search (text) box, **72**, 74, **109**
Search command, 152
Search engines, **75**, **109**
    blog, 96
    desktop, 79–81
    multimedia, 77–79
    web, 75, 76, 77–79
Search hijackers, **102**, **110**
Searching the web, 74–79
    multimedia content and, 77–79
    serious techniques for, 80
    strategies for, 76–77
    tools for, 75–76
Search services, **75**, **110**
Secondary storage, 26, **43**, 220–232
    future developments in, 236–237
    longevity of data and, 415
    online, 232
Secondary storage hardware, 30–31, **220–232**, **246**
    flash memory cards, **232**
    floppy disks, **221**–222
    hard disks, **222–225**
    magnetic tape, **229–230**
    optical disks, **225–229**
    smart cards, **230–231**
    Zip disks, **222**
Second-generation (2G) technology, 338–339
Sectors, **222**, **246**
Security, 344–354, 468–482, **478**, **494**
    access rights and, 479–480
    biometric, **277**, **351**, 353, **480**
    broadband connections and, 56
    components of, 478
    computer crime and, **471–476**, 478–479
    database, 414

Security—*Cont.*
 denial-of-service attacks and, **345**
 disaster-recovery plans and, 481–482
 employee monitoring and, 477
 encryption and, **353**–354, 480–481
 errors/accidents and, 468–470
 free online check of, 346
 general procedures for, 481
 hackers/crackers and, 348–350
 identification systems and, 479–480
 Macintosh OS and, 140
 natural hazards and, 470–471
 OS management of, 124
 password, 350–351, 352–353
 review questions/exercises on, 495–496
 service packs and, 140
 software for, 350–354
 terrorism and, 349, 476
 Trojan horses and, **346**
 types of threats to, 344, 468, 469
 viruses and, **346**
 websites and, 347, 478
 WiFi technology and, 341
 worms and, **345**–346
Selection control structure, 510, 512
Self-scanning checkout, 267, 268
Semiconductors, **192**, **246**
Sensors, **275**, 291–292, 293, **302**
Sequence control structure, 510, 512
Serial ports, **213**, **246**
Servers, **24**, **43**, **60**, **110**
Service packs, 138, 140
Service programs, 124
Services, theft of, 472
Set-top box, 257
Sexual predators, 486–487
SGML (Standard Generalized Markup
 Language), 528–529
Shareware, **146**, **183**
Sheet-fed scanners, 266
Sheet music, 9
Shells, 141
Shop bots, 429
Short-range wireless communications,
 340–344
Silicon, 34, **191**, **246**
SIMMs, 210–211
Simpson, O. J., 465–466
Simulators, **443**, 444, **458**
Sinatra, Frank, 465
Single-lens reflex (SLR) cameras, **380**, **403**
Single-user license, 146
Sirius Satellite Radio, 376
Site license, 145
Slide shows, 164
Slippery slope, 532
Smart cards, 230–231, **246**
Smart Display, 280
Smart houses, 2
Smart mobile devices, 2
Smart mobs, 2–3
Smartphones, 14–15, **391**–397, **403**
 how they work, 392
 services and features, 392–396
 societal effects of, 397
 See also Cellular telephones
Smart tags, 268
Smart TVs, **388**, **403**
SMS (Short Message Service), **393**, **403**
SMTP server, 81, 82
Snopes.com website, 467
Social Security number (SSN), 411, 454
Softcopy output, **277**, **302**
Soft goods, 97
Software, **25**, 32–33, **43**, 118
 abandonware, 147
 antivirus, 126, **350**
 blocking, 485
 commercial, 145–146
 converting, 504–505
 creating, 505–515
 custom, 147
 documentation, 147

 errors in, 470
 firewall, **350**
 freeware, **146**
 IT timeline and, 162–169
 make-or-buy decision for, 503
 open-source, **142**
 origins of term, 118
 pirated, **147**, 471
 productivity, **148**–149
 protecting, 296, 481
 public-domain, **146**
 rentalware, **147**
 security, 350–354
 shareware, **146**
 tech support, 176–177
 theft of, 471
 tutorials, 147
 utility, 124–127
 versions/releases, 146
 See also Application software; System
 software
Software-defined radio, 356
Software engineering, 505–506
 See also Programming
Software license, **145**–146, **183**
Software Publishers Association (SPA), 478
Solaris, 141
Solid-state devices, **191**, **246**
Sony PlayStation3, 398, 399
Sorting
 database records, 161
 email messages, 85
Sound board, 218, **270**, **303**
Sound card, **31**, **43**, 219, **246**, 288
Sound manipulation, 465
Sound-output devices, **288**, **303**
Source code, 519
Source data-entry devices, **265**–270,
 292–294, **303**
Source program files, 411
Spacing, text, 154
Spaghetti code, 510
Spam, **99**–100, **110**
Speakers, **32**, **44**
Spear-phishing, 101
Specialized search engines, 76, 77
Special-purpose keys, **127**, **183**
Specialty printers, 287
Specialty software, 149, 163–175
Specific Absorption Rate (SAR), 290–291
Speech-recognition systems, **274**–275,
 292–293, **303**, 442
Speed
 data transmission, 52–53
 processing, 35, 206–207, 212–213
Spelling checker, 152
Spiders, **75**, **110**
Spoofing, **101**, **110**
Spooling, 121
Sports, data mining used in, 426
Spotlight search engine, 136–137
Spreadsheets, **156**–159, **183**
 creating charts from, 158–159
 features of, 156–158
 illustrated overview of, 157
Spyware, **102**–103, **110**
SQ3R reading method, 38–39
SRAM chips, 210
Standard-definition television (SDTV),
 389, **403**
Star network, **324**, **362**
Start button, 132, 133
Static IP address, 63
Statistics, questionable, 532
Stealth virus, 347
Stern, Howard, 376
Stewart, Martha, 465
Stiffy disks, 221
Storage, 25–26
 digital camera, 381–382
 future of, 236–237
 IT timeline and, 192–201
 online, 232

PDA, 386
 primary, **25**–26, 28
 secondary, **26**, 30–31, 220–232
 smartphone, 392
 volatile, **210**
Strategic-level management, 434
Straw man argument, 532
Streaming audio, **95**, **110**
Streaming video, **94**–95, **110**
Stress, 484–485
Strong AI, 446, 447
Stroustrup, Bjarne, 527
Structured programming, **508**–513, **537**
Structured query language (SQL), **420**, **458**
Students
 classroom computer use by, 493
 distance learning for, **5**–6
 infotech challenges for, 464
 internet addiction among, 489
 MP3 players used by, 374–375
 online evaluations by, 493
 plagiarism issues for, 104
 study and learning tips for, 38–39
Study skills, 38–39
Stulman, Nate, 489
Stylus, 262
Subdirectories, 122
Subject directories, **75**, 76, **110**
Subject guide, 74
Subprograms/subroutines, 508
Subscription services, 95
Suitor eye-gaze technology, 294
Summary reports, 437
Supercomputers, **21**–22, **44**
Superscalar architecture, 213
Supervisor, **121**, **183**
Supervisory managers, 434
Surfing the web, **64**, **110**
Surge protector, **201**, 203
SVGA (super video graphics array), **282**, **303**
Switch, **322**, **362**
Switched-network telecommunications
 model, **371**, **403**
SXGA (super extended graphics array),
 282, **303**
Symbian operating system, 144
Syntax, **513**, **537**
Syntax errors, **514**, **537**
System, **498**, **537**
System board. *See* Motherboard
System cabinet, 28, 201
System clock, **206**, **246**
Systems analysis, **500**–502, **537**
Systems analysis and design, **499**, **537**
 participants in, 499
 purpose of, 498
 review questions/exercises on, 539–540
 six phases of, 499–505
 See also Systems development life cycle
Systems analyst, **499**, **537**–538
Systems design, **502**–503, **538**
Systems development, **503**–504, **538**
Systems development life cycle (SDLC),
 499–505, **538**
 analyzing the system, 500–502
 designing the system, 502–503
 developing/acquiring the system,
 503–504
 implementing the system, 504–505
 maintaining the system, 505
 overview of phases in, 499
 participants in, 499
 preliminary investigation, 500
Systems implementation, **504**–505, **538**
Systems maintenance, **505**, **538**
System software, **32**, **44**, **118**–144, **183**
 components of, 119
 device drivers, 119, **124**
 Help command, **134**–135
 IT timeline and, 162–169
 language translators, 119, 199
 operating systems, **119**–124, 135–144
 review questions/exercises on, 185–188

service packs, 138, 140
tech support for, 176–177
user interface, **127**–135
utility programs, 119, **124**–127
System testing, 504
System unit, **28**, 194–220

T1 line, **56**, 59, **110**
Tablet PCs, 263, 387
Tactical-level management, 434
Tags, **81**, **110**
Tape cartridges, **229**, **246**
Taskbar, **132**, 134, **183**
Task management, 122–123
Taxonomy of Educational Objectives (Bloom), 45
Tax software programs, 168
Technical staff, 499
Techno-blight, 483, 484
Tech support, 176–177
Telecommunications, 5
    historical emergence of, 50–55
    switched-network model of, **371**
    tree-and-branch model of, **371**
    See also Communications technology
Telecommuting, 327
Teleconferencing, 357
    See also Videoconferencing
Telemedicine, **6**, **44**
Telephones
    cellular, 14–15, 337–339, 355
    internet telephony and, 92–93, **107**
    modems connected to, 53–54, 313–314
    smartphones, 14–15
Television, 387–391
    closed-circuit, 357
    contemporary technologies for, 388
    manipulation of content on, 467
    smartphones and, 395–396
    societal effects of new, 390–391
    three types of, 388–389
Telework, 327–328
Temp file removal, 130
Templates, **153**, **183**
    document, 153
    presentation graphics, 164, 166
    worksheet, 158
Terabyte (T, TB), 30, **197**, **246**
Terminals, **22**, **44**, 256
Term papers, 104
Terrorists, 476
Testing
    information systems, 503–504
    internet connection speed, 57
    program code, 514
Text messaging, **392**–393, **403**
Text-to-speech (TTS) systems, 288
TFT display, 281
Theft
    hardware, 296, 471
    identity, **452**, 453–454
    information, 472–473
    music and movie, 471–472
    software, 471
    time and services, 472
Thermal printers, **285**–286, **303**
Thermal wax-transfer printers, 286
Thesaurus, 153
Thinking skills, 45
Third-generation (3G) technology, 339
Thomas, Christie, 327
3-D display systems, 295
3-D printers, 295
3G technology, **58**, 59, **110**
Thrill-seeker hackers, 349
Tiger operating system, 136–137
Time, theft of, 472
Time-sharing, 123
Time slicing, 123
Title bar, **132**, **183**
TiVo systems, 390
Token Ring technology, 324–325

Tomlinson, Ray, 85
Toolbar, **132**, **183**
Top-down program design, **508**, **538**
Top-level domain, 67, 83
Top managers, 434
Topology of networks, **322**–324, **362**
Torvalds, Linus, 142
Touchpad, **261**, **303**
Touch screens, **261**–262, 292, **303**
Tower PCs, **23**, **44**, 201
Tracer Mouse, 260
Trackball, **260**, **303**
Tracking document changes, 155
Tracks, **221**–222, **246**
Trading agents, 429
Transaction processing system (TPS), **435**–436, **458**
Transceivers, 330
Transcoding, 355
Transferring files
    digital cameras, 384–385
    MP3 players, 373
    PDA devices, 386
Transistors, 34, **191**, 204, **246**
Transmission Control Protocol/Internet Protocol (TCP/IP), **62**, **110** 111
Tree-and-branch telecommunications model, **371**, **403**
Trojan horse, **346**, 347, **362**, 473
Tucker, Lew, 344
Tukey, John, 118
Turing, Alan, 448, 449
Turing test, 448–449
Tutorials, 147
TWAIN technology, 266
Twisted-pair wire, **325**, **363**
Two-way communications, 333, 337–344
Two-way pagers, 337

UltraCard technology, 230
Ultra wideband (UWB) technology, **342**–343, **363**
UMTS wireless standard, 339
Undo command, 152
Unicode, **197**, **247**
Uninstall utility, 134
Unit testing, 503
Unix, **141**, **183**, 525
Upgrading, **202**, 217, **247**
Uplinking, 332
Uploading, **53**, **111**
UPS (uninterruptible power supply), 202, 203
URL (Uniform Resource Locator), **66**–67, **111**
USA PATRIOT Act (2001), 450
USB hubs, 215, 216
USB OTG, 216
USB ports, **214**–216, **247**
Usenet, **89**, **111**
User ID, 83
User interface, **127**–135, **183**–184
    expert system, 441
    GUI features, 130–134
    Help command, 134–135
    keyboard, 127–128
    mouse, 128, 129
User name, 83
Users
    participation in systems development by, 499
    preparing documentation for, 515
    training on new systems, 505
Utility programs, 119, **124**–127, **184**
    DBMS utilities, **416**
    open-source utilities, 143
UXGA (ultra extended graphics array), **282**, **303**

Vacuum tubes, 190
Values, spreadsheet, **158**, **184**
Vandalism, computer, 474–475
Varkey, Dax, 400

V-chip technology, 486
Vector images, 171
Versions, software, 146
Vertical portals, 73
Very-high-level programming languages, **520**, **538**
Video, **288**, **304**
    analog to digital, 272
    digital cameras and, 383–384
    editing software for, 171
    future developments in, 295
    manipulation of, 467
    search tools for, 78, 79
    streaming, **94**–95
    web-based, 78, 79, 94–95
Video blogs, 92, 96
Video cards, **31**, **44**, 271, 272, 281
Videoconferencing, **288**, **304**, 357
Video display terminals (VDTs), 256
    See also Display screens
Video on demand (VOD), **390**–391, **403**
Video files, 412, 413
Videogame ratings, 486
Videogame systems, 397–399
    Microsoft Xbox 360, 397, 398–399
    Nintendo Revolution, 398, 399
    Sony PlayStation3, 398, 399
Videophones, 357
Viewable image size (vis), 278
Virilio, Paul, 470
Virtual, meaning of, **8**, **44**
Virtual File Allocation Table (VFAT), 223
Virtual Keyboard, 255
Virtual Linux Service, 235
Virtual meetings, 357
Virtual memory, 121, **212**, **247**
Virtual office, 327–328
Virtual private networks (VPNs), **320**, **363**
Virtual reality (VR), **443**, 444, **458**
Virtual Reality Modeling Language (VRML), **529**–530, **538**
Viruses, 126, **346**, **363**
    as attached files, 86, 346–347
    free online check for, 346
    how they're spread, 346–348
    software for protecting against, 126, 346, 348, 350
    tips on preventing, 127, 348
    types of, 347
Visual BASIC, 528
Visual programming, **528**, **538**
Vocal references, 274
Voiceband communications, 329
Voice mouse, 258
Voice-output devices, **288**, **304**
Voice-recognition systems, 353
VoIP phoning, 92–93, 394–395
Volatile storage, **210**, **247**
Voltage regulator, 202, 203
Von Neumann, John, 20
Vortals, 73
Voting technology, 10
VRML (Virtual Reality Modeling Language), **529**–530, **538**

Wafers, 193
Wall, Larry, 531
Wallace, William A., 449
Warm boot, 121
Warnock, John, 168
Warwick, Ken, 6
Watermark, 152
Watson, Ken, 345
Wavetable synthesis, 219
Weak AI, 446
Weatherbee, Danielle, 289
Weather meters, 387
Web-based mail, **82**, **111**
Web browsers, **64**, **111**
    development of, 67
    features of, 69, 70, 71
    navigating the Web with, 69–72

Webcams, **271**, **304**, 357
Webcasting, **95**, **111**
Web conferencing, 357
Web database, 422
Web documents, 155
Web exercises
    on application software, 188
    on artificial intelligence, 461
    on communications technology, 365–366
    on databases, 461
    on hardware, 250, 306–307
    on information technology, 48
    on the internet, 114–116
    on personal technology, 405–406
    on security, 496
    on systems analysis and design, 540
    on system software, 188
    See also Review questions/exercises
Web page design software, **173**, **184**
Web pages, **66**, **111**
    hyperlinks on, 69, 72
    interactivity of, 72
    multimedia effects on, 93–95
    resources for building, 174
    software for creating, 173
    See also Websites
Web portals, **73**–74, **111**
Web servers, 319
Websites, **66**, **111**
    bookmarking, 72
    history of visited, 71
    secure, 478
    See also Web pages; World Wide Web
Web terminal, 257
What-if analysis, **158**, **184**
White-hat hackers, 349
Whittler, David, 17
Wide area networks (WANs), **316**–317, **363**
Wi-Fi phishing, 474
Wi-Fi Protected Access (WPA), 354
Wi-Fi technology, **58**, 59, **111**, 340–341, 394
Wikipedia, 78, 408
Wikis, 37, 78
Wildcards, 80
Wilde, Melissa, 5
Wildstrom, Stephen, 79
Williams, Penny, 467
WiMax wireless standard, 339–340
Window (computer display), **86**, **111**, **132**, **184**
Windows Media Player, 144
Windows Mobile, **144**
Windows operating systems, 137–139
    Windows 2000, **139**, 141
    Windows CE, **144**

Windows NT, **139**–140
Windows Server 2003, 141
Windows Vista, 139
Windows XP, **137**–138, 139, 143
Windows XP Media Center Edition, 138
Wired communications media, 325–326
Wired Equivalent Privacy (WEP), 354
Wireless Application Protocol (WAP), **329**, **363**
Wireless communications, 328–344
    bandwidth and, 329
    electromagnetic spectrum and, **329**
    frequencies used for, 330–331
    internet connections and, 57–58
    long-distance, 333–340
    radio-frequency spectrum and, **329**, 330–331
    short-range, 340–344
    software-defined, 356
    types of, 330–333
Wireless email devices, 337, 394
Wireless internet service, 58
Wireless keyboards, 255
Wireless LANs, 340–341
Wireless laptops, 239
Wireless networks, **58**, **111**
Wireless pocket PC, 257
Wireless portals, 73
Wireless USB, **343**, **363**
Wirth, Niklaus, 524
Wizards, **153**, **184**
Word games, 39
Word processing software, **149**–155, **184**
    creating documents, 150
    editing documents, 150, 152–153
    formatting documents, 153–155
    illustrated overview of, 151
    printing documents, 155
    saving documents, 155
    tracking document changes, 155
    web documents and, 155
Word size, **208**, **247**
Word wrap, **150**, **184**
Workgroup computing, 357
Workplace
    monitoring employees in, 477
    quality-of-life issues and, 487–488
    See also Employment
Worksheet files, 148, 156
Worksheet templates, 158
Workstations, **22**, **44**
World Wide Web (WWW), **18**, **44**, 64–81
    addiction to, 489
    addresses used on, 66–68
    blogs on, 96

browser software, **64**, 69–72
business conducted on, 97–98
citing sources from, 78
domain abbreviations on, 68
evaluating information on, 75, 78, 79
government resources on, 10–11
HTML and, 68
hyperlinks on, 69
intrusiveness of, 98–103
job searching on, 12–13
lecture notes posted on, 493
multimedia on, 64, 77–79, 93–95
navigating, 69–72
plagiarism issues, 104
podcasting and, 96–97
portal sites on, 73–74
push technology and, **95**
review questions/exercises on, 112–116
RSS newsreaders on, 96
searching, 74–79
telephone calls via, 92–93
terminology associated with, 64, 66
tips for searching, 80
workplace issues and, 477
    See also Internet; Websites
World Wide Web Consortium (W3C), 65
Worms, **345**–346, 347, **363**
Write-protect notch, 221
Writing data, **211**, 221, **247**
Writing program documentation, **514**–515, **538**

Xbox 360, 397, 398–399
Xen software, 140
XGA (extended graphics array), **282**, **304**
XML (extensible markup language), **96**, **111**, **530**, **538**
XM Satellite Radio, 376

Yahoo!, 73, 74, 174
Yang, Jerry, 368
Yazykov, Viktor, 6
Yu, Stella, 489

Zemeckis, Robert, 467
ZigBee technology, 343
Zip-disk drive, **30**, **44**, 222
Zip disks, **222**, **247**
Zombie computers, **474**, **494**
Z-wave technology, 343

CL

004
WIL